Lecture Notes in Computer Science 936
Edited by G. Goos, J. Hartmanis and J. van Leeuwen

Springer
Berlin
Heidelberg
New York
Barcelona
Budapest
Hong Kong
London
Milan
Paris
Tokyo

V.S. Alagar Maurice Nivat (Eds.)

Algebraic Methodology and Software Technology

4th International Conference, AMAST '95
Montreal, Canada, July 3-7, 1995
Proceedings

Springer

Series Editors

Gerhard Goos
Universität Karlsruhe
Vincenz-Priessnitz-Straße 3, D-76128 Karlsruhe, Germany

Juris Hartmanis
Department of Computer Science, Cornell University
4130 Upson Hall, Ithaca, NY 14853, USA

Jan van Leeuwen
Department of Computer Science, Utrecht University
Padualaan 14, 3584 CH Utrecht, The Netherlands

Volume Editors

V.S. Alagar
Department of Computer Science, Concordia University
1455 de Maisonneuve Bvld. West, Montreal, Quebec H3G 1M8, Canada

Maurice Nivat
LITP, Université Paris 7- Denis Diderot
2 Place Jussieu, F-75005 Paris, France

Die Deutsche Bibliothek - CIP-Einheitsaufnahme

Algebraic methodology and software technology : 4th international conference ; proceedings / AMAST '95, Montreal, Canada, July 3 - 7, 1995. V. S. Alagar ; Maurice Nivat (ed.). - Berlin ; Heidelberg ; New York : Springer, 1995
(Lecture notes in computer science ; Vol. 936)
ISBN 3-540-60043-4
NE: Alagar, Vangalur S. [Hrsg.]; AMAST <4, 1995, Montreal>; GT

CR Subject Classification (1991): F.3-4, D.2, C.3, D.1.6, I.2.3, H.2.1-4

ISBN 3-540-60043-4 Springer-Verlag Berlin Heidelberg New York

Typesetting: Camera-ready by author
SPIN: 10486305 06/3142-543210 - Printed on acid-free paper

Preface

This is the fourth biennial conference on Algebraic Methodology and Software Technology (AMAST). The previous three editions of AMAST were held at the University of Iowa, USA in 1989 and 1991, and at the University of Twente, The Netherlands in 1993. Starting next year it is planned to hold it annually.

Algebra and logic are the two branches of mathematics on which formalisms for computer software and hardware systems have been successfully built. Over the last few years formal methods have been increasingly accepted and applied in remarkably diverse areas such as concurrent and reactive systems with applications in telephony, and atomic power plants. The adequacy of the theory determines and is determined by the suitability of application in practice. The AMAST series of conferences and workshops provide the forum for creative discussion and dissemination of results arising in the context of such an adequate theory and practice in algebraic and logical foundations.

Since its inception, the conference has continually evolved, broadened its scopes and widened its activities, and has now become truly international in its organization and attendance. Starting 1993, the first day of the conference has been dedicated to mathematics education of software engineers. The theme this year takes a step further to encourage discussion on the role of mathematics education and training in the industrial application of formal methods. The broadening scope of the conference is also reflected in the industry sponsorships received, the variety of research papers submitted, and the research prototype demonstrations to be held during the conference.

We received in all some 100 submissions from 26 countries, with a strong representation from Europe and North America. We invited four speakers for the education day, four speakers for the conference days, one speaker for the banquet, and accepted 29 refereed papers, complemented by 6 research prototype tools.

This volume has three parts to it: the *contributions of invited speakers*; *refereed papers*, and *research prototype reports*. The distinguished invited speakers are David Gries, Jeannette Wing, Dan Craigen, Ted Ralston, Ewa Orłowska, Krzysztof Apt, Joseph Goguen, and Rohit Parikh. We were also fortunate to have Pamela Zave as our banquet speaker. The accepted papers cover the four major conference themes: algebraic and logical foundations, concurrent and reactive systems, software technology, and logic programming and databases. The subjects of these papers include a wide variety of methods, notations, semantics, and proofs contributing directly to an advancement and rigorous development of software systems. The research prototype reports describe on-going experimentation based on theoretical issues in logic, algebra and proof theory.

Montreal, April 1995
V.S. Alagar

Acknowledgements

The following people have contributed to the success of AMAST'95:

General Chair: Maurice Nivat

Program Chair: V.S. Alagar

Program Committee
Martin Abadi (USA)
Gregor Bochmann (CAN)
Chris Brink (South Africa)
Pierre Deransart (France)
Michael Ferguson (CAN)
Kokichi Futatsugi (Japan)
Armando Haeberer (Brazil)
Nicolas Halbwachs (France)
Jiawei Han (CAN)
Michael Johnson (Australia)
Deepak Kapur (USA)
Hélène Kirchner (France)
Laks V.S. Lakshmanan (CAN)
Giorgio Levi (Italy)
Luigi Logrippo (CAN)
Silvio Lemos Meira (Brazil)
José Meseguer (USA)
Hafedh Mili (CAN)
Rokia Missaoui (CAN)
Peter Mosses (Denmark)
Istvan Németi (Hungary)
Rocco De Nicola (Italy)
Prakash Panangadan (CAN)
Don Pigozzi (USA)
R. Ramanujam (India)
R.K. Shyamasundar (India)
Andrzej Tarlecki (Poland)
Frits Vaandrager (Netherlands)
Martin Wirsing (Germany)

Organizing Committee
Chair: Teodor Rus (USA)

Tools and Demos Chair:
Peter Grogono (CAN)

Finance Chair:
T. Radhakrishnan (CAN)

Publicity
Chair: Charles Rattray (UK)
Michel Bidoit (France)
Pankaj Goyal (USA)
Giuseppe Scollo (Netherlands)
Ralph Wachter (USA)

Local Arrangements
Chair: Rokia Missaoui (CAN)
A. Abran (CAN)
A. Das (CAN)
B. Kerherve (CAN)

This conference was financially supported by Concordia University, Montreal, Université du Québec à Montréal (UQAM), Bell-Northern Research, Nun's Island, Montreal, Natural Sciences and Engineering Research Council of Canada and the Office of Naval Research, USA.

Excellent secretarial help was provided by Mrs. Terry Czernienko and Ms. Angie De Benedictis.

Alfred Hofmann of Springer-Verlag provided valuable support in publishing these proceedings.

External Referees

The program committee members and a number of external referees refereed all submitted papers and helped in the selection of papers from among the many excellent submissions received. Their invaluable timely reports and assistance during the selection process made it possible to bring the proceedings to its present form.

L. Aceto
R. Achuthan
I. Alouini
R. Alur
J. Andrews
D. Aspinall
R. Backhouse
M.A. Bednarczyk
S. Bensalem
C. Bergman
A. Bernardi
J. Bicarregui
M. Boreale
G. Boudol
A. Bouhoula
E. Brinksma
R. Buckland
G. Butler
M.V. Cengarle
M. Cerioli
E.M. Clarke
R. Cockett
B. Courcelle
L. De Baets
P. Degano
P. Devienne
J. Dingel
E.J. Doedel
F.Q. Dong
R. Dssouli
B. Dumant
J.M. Dunn
M. Falaschi
K. Farooqui
J.-C. Fernandez
J.L.L. Fiadeiro
W. Fokkink
M.F. Frias
R. Giacobazzi
R. Gotzhein
J.-C. Grégoire
G. Grudziński
M. Gyssens
E. Haghverdi
E.H. Haeusler
L. Heerink
R. Hennicker
T. Henzinger
R. Jagdeesan
H. Jamil
R. Janicki
L. Jategoenkar
S. Jha
B.v. Karger
J.-P. Katoen
R. K. Keller
B. Kerherve
F. Khendek
S. Klusener
K. Koperski
R.L.C. Koymans
G. Konstantinos
H. Korver
M. Koutny
A. Labella
L. Laforest
Y. Lakhnech
R. Langerak
G.T. Leavens
U. Lechner
N. Leone
B. Lefebvre
F. Leßke
H.F. Li
T. Li
J.-Y. Lin
P. Lincoln
L. Liu
H.C.R. Lock
K. Lodaya
R.D. Maddux
F. Maraninchi
L.S. Marshall
N. Marti-Oliet
A. Masini
D. Mery
M. Mukund
D. Murphy
L. Narayanan
F. Nickl
M. Okada
F. Orejas
P.K. Pandya
A. Petrenko
F. Parisi-Presicce
K. Periyasamy
A. Piperno
D.K. Probst
N. Raja
Y.S. Ramakrishna
C.R. Ramakrishnan
G. Ramanathan
S. Ramesh
P. Raymond
G. Richard
B. Ritchie
R. Robbana
W.-P. de Roever
I. Sain
A. Salibra
D. Sannella
D.R. Schmidt
G. Schmidt
P. Scott
J.P. Seldin
G. Sereny
R. Shinghal
N. Shiri
J. Sifakis
J. Sincennes
D.R. Smith
C.P. Stirling
I.N. Subramanian
P.S. Subramanian
P.S. Thiagarajan
G. Tremblay
E. Tronci
F. Turini
D.N. Turner
P.A.S. Veloso
R.Villemaire
B. Virot
R.F.C. Walters
J. Wei
J. Winkowski
J.C.S. van der Woude
Z. Xie
G.M. Yee
E. Zaffanella
W. Ziarko
A. Zwarico

AMAST Sponsors

Concordia University, Montreal
Université du Québec à Montréal (UQAM)
Bell-Northern Research, Nun's Island, Montreal
Natural Sciences and Engineering Research Council of Canada
Office of Naval Research, USA.

Table of Contents

Invited Lectures

Refereed Contributions

Algebraic and Logical Foundations

Equational Logic as a Tool

David Gries

Computer Science, Cornell University
Ithaca, New York 14853 USA

Abstract. Software tools and methods that approach being formal are not readily used by programmers, software engineers, and even most computer scientists. (There are avid users of mechanical verifiers and proof checkers, but they are a small the minority.) One reason for this is that the foundation of many formalisms —propositional and predicate logic— has been viewed and taught more as an object of study than as a useful tool.

We believe that formal logic *can* a useful mental tool. In fact, logic is the glue that binds together methods of reasoning, in all domains. Further, logic can be taught in a way that imparts appreciation for logic and rigorous proof, as well as some skill in formal manipulation. This is most easily done using an *equational* logic —a logic based on substitution of equals for equals and the kinds of manipulations people in scientific disciplines already perform. We outline this logic, explain its pedagogical advantages, and discuss teaching it.

1 Introduction

A major goal of the AMAST conferences is to put software-development methodology on firm, mathematical foundations. Emphasis is being placed on algebraic and logical foundations, with the hope of establishing algebraic and logical methodologies that are viable alternatives to the prevailing ad-hoc approaches to software engineering.

However, placing software-development on firm, mathematical foundations won't happen as long as practitioners eschew those foundations. Most practitioners are afraid of formalisms, of mathematical notation, of formal manipulation, even of proof. They have difficulty formalizing informal statements, and they don't see the need for it. They would rather write all their specifications and implementations —as well as their reasoning about them— in a largely informal manner.

Further, this viewpoint is not restricted to programmers and software engineers. Many, if not most, computer scientists feel the same way. If you don't believe this, just look at how few computer science departments teach formal methods in their core courses. Look at how few computer science departments teach the formal development of programs even as a senior elective, let alone do a good job of teaching about good specifications or loop invariants in their first courses. And the vast number of introductory programming texts say almost nothing about formal notation, precise specifications, or proofs of correctness —even though the ideas are now 25 years old.

We believe that people are afraid of formalism and proof because they have never been taught formalism and proof in a way that develops skill. Even math courses don't talk about proof. Of course, a math teacher will present a proof and then expect students to be able to develop similar proofs, but the topic of proof itself is rarely discussed. It is almost as if students are just supposed to *know* about proof, and if they don't, then they are not born mathematicians and will never become mathematicians.

We computer scientists, as well as the mathematicians, do a bad job of teaching formalism and its use. Lower-level college courses usually avoid using formalism, in both definitions and proofs. Later, when students have mastered definitions and proofs written largely in English, they may be shown how informal reasoning could be formalized, but the impression is left that such formalization would not be worth the effort. And because proofs are largely informal, the design of proofs is not taught —there is little or no discussion of principles or strategies for designing proofs.

Few are happy with the results of these courses. Generally, students' reasoning abilities are poor, even after several math courses. Many students still fear math and notation, and the development of proofs remains a mystery to most. Students are not equipped with the tools needed to employ mathematics in solving new problems.

In this article, we discuss ways of improving this state of affairs. We:

1. Introduce equational propositional and predicate logic and relate it to other logics.
2. Discuss the advantages of equational logic, showing how it can be a useful formal tool in many different domains.
3. Explain the pedagogical advantages of equational logic and give some hints on teaching formal logic as a tool.

The approach outlined in this article is embodied in the text *A Logical Approach to Discrete Math* [2] and the accompanying instructor's manual [3]. We believe the approach is exciting and revolutionary, and we hope that others will write new and better texts embodying the approach in the coming years.

2 Equational propositional logic

The goal in introducing a new logic is to make the manipulation of formulas as easy and effective as possible. This involves more than just the inference rules and axioms; various notational aspects have to be considered, such as the choice of precedences of operators (to reduce the need for parentheses) and the choice of notation for quantification. We intersperse our presentation of propositional and predicate logic with discussions of our various choices.

Our equational propositional logic **E** has the same theorems as any conventional propositional logic; the difference is in the inference rules and the axioms.

Table 1. Table of Precedences

(a)	$[x := e]$ (textual substitution)	(highest precedence)
(b)	. (function application)	
(c)	unary prefix operators: $+ \; - \; \neg \; \sim \; \mathcal{P}$	
(d)	$\cdot \; / \; \div$ mod	
(e)	$+ \; - \; \cup \; \cap \; \times$	
(f)	$= \; < \; > \; \in \; \subset \; \subseteq \; \supset \; \supseteq$	(conjunctional, see page 3)
(g)	$\vee \; \wedge$	
(h)	$\Rightarrow \; \Leftarrow$	
(i)	$\equiv$	(lowest precedence)

All nonassociative binary infix operators associate to the left, except $\Rightarrow$, which associate to the right.

The operators on lines (f), (h), and (i) may have a slash / through them to denote negation —e.g. $b \not\equiv c$ is an abbreviation for $\neg(b \equiv c)$.

For example, equivalence plays a prominent role in logic **E**, but it is a second-class citizen in other logics, where implication dominates.

We profit from using two different symbols for equality. Generally speaking, $b = c$ is defined as long as b and c have the same type —e.g. both booleans, both integers, both set of integers. Further, equality $=$ is treated conjunctionally, so $b = c = d$ is an abbreviation for $b = c \wedge c = d$.

Symbol $\equiv$ is used only for equality over the booleans: $b \equiv c$ and $b = c$ are the same, for boolean expressions b and c. Symbol $\equiv$ is assigned a lower precedence than $=$; this allows us to eliminate some parentheses[1]. For example, we can write

$$x = y \vee x < y \;\equiv\; x \leq y \quad .$$

Having two different symbols for equality allows us to make use of the associativity of equality over the booleans. Thus, we write $b \equiv c \equiv d$ for either $b \equiv (c \equiv d)$ or $(b \equiv c) \equiv d$, since they are equivalent. Had we used only $=$ for equality, we could not have benefited from associativity because $a = b = c$ already means $a = b \wedge b = c$. As with arithmetic manipulations, we often use symmetry (commutativity) and associativity of operators without explicit mention. Logicians have not used associativity of $\equiv$ much. For example, Rosser [4] uses $\equiv$ conjunctionally instead of associatively. Perhaps this is because logicians have been more interested in *studying* rather than *using* logic.

Our precedences for propositional calculus (and integers and set theory) are given in Table 1.

The four inference-rule schemas of logic **E** are given below, using the notation $E[v := P]$ to denote textual substitution of expression P for free occurrences

[1] As another simplification, we write application of function f to a simple argument b as $f.b$. Function-application operator . has highest precedence.

Table 2. Axioms of Equational Propositional Logic

Associativity of ≡: $((p \equiv q) \equiv r) \;\equiv\; (p \equiv (q \equiv r))$	(1)
Symmetry of ≡: $p \equiv q \equiv q \equiv p$	(2)
Identity of ≡: $true \equiv q \equiv q$	(3)
Definition of *false*: $false \equiv \neg true$	(4)
Distributivity of ¬ over ≡: $\neg(p \equiv q) \;\equiv\; \neg p \equiv q$	(5)
Definition of ≢: $(p \not\equiv q) \;\equiv\; \neg(p \equiv q)$	(6)
Symmetry of ∨: $p \vee q \;\equiv\; q \vee p$	(7)
Associativity of ∨: $(p \vee q) \vee r \;\equiv\; p \vee (q \vee r)$	(8)
Idempotency of ∨: $p \vee p \;\equiv\; p$	(9)
Distributivity of ∨ over ≡: $p \vee (q \;\equiv\; r) \;\equiv\; p \vee q \;\equiv\; p \vee r$	(10)
Excluded Middle: $p \vee \neg p$	(11)
Golden rule: $p \wedge q \;\equiv\; p \;\equiv\; q \;\equiv\; p \vee q$	(12)
Definition of Implication: $p \Rightarrow q \;\equiv\; p \vee q \;\equiv\; q$	(13)
Consequence: $p \Leftarrow q \;\equiv\; q \Rightarrow p$	(14)

of variable v in expression E:

$$\textbf{Leibniz:} \quad \frac{P = Q}{E[v := P] = E[v := Q]}$$

$$\textbf{Transitivity:} \quad \frac{P = Q,\; Q = R}{P = R}$$

$$\textbf{Substitution:} \quad \frac{P}{P[v := Q]}$$

$$\textbf{Equanimity:} \quad \frac{P,\; P \equiv Q}{Q}$$

The axioms of **E** are given in Table 2, ordered and grouped as we teach them. Equivalence is introduced first. And, because the first axiom says that equivalence is associative, thereafter we eliminate parentheses from sequences of equivalences.

As an example of the use of associativity of $\equiv$, note that we can parse (2) in five ways, thus reducing the number of axioms and theorems that have to be listed: $((p \equiv q) \equiv q) \equiv p$, $(p \equiv q) \equiv (q \equiv p)$, $(p \equiv (q \equiv q)) \equiv p$, $p \equiv ((q \equiv q) \equiv p)$, and $p \equiv (q \equiv (q \equiv p))$.

To see that the definition of conjunction, Golden rule (12), is valid, check its truth table or else use associativity and symmetry to rewrite it as

$$p \equiv q \quad \equiv \quad p \wedge q \equiv p \vee q$$

Now, it may be recognized as the law that says that two booleans are equal iff their conjunction and disjunction are equal.

Operator $\Leftarrow$ (see (14)) is included because some proofs are more readily developed or understood using it rather than $\Rightarrow$. Its use eliminates many "rabbits" from proofs —we mention this topic again, later.

As usual, a *theorem* of the logic is either (i) an axiom or (ii) the conclusion of an inference rule whose premises are theorems.

In general, we write proofs in an equational or calculational style, as shown below in a proof of $p \vee q \;\equiv\; p \vee \neg q \;\equiv\; p$. This style is reminiscent of how proofs are written in high school algebra. Leibniz is the dominant inference rule.

$$
\begin{array}{ll}
 & p \vee q \equiv p \vee \neg q \\
= & \quad \langle \text{Distributivity of } \vee \text{ over } \equiv \text{ (10), with } r := \neg q \rangle \\
 & p \vee (q \equiv \neg q) \\
= & \quad \langle q \equiv \neg q \equiv \mathit{false} \text{ (previously proved)} \rangle \\
 & p \vee \mathit{false} \\
= & \quad \langle \text{Identity of } \vee,\ q \vee \mathit{false} \equiv q \text{ (previously proved)} \rangle \\
 & p
\end{array}
$$

Each step of the proof consists of a substitution of equals for equals, using inference rule Leibniz. For example, the middle step can be written in inference-rule form as (using f for *false*)

$$
\frac{(q \equiv \neg q) \equiv f}{p \vee (q \equiv \neg q) \equiv p \vee f} \quad \text{or} \quad \frac{(q \equiv \neg q) \equiv f}{(p \vee (p \vee z)[z := q \equiv \neg q] \equiv (p \vee z)[z := f]}
$$

The hint following $=$ on a line of a proof must contain the premise of the inference rule. In the first step, the hint is axiom (10) with the substitution $r := \neg q$, or $(p \vee (q \equiv r) \equiv p \vee q \;\equiv\; p \vee r)[r := \neg q]$. So, inference rule Substitution is being used to generate the theorem that is the necessary premise. This illustrates how inference rule Substitution is used in this proof format. Often, when its use is obvious, the substitution itself is not mentioned. For example, in the last step, the hint is really "Identity of $\vee$, with $q := p$".

Inference rule Transitivity is used implicitly to conclude that the first line equals the fifth and then that the first line equals the last.

Finally, if the last line is a known theorem, we would append to it a hint that indicates which theorem it is; the reader would then implicitly use inference rule Equanimity to conclude that the first line is a theorem as well.

With this proof style, the inference rules need not mentioned explicitly; instead, the proof format indicates where each is being used.

We contrast calculational proof style with the Hilbert proof style. In the Hilbert style, a proof consists of a sequence of lines, each containing a theorem and the reason why it is a theorem —either it is an axiom or it is the conclusion of an inference rule. In the latter case, the inference rule is named and its premises are indicated —the premises must be axioms, be previously proved theorems, or appear on previous lines. Here is a proof of $p \vee q \equiv p \vee \neg q \equiv p$ in the Hilbert style.

1. $p \vee (q \equiv \neg q) \equiv p \vee q \equiv p \vee \neg q$	Substitution, (10)
2. $p \vee (q \equiv \neg q) \equiv p \vee \mathit{false}$	Leibniz, $q \equiv \neg q \equiv \mathit{false}$
3. $p \vee q \equiv p \vee \neg q \equiv p \vee \mathit{false}$	Transitivity, 1, 2
4. $p \vee \mathit{false} \equiv p$	Substitution, $q \vee \mathit{false} \equiv q$
5. $p \vee q \equiv p \vee \neg q \equiv p$	Transitivity, 2, 4

The calculational style has several advantages over the Hilbert style. First, in the Hilbert style, all inference rules must be named, so there is more writing. Second, the Hilbert style requires repeating formulas again and again, thus making the proof appear longer and harder to read. For example, in the calculational style, the LHS of the theorem to be proved appears once —on the first line— while in the Hilbert proof it appears within lines 1, 3, and 5.

More importantly, the structure of a proof is not readily apparent. Proofs are kind of bottom-up, building new theorems in (seemingly) no particular order. In fact, some lines could be interchanged without hurting the proof. One sees no particular *reason* for each line.

The calculational style, on the other hand, leads to more "goal-directed" proofs. At each step, the shapes of the current expression and the goal help determine the next substitution to make. Making this determination is largely a pattern-matching process —matching the current expression with equivalences that could be used in effecting the substitution. Often, the shapes of the formulas involved are such that *one* possibility for a substitution stands out.

In teaching about proofs and their development, we stress that as few rabbits should be pulled out of a hat as possible. If a proof is written well, the reader will think at each step "I would have done the same thing if I were developing the proof". It is possible to get this idea across to students using the calculational style; it is much much harder with the Hilbert style. In fact, we can teach proof strategies and principles, a few of which appear in Table 3. These strategies and principles are not particularly deep, but their conscious application does indeed help, and they enable the teaching of proof development.

Finally, a word should be said about using Gentzen's *natural deduction* logic [1]. In this logic, modus ponens plays the major role; *inference* is stressed rather than *equivalence*.

In our experience, and that of many others, natural deduction proofs are usually longer than corresponding equational proofs and harder to develop and understand. The general idea seems to be first to break a formula into its many constituents and then to build up a desired formula from these constituents. The calculational style, on the other hand, allows the replacement or introduction of large sub-formulas. Further, practically *no one* writes formal natural-deduction proofs as a matter of course in their work, so natural deduction can not be taught as a system that people actually use.

Gerhard Gentzen developed natural deduction in order to formalize how mathematicians actually wrote their informal proofs. This, perhaps, is a good reason for eschewing natural deduction. The last thing we want to teach is how to

Table 3. Some Proof Strategies and Principles
Strategy. Identify applicable theorems by matching the structure of expressions or subexpressions. The operators that appear in an expression and the shape of its subexpressions can focus the choice of theorems to be used in manipulating it.
Strategy. To prove $P \equiv Q$, start with the side that has the most structure or complexity, because that additional structure can help guide the development.
Principle. Structure proofs to avoid repeating the same expression on many lines.
Strategy. To prove a theorem about an operator, first eliminate it using its definition, then manipulate, and finally reintroduce it using its definition (if necessary).
Principle. Lemmas can provide structure, bring to light interesting facts, and ultimately shorten a proof.

formalize clumsy, informal arguments. Much better is to teach a complementary style that leads to more elegant proofs.

Of course, using metatheorems and derived inference rules, one can graft different proof styles onto any propositional logic. However, the first logic one learns has a serious impact on how one thinks about logic and proof later on, so we should be careful about what logic we teach first.

After teaching a very strict use of equational logic, in which more than 80 theorems can be proved in the strict equational style, we relax the style in several ways. We show how the conventional informal methods of proof —assuming the antecedent of an implication and proving the consequent, case analysis, contradiction, mutual implication, and contrapositive— are rooted in propositional logic. Suddenly, the kinds of informal proof that students have seen earlier in their education begin to make sense.

We also explain clearly a few proof methods that deal with quantification. For example, consider proving cancellation theorem $a+b=a+c \Rightarrow b=c$ in the theory of integers, using axiom $(\exists x{:}nat \mid: x+a=0)$. Usually, this proof uses (because of the axiom) the existence of a *witness* $\hat{a}$ that satisfies $\hat{a}+a=0$. But the idea that one can actually use such a witness is rarely substantiated. In *A Logical Approach*, however, we state and prove the following Metatheorem, which can be used to substantiate the use of a witness.

Metatheorem Witness. Given that x doesn't occur free in P, Q, R,

$(\exists x \mid R : P) \Rightarrow Q$ is a theorem iff
$(R \wedge P)[x := \hat{x}] \Rightarrow Q$ is a theorem.

But we are getting ahead of ourselves; let us talk in more detail about quantification.

3 Quantification and the predicate calculus

The treatment of quantification in our course unifies what has previously been a rather chaotic topic. In math, there are several different notations for quantification, for example:

$$\begin{aligned} \Sigma_{i=1}^{3} i^2 &= 1^2+2^2+3^2 \\ (\forall x).1 \le x \le 3 \Rightarrow b[x]=0 &\equiv b[1]=0 \wedge b[2]=0 \wedge b[3]=0 \\ (\exists x).1 \le x \le 3 \wedge b[x]=0 &\equiv b[1]=0 \vee b[2]=0 \vee b[3]=0 \end{aligned}$$

There appears to be no consistency of concept or notation here. Compounding the problem is that students are not taught rules for manipulating specific quantifiers —much less general rules that hold for all.

We use a single notation for all quantifications. Let $\star$ be any binary, associative, and symmetric operator that has an identity.[2] The notation

$$(\star i{:}T \mid R.i : P.i) \tag{15}$$

denotes the accumulation of values $P.i$, using operator $\star$, over all values i for which range predicate $R.i$ holds. T is the type of dummy i; it can be left out in contexts where the type is obvious.[3] If range $R.i$ is *true*, we may write the quantification as $(\star i \mid: P.i)$.

In the examples below, the type is omitted; also, **gcd** is the greatest common divisor operator.

$$\begin{aligned} (+i \mid 1 \le i \le 3 : i^2) &= 1^2+2^2+3^2 \\ (\wedge x \mid 5 \le x \le 10 \wedge prime.x : b[x]=0) &\equiv b[5]=0 \wedge b[7]=0 \\ (\,\mathbf{gcd}\ i \mid 2 \le i \le 4 : i^2) &= 2^2 \ \mathbf{gcd}\ 3^2 \ \mathbf{gcd}\ 4^2 \end{aligned}$$

Having a single notation (15) for quantification lets us discuss bound variables, scope of variables, free variables, and textual substitution for all quantifications, once and for all. Further, we can give axioms that hold for all such quantifications (see Table 4).

Having discussed quantification in detail, we then turn to pure predicate logic itself. This calls for just three more axioms that deal specifically with universal and existential quantification —see Table 5. We do follow convention and use $\forall$ and $\exists$ instead of $\wedge$ and $\vee$.

Range $R.i$ in notation (15) for quantification can be any predicate. For universal and existential quantification, however, a range is not necessary. Nevertheless, consistency of notation encourages us to use a single notation, even for universal and existential quantification. Furthermore, in many manipulations, range $R.i$ may remain unchanged while term $P.i$ is modified, and the separation of the range and term makes this easier to see. Here, the desire for ease of formal manipulation has dictated the choice of notations.

---

[2] An associative and symmetric operator without an identity can be used, but no instance of an axiom or theorem with a *false* range can be used.

[3] Space limitations preclude a discussion of types in a typed calculus here.

Table 4. Axioms for Quantification

Identity accumulation: $(\star x \mid R : u) = u$	(16)
Empty range: $(\star x \mid false : P) =$ (the identity of $\star$)	(17)
One-point rule: $(\star x \mid x = E : P) = P[x := E]$	(18)
Distributivity: $(\star x \mid R : P) \star (\star x \mid R : Q) = (\star x \mid R : P \star Q)$	(19)
Range split: $(\star x \mid R \vee S : P) \star (\star x \mid R \wedge S : P) = (\star x \mid R : P) \star (\star x \mid S : P)$	(20)
Interchange of dummies: $(\star x \mid R : (\star y \mid Q : P)) = (\star y \mid Q : (\star x \mid R : P))$	(21)
Nesting: $(\star x, y \mid R \wedge Q : P) = (\star x \mid R : (\star y \mid Q : P))$	(22)
Dummy renaming: $(\star x \mid R : P) = (\star y \mid R[x := y] : P[x := y])$	(23)

(The usual caveats concerning the absence of free occurrences of dummies in some expressions are needed to avoid capture of variables. Further, each axiom holds only if the quantifications mentioned in it are defined.)

Table 5. Additional Axioms for Predicate Calculus

Trading: $(\forall x \mid R : P) \equiv (\forall x \mid : R \Rightarrow P)$	(24)
Distributivity of $\vee$ over $\forall$: $P \vee (\forall x \mid R : Q) \equiv (\forall x \mid R : P \vee Q)$	(25)
(Generalized) De Morgan: $(\exists x \mid R : P) \equiv \neg(\forall x \mid R : \neg P)$	(26)

Issues of scope, bound variables, etc., make quantification and predicate calculus more complicated than propositional calculus, and some think that quantification is too complicated for freshmen and sophomores. However, many courses in math, computer science, physics, and engineering require quantification in one form or another, so not teaching quantification means that students are unprepared for those classes. In fact, the lack of knowledge of basic tools like quantification explains partly why students are apprehensive about mathematics.

Thus, we advocate teaching quantification carefully, completely, and rigorously, but in a manner that instills confidence. We have found that this can be done.

4 Using logic throughout the discrete-math curriculum

Most discrete math texts have a chapter on logic, but the concepts and notation of that chapter are not fully integrated into the rest of the text. (In fact, logic may be the subject of the last chapter.) Logic is usually taught in one or two weeks, and the students don't have time to digest the material and gain a skill in formal manipulation. Hence, many students see the study of logic as an academic exercise and not as something useful.

To change this view of logic, we advocate the following.

- Spend at least 6 weeks on logic.
- Emphasize the development of formal proofs and give the students enough practice that they gain a skill in developing formal proofs.
- Use formal proofs wherever possible in teaching discrete math, so that students see that logic is indeed the glue that binds together reasoning in all domains.
- Show the students that the use of logic is effective, in that it clarifies and shortens arguments.

In teaching proof, it does not suffice to develop arguments in English and then formalize them. Many times, informal arguments are by nature long-winded and obtuse, and formalizing them doesn't help. Equational logic provides an alternative, complementary method of proof, which is often better.

Drastic examples of the difference between English and equational-style formal proofs can be seen in many word problems. To illustrate, we use one problem from Smullyan's book [5] (We thank Michael Barnett for bringing this problem to our attention).

> Three people are talking together. Each is either a knight (always tells the truth) or a knave (always lies). Person B says, "C and D are of the same type (both knights or both knaves)." A stranger walking by asks D, "Are B and C the same type?" What is D's answer?

To solve this riddle, we first formalize. Use b for "B is a knight" and c and d similarly. B's statement is equivalent to $C \equiv D$. Since B said it, this statement is true iff B is a knight, so we formalize it as

$$B \equiv (C \equiv D) \tag{27}$$

An answer "yes" by D is equivalent to $B \equiv C$; "no" to $B \not\equiv C$. We write D's answer as $B ? C$, where we have to determine whether ? is $\equiv$ or $\not\equiv$. Since D gives the answer, we have

$$D \equiv (B ? C) \tag{28}$$

Taking (27) and (28) as axioms that describe the situation, we manipulate. We write the formulas without parentheses because $\equiv$ is associative and $\equiv$ and $\not\equiv$ are mutually associative.

$$\begin{array}{ll} & D \equiv B ? C \quad \text{—(28)} \\ = & \quad \langle (27) \rangle \\ & B \equiv C \equiv B ? C \end{array}$$

Since the first line is a theorem, by inference rule Equanimity so is the third. But this is the case only if $\equiv$ is used for ?. Therefore, D says "yes".

The development of this simple proof is within the abilities of most students who have studied equational logic for one week. Compare it with Smullyan's proof in [5], *half* of which is given in Table 6.

The demonstration of such remarkably simple solutions to seemingly difficult word problems after just one week of teaching logic goes a long way to providing

Table 6. Half of Smullyan's Proof of the Knight-Knave Problem

I am afraid we can solve this one only by analysis into cases.

Case One. B is a knight. Then C and D are really the same type. If D is a knight, then C is also a knight, hence of the same type as B, so D, being truthful, must answer "yes". If D is a knave, then C is a knave (since he is the same type as D), hence is of different type from B. So D, being a knave, must lie and answer "yes".

Case Two. ...

the motivation that students want. However, one can do much more. In the next few pages, we show briefly how propositional and predicate calculus form the underpinnings of other subjects of discrete math. (We omit discussion of the obvious application of software specifications and the development of proofs of correctness of programs.)

Set theory

Set theory is the first example of an extension of the pure predicate calculus. We first introduce the notation $\{d{:}t \mid R : E\}$ for set comprehension, the set of values of expression E for values of dummy d (of type t) for which R is true. For example, $\{i{:}\,nat \mid 1 \leq i < 100 \wedge prime.i : i^2\}$ is the set of squares of all primes less than 100.

We sometimes use the abbreviation $\{x \mid R\}$ for $\{x \mid R : x\}$

This informal definition of set comprehension can be explored with many examples, until students get a good feel for it. Then, it is time for a formal definition. This is done by defining set membership and set equality.

A value v is in $\{d{:}t \mid R : E\}$ exactly when $v = E$ for some value of the dummy for which R is true. This we formalize as follows: Provided d does not occur free in expression V, we define

$$V \in \{d{:}t \mid R : E\} \equiv (\exists d{:}t \mid R : V = E\} \quad .$$

Then we define equality of sets S and T:

$$\textbf{Extensionality:}\ S = T \equiv (\forall v \mid : v \in S \equiv v \in T) \quad .$$

We can also formally define all the conventional set operations, e.g.

$$\textbf{Union:}\ v \in S \cup T = v \in S \vee v \in T \quad .$$

Now, one can prove formally many theorems of set theory. For example, below, we prove $(\forall x \mid : P \Rightarrow Q) \equiv \{x \mid P\} \subseteq \{x \mid Q\}$, which shows the relationship between implication and subset. Actually, the development of this theorem helped *us*. We had always equated $P \Rightarrow Q$ with $\{x \mid P\} \subseteq \{x \mid Q\}$, and developing the proof forced us to correct our error. The value of formalism in correcting one's thoughts and making them precise is one point that we try to get across to students.

$$\begin{array}{ll} & \{x \mid P\} \subseteq \{x \mid Q\} \\ = & \quad \langle \text{Definition of } \subseteq \rangle \\ & (\forall x \mid x \in \{x \mid P\} : x \in \{x \mid Q\}) \\ = & \quad \langle x \in \{x \mid P\} \equiv P \text{, twice} \rangle \\ & (\forall x \mid P : Q) \\ = & \quad \langle \text{Trading} \rangle \\ & (\forall x \mid : P \Rightarrow Q) \end{array}$$

One can also use formalism to illustrate Russell's paradox in a beautifully clear way, far better than the usual informal pingpong argument. The set S of all sets that do not contain themselves, if definable, can be defined as

$$(\forall x \mid : x \in S \equiv \neg(x \in x)) \quad .$$

Since S is a set, we can instantiate the definition with S itself, yielding the contradiction $S \in S \equiv \neg(S \in S)$. What can be simpler!

Mathematical induction

Even though discussion of induction is restricted to induction over the natural numbers, this topic is hard for many students. We believe that formalism can make this topic easier. First, note that many illustrative problems deal with summations, e.g. proving $(\Sigma i \mid 0 \le i < n : i^2) = n^2$. Solving these problems requires skill in manipulating quantifications, which the usual discrete math course does not give students. Our approach, on the other hand, gives them all the necessary skills in manipulating quantifications *before* teaching them induction.

Second, the formalism can clarify in a way that informalism cannot. We illustrate this with an example. Our definition of math induction over the natural numbers is

$$(\forall n{:}\, nat \mid : P.n) \equiv (\forall n{:}\, nat \mid : (\forall i \mid 0 \le i < n : P.i) \Rightarrow P.n) \quad .$$

Hence, we force the students to put any statement to be proved by induction in the form $(\forall n{:}\, nat \mid : P.n)$. Define

$$\begin{array}{ll} b^0 = 1 \quad , & \\ b^n = b \cdot b^{n-1} & , (\text{for } 1 \le n) \end{array} \tag{29}$$

and consider proving $b^{m+n} = b^m \cdot b^n$ for all natural numbers n and m. Informal methods have a very difficult time with this problem, because they cannot explain well the different roles of m and n in the proof. With our formal approach, we require that the statement be written in the form $(\forall n{:}\, nat \mid : P.n)$. We formalize the statement to be proved and manipulate the formalization.

$$\begin{array}{ll} & (\forall m, n \mid 0 \le m \wedge 0 \le n : b^{m+n} = b^m \cdot b^n) \\ = & \quad \langle \text{Nesting (22)} \rangle \\ & (\forall n \mid 0 \le n : (\forall m \mid 0 \le m : b^{m+n} = b^m \cdot b^n)) \\ = & \quad \langle \text{Defining } P.n \equiv (\forall m \mid 0 \le m : b^{m+n} = b^m \cdot b^n) \rangle \\ & (\forall n \mid 0 \le n : P.n) \end{array}$$

Table 7. Proof of equivalence of induction and well-foundedness

The proof rests on the fact that for any subset S of U we can construct the expression $P.z \equiv z \notin S$, and for any boolean expression $P.z$ we can construct the set $S = \{z \mid \neg P.z\}$. For any subset S of U and corresponding expression $P.z \equiv z \notin S$, we change (31) into (30):

$$
\begin{array}{ll}
 & S \neq \emptyset \;\equiv\; (\exists x \mid : x \in S \wedge (\forall y \mid y \prec x : y \notin S)) \\
= & \quad \langle X \equiv Y \equiv \neg X \equiv \neg Y \text{; Double negation} \rangle \\
 & S = \emptyset \;\equiv\; \neg(\exists x \mid : x \in S \wedge (\forall y \mid y \prec x : y \notin S)) \\
= & \quad \langle \text{De Morgan; twice} \rangle \\
 & S = \emptyset \;\equiv\; (\forall x \mid : x \notin S \vee \neg(\forall y \mid y \prec x : y \notin S)) \\
= & \quad \langle P.z \equiv z \notin S \text{ —replace occurrences of } S \rangle \\
 & (\forall x \mid : P.x) \;\equiv\; (\forall x \mid : P.x \vee \neg(\forall y \mid y \prec x : P.y)) \\
= & \quad \langle \text{Law of implication} \rangle \\
 & (\forall x \mid : P.x) \;\equiv\; (\forall x \mid : (\forall y \mid y \prec x : P.y) \Rightarrow P.x)
\end{array}
$$

Now the problem becomes clear: $(\forall m \mid 0 \leq m : b^{m+n} = b^m \cdot b^n)$ is to be proved for natural numbers n by induction on n.

Because of the use of formalism, we can teach much more about induction. The induction principle for an arbitrary universe U and relation $\prec$ can be defined as follows:

$$(\forall x \mid : P.x) \equiv (\forall x \mid : (\forall y \mid y \prec x : P.y) \Rightarrow P.x) \quad \text{(for all } P\text{)} \qquad (30)$$

Secondly, $\langle U, \prec \rangle$ is called well founded iff every every nonempty subset of U has a minimal element (with respect to $\prec$):

$$S \neq \emptyset \;\equiv\; (\exists x \mid : x \in S \wedge (\forall y \mid y \prec x : y \notin S)) \quad \text{(for all } S\text{).} \qquad (31)$$

We show students the proof in Table 7 that $\langle U, \prec \rangle$ admits induction precisely when it is well founded. Further, we require that students to prove this theorem on a test, and, with previous warning that it will be on the test, 95% of them get it right.

On the other hand, few students understand an informal proof of this theorem, let alone are able to prove it themselves later on.

Solving recurrence relations

One of the concepts used in solving recurrence relations is the *generating function*. The generating function $G(z)$ for a sequence $x = x_0, x_1, x_2, \ldots$ of real numbers is the infinite polynomial

$$G(z) = (\Sigma i \mid 0 \leq i : x_i \cdot z^i)$$

For example, the generating function for the sequence $2, 4, 6, \ldots$ is

$$2 + 4 \cdot z + 6 \cdot z^2 + 8 \cdot z^3 + \ldots = (\Sigma i \mid 0 \leq i : 2 \cdot (i+1) \cdot z^i)$$

The generating function is not used to evaluate the sum at a given point z. Think of it instead as a new kind of mathematical entity, whose main purpose is to give a different representation for a sequence (or a function over the natural numbers), a representation that can be manipulated.

Consider the sequence $c^0, c^1, c^2, \ldots$ for some nonzero constant c. Let us try to find a simple representation of its generating function. We manipulate its generating function:

$$\begin{array}{ll} & G(z) = (\Sigma i \mid 0 \leq i : c^i \cdot z^i) \\ = & \langle \text{Split off first term; } c^0 \cdot z^0 = 1\text{; Arithmetic} \rangle \\ & G(z) - 1 = (\Sigma i \mid 1 \leq i : c^i \cdot z^i) \\ = & \langle \text{Change of dummy} \rangle \\ & G(z) - 1 = (\Sigma i \mid 0 \leq i : c^{i+1} \cdot z^{i+1}) \\ = & \langle \text{Distributivity of } \cdot \text{ over } \Sigma \text{ —factor out } c \cdot z \rangle \\ & G(z) - 1 = c \cdot z \cdot (\Sigma i \mid 0 \leq i : c^i \cdot z^i) \\ = & \langle \text{Definition of } G(z) \rangle \\ & G(z) - 1 = c \cdot z \cdot G(z) \\ = & \langle \text{Arithmetic} \rangle \\ & G(z) = 1/(1 - c \cdot z) \end{array}$$

Hence, we have $G(z) = 1/(1 - c \cdot z)$!

Informal descriptions of why $G(z) = 1/(1 - c \cdot z)$ just don't work. The *only* way to understand this result is to derive it, as we have just done. However, if the student does not have manipulative ability, which is the case in most discrete math courses, then they have great difficulty with this proof. On the other hand, with previous study of quantification, the above manipulation is easy.

Combinatorics and the pigeonhole principle

The pigeonhole principle is usually stated as follows.

> **Pigeonhole principle:** If more than n pigeons are placed in n holes, at least one hole will contain more than one pigeon.

This principle has been used in this informal way for at least 100 years. It has been formalized by logicians, but in order to study rather than use it. Several years ago, Dijkstra provided a different formalization of it, aimed at its use rather than its study, resulting in a remarkable discovery.

The first point to note is that with more than n pigeons and with n holes, the average number of pigeons per hole is greater than one. The second point to note is that the statement "at least one hole contains more than one pigeon" is equivalent to "the maximum number of pigeons in any hole is greater than one".

Thus, we abstract away from pigeons and holes and just talk about a bag S of real numbers (the number of pigeons in each hole) and restate the pigeonhole principle more mathematically. Let $av.S$ denote the average of the elements of bag S and $max.S$ denote the maximum. Then the pigeonhole principle is:

Pigeonhole principle: $\text{av}.S > 1 \Rightarrow \text{max}.S > 1$.

But this form of the principle can be generalized to the following. Provided S is nonempty,

Genralized Pigeonhole principle: $\text{av}.S \leq \max.S$.

It is easy to prove that the generalized pigeonhole principle implies the pigeonhole principle, but the implication in the other direction does not hold. Hence, the generalized pigeonhole principle is indeed more general. Second, we do not have to accept the principle as intuitively true, for it can be proved rather easily, given formal definitions of the average and the maximum of a bag of numbers.

The point to make here —a point that we should be making continually to our students— is that good use of formalism and notation can result in enhanced understanding and better mental tools. We don't formalize for the sake of formalizing; we formalize in order to clarify, to expand our knowledge, to increase our skills, to provide newer and better mental tools.

5 Conclusion

From time to time, we having jokingly (but with some seriousness) characterized students in the conventional discrete math course as follows.

> Students enter the conventional discrete course with a lack of understanding of proof and fear of math and math notation.
>
> Students leave the conventional discrete course with a lack of understanding of proof, fear of math and math notation, and a hatred of discrete math.

This is because the conventional approach to teaching discrete math is too informal,too fast, and too disjoint.

We believe that our new approach can change this view of discrete math. However, the approach requires that the instructor change as well. They must realize that much in mathematics has never been explained to the student. So the they must go out of their way to make *everything* absolutely clear.

The instructor must realize that informal arguments are often quite obscure, cannot be digested enough to explain later to others, and rarely lead to an understanding of proof methods. So the instructor must use the formalism throughout to teach about proofs and their development.

The instructor must realize that it takes time to digest new concepts and notations enough to gain a skill in their use. So the instructor must go slowly enough through the development of logic and proof methods —the glue that binds together reasoning in all domains— that students have a chance to gain that skill.

The instructor must realize that at first, students will not be inclined toward the use of formalism. So the instructor must take every opportunity to show the use of formalism in a good light, to motivate with suitable examples, and to show how the use of formality leads to more understanding and better mental tools. Students in the course will have no other experience to relate to, so it makes sense to show them from time to time the difference between the old and the new approach.

Our experience is that teaching using our approach does indeed produce students who are less afraid of math and notation, who have more confidence in their mathematical ability, and who appreciate to some extent the use of formalism.

You may think that this is only because we teach at an elite, research-oriented university. Others tell a different story. At a panel on using this approach at the SIGCSE conference in Nashville, March 1995, three instructors from institutions whose major role is teaching discussed their experiences in using our approach: Stan Warford of Pepperdine, Joan Krone of Denison, and Pete Weston of Daniel Webster College. They said that they would not go back to teaching in the more traditional style. They said that students lost (at least some of) their fear of proof, that proof turned out to be the *easy* part of the course. Weston mentioned that the approach helped the *weaker* student more than the stronger one, because it made everything so patently clear. (Weston used the approach in teaching incoming freshman in a course that paralleled the first programming course.)

Thus, we believe there is enough success in the use of our approach for the more doubtful people to think seriously about using it.

The people at the AMAST conference believe in the use of formal methods in software engineering and related fields and are involved in developing the methods of the future. However, those methods will be used only if practitioners believe in them as well. Hence, we must prepare practitioners to be able to accept them. We offer our new and somewhat revolutionary approach to teaching logic and discrete math for this purpose.[4]

Acknowledgements

I thank first of all my coauthor of *A Logical Approach to Discrete Math*, Fred B. Schneider, whose work on this text was crucial to its development. Thanks also to all the people in the field of formal development of programs. It is their persistence in coming to grips with formality and calculation that led to our text, which just attempts to explain what the people in the field have known for some time.

---

[4] Email the author at gries@cs.cornell.edu for information on obtaining the instructor's manual for *A Logical Approach to Discrete Math*. The instructor's manual has several essays about teaching using this approach, hints on teaching from all the chapters, and answers to the over-900 exercises, most of which have an equational proof in them.

References

1. Gentzen, G. Untersuchungen über das logische Schliessen. *Mathematische Zeitchrift 39* (1935), 176-210, 405-432. (An English translation appears in M.E. Szabo (ed.). *Collected Papers of Gerhard Gentzen.* North Holland, Amsterdam, 1969).
2. Gries, D., and F.B. Schneider. *A Logical Approach to Discrete Math.* Springer-Verlag, New York, 1993.
3. Gries, D., and F.B. Schneider. *Instructor's Manual for "A Logical Approach to Discrete Math".* Gries and Schneider, Ithaca, 1993. (To discuss obtaining a copy, email gries@cs.cornell.edu.)
4. Rosser, B. *Logic for Mathematicians.* McGraw-Hill, New York, 1953.
5. Raymond M. Smullyan. *What Is the Name of This Book?* Prentice- Hall, Englewood Cliffs, 1978.

Teaching Mathematics to Software Engineers

Jeannette M. Wing*

School of Computer Science, Carnegie Mellon University,
Pittsburgh, PA 15213, USA

Abstract. Based on my experience in teaching formal methods to practicing and aspiring software engineers, I present some of the common stumbling blocks faced when writing formal specifications. The most conspicuous problem is learning to abstract. I address all these problems indirectly by giving a list of hints to specifiers. Thus this paper should be of interest not only to teachers of formal methods but also to their students.

1 Context: Teach *What* to *Whom*?

Let me explain this paper's title, and in particular what I mean by "mathematics" and "software engineers." By "mathematics" I mean the mathematical foundations and techniques that underlie a wide range of *formal methods*. The foundations are mathematical logic and discrete mathematics. The techniques are specification and reasoning. Specification is a synthesis process involving the construction of a mathematical model or theory from an informally described set of concepts (e.g., an English-language description of a system's requirements). In this paper, when I say "specify" ("specification") I mean "*formally* specify" ("*formal* specification"). Reasoning is an analysis process involving proving properties about a system from a formal specification; the proofs themselves may be informal or formal, but their logical basis is formal. This paper focuses more on specification than reasoning.
By "software engineer" I mean to include practicing software engineers in industry, who may be furthering their education by taking courses while on the job, or taking a leave from work to get an advanced (typically, Master's) degree; graduate students who design and build large software systems, typically as an integral component of their Ph.D. thesis work;.

* This research is sponsored by the Wright Laboratory, Aeronautical Systems Center, Air Force Materiel Command, USAF, and the Advanced Research Projects Agency (ARPA) under grant number F33615-93-1-1330. Views and conclusions contained in this document are those of the authors and should not be interpreted as necessarily representing official policies or endorsements, either expressed or implied, of Wright Laboratory or the United States Government.

and undergraduate students who want to learn some basic software engineering principles before working at their first job.
My target teaching audience is not students of mathematics but students who are or will be software engineers. What and how to teach mathematical foundations to them differs from the contents and methods taught in traditional courses on logic and discrete mathematics. Indeed this observation led a group of us (Alan Brown, David Garlan, Daniel Jackson, James Tomayko, and me) to redesign the core curriculum of the Carnegie Mellon Master's of Software Engineering Program to include a course called "Models of Software Systems" [7, 6]. This course aims exactly at teaching the mathematical foundations and techniques relevant to the practice of software engineering.
This paper is based on the following experiences, starting with the most influential:

- Co-teaching (with David Garlan) the CMU MSE "Models" course during the Fall 1993 and Fall 1994 semesters. A total of thirty-six Master's and advanced undergraduate students enrolled. Z [19] and CSP [8] are the predominate formal methods used.
- Working one-on-one with CMU Ph.D students, mine and those of other faculty members, in the "Programming Systems" area. I have worked closely with or served as specification consultant for over twenty students in the past ten years.
- Working one-on-one with a Japanese industrial visitor, a software engineer in control systems, this past year. He is using the Larch/C Interface Language (LCL).
- Co-teaching (with Daniel Jackson) the CMU MSE "Analysis of Software Artifacts" course during the Spring 1995 semester. Students use the formalisms introduced in "Models" and analyze specifications of pieces of real systems. Fifteen students are enrolled.
- Teaching a Ph.D.-level graduate course at CMU, "Reasoning about Concurrent and Distributed Systems," Spring 1994, based on the MIT course, "Principles of Computer Systems," developed by Butler Lampson and William Weihl [12]. Students use the specification language *Spec* (roughly a combination of Modula-3 and Dijkstra's guarded command language) to describe system interfaces. About ten students either were enrolled or audited during the course of the term.
- Co-teaching (with John Guttag) a graduate course at MIT, "Specifications in Software Development" during the Fall 1992 semester. We covered Larch [9] in some detail. Ten Master's and Ph.D. students enrolled.

- Working as a laboratory assistant (while I was a graduate student) for the industrial short course, "Design and Implementation of Modular Software." Barbara Liskov and John Guttag were the instructors. We worked together on four incarnations of this course during 1980-1982. We briefly covered an algebraic specification technique.
- Teaching (Spring 1993) and co-teaching (with Bernd Bruegge, Spring 1991) the CMU undergraduate software engineering course. The major part of this course is an integrated class project whose subcomponents are developed by separate teams. Students are expected to write stylized, informal interface specifications.

Over the years I have been accumulating specification hints that I give to students in response to common problems and recurrent questions. So, I will use an indirect way to explain to teachers of formal methods the lessons I have learned; in the following four sections I speak directly to students of formal methods and exemplify the kinds of stumbling blocks they encounter when first using formal methods. Herein I use "you" to refer to the reader who is the software engineer attempting to write a formal specification.

I've broadly categorized the issues along the following dimensions:

- Figuring out *why* you are going through this specification effort (Section 2). What do you hope to get out of using formalism?
- Figuring out *what* of the system you want *to specify* (Section 3).
- Figuring out *how* to specify (Section 4). The most important hurdle to overcome is learning to abstract. I also give specific suggestions on how to make incremental progress when writing a specification.
- Figuring out *what to write down* (Section 5). Learn a formal method's set of conventions but do not feel constrained by them. Also, we all make logical errors sometimes; I point out some common troublespots in getting the details of a specification right.

I will illustrate my points with examples, usually in Z or Larch. Many actually make more than one point.

I close with a discussion of challenges to the technical community in Section 6.

2 Why Specify?

You should first ask yourself this question, "Why specify?" You might choose to specify because you want additional documentation of your system's interfaces, you want a more abstract description of your system design, or you want to perform some formal analysis of your system. What you write should be determined by what it is you want to do with your specification.

You should then ask yourself "Why *formally* specify?" Your answer determines what is to be formalized, what formal method to use, and what benefits you expect from a formal specification not attainable from just an informal one. When I have asked this question of a software engineer, here are the kinds of responses I have heard:

- Showing that a property holds globally of the entire system.
  - I want to characterize the "correctness condition" I can promise the user of my system.
  - I want to show this property is really a system invariant.
  - I want to show my system meets some high-level design criteria.
- Error handling
  - I want to specify what happens if an error occurs.
  - I want to specify the right thing happens if an error occurs.
  - I want to make sure this error never occurs.
- "Completeness"
  - I want to make sure that I've covered all the cases, including error cases, for this protocol.
  - I'd like to know that this language I've designed is computationally complete.
- Specifying interfaces.
  - I'd like to specify a hierarchy of C++ classes.
  - I'd like a more formal description of this system's user interface.
- Getting a handle on complexity.
  - The design is getting too complicated. I can't fit it all in my head. I need a way to think about it in smaller pieces.
- Change control.
  - Everytime I change one piece of code I need to know what other pieces are affected. I'd like to know where else to look without looking at all modules, without looking at all the source code.

Judicious use of mathematics, e.g., by applying formal methods, can help address all these problems to varying levels of detail and rigor.

3 What to Specify?

Formal methods are not to the point where an entire large, software system can be formally specified. You may be able to specify one aspect of it, e.g., its functionality or its real-time behavior; you may be able to specify many aspects of a part of it, e.g., specifying both functionality and real-time behavior of its safety-critical part. In practice, you may care to specify only one aspect of a part of a system anyway.

In writing a specification, you should know whether the specification is describing *required* or *permitted* behavior. Must or may? Since a specification can be viewed as an abstraction of many possible, legitimate implementations, you might most naturally associate a specification with describing permitted behavior. An implementation may have any of the behaviors permitted by the specification, but the implementor is not required to realize all. For example, a nondeterministic *choose* operation specified for sets will have a deterministic implementation. However, the expression "software system requirements" suggests that a customer may in fact *require* certain behavior. For example, in specifying an abstract data type's interface, the assumption is that all, not some subset, of the operations listed must be implemented.

Once it is clear what you hope to gain from the specification process, you can turn to determining exactly what should be formalized.

In increasing order of level of detail, you might want to formalize a global *correctness condition* for the system, one or more system *invariants*, the *observable behavior* of a system, or *properties* about entities in a system.

Correctness Conditions

You usually have some informal notion of a global *correctness condition* that you expect your system to uphold. It might be something as standard as serializability, cache coherence, or deadlock freedom. Or, it might be very specific to the protocol or system at hand. If it is standard, then very likely someone else has developed a formal model for characterizing a system and a logic such that the correctness condition can be formally stated and proven. E.g., serializability has been thoroughly studied by the database community from all angles, theoretical to practical. If your correctness condition can be cast in terms of a well-known theory, it pays to reuse that work and not invent from scratch.

If it is not standard then an informal statement of the correctness condition should drive the formalization of the system model and expression of the correctness condition. For example, in work by Mummert et al. [16], the authors started with this informal statement of cache coherence for a distributed file system:

> If a client believes that a cached file is valid then the server that is the authority on that file had better believe that the file is valid.

They developed a system model (a state machine model) and logic (based on the logic of authentication [4]) that enabled them to turn the informal statement into the following formal statement:

For all clients C, servers S, and objects d for which S is the repository, if C believes $valid(d_C)$ then S believes $valid(d_C)$

where clients, servers, objects, repository, believes and *valid* are formally defined concepts. The point is that the formal statement does not read too differently from the informal one.

Keep in mind this rule-of-thumb when formalizing from an informal statement: Let the things you want to describe formally drive the description of the formal model. There is a tendency to let the formal method drive the description of the formal model; you end up specifying what you can easily specify using that method. That is fine as far as it goes, but if there are things you cannot say or that are awkward to express using that method, you should not feel bound by the method. Invent your own syntax (to be defined later), add auxiliary definitions, or search for a complementary method.

The process of constructing a formal model of a system and formally characterizing the intended correctness condition can lead to surprises. More than once in my work with Ph.D. students who were formalizing the systems they were building, they had to back off from their expected and desired correctness condition by realizing that it was too strong, not always guaranteed (e.g., not guaranteed for some failure case or for a "fast-path" case), or only locally true (holds for a system component but not the entire system). Correctness conditions for distributed systems are likely to be weaker than expected or desired because of the presence of failures (nodes or links crashing) and transmission delays; it is likely that there are "windows of vulnerability" during which the correctness condition cannot be guaranteed.

Invariants

The most common way to characterize certain kinds of correctness conditions is as a *state invariant*. An invariant is a property that does not change as the system goes from state to state. Remember also:

- An invariant is just a predicate. If you define some appropriate assertion language, it is usually not a big deal to express an invariant formally.
- "True" is an invariant of any system. It's the weakest invariant and hence not a very satisfying one; you probably want to say something more interesting about your system. If "true" ends up being your strongest invariant, revisit your system design.
- An invariant can serve multiple purposes. It is usually used to pare down a state space to the states of interest. For example, it can be

used to characterize the set of *reachable* states or the set of *acceptable/legal* ("good") states. (These two sets are not always the same. For example, you might want the set of acceptable states to be a subset of the set of reachable ones.) *Representation invariants* are used to define the domain of an abstraction function, used when showing that one system "implements" another [13].

- Different formal methods treat invariants differently. (See *Implicit versus Explicit* in Section 5.1 for an elaboration of this point.) Make sure you understand invariants in the context of the formal method you are using.
- Hard questioning of system invariants can lead to radically new designs.

To illustrate the last point, consider this example from the garbage collection community. One class of copying garbage collection algorithms relies on dividing the heap into two semi-spaces, *to-space* and *from-space*; in one phase of these algorithms, objects are copied from from-space to to-space [2]. Traditional copying garbage collection algorithms obey a "to-space invariant": The user accesses objects only in to-space. Nettles and O'Toole observed that breaking this invariant and maintaining an alternative "from-space invariant" (the user accesses objects only in from-space) leads to simpler designs that are much easier to implement, analyze, and measure [17]. This observation led to a brand new class of garbage collection algorithms.

Observable Behavior

Given that state invariants are a good way to characterize desired system properties, formalizing the state transitions will allow you to prove that the invariants are maintained. When you specify state transitions, what you are specifying is the behavior of the system as it interacts with its environment, i.e., the system's observable behavior.

It might seem obvious that what you want to specify is the observable behavior of a system, but sometimes when you are buried in the details of the task of specifying, you forget the bigger picture. Suppose you take a state machine approach to modeling your system. Here is a general approach to specifying observable behavior:

1. Identify the level of abstraction (see Section 4.1) at which you are specifying the system. This level determines the interface boundary that you are specifying; it determines what is or is not observable. For example, a bus error at the hardware level is not expected to be an observable event in the execution of an text formatter like Word, but `core dumped` is certainly an observable event when using a text editor like `emacs`.

2. Characterize the observable entities in a state at that given level of abstraction. These are your state variables or objects. This step forces you to identify the relevant abstract *types* of your system (See the section on *Properties of State Entities* below.)
3. Characterize a set of initial states, and if appropriate, a set of final states.
4. Identify the operations that can access or modify the observable entities. These define your state transitions.
5. For each operation, characterize its observable effect on the observable state entities. For example, use Z schemas, Larch interfaces, or VDM pre/post-conditions.

Observable behavior should include any change in state that is observable to the user. If you are specifying an operation, then the kinds of observable state changes include changes in value to state objects, observable changes in the store (new objects that appear and old objects that disappear), objects returned by the operation, and signaled exceptions or errors.

Another way to think about observable behavior is to think about *observable equivalence* [15, 11, 10]. Ask "Can I distinguish between these two things?" where "things" might be states, individual entities in a state, traces of a process, or behavior sets of a process, depending on what you are specifying. If the answer is "yes," then there must be way to tell them apart (perhaps by using unique names or perhaps by defining an *equal* operation); if "no," then there must not be any way for the observer to tell them apart.

Properties of State Entities

The most important property to state of any entity in a system is its *type*. This statement is true regardless of the fine distinctions between the different type systems that different formal methods and specification (and programming) languages have. Since for a specification we are not concerned about compile-time and/or run-time costs of checking types, there is never a cost incurred in documenting the type of an entity in a specification.

Since a type can be viewed as an abbreviation for a little theory, declaring an entity's type is a succinct way of associating a possibly infinite set of properties with the entity in one or two words. A truly powerful abstraction device!

For entities that are "structured" objects (e.g., an object that is a collection of other objects), when determining its type, the kinds of distinguishing properties include:

- Ordering. Are elements ordered or unordered? If ordering matters, is the order partial or total? Are elements removed FIFO, LIFO, or by priority?

- Duplicates. Are duplicates allowed?
- Boundedness. Is the object bounded in size or unbounded? Can the bound change or it is fixed at creation time?
- Associative access. Are elements retrieved by an index or key? Is the type of the index built-in (e.g., as for sequences and arrays) or user-definable (e.g., as for symbol tables, hash tables)?
- Shape. Is the structure of the object linear, hierarchical, acyclic, n-dimensional, or arbitrarily complex (e.g., graphs, forests)?

For entities that are relations, the kinds of distinguishing properties include whether the relation is a function (many-to-one), partial, finite, defined for only a finite domain, surjective, injective, bijective, and/or any (meaningful) combination of these.

Finally, algebraic properties help characterize any relational entity or any function or relation defined on a structured entity. The common algebraic properties include: idempotency, reflexivity, symmetry, transitivity, commutativity, associativity, distributivity, existence of an identity element, and existence of an inverse relation or function. Algebras are a well-known mathematical model for both abstract data types and processes [8, 15, 1]. For example, this algebraic equation characterizes the idempotency of inserting the same element into a set multiple times:

$$insert(insert(s, e), e) = insert(s, e)$$

and this characterizes *insert*'s commutativity property:

$$insert(insert(s, e_1), e_2) = insert(insert(s, e_2), e_1)$$

And, since processes can be viewed as denoting structured objects (e.g., sets of traces), it makes sense to ask about whether algebraic properties hold for operations on processes. For example, for CSP processes, parallel composition is both commutative and associative:

$$P \parallel Q = Q \parallel P$$

$$P \parallel (Q \parallel R) = (P \parallel Q) \parallel R$$

4 How to Specify?

Given that you understand why you are specifying and what it is you want to specify, in what ways should you try to think about the system so that you can begin to specify and then make progress in writing your specification? Not surprisingly, we rely on the tried and true techniques: ***abstraction*** and ***decomposition***. In specifying large, complex systems, abstraction is useful for focusing your attention to one level of detail at a time; decomposition, for one small piece of the system (at a given abstraction level) at a time. Both enable local reasoning.

4.1 Learn to Abstract: Try Not to Think Like a Programmer

The skill that people find the most difficult to acquire is the ability to abstract. One aspect of learning to abstract is being able to think at a level higher than programmers are used to.

Try to think definitionally not operationally.
A student said the following to me when trying to explain his system design:

> If you do this and then that and then this and then that, you end up in a good state. But if you do this and then that and then this, you end up in a bad state.

Rather than thinking of what characterizes all good states, people find it easier to think about whether a particular sequence of operations leads to a good or bad state. This operational approach means ending up trying to enumerate all possible interleavings; this enumeration process quickly gets out of control, which is typically when a student will come knocking at my door for help. This problem is related to understanding invariants (see *Invariants* in Section 3). Invariably, the very first thing I need to teach students when I work with them one-on-one is what an invariant is.

Try not to think computationally.[2]
When writing specifications abstraction is intellectually liberating because you are not bound to think in terms of computers and their computations.
The following predicate

$$s = s' \frown \langle e \rangle$$

might appear in the post-condition of the specification of a *remove* operation on sequences. Here, s stands for the sequence's initial value; s', its final value; e, the element removed and returned. You most naturally might read "=" as assignment (especially if you are a C programmer) and not as a predicate symbol used here to relate values of objects in two different states; you may need to stare at these kinds of predicates for a while before realizing the assertional nature (and power) of logic.

Try constructing theories, not just models.
Building models is an abstraction process; but defining a theory takes a different kind of abstraction skill. When you construct a model of a system in terms of mathematical structures like sets, sequences, and relations, you get all properties of sets and relations "for free." This has

[2] Another way of saying the same thing as above.

the advantage that you do not have to spell them out every time you specify a system, but the disadvantage that some of those properties are irrelevant to your system. Thus, in a model-based constructive approach, you also need to provide a way to say which properties about the standard mathematical structures are irrelevant. (You might do this stripping away of properties in terms of invariants.) For example, you might specify a stack in terms of a sequence, where the top of the stack corresponds to one end of the sequence. Then, you need not only to state which end of the sequence serves as the top of the stack, but also to eliminate some sequence properties, e.g., being able to index into a sequence or concatenate two sequences, because they have no relevance for stacks. Alternatively, in a theory-based approach you state explicitly exactly what properties you want your system to have. Any model that satisfies that theory is deemed to be acceptable. For example, the essence of stacks is captured by the well-known equations:

$$pop(push(s,e)) = s$$
$$top(push(s,e)) = e$$

Sequences, or any other particular mathematical structure, do not enter the picture at all.

Like many, you may find methods like Z and VDM appealing because they encourage a model-based rather than theory-based approach to specification. You can build up good intuition about your system if you have a model in hand. However, to practice learning how to abstract, try writing algebraic or axiomatic assertions about the model.

4.2 How To Proceed: Incrementally

At any given level of abstraction, we ignore some detail about the system below. You might feel anxious to specify everything for fear of being "incomplete." Learning to abstract means learning when it is okay to leave something unspecified. This aspect of the abstraction process also allows incremental specification. In any case, it is probably better to specify something partially than not at all.

Here are four common and important examples of incremental abstraction techniques: (1) first assume something is true of the input argument and capture this assumption in a pre-condition, then weaken the precondition; (2) first handle the normal case, then the failure case; (3) first ignore the fact that ordering (or no duplicates, etc.) matters, then strengthen the post-condition; (4) first assume the operation is atomic, then break it into smaller atomic steps. Let's look in turn at each of these examples in their generality and in more detail.

Use pre-conditions.
Putting your "programmers" cap on, think of pre-conditions in the context of procedure call. A pre-condition serves two purposes: an obligation on the caller to establish before calling the procedure and an assumption the implementor can make when coding the procedure.
More generally, pre-conditions are a way of specifying assumptions about the environment of a system component. Thus, such assumptions can and should be spelled out and written down explicitly, and in so doing, you can specify and reason about a piece of the system without having to think about the entire system all at once. Thus pre-conditions enable partial specification, incremental design, and local reasoning—all attractive means of dealing with the complexity of large software systems.
One technical difficulty that trips some people is what it means or what happens if a pre-condition is not met. In many specification techniques (like Z and Larch), when an operation's pre-condition is not met, the interpretation is "all bets are off." The interpretation is that the pre-condition is a *disclaimer*.[3] In other words, the operation is free to do anything, including not terminate, if the pre-condition does not hold. The technical justification is that when an operation is specified using pre- and post-conditions, the logical interpretation of the specification is an implication:

$pre \Rightarrow post$

When the pre-condition is "false" then the implication is vacuously true, so any behavior should be allowed.
However, a stronger interpretation (e.g., taken by InaJo [18] and I/O automata [14]) is that a pre-condition should be interpreted as a *guard*. No state transition should occur if the pre-condition/guard is not met. Here the interpretation of a pre/post-condition specification is conjunction:

$pre \wedge post$

The difference is that under the disclaimer interpretation, for any state s in which the pre-condition does not hold, the state pair, $\langle s, s' \rangle$, for any state s', would be in the state transition relation; under the guard interpretation, no such state pair would be in the relation.[4]
There are other possible interpretations: For example, if the pre-condition is not met, it could mean that the state transition always goes to a special "error" state and termination is guaranteed, or it could mean the state transition leads to either an "error" state or non-termination. The

[3] Thanks to Daniel Jackson for this term.

[4] There is further confusion in interpreting pre-conditions in Z because even though you might write explicitly in your schema the conjunction, $xpre \wedge post$, where $xpre$ is the "explicit" pre-condition, the meaning is the implication, $pre \Rightarrow post$, where pre is the calculated pre-condition and usually not identical to $xpre$ [5].

point is that for the formal method you are using you should understand what it means both when a pre-condition is met and not met.

Finally, in the presence of concurrency, an operation's pre-condition is usually interpreted as a guard. More subtly, the predicate is evaluated in the state in which the operation begins executing, not in the state in which the operation is called. Because of concurrency, a scheduler may delay the start of the execution of an operation to some time after the call of the operation; since there is time between the state in which the operation is called and the state in which it starts executing, an intervening operation (executed by some other process) may change the system's state. Thus, a predicate that holds in the state when the operation is called may no longer hold in the state when the operation begins to execute. The point is to realize that in the presence of concurrency, there may be a change in interpretation of pre-conditions. [5]

Specify errors/exceptions/failures.

It is as important to specify erroneous or exceptional behavior as it is to specify normal behavior. If an operation can lead to an undesired state, you should specify the conditions under which this state is reachable. If you are lucky the specification language has some notational convenience (e.g., Larch's **signals** clause) or prescribed technique (e.g., Z's schema calculus) to remind you to describe error conditions; otherwise, handling errors may have to be disguised in terms of input or output arguments that serve as error flags.

There is a close correlation between pre-conditions and handling errors. Z draws this connection by advocating this convention using schema disjunction:

$$TotalOp = NormalOp \vee ErrorOp$$

where NormalOp is the specification (schema) of the Op operation under normal conditions, and ErrorOp is the specification of Op under the condition in which the pre-condition (which must be *calculated* [19] from NormalOp) does not hold. Thus, TotalOp gives the specification of Op under all possible conditions.

Larch, on the other hand, draws the connection by advocating weakening the pre-condition, e.g., defining it to be equivalent to "true," and correspondingly strengthening the post-condition. Thus,

```
Op = op()
        requires P
        ensures  Q
```

[5] Larch calls the guard a *when-condition* to distinguish it from the standard pre-condition written in a requires clause.

turns into:

Op = **op**() **signals** (error)
  requires true
  ensures if P **then** Q **else signal** error

For interfaces to distributed systems, you cannot ignore the possibility of failure due to network partitions or crashed nodes. You could abstract from the different kinds of failures by introducing a generic "failure" exception that stands for errors arising from the distributed nature of your system.

The two main points to remember are (1) in support of incremental specification, specify the normal case and then handle the error cases, but (2) do not forget to handle the error cases!

Use nondeterminism.

Introducing nondeterminism is an effective abstraction technique. Nondeterminism permits design freedom and avoids implementation bias.

Nondeterminism may show up in many ways. First, it may be inherent to the behavior of an operation or object. Consider the *choose* operation on sets:

choose = **op** (s: set) **returns** (e: elem)
  requires $s \neq \emptyset$
  ensures $e \in s$

The post-condition says that the element returned is a member of the set argument; it does not specify exactly which element is returned.

Nondeterminism may result from explicit use of disjunction in a post-condition:

light = **op**() **returns** (c: color)
  ensures $c = \mathit{red} \vee c = \mathit{amber} \vee c = \mathit{green}$

or more generally, from the explicit use of an existential quantifier:

positively_random = **op** () **returns** (i: int)
  ensures $\exists\, i \,.\, i > 0$

From a state machine model viewpoint (for instance when discussing deterministic and nondeterministic finite state automata), nondeterminism should not be confused with choice. Suppose δ is a state transition relation,

$$\delta : \mathit{State}, \mathit{Action} \rightarrow 2^{\mathit{State}}$$

Then an example of choice is:

$$\delta(s, a_1) = \{t\}$$
$$\delta(s, a_2) = \{u\}$$

which says from state s you can either do the action a_1 (and go to the next state, t), or do the action a_2 (and go to the next state, u). However, an example of nondeterminism is:

$$\delta(s, a_1) = \{t, u\}$$

which says from state s you can do action a_1 and go to either state t or u.

Some formal methods for concurrent systems introduce their own notions of nondeterminism/choice; for example, CSP has two operators, one for internal choice ($\sqcap$) made by the machine and the other for external choice ($\Box$) made by the environment[6]. CCS has yet a different way to model nondeterminism.

The two main points are that (1) nondeterminism is a useful and important way to abstract, but (2) be careful to understand any given method's way of modeling nondeterminism and/or choice to use it properly.

Use Atomic Operations

For any system it is important to identify what the *atomic* operations are. At any level of abstraction an atomic operation may be implemented in terms of sequences of lower-level atomic operations (e.g., a *write* operation to a file on disk might be implemented in terms of a sequence of *write* operations to individual disk blocks). Even assignment can be broken down into sequences of loads and stores to/from memory and registers.

For a sequential program, usually it is assumed that each procedure is executed atomically; this assumption is rarely stated explicitly.

For a concurrent system, it is critical to state explicitly what operations are assumed to be atomic. The atomicity of an operation, Op, guarantees that no other operation can interfere with the execution of Op and that you can abstract away from any intermediate (lower-level) state that the operation might actually pass through.

5 What to Write?

With your pen poised over a blank sheet of paper or fingers over your keyboard, you now face the problem of what to write down. If you are using a specific formal method like Z, VDM, or Larch, you must know the syntax and semantics of its specification language. It is not enough to know what the syntactic features are; you need to understand what each means.

Also, embedded within any formal method is an *assertion language*, usually based on some variation of first-order predicate logic. With assertions you nail down precisely your system's behavior. It is in your assertions where the smallest change in syntax can have a dramatic change in semantics. Getting the details of your assertions right is typically when

[6] Hoare calls the former "nondeterministic or" and the latter "general choice."

you discover most of the conceptual misunderstandings of your system's design.

It is important to understand the difference between syntax and semantics. For example, the typical assertion language for algebraic specification languages gives grammatical rules for formulating syntactically legal terms out of function and variable symbols. Each syntactically legal term denotes a value in some underlying algebraic model. For example, the term, $insert(insert(\emptyset, e_1), e_2)$ is a syntactic entity that denotes the set value, $\{e_1, e_2\}$, which is a semantic entity. And so for a standard model of sets, the syntactically different term $insert(insert(\emptyset, e_2), e_1)$ denotes the same semantic set value.

5.1 General Rules-of-Thumb

What typically distinguishes a formal method from mathematics is its methodological aspects, e.g., stylistic conventions. A specification written in the style of a given formal method is usually not just an unstructured set of formulae. Syntactic features make it easier to read the specification (e.g,. the lines in a Z schema), remind the specifier what to write (e.g., the **modifies** clause in Larch), and aid in structuring a large specification into smaller, more modular pieces (e.g., Z schemas, Larch traits).

Implicit vs. Explicit

Most formal methods have well-defined specification languages so the choice of what you explicitly write down is guided by the grammar and constructs of the language.

However, there is a danger of forgetting the power of the unsaid. What is *not* explicitly stated in a specification often has a meaning. A naive specifier is likely to be unaware of these implicit consequences, thereby be in danger of writing nonsense. Here are three examples.

The first example is the *frame* issue. If you are specifying the behavior of one piece of the system in one specification module, you should say what effects that piece has *on the rest* of the system. In some formal methods (e.g., InaJo), you are forced to say explicitly what other pieces of the system do not change (NC''):

$$NC''(x_1, \ldots, x_n)$$

This is sometimes impractical if n is large, or worse, if you do not or cannot know what the $x_1, \ldots, x_n$ are in advance.

In some methods (e.g., Larch), you say only what may (but is not required to) change; anything not listed explicitly is required *not* to change:

modifies $y_1, \ldots, y_m$

This says $y_1 \ldots y_m$ may change but the rest of the system stays the same.

A subtler point about the Larch **modifies** clause is that there is significance to the omission of the clause. The absence of a **modifies** clause says that no objects may change. Thus, you might write a post-condition that asserts some change in value to an input argument or global; this assertion would be inconsistent with an omitted **modifies** clause.

Z's Δ and Ξ operators on schemas are similar to InaJo's *NC* construct; they allow you to make statements local to individual operations about whether they change certain state variables or not. Use of these schema operators on say the schemas, S_i, leaves implicit the invariant properties of the system captured in S_i. These properties can be made explicit by "unrolling" the schemas S_i.

This feature of Z is related to my second example of implicit vs. explicit specification, which has to do with invariants. In some formal methods like Z, state invariants are stated explicitly. They are a critical part of the specification, i.e., "property" component of a Z schema, and used to help calculate operation pre-conditions. In others like Larch, they are implicit and must be proven, usually using some kind of inductive proof rule. Finally, in others like VDM, they are redundant. They are stated explicitly and contribute to the checklist of proof obligations generated for each operation.

Finally, the third example has to do with implicit quantification. In many algebraic specification languages the i equations in this list

e_1

...

e_i

are implicitly conjoined and quantified as follows:

$$\exists f_1 \ldots \exists f_n \,.\, \forall x_1 \ldots \forall x_m \,.\, e_1 \wedge \ldots \wedge e_i$$

where $f_1 \ldots f_n$ are the function symbols and $x_1 \ldots x_m$ are the variables that appear in $e_1 \ldots e_i$.

This kind of implicit quantification has subtle consequences. Consider the following (incorrect) equational specification of an operation that determines whether one set is a subset of another:

$$s_1 \subseteq s_2 = (e \in s_1 \Rightarrow e \in s_2)$$

What you really mean is:

$$s_1 \subseteq s_2 = \forall e.(e \in s_1 \Rightarrow e \in s_2)$$

but in most algebraic specification languages, writing a quantifier in the equation is syntactically illegal; the tipoff to the error is the occurrence of the free variable e on the righthand side of the first equation.

Auxiliary Definitions
Do not be afraid to use auxiliary definitions:

- To shorten individual specification statements. For example, when argument lists to functions get too long (say, greater than four), then it probably means the function being defined is "doing too much."
- To "chunkify" and enable reuse of concepts. When a long expression (say, involving more than two logical operators and three function symbols) appears multiple (say, more than two) times, then it probably means that chunk of information can be given a name and the name reused accordingly.
- To postpone specifying certain details. When you find yourself going into too much depth while specifying one component of the system in neglect of specifying the rest of the system, then introduce a placeholder term to be defined later.

Notation
You should not feel overly constrained by the specification notation. If there is a concept you want to express and you cannot express it easily in the given notation, invent some convenient syntax, say what you want, and defer giving it a formal meaning till later. Don't get stuck just because your notation is restrictive. On the other hand, don't forget to define your inventions. It may be at odds with the rest of the semantics. If you're lucky, however, you will have thought of a new specification language construct that is more generally useful than for just your problem at hand.

Proofs
Most likely you will not be proving theorems about your system from your specifications, but if you are, the first difficult aspect about doing proofs is knowing how formal to be. For realistic systems and/or large examples, it's impractical to do a completely formal proof, in the strictest sense of "formal" as used in mathematical logic. What you should strive for when writing out an informal proof is to justify each proof step where in principle the step could be formalized.
Given that you are doing only informal proofs, the second difficult aspect is knowing when you can skip steps. Some steps are "obvious" but others are not. Also, what may seem "obvious" often reflects a hole in your argument.
It is possible to do large formal proofs using machine aids like proof checkers and theorem provers. There is of course a tradeoff between the effort needed to learn to use one of these tools and its input language and underlying logic and the benefit gained by doing the more formal

proof. If what you are trying to prove is critical, it may pay to invest the time and energy; moreover, this cost need be done only once, the first time. If you plan to do more than one (critical) proof, it may be worth your while. Finally, using machine aids keeps you honest because they do not let you skip steps.

Choosing the degree of formality and how much proof detail to give takes experience and practice, gained by both reading other people's proofs and constructing your own. A background in mathematics usually helps.

There are common proof techniques that you should have in your arsenal: proof by induction, case analysis, proof by contradiction, and equational reasoning (substituting equals for equals). You should be familiar with natural deduction though you probably would use it for only small, local proofs.

Finally, the familiar commuting diagram from mathematics plays a central role in proofs of correctness for software systems. For example, an interpretation for Fig. 1 in the context of state machines is to suppose that f is an action of a concrete machine on the concrete state x. If $\langle x, f, x' \rangle$ is a state transition of the concrete machine, then there exists an abstract action g such that $\langle A(x), g, A(x') \rangle$ is a state transition of the abstract machine.

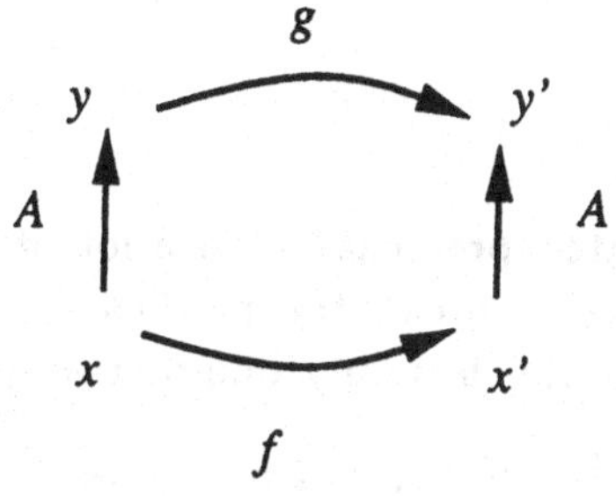

Fig. 1. A Commuting Diagram

In the context of abstract data types, the interpretation is that given that x is a concrete representation for y, the concrete function f *implements* the abstract function g under the abstraction function A. That is,

$$A(f(x)) = g(A(x))$$

More elaborate diagrams, for example, that allow sequences of actions rather than single actions generalize this basic idea. The "CLInc Stack" case study [3] of proving the correctness of the implementation of a small programming language down to the hardware level relies fundamentally on a stack of commuting diagrams.

5.2 The Details

I now turn to the nitty gritty of specification: getting the technical details right.

Logical Errors

Common logical errors that I have seen specifiers (including myself) make involve implication and quantification.

Implication. Remember that *false* implies anything so that

$false \Rightarrow \ldots$

is vacuously true, and that anything implies *true* so that

$\ldots \Rightarrow true$

reduces to true.

Quantifiers. Problem spots include nested quantifiers, ordering of quantifiers (especially modal operators for a temporal logic), and combining quantifiers and implication (e.g., what happens to a formula when bringing a quantifier outside an implication). Another confusion arises when qualifying a quantified variable with set membership, $\in$. That is,

$\forall x \in T \,.\, P(x)$

translates to

$\forall x \,.\, x \in T \Rightarrow P(x)$

but

$\exists x \in T \,.\, P(x)$

translates to

$\exists x \,.\, x \in T \wedge P(x)$

If you have a complicated predicate with a lot of embedded quantifiers, you may find it helpful to break the predicate into pieces, where each piece is in prenex form and has only one or two quantifiers.

Properties about sets, functions, and relations

When specifying objects such as sets, bags, and sequences that are collections of objects you may be prone to making the following common errors.

Saying

$x \in s'$

in the post-condition of an *insert* operation on sets may be too weak. It does not say that elements in the set that were originally in s are still there.

Saying

$s' = s - \{x\}$

in the post-condition of a *remove* operation on sets is too weak. You need to say

$s' = s - \{x\} \wedge x \in s$

since in the first case x may not be a member of s and the post-condition could hold by returning an arbitrary value; moreover, the set would not change in value, probably not the intended behavior for a *remove* operation.

Saying

$$s - s' = \{x\}$$

is also not strong enough. Here you need to add that s' is a proper subset of s:

$$s - s' = \{x\} \wedge s' \subset s$$

since the first case allows s' to have extra elements.

Saying something like

$$f(choose(s)) \wedge g(choose(s))$$

in the post-condition of an operation where *choose* is not deterministic (i.e., not a function) is weaker than saying

$$\exists\, x\,.\, x = choose(s) \wedge f(x) \wedge g(x)$$

since in the first case the different occurrences of *choose* could return different values. Of course if *choose* is a function, then it is guaranteed to return the same value.

Recursive definitions, commonly found in algebraic specifications, may at first look puzzling. For example, in specifying the *delete* operation for sets

$$delete(insert(s, e_1), e_2) = \text{if } e_1 = e_2 \text{ then } delete(s, e_2) \\ \text{else } insert(delete(s, e_2), e_1)$$

a common error is to forget to reapply *delete* recursively if e_1 and e_2 are equal or to forget to "reinsert" e_1 if they are not.

6 Challenges for the Technical Community

There exist expressive enough logics to support most kinds of properties that software engineers want to state of their systems, including temporal behavior (safety and liveness properties), real-time behavior (hard and soft time constraints), and erroneous behavior (failures, exceptions, etc.). So we are not lacking in the number and kinds of logics.

However, since no one logic is going to be the most appropriate for specifying all aspects of a software system, one technical challenge for the mathematical logic community is to be able to use different logics together. We might combine logics at the surface (syntactic) level or at the deeper semantic level or at some level in between. We might translate different logics into a common logic.

A more pragmatic need is better and more sophisticated tools to help in the semantic analysis of specifications. As tools increase in logical power, we can do more reasoning of more complex and larger systems. For example, rewrite rule engines and similar tools that manipulate algebraic (equational) specifications can only go so far; since more sophisticated logics are needed to describe real systems, more sophisticated reasoning tools are needed. Analogous to the need for ways to combine logics is the need to combine tools that support different logics and/or different proof techniques. For example, investigating the combination of model checking and theorem proving seems particularly promising and fruitful. Finally, considering the subject of this paper, we still face the educational challenge of teaching mathematical foundations like logic and discrete mathematics to practicing or aspiring software engineers. We need to go beyond giving the traditional courses and think about who the target students are (usually computer scientists and engineers); these new kinds of courses must make direct connections between mathematical concepts and real software systems.

Acknowledgments

I thank David Garlan and Daniel Jackson for their feedback on an earlier draft of this paper, for the fun discussions we have had on this topic, and for their collaboration on designing and teaching the Models, Methods, and Analysis courses for the MSE program.
I thank all the students with whom I have worked over the years for their interest and patience in trying their hand at formal specification.

References

1. DIS 8807. Information systems processing–open systems interconnection–lotos. Technical report, International Standards Organization, 1987.
2. H. G. Baker. List Processing in Real Time on a Serial Computer. *Communications of the ACM*, 21(4):280–294, 1978.
3. W.R. Bevier, W.A. Hunt, Jr., J S. Moore, and W.D. Young. An approach to systems verification. *Journal of Automated Reasoning*, 5:411–428, 1989.
4. M. Burrows, M. Abadi, and R. Needham. A logic of authentication. *ACM Transactions on Computer Systems*, 8(1):18–36, February 1990.

5. D. Garlan. Preconditions for understanding. In *Proceedings of the Sixth Int'l Conf. on Software Specification and Design*, pages 242–245, October 1991.
6. D. Garlan, G. Abowd, D. Jackson, J. Tomayko, and J.M. Wing. The CMU Master of Software Engineering Core Curriculum. In *Proceedings of the Eighth SEI Conference on Software Engineering Education (CSEE)*. Springer-Verlag, 1995.
7. D. Garlan, A. Brown, D. Jackson, J. Tomayko, and J. Wing. The CMU Masters in Software Engineering Core Curriculum. Technical Report CMU-CS-93-180, Carnegie Mellon Computer Science Department, August 1993.
8. C.A.R. Hoare. *Communicating Sequential Processes*. Prentice-Hall International, 1985.
9. J.J. Horning, J.V. with S.J. Garland Guttag, K.D. Jones, A. Modet, and J.M. Wing. *Larch : Languages and Tools for Formal Specification*. Springer-Verlag, New York, 1993.
10. C.B. Jones. *Systematic Software Development Using VDM*, chapter 15. Prentice-Hall International, 1986.
11. Deepak Kapur. Towards a theory of abstract data types. Technical Report 237, MIT LCS, June 1980. Ph.D. Thesis.
12. B. Lampson, W. Weihl, and U. Maheshwari. Principles of computer systems. Technical Report MIT/LCS/RSS-22, MIT Lab. for Comp. Science, 1993. Lecture Notes for 6.826, Fall 1992.
13. B. Liskov and J. Guttag. *Abstraction and Specification in Program Development*. McGraw-Hill/MIT Press, 1986.
14. N. Lynch and M. Tuttle. Hierarchical correctness proofs for distributed algorithms. Technical report, MIT Laboratory for Computer Science, Cambridge, MA, April 1987.
15. A.J.R.G. Milner. *A Calculus of Communicating Systems*, volume 92 of *Lecture Notes in Computer Science*. Springer-Verlag, 1980.
16. L. Mummert, J.M. Wing, and M. Satyanarayanan. Using belief to reason about cache coherence. In *Proceedings of the Symposium on Principles of Distributed Computing*, pages 71–80, August 1994. Also CMU-CS-94-151, May 1994.
17. Scott M. Nettles and James W. O'Toole. Real-Time Replication Garbage Collection. In *SIGPLAN Symposium on Programming Language Design and Implementation*. ACM, June 1993.
18. John Scheid and Steven Holtsberg. Ina Jo specification language reference manual. Technical Report TM-6021/001/06, Paramax Systems Corporation, A Unisys Company, June 1992.
19. J.M. Spivey. *Introducing Z: a Specification Language and its Formal Semantics*. Cambridge University Press, 1988.

The Role of Education and Training in the Industrial Application of Formal Methods

Ted Ralston,[1] Susan Gerhart,[2] Dan Craigen[3]

[1] Odyssey Research Associates, Ithaca NY 14850, USA
[2] Research Institute for Computing and Information Systems
University of Houston, Houston Texas 77058, USA
[3] ORA Canada, Ottawa Ontario K1Z 6X3, Canada

Abstract. During two one-year studies [11, 7], data was collected on education and training background and requirements for the successful transfer of formal methods to industry. This paper reports our observations and conclusions from both studies. One of the primary purposes of the second study was to provide a more systematic record of industrial experiences on a number of alleged deficiencies in the application of formal methods. One of these deficiencies is the contention that formal methods require prodigious mathematical education and talent for their successful use. Our interviews explored such issues as in-house company education; external education; differences in curricula between North America and Europe; profiles of personnel working on formal methods projects; and the role of tools as educational media.

1 Introduction

This paper reports on results from two separate and independent, one-year studies [11, 7] of the transfer of formal methods to industry, with particular reference to extracting observations on education and training issues from these studies. The first study was conducted at the Microelectronics and Computer Technology Corporation (MCC Formal Methods Transition Study or FMTS) [11]. The purpose of the FMTS was to give the thirteen sponsors (twelve commercial companies and one U.S. government agency) information about a broad range of issues related to formal methods—including requirements for staff skills and the education and training materials and means to obtain them—for the purpose of helping these companies assess the applicability and suitability of formal methods to meet their software engineering needs.

The second study, sponsored by the U.S. National Institute of Standards and Technology, the U.S. Naval Research Laboratory, and the Atomic Energy Control Board of Canada [7], surveyed twelve cases of significant industrial application of formal methods in North America and Europe.[4] These twelve cases were analyzed using a set of criteria covering business and software engineering

[4] Project resources were insufficient for the inclusion of Asian-based industrial cases.

issues, including education and skill levels of the practitioners. The purpose of the Survey was to provide a systematic appraisal of the performance of formal methods in industrial applications and examine the validity of a number of general observations and claims (both positive and negative) about formal methods that had been largely based on anecdotes.

2 Education as a Transfer Vector

Education is of particular importance to the transfer of formal methods because it is often alleged that staff skills and capabilities are an impediment to the transfer and successful adoption of formal methods. The use of formal notations and languages is widely perceived by software professionals to require staff with prodigious mathematics experience and skill in logic and mathematical analysis. This level of capability is made more problematic because the extent of mathematics content in undergraduate and graduate computer science courses has been declining. This allegation is particularly telling because it is a corollary to another—that formal methods is too costly—since a major part of the alleged high cost is in (i) training and (ii) salary requirements of recruits with advanced degrees.

We examine this issue from two perspectives: first, the educational backgrounds of the practitioners in the twelve surveyed cases and the staff of the MCC FMTS; and, second, the perspective of the various approaches and levels of effort used by companies to provide education to their staff.

2.1 Educational Backgrounds

The Survey included questions that were designed to provide skills profiles of the practitioners. Not surprisingly, the skill base was varied, ranging from one individual who had no university education (and had taught himself formal methods) to PhDs in number theory and logic.

As an overall observation, we noted that skills appear to be building slowly within companies that are attempting experimental use of formal methods on industrial-strength projects. This latter qualifier is important, because in companies that are *not* attempting experimental use of formal methods, there does not seem to be a concomitant buildup of relevant skills (with the possible exception of "process" awareness). We believe this is important because it reflects a fairly widespread industrial lack of interest in terms of job hiring criteria and such a lack of interest has an influence on university and technical training curricula.

The dearth of mathematical background in the general software workforce follows through in these cases. However, we observed in several of the cases that trained engineers were able to adapt to formal notations with relative ease when they were presented in a systematic and project-focused manner. Given the elementary nature of the mathematics of most formal methods, the greater problem may be in the ability of formal methods users to model their systems properly and carry through designs.

Contrasting these findings with the experience in the MCC FMTS, in which we found little or no expertise in the relevant subjects in the sponsor companies' software engineering groups, it would indicate some (albeit slight) degree of improvement.

We also concluded that formal methods are steadily maturing (over a 10–15 year time frame). In explaining what this conclusion means, we observed that both technical and non-technical barriers had to be overcome for formal methods to be used industrially. These barriers are summarized (with putative solutions) in Figures 1 and 2. In both cases, education issues played a strong role.

---

1. Expressing system functions: Abstract state machine modelling techniques predominate.
2. Following system architectures: Language modularization techniques were adequate, although most could be improved.
3. Providing the basic tool support: Standard text processors were adapted, although only as interim measures.
4. Utilizing domain expertise: Definitions, high-level modelization, and hard intellectual work extracted the needed domain information.

Fig. 1. Technical Barriers

---

1. Incomprehensible notations. (Solution: Iterating with developers until notations become readable to non-formalists.)
2. Required mathematics. (Solution: Identifying the necessary background mathematics and then developing and offering courses.)
3. Inadequate metrics and assessment criteria. (Solution: Starting to collect data, formulating models for calibrating processes using formal methods.)
4. Applying formal methods without tool support. (Solution: Integrate more CASE and SEE tools from the beginning, including making these tools available to universities.)

Fig. 2. Non-technical Barriers.

---

3 In-house Company Education

In both the MCC FMTS and the Survey, we found (somewhat surprisingly) that companies tend to use one mode of education and training: short duration courses. With respect to formal methods, the most systematic approach we encountered in the twelve cases surveyed was carried out at IBM's Hursley Laboratory through their adaptation of the IBM-wide Software Engineering

Workshops [18] to teach formal methods. These workshops offered essentially university level instruction on a weekly basis, over nine weeks, that covered abstraction, encapsulation, predicate calculus, first-order logic, sets and state transition models.

More typically, we found that companies offered courses of substantially shorter duration (on the order of one to three weeks) as a precursor to the actual development. These shorter courses consisted of a brief general introduction to logic and notation, the syntax and semantics of the chosen language, and some exercises using the method/language and (if available) a rudimentary tool (usually an editor and type checker). In some of the surveyed companies, actual examples in the domain of the intended project were used both to familiarize the developers with domain knowledge but also to "customize" the formal method to the domain.

4 External Education

One fairly consistent feature of all the cases surveyed was the availability of expertise in a nearby academic institution, and the willingness of the company to draw on these experts by means of contract consulting. In some cases, the academic experts were initially brought into the company to teach courses, and were retained to help with the development project (sometimes as mentors and sometimes as a direct part of the development team). In others, the experts were used as external reviewers. Regardless of the role as formally contracted, the effect of these experts was to increase the capabilities of the practitioners.

In much the same way, the MCC FMTS staff acted as experts to the thirteen sponsor organizations. The FMTS staff assimilated and transmitted information gained from both study and experimentation covering six formal methods.

5 Differences in Curricula Between North America and Europe

It is often noted that computer science curricula in North America and Europe differ significantly. European undergraduate computer science programs contain a greater degree of discrete mathematics (e.g., logic, set theory, graph theory, automata theory) than their North American counterparts. North American programs tend to stress high-level programming languages and software engineering process (with an emphasis on immediate coding or "hacking" in the sense of a free-form unstructured intuitive approach to writing programs). European universities tend to stress rigour, formalization, and abstraction at a theoretical level to a greater extent than the majority of North American undergraduate programs in computer science.

In the surveyed cases, we found definite evidence of this difference. Practitioners in Europe had a more immediate familiarity with logic notation and formal specification methods and languages, than their North American counterparts (especially those North American practitioners that came from a more traditional science field such as physics or engineering). Having said that, there were a couple of instances in which engineers without training or education in discrete mathematics were able learn and apply formal specification notations and languages without a long time or cost delay. Since only four of the twelve surveyed cases used formal proofs, it is difficult to reach any general conclusions regarding a difference in accessibility and understanding of proving techniques because of educational background. On each of the four cases that did use proofs, only one or two individuals in each team actually developed proofs and these individuals were mathematicians.

6 Profiles of Personnel Working on Formal Methods Projects

Some useful insights into education/training needs can be seen in short descriptions of the backgrounds of each team in the twelve industrial cases we examined in the Survey.

While it is difficult to extract many broad comprehensive insights into the effect of educational background on the project results, it is possible to make some observations which seem to cut across most of surveyed projects. First, the successful projects tended to have one or two individuals highly skilled in formal methods who acted as teachers or facilitators to others in the project. Second, the level of the knowledge of discrete mathematics required in all the projects tended to be at the undergraduate university level. Third, it appears to be more effective to teach formal methods to engineers and other domain specialists than it is to teach the domain knowledge to the software engineers.

Summary background descriptions for the educational backgrounds of the Survey case studies follow. For each case study, a citation to a published paper is provided. Further citations can be found in the Survey report [7].

SSADM (a CASE tool developed for the SSADM method by Praxis Systems Ltd). A team of six individuals, most of whom had formal methods training in university and at their company prior to working on this project. Praxis emphasizes the use of formal methods for high quality software systems as a major business differentiator. (Reference: [4].)

Customer Information Control System [CICS] (a re-engineering of legacy code). Expertise from Oxford University was brought in to train and provide expert consulting to some thirty-five people. Of this number, about five worked more or less full time on the CICS project, with others working on a part-time basis as needed. All thirty-five had been through the IBM Software Engineering Workshop courses on formal methods and had diverse educational backgrounds. (Reference: [14].)

Cleanroom/COBOL SF (restructuring COBOL code). There is a wide variation in the educational backgrounds of those working on Cleanroom projects, ranging from re-trainees to PhDs. In general, however, they are well educated with Bachelor's or Master's degrees in computer science and mathematics and five to ten years of industrial experience. (Reference: [17].)

Darlington Nuclear Reactor Control Software (a certification effort requiring high assurance achieved using post-development mathematical analysis of requirements and code using the Software Cost Reduction method). Most of the industrial and regulatory personnel involved in this project had engineering degrees with limited software engineering training. The two consultants involved are university professors of computer science, and one also has an engineering degree. Both are active researchers in computing science, specializing in the use of computers in safety-critical systems. (Reference: [1].)

LaCoS (a European government technology transfer project seeking to transfer the RAISE formal method to industry). This project is composed of seven individual projects, each involving a different company. In the two projects we surveyed, the personnel's educational backgrounds consisted of individuals with graduate degrees (two PhDs and several Master's degrees with at least half the teams having extensive training in formal methods). (Reference: [19].)

Multinet Gateway System (development of a secure datagram network service). Most of the personnel at Ford Aerospace that carried out this project had Bachelor's degrees. One individual did not have a college degree and was self-taught in logic and mathematics. (Reference: [9].)

SACEM Train Control System (certification of a new software control system for trains running on the Paris Metro system). The team involved in developing the SACEM software consisted of transportation and safety engineers and computer scientists. The latter had graduate degrees with one PhD (a specialist on the B method). (Reference: [5].)

TBACS (a smartcard authentication system for non-classified U.S. government systems). The two key individuals on this project both have Master's degrees in computer science and have had training in software engineering practice. The individual who carried out the formal verification had taken one university level course in discrete mathematics as a part of his Master's degree. (Reference: [15].)

Tektronix Oscilloscope software (formal specification of software infrastructure for a new family of oscilloscopes). The team of some ten to twelve individuals, consisted of oscilloscope engineers, production programmers and software engineering researchers. The educational backgrounds were Bachelor's degrees and PhDs in electrical engineering, physics and computer science, and two of the researchers had extensive experience in the study and teaching of formal methods. (Reference: [10].)

Traffic Alert and Collision Avoidance System (TCAS is a computer-controlled instrument system installed on commercial aircraft). This project involved a large number of individuals with a wide variety of backgrounds, the majority of whom served on a committee to develop and review the specification of the TCAS system. The team developing the TCAS specification was a group of computer science graduate students led by their professor, a noted expert in safety engineering. (Reference: [16].)

Inmos Transputer (a hardware specification and verification of a production 32-bit VLSI circuit). The team was made up of seven people: a computer architect, a designer, two mathematicians, two electrical engineers, and one programmer. The key individual on the project was one of the mathematicians, whose PhD covered applying formal specification and verification to a scheduler for an earlier generation of the circuit. However, all the members of the team had received education and training in formal methods either at university or as employees of Inmos. (Reference: [2].)

Hewlett-Packard Medical Instruments (specification of a real-time database component to an HP patient monitoring system). The personnel involved in this project were product engineers, quality assurance engineers, and software engineering researchers. The production engineering division personnel had Master's degrees in computer science and computer engineering, respectively. The quality assurance person had a Master's degree in computer science, and the two researchers had a PhD in mathematics and Master's degrees in mathematics and physics, respectively. (Reference: [3].)

7 The Role of Tools as Educational Media

While our overall observation regarding tools concluded that tool support was neither necessary nor sufficient for the successful application of formal methods, we noted that tools can play two types of role in education/training. The first is as a means of helping introduce formal methods newcomers to rudimentary (but often new or unfamiliar) procedures, such as type checking, or notations (most formal methods use notation systems that involve characters and symbols which are unfamiliar to non-mathematically trained individuals). The second role is as a low cost way of facilitating interest at the early stages of exploration and experimentation.

The majority of formal methods tools currently available are rather primitive (when compared to available functionality in PC or CASE products) and consist of rudimentary editing and text processing utilities, factors which make them difficult to use in production or educational environments. There is also a regrettable lack of documentation, which hampers the use of these tools in general but, in particular, for education.

Formal methods tools that are available fall into two classes: those that support formal specification languages and an attendant process, and those that are used for formal verification because of the high degree of safety or security required. The former tools are largely editors, parsers, syntax analysis and type checkers that help in the reading and writing of specifications. The second class of tools are automated theorem provers, which require a good understanding of structuring and guiding proof attempts, and are difficult to use without extensive human-computer interactions. Despite these difficulties, the availability of tools was seen to assist the teaching of a systematic and rigorous approach by facilitating the combination of practice and theory in an automated fashion.

Tools can also reflect the "pace of technology" improvement, and hence help maintain currency. However, few if any of the formal methods tools surveyed reflected any attempt to keep pace with new trends or functionality emerging from other disciplines in computer engineering or science.

8 Conclusions

The pedagogical material collected in the Survey suggests that the successful application of formal methods depends in part on enhancing the teaching of discrete mathematics in university and continuing education programs.

There is a clear difference between the discrete mathematics content in European and North American curricula. However, there is some evidence suggesting that Canadian computer science curricula are somewhat closer to European in this respect.

The availability of systematic and comprehensive continuing education programs in industry is mixed. In many companies, the norm is to support limited, if any, in-house programs, with a marked preference to send staff to outside instructors. The amount of time that staff can take for such outside continuing education has been shrinking (from two or three weeks to one week), as has the amount of time for educational leave. There is also a tendency to rely on conference tutorials to help provide continuing education.

The availability of tools, in and of themselves, did not significantly enhance or retard the results of the projects. The rudimentary formal methods tools did contribute to a somewhat faster uptake of the concepts when utilized in educational experiments. For the most part, these tools are either free or cost a few hundred dollars, which would not unduly burden even the most resource-poor universities or in-house company education departments.

References

1. G. Archinoff, et al. "Verification of the Shutdown System Software at the Darlington Nuclear Generating Station." *International Conference on Control Instrumentation and Nuclear Installations*, Glasgow, Scotland, May 1990.
2. G. Barrett. "Formal Methods Applied to a Floating Point Number System." *IEEE Transactions on Software Engineering*, 1989.
3. S. Bear. "An Overview of HP-SL." *Proceedings of VDM'91: Formal Development Methods*, Volume 551, Lecture Notes in Computer Science, Springer-Verlag, 1991.

4. D. Brownbridge. "Using Z to Develop a CASE Toolset." *1989 Z User Meeting,* Workshops in Computing, Springer-Verlag 1989.
5. M. Carnot, C. DaSilva, B. Dehbonei and F. Meija. "Error-free Software Development for Critical Systems using the B methodology." *Third International Symposium on Software Reliability Engineering,* IEEE, 1992.
6. Dan Craigen, Susan Gerhart, Ted Ralston. "Formal Methods Reality Check: Industrial Usage." *IEEE Transactions on Software Engineering,* 21(2), February 1995.
7. Dan Craigen, Susan Gerhart, Ted Ralston. "An International Survey of Industrial Applications of Formal Methods." U.S. National Institute of Standards and Technology, March 1993, Technical Report NIST GCR 93/626 (Volumes 1 and 2). Also published by the U.S. Naval Research Laboratory (Formal Report 5546-93-9582, September 1993) and the Canadian Atomic Energy Control Board reports INFO-0474-1 (vol 1) and INFO-0474-2 (vol 2), January 1995. Also available at http://www.ora.on.ca/.
8. Dan Craigen, Susan Gerhart and Ted Ralston. "Formal Methods Technology Transfer: Impediments and Innovation." In *Applications of Formal Methods,* M.G. Hinchey and J.P. Bowen, Editors. Prentice-Hall International Series in Computer Science, September 1995.
9. G. Dinolt, et al. "Multinet Gateway–Towards A1 Certification." *Symposium on Security and Privacy,* IEEE 1984.
10. D. Garlan and N. Delisle. "Formal Specifications as Reusable Frameworks." *Proceedings of VDM'90: VDM and Z!,* Vol. 428, Lecture Notes in Computer Science, Springer-Verlag 1990.
11. Susan Gerhart, Kevin Greene, Damir Jamsek, Mark Bouler, Ted Ralston, David Russinoff. "MCC Formal Methods Transition Study," MCC Technical Report FTP-FT-200-91, August 31, 1991.
12. Susan Gerhart, Dan Craigen and Ted Ralston. "Observations on Industrial Practice Using Formal Methods." In *Proceedings of the 15th International Conference on Software Engineering,* Baltimore, Maryland, May 1993.
13. Susan Gerhart, Dan Craigen and Ted Ralston. "Experiences with Formal Methods in Critical Systems." *IEEE Software,* January 1994.
14. I. Houston, S. King. "CICS Project Report: Experiences and Results from the use of Z." *Proceedings of VDM'91: Formal Development Methods,* Volume 551, Lecture Notes in Computer Science, Springer-Verlag, 1991.
15. D. Kuhn and J. Dray. "Formal Specification and Verification of Control Software for Cryptographic Equipment." *Sixth Computer Security Applications Conference,* 1990.
16. N. Leveson, et al. "Experiences using Statecharts for a System Requirements Specification." UC Irvine technical report, TR-92-106. Submitted for publication.
17. R. Linger and H. Mills. "A Case Study in Cleanroom Software Engineering: the IBM COBOL Structuring Facility." *COMPSAC,* IEEE 1988.
18. John Wordsworth. *Software Development with Z.* Addison-Wesley, 1992.
19. "Experiences from Applications of RAISE." LaCoS Project Reports, dated June 1991 and March 1992.

Information Algebras

Ewa Orlowska

Institute of Theoretical and Applied Computer Science
Polish Academy of Sciences
E–mail: orlowska@plearn.bitnet

1. Overview

Modern logic has developed an important paradigm of relationship between logical and algebraic systems. Along these lines we put forward a general perspective on modeling incomplete information and we propose an algebraic and logical framework for developing representational mechanisms which are capable of dealing with uncertainty. In section 2 we explicate two types of incompleteness of information that are grounded in a nonnumerical view of imprecision. In section 3 we discuss manifestation of these types of incompleteness in information systems. Next, we show how information presented in the form of an information system gives rise to classes of frames and classes of algebras. The focus of section 4 is on characterisation of these classes in an abstract way. The classes of information frames and information algebras are introduced that capture the relevant features of incompleteness. We propose a classification of information frames and a classification of information algebras that exhibit the main differences between the classes of frames and algebras, respectively, derived from information systems. In section 5 we consider information systems with null values and we apply the methodology elaborated in sections 3 and 4 to defining the relevant information frames and algebras. In section 6 we suggest a framework for duality theory for information frames and algebras. In section 7 we introduce a concept of informational representability of algebras.

2. Indiscernibility and orthogonality: two paradigms of incompleteness

Uncertainty of knowledge is a subject of studies in logic and computer science, and many theories of uncertainty and formalisms for quantifying uncertainty have been developed over years. We increasingly accept the fact that uncertainty is a part of reality, and that our knowledge is generally very incomplete. We also realise that uncertainty should be included in the formal models, and methods of reasoning in these models must be able to cope with uncertainty phenomenon in order to correspond more closely to reality and human common sense reasoning.

Numerous formalisms for reasoning under uncertainty have been studied in logic and computer science. They can be classified into two categories:

Numerical approaches: probability theory, fuzzy set theory, Dempster–Shafer theory, possibility theory, etc.,

Nonnumerical approaches: multivalued logics, nonmonotonic logics, information logics, rough set theory, etc.

Each theory of uncertainty is based on a particular view of incompleteness of information that leads to imprecision and vagueness of concepts discovered from that information. For instance, in probability theory incompleteness is reflected by a probability of information items, in multivalued logics every piece of information is assigned a degree of truth, in rough set theory every entity has its lower and upper approximation, etc. Each of these theories develops formal techniques for representation and reasoning with information that is laden with incompleteness of the underlying type.

Incompleteness of information is a cause of uncertainty of knowledge acquired from that information. A proper representation of incomplete information should allow a user formation of concepts even if they will not always be crisp and precise. The two basic carriers of concepts are their extensions and intensions. In the presence of incomplete information the extension and the intension of a concept discovered from that information might not be fully compatible. In order to isolate and to reflect those aspects of incompleteness that are most essential for that incompatibility we need a representational means with explicit access to the two levels of information: the level of individual entities (objects), and the higher level of properties (sets of objects). A representation formalism should provide means for expressing the extensions and the intensions of concepts and should allow for exhibiting the relevant type of incompleteness of primary information that the concepts are acquired from. Information frames and algebras considered in this paper are among the formalisms that satisfy these requirements.

The indiscernibility paradigm emerged from the observation that characterization of objects in terms of their properties might result in indistinguishability: some objects are 'the same' as far as the admitted properties are concerned. Hence, instead of crisp entities we rather grasp classes of objects such that each class contains those objects that cannot be distinguished one from another in terms of the given properties. It follows that also membership of an object in a set can only be defined modulo the properties of objects, and we might not be able to determine a sharp boundary between a set and its complement. As a consequence the problem arises of extension–intension incompatibility. The properties that form the intension of a concept do not necessarily enable us to define its extension in a precise way, in general the extension is determined with a tolerance. To model indiscernibility–type incompleteness of information we introduce a family of relations that reflect impossibility of discerning all the individual objects. The information relations from this group model degrees of indistinguishability in a nonnumerical way and they enable us to define a hierarchy of definability classes of concepts and to express extension–intension relationships.

However, in many situations it is more suitable to ask not for indistinguishability but for its opposite. To model in a nonnumerical way degrees of distinguishability we consider a family of orthogonality–type relations. We illustrate applications of relations from this class to representation of cores of sets of attributes in information systems. The concept of orthogonality plays a crucial role in quantum logic and various classes of logical and algebraic systems have been introduced in this connection (Cattaneo and Nistico 1989, Cattaneo et al. 1993, Goldblatt 1991). In those systems orthogonality is a semantical counterpart of negation or complement. In this paper we propose modeling of orthogonality that is related to representation of incomplete information..

3. Motivation for information frames and algebras: Frames and algebras derived from information systems

Let an information system (OB, AT, $\{VAL_a: a \in AT\}$) be given such that OB is a nonempty set of objects, AT is a nonempty set of attributes, each VAL_a is a nonempty set of values of attribute a. Each attribute is a function a: $OB \rightarrow Sb(VAL)$, where VAL is the union of all VAL_a, that assigns subsets of values to objects such that $a(x) \neq \emptyset$ for all a and x. Any set a(x) can be viewed as a property of object x. For example, if attribute a is 'colour' and a(x)={green}, then x possesses property 'to be green', if a is 'age' and x is 25 years old, then a(x)={25}. However, we might know the age of x only approximately, say between 20 and 28, and then a(x)={20,...,28}. If a(x) is a singleton set, then we have a deterministic information about x, otherwise information is nondeterministic.

Let A be a subset of AT. The following classes of information relations derived from an information system reflect incompleteness of information manifested by approximate (relative to properties) indistinguishabilty:

Strong indiscernibility: $(x,y) \in ind(A)$ iff $a(x)=a(y)$ for all $a \in A$
Weak indiscernibility: $(x,y) \in wind(A)$ iff $a(x)=a(y)$ for some $a \in A$.
Strong similarity: $(x,y) \in sim(A)$ iff $a(x) \cap a(y) \neq \emptyset$ for all $a \in A$
Weak similarity: $(x,y) \in wsim(A)$ iff $a(x) \cap a(y) \neq \emptyset$ for some $a \in A$

Intuitively, two objects are A–indiscernible whenever all the A–properties that they possess are the same, in other words up to discriminative resources from A these objects are the same. Objects are weakly A–indiscernible whenever some of their A–properties are the same. Objects are A–similar (weakly similar) whenever all (some) their A–properties are similar in the sense that they share some values. In the case of deterministic information strong (weak) similarity coincides with strong (weak) indiscernibility, respectively. All the relations defined above reflect indiscernibility–type incompleteness of information that is provided in a form of an information system. This type of incompleteness is the subject of investigations in the rough set theory (Pawlak 1991). Logics for reasoning about information relations defined above are investigated in Orlowska (1983, 1984, 1985, 1987), Vakarelov (1987, 1989, 1991a,b).

Important applications of information relations from the indiscernibility group are related to representation of approximations of subsets of objects in information systems. Let X be a subset of OB, then the lower R(A)–approximation of X (L(R(A))X) and the upper R(A)–approximation of X (U(R(A))) are defined as follows:

$L(R(A))X=\{x\in OB\text{: for all } y\in OB \text{ if } (x,y)\in R(A)\text{, then } y\in X\}$

$U(R(A))X=\{x\in OB\text{: there is } y\in OB \text{ such that } (x,y)\in R(A) \text{ and } y\in X\}$.

In the classical rough set setting, if relation R(A) is a strong indiscernibility relation, then we obtain the following hierarchy of definability of sets. A subset X of OB is said to be:

A–definable iff $L(ind(A))X=X=U(ind(A))X$

roughly A–definable iff $L(ind(A))X\neq\emptyset$ and $U(ind(A))X\neq OB$

internally A–indefinable iff $L(ind(A))X=\emptyset$

externally A–indefinable iff $U(ind(A))X=OB$

totally A–indefinable iff internally A–indefinable and externally A–indefinable.

The other application of the above information relations is related to modeling uncertain knowledge acquired from information about objects provided in an information system (Orlowska 1983, 1987, 1989). Let X be a subset of OB, we define A–positive (POS(A)X), A–borderline (BOR(A)X) and A–negative (NEG(A)X) instances of X as follows:

$POS(A)X=L(ind(A))X$

$BOR(A)X=U(idn(A))X-L(ind(A))X$

$NEG(A)X=OB-U(ind(A))X$.

Knowledge about a set X of objects that can be discovered from information provided in an information system can be modeled in the following way:

$K(A)X=POS(A)X\cup NEG(A)X$.

Intuitively, A–knowledge about X consists of those objects that are either A–positive instances of X (they are members of X up to properties from A) or A–negative instances of X (they are not members of X up to properties from A). We say that A–knowledge about X is:

complete iff $K(A)X=OB$, otherwise incomplete

rough iff $POS(A)X, BOR(A)X, NEG(A)X\neq\emptyset$

pos–empty iff $POS(A)X=\emptyset$

neg–empty iff $NEG(A)X=\emptyset$

empty iff pos–empty and neg–empty.

Clearly, the above applications can be extended to the other information relations from the indiscernibility group. The applications of similarity relations to modeling approximations can be found in Pomykala (1987, 1988).

Incompleteness of information manifested by approximate distinguishability can be modeled with orthogonality–type information relations derived from information systems. We define the following classes of these relations:

Strong orthogonality: $(x,y) \in ort(A)$ iff $a(x) \subseteq -a(y)$ for all $a \in A$, where $-$ is the complement with respect to set VAL_a

Weak orthogonality: $(x,y) \in wort(A)$ iff $a(x) \subseteq -a(y)$ for some $a \in A$

Strong diversity: $(x,y) \in div(A)$ iff $a(x) \neq a(y)$ for all $a \in A$

Weak diversity: $(x,y) \in wdiv(A)$ iff $a(x) \neq a(y)$ for some $a \in A$

Objects are A–orthogonal (weakly orthogonal) whenever all (some) of their A–properties are disjoint, and they are A–diverse (weakly diverse) if all (some) of their A–properties are different. In the case of deterministic information strong (weak) orthogonality coincides with strong (weak) diversity, respectively.

Applications of diversity relations are related to algorithms for finding cores of sets of attributes. Let an information system be given, and let A be a subset of AT. We say that an attribute $a \in A$ is indispensable in A iff $ind(A) \neq ind(A-\{a\})$, that is there are some objects such that a is the only attribute from A that can distinguish between them. A reduct of A is a minimal subset A' of A such that every $a \in A'$ is indispensable in A' and $ind(A')=ind(A)$. The core of A is defined as $CORE(A)=\cap\{A' \subseteq AT\colon A' \text{ is a reduct of } A\}$. For any pair x,y of objects we define the discernibility set $D_{xy}=\{a \in AT\colon (x,y) \in div(\{a\})\}$. It is proved in Rauszer and Skowron (1992) that $CORE(A)=\{a \in A\colon \text{there are } x,y \in OB \text{ such that } D_{xy}=\{a\}\}$.

Proposition 1

For any set A of attributes information relations satisfy the following conditions:

(a) $ind(A) \subseteq sim(A)$, $wind(A) \subseteq wsim(A)$, $ort(A) \subseteq div(A)$, $wort(A) \subseteq wdiv(A)$

For R=ind, sim, ort, div we have

(b) If $a \in AT$, then $R(\{a\})=wR(\{a\})$

(c) If $A \neq \varnothing$, then $R(A) \subseteq wR(A)$

(d) $R(\varnothing)=OB \times OB$, $wR(\varnothing)=\varnothing$

(e) $R(A \cup B)=R(A) \cap R(B)$, $wR(A \cup B)=R(A) \cup R(B)$

(f) If $A \subseteq B$, then $R(B) \subseteq R(A)$, $wR(A) \subseteq wR(B)$.

By a frame derived from information system $S=(OB, AT, \{VAL_a\colon a \in AT\})$ we mean a relational system $K_{S,R}=(OB, \{R(A)\colon A \subseteq AT\})$, where $\{R(A)\colon A \subseteq AT\}$ is any of the families of information relations defined above. Observe that relations in these frames depend on subsets of set AT. These subsets play the role of indices, that is we deal with families of relations indexed with subsets of a set.

Information relations determine families of operators in set OB of objects. Let R be any of information relations in OB, and let X be a subset of OB. We define:

$[R]X=\{x\in OB: \text{for all } y \text{ if } (x,y)\in R, \text{ then } y\in X\}$
$<R>X=\{x\in OB: \text{there is } y \text{ such that } (x,y)\in R \text{ and } y\in X\}$
$[[R]]X=\{x\in OB: \text{for all } y \text{ if } y\in X, \text{ then } (x,y)\in R\}$
$<<R>>X=\{x\in OB: \text{there is } y \text{ such that } y\notin X \text{ and } (x,y)\notin R\}$

Let $Rx=\{y: (x,y)\in R\}$ be the set of all the R–successors of x. We clearly have the following: $x\in[R]X$ iff $Rx\subseteq X$, $x\in<R>X$ iff $Rx\cap X\neq\emptyset$, $x\in[[R]]X$ iff $-Rx\subseteq -X$, $x\in <<R>>X$ iff $-Rx\cap -X\neq\emptyset$. It follows that if R is a relation from the indiscernibility group, then $[R]X$ and $<R>X$ are lower and upper approximation of X, respectively. From the logical perspective operators $[\ ]$ and $<>$ are necessity and possibility operators, respectively, and $[[\ ]]$, $<<>>$ are sufficiency operators.

Proposition 2

For R=ind, sim, ort, div the above operators satisfy the following conditions:

(a) If $X\neq\emptyset$, then $<R(\emptyset)>X=OB$
(b) $<R(A)>\emptyset=\emptyset$, $<wR(A)>\emptyset=\emptyset$
(c) $<wR(\emptyset)>X=\emptyset$
(d) $<>$ and $[\ ]$ are monotonic
(e) If $X\neq OB$, then $<<wR(\emptyset)>>X=OB$
(f) $<<R(A)>>OB=\emptyset$, $<<wR(A)>>OB=\emptyset$
(g) $<<R(\emptyset)>>X=\emptyset$
(h) $<<>>$ and $[[\ ]]$ are antimonotonic.

We say that operator f is dual to operator g iff $f(R)X=-g(R)-X$ for any X and R. It is easy to see that we have the following pairs of dual operators:

$<>$ $[\ ]$
$<<>>$ $[[\ ]]$

Operators f and g are said to be conjugated iff $f(R)X=g(-R)-X$ for any X and R. The following are the conjugated pairs of operators:

$<>$ $<<>>$
$[\ ]$ $[[\ ]]$

Any information system $S=(OB, AT, \{VAL_a: a\in AT\})$ determines derived algebras in the following way. Let $(Sb(OB), -,\cup,\cap,OB,\emptyset)$ be the Boolean algebra of all the subsets of the set OB of objects. Each set $A\subseteq AT$ determines a family of information relations defined as above, and these relations in turn determine the respective operators. Adjoining these operators to the Boolean algebra we form the following algebras, for R=ind, sim, ort, div:

$B_{S,<R>}=(Sb(OB), -,\cup,\cap,OB,\emptyset,\{<R(A)>: A\subseteq AT\})$
$B_{S,<wR>}=(Sb(OB), -,\cup,\cap,OB,\emptyset,\{<wR(A)>: A\subseteq AT\})$

$B_{S,<<R>>}=(Sb(OB), -, \cup, \cap, OB, \varnothing, \{<<R(A)>>: A \subseteq AT\})$
$B_{S,<<wR>>}=(Sb(OB), -, \cup, \cap, OB, \varnothing, \{<<wR(A)>>: A \subseteq AT\})$.

Algebras $B_{S,<R>}$ and $B_{S,<wR>}$ are very close to the normal Boolean algebras with operators introduced in Jonsson and Tarski (1951, 1952). A normal Boolean algebra with operators (normal BAO) is an algebra (B, $\{f_i: i \in I\}$) such that B is a Boolean algebra and each f_i is an operation of some finite rank that is additive in each of its arguments, and normal that is it takes on the value 0 whenever one of the arguments is 0. Algebras $B_{S,<R>}$ and $B_{S,<wR>}$ satisfy these conditions, however the family of operators adjoined to the Boolean algebra is indexed with subsets, and not with elements of a set. It is a consequence of the fact that the information relations are parametrized with subsets of attributes. The remaining algebras apparently do not belong to BAO: $<<>>$ are neither normal nor additive. Observe that within the family of the normal BAO's we cannot distinguish between operators determined by strong and weak relations. In the following section we propose families of frames and families of algebraic systems that characterise all the relevant features of the relations and the operators from frames derived from information systems and the algebras derived from information systems, respectively.

4. Information frames and algebras

In the present section we define classes of frames and algebras that provide an abstract framework for modeling incomplete information.

Let $K=(U, \{R(P): P \subseteq A\})$ be a relational system such that U is a nonempty set, A is a nonempty finite set, and each R(P) is a binary relation in U.

K is an information frame with strong relations of type A (FS) iff for every $P \subseteq A$ the relations R(P) satisfy the following conditions:

$R(P \cup Q)=R(P) \cap R(Q)$
$R(\varnothing)=U \times U$.

K is an information frame with weak relations of type A (FW) iff for every $P \subseteq A$ the relations R(P) satisfy the following conditions:

$R(P \cup Q)=R(P) \cup R(Q)$
$R(\varnothing)=\varnothing$

We say that an information frame $K=(U, \{R(P): P \subseteq A\})$ is a reflection of frame $K'=(U', \{R'(P): P \subseteq A\})$ iff $U=U'$ and for all $P \subseteq A$ it holds $R(P)=-R'(P)$, and we write $K=-K'$. Observe, that K is a frame with strong relations iff $-K$ is the frame with weak relations.

We define more specific classes of frames postulating some conditions that the relations are supposed to satisfy.

Indiscernibility frame (FIND): an information frame with strong relations R(P) that are equivalence relations

Similarity frame (FSIM): strong tolerance relations

Orthogonality fame (FORT): strong relations whose complements are tolerances

Diversity frame (FDIV): strong relations whose complements are tolerances and complement of R({a}) is transitive for all $a \in A$.

Weak indiscernibility frame (FWIND): relations R(P) are weak tolerances and R({a}) is transitive for all $a \in A$

Weak similarity frame (FWSIM): weak tolerances

Weak orthogonality frame (FWORT): weak relations whose complements are tolerances

Weak diversity frame (FWDIV): weak relations whose complements are equivalences.

We clearly have:

Proposition 3

(a) $K_{S,ind} \in FIND$, $K_{S,sim} \in FSIM$, $K_{S,ort} \in FORT$, $K_{S,div} \in FDIV$,

(b) $K_{S,wind} \in FWIND$, $K_{S,wsim} \in FWSIM$, $K_{S,wort} \in FWORT$, $K_{S,wdiv} \in FWDIV$.

(c) If $K \in FIND$ (FWIND, FSIM, FWSIM), then $-K \in FWDIV$ (FDIV, FWORT, FORT)

(d) If $K \in FORT$ (FWORT, FDIV, FWDIV), then $-K \in FWSIM$ (FSIM, FWIND, FIND).

As usual frames determine Kripke–style models of modal logics. Various classes of information logics have been studied in the literature. Kripke models based on frames with parametrised relations are considered in (Orlowska 1988). Logical systems with modal operators determined by parametrised strong indiscernibility and similarity relations are investigated in Konikowska (1987, 1994). Modal logics with bimodal base (R, –R) are considered in Gargov et al. (1987), Goranko (1988, 1990), they are strongly related to logics with operators <<>> and [[ ]] . The operators similar to <<>> and [[ ]] are investigated in Humberstone (1983). However, in all those papers the logics under consideration have models based on frames with single relation and its complement, not on frames with families of parametrised relations.

Information algebras are extensions of complete atomic Boolean algebras obtained by adjoining families of unary parametrised operators and by postulating some new axioms. We define four classes of information algebras.

Let $\Delta=(B, (f(P): P \subseteq A\})$ be an algebraic system such that B is a complete and atomic Boolean algebra $(B,-,+,\cdot,1,0)$, A is a nonempty finite set, and each f(P) is an unary operator in B.

Δ is an information algebra with strong normal operators of type A (SN) iff for every $P \subseteq A$ the operators f(P) satisfy the following conditions:

$f(P)0=0$

$f(P)(x+y)=f(P)(x)+f(P)(y)$

If $x \neq 0$, then $f(\varnothing)x=1$

If x is an atom of B, then $f(P \cup Q)x=f(P)x \cdot f(Q)x$

Δ is an information algebra with weak normal operators of type A (WN) iff for every $P \subseteq A$ the operators f(P) satisfy the following conditions:

$f(P)0=0$

$f(P)(x+y)=f(P)(x)+f(P)(y)$

$f(\varnothing)x=0$

$f(P \cup Q)x=f(P)x+f(Q)x$

Δ is an information algebra with strong conormal operators of type A (SCN) iff for every $P \subseteq A$ the operators f(P) satisfy the following conditions:

$f(P)1=0$

$f(P)(x \cdot y)=f(P)x+f(P)y$

$f(\varnothing)x=0$

$f(P \cup Q)x=f(P)x+f(Q)x$

Δ is an information algebra with weak conormal operators of type A (WCN) iff for every $P \subseteq A$ the operators f(P) satisfy the following conditions:

$f(P)1=0$

$f(P)(x \cdot y)=f(P)x+f(P)y$

If $x \neq 1$, then $f(\varnothing)x=1$

If x is a coatom of B, then $f(P \cup Q)x=f(P)x \cdot f(Q)x$

We clearly have:

$B_{S,<R>} \in ASN$, $B_{S,<wR>} \in AWN$, $B_{S,<<R>>} \in ASCN$, and $B_{S,<<wR>>} \in AWCN$.

The classes defined above enable us to distinguish between weak and strong operators and between normal and conormal operators. To model the operators determined by information relations from a particular class, we need more specific classes of algebras.

An indiscernibility algebra of type A (IND) is an algebra from class SN such that the following conditions are satisfied for all $P \subseteq A$:

$x \leq f(P)x$

$f(P)(x \cdot f(P)y)=f(P)x \cdot f(P)y$.

A weak indiscernibility algebra of type A (WIND) is an algebra from class WN such that the following conditions are satisfied for all $P \subseteq A$:

$x \le f(P)x$
$x \cdot f(P)y = 0$ iff $y \cdot f(P)x = 0$.
$f(a)(x \cdot f(a)y) = f(a)x \cdot f(a)y$ where $a \in A$

A similarity (weak similarity) algebra od type A (SIM, WSIM, respectively) is an algebra from class SN (WN) such that the following conditions are satisfied for all $P \subseteq A$:

$x \le f(P)x$
$x \cdot f(P)y = 0$ iff $y \cdot f(P)x = 0$.

An orthogonality (weak orthogonality) algebra of type A (ORT, WORT, respectively) is an algebra from class SCN (WCN) such that the following conditions are satisfied for all $P \subseteq A$:

$x \le f(P) - x$
$x \cdot f(P)y = 0$ iff $y + f(P)x = 1$.

A diversity algebra of type A (DIV) is an algebra from class SCN such that the following conditions are satisfied for all $P \subseteq A$:

$x \le f(P) - x$
$x \cdot f(P)y = 0$ iff $y + f(P)x = 1$.
$f(a)(x + -f(a)y) = f(a)x \cdot f(a)y$ where $a \in A$

A weak diversity algebra of type A (WDIV) is an algebra from class WCN such that the following conditions are satisfied for all $P \subseteq A$:

$x \le f(P) - x$
$f(P)(x + -f(P)y) = f(P)x \cdot f(P)y$

Algebraic systems related to indiscernibility–type incompleteness have been originated in Iwinski (1987). The classes of algebras introduced by Iwinski have been an inspiration for the studies of algebras derived from approximation spaces (Duentsch 1992, Pomykala and Pomykala 1988, Pagliani 1994, Novotny 1994). Indiscernibility algebras in a slightly different form have been introduced and investigated in Comer (1991). The intended interpretation is that in these algebras set A is a counterpart of a set of attributes, and operators f(P) are the algebraic versions of operators <ind(P)>. An attempt to define similarity algebras has been made in Pomykala (1993).

Proposition 4
Information algebras satisfy the following conditions:
(a) $B_{S<ind>} \in IND$, $B_{S,<wind>} \in WIND$, $B_{S,<sim>} \in SIM$, $B_{S,<wsim>} \in WSIM$,
(b) $B_{S,<<ort>>} \in ORT$, $B_{S,<<wort>>} \in WORT$, $B_{S,<<div>>} \in DIV$, $B_{S,<<wdiv>>} \in WDIV$,
(c) If Δ is an algebra with normal operators, then the operators are monotonic,
(d) If Δ is an algebra with conormal operators, then the operators are antimonotonic,

(e) If Δ is an algebra with strong operators of type A, then for every $P,Q \subseteq A$ if $P \subseteq Q$ then $f(Q)x \leq f(P)x$,
(f) If Δ is an algebra with weak operators of type A, then for every $P,Q \subseteq A$ if $P \subseteq Q$ then $f(P)x \leq f(Q)x$.

We say that an information algebra $\Delta=(B, \{f(P): P \subseteq A\})$ of type A is a reflection of an information algebra $\Delta'=(B', \{f'(P): P \subseteq A\})$ of type A iff $B=B'$ and $f'(P)x=f(P)-x$ for all x.

Proposition 5
(a) Every algebra from class SN is a reflection of an algebra from class WCN
(b) Every algebra from class WN is a reflection of an algebra from class SCN.

5. Modeling information with null values

Incompleteness of information reflected by information relations considered in section 3 is a kind of inherent incompleteness that appears in any collection of information items that provide description of objects in terms of their properties. However, it is often the case that information is directly incomplete in the following sense: properties of some objects are not known. In this section we consider information systems $(OB, AT, \{VAL_a: a \in AT\})$ such that for some attributes a and objects x set a(x) might be empty. Information systems with this property are referred to as systems with null values. We discuss relationships between this direct incompleteness and the types of incompleteness introduced in the preceding sections. We propose indiscernibility–type relations and orthogonality–type relations adequate for modeling the underlying types of incompleteness in information systems with null values.

We define the following information relations derived from an infromation system with null values:

Strong (weak) proper indiscernibility:

$(x,y) \in pind(A)$ $(wpind(A))$ iff $a(x)=\emptyset$ or $a(y)=\emptyset$ or $a(x)=a(y)$ for all (some) $a \in A$

Strong (weak) proper diversity:

$(x,y) \in pdiv(A)$ $(wpdiv(A))$ iff $a(x) \neq \emptyset$ and $a(y) \neq \emptyset$ and $a(x) \neq a(y)$ for all (some) $a \in A$

Proper similarity relations and proper orthogonality relations are defined in the same way as in section 3, however, their properties in information systems with null values are different. We say that a binary relation R in a set U is weakly reflexive iff for all $x,y \in U$ if $(x,y) \in R$, then $(x.x) \in R$.

We define the following classes of information frames of the form $(U, \{R(P): P \subseteq A\})$:

Proper similarity frame (FPSIM): strong, weakly reflexive, and symmetric relations
Proper weak similarity frame (FPWSIM): weak, weakly reflexive, and symmetric relations
Proper orthogonality frame: (FPORT): strong relations whose complements are weakly reflexive and symmetric
Proper weak orthogonality frame (FPWORT): weak relations whose complements are weakly reflexive and symmetric

The following classes of information algebras of the form (U, $\{f(P): P \subseteq A\}$) are the counterpatrs of these classes of frames:

Proper similarity (weak similarity) algebra (PSIM, PWSIM): information algebra with strong (weak) normal operators of type A such that the following conditions are satisfied for all $P \subseteq A$:

If $f(P)1 \neq 0$, then for all $x \in U$, $x \leq f(P)x$
$x \cdot f(P)y = 0$ iff $y \cdot f(P)x = 0$

Proper orthogonality (weak orthogonality) algebra (PORT, PWORT): information algebra with strong (weak) conormal operators such that the following conditions are satisfied for all $P \subseteq A$:

If $f(P)0 \neq 1$, then for all $x \in U$, $x \leq f(P)-x$
$x \cdot f(P)y = 0$ iff $y + f(P)x = 1$

Abstract characterization of proper indiscernibility and proper diversity frames and algebras is an open problem. Proper indiscernibility relations derived from an information system are reflexive and symmetric, but in order to distinguish this class of frames from similarity frames we need more sophisticated characterisation.

6. Preliminaries of duality

In the theory of standard BAO's the problem of duality between relational structures and Boolean algebras with operators is an important issue. We recall the basic notions in the simple case of algebras with a single operator. Let K=(U,R) be a frame such that R is a binary relation in a nonempty set U, the complex algebra of K is the algebra $K^+=(Sb(U),-,\cup,\cap,U,\emptyset, f_R)$ where $f_R X=\{x \in U$: there is $y \in U$ such that $(x,y) \in R$ and $y \in X\}$. K^+ is a complete, atomic, normal BAO. Given a Boolean algebra (B, f)$\in$BAO, the canonical relational structure of B is the frame $B_+=(F_B, R_f)$ such that F_B is the family of all the maximal filters of B, and $(u,v) \in R_f$ iff for all $x \in B$, if $x \in v$ then $fx \in u$. It is known that B is isomorphic to $(B_+)^+$, but K is not necessarily isomorphic to $(K^+)_+$. In this section we introduce basic notions of duality theory between information algebras and information frames.

Let $K=(U, \{R(P):P\subseteq A\})$ be an information frame, we define the normal (conormal) complex algebra K^{+N} (K^{+CN}) of K as follows:

$K^{+N}=(Sb(U),-,\cup,\cap,U,\emptyset, \{f_{R(P)}: P\subseteq A\})$
where $f_{R(P)}X=\{x\in OB$: there is y such that $(x,y)\in R(P)$ and $y\in X\}$

$K^{+CN}=(Sb(U),-,\cup,\cap,U,\emptyset, \{f_{R(P)}: P\subseteq A\})$
where $f_{R(P)}=\{x\in OB$: there is y such that $y\notin X$ and $(x,y)\notin R(P)\}$.

Proposition 9

Complex algebras satisfy the following conditions:
(a) If frame $K\in FS$, then $K^{+N}\in SN$ and $K^{+CN}\in SCN$,
(b) If frame $K\in FW$, then $K^{+N}\in WN$ and $K^{+CN}\in WCN$.

By way of example we show a construction of the canonical structure for algebras from class WSIM. Let an information algebra $\Delta=(B, (f(P)$: $P\subseteq A\})\in WSIM$ be given, we define the canonical structure Δ_+ for Δ in the following way:

$\Delta_+=(F_B, \{R_f(P): P\subseteq A\})$ where F_B is the family of all the maximal filters of B, and $(u,v)\in R_f(P)$ iff for all $x\in B$ if $x\in v$, then $f(P)x\in u$.

Proposition 6

If $\Delta\in WSIM$, then $\Delta_+\in SIM$

Proof: We show that in Δ_+ we have $R_f(P)\cup R_f(Q)\subseteq R_f(P\cup Q)$. If for all x, $x\in v$ implies $f(P)x\in u$ or for all x, $x\in v$ implies $f(Q)x\in u$, then since $f(P\cup Q)x=f(P)x\cup f(Q)x$ for weak operator f, and u is a maximal filter, we also have for all x, $x\in v$ implies $f(P\cup Q)x\in u$. To show $R_f(P\cup Q)\subseteq R_f(P)\cup R_f(Q)$, assume that for every x, if $x\in v$ then either $f(P)x\in u$ or $f(Q)x\in u$. Suppose that there is $y\in v$ such that $f(P)x\notin u$, and there is $z\in v$ such that $f(Q)z\notin u$. Take $x=y\cdot z$, clearly $x\in v$. We have $y\cdot z\leq y$, and since f is a normal operator, we have $-f(P)y\leq -f(P)(y\cdot z)$. Since $-f(P)y\in u$, we also have $-f(P)(y\cdot z)$ $\in u$, and hence $f(P)(y\cdot z)\notin u$. In a similar way we show that $f(Q)(y\cdot z)\notin u$, and these two facts contradict the assumption. To show that $R_f(\emptyset)=\emptyset$, observe that if operator f is a weak normal operator, then $f(\emptyset)x=0$ for all x, and $0\notin u$ for any maximal filter u, which completes the proof of the fact that relations $R_f(P)$ are weak.
Now we show that in frame Δ_+ relations $R_f(P)$ are tolerance relations. Reflexivity follows from axiom $x\leq f(P)x$. To prove symmetry, observe that condition $x\cdot f(P)y=0$ iff $y\cdot f(P)x=0$ holds iff condition $x\leq -f(P)y$ iff $y\leq -f(P)x$ holds. Since $-f(P)x\leq -f(P)x$, we obtain $x\leq -f(P)-f(P)x$. Assume that $(u,v)\in R_f(P)$, and let $x\in u$. It follows that $-f(P)-f(P)x\in u$. Suppose that $f(P)x\notin v$, then $-f(P)x\in v$ and hence by the assumption $f(P)-f(P)x\in u$, a contradiction.

7. Informational representability

The standard representation theorem for the algebras from class BAO says that every algebra from this class is isomorphic to a subalgebra of the complex algebra of some frame. The analogous theorems can be obtained for information algebras. However, in the informational framework more interesting are the theorems that establish informational representability of algebras.

We say that an algebra Δ from a class C of information algebras is informationally representable iff there is an information system S and an algebra B_S $\in C$ such that Δ is isomorphic to B_S.

Informational representability of algebras is analogous to informational representability of models of information logics (Vakarelov 1987, 1991a,b, Orlowska1993). An example of algebraic informational representability theorem is a representation theorem presented in Comer (1991) for a subclass of indiscernibility algebras defined there. However, that class of algebras is very restrictive, namely the algebras satisfy the following condition: if x is an atom, then f(P)x=x. In information terms it corresponds to the fact that every subset of objects is definable in terms of indiscerniscernibility relation that determines operator f. In other words, the class of these algebras enables us modeling of sharp concepts only, such that their extensions and intensions are compatible.

The general conclusion of this work is that within the two paradigms of incompleteness manifested by indiscernibility or orthogonality there is a rich diversity of forms that can be isolated and exibited both in algebraic and logical terms. The systematic study of the modal logics of the classes of information frames introduced in this paper is needed, in particular development of a proof theory for these logics and investigation of informational representability of their models is of practical interest. It would be also interesting to investigate the multi–modal versions of the logics whose semantics is determined by frames with several families of parametrised relations, and moreover some relationships exist between relations from those families.

References

Cattaneo,G. and Nistico,G. (1989) Brouver–Zadeh posets and three–valued Lukasiewicz posets. Fuzzy Sets and Systems 33, 165–190.

Cattaneo,G., Dalla Chiara,M.L. and Giuntini,R. (1993) Fuzzy–intuitionistic quantum logics. Studia Logica 52, 1–24.

Comer,S. (1991) An algebraic approach to the approximation of information. Fundamenta Informaticae 14, 492–502.

Duentsch, I. (1992) Rough relation algebras. Fundamenta Informaticae.

Gargov,G., Passy,S. and Tinchev,T. (1987) Modal environment for Boolean speculations. In: Skordev,D. (ed) Mathematical Logic and Applications. Plenum Press, New York, 253–263.

Goldblatt,R. (1991) Semantic analysis of orthologic. In: Goldblatt,R. Mathematics of Modality. CSLI Lecture Notes No. 43, Stanford, California, 81–97.

Goranko,V. (1987) Completeness and incompleteness in the bimodal base L(R,–R). Proceedings of the Conference on Mathematical Logic Heyting'88, Chaika, Bulgaria. Plenum Press, New York.

Goranko,V. (1990) Modal definability in enriched languages. Notre Dame Journal of Formal Logic 31, 81–105.

Humberstone,I. (1983) Inaccessible worlds. Notre Dame Journal of Formal Logic 24, 346–352.

Iwinski,T. (1987) Algebraic approach to rough sets. Bulletin of the PAS, Mathematics, vol 35, 673–683.

Jonsson,B. and Tarski,A. (1951) Boolean algebras with operators. Part I. American Journal of Mathematics 73, 891–939.

Jonsson,B. and Tarski,A. (1952) Boolean algebras with operators. Part II. American Journal of Mathematics 74, 127–162.

Konikowska,B. (1987) A formal language for reasoning about indiscernibility. Bulletin of the PAS, Mathematics, vol 35, 239–249.

Konikowska,B. (1994) A logic for reasoning about similarity. In: Orlowska,E. (ed) Reasoning with incomplete information. In preparation.

Novotny,M. (1994) Dependence spaces of information systems. In: Orlowska,E. (ed) Modeling Incomplete Information. Fundamentals and Applications. In preparation.

Orlowska,E. (1983) Semantics of vague concepts. In: Dorn,G. and Weingartner,P. (eds) Foundations of Logic and Linguistics. Problems and Solutions. Selected contributions to the 7th International Congress of Logic, Methodology, and Philosophy of Science, Salzburg 1983, London–New York, Plenum Press, 465–482.

Orlowska,E (1984) Logic of indiscernibility relations. Lecture Notes in Computer Science 208, Springer, Berlin–Heidelberg–New York, 177–186.

Orlowska,E. (1985) Logic of nondeterministic information. Studia Logica XLIV, 93 –102.

Orlowska,E. (1987) Logic for reasoning about knowledge. Bulletin of the Section of Logic 16, No 1, 26–38. Also Zeitschrift fuer Mathematische Logik und Grundlagen der Mathematik 35, 1989, 559–572.

Orlowska,E. (1988) Kripke models with relative accessibility and their application to inferences from incomplete information. In: Mirkowska,G. and Rasiowa,H. (eds) Mathematical Problems in Computation Theory. Banach Center Publications 21, 329 –339.

Orlowska,E. (1993) Rough set semantics for nonclassical logics. Proceedings of the Workshop on Rough Sets and Knowledge Discovery, Banff, Canada, 143–148.

Pagliani,P. Rough sets theory and logic–algebraic structures. In: Orlowska,E. (ed) Modeling Incomplete Information. Fundamentals and Applications. In preparation.

Pawlak,Z. (1991) Rough sets. Kluwer, Dordrecht.

Pomykala,J.A. (1987) Approximation operations in approximation space. Bulletin of the PAS, Mathe–matics, vol 35, 653–662.

Pomykala,J.A. (1988) On definability in the nondeterministic information system. Bulletin of the PAS, Mathematics, vol 36, 193–210.

Pomykala,J.A. (1993) Approximation, similarity and rough constructions. ILLC Prepublication Series, University of Amsterdam, CT–93–07.

Pomykala, J.A. and Pomykala,J.M. (1988) The Stone algebra of rough sets. Bulletin of the PAS, Mathematics vol 36, 495–508.

Rauszer,C. and Skowron,A. (1992) The discernibility matrices and functions in information systems. In: Slowinski,R. (ed) Intelligent decision support. Handbook of applications and advances in the rough set theory. Kluwer, Dordrecht, 331–362.

Vakarelov,D. (1987) Abstract characterization of some knowledge representation systems and the logic NIL of nondeterministic information. In: Ph.Jorrand and V.Sgurev (eds) Artificial Intelligence II, Methodology, Systems, Applications. North Holland, Amsterdam, 255–260.

Vakarelov,D. (1989) Modal logics for knowledge representation. Lecture Notes in Computer Science 363, Springer, Berlin–Heidelberg–New York, 257–277.

Vakarelov,D. (1991a) Logical analysis of positive and negative similarity relations in property systems. Proceedings of the First World Conference on the Fundamentals of Artificial Intelligence, Paris, France, 1991, 491–500.

Vakarelov,D. (1991b) A modal logic for similarity relations in Pawlak knowledge representation systems. Fundamenta Informaticae 15, 61–79.

Verification of Logic Programs with Delay Declarations

Krzysztof R. Apt[1,2] and Ingrid Luitjes[2]

[1] CWI, P.O. Box 94079, 1090 GB Amsterdam, The Netherlands
[2] Department of Mathematics and Computer Science
University of Amsterdam, Plantage Muidergracht 24
1018 TV Amsterdam, The Netherlands

Abstract. Logic programs augmented with delay declarations form a higly expressive programming language in which dynamic networks of processes that communicate asynchronously by means of multiparty channels can be easily created. In this paper we study correctness these programs. In particular, we propose proof methods allowing us to deal with occur check freedom, absence of deadlock, absence of errors in presence of arithmetic relations, and termination. These methods turn out to be simple modifications of the corresponding methods dealing with Prolog programs. This allows us to derive correct delay declarations by analyzing Prolog programs. Finally, we point out difficulties concerning proofs of termination.

Note. The research of the first author was partly supported by the ESPRIT Basic Research Action 6810 (Compulog 2).

1 Introduction

In Kowalski [Kow79] the slogan "Algorithm = Logic + Control" was coined. This paper suggested logic programming as a formalism for a systematic development of algorithms. The idea was to endow a logic program with a control mechanism to obtain an executable program. Prolog is a realization of this idea where the control consists of the leftmost selection rule combined with the depth-first search in the resulting search tree.

But this idea of a control is in many cases overly restrictive. As an extreme example consider a theorem prover written in Prolog. A proof rule

$$\frac{A_1, \ldots, A_n}{B}$$

naturally translates into a Prolog clause

$$\texttt{prove(B)} \leftarrow \texttt{prove(A}_1\texttt{)}, \ldots, \texttt{prove(A}_n\texttt{)}$$

according to which the premises $A_1, \ldots, A_n$ have to be proved in the abovementioned order. In contrast, in the underlying logic the order in which the premises are to be proved is usually arbitrary.

In general, however, some order between the actions of a logic program is necessary. As an example consider the `QUICKSORT` program:

```
qs(Xs, Ys) ← Ys is an ordered permutation of the list Xs.
qs([], []).
qs([X | Xs], Ys) ←
    part(X, Xs, Littles, Bigs),
    qs(Littles, Ls),
    qs(Bigs, Bs),
    app(Ls, [X | Bs], Ys).

part(X, Xs, Ls, Bs) ← Ls is a list of elements of Xs which are < X,
                      Bs is a list of elements of Xs which are ≥ X.
part(_, [], [], []).
part(X, [Y | Xs], [Y | Ls], Bs) ← X > Y, part(X, Xs, Ls, Bs).
part(X, [Y | Xs], Ls, [Y | Bs]) ← X ≤ Y, part(X, Xs, Ls, Bs).
```

augmented by the following APPEND program:

```
app(Xs, Ys, Zs) ← Zs is the result of concatenating the lists Xs and Ys.
app([], Ys, Ys).
app([X | Xs], Ys, [X | Zs]) ← app(Xs, Ys, Zs).
```

It is easy to see that starting with a query qs(s, Ys), where s is a list of integers, QUICKSORT diverges when the rightmost atom is repeatedly selected. Also, a run time error results if the arithmetic atoms X > Y and X ≤ Y are selected "too early". To prevent this type of undesired behaviour some coordination between the actions of the program is necessary.

For this purpose in Naish [Nai82] delay declarations were proposed. The idea is to replace the Prolog selection rule by a more flexible selection mechanism according to which atoms are delayed until they become "sufficiently" instantiated. This is achieved by adding to the program so-called delay declarations. Then at each stage of the execution of a logic program only atoms satisfying the delay declarations can be selected. In presence of trivial delay declarations any atom can be selected and a nondeterministic logic program without any control declaration is then obtained.

More recently, the delay declarations were studied in Lüttringhaus-Kappel [LK93] and also incorporated in various versions of Prolog, notably Sicstus Prolog, and in Gödel, the programming language proposed by Hill and Lloyd [HL94]. Actually, in all these references a more restricted selection rule is employed, according to which the leftmost non-delayed atom is selected. This selection rule allows us to model within the logic programming the coroutine mechanism and lazy evaluation. In contrast, the selection rules here studied. and originally considered in Naish [Nai88], allow us to model parallel executions.

Returning to the above QUICKSORT program we note that the coordination between the program actions has to do with the need to produce the appropriate "inputs" before executing the "calls" which use them. The requirement to compute these inputs can be expressed by means of the following delay declarations, where the last two declarations are meant to ensure that the arithmetic relations are called with the right inputs:

```
DELAY qs(X, _) UNTIL nonvar(X).
DELAY part(_, Y, _, _) UNTIL nonvar(Y).
DELAY app(X, _, _) UNTIL nonvar(X).
DELAY X > Y UNTIL ground(X) ∧ ground(Y).
DELAY X ≤ Y UNTIL ground(X) ∧ ground(Y).
```

The behaviour of the resulting program is highly non-trivial, since during its executions dynamic networks of asynchronously communicating processes are created.

Delay declarations form a powerful control mechanism. In general, we can identify three natural uses of them:

- to enforce termination,
- to prevent absence of errors in presence of arithmetic operations,
- to impose a synchronisation between various actions of the program; this makes it possible to model parallel executions.

In this paper we illustrate each of these uses of delay declarations and provide formal means of justifying them. More specifically, we study here various correctness aspects of logic programs in presence of delay declarations, visibly occur check freedom, absence of deadlock, absence of errors in presence of arithmetic operations, and termination. In each case we propose a simple method which can be readily applied to several well-known programs.

These results imply that for the query `qs(s, Ys)`, where `s` is a list of integers, `QUICKSORT` augmented by the above delay declarations is occur-check free and deadlock free. Moreover, no errors due to the presence of arithmetic operations arise and under some additional assumptions all derivations terminate.

Interestingly, the suggested proof methods turn out to be simple modifications of the corresponding methods dealing with Prolog programs. So the transition from Prolog to programs with delay declarations is quite natural, even though the latter ones permit more execution sequences and more complex "intermediate situations". This observation is further substantiated by showing how "correct" delay declarations can be derived by analyzing Prolog programs so that the given Prolog program can be executed in a more flexible way.

2 Preliminaries

In what follows we use the standard notation of Lloyd [Llo87] and Apt [Apt90], though we work here with *queries*, that is sequences of atoms, instead of *goals*, that is constructs of the form $\leftarrow Q$, where Q is a query. In particular, given a syntactic construct E (so for example, a term, an atom or a set of term equations) we denote by $Var(E)$ the set of the variables appearing in E. Recall that an mgu θ of a set of term equations E is called *relevant* if $Var(\theta) \subseteq Var(E)$.

The following lemma (see e.g. Apt and Pellegrini [AP94]) will be needed in Section 6.

Lemma 1 (Iteration). *Let E_1, E_2 be two sets of term equations. Suppose that θ_1 is a relevant mgu of E_1 and θ_2 is a relevant mgu of $E_2\theta_1$. Then $\theta_1\theta_2$ is a relevant mgu of $E_1 \cup E_2$. Moreover, if $E_1 \cup E_2$ is unifiable then a relevant mgu θ_1 of E_1 exists, and for any mgu θ_1 of E_1 a relevant mgu θ_2 of $E_2\theta_1$ exists, as well.* □

We now define the syntax of delay declarations. We loosely follow here Hill and Lloyd [HL94]. First, we define inductively a set of *conditions*:

- true is a condition,
- given a variable X, ground(X) and nonvar(X) are conditions,
- if c_1 and c_2 are conditions, then $c_1 \wedge c_2$ and $c_1 \vee c_2$ are conditions.

Next, we define inductively when an instance of a condition *holds*:

- true holds,
- ground(s) holds if s is a ground term,
- nonvar(s) holds if s is a non-variable term,
- $c_1 \wedge c_2$ holds if c_1 and c_2 hold,
- $c_1 \vee c_2$ holds if c_1 or c_2 holds.

Call an atom a *p-atom* if its relation symbol is *p*. A delay declaration associated for a relation symbol *p* has the form DELAY A UNTIL COND, where A is a *p*-atom and COND is a condition. From now on we consider logic programs augmented by the delay declarations, one for each of its relation symbols. In the presentation below we drop the delay declarations of the form DELAY A UNTIL true.

The following simple definition explains the use of delay declarations.

Definition 2.

- We say that an atom B *satisfies a delay declaration* DELAY A UNTIL COND if for some substitution θ both B = Aθ and CONDθ hold.
  In particular, if $X_1, \ldots, X_n$ are different variables, then $p(s_1, \ldots, s_n)$ satisfies a delay declaration DELAY $p(X_1, \ldots, X_n)$ UNTIL COND if COND$\{X_1/s_1, \ldots, X_n/s_n\}$ holds.
- An SLD-derivation *respects the delay declarations* if all atoms selected in it satisfy their delay declarations. □

Intuitively, in presence of delay declarations only atoms which satisfy their delay declarations can be selected. So in presence of delay declarations we allow only those selection rules which generate SLD-derivations respecting the delay declarations. Note that in presence of delay declarations a query can be generated in which no atom can be selected because none of them satisfies its delay declaration. We view such a fragment of an SLD-derivation as a finite SLD-derivation.

In what follows we shall study correctness of logic programs augmented with the delay declarations. To show the usefulness of the obtained results and to see their limitations, we shall analyse in this paper three example programs: **QUICKSORT** from the introduction, and the following two.

The program **IN_ORDER** constructs the list of all nodes of a binary tree by means of an in-order traversal:

```
in_order(Tree, List) ← List is a list obtained by the in-order
                       traversal of the tree Tree.

in_order(void, []).
in_order(tree(X, Left, Right), Xs) ←
  in_order(Left, Ls),
  in_order(Right, Rs),
  app(Ls, [X | Rs], Xs).
```

augmented by the **APPEND** program.

together with the following delay declarations:

```
DELAY in_order(X, _) UNTIL nonvar(X).
DELAY app(X, _, _) UNTIL nonvar(X).
```

Finally, the program **SEQUENCE** (see Coelho and Cotta [CC88, page 193]) solves the following problem: arrange three 1's, three 2's, ..., three 9's in sequence so that for all $i \in [1, 9]$ there are exactly i numbers between successive occurrences of i.

```
sublist(Xs, Ys) ← app(_, Zs, Ys), app(Xs, _, Zs).

sequence([_,_,_,_,_,_,_,_,_,_,_,_,_,_,_,_,_,_,_,_,_,_,_,_,_,_,_]).

question(Ss) ←
   sequence(Ss),
   sublist([1,_,1,_,1], Ss),
   sublist([2,_,_,2,_,_,2], Ss),
   sublist([3,_,_,_,3,_,_,_,3], Ss),
   sublist([4,_,_,_,_,4,_,_,_,_,4], Ss),
   sublist([5,_,_,_,_,_,5,_,_,_,_,_,5], Ss),
   sublist([6,_,_,_,_,_,_,6,_,_,_,_,_,_,6], Ss),
   sublist([7,_,_,_,_,_,_,_,7,_,_,_,_,_,_,_,7], Ss),
   sublist([8,_,_,_,_,_,_,_,_,8,_,_,_,_,_,_,_,_,8], Ss),
   sublist([9,_,_,_,_,_,_,_,_,_,9,_,_,_,_,_,_,_,_,_,9], Ss).
```

augmented by the **APPEND** program.

together with the following the delay declaration:

```
DELAY app(_, _, Z) UNTIL nonvar(Z).
```

3 Occur-check Freedom

In most Prolog implementations for the efficiency reasons so-called occur-check is omitted from the unification algorithm. It is well-known that this omission can lead to incorrect results. The resulting difficulties are usually called the *occur-check problem.* They have motivated a study of conditions under which the occur-check can be safely omitted. In this section we study this problem for logic programs augmented with delay declarations. To this end we recall the relevant definitions. We follow here Apt and Pellegrini [AP94] though we adapt its framework to arbitrary SLD-derivations. The following notion is due to Deransart, Ferrand and Téguia [DFT91].

Definition 3. A system of term equations E is called *not subject to occur-check* (NSTO in short) if no execution of the Martelli-Montanari unification algorithm (see Martelli and Montanari [MM82]) started with E ends with a system that includes an equation of the form $x = t$, where x is a variable and t a term different from x, but in which x occurs. □

We can now introduce the crucial notion of this section.

Definition 4.

- Consider an SLD-derivation ξ. Let A be an atom selected in ξ and H the head of the input clause selected to resolve A in ξ. Suppose that A and H have the same relation symbol. Then we say that the system $A = H$ *is considered in* ξ.
- An SLD-derivation is called *occur-check free* if all the systems of equations considered in it are NSTO. □

Recall that in presence of delay declarations selection of an atom implies that it satisfies its delay declaration.

In what follows we identify some syntactic conditions that allow us to draw conclusions about the occur-check freedom. To this end we use modes. Informally, modes indicate how the arguments of a relation should be used.

Definition 5. Consider an n-ary relation symbol p. By a *mode* for p we mean a function m_p from $\{1, \dots, n\}$ to the set $\{+, -\}$. If $m_p(i) =$ '+', we call i an *input position* of p and if $m_p(i) =$ '−', we call i an *output position* of p (both w.r.t. m_p). By a *moding* we mean a collection of modes, each for a different relation symbol. □

We write m_p in a more suggestive form $p(m_p(1), \dots, m_p(n))$. For example, `member(-,+)` denotes a binary relation `member` with the first position moded as output and the second position moded as input.

The definition of moding assumes one mode per relation in a program. Multiple modes may be obtained by simply renaming the relations. In this paper we adopt the following

Assumption *Every considered relation* has a fixed mode associated with it.

This assumption will allow us to talk about input positions and output positions of an atom.

Definition 6.

- A family of terms is called *linear* if every variable occurs at most once in it.
- An atom is called *input* (resp. *output*) *linear* if the family of terms occurring in its input (resp. output) positions is linear.
- An atom is called *input-output disjoint* if the family of terms occurring in its input positions has no variable in common with the family of terms occurring in its output positions. □

In the sequel we shall use the following lemma of Apt and Pellegrini [AP94].

Lemma 7. *Consider two atoms A and H with the same relation symbol. Suppose that*

- *they have no variable in common,*
- *one of them is input-output disjoint,*
- *one of them is input linear and the other is output linear.*

Then $A = H$ *is NSTO.* □

The first result is an immediate consequence of Lemma 7. The idea is that when the head of every clause of P is output linear, it suffices to delay all the atoms until their input positions become ground. To this end we introduce the following notion that relates the delay declarations to modes.

Definition 8. We say that the delay declarations *imply the moding* if every atom which satisfies its delay declaration is ground in its input positions. □

Theorem 9 (Occur-check Freedom 1). *Suppose that*

- *the head of every clause of P is output linear,*
- *the delay declarations imply the moding.*

Then all SLD-derivations of $P \cup \{Q\}$ *which respect the delay declarations are occur-check free.*

Proof. By Lemma 7. □

This result allows us to deal only with the delay "until ground" declarations. So for example, we cannot draw yet at this stage any conclusions concerning the `QUICKSORT` program from the introduction. To deal with other forms of delay declarations we use the following notion introduced in Chadha and Plaisted [CP94] and further studied in Apt and Pellegrini [AP94].

To simplify the notation, when writing an atom as $p(\mathbf{u}, \mathbf{v})$, we now assume that $\mathbf{u}$ is a sequence of terms filling in the input positions of p and that $\mathbf{v}$ is a sequence of terms filling in the output positions of p.

Definition 10.

- A query $p_1(\mathbf{s_1}, \mathbf{t_1}), \ldots, p_n(\mathbf{s_n}, \mathbf{t_n})$ is called *nicely moded* if $\mathbf{t_1}, \ldots, \mathbf{t_n}$ is a linear family of terms and for $i \in [1, n]$

$$Var(\mathbf{s_i}) \cap (\bigcup_{j=i}^{n} Var(\mathbf{t_j})) = \emptyset.$$

- A clause

$$p_0(\mathbf{s_0}, \mathbf{t_0}) \leftarrow p_1(\mathbf{s_1}, \mathbf{t_1}), \ldots, p_n(\mathbf{s_n}, \mathbf{t_n})$$

is called *nicely moded* if $\leftarrow p_1(\mathbf{s_1}, \mathbf{t_1}), \ldots, p_n(\mathbf{s_n}, \mathbf{t_n})$ is nicely moded and

$$Var(\mathbf{s_0}) \cap (\bigcup_{j=1}^{n} Var(\mathbf{t_j})) = \emptyset.$$

In particular, every unit clause is nicely-moded.
- A program is called *nicely moded* if every clause of it is. □

Intuitively, the concept of being nicely moded prevents a "speculative binding" of the variables which occur in output positions — these variables are required to be "fresh". Note that a query with only one atom is nicely moded iff it is output linear and input-output disjoint.

The following lemma is crucial. It shows persistence of the notion of nice modedness in presence of a natural assumption.

Lemma 11 (Nice modedness). *Every SLD-resolvent of a nicely moded query and a nicely moded clause with an input-linear head, that is variable-disjoint with it, is nicely moded.*

Proof. The proof is quite long and can be found in Luitjes [Lui94]. It is similar to the proof of an analogous lemma, 5.3, in Apt and Pellegrini [AP94, pages 719-724]. □

Corollary 12 (Nice modedness). *Let P and Q be nicely moded. Suppose that the head of every clause of P is input linear. Then all queries in all SLD-derivations of $P \cup \{Q\}$ are nicely moded.* □

This corollary brings us to the following conclusion.

Theorem 13 (Occur-check Freedom 2). *Let P and Q be nicely moded. Suppose that*

- *the head of every clause of P is input linear.*

Then all SLD-derivations of $P \cup \{Q\}$ are occur-check free.

Proof. By the Nice modedness Corollary 12 all queries in all SLD-derivations of $P \cup \{Q\}$ are nicely moded. But every atom of a nicely moded query is input-output disjoint and output linear. So the claim follows by Lemma 7. □

This result shows that for nicely moded programs and queries occur-check freedom can be ensured inpendendently of the selection rule, so without taking into account the delay declarations. In Chadha and Plaisted [CP94] and Apt and Pellegrini [AP94] it was shown that the **QUICKSORT** program with the moding qs(+,-), partition(+,+,-,-), app(+,+,-), +>+, +≤+ satisfies the condition of the above theorem, so this result applies to any query of the form qs(s, Ys), where s is a list of integers.

To apply the above result to the **IN_ORDER** program consider the following moding: in_order(+,-), app(+,+,-). It is straightforward to check that then **IN_ORDER** is nicely moded and the head of every clause of **IN_ORDER** is input linear.. So by the Occur-check Freedom 2 Theorem 13 we conclude that for any term t and a variable Ys that does not occur in t, all SLD-derivations of **IN_ORDER** ∪{in_order(t, Ys)} are occur-check free.

Finally, to deal with the **SEQUENCE** program take the following moding: sublist(-,+), sequence(+), question(+), app(-,-,+). Thanks to the use of anonymous variables it is easy to check that **SEQUENCE** is then indeed nicely moded and that the heads of all clauses are input linear. So by the Occur-check Freedom 2 Theorem 13 all SLD-derivations of **SEQUENCE** ∪{question(Ss)} are occur-check free.

4 Absence of Deadlock

In presence of delay declarations a query can be generated in which no atom can be selected, because none of them satisfies its delay declaration. Then the computation cannot proceed. Such a situation is obviously undesirable. We call it a deadlock and study here means to avoid it. Let us begin with a formal definition.

Definition 14. An SLD-derivation *flounders* if it contains a query no atom of which satisfies its delay declaration. We say that $P \cup \{Q\}$ *deadlocks* if an SLD-derivation of $P \cup \{Q\}$ which respects the delay declarations flounders. □

We now propose syntactic conditions which allow us to prove absence of deadlock. The main tool is the notion of a well-moded program and query. Let us recall the definition (see e.g. Dembinski and Maluszynski [DM85] and Drabent [Dra87]).

4.1 Well-moded Queries and Programs

Definition 15.

– A query $p_1(\mathbf{s_1}, \mathbf{t_1}), \ldots, p_n(\mathbf{s_n}, \mathbf{t_n})$ is called *well-moded* if for $i \in [1, n]$

$$Var(\mathbf{s_i}) \subseteq \bigcup_{j=1}^{i-1} Var(\mathbf{t_j}).$$

- A clause

$$p_0(\mathbf{t_0}, \mathbf{s_{n+1}}) \leftarrow p_1(\mathbf{s_1}, \mathbf{t_1}), \ldots, p_n(\mathbf{s_n}, \mathbf{t_n})$$

  is called *well-moded* if for $i \in [1, n+1]$

$$Var(\mathbf{s_i}) \subseteq \bigcup_{j=0}^{i-1} Var(\mathbf{t_j}).$$

- A program is called *well-moded* if every clause of it is. □

The following lemma shows the persistence of the notion of well-modedness. It strengthens the version given in Apt and Pellegrini [AP94] to arbitrary SLD-resolvents.

Lemma 16 (Well-modedness). *Every SLD-resolvent of a well-moded query and a well-moded clause, that is variable-disjoint with it, is well-moded.*

Proof. An SLD-resolvent of a query and a clause is obtained by means of the following three operations:

- instantiation of a query,
- instantiation of a clause,
- replacement of an atom, say H, of a query by the body of a clause whose head is H.

So we only need to prove the following two claims.

Claim 1 *An instance of a well-moded query (resp. clause) is well-moded.*

Proof. It suffices to note that for any sequences of terms $\mathbf{s}, \mathbf{t}_1, \ldots, \mathbf{t}_n$ and a substitution θ, $Var(\mathbf{s}) \subseteq \bigcup_{j=1}^{n} Var(\mathbf{t}_j)$ implies $Var(\mathbf{s}\theta) \subseteq \bigcup_{j=1}^{n} Var(\mathbf{t}_j\theta)$. □

Claim 2 *Suppose that* $\mathbf{A}, H, \mathbf{C}$ *is a well-moded query and* $H \leftarrow \mathbf{B}$ *is a well-moded clause. Then* $\mathbf{A}, \mathbf{B}, \mathbf{C}$ *is a well-moded query.*

Proof. Let

$\mathbf{A} := p_1(\mathbf{s}_1, \mathbf{t}_1), \ldots, p_k(\mathbf{s}_k, \mathbf{t}_k),$
$H := p(\mathbf{s}, \mathbf{t}),$
$\mathbf{B} := p_{k+1}(\mathbf{s}_{k+1}, \mathbf{t}_{k+1}), \ldots, p_{k+l}(\mathbf{s}_{k+l}, \mathbf{t}_{k+l}),$
$\mathbf{C} := p_{k+l+1}(\mathbf{s}_{k+l+1}, \mathbf{t}_{k+l+1}), \ldots, p_n(\mathbf{s}_{k+l+m}, \mathbf{t}_{k+l+m}).$

Fix now $i \in [1, k+l+m]$. We need to prove that $Var(\mathbf{s}_i) \subseteq \bigcup_{j=1}^{i-1} Var(\mathbf{t}_j)$.
Case 1 $i \in [1, k]$.
Note that $\mathbf{A}$ is well-moded, since $\mathbf{A}, H, \mathbf{C}$ is well-moded. Hence the claim follows.
Case 2 $i \in [k+1, k+l]$.
$H \leftarrow \mathbf{B}$ is well-moded, so $Var(\mathbf{s_i}) \subseteq Var(\mathbf{s}) \cup \bigcup_{j=k+1}^{i-1} Var(\mathbf{t}_j)$. Moreover, $\mathbf{A}, H, \mathbf{C}$ is well-moded, so $Var(\mathbf{s}) \subseteq \bigcup_{j=1}^{k} Var(\mathbf{t}_j)$. This implies the claim.

Case 3 $i \in [k+l+1, k+l+m]$.
$\mathbf{A}, H, \mathbf{C}$ is well-moded, so $Var(\mathbf{s}_i) \subseteq \bigcup_{j=1}^{k} Var(\mathbf{t}_j) \cup Var(\mathbf{t}) \cup \bigcup_{j=k+l+1}^{i-1} Var(\mathbf{t}_j)$ and $Var(\mathbf{s}) \subseteq \bigcup_{j=1}^{k} Var(\mathbf{t}_j)$. Moreover, $H \leftarrow \mathbf{B}$ is well-moded, so

$$Var(\mathbf{t}) \subseteq Var(\mathbf{s}) \cup \bigcup_{j=k+1}^{k+l} Var(\mathbf{t}_j).$$

This implies the claim. □

□

Corollary 17 (Well-modedness). *Let P and Q be well-moded. Then all queries in all SLD-derivations of* $P \cup \{Q\}$ *are well-moded.* □

The following definition provides a link between the delay declarations and moding.

Definition 18. We say that the delay declarations *are implied by the moding* if every atom which is ground in its input positions satisfies its delay declaration. □

We can now state and prove the desired result.

Theorem 19 (Absence of Deadlock 1). *Let P and Q be well-moded. Suppose that the delay declarations are implied by the moding. Then* $P \cup \{Q\}$ *does not deadlock.*

Proof. By the Well-modedness Corollary 17 all queries in all SLD-derivations of $P \cup \{Q\}$ are well-moded. But the first atom of a well-moded query is ground in its input positions. Hence, by the assumption, in every SLD-derivation of $P \cup \{Q\}$ the first atom of every query satisfies its delay declaration. Consequently, no SLD-derivation of $P \cup \{Q\}$ flounders. □

Intuitively, the above result states that under the abovementioned conditions, at every stage of a computation the first atom can always be selected.

The above result can be applied to the `QUICKSORT` program. Indeed, with the moding considered in Section 3 `QUICKSORT` is easily seen to be well-moded, the query `qs(s, Ys)`, where `s` is a list of integers, is well-moded and the delay declarations considered in the introduction are clearly implied by this moding. In fact, the same reasoning applies when the original delay declarations are strengthened by replacing everywhere "nonvar" by "ground".

The Absence of Deadlock 1 Theorem 19 can also be used for the `IN_ORDER` program. Indeed, in the moding considered in the previous section `IN_ORDER` is clearly well-moded. So the above theorem is applicable to any query of the form `in_order(t, Ys)`, where `t` is a ground term. However, no conclusion can be drawn if `t` is not ground. Below we shall see how to draw such stronger conclusions.

Finally, the above theorem cannot be applied to the program SEQUENCE. Indeed, it is easy to see that no moding exists for which both SEQUENCE and the query question(Ss) are well-moded. So the above theorem cannot be applied to the SEQUENCE program *no matter* what delay declarations are used.

4.2 Well-typed Queries and Programs

To overcome these difficulties we generalise the above approach by using the notion of a type. The presentation below of well-typed queries and programs is taken from Apt and Etalle [AE93]. We begin by adopting the following general definition.

Definition 20. A *type* is an non-empty set of terms closed under substitution. □

We now fix a specific set of types, denoted by *Types*, which includes:

U — the set of all terms,
List — the set of lists,
Gae — the set of of all ground arithmetic expressions (gae's, in short),
ListGae — the set of lists of gae's.
Tree — the set of binary trees, defined inductively as follows: void is a tree and if s, t are trees, then for any term u, tree(u,s,t) is a tree.

Of course, the use of the type *List* assumes the existence of the empty list [] and the list constructor [.|.] in the language, etc.

We call a construct of the form $s : S$, where s is a term and S is a type, a *typed term*. Given a sequence $\mathbf{s} : \mathbf{S} = s_1 : S_1, \ldots, s_n : S_n$ of typed terms, we write $\mathbf{s} \in \mathbf{S}$ if for $i \in [1, n]$ we have $s_i \in S_i$. Further, we abbreviate the sequence $s_1\theta, \ldots, s_n\theta$ to $\mathbf{s}\theta$. Finally, we write

$$\models \mathbf{s} : \mathbf{S} \Rightarrow \mathbf{t} : \mathbf{T},$$

if for all substitutions θ, $\mathbf{s}\theta \in \mathbf{S}$ implies $\mathbf{t}\theta \in \mathbf{T}$.

Next, we define types for relations.

Definition 21. By a *type* for an n-ary relation symbol$p$ we mean a function t_p from $[1, n]$ to the set *Types*. If $t_p(i) = T$, we call T *the type associated with the position i of p*. □

In the remainder of this paper we consider a combination of modes and types and adopt the following

Assumption *Every considered relation* has a fixed mode and a fixed type associated with it.

This assumption will allow us to talk about types of input positions and of output positions of an atom. An n-ary relation p with a mode m_p and type t_p will be denoted by

$$p(m_p(1) : t_p(1), \ldots, m_p(n) : t_p(n)).$$

For example, **member**$(- : U, + : List)$ denotes a binary relation **member** with the first position moded as output and typed as U, and the second position moded as input and typed as *List*.

To simplify the notation, when writing an atom as $p(\mathbf{u} : \mathbf{S}, \mathbf{v} : \mathbf{T})$ we now assume that $\mathbf{u} : \mathbf{S}$ is a sequence of typed terms filling in the input positions of p and $\mathbf{v} : \mathbf{T}$ is a sequence of typed terms filling in the output positions of p. We call a construct of the form $p(\mathbf{u} : \mathbf{S}, \mathbf{v} : \mathbf{T})$ a *typed atom* and a sequence of typed atoms a *typed query*. We say that a typed atom $p(s_1 : S_1, \ldots, s_n : S_n)$ is *correctly typed in position i* if $s_i \in S_i$ and use an analogous terminology for typed queries.

The following notion is due to Bronsard, Lakshman and Reddy [BLR92].

Definition 22.

- A query
$$p_1(\mathbf{i_1} : \mathbf{I_1}, \mathbf{o_1} : \mathbf{O_1}), \ldots, p_n(\mathbf{i_n} : \mathbf{I_n}, \mathbf{o_n} : \mathbf{O_n})$$
is called *well-typed* if for $j \in [1, n]$
$$\models \mathbf{o_1} : \mathbf{O_1}, \ldots, \mathbf{o_{j-1}} : \mathbf{O_{j-1}} \Rightarrow \mathbf{i_j} : \mathbf{I_j}.$$
- A clause
$$p_0(\mathbf{o_0} : \mathbf{O_0}, \mathbf{i_{n+1}} : \mathbf{I_{n+1}}) \leftarrow p_1(\mathbf{i_1} : \mathbf{I_1}, \mathbf{o_1} : \mathbf{O_1}), \ldots, p_n(\mathbf{i_n} : \mathbf{I_n}, \mathbf{o_n} : \mathbf{O_n})$$
is called *well-typed* if for $j \in [1, n+1]$
$$\models \mathbf{o_0} : \mathbf{O_0}, \ldots, \mathbf{o_{j-1}} : \mathbf{O_{j-1}} \Rightarrow \mathbf{i_j} : \mathbf{I_j}.$$
- A program is called *well-typed* if every clause of it is. □

Note that a query with only one atom is well-typed iff this atom is correctly typed in its input positions. The following lemma shows persistence of the notion of well-typedness. It strenghtens a result mentioned in Bronsard, Lakshman and Reddy [BLR92] to arbitrary SLD-resolvents.

Lemma 23 (Well-typedness). *Every SLD-resolvent of a well-typed query and a well-typed clause, that is variable-disjoint with it, is well-typed.*

Proof. The proof is analogous to that of the Well-modedness Lemma 16 and is omitted. □

Corollary 24 (Well-typedness). *Let P and Q be well-typed. Then all queries in all SLD-derivations of $P \cup \{Q\}$ are well-typed.* □

Finally, we link the delay declarations with types.

Definition 25. We say that the delay declarations *are implied by the typing* if every atom which is correctly typed in its input positions satisfies its delay declaration. □

We can now state and prove the desired result.

Theorem 26 (Absence of Deadlock 2). *Let P and Q be well-typed. Suppose that the delay declarations are implied by the typing. Then $P \cup \{Q\}$ does not deadlock.*

Proof. By the Well-typedness Corollary 24 all queries in all SLD-derivations of $P \cup \{Q\}$ are well-typed. But the first atom of a well-typed query is correctly typed in its input positions. Hence, by the assumption, in every SLD-derivation of $P \cup \{Q\}$ the first atom of every query satisfies its delay declaration. Consequently, no SLD-derivation of $P \cup \{Q\}$ flounders. □

Let us see how to apply this theorem to the `IN_ORDER` program. Consider the following typing:
`in_order`$(+ : Tree, - : List)$,
`app`$(+ : List, + : List, + : List)$.

We leave to the reader the task of checking that `IN_ORDER` is then well-typed and that the delay declarations are implied by the typing. We conclude that for any term `t` which is a tree, `IN_ORDER` $\cup${`in_order(t, Ys)`} does not deadlock, which strengthens the conclusion drawn by means of the Absence of Deadlock 1 Theorem 19.

Finally, note that this theorem can also be applied to the `SEQUENCE` program. Indeed, consider the following typing:
`question`$(- : List)$,
`sequence`$(- : List)$,
`sublist`$(+ : List, + : List)$,
`app`$(- : List, - : List, + : List)$.

Again, it is easy to see that `SEQUENCE` and the query `question(Ss)` are then well-typed and that the delay declaration is implied by the typing. It is worthwhile to note that if we change in this declaration "nonvar" to "ground", then `SEQUENCE` $\cup${`question(Ss)`} does deadlock.

5 Absence of Errors

One of the natural uses of the delay declarations is to prevent run time errors in presence of arithmetic relations. This is for example the idea behind the delay declarations

```
DELAY X > Y UNTIL ground(X) ∧ ground(Y).
DELAY X ≤ Y UNTIL ground(X) ∧ ground(Y).
```

used in the introduction which ensure that both relations are called only with ground arguments. Now, to prove absence of errors a stronger property is needed, namely that the arguments of these relations are ground arithmetic expressions. However, the syntax of the delay declarations does not allow us to express this stronger information.

The aim of this section is to provide means to deduce this stronger property and thus to prove absence of errors in presence of arithmetic relations. To this end we shall use the notions of a well-typed query and a well-typed program introduced in the previous section.

The following simple observation provides us with some means to prove that if an atom is ground in its input positions, then it is correctly typed in its input positions.

Lemma 27. *Suppose that* $\mathbf{A}, B, \mathbf{C}$ *is a well-typed query such that*

- B *is ground in its input positions,*
- *for some substitution* θ, $\mathbf{A}\theta$ *is correctly typed in its output positions.*

Then B *is correctly typed in its input positions.*

Proof. Let $\mathbf{A} = p_1(\mathbf{i_1} : \mathbf{I_1}, \mathbf{o_1} : \mathbf{O_1}), \ldots, p_n(\mathbf{i_n} : \mathbf{I_n}, \mathbf{o_n} : \mathbf{O_n})$ and $B = p_{n+1}(\mathbf{i_{n+1}} : \mathbf{I_{n+1}}, \mathbf{o_{n+1}} : \mathbf{O_{n+1}})$. By the definition of well-typedness

$$\models \mathbf{o_1} : \mathbf{O_1}, \ldots, \mathbf{o_n} : \mathbf{O_n} \Rightarrow \mathbf{i_{n+1}} : \mathbf{I_{n+1}}.$$

But by the assumption $\mathbf{o_i}\theta \in \mathbf{O_i}$ for $i \in [1, n]$, so $\mathbf{i_{n+1}}\theta \in \mathbf{I_{n+1}}$ which implies the claim since by assumption $\mathbf{i_{n+1}}\theta = \mathbf{i_{n+1}}$. □

We need to apply this lemma at every stage of the SLD-derivations. To prove the second assumption we shall use the following notion introduced in Apt and Etalle [AE93].

Definition 28.

- A query $p_1(\mathbf{s_1}, \mathbf{t_1}), \ldots, p_n(\mathbf{s_n}, \mathbf{t_n})$ is called *simply moded* if $\mathbf{t_1}, \ldots, \mathbf{t_n}$ is a linear family of variables and for $i \in [1, n]$

$$Var(\mathbf{s_i}) \cap (\bigcup_{j=i}^{n} Var(\mathbf{t_j})) = \emptyset.$$

- A clause

$$p_0(\mathbf{s_0}, \mathbf{t_0}) \leftarrow p_1(\mathbf{s_1}, \mathbf{t_1}), \ldots, p_n(\mathbf{s_n}, \mathbf{t_n})$$

  is called *simply moded* if $p_1(\mathbf{s_1}, \mathbf{t_1}), \ldots, p_n(\mathbf{s_n}, \mathbf{t_n})$ is simply moded and

$$Var(\mathbf{s_0}) \cap (\bigcup_{j=1}^{n} Var(\mathbf{t_j})) = \emptyset.$$

  In particular, every unit clause is simply moded.
- A program is called *simply moded* if every clause of it is. □

So simple modedness is a special case of nice modedness. The sole difference lies in the new assumption that each output position of a query or of a body of a clause is filled by a variable. The following lemma clarifies our interest in the notion of simple modedness.

Lemma 29. *Let* $\mathbf{A}$ *be a typed query which is simply moded. Then for some substitution* θ*,* $\mathbf{A}\theta$ *is correctly typed in its output positions.*

Proof. By the definition the output positions of $\mathbf{A}$ are filled by different variables. Choose for each variable occurring in an output position of $\mathbf{A}$ a term from the type associated with this position. So obtained bindings form the desidered substitution. □

The next step is to prove persistence of the notion of simple modedness. In Apt and Etalle [AE93] this property is established for the SLD-derivations w.r.t. the leftmost selection rule. But the generalization to arbitrary SLD-derivations does not work, as the following example shows. Consider the moding $p(-), r(+)$ and let $P = \{p(a) \leftarrow\}$ and $Q = p(x), r(x)$. Then $p(a)$ is an SLD-resolvent of Q and $p(a)$ is not simply moded.

However, a simple additional condition does ensure persistence. Namely, we have the following lemma.

Lemma 30 (Simple modedness). *Every SLD-resolvent of a simply moded query and a simply moded clause, that is variable-disjoint with it, is simply moded, when the input part of selected atom is an instance of the input part of the head of the clause.*

Proof. Omitted. □

Corollary 31 (Simple modedness). *Let* P *and* Q *be simply moded. Consider an SLD-derivation* ξ *of* $P \cup \{Q\}$ *such that the input part of each selected atom is an instance of the input part of the head of the used clause. Then all queries in* ξ *are simply moded.* □

To use this result we now link the delay declarations with matching.

Definition 32. We say that the delay declarations *imply matching* if for every atom $p(\mathbf{u}, \mathbf{v})$ which satisfies its delay declaration and for every head $p(\mathbf{u}', \mathbf{v}')$ of a clause if $p(\mathbf{u}, \mathbf{v})$ and $p(\mathbf{u}', \mathbf{v}')$ unify, then $\mathbf{u}$ is an instance of $\mathbf{u}'$. □

In particular, if the delay declarations imply the moding, then they imply matching, but not conversely. This brings us to the following conclusion.

Theorem 33 (Correct Typing). *Suppose that*

- *P and Q are well-typed and simply moded,*
- *the delay declarations imply matching.*

Then in all SLD-derivations of $P \cup \{Q\}$ *which respect the delay declarations every selected atom which is ground in its input positions is correctly typed in its input positions.*

Proof. It is a direct consequence of the Lemmata 27, 29, the Well-typedness Corollary 24 and the Simple modedness Corollary 31. □

This result can be used to prove absence of errors in presence of arithmetic relations by using for each arithmetic relation p a delay declaration

```
DELAY p(X, Y) UNTIL ground(X) ∧ ground(Y)
```

and a typing $p(+ : Gae, + : Gae)$. In the case of QUICKSORT take the typing
qs$(+ : ListGae, - : ListGae)$,
part$(+ : Gae, + : ListGae, - : ListGae, - : ListGae)$,
app$(+ : ListGae, + : ListGae, - : ListGae)$
$> (+ : Gae, + : Gae)$,
$\leq (+ : Gae, + : Gae)$,
Then QUICKSORT is well-typed (see Apt [Apt93]). Also, it is clearly simply moded. Moreover, the delay declarations from the introduction imply matching. Indeed, if a non-variable term unifies with [X | Xs], then it is an instance of [X | Xs].

We conclude that for s a list of integers, all SLD-derivations of QUICKSORT $\cup${qs(s, X)} which respect the delay declarations from the introduction do not end in an error.

6 Termination

Finally, we study termination of logic programs with delay declarations. The key idea in our approach is the restriction to a specific class of SLD-derivations.

6.1 Termination via Determinacy

Definition 34. We say that an SLD-derivation is *determinate* if every selected atom unifies with a variant of at most one clause head, that is, every selected atom can be resolved using at most one clause. □

The following simple observation, which is of independent interest, forms the basis of our approach.

Lemma 35. *Suppose that*

– *an SLD-derivation of $P \cup \{Q\}$ is successful.*

Then all determinate SLD-derivations of $P \cup \{Q\}$ are successful, hence finite.

Proof. Consider a determinate SLD-derivation ξ of $P \cup \{Q\}$. ξ is a branch in an SLD-tree $\mathcal{T}$ for $P \cup \{Q\}$. By the strong completeness of the SLD-resolution $\mathcal{T}$ is successful and since ξ is determinate, $\mathcal{T}$ has just one branch, namely ξ. So ξ is successful, and a fortiori finite. □

To use this result, we link the delay declarations with the notion of determinacy.

Definition 36. We say that the delay declarations *imply determinacy* if every atom which satisfies its delay declaration unifies with a variant of at most one clause head. □

This brings us to the following result.

Theorem 37 (Termination 1). *Suppose that*

- *an SLD-derivation of $P \cup \{Q\}$ is successful,*
- *the delay declarations imply determinacy.*

Then all SLD-derivations of $P \cup \{Q\}$ which respect the delay declarations are successful, hence finite.

Proof. It is an immediate consequence of Lemma 35, because if the delay declarations imply determinacy, then every SLD-derivation which respects these delay declarations is determinate. □

Let us see now how to apply this result to the `IN_ORDER` program. Using the approach of Apt [Apt93] it is straightforward to prove that for a tree `t`, `IN_ORDER` $\cup$`{in_order(t, Ys)}` satisfies the first condition of the above theorem. Also, the assumed delay declarations clearly imply determinacy. We conclude that all SLD-derivations of `IN_ORDER` $\cup$`{in_order(t, Ys)}` which respect the delay declarations are finite.

However, the above theorem cannot be directly applied to the `QUICKSORT` program because the delay declarations from the introduction do not imply determinacy in the case of the `part` relation. On the other hand, it is possible to adjust these delay declarations and slightly modify the execution of the program so that for the query `qs(s, Ys)`, where `s` is a list of integers, termination can be established.

Namely, consider the following alternative *set* of delay declarations for the `part` relation, given in Naish [Nai88]:

```
DELAY part(_, [], _, _) UNTIL true.
DELAY part(X, [Y | _], _, _) UNTIL ground(X) ∧ ground(Y).
```

We now say that an atom *satisfies a set of delay declarations* if it satisfies at least one of them. The idea behind these new delay declarations is to enforce that the arguments of the arithmetic atoms `X` $>$ `Y` and `X` $\leq$ `Y` are ground once they are introduced through the selection of a part-atom. Note that now the delay declarations for the arithmetic relations become superfluous in the sense that they are always satisfied.

Also, observe that for this new set of delay declarations absence of deadlock is now ensured by the Absence of Deadlock 2 Theorem 26 and not anymore by the Absence of Deadlock 1 Theorem 19. Indeed, these new delay declarations are clearly implied by the typing given in Section 5 but are not any longer implied by the moding given in Section 3, as not all ground terms are of the form `[y|ys]`.

Further, note that this new set of delay declarations still implies matching. So the proof of the absence of errors in all SLD-derivations of `QUICKSORT` $\cup$`{qs(s, X)}` respecting the delay declarations and given in the previous section remains valid.

However, the determinacy is still not ensured. To deal with this problem we now modify the execution of the programs by viewing the arithmetic atoms as guards in the sense of e.g. Shapiro [Sha89].

6.2 Termination for Guarded Programs

By treating atoms as guards we mean the following generalization of the SLD-resolution, which we call the *SLDG-resolution*. By a *guarded clause* we refer to a construct of the form $H \leftarrow \mathbf{G} \mid \mathbf{B}$, where H is an atom and $\mathbf{G}$ and $\mathbf{B}$ are sequences of atoms. A *guarded program* is a set of guarded clauses. If $\mathbf{G}$ is empty, then we drop the vertical bar "|". The atoms in $\mathbf{G}$ are called *guards*. Note that we do not insist that a guarded program is a finite set of clauses. The reason is that we wish to have the possibility of defining the relations used in the guards by a possibly infinite set of ground facts. It is clear how to extend the notions of well-modedness etc. to guarded programs.

We now view `QUICKSORT` as a guarded program by rewriting the last two clauses defining the part relation as follows:

```
part(X, [Y | Xs], [Y | Ls], Bs) ← X > Y | part(X, Xs, Ls, Bs).
part(X, [Y | Xs], Ls, [Y | Bs]) ← X ≤ Y | part(X, Xs, Ls, Bs).
```

We call the resulting program `QUICKSORT-G`. We assume that both in `QUICKSORT` and in `QUICKSORT-G` the arithmetic relations are defined by the infinite set of ground facts.

Consider now a query $\mathbf{A}, B, \mathbf{C}$ and a variable disjoint with it guarded clause $H \leftarrow \mathbf{G} \mid \mathbf{B}$. We say that B *guardedly unifies* with H if for some mgu θ of B and H the query $\mathbf{G}\theta$ is ground and succeeds. We call then $(\mathbf{A}, \mathbf{B}, \mathbf{C})\theta$ an *SLDG-resolvent* of $\mathbf{A}, B, \mathbf{C}$ and $H \leftarrow \mathbf{G} \mid \mathbf{B}$. So, intuitively, the test that the guards are ground and succeed forms now a part of the unification process. If $\mathbf{G}$ is empty, then the SLDG-resolvents coincide with the SLD-resolvents. So SLDG-resolution is indeed a generalization of the SLD-resolution. It is now clear how to define the SLDG-derivations and the SDLG-trees.

Given a guarded program, we now say that the delay declarations *imply determinacy* if every atom which satisfies its set of delay declarations guardedly unifies with a variant of at most one guarded clause head.

Note that the new set of delay declarations implies determinacy for the guarded program `QUICKSORT-G` because in the case of the part relation for any two ground terms `s`,`t` at most one of the queries `s` $>$ `t` and `s` $\leq$ `t` succeeds. The remaining delay declarations imply determinacy because if a non-variable term unifies with `[X | Xs]`, then it is does not unify with `[ ]`.

Definition 38. We say that a guarded program is *regular* if

– every relation used in the guards is defined by a set of ground facts,
– for every other relation symbol p either
  • in all clauses defining p all guards are empty, or
  • for some sequence of terms $\mathbf{s}$ and a set of variables $V \subseteq Var(\mathbf{s})$, every guarded clause defining p is of the form

$$p(\mathbf{s}, \mathbf{t}) \leftarrow \mathbf{G} \mid \mathbf{B}$$

for some $\mathbf{t}$, $\mathbf{G}$, $\mathbf{B}$, such that $Var(\mathbf{G}) = V$. □

Note that `QUICKSORT-G` is regular. In fact, we observed that most of the programs that use arithmetic comparison relations are regular when viewed as guarded programs.

The following theorem explains our approach to the proofs of termination for guarded programs. It is a modification of the Termination 1 Theorem 37.

Theorem 39 (Termination 2). *Suppose that*

– *P is a regular guarded program,*
– *P and Q are simply moded,*
– *the delay declarations imply determinacy and matching,*
– *an SLDG-derivation of $P \cup \{Q\}$ is successful.*

Then all SLDG-derivations of $P \cup \{Q\}$ which respect the delay declarations are successful, hence finite.

Proof. Omitted. □

Let us return now to `QUICKSORT-G`. In the previous sections we already checked before that `QUICKSORT-G` with the query `qs(s, Ys)`, where `s` is a list of integers, satisfies the first three condtions of the above theorem. Now, using the approach of Apt [Apt93] it is easy to show that `QUICKSORT-G` ∪{`qs(s, Ys)`} satisfies the last condition. We conclude by the Termination 2 Theorem 39 that all SLDG-derivations of `QUICKSORT` ∪{`qs(s, Ys)`} which respect the modified delay declarations are finite.

It is worthwhile to point out that the SLDG-resolution is very meaningful from the operational point of view. In the case of `QUICKSORT-G` it prevents a choice of a "wrong" alternative in the definition of the `part` relation and consequently, obviates a backtracking.

Finally, note that we cannot apply either the Termination 1 Theorem 37 or the Termination 2 Theorem 39 to the `SEQUENCE` program, because the delay declaration

```
DELAY app(_, _, Z) UNTIL nonvar(Z)
```

does not imply determinacy. Currently we are working on techniques allowing us to deal with termination in absence of determinacy.

The above two theorems are applicable only to the queries which have successful SLD-derivations. At this moment we know how to deal with termination for arbitrary queries at the cost of restricting attention to fair SLD-derivations. Recall that an SLD-derivation is *fair* if it is finite or if every atom occurring in it is eventually selected (after some possible instantiations).

Below, by an *LD-derivation* we mean an SLD-derivation via the leftmost selection rule. So LD-derivations are SLD-derivations generated by the Prolog selection rule. In the literature a lot of attention has been devoted to the study of termination with respect to the Prolog selection rule (see for example the survey article of De Schreye and Decorte [SD94]). This work can be use to deal with termination in presence of delay declarations. Namely, we have the following result.

Theorem 40 (Termination 3). *Suppose that*

– *all LD-derivations of $P \cup \{Q\}$ are finite.*

Then all fair SLD-derivations of $P \cup \{Q\}$ are finite.

Proof. The proof uses a generalization of the Switching Lemma (see Lloyd[Llo87, pages 50-51] to infinite SLD-derivations. Using it one can prove that if an infinite fair SLD-derivation exists, then an infinite LD-derivation exists. We omit the details. □

6.3 Termination of APPEND

It is important to realize that in general termination in presence of delay declarations is very subtle. As an example of the difficulties consider the APPEND program augmented with the delay declaration

```
DELAY app(X, _, _) UNTIL nonvar(X).
```

It seems that APPEND then terminates for all queries, that is, for all queries Q, all SLD-derivations of APPEND $\cup\{Q\}$ which respect this delay declaration are finite. The informal argument goes as follows. When the second clause of APPEND can be used to resolve a query of the form app(s, t, u), then s is of the form [x|xs], so in the next resolvent the first argument of app is xs, thus shorter. So eventually, the first argument is either a constant, in which case the second clause cannot be used, or else a variable and in this case the SLD-derivation terminates due to the delay declaration.

However, this reasoning is incorrect. Namely, consider the query app([X | U], Ys, U). Using the second clause it resolves to itself and consequently an infinite SLD-derivation for app([X | U], Ys, U) can be generated with the first argument of app always being non-variable. (This fact was noticed in Naish [Nai92] and Lüttringhaus-Kappel [LK93]).

Thus the query app([X | U], Ys, U) does not terminate in presence of the delay declaration DELAY app(X, _, _) UNTIL nonvar(X). Similarly, the query app(U, Ys, [X | U]) does not terminate in presence of the delay declaration DELAY app(_, _, Z) UNTIL nonvar(Z).

It seems that the problem has to do with the fact that the first and the third arguments of app share a variable. However, limiting one's attention to the queries of the form app(s, t, u) where s,t,u are pairwise variable disjoint does not help, as the query app([(U,X1,X1)|U], Ys, [([X2|V1],W,V1)|W]) resolves to the above query app([X | U], Ys, U).

What does hold is the following limited property.

Theorem 41. *Suppose that* s,t,u *are terms such that* s *and* u *are variable disjoint.*

(i) If u *is linear, then all SLD-derivations of* APPEND ∪ {app(s, t, u)} *that respect the delay declaration* DELAY app(X, _, _) UNTIL nonvar(X) *terminate.*

(ii) If t *is linear, then all SLD-derivations of* APPEND ∪ {app(s, t, u)} *that respect the delay declaration* DELAY app(_, _, Z) UNTIL nonvar(Z) *terminate.*

(iii) If s *and* t *are linear, then all SLD-derivations of* APPEND ∪ {app(s, t, u)} *that respect the delay declaration*

DELAY app(X, _, Z) UNTIL nonvar(X) ∨ nonvar(Z)

terminate.

Proof. (i) Let l(v) denote the number of symbols of the term v. Suppose that app(s, t, u) satisfies the delay declaration and resolves to app(s', t', u') using the second clause of APPEND. We prove that then s' and u' are variable disjoint, u' is linear and the pair (l(u'), l(s')) is smaller than (l(u), l(s)) in the lexicographic ordering.

s is not a variable, so it is of the form [x|xs], and u is not a constant. Two cases arise where we assume that app(s, t, u) and app([X | Xs], Ys, [X | Zs]) are variable disjoint.

Case 1 u is a variable.
Then {X/x, Xs/xs, Ys/t, u/[x|Zs]} is an mgu of app(s, t, u) and app([X | Xs], Ys, [X | Zs]), so app(s, t, u) resolves to app(xs, t, Zs). Now xs and Zs are variable disjoint, Zs is linear and the pair (l(Zs), l(xs)) is smaller than (l(u), l([x|xs])) in the lexicographic ordering.

Case 2 u is a compound term.
Then u is of the form [z|zs]. By the Iteration Lemma 1 the substitution {X/x, Xs/xs, Ys/t, Zs/zs}θ, where θ is a relevant mgu of x and z, is an mgu of app(s, t, u) and app([X | Xs], Ys, [X | Zs]). By assumption x is variable disjoint with zs. Moreover, [z|zs] is linear, so z is variable disjoint with zs, as well. Thus by the relevance of θ, zsθ = zs, so app(s, t, u) resolves to app(xsθ, tθ, zs). Now, zs is linear, so xsθ and zs are variable disjoint, because *Var*(xsθ) ⊆ *Var*(s) ∪ *Var*(z). Moreover, the pair (l(zs), l(xsθ)) is smaller than (l([z|zs]), l(s)) in the lexicographic ordering.

This proves the claim.

(ii) By symmetry with (i).

(iii) Suppose that an infinite SLD-derivation ξ of APPEND $\cup$\{app(s, t, u)\} exists that respects the delay declaration

DELAY app(X, _, Z) UNTIL nonvar(X) $\vee$ nonvar(Z).

Then either s or u is not a variable. Assume without loss of generality that s is not a variable. By (i) a descendant of app(s, t, u) in ξ exists which is of the form app(Xs, t', u') for a variable Xs and some terms t' and u'. Since ξ is infinite, u' is not a variable.

In the moding app(+,+,-) APPEND and the query app(s, t, u) are nicely moded and the head of every clause of APPEND is input linear. Thus by the Nice modedness Corollary 12 u' is linear and Xs and u' are variable disjoint.

By (ii) a descendant of app(Xs, t', u') in ξ exists which is of the form app(s'', t'', Zs) for some terms s'' and t'' and a variable Zs. Moreover, in every descendant of app(Xs, t', u') in ξ the first argument of app remains a variable and consequently app(s'', t'', Zs) does not satisfy the delay declaration DELAY app(X, _, Z) UNTIL nonvar(X) $\vee$ nonvar(Z). A contradiction. □

This isolated result shows how elaborate arguments are sometimes needed to prove simple termination results in presence of delay declarations. In this connection let us mention a related work of Naish [Nai86] and Lüttringhaus-Kappel [LK93] who automatically generate delay declarations which ensure that a given program terminates with respect to a selection rule according to which the leftmost non-delayed atom is selected.

7 Conclusions

In this paper we dealt with the correctness of logic programs augmented by the delay declarations. To this end we strengthened and adapted methods that were originally developed for the study of Prolog programs.

Interestingly, we can reverse the situation and derive the appropriate delay declarations by analysing the Prolog programs. Consider for example the QUICKSORT program with the query qs(s, Ys), where s is a list of integers. Once we have shown in Section 4 that it is well-moded with the moding qs(+,-), partition(+,+,-,-), app(+,+,-), +>+, +$\leq$+ we can augment it by arbitrary delay declarations which are implied by this moding and conclude by virtue of the Absence of Deadlock 1 Theorem 19 that for the query in question no deadlock arises.

Once we have shown in Section 4 that it is well-typed and simply moded with the typing

qs$(+ : \mathit{ListGae}, - : \mathit{ListGae})$,

part$(+ : \mathit{Gae}, + : \mathit{ListGae}, - : \mathit{ListGae}, - : \mathit{ListGae})$,

app$(+ : \mathit{ListGae}, + : \mathit{ListGae}, - : \mathit{ListGae})$

$> (+ : \mathit{Gae}, + : \mathit{Gae})$,

$\leq (+ : \mathit{Gae}, + : \mathit{Gae})$,

we can augment it by arbitrary delay declarations which imply matching and conclude by virtue of the Correct Typing Theorem 33 that for the query in question no run time errors due to the use of arithmetic relations arise.

Finally, once we have shown in Section 6 that some SLDG-derivation of `QUICKSORT-G` ∪{`qs(s, Ys)`} is successful, we can augment `QUICKSORT-G` by arbitrary delay declarations which imply determinacy and matching, and conclude by virtue of the Termination 2 Theorem 39 that all SLDG-derivations of `qs(s, Ys)` which respect these delay declarations are finite. This shows that it is possible to derive correct parallel logic programs by analysing Prolog programs.

We conclude by noticing that the "stronger" the delay declarations are the bigger the chance that a deadlock arises, but the smaller the chance that divergence can result. So deadlock freedom and termination seem to form two boundaries within which lie the "correct" delay declarations.

Acknowledgements

The first author would like to thank Elena Marchiori and Frank Teusink for helpful discussions on the subject of the delay declarations.

References

[AE93] K. R. Apt and S. Etalle. On the unification free Prolog programs. In A. Borzyszkowski and S. Sokolowski, editors, *Proceedings of the Conference on Mathematical Foundations of Computer Science (MFCS 93)*, Lecture Notes in Computer Science, pages 1–19, Berlin, 1993. Springer-Verlag.

[AP94] K. R. Apt and A. Pellegrini. On the occur-check free Prolog programs. *ACM Toplas*, 16(3):687–726, 1994.

[Apt90] K. R. Apt. Logic programming. In J. van Leeuwen, editor, *Handbook of Theoretical Computer Science*, pages 493–574. Elsevier, 1990. Vol. B.

[Apt93] K. R. Apt. Declarative programming in Prolog. In D. Miller, editor, *Proc. International Symposium on Logic Programming*, pages 11–35. MIT Press, 1993.

[BLR92] F. Bronsard, T.K. Lakshman, and U.S. Reddy. A framework of directionality for proving termination of logic programs. In K. R. Apt, editor, *Proceedings of the Joint International Conference and Symposium on Logic Programming*, pages 321–335. MIT Press, 1992.

[CC88] H. Coelho and J. C. Cotta. *Prolog by Example*. Springer-Verlag, Berlin, 1988.

[CP94] R. Chadha and D. A. Plaisted. Correctness of unification without occur check in Prolog. *Journal of Logic Programming*, 18(2):99–122, 1994.

[DFT91] P. Deransart, G. Ferrand, and M. Téguia. NSTO programs (not subject to occur-check). In V. Saraswat and K. Ueda, editors, *Proceedings of the International Logic Symposium*, pages 533–547. The MIT Press, 1991.

[DM85] P. Dembinski and J. Maluszynski. AND-parallelism with intelligent backtracking for annotated logic programs. In *Proceedings of the International Symposium on Logic Programming*, pages 29–38, Boston, 1985.

[Dra87] W. Drabent. Do Logic Programs Resemble Programs in Conventional Languages? In *International Symposium on Logic Programming*, pages 389–396. San Francisco, IEEE Computer Society, August 1987.

[HL94] P. M. Hill and J. W. Lloyd. *The Gödel Programming Language.* The MIT Press, 1994.

[Kow79] R. Kowalski. Algorithm = Logic + Control. *Communications of ACM*, 22:424–431, 1979.

[LK93] S. Lüttringhaus-Kappel. Control generation for logic programs. In D. S. Warren, editor, *Proceedings of the 10th Int. Conf. on Logic Programming, Budapest*, pages 478–495. MIT, July 1993.

[Llo87] J. W. Lloyd. *Foundations of Logic Programming.* Springer-Verlag, Berlin, second edition, 1987.

[Lui94] I. Luitjes. Logic programming and dynamic selection rules. Scriptie (Master's Thesis), University of Amsterdam, 1994.

[MM82] A. Martelli and U. Montanari. An efficient unification algorithm. *ACM Transactions on Programming Languages and Systems*, 4:258–282, 1982.

[Nai82] L. Naish. An Introduction to MU-PROLOG. Technical Report TR 82/2, Dept. of Computer Science, Univ. of Melbourne, 1982.

[Nai86] L. Naish. *Negation and Control in Prolog.* Number 238 in Lecture Notes in Computer Science. Springer-Verlag, 1986.

[Nai88] L. Naish. Parallelizing NU-Prolog. In *Proceedings of the Fifth Annual Symposium on Logic in Computer Science*, pages 1546–1564. The MIT Press, 1988.

[Nai92] L. Naish. Coroutining and the construction of terminating logic programs. Technical Report 92/5, Department of Computer Science, University of Melbourne, 1992.

[SD94] D. De Schreye and S. Decorte. Termination of logic programs: the never-ending story. *Journal of Logic Programming*, 19-20:199–260, 1994.

[Sha89] E. Shapiro. The family of concurrent logic programming languages. *ACM Computing Surveys*, 21(3):412–510, 1989.

An Introduction to Category-based Equational Logic*

Joseph A. Goguen[1] and Răzvan Diaconescu[2]

[1] Programming Research Group, Oxford University Computing Lab, Oxford OX1 3QD, United Kingdom
[2] Institute of Mathematics of the Romanian Academy, PO Box 1-764, 70700 Bucharest, Romania

Abstract: This paper surveys *category-based equational logic*, which generalises both the theoretical and computational aspects of equational logic and its model theory (general algebra) far beyond terms, so as to include: Horn clause logic, with and without equality; all variants of order and many sorted equational logic, including working modulo a set of axioms; constraint logic programming over arbitrary user-defined data types; and any combination of the above. This unifies several important computational paradigms, and opens the door to still further generalisations. Results include completeness of deduction, a Herbrand theorem, completeness of paramodulation, generic modularisation techniques, and a model theoretic semantics for extensible constraint logic programing.

1 Introduction

Category-based equational logic (abbreviated **CBEL**) abstracts out the essential ingredients of equational logic. Equations, deduction, models (algebras), congruences, satisfaction, etc. are treated in an arbitrary category of models satisfying certain mild conditions, including a forgetful functor to a category of *domains*. This encodes the principle that a model interprets a signature (often called a "vocabulary" in classical logic) into a domain, usually a set, or for typed systems, a collection of sets. Category-based equational logic programming (abbreviated **CB** and **ELP**, respectively) extends this generalisation to computation, including rewriting, paramodulation, and constraint solving. By viewing terms semantically as elements of a free model rather than as syntactic constructs, the universal quantifiers in equations are generalised to (usually free) models, and valuations are generalised to model morphisms.

* The research in this paper was supported in part by the Science and Engineering Research Council, the CEC under ESPRIT-2 BRA Working Groups 6071, IS-CORE (Information Systems COrrectness and REusability) and 6112, COMPASS (COMPrehensive Algebraic Approach to System Specification and development), Fujitsu Laboratories Limited, and a contract managed by the Information Technology Promotion Agency (IPA), Japan, as part of the Industrial Science and Technology Frontier Program "New Models for Software Architectures," sponsored by NEDO (New Energy and Industrial Technology Development Organization).

We introduce concepts and results at the highest appropriate level of abstraction, and then gradually make them more concrete. For example, the completeness of category-based equational deduction is proved at a very abstract level which applies to Horn clause logic and extensible constraint logic programming just as well as to many and order sorted equational logics. Completeness theorems for particular logics appear by specialising the abstract rules of deduction to exploit more specific assumptions and reveal the usual syntactic formulations.

The completeness of category-based equational deduction leads to a generic Herbrand theorem in the style of [21], characterising Herbrand models as initial models of the program regarded as an equational theory. When applied to constraint logics (in Section 6), this gives a Herbrand theorem for extensible constraint logic programming. This rather sophisticated new result requires minimal effort using the DB machinery. We also treat very general parameterised modules for ELP. This paper surveys results in [13] and [11], but differs in taking the more concrete view that domains are many sorted sets.

1.1 Applications to Computing

Category-based equational logic yields a semantic approach to completeness of the operational semantics of various programming paradigms, independent of the logic involved, as long as it is equational in our very broad sense. Rewriting is defined on algebraic entities more abstract than terms by abstracting the properties of contexts from the standard cases, and paramodulation is generalised to ***semantic paramodulation***, defined by an inference rule over an arbitrary fixed model. This contrasts with the usual combinatorial paramodulation-based operational approaches.

Equational deduction modulo a theory arises when some equations in a program are non-orientable, making them useless as rules for rewriting or narrowing. The most familiar cases are associativity (A), commutativity (C) and their combination (AC), but graph rewriting [4] is also an example. Our semantic perspective motivates ***paramodulation modulo a model morphism***, which we use to show that computing in the quotient model of a theory is the same as computing modulo that theory.

Integrating our framework with that of institutions yields a foundation for ELP modules in the style of OBJ [24], where modules have interface theories, and are parameterised by other modules [16]. This uses the institution of all category-based equational logics, which also gives a semantics for queries and solution forms in the context of modularisation. In the institution of all CBELs, signatures are functors, abstracting the fact that in any equational logic, a signature determines a category of models, a category of domains, and a forgetful functor between them.

Extensible constraint logic programming fits the ELP paradigm as a CBEL whose models form a comma category over a fixed **built-in model** of data types over which constraints are solved, following the generalisation in [21] from initial algebras to initial model morphisms from built-ins. This allows ELP to benefit from the maturity of equational logic, including its implementation technology.

1.2 Beyond Conventional Abstract Model Theory

The approach of this paper is ***abstract model theory*** in roughly the same spirit as the "Hungarian School" (e.g., [1]), which means much more than the familiar logical tradition of abstracting Tarskian semantics to extend classical first order model theory towards other logical systems [3, 2]. The DB framework is closer to the theory of institutions [18] (see Section 5.2) in that:

- it parameterises over signatures, rather than assuming a fixed signature given in advance;
- it abstracts Tarski's semantic definition of truth [42] to a (functorial) relation of satisfaction between models and sentences; and
- it uses category theory to achieve generality and simplicity, for example, by letting models be objects in an arbitrary category.

Two differences between our approach and that of institutions are:

- our satisfaction is significantly less abstract than institutional satisfaction, because it captures the essence of equational logic satisfaction; and
- our framework does not have any direct formulation of the axiom for institutions that "truth is invariant under change of notation."

Nonetheless, Section 5 shows how CBEL naturally gives rise to an institution.

2 Preliminaries

This section presents notation, introduces the category-based framework, and discusses some examples.

2.1 Basic Definitions and Notations

We assume familiarity with the basics of universal algebra and category theory, and generally use the same notation as Mac Lane [36], except that composition is denoted ";" and written in diagrammatic order. The application of functions (and functors) to arguments may be written either normally using parentheses, or else in diagrammatic order without parentheses. Categories usually have a name with first letter in capital bbold font; for example, the category of sets is $\mathbb{S}et$. The class of objects of $\mathbb{C}$ is denoted $|\mathbb{C}|$. Functors are usually denoted by capital caligraphic letters. The empty string is denoted $[]$.

Binary Relations. A binary relation $\{\langle s_i, t_i\rangle \mid i \in I\} \subseteq A \times A$ on a (possibly many sorted) set A may be written $\langle s, t\rangle$ where $s, t\colon I \to A$. A binary relation in the ordinary sense may be represented by many different parallel pairs of functions, corresponding to different **index sets** I. When I is a singleton, we may identify the ordered pair $\langle s, t\rangle$ with the corresponding singleton relation. A singleton relation is called **atomic**. Given a map $h\colon A \to B$, we extend the notation of map application from elements to tuples, so that $h(\langle s,t\rangle)$ or $\langle h(s), h(t)\rangle$ actually means $\langle s;h, t;h\rangle$ for any relation $\langle s,t\rangle$ on A. We also extend the notation sQt from ordered pairs of elements to binary relations, so that sQt means $\langle s,t\rangle \subseteq Q$ for Q and $\langle s,t\rangle$ arbitrary binary relations. The indexed set version of relational inclusion is: given $s,t\colon I \to A$ and $s',t'\colon I' \to A$, then $\langle s,t\rangle \subseteq \langle s',t'\rangle$ iff there is a map $j\colon I \to I'$ such that $j;s' = s$ and $j;t' = t$. Composition of binary relations is denoted $\circ$, so that

$$Q \circ R = \{\langle a,d\rangle \mid \text{ there exists } b \text{ such that } aQb \text{ and } bRd\}\,.$$

This approach to relations will help our treatment of congruences.

Categories. Here we review material found many places, e.g., [36]. Given a functor $\mathbf{C} \xrightarrow{C} \mathbf{E}$ and an object $a \in |\mathbf{E}|$, the **comma category** $(a{\downarrow}C)$ has arrows $a \xrightarrow{t} cC$ as objects and arrows f as morphisms such that the following diagram commutes:

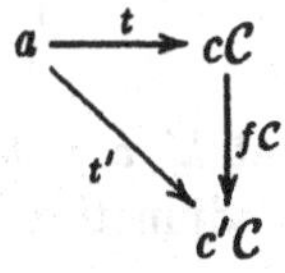

A **diagram** in a category $\mathbf{C}$ is a functor $J \xrightarrow{C} \mathbf{C}$. A **cone** $\gamma\colon d \to C$ consists of an object $d \in |\mathbf{C}|$ (called the **apex**) and a $|J|$-indexed family of arrows $\{d \xrightarrow{\gamma_i} C(i)\}_{i\in|J|}$ such that $\gamma_j; C(u) = \gamma_i$ for any $u\colon i \to j$ in J. A **limit** of C is a "minimal" cone $\mu\colon c \to C$ over C, in the sense that for any other cone $\gamma\colon d \to C$ there is a unique $f\colon d \to c$ in $\mathbf{C}$ such that $f;\mu = \gamma$.

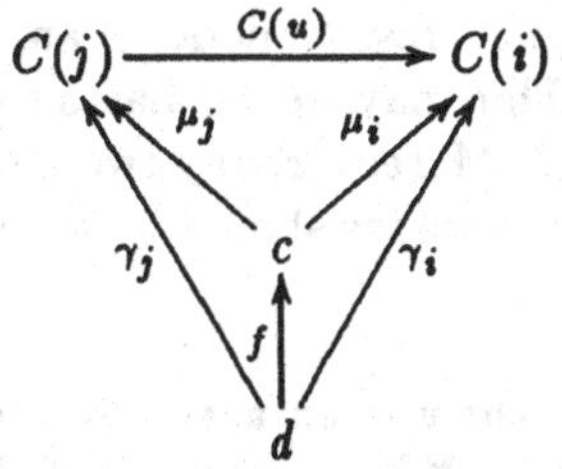

Co-cone and **colimit** are dual to cone and limit, i.e., their defintions are obtained by reversing the arrows in the definition of limits, as in:

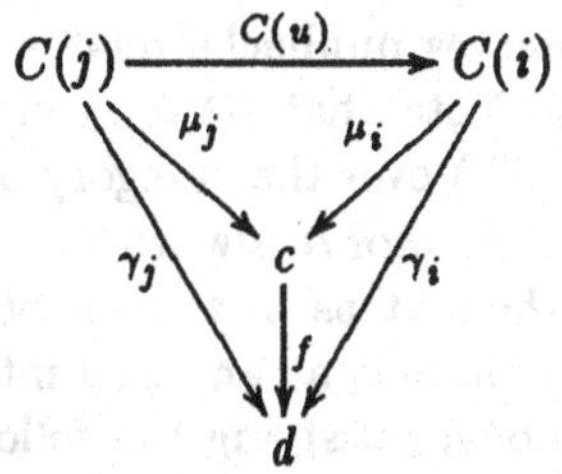

Particular limits and colimits are obtained by fixing the shape of diagrams, i.e., the category J. When J is *discrete* (i.e., consists only of identity arrows) we get **products** and **coproducts**, respectively. When J consists only of two objects and a parallel pair of arrows between them, we get **equalisers** and **coequalisers** (denoted *coeq*), respectively. When J is

we get **pullbacks** and **pushouts**. The **kernel** (denoted *ker*) of an arrow is the pullback of that arrow with itself.

An object A in a category $\mathbf{C}$ is **projective** iff for any coequaliser c and any arrow h given as in the following diagram,

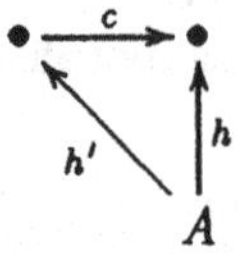

there exists an arrow h' such that $h'; c = h$. Projectivity is an abstraction of freeness that does not require an adjunction.

2.2 Models and Domains

The semantics of a logical system is given by its *models*. In general, soundness of the inference rules of a logical system is checked against its models using a satisfaction relation, in the style of Tarski [42]. We assume that models and their morphisms form a category. As in institutions [18], category-based equational logics are "localised" to signatures. A model is an interpretation of a particular signature into a *domain*. Thus any model has an underlying domain, and this correspondence is functorial. Moreover, any two parallel model morphisms identical as maps between their domains should be the same. These assumptions are summed up in the following:

> **[Basic Framework]:** There is an abstract category of *models* $\mathcal{A}$ and a *forgetful functor* $\mathcal{U} : \mathcal{A} \to \mathbf{X}$ to a category of *domains* $\mathbf{X}$ that is faithful and preserves pullbacks.

The simplicity of these assumptions reflects the simplicity of equational logic. The condition that $\mathcal{U}$ preserves pullbacks relates to congruences being equivalences, as discussed later. Note that $(\mathcal{A}, \mathcal{U})$ can be regarded as a **concrete category** (in the sense of [32]) over the category of domains.

In practice, the forgetful functor $\mathcal{U}$ always has a left adjoint $\mathcal{F}$, which means that for every $X \in |\mathbf{X}|$, thought as a domain of variables, there is a model $X\mathcal{F}$, free in the sense that there is a canonical interpretation $X\eta : X \to X\mathcal{F}\mathcal{U}$ of the variables into the model satisfying the following universal property: for each $f\colon X \to A\mathcal{U}$ interpreting variables in a model A, there is a unique model morphism $f^\sharp\colon X\mathcal{F} \to A$ extending f, in the sense that $X\eta; f^\sharp\mathcal{U} = f$.

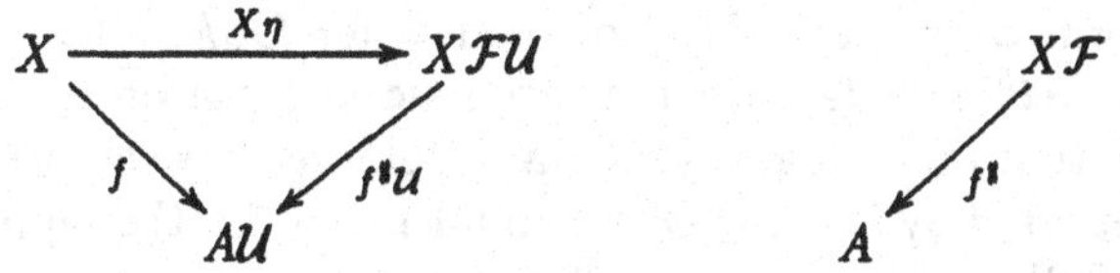

Note that $\mathcal{U}$ preserves pullbacks when it has a left adjoint (e.g., see [36]).

The signatures of computing science logics usually involve a set of sorts. Then the categories of domains are categories of many sorted sets, i.e., $\mathbf{X} = \mathbf{Set}^S$ for some set S of sorts. Although CBEL was originally developed more abstractly [13, 11], here we usually simplify by assuming that domains are many sorted sets, i.e., that the following

[Simplifying Assumption]: $\mathbf{X} = \mathbf{Set}^S$

holds unless explicitly stated otherwise. This avoids the difficult technical details of categorical relations and their finiteness properties found in [11] and [13].

2.3 Examples

This subsection briefly sketches several examples, assuming familiarity with their basic concepts, and showing how they fall under the **Basic Framework**.

Many Sorted Algebra. This example is probably the best known. Expositions of many sorted algebra (abbreviated **MSA**) including the standard definitions for signature, algebra, homomorphism, etc., can be found many places, e.g., [23, 17], and are assumed here. Given a many sorted signature (S, Σ), let $\mathcal{A}lg_\Sigma$ denote the category of Σ-algebras with Σ-homomorphisms. There is a forgetful functor $\mathcal{U}_\Sigma\colon \mathcal{A}lg_\Sigma \to \mathbf{Set}^S$ from Σ-algebras to S-sorted sets, forgetting the interpretations of the operation symbols in Σ. This functor has a left adjoint. Given a set X of variable symbols, let $T_\Sigma(X)$ denote the (S-sorted) **term algebra** with operation symbols from Σ and variable symbols from X; it is the **free** Σ-algebra **generated** by X, in the sense that if $v\colon X \to A$ is an **assignment**, i.e., a (many sorted) function to a Σ-algebra A, then there is a unique extension of v to a Σ-homomorphism $v^\sharp\colon T_\Sigma(X) \to A$. We let T_Σ denote the **initial** term Σ-algebra, $T_\Sigma(\emptyset)$, recalling that this means there is a unique Σ-homomorphism $!_A\colon T_\Sigma \to A$ for any Σ-algebra A. Call $t \in T_\Sigma$ a **ground Σ-term**. Given a ground Σ-term t, let t_A denote the element $!_A(t)$ in A. Call A **reachable** iff $!_A$ is surjective, i.e., iff each element of A is "named" by some ground term.

Order Sorted Algebra. Order sorted algebra (abbreviated **OSA**) adds to MSA a partial ordering on sorts, which is interpreted as inclusion among the corresponding carriers; all approaches to OSA share this essential idea. See [19] for a recent survey, including all basic OSA definitions (signature, algebra, homomorphism, regularity, etc.).

Given an order sorted signature $(S, \leq, \Sigma)$, the Σ-algebras and their homomorphisms form a category $\mathcal{A}lg_\Sigma$ of models for OSA. The forgetful functor $\mathcal{U}_\Sigma \colon \mathcal{A}lg_\Sigma \to \mathbf{Set}^S$ forgets both the algebraic *and* the order sorted structure. We emphasise that the domains for OSA should *not* have an order sorted structure, as is supported by the way OSA is implemented. Other approaches to OSA mentioned in [19] can be treated similarly.

Horn Clause Logic. Theorem 1 below describes an embedding of the category of models of a first order signature as a retract of the category of algebras of an MSA signature obtained from the original first order signature by turning predicates into operations. Interpreting predicates as boolean valued operations is hardly new; it has even been used to lift narrowing to an operational semantics for logic programming [7]. However, this approach (from [10]; see also [13, 11]) is somewhat different, because it does not assume a full boolean structure on the new sort of truth values. Moreover, the model theoretic aspect is emphasised.

A first order signature (S, Σ, Π) consists of a many sorted signature (S, Σ) of function symbols, plus a family $\Pi = \{\Pi_w \mid w \in S^*\}$ of predicate symbols. Given a first order signature (S, Σ, Π), let $\mathbf{Mod}_{\Sigma,\Pi}$ denote the category of (S, Σ, Π)-models and their morphisms.

Theorem 1. *Given a many sorted first order signature (S, Σ, Π), construct an MSA signature $(S^b, \Sigma^b \cup \Pi^b)$ as follows:*

- $S^b = S \cup \{b\}$, *for* $b \notin S$,
- $\Sigma^b_{[],b} = \{\mathtt{t}\}$ *and* $\Sigma^b_{w,s} = \Sigma_{w,s}$ *for* $w, s \in S^* \times S$ *and* $\mathtt{t}$ *a new symbol, and*
- $\Pi^b_{s_1 \ldots s_n, b} = \{\pi^b \mid \pi \in \Pi_{s_1 \ldots s_n}\}$.

Then:

1. *There is a forgetful functor* $\mathcal{H}_{\Sigma,\Pi} \colon \mathcal{A}lg_{\Sigma^b \cup \Pi^b} \to \mathbf{Mod}_{\Sigma,\Pi}$ *such that for all* $\pi \in \Pi$ *and all* $(\Sigma^b \cup \Pi^b)$*-algebras* A, *we have* $a \in \pi_{\mathcal{H}_{\Sigma,\Pi}(A)}$ *iff* $\pi^b_A(a) = \mathtt{t}_A$.
2. *The functor* $\mathcal{H}_{\Sigma,\Pi}$ *has a left adjoint left inverse* $\mathcal{E}_{\Sigma,\Pi}$.
3. *There is a "translation"* $\alpha_{\Sigma,\Pi}$ *of* (S, Σ, Π)*-Horn clauses to conditional* $(\Sigma^b \cup \Pi^b)$*-equations that regards every* Σ*-equation as a* Σ^b*-equation in the canonical way and maps an atom* $\pi(x)$ *to the* $(\Sigma^b \cup \Pi^b)$*-equation* $\pi^b(x) = \mathtt{t}$.
4. *For any Horn clause* φ *and any* $(\Sigma^b \cup \Pi^b)$*-algebra* A,
   $A \models_{\Sigma^b \cup \Pi^b} \alpha_{\Sigma,\Pi}(\varphi)$ *iff* $\mathcal{H}_{\Sigma,\Pi}(A) \models_{\Sigma,\Pi} \varphi$.

□

A consequence of this result is that given a first order signature (S, Σ, Π), the category of models for Horn clause logic (abbreviated **HCL**) can be taken as $\mathcal{A}lg_{\Sigma^b \cup \Pi^b}$ instead of $\mathbf{Mod}_{\Sigma,\Pi}$, and its sentences as conditional equations instead of Horn clauses. Notice that in HCL, unlike MSA, the forgetful functor from models to domains, $\mathcal{A}lg_{\Sigma^b \cup \Pi^b} \to \mathbf{Set}^S$, is *not* monadic.

Equational Logic Modulo Axioms. Equational deduction modulo a set of axioms (abbreviated **ELM**) is needed for rewriting when there are non-orientable equations; detailed expositions are given in [17, 30, 8, 35]. Although in practice non-orientable rules are mostly unconditional, there is no theoretical reason to exclude equational deduction modulo a set of conditional equations. Idempotence is a non-orientable conditional axiom, when given in the form `x + y = x if x = y`.

Definition 2. Given an MSA signature (S, Σ) and a set E of Σ-equations, a **Σ-term modulo E** is an element t of $T_{\Sigma,E}(X)$, the quotient of the term algebra $T_\Sigma(X)$ by E. □

Equational deduction modulo E generalises the usual concepts of MSA to "concepts modulo E", including the inference rules [17]. A model theory for equational logic modulo E requires an adequate notion of model, and it is natural to use $\mathcal{A}lg_{\Sigma,E}$, which gives "algebras modulo axioms." The category of domains is the category $\mathbf{Set}^S$ of S-sorted sets and functions, and the forgetful functor $\mathcal{U}_{\Sigma,E} : \mathcal{A}lg_{\Sigma,E} \to \mathbf{Set}^S$ forgets both the axioms and the MSA structure.

Example 1. The logic of Mosses's unified algebras [38] can be regarded as equational logic modulo a conditional theory. All unified specifications of a given unified signature contain a core of Horn clauses. Unified algebras appear as models of this specification [13, 11]. □

Summary of Examples. The following summarises the examples discussed above:

	$\mathcal{A}$ (category of models)	$\mathcal{U}$ forgets:
MSA	$\mathcal{A}lg_\Sigma$	algebraic structure
OSA	$\mathcal{A}lg_\Sigma$	algebraic structure + order sortedness
HCL	$\mathcal{A}lg_{\Sigma^b \cup \Pi^b}$	algebraic structure + sort b
ELM	$\mathcal{A}lg_{\Sigma,E}$	algebraic structure + axioms

Any combination of these logical systems is possible, e.g., order sorted Horn clause logic with equality. The logic underlying Eqlog combines all these systems, plus constraint logic programming.

3 Category-based Equational Deduction

This section develops a categorical proof theory for category-based equational logic and proves its completeness with respect to its model theory. The following technical assumption underlies all of Section 3:

> **[Deduction Framework]: Basic Framework + the category $\mathcal{A}$ of models has pullbacks and coequalisers.**

The proof theory of CBEL uses categorical abstractions of basic concepts from equational logic and universal algebra, including congruence, term algebra, substitution, equation and satisfaction. These are all defined below.

3.1 Congruences

Quotient models and the complete system of inference rules for CBEL both require a suitable notion of congruence.

Definition 3. Let A be an arbitrary model. Then a binary relation Q on the domain of A is a **congruence** iff it is a kernel of a model morphism, i.e., iff there is a morphism φ in $\mathbb{A}$ such that $Q = \mathcal{U}(ker\varphi)$. The **quotient** of A by Q is the coequaliser of $ker\varphi$. Its target model is denoted A/Q and is also called the **quotient** of A. □

Kernels and coequalisers play a complementary rôle: coequalisers construct the quotient coresponding to a congruence (which can be a kernel of *any* model morphism), while kernels recover the congruence (which is implicit for quotients) as a binary relation.

Definition 4. Let Q be a binary relation on the domain of a model A. Then the **congruence closure of** Q is the least congruence on A containing Q, denoted $\mathsf{C}(Q)$. □

Definition 5. Suppose that congruence closures of binary relations exist in $\mathbb{A}$. Then a forgetful functor $\mathcal{U} : \mathbb{A} \to \mathbf{Set}^S$ is **finitary** iff

$$\mathsf{C}(Q) = \bigcup\{\mathsf{C}(Q_0) \mid Q_0 \subseteq Q \text{ finite}\}$$

for any model A and any binary relation Q on the domain of A. □

All forgetful functors in the examples of Section 2.3 are finitary. This is because all operation and relational symbols involved take only a finite number of arguments, as will be explained in Section 3.3.

3.2 Equations, Satisfaction and Completeness

Equations are traditionally pairs of terms constructed from the symbols of a signature plus some variables. Goguen and Meseguer [20] first made quantifiers part of the concept of equation, for MSA. Although terms are syntactic constructs, from a model theoretic perspective they are just elements of the free term model over the set of quantified variables. Any valuation of the variables into a model extends uniquely to a model morphism evaluating both sides of the equation. Thus, a more semantic treatment of quantification regards quantifiers as models rather than sets of variables, and regards valuations as model morphisms rather than functions; this was already done in [6] for MSA. This non-trivial generalisation of equation and satisfaction extends naturally to equational deduction, and in our opinion gives a pleasing unity and generality to the whole area.

Definition 6. Let A be any model. Then a **$\mathcal{U}$-identity** on A is a binary relation $\langle s, t\rangle$ on the domain of A. A $\mathcal{U}$-identity $\langle s, t\rangle$ on A is **satisfied** by a model B with respect to a model morphism $h\colon A \to B$ iff[3] $h(s) = h(t)$.

A **$\mathcal{U}$-equation** is a universally quantified expression $(\forall A)\langle s, t\rangle$ where A is a model representing the quantifier and $\langle s, t\rangle$ is a $\mathcal{U}$-identity on A. A model B **satisfies** $(\forall A)\langle s, t\rangle$ iff B satisfies the $\mathcal{U}$-identity $\langle s, t\rangle$ for all morphisms $h\colon A \to B$. This is written $B \models (\forall A)\langle s, t\rangle$. We feel free to drop the prefix $\mathcal{U}$- when it is clear from context. □

Representing an equation as a parallel pair of arrows is hardly new, e.g., see [26, 27]. Our formulation gives *families of equations* as sentences, rather than single equations, in agreement with Rodenburg's work [39] showing that equational logic with conjunction satisfies the Craig interpolation property whereas the usual formulation does not. When the quantifiers of equations are free models, i.e., $A = X\mathcal{F}$ for some S-sorted set X, we may write $(\forall X)\langle s, t\rangle$ instead of $(\forall X\mathcal{F})\langle s, t\rangle$.

Definition 7. A **conditional $\mathcal{U}$-equation** is an expression having the form $(\forall A)\ \langle s', t'\rangle$ **if** $\langle s, t\rangle$ where A is a model representing the quantifier, $\langle s', t'\rangle$ is a $\mathcal{U}$-identity on A, and $\langle s, t\rangle$ is a finite binary relation on the domain of A representing the hypotheses (i.e, the condition) of the equation. A model B **satisfies** $(\forall A)\langle s', t'\rangle$ **if** $\langle s, t\rangle$ iff for any morphism $h\colon A \to B$, $h(s) = h(t)$ implies $h(s') = h(t')$. This is written $B \models (\forall A)\langle s', t'\rangle$ **if** $\langle s, t\rangle$. If e is a (possibly conditional) $\mathcal{U}$-equation and Γ is a set of (possibly conditional) $\mathcal{U}$-equations, then $\Gamma \models e$ means that $A \models \Gamma$ implies $A \models e$ for all models A. A conditional **$\mathcal{U}$-rule** is an oriented conditional $\mathcal{U}$-equation whose left side is atomic[4], usually written $(\forall A)\ l \to r$ **if** $\langle s, t\rangle$. We feel free to drop $\mathcal{U}$- when the context permits. □

Our approach to the completeness of category-based equational deduction is traditional in that the central concept is the congruence determined by a set Γ of (possibly conditional) equations on a model A (e.g., see [5]). The most abstract completeness result states the equivalence of two versions of this congruence: all unconditional equations quantified by A that can be *syntactically inferred* from Γ; and all unconditional equations quantified by A that are *semantic consequences* of Γ. Our semantic treatment of equation and satisfaction allows the congruences determined by Γ on free models and on other models to be treated the same way. Despite the generality and abstraction, the rules of inference for CB equational deduction can be made explicit for concrete examples, and can be recognised even in the most abstract formulation. The following subsections introduce more specific assumptions that reveal the usual syntactic formulations, which require finiteness of the hypotheses of conditional equations and the arities of operations.

Note that unconditional $\mathcal{U}$-equations are a special case of conditional $\mathcal{U}$-equations.

---

[3] Or more precisely $(h\mathcal{U})(s) = (h\mathcal{U})(t)$, but we prefer the simpler notation.

[4] This is because it is more convenient to work with singleton rules.

Definition 8. Let Γ be a set of conditional equations. Then a congruence $\equiv$ on A is **closed under Γ-substitutivity** iff for any $(\forall B)\langle s', t'\rangle$ **if** $\langle s, t\rangle$ in Γ and any morphism $h\colon B \to A$, $h(s) \equiv h(t)$ implies $h(s') \equiv h(t')$. The least congruence on A closed under Γ-substitutivity is denoted $\equiv_\Gamma^A$. □

In the usual concrete examples, a relation is closed under Γ-substutivity iff it contains all the pairs generated as substitution instances of the equations in Γ.

Theorem 9. Completeness Theorem *If the forgetful functor $\mathcal{U}$ is finitary and all equations in Γ have projective quantifiers, then:*

1. *the least congruence closed under Γ-substitutivity, denoted $\equiv_\Gamma^A$, exists;*
2. *$A/{\equiv_\Gamma^A}$ is the free Γ-model over A; and*
3. *$\Gamma \models (\forall A)\langle s, t\rangle$ iff $s \equiv_\Gamma^A t$.*

□

The proof may be found in [11, 13]. Note that $A/{\equiv_\Gamma^A}$ being the free Γ-model over A refers to freeness with respect to the forgetful functor from Γ-models to A; in a sense this is already an abstract statement of completeness. Recall from Section 2.1 that $s \equiv_\Gamma^A t$ means $\langle s, t\rangle \subseteq {\equiv_\Gamma^A}$ (as S-sorted sets). If we instead write this as $\Gamma \vdash (\forall A)\langle s, t\rangle$, then 3. above takes the familiar form

$$\Gamma \models (\forall A)\langle s, t\rangle \ \text{ iff } \ \Gamma \vdash (\forall A)\langle s, t\rangle \ .$$

The proof of Theorem 9 brings out the syntactic character of $\equiv_\Gamma^A$, showing it is the closure under the syntactic consequences of Γ using the inference rules

$$\text{[congruence]} \quad \frac{(\forall A)\langle s, t\rangle}{(\forall A)\mathrm{C}\langle s, t\rangle}$$

$$\text{[substitutivity]} \quad \frac{(\forall A)\langle s; h\mathcal{U},\ t; h\mathcal{U}\rangle}{(\forall A)\langle s'; h\mathcal{U},\ t'; h\mathcal{U}\rangle}$$

where $(\forall B)\langle s', t'\rangle$ **if** $\langle s, t\rangle$ is in Γ and $h: B \to A$ is any model morphism. More technically, $\equiv_\Gamma^A$ is the union of an infinite chain of relations $\{Q_n\}_{n\in\omega}$ on the domain of A, obtained by alternating application of the above inference rules, where Q_{n+1} is obtained by taking the congruence closure of Q_n, or else the union of all conclusions of sentences in Γ for which the hypotheses belong to Q_n. See [13, 11] for details.

3.3 Consequences of Freeness

So far, we have not assumed freeness of quantifing models, or even the existence of free models. This assumption allows us to give more concrete inference rules for equational deduction, by splitting the congruence rule into equivalence (i.e., reflexivity, symmetry, and transitivity) and closure under operations. It also lets us see how finitarity of the forgetful functor amounts to finiteness of the model operation arities. This subsection assumes the following:

[Adjointness Framework]: Deduction Framework + the forgetful functor $\mathcal{U}$ has a left adjoint $\mathcal{F}$.

The congruence closure of a binary relation can be constructed in steps that are very similar to the usual rules for equivalence (i.e., reflexivity, symmetry and transitivity) plus congruence (i.e., closure under "model operations") in equational logic [20, 22, 21].

Proposition 10. *Let $I \xrightarrow{\langle s,t\rangle} A\mathcal{U}$ be a relation on the domain of a model A. Then the congruence closure of $\langle s,t\rangle$ exists, and is constructed as follows:*

- operations: *let $s^\natural$ and $t^\natural$ to the unique extensions of s and t, respectively, to model morphisms $I\mathcal{F} \to A$, and*
- equivalence: *let $\langle S, T\rangle$ be the kernel of $coeq\langle s^\natural, t^\natural\rangle$.*

Then the congruence closure $\mathbf{C}\langle s,t\rangle$ *is* $\langle S\mathcal{U}, T\mathcal{U}\rangle$. □

The proof may be found in [11, 13]. The operations step closes the original relation $\langle s,t\rangle$ under model operations; this is achieved categorically by using the universal property of the free model over the indices of the relation. The equivalence step constructs the equivalence generated by the closure under operations. Because this is done at the level of model morphisms, it is a congruence, and therefore (it is not hard to see) closed under operations.

Definition 11. Given a binary relation $\langle s,t\rangle$ on the domain of a model A, then $\langle s,t\rangle$ is **closed under operations** iff $\langle s^\natural\mathcal{U}, t^\natural\mathcal{U}\rangle \subseteq \langle s,t\rangle$. The **closure of** $\langle s,t\rangle$ **under operations** is the least relation closed under operations that contains $\langle s,t\rangle$; it is denoted $\mathbf{Op}\langle s,t\rangle$. □

Fact 12. *Let $\langle s,t\rangle$ be a binary relation on the domain of a model A. Then its closure under operations exists and is given by $\langle s^\natural\mathcal{U}, t^\natural\mathcal{U}\rangle$.* □

The following may help clarify the situation:

Example 2. Let (S,Σ) be a many sorted signature and $\langle s,t\rangle$ an S-sorted binary relation on the carrier of an S-sorted Σ-algebra A. Then

- $\langle s^\natural\mathcal{U}, t^\natural\mathcal{U}\rangle$ is obtained by taking the union of the increasing chain of S-sorted relations $\langle s^n, t^n\rangle_{n\in\omega}$ where $\langle s^0, t^0\rangle = \langle s,t\rangle$ and $\langle s^{n+1}, t^{n+1}\rangle = \langle s^n, t^n\rangle \cup \{\langle\sigma_A(s^n), \sigma_A(t^n)\rangle \mid \sigma \in \Sigma\}$. $\langle\sigma_A(s^n), \sigma_A(t^n)\rangle$ is obtained by relating the results of all applications of the operation σ_A to all pairs of elements related by $\langle s^n, t^n\rangle$. The union $\bigcup_{n\in\omega}\langle s^n, t^n\rangle$ is the same as relating all results of applications of derived operators to pairs of elements related by $\langle s,t\rangle$.
- Closing $\langle s^\natural\mathcal{U}, t^\natural\mathcal{U}\rangle$ under equivalence produces the congruence coequalising the S-sorted Σ-morphisms $s^\natural$ and $t^\natural$. The congruence is recovered categorically as the kernel of the coequaliser of $s^\natural$ and $t^\natural$.

□

In most cases, the congruence closure of a binary relation can also be constructed by reversing these steps, i.e., first closing under equivalence and then under model operations. Although our present framework is too abstract for showing validity of this alternative construction, half of it still holds, namely $\mathbf{Op}\langle\bar{s},\bar{t}\rangle \subseteq \mathbf{C}\langle s,t\rangle$, where $\langle\bar{s},\bar{t}\rangle$ denotes the equivalence closure of $\langle s,t\rangle$ on the domain of a model A. The following definition captures equality of these two constructions as an assumption:

Definition 13. We say that **congruences are concrete** iff any equivalence closed under operations is a congruence. □

In the concrete cases, congruences are equivalences closed under operations, and in fact, congruences are usually defined this way.

Corollary 14. *If congruences are concrete, then the relation obtained by closure under the following rules is* $\equiv^A_\Gamma$:

[reflexivity] $$\frac{}{(\forall A)\langle s,s\rangle}$$

[symmetry] $$\frac{(\forall A)\langle s,t\rangle}{(\forall A)\langle t,s\rangle}$$

[transitivity] $$\frac{(\forall A)\langle s,t\rangle \quad (\forall A)\langle t,u\rangle}{(\forall A)\langle s,u\rangle}$$

[operations] $$\frac{(\forall A)\langle s,t\rangle}{(\forall A)\langle s^{\mathfrak{l}}\mathcal{U}, t^{\mathfrak{l}}\mathcal{U}\rangle}$$

[substitutivity] (as given earlier)

□

Thus we can say that category-based equational logic is complete under the above inference rules, which are essentially the same as the usual concrete rules, except of course that quantifiers are models instead of variables.

The following result shows how finitarity of $\mathcal{U}$ (Definition 5) reduces in practice to finiteness of model operations. The CB formulation of finitary model operations is that the forgetful functor $\mathcal{U}$ preserves filtered colimits.

Proposition 15. *The forgetful functor $\mathcal{U}$ is finitary if it preserves filtered colimits.* □

All of the forgetful functors from categories of models to categories of domains presented in Section 2.3 are finitary.

Variable based quantifiers, i.e., quantifiers of the form $X\mathcal{F}$ for a many sorted set X, are projective whenever coequalisers of model morphisms are surjections, which is almost always true in practice.

Proposition 16. *If each coequaliser in the category of models is a surjective many sorted function at the domain level, then each free model is projective.* □

It is worth mentioning that for arbitrary categories of domains (as in [11, 13]), the condition on the surjectivity of the model coequalisers has a formulation very similar to categorical formulations of the Axiom of Choice.

3.4 A Herbrand Theorem

An important consequence of the most abstract completeness theorem (9) is a category-based Herbrand theorem. Our approach uses the categorical characterisation of Herbrand universes as initial models suggested in [21]. This section uses only the **Deduction Framework**.

Definition 17. A **$\mathcal{U}$-query** is an existentially quantified expression $(\exists A)\langle s,t\rangle$, where A is a model representing the quantifier and $\langle s,t\rangle$ is a $\mathcal{U}$-identity on A. We feel free to drop the prefix $\mathcal{U}$- when it is clear from context. A **solution of** $(\exists A)\langle s,t\rangle$ in a model B is a morphism $h\colon A \to B$ such that $\langle s,t\rangle$ is satisfied in B with respect to h. When B is a free model, h is called a **solution form**. We write $B \models q$ if there is a solution to q in B. □

Corollary 18. *Herbrand Theorem* *If* $\mathbb{A}$ *has an initial model and all quantifiers of equations in* Γ *are projective, then:*

1. *the initial model of* Γ *exists; let us denote it* 0_Γ*; and*
2. $\Gamma \models (\exists A)q$ *iff* $0_\Gamma \models (\exists A)q$*, for any* $\mathcal{U}$*-query* $(\exists A)q$ *and any model* A*.*

□

For the proof, see [13] and [11], which also prove a non-empty sorts Herbrand theorem from the above, for which the domain of the intial model has all carriers non-empty. This provides foundations for solving queries using techniques like resolution and paramodulation.

4 Semantic Paramodulation

This section extends paramodulation to semantic paramodulation in the framework of CBEL, and studies the relationship between the paramodulation relation induced by a program Γ on a model A and the least congruence on A closed under Γ-substitutivity. This goes to the heart of CB operational semantics for ELP, because the completeness of paramodulation is explained by the identity of these two relations. This identity occurs exactly when the paramodulation relation is transitive. Paramodulation modulo some axioms is generalised to paramodulation modulo a model morphism. Finally rewriting and confluence are treated. The technical assumptions for this section are given by the:

> **[Paramodulation Framework]: Adjointness Framework** + preservation of filtered colimits by $\mathcal{U}$ + concrete congruences.

4.1 Contexts

Contexts play a basic rôle paramodulation and rewriting. This section gives a CB definition of context, abstracted from properties of contexts in MSA; this allows defining rewriting on entities more abstract than terms. A characteristic property of contexts is that they are *unary*:

Definition 19. Let Σ be a MSA signature. Then a **Σ-context** is a Σ-term with one variable symbol having a single occurrence of that variable symbol. □

Fact 20. *Given a Σ-algebra A, a Σ-context c determines a map $c_A\colon A \to A$ that evaluates the context for any given value a in A of the variable symbol z in c. This is represented by the following diagram,*

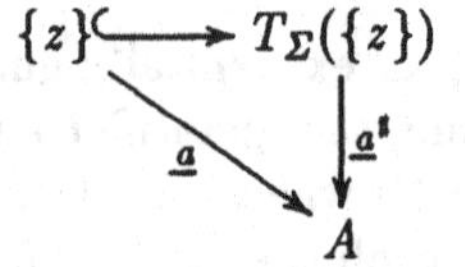

where for each $a \in A$, $\underline{a}\colon \{z\} \to A$ satisfies $\underline{a}(z) = a$. Then $c_A\colon A \to A$ is defined by $c_A(a) = \underline{a}^!(c)$ where $\underline{a}^!$ is the unique extension of $\underline{a}$ to a Σ-homomorphism. We usually write $c_A[t]$ instead of $c_A(t)$ □

In general c_A is *not* homomorphic. However, contexts do form a monoid under composition (i.e., plugging one context into another), and evaluation of contexts commutes with homomorphisms:

Proposition 21. *Given a context c over an MSA signature Σ and a Σ-homomorphism $h\colon A \to B$, then $c_A; h = h; c_B$.* □

This suggests the *natural transformational* nature of contexts:

Definition 22. Given $\mathcal{U}\colon \mathsf{A} \to \mathbf{Set}^S$, then a **$\mathcal{U}$-pre-context** is a natural transformation $c\colon \mathcal{U} \to \mathcal{U}$. **Composition** of $\mathcal{U}$-pre-contexts is the usual composition of natural transformations. The **evaluation** of a pre-context c on $t \in \mathcal{U}(A)_s$ is $(c_A)_s(t)$. More generally, given $t\colon I \to A\mathcal{U}$, let $c_A[t] = t; c_A$. This extends to relations $\langle s, t\rangle\colon I \to A\mathcal{U}$ by defining $c_A[\langle s, t\rangle] = \langle s; c_A, t; c_A\rangle$.

A binary relation $\langle s, t\rangle$ on the domain of a model A is **closed under context evaluation** for a class C of pre-contexts iff $\langle c_A[s], c_A[t]\rangle \subseteq \langle s, t\rangle$ for each $\mathcal{U}$-pre-context $c \in C$. The least relation closed under context evaluation for C and containing $\langle s, t\rangle$ is called the **context closure** of $\langle s, t\rangle$ for C. □

Proposition 23. *Let Q be a binary relation on the domain of a model A. If Q is closed under operations, then it is also closed under context evaluation for any class C of pre-contexts.* □

The converse of the above holds for transitive relations on reachable algebras in MSA. This property seems closely related to finiteness of the arities of the model operations, and is central to the CB definition of context:

Definition 24. A **context monoid** for $\mathcal{U}\colon \mathsf{A} \to \mathbf{Set}^S$ is a submonoid $\mathcal{C}$ of the monoid of all $\mathcal{U}$-pre-contexts such that any transitive relation on a reachable model that is closed under context evaluation for $\mathcal{C}$ is also closed under operations. □

Some context monoids for a given CBEL could be very different from the standard ones, giving rise to novel notions of rewriting and paramodulation.

Corollary 25. *Let $\mathcal{C}$ be a fixed context monoid for $\mathcal{U}$. Then an equivalence on A is a congruence iff it is closed under context evaluation.* □

4.2 Semantic Paramodulation

To motivate our category-based treatment of paramodulation, we briefly review ordinary paramodulation for MSA, and show how resolution can be regarded as a restricted form of paramodulation. Then we develop CB paramodulation, and paramodulation modulo a model morphism.

MSA Paramodulation. This subsection reviews paramodulation and narrowing for ordinary MSA equational logic programming. Recall that an *occurrence* in a term is a pointer to a subterm. Given a term t and an occurrence π in t, let $t|_\pi$ denote the subterm of t whose root is at π, and let $t|_{\pi \leftarrow t'}$ denote the term obtained from t by replacing $t|_\pi$ with t' in t. An **equational goal** is a pair $\langle t_1, t_2 \rangle$ of terms, and we extend occurrences from terms to goals by regarding a goal $\langle t_1, t_2 \rangle$ as a term having two subterms, t_1 and t_2. The instantiation of a term by a substitution θ is denoted $t\theta$, and the composition of substitutions is indicated simply by concatenation, with application performed left to right. The empty substitution is denoted ϵ. Given terms t_1 and t_2, a substitution θ is a **unifier** iff $t_1\theta = t_2\theta$, and θ is the **most general unifier** of t_1 and t_2 iff for any other unifier γ of t_1 and t_2 there exists a (necessarily unique) substitution γ' such that $\theta\gamma' = \gamma$.

Definition 26. Let Σ be an MSA signature. Then a **program** over Σ is a set Γ of Σ-rules, and the **paramodulation** rule is

$$\frac{\mathbf{G} \cup \{\langle t_1, t_2 \rangle\}}{\mathbf{G}\theta \cup \langle s\theta, t\theta \rangle \cup \{(\langle t_1, t_2 \rangle|_{\pi \leftarrow r})\theta\}}$$

where $(\forall X)\ l \to r$ **if** $\langle s, t \rangle$ is a new variant[5] of a rule in Γ, $\mathbf{G}$ is a list of goals, and θ is the most general unifier of l and $\langle t_1, t_2 \rangle|_\pi$. A single inference step of this rule is indicated with $\longrightarrow\!\!\!\!\to_p$. A **narrowing** step is a paramodulation step such that $\langle t_1, t_2 \rangle|_\pi$ is *not* a variable. Let □ denote the empty list of goals. Then a chain of inference steps is called a **refutation** iff it ends in □. □

[5] Obtained by renaming all variables in the rule with names not used elsewhere.

Most implementations of ELP use a restricted form of paramodulation. The best known are narrowing and its refinements. Narrowing was introduced by Slagle [40], and later used as a basis for ***semantic unification*** algorithms (i.e., for unification modulo a set of rules [14]). ***Basic narrowing*** was introduced by Hullot [31]. The completeness of innermost narrowing for canonical term rewriting systems is due to Fribourg [15]. Hölldobler's thesis [28] gives a systematic account of this field, with interesting historical references.

Resolution as Restricted Paramodulation. Resolution can be regarded as restricted paramodulation, via the embedding of Horn clause logic into equational logic in Section 2.3.

Definition 27. Let (S, Σ, Π) be a first order signature and Γ a set of (S, Σ, Π)-clauses. Then the **resolution** rule is

$$\frac{\mathbf{G} \cup \{p(t)\}}{\mathbf{G}\theta \cup C\theta}$$

where $(\forall X)p(s)$ **if** C is a new variant of a clause in Γ, p is a relation symbol in Π, and θ is the most general unifier of s and t. The **reflection** rule is

$$\frac{\mathbf{G} \cup \{\langle t_1, t_2 \rangle\}}{\mathbf{G}\theta}$$

where **G** is a list of goals and θ is the most general unifier of t_1 and t_2. □

Fact 28. *Under the transformations in Section 2.3, a resolution step is a narrowing step followed by a reflection step.* □

This result is primarily of theoretical interest, because it would yield an inefficient operational semantics.

The Paramodulation Relation. We now introduce semantic paramodulation as a binary relation induced by a program on an arbitrary model. We assume a fixed context monoid C for $\mathcal{U}$. Recall that for simplicity we often omit the forgetful functor $\mathcal{U}$ from domain maps that underlie model morphisms, e.g., we write $s; h$ rather than $s; h\mathcal{U}$.

Definition 29. Given a set Γ of conditional $\mathcal{U}$-rules and a model A, then a binary relation $\sim$ on A is **closed under Γ-paramodulation** iff for any rule $(\forall B)\ l \rightarrow r$ **if** $\langle s, t \rangle$ in Γ, any model morphism $h : B \rightarrow A$, and any context c,

$$c_A[h(l)] \sim b \ \text{ if } \ h(s) \sim h(t) \ \text{ and } \ c_A[h(r)] \sim b\ ,$$

for any b in the domain of A. The least binary relation on A closed under reflexivity, symmetry and Γ-paramodulation is denoted $\sim^A_\Gamma$ and called the **paramodulation relation** induced by Γ. □

The least binary reflexive-symmetric relation closed under paramodulation is an algebraic abstraction of the relation on terms induced by paramodulation as a refutation rule:

Fact 30. *Let T_Σ be the initial Σ-algebra for an MSA signature Σ, i.e., the algebra of ground terms. Then for any set Γ of conditional Σ-rules,*

$$\sim_\Gamma^{T_\Sigma} = \{\langle t_1, t_2\rangle \mid \langle t_1, t_2\rangle \xrightarrow{\Gamma}{\!\!\to}_p^* \Box\} ,$$

i.e., the least relation on T_Σ closed under reflexivity, symmetry and Γ-paramodulation consists exactly of those pairs of terms for which there is a refutation over Γ using paramodulation and reflexivity[6]. □

Given a program Γ, we define **semantic paramodulation** for a model A as an inference rule on A-**goals**, which are pairs of elements from A, as follows:

$$\text{[sp]} \quad \frac{\langle h(s), h(t)\rangle \quad \langle c_A[h(r)], b\rangle}{\langle c_A[h(l)], b\rangle}$$

for any rule $(\forall B)\ l {\to} r$ **if** $\langle s, t\rangle$ in Γ, any model morphism $h : B \to A$, and any context c. This rule is considered *symmetrical*, in the sense that computation steps (e.g., narrowing) can be applied to either side of a goal.

Proposition 31. *For any model A, the least relation on A closed under reflexivity, symmetry and Γ-paramodulation exists, and is given by*

$$\sim_\Gamma^A = \bigcup_{n\in\omega} \sim_{\Gamma,n}^A$$

where $\sim_{\Gamma,n}^A$ is the reflexive-symmetric relation generated by applying at most n *semantic paramodulation steps.* □

Theorem 32. *Let Γ be a set of conditional $\mathcal{U}$-rules and fix a model A. Then:*

1. soundness: $\sim_\Gamma^A \subseteq \equiv_\Gamma^A$; *and*
2. completeness: *if A is reachable, then $\sim_\Gamma^A = \equiv_\Gamma^A$ iff $\sim_\Gamma^A$ is transitive.*

□

The proof may be found in [11, 13]. This result means that any pair of elements of a model that can be refuted with paramodulation, can also be proved with standard equational deduction, and *vice versa* if the paramodulation relation is transitive. We sum this up with the following:

> Completeness of semantic paramodulation = transitivity of the paramodulation relation.

[6] This means eliminating trivial goals, which differs from the way reflexivity is used in equational deduction, because refutations run in the opposite direction.

We now show that $\sim_\Gamma^A$ is transitive when backward application of rules in Γ is allowed. Let $\overline{\Gamma}$ denote the set of conditional $\mathcal{U}$-rules obtained by reversing the orientation of rules in Γ, i.e.,

$$\overline{\Gamma} = \{(\forall B) r \rightarrow l \textbf{ if } \langle s,t\rangle \mid (\forall B)\ l \rightarrow r \textbf{ if } \langle s,t\rangle \in \Gamma\}\ .$$

Proposition 33. *Let Γ be a set of conditional $\mathcal{U}$-rules. Then for any model A, $\sim^A_{\Gamma\cup\overline{\Gamma}}$ is transitive.* □

The completeness of semantic paramodulation under bidirectional applications is given by the following:

Corollary 34. *Let Γ be a set of conditional $\mathcal{U}$-rules. Then for any reachable model A, we have $\sim^A_{\Gamma\cup\overline{\Gamma}} = \equiv^A_\Gamma$.* □

Paramodulation Modulo a Model Morphism. This subsection studies the relationship between provability by paramodulation in a model and provability by paramodulation in the quotient model. A standard example is the quotient of an initial model (of ground terms) modulo axioms (see Section 4.3). The following inference rule gives a way to integrate quotienting into the proof system:

Definition 35. Let $\sim_\Gamma^{A,f}$ be the least reflexive-symmetric relation closed under Γ-paramodulation and under the rule

$$[\mathsf{modf}]\quad \frac{Q}{ker(f)\circ Q\circ ker(f)}$$

where Q is any binary relation on the domain of A. □

The relation $\sim_\Gamma^{A,f}$ exists and can be obtained in the manner of Proposition 31 by an alternation of Γ-paramodulation steps with modf steps. The theorem below [11, 13] relates this relation with plain paramodulation. It is not hard to see that any model morphism preserves provability under paramodulation, i.e., that $\sim_\Gamma^A;f \subseteq \sim_\Gamma^{A'}$, for Γ a set of conditional $\mathcal{U}$-rules and $f\colon A \rightarrow A'$ a model morphism. But equality doesn't hold in general because this formulation ignores the rôle of quotienting in proofs.

Theorem 36. *Let $f\colon A \rightarrow A'$ be a model morphism and Γ a set of conditional $\mathcal{U}$-rules. Then*

1. $\sim_\Gamma^{A,f};f \subseteq \sim_\Gamma^{A'}$, *and*
2. $\sim_\Gamma^{A,f};f = \sim_\Gamma^{A'}$ *iff f is a coequaliser and all quantifiers in Γ are projective.*

□

Semantic paramodulation together with the rule modf define **paramodulation modulo a model morphism**. The above theorem shows that

Paramodulation modulo a model morphism is paramodulation in the quotient model.

Paramodulation modulo axioms can be seen as an example of paramodulation modulo a model morphism, because any theory determines a quotient morphism for each model A (Theorem 9) by constructing the free model over A satisfying the theory. This quotient map can be considered a semantic expression of the theory. Under the semantic approach to equational theories in Definition 6, these appear as two sides of the same concept, by regarding the kernel of a model morphism as a theory, or better as the consequences of a theory in the source of the morphism.

4.3 Confluence

Using the rules of a program as non-oriented equations can lead to very inefficient search through the space of paramodulation chains. A key first step in reducing the search space is to use the orientation of rules; this adds direction to refutation. Completeness of paramodulation with oriented rules requires confluence of the program. This section explains the relationship between transitivity of the paramodulation relation determined by a program Γ on a model A, and the confluence of Γ as a set of oriented rules.

Confluence (also called the Church-Rosser property) is central to the theory of rewriting. Confluence and termination are essential properties of rewriting systems as models of computation. Confluent and terminating rewriting systems can be used as decision procedures for equality [8, 35, 17]. Our concept of confluence for a program generalises the traditional one in that it allows any model, rather than being fixed on a model of ground terms.

Semantic Rewriting. A program determines a rewriting relation on the domain of any model:

Definition 37. Given Γ a set of conditional $\mathcal{U}$-rules, then a binary relation $\gg$ on a model A is **closed under Γ-rewriting** iff for any rule $(\forall B)\ l \to r$ **if** $\langle s, t\rangle$ in Γ and any morphism $h\colon B \to A$,

$$c_A[h(l)] \gg c_A[h(r)] \text{ if } h(s) \sim^A_\Gamma h(t)$$

for any context c. The least relation on A closed under reflexivity, transitivity and Γ-rewriting is denoted $\gg^A_\Gamma$. □

Fact 38. *Let Γ be a set of conditional $\mathcal{U}$-rules. Then for any model A, $\gg^A_\Gamma$ exists and is given by*

$$\gg^A_\Gamma = (\rho^A_\Gamma)^* \quad \textit{where} \quad \rho^A_\Gamma = \{\langle c_A[h(l)], c_A[h(r)]\rangle \mid h(s) \sim^A_\Gamma h(t)\}\ ,$$

i.e., $\gg^A_\Gamma$ is the transitive-reflexive closure of the least relation closed under Γ-rewriting. □

In Definition 37, h is a "matcher" for the left side of a rule to an element of the model. For OBJ, the model A is the initial algebra of ground terms (or the initial algebra of a theory for rewriting modulo a theory). In this case, h matches the left side of a rule in the program with a subterm of the term to be rewritten. But the rewriting is done *only after* the system proves the validity of the hypotheses instantiated by the matcher h. The algebraic formulation of this last condition is given by $h(s) \sim_{\Gamma}^{A} h(t)$, since $\sim_{\Gamma}^{A}$ contains exactly the identities in A that can be proved from Γ with paramodulation.

It is not hard to prove that the rewriting relation is preserved under model morphisms, i.e., that $\gg_{\Gamma}^{A};f \subseteq \gg_{\Gamma}^{A'}$, for any set Γ of conditional $\mathcal{U}$-rules and any model morphism $f\colon A \to A'$. But as with paramodulation, equality does not hold in general.

Definition 39. Let Γ be a set of conditional $\mathcal{U}$-rules and $f\colon A \to A'$ a model morphism. Then a binary relation $\gg$ on A is closed under **Γ-rewriting modulo** f iff for any rule $(\forall B)\ l \to r$ **if** $\langle s, t\rangle$,

$$c_A[h(l)] \gg c_A[h(r)] \text{ iff } h(s) \sim_{\Gamma}^{A,f} h(t)\,.$$

The least relation on A closed under reflexivity, transitivity, Γ-rewriting modulo f, and modf is denoted $\gg_{\Gamma}^{A,f}$. □

The following result [11, 13] in the spirit of Section 4.2 shows that rewriting modulo a model morphism is the same as rewriting in the quotient model:

Theorem 40. *Let Γ be a set of conditional $\mathcal{U}$-rules with projective quantifiers and let f be a coequaliser. Then $\gg_{\Gamma}^{A,f};f = \gg_{\Gamma}^{A'}$.* □

Transitivity and Confluence.

Definition 41. Given a model A and a set Γ of conditional $\mathcal{U}$-rules, then Γ is **A-confluent** iff the rewriting relation $\gg_{\Gamma}^{A}$ is confluent. □

A-confluence generalises the usual notion of confluence in term rewriting[7]. The simplest and best case is where A is the initial algebra T_{Σ} of ground terms over an MSA signature Σ. Section 4.3 explains the relationship between A-confluence and confluence modulo an equivalence. The following establishes the crucial link between confluence of Γ and transitivity of the paramodulation relation induced by Γ:

Proposition 42. *Given a model A and a set Γ of conditional $\mathcal{U}$-rules, then Γ is A-confluent iff $\sim_{\Gamma}^{A}$ is transitive.* □

The following shows that semantic paramodulation is complete for confluent oriented rules; a proof may be found in [11, 13].

Corollary 43. *Let A be a reachable model and Γ a set of A-confluent conditional atomic $\mathcal{U}$-rules. Then $\sim_{\Gamma}^{A} = \equiv_{\Gamma}^{A}$.* □

[7] See [17] for a detailed exposition of confluence for term rewriting systems; important surveys include [8] and [30].

Confluence Modulo a Model Morphism. This subsection shows that A-confluence (Definition 41) corresponds to confluence of rewriting on equivalence classes under the quotienting morphism defined by a theory. Rewriting on congruence classes (called **class rewriting** in [8]) was introduced by Lankford and Ballantyne [37] for *permutative* congruences, where each congruence class is finite, e.g., for associative and/or commutative axioms.

Let Σ be an MSA signature and E a set of Σ-equations. In the context of the definitions in Section 4.2, let A be the algebra T_Σ of ground terms, let A' be the initial (Σ, E)-algebra $T_{\Sigma,E}$, and let $f\colon T_\Sigma \to T_{\Sigma,E}$ be the quotient morphism. Then rewriting [paramodulation] modulo f is the same as rewriting [paramodulation] modulo E. Given a set Γ of conditional Σ-rules, the class rewriting relation defined by Γ and E (denoted Γ/E in [8, 34]) is $\gg_\Gamma^{T_\Sigma,f}$. By Theorem 40 we have:

Corollary 44. *A term rewriting system Γ is confluent modulo axioms E iff it is $T_{\Sigma,E}$-confluent.* □

Several papers [29, 34] and surveys [30, 8] study various notions of confluence modulo axioms and their relationship with confluence of class rewriting.

5 The Institution of Category-based Equational Logics

Although modularisation is basic to modern computing, it has been little studied for logic-based programming. We treat modularisation for ELP using the institution of CBELs in three different ways: (1) to provide a generic satisfaction condition for equational logics; (2) to give a CB semantics for queries and their solutions; and (3) as an abstract definition of compilation from one (equational) logic programming language to another.

Regarding (2), we study soundness and completeness for ELP queries and their solutions. This can be understood as ordinary soundness and completeness in a suitable institution. Soundness holds for all module imports, but completeness only holds for conservative module imports. Category-based equational signatures are seen as modules, and morphisms of such signatures as module imports. Regarding (3), completeness corresponds to compiler correctness. This paper presents (1) in some detail, but just outlines (2). A full development of (1) and (2) can be found in [11], while (3) is explained in [12].

5.1 Categorical Relations

This section assumes the **Basic Framework** in full generality, i.e., without the **Simplifying Assumption** that domains are many sorted sets. We therefore need a categorical definition of binary relations:

Definition 45. Let A be an object of a category $\mathbf{X}$. Then a **binary relation representation on** A is a parallel pair of arrows $s, t \in \mathbf{X}(I, A)$, denoted $I \xrightarrow{\langle s,t \rangle} A$

or just $\langle s,t\rangle$. Let $I \xrightarrow{\langle s,t\rangle} A$ and $I' \xrightarrow{\langle s',t'\rangle} A$ be binary relation representations on the same object A. Then $\langle s,t\rangle$ is **included in** $\langle s',t'\rangle$ (denoted $\langle s,t\rangle \subseteq_A \langle s',t'\rangle$, or just $\langle s,t\rangle \subseteq \langle s',t'\rangle$) iff there is a map $h: I \to I'$ between the objects of indices such that $s = h; s'$ and $t = h; t'$. Two relation representations Q and Q' on the same object A are **equivalent** (denoted $Q \equiv_A Q'$, or just $Q \equiv Q'$) iff $Q \subseteq Q'$ and $Q' \subseteq Q$. Then a **binary relation** on A is an equivalence class of $\equiv_A$. □

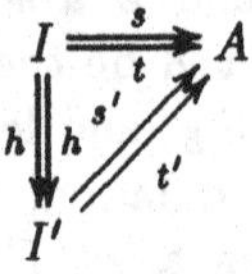

Although binary relations are classes of equivalent representations, for simplicity we often use representations instead of classes. The concept of inclusion between binary relation representations extends to binary relations proper without difficulty. The definitions of sentence and equational satisfaction (i.e., Definition 6) also extend to categorical relations without difficulty.

5.2 Institutions

Institutions are much more abstract than Tarski's model theory, and also have another basic ingredient, namely signatures and the possibility of translating sentences and models across signature morphisms. A special case of this translation is familiar in first order model theory: if $\Sigma \to \Sigma'$ is an inclusion of first order signatures and M is a Σ'-model, then we can form the *reduct* of M to Σ, denoted $M\restriction_\Sigma$. Similarly, if e is a Σ-sentence, we can always view it as a Σ'-sentence (but there is no standard notation for this). The key axiom, called the satisfaction condition, says that ***truth is invariant under change of notation***, which is surely a very basic intuition for traditional logic.

Definition 46. An **institution** $\Im = (\mathbf{Sign}, Sen, \text{Mod}, \models)$ consists of

1. a category $\mathbf{Sign}$, whose objects are called **signatures**,
2. a functor $Sen: \mathbf{Sign} \to \mathbf{Set}$, giving for each signature a set whose elements are called **sentences** over that signature,
3. a functor $\text{Mod}: \mathbf{Sign}^{op} \to \mathbf{Cat}$ giving for each signature Σ a category whose objects are called **Σ-models**, and whose arrows are called **Σ-(model) morphisms**, and
4. a relation $\models_\Sigma \subseteq |\text{Mod}(\Sigma)| \times Sen(\Sigma)$ for each $\Sigma \in |\mathbf{Sign}|$, called **$\Sigma$-satisfaction**,

such that for each morphism $\varphi: \Sigma \to \Sigma'$ in $\mathbf{Sign}$, the **satisfaction condition**

$$M' \models_{\Sigma'} Sen(\varphi)(e) \quad \text{iff} \quad \text{Mod}(\varphi)(M') \models_\Sigma e$$

holds for each $M' \in |\text{Mod}(\Sigma')|$ and $e \in Sen(\Sigma)$. We may denote the reduct functor $\text{Mod}(\varphi)$ by $\_\restriction_\varphi$ and the sentence translation $Sen(\varphi)$ by $\varphi(\_)$ or even $\_\varphi$. □

Definition 47. A **theory** (Σ, E) in an institution $\Im = (\mathbf{Sign}, Sen, \textsc{Mod}, \models)$ consists of a signature Σ and a set E of Σ-sentences closed under semantic entailment, i.e., $e \in E$ if $E \models_\Sigma e$[8].

A **theory morphism** $\varphi\colon (\Sigma, E) \to (\Sigma', E')$ is a signature morphism $\varphi\colon \Sigma \to \Sigma'$ such that $Sen(\varphi)(E) \subseteq E'$. Let $\mathbf{T}h(\Im)$ denote the category of all theories in $\Im$. □

For any institution $\Im$, the model functor $\textsc{Mod}$ extends to $\mathbf{T}h(\Im)$, by mapping a theory (Σ, E) to the full subcategory $\textsc{Mod}(\Sigma, E)$ of $\textsc{Mod}(\Sigma)$ formed by the Σ-models that satisfy E.

The principle of "initial algebra semantics" is formalised at the level of institutions using liberality [18] as follows:

Definition 48. Let $\Im = (\mathbf{Sign}, Sen, \textsc{Mod}, \models)$ be an institution. Then a theory morphism φ is **liberal** iff the reduct functor $\textsc{Mod}(\varphi)$ has a left adjoint, and it is **persistent** iff it is liberal and $\textsc{Mod}(\varphi)$ has a left inverse. An institution $\Im$ is **liberal** iff every theory morphism in $\mathbf{T}h(\Im)$ is liberal. □

Equational logics tend to be liberal, while first order logics tend not to be. Tarlecki [41] relates the liberality of an institution to the quasi-variety property satisfied by the class of models of every theory in the institution, namely closure under products and submodels[9].

5.3 The Satisfaction Condition

To get an institution for CBEL, we need to define signature morphisms for CBEL, and define how models and sentences translate along signature morphisms; in particular, we need to know how quantifiers translate along signature morphisms. Then we must check that the satisfaction relation between CBEL models and the sentences in Definition 6 satisfies the satisfaction condition. It will help to first look closer at the MSA case:

Definition 49. An MSA **signature morphism** $\varphi\colon (S, \Sigma) \to (S', \Sigma')$ is a pair $\langle f, g\rangle$, where $f\colon S \to S'$ is a map on sorts and g is an $S^* \times S$-indexed family of maps $g_{u,s}\colon \Sigma_{u,s} \to \Sigma'_{f^*(u), f(s)}$ on operation symbols. □

Any MSA signature morphism φ determines a forgetful functor $\mathcal{A}lg(\varphi)\colon \mathcal{A}lg_{\Sigma'} \to \mathcal{A}lg_\Sigma$ on models and another $\mathbf{Set}^f\colon \mathbf{Set}^{S'} \to \mathbf{Set}^S$ on domains, such that

$$\begin{array}{ccc} \mathcal{A}lg_{\Sigma'} & \xrightarrow{\;\mathcal{U}'\;} & \mathbf{Set}^{S'} \\ {\scriptstyle \mathcal{A}lg(\varphi)}\downarrow & & \downarrow{\scriptstyle \mathbf{Set}^f} \\ \mathcal{A}lg_{\Sigma} & \xrightarrow{\;\mathcal{U}\;} & \mathbf{Set}^{S} \end{array}$$

[8] Meaning that $M \models_\Sigma e$ for any Σ-model M that satisfies all sentences in E.

[9] For the usual logical systems, this corresponds exactly to the power of Horn clause axiomatisations.

commutes, where $\mathcal{U}$ and $\mathcal{U}'$ are the corresponding forgetful functors from many sorted algebras to many sorted sets. Each forgetful functor has a left adjoint, so that any model has a free extension along a signature morphism[10]. Forgetting model structure first along a signature morphism and then to domains is the same as forgetting first to domains and then along domain structure. These ideas are formalised by the following:

Definition 50. **A category-based equational signature** is a functor $\mathcal{U} : \mathbb{A} \to \mathbb{X}$, and a **morphism of category-based equational signatures** is a pair of functors $\langle \mathcal{M}, \mathcal{D} \rangle : \mathcal{U} \to \mathcal{U}'$ such that $\mathcal{M};\mathcal{U} = \mathcal{U}';\mathcal{D}$ and $\mathcal{D}$ has a left adjoint. □

$$\begin{array}{ccc} \mathbb{A}' & \xrightarrow{\mathcal{U}'} & \mathbb{X}' \\ {\scriptstyle \mathcal{M}}\downarrow & & \downarrow{\scriptstyle \mathcal{D}} \\ \mathbb{A} & \xrightarrow{\mathcal{U}} & \mathbb{X} \end{array}$$

Notice (see Definition 48) that a morphism of CB equational signatures is liberal iff $\mathcal{M}$ has a left adjoint.

The following shows the analogy of concepts in MSA and CBEL:

MSA	CBEL
signature (S, Σ)	functor $\mathcal{U} : \mathbb{A} \to \mathbb{X}$
S	$\mathbb{X}$
Σ	$\mathbb{A}$
$\varphi = \langle f, g \rangle$	$\langle \mathcal{M}, \mathcal{D} \rangle$
f	$\mathcal{D}$
g	$\mathcal{M}$
$\mathbb{S}et^f$	$\mathcal{D}$
$\mathbb{A}lg(\varphi)$	$\mathcal{M}$
Σ-equation	$\mathcal{U}$-equation

Before defining translations of equations along CB equational signature morphisms, we look again at the many sorted case. Some readers may first wish to review Kan extensions, e.g., in [36].

Example 3. A function $f : S \to S'$ translates an S-sorted set X into the S'-sorted set $X^{\sim}$ by taking the (pointwise) left Kan extension of f along X:

$$X^{\sim}_{s'} = \coprod_{f(s)=s'} X_s \quad \text{for any sort} \quad s' \in S' .$$

$$\begin{array}{ccc} S & \xrightarrow{f} & S' \\ & {\scriptstyle X}\searrow & \\ & & \mathbb{S}et \end{array} \qquad \begin{array}{c} S' \\ \downarrow{\scriptstyle X^{\sim}} \\ \mathbb{S}et \end{array}$$

[10] While free extensions along theory morphisms are problematic in many logical systems, most still support free extensions along signature morphisms, including first order logic.

An MSA signature morphism $\varphi = \langle f, g\rangle\colon (S, \Sigma) \to (S', \Sigma')$ defines an S-sorted map $\varphi^*_X\colon T_\Sigma(X) \to T_{\Sigma'}(X^\sim)\restriction_\varphi$ as follows:

$$\begin{array}{ccc} X & \xrightarrow{X\eta} & \mathcal{U}(T_\Sigma(X)) \\ & \searrow^{j} & \downarrow{\scriptstyle j^!\mathcal{U}=\varphi^*_X} \\ & & \mathcal{U}(T_{\Sigma'}(X^\sim)\restriction_\varphi) \end{array}$$

First note that $X \subseteq \mathcal{U}(T_{\Sigma'}(X^\sim)\restriction_\varphi)$ because if $x \in X_s$ then $x \in X^\sim_{f(s)}$ and $X^\sim_{f(s)} \subseteq T_{\Sigma'}(X^\sim)_{f(s)} = (T_{\Sigma'}(X^\sim)\restriction_\varphi)_s$; let $j\colon X \to \mathcal{U}(T_{\Sigma'}(X^\sim)\restriction_\varphi)$ denote this inclusion. Now define $\varphi^*_X = j^!$, where $j^!$ is the unique extension of j to a Σ-homomorphism $T_\Sigma(X) \to T_{\Sigma'}(X^\sim)\restriction_\varphi$. Then a Σ-equation $(\forall X)\langle t_1, t_2\rangle$ translates to the Σ'-equation $(\forall X^\sim)\langle \varphi^*_X(t_1), \varphi^*_X(t_2)\rangle$. □

Notice that the term algebra $T_{\Sigma'}(X^\sim)$ is exactly the free extension of $T_\Sigma(X)$ along φ. From this, we conclude that:

Translations of quantifiers are free extensions along signature morphisms.

This also covers quantifiers that are not free models. The translation of equations along signature morphisms in MSA is a particular case of the following:

Definition 51. Let $\langle \mathcal{M}, \mathcal{D}\rangle$ be a liberal morphism of category-based equational signatures $(\mathbb{A} \xrightarrow{\mathcal{U}} \mathbb{X}) \to (\mathbb{A}' \xrightarrow{\mathcal{U}'} \mathbb{X}')$. Then the $\mathcal{U}$-equation $(\forall A)\langle s, t\rangle$ translates to the $\mathcal{U}'$-equation $(\forall A^{\mathcal{M}})\langle s^*, t^*\rangle$,

$$\begin{array}{ccc} I & \xrightarrow{I\theta} & I^{\mathcal{D}}\mathcal{D} \\ {\scriptstyle t}\downarrow\downarrow{\scriptstyle s} & & {\scriptstyle t^*\mathcal{D}}\downarrow\downarrow{\scriptstyle s^*\mathcal{D}} \\ A\mathcal{U} & \xrightarrow[A\alpha\mathcal{U}]{} & A^{\mathcal{M}}\mathcal{M}\mathcal{U} = A^{\mathcal{D}}\mathcal{U}'\mathcal{D} \end{array}$$

where $\_^{\mathcal{D}}$ denotes the left adjoint of $\mathcal{D}$, $\_^{\mathcal{M}}$ denotes the left adjoint of $\mathcal{M}$, α and θ denote the units of the adjunctions determined by $\mathcal{M}$ and $\mathcal{D}$, and s^* and t^* denote the unique "extensions" of $s; A\alpha\mathcal{U}$ and $t; A\alpha\mathcal{U}$ to maps in $\mathbb{X}'$. Similarly, the $\mathcal{U}$-query $(\exists A)\langle s, t\rangle$ translates to the $\mathcal{U}'$-query $(\exists A^{\mathcal{M}})\langle s^*, t^*\rangle$. □

The following result (from [11, 13]) is the satisfaction condition for equational logic systems; it extends to conditional equations without difficulty.

Theorem 52. *Let $\langle \mathcal{M}, \mathcal{D}\rangle$ be a liberal morphism of category-based equational signatures $(\mathbb{A} \xrightarrow{\mathcal{U}} \mathbb{X}) \to (\mathbb{A}' \xrightarrow{\mathcal{U}'} \mathbb{X}')$. Then for any model $B \in |\mathbb{A}'|$ and any sentence $(\lambda A)\langle s, t\rangle$ with $\lambda \in \{\forall, \exists\}$,*

$$B \models_{\mathcal{U}'} (\lambda A^{\mathcal{M}})\langle s^*, t^*\rangle \quad \textit{iff} \quad B\mathcal{M} \models_{\mathcal{U}} (\lambda A)\langle s, t\rangle\ .$$

□

A proof of institutionality for each example in Section 2.3 can be obtained by specialising the proof of the above.

5.4 Many Sorted Institutions

This subsection introduces a class of institutions where the notion of sort is made explicit in signatures. Then the category of domains for a theory is in fact the category of models for the simple signature having just the sorts of the signature of the theory. Assuming some liberality of the institution, the forgetful functor from the models of the theory to domains has a left adjoint. Any such institution can be regarded as a fragment of the institution of category-based equational logics in a natural way. This gives us a generic way to show that MSA, OSA, HCL, ELM, etc. are CBELs.

Definition 53. A many sorted institution is a tuple of the form

$$(\Im = (\mathbf{Sign}, \mathcal{S}ort, \mathrm{MOD}, Sen, \models)$$

such that

- $(\mathbf{Sign}, \mathrm{MOD}, Sen, \models)$ is an institution,
- $\mathcal{S}ort\colon \mathbf{Sign} \to \mathbf{Set}$ is a functor having a left adjoint left inverse $\mathcal{Q}$, and
- $\Im$ is liberal on signature morphisms.

A **domain** in $\Im$ is a signature of the form $\mathcal{Q}(S)$ for S an arbitrary set. □

Now the main result:

Proposition 54. *Let $\Im = (\mathbf{Sign}, \mathcal{S}ort, \mathrm{MOD}, Sen, \models)$ be a many sorted institution with ε the co-unit of the persistent adjunction $\mathcal{Q} \dashv \mathcal{S}ort\colon \mathbf{Set} \to \mathbf{Sign}$. Then any signature morphism $\varphi\colon \Sigma \to \Sigma'$ determines a liberal morphism of category-based equational signatures*

$$\langle \mathrm{MOD}(\varphi), \mathrm{MOD}(\mathcal{Q}(\mathcal{S}ort(\varphi)))\rangle : \mathcal{U}_\Sigma \to \mathcal{U}_{\Sigma'} \ ,$$

where $\mathcal{U}_\Sigma = \mathrm{MOD}(\varepsilon_\Sigma)$ is the forgetful functor from the category $\mathrm{MOD}(\Sigma)$ of Σ-models to the category $\mathrm{MOD}(\mathcal{Q}(\mathcal{S}ort(\Sigma)))$ of domains. □

The proof may be found in [11, 13]. The liberality condition in Definition 53 is very mild in practice. Even institutions notorious for not being liberal, like first order logic, are still liberal on signature morphisms.

Corollary 55. *The signature morphisms in MSA, OSA, HCL, ELM are all liberal morphisms of CB equational signatures.* □

5.5 Queries, Solutions and Modularisation

This section gives a categorical semantics for ELP queries and their solutions in the context of modularisation in the style of OBJ and Eqlog, and discusses the soundness and completeness of module imports. We take the view of [9] that modules are presentations and that module imports are morphisms of presentations. A denotational semantics for ELP is given in [21] based on initial algebra semantics. Due to the presence of logical variables, the denotation of an ELP module is given by an adjunction rather than an initial model, in fact the adjunction determined by the forgetful functor from the category of models of the given module to the category of domains representing the mathematical structure for collections of logical variables. Our approach exploits the fact that CB equational signatures are abstract enough to include ELP modules[11]. The principle underlying our semantics for ELP queries and their solutions is formulated as follows:

> **Module denotation is abstracted to category-based equational signatures with left adjoints.**

Definition 56. Let P be an ELP module. Then its **denotation** $[\![P]\!]$ is the forgetful functor $[\![P]\!]\colon \mathrm{MOD}(P) \to \mathrm{DOM}(P)$ from its models to its domains. □

For the purposes of this paper, it is not necessary to define the denotation of a module import $P \xrightarrow{\iota} P'$; we only need it to be a morphism of CB equational signatures, $[\![\iota]\!]\colon [\![P]\!] \to [\![P']\!]$.

Definition 57. A **query** for an ELP module P is a $[\![P]\!]$-query. A **solution** for a query $(\exists B)\langle t_1, t_2\rangle$ in a P-model A is a morphism $h\colon B \to A$ such that $t_1; h[\![P]\!] = t_2; h[\![P]\!]$.

Let $P \xrightarrow{\iota} P'$ be a module import. Then the **translation** of queries along ι (i.e., from P-queries to P'-queries) is given by translation along the morphism of category-based equational signatures $[\![\iota]\!]$ as described in Definition 51. We denote this translation by $\_\iota$. □

The interpretation of the satisfaction condition (Theorem 52) in this context is that for any P-query q, any module import $\iota\colon P \to P'$, and any P'-model A, there is a canonical one-to-one correspondence between the solutions of $q\iota$ in A and the solutions of q in $A\mathcal{M}$, where $\mathcal{M}$ is the model reduct component of $[\![\iota]\!]$.

Definition 58. Let $\iota\colon P \to P'$ be a module import. Then ι is **sound** iff for any P-query q and any solution form s for q, $s\iota$ is a solution form for $q\iota$; also ι is **complete** iff for any P-query q and any solution form s' for $q\iota$ there exists a solution form s for q such that $s' = s\iota$. □

A sound and complete module import $P \to P'$ "protects"[12] solution forms, in the sense that any P-query has the same solutions in P' as in P.

Fact 59. *The composition of sound [complete] module imports is sound [complete].* □

There is great similarity between soundness and completeness for module imports and for logical systems. In fact, [11, 12] show that both are instances of a categorical formulation of soundness and completeness for institutions.

---

[11] This is technically similar to the way ELM is encoded as a CBEL.

[12] Compare this with "protecting imports" in the sense of OBJ [24].

Definition 60. A morphism $\langle \mathcal{M}, \mathcal{D} \rangle : (\mathsf{A} \xrightarrow{u} \mathsf{X}) \to (\mathsf{A}' \xrightarrow{u'} \mathsf{X}')$ of category-based equational signatures is **essentially persistent** iff it is liberal and the adjunctions corresponding to both $\mathcal{M}$ and $\mathcal{D}$ are persistent. A module import ι is **essentially persistent** iff its denotation $[\![\iota]\!]$ is an essentially persistent morphism of category-based equational signatures. □

When domains are many sorted sets, persistence of the adjunction on domains corresponds exactly to injectivity on sorts of the module import; this relates to the use of persistency for the *protecting extensions* of built-ins in Eqlog [21]. The proof of the following may be found in [11, 12]:

Theorem 61. Completeness *Let $P \xrightarrow{\iota} P'$ be a module import. Then*

1. *ι is sound, and*
2. *ι is complete if it is essentially persistent.*

□

6 Extensible Constraint Logic Programming

Constraint programming is a powerful paradigm with many applications. In general, constraint programming systems solve constraints over a fixed collection of data types or computational domains[13]; here we treat constraint programming over arbitrary user-definable data types, and call this *extensible* CLP.

This section presents a model theoretic semantics for constraint logic programming, without directly addressing the computational aspect. Our approach departs from the usual by following [21] in proving a Herbrand theorem for *constraint logic*, which is the logic underlying extensible constraint logic programming. As with the CLP approach of Jaffar and Lassez [33], both constraint relations and programs are (sets of) sentences in the same logical system. But our constraint logics are much more general than Horn clause logic. Also, the computational domain plays a central rôle in our definition of constraint logic, rather than being axiomatised in Horn clause logic, as in [33].

When regarded as a model in constraint logic, the computational domain is an *initial* model. This is mathematically linked to the semantics of OBJ-like module systems, the fundamental idea being to regard the models of extensible constraint logic programming as expansions of an appropriate ***built-in model** A* along a signature inclusion $\iota : \Sigma \hookrightarrow \Sigma'$, where Σ is the signature of built-in sorts, operations and relations, and Σ' adds new "logical" symbols. In practice, the constraint relations (i.e., the logical relations one wishes to impose on potential solutions) are limited to atomic sentences involving both Σ-symbols and elements of the built-in model A. However, at the theory level there is no reason to restrict constraint relations to be atomic formulae. The models for constraint

[13] A computational domain can be regarded as a model (not necessarily the standard one) of a certain data type specification.

logic programming are expansions of the built-in model to the larger signature Σ', and morphisms of constraint models must preserve the built-ins. Thus the constraint models form a comma category, $(A{\downarrow}\mathrm{MOD}(\iota))$.

6.1 Generalised Polynomials and Constraint Satisfaction

Consistent with our previous notation, let A^{ι} denote the free expansion of the built-in model A along the inclusion (of the the built-ins' signature) $\iota\colon \Sigma \to \Sigma'$, and let $\mathcal{F}'$ be a left adjoint to the forgetful functor $\mathcal{U}'\colon \mathrm{MOD}(\Sigma') \to \mathrm{DOM}(\Sigma')$[14]. The rôle played by terms in ordinary logic is played by ***generalised polynomials***[15] in our constraint logic. They are term-like structures involving both operator symbols and elements of the built-in model. Generalised polynomials can be regarded as elements of models in the same way that ordinary terms were regarded as elements of (free) models in our semantic approach to sentence and satisfaction in CBEL.

Given a domain X (in practice a domain of variables), the Σ'-model of polynomials over X is the coproduct $A^{\iota} \coprod X\mathcal{F}'$ of A^{ι} with the free Σ'-model $X\mathcal{F}'$, denoted $A[X]$. When $\Sigma = \Sigma'$ are unsorted algebraic signatures, this is a well known construction in universal algebra [25]. But the best known example is the usual polynomial rings of linear algebra. The universal property of the model of generalised polynomials allows generalising from collections of variables to models, and from valuation maps to model morphisms.

Definition 62. Let B be any Σ'-model. Then the **model of generalised polynomials over** B is the coproduct $A^{\iota} \coprod B$, denoted $A[B]$. □

Constraint logic can be defined for any CBEL, by abstracting the signature of built-ins Σ to a CB equational signature $\mathcal{U}$, Σ' to $\mathcal{U}'$, and the inclusion $\iota\colon \Sigma \to \Sigma'$ to a morphism of CB equational signatures $\mathcal{U} \to \mathcal{U}'$. In this way, we extend the extensible constraint logic programming paradigm to any CBEL.

Definition 63. Let $\langle \mathcal{M}, \mathcal{D}\rangle\colon (\mathbb{A} \xrightarrow{\mathcal{U}} \mathbb{X}) \to (\mathbb{A}' \xrightarrow{\mathcal{U}'} \mathbb{X}')$ be a liberal morphism of category-based equational signatures. Fix a model $A \in |\mathbb{A}|$ (for the **built-ins**). Then a **constraint model** is a model in $\mathbb{A}'$ whose reduct to the signature of the built-ins contains an image of A, i.e., a map $c\colon A \to C\mathcal{M}$ with $C \in |\mathbb{A}'|$. A **model morphism** $c \to c'$ is a map $h\colon C \to C'$ in $\mathbb{A}'$ such that the diagram below commutes:

$$\begin{array}{ccc} A & \xrightarrow{c} & C\mathcal{M} \\ & \searrow^{c'} & \downarrow h\mathcal{M} \\ & & C'\mathcal{M} \end{array}$$

[14] From the category of the models of the signature Σ' to the category of domains of Σ'.

[15] The ordinary polynomials from linear algebra are a special case. The word "generalised" here plays the same rôle as "general" in the phrase "general algebra."

A constraint identity in $B \in |\mathbb{A}'|$ is a binary relation $I \xrightarrow{\langle s,t \rangle} (A[B])\mathcal{U}'$. An identity $\langle s,t \rangle$ in B is satisfied by a model $A \xrightarrow{c} C\mathcal{M}$ with respect to a model morphism $f\colon B \to C$ iff $s; [f, c^\dagger]\mathcal{U}' = t; [f, c^\dagger]\mathcal{U}'$, where $c^\dagger$ is the unique extension of c to a model morphism $A^{\mathcal{M}} \to C$.

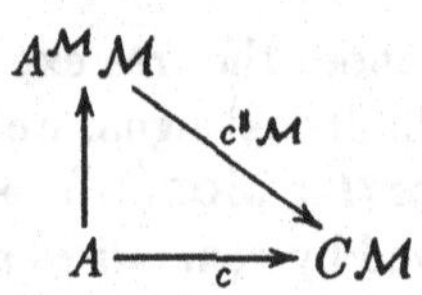

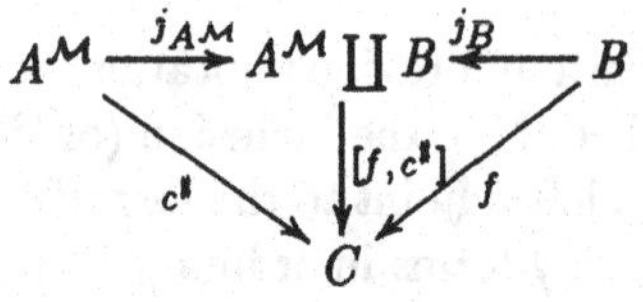

This definition extends to **constraint equations, queries and their satisfaction** by constraint models in the same manner as in Definition 6. □

The key technical idea of our approach to the semantics of constraint logic programming is to see constraint logic as a CBEL. The usual concrete algebraic and model theoretic approaches cannot accommodate this logic because their notions of signature are not abstract enough. We consider this a good example of the benefits our category-based approach. We first summarize this, and then formulate it precisely:

> Constraint logic = equational logic in a special category-based equational signature.

Definition 64. Let $\langle \mathcal{M}, \mathcal{D} \rangle \colon (\mathbb{A} \xrightarrow{\mathcal{U}} \mathbf{X}) \to (\mathbb{A}' \xrightarrow{\mathcal{U}'} \mathbf{X}')$ be a liberal morphism of category-based equational signatures. Then any model $A \in |\mathbb{A}|$ determines a forgetful functor $\mathcal{U}'_A \colon (A {\downarrow} \mathcal{M}) \to \mathbf{X}'$ such that $\mathcal{U}'_A = \mathcal{M}_A ; \mathcal{U}'$, where $\mathcal{M}_A$ is the forgetful functor $(A {\downarrow} \mathcal{M}) \to \mathbb{A}'$. □

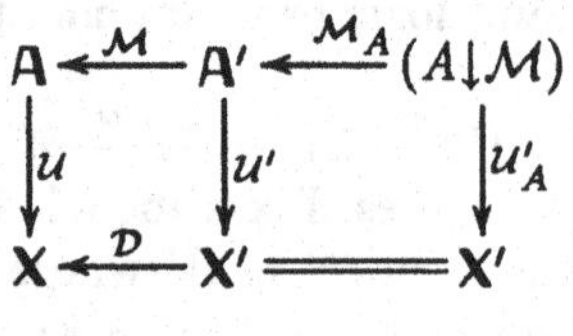

In this way the constraint logic introduced in Definition 63 is the CBEL determined by the forgetful functor $\mathcal{U}'_A$.

6.2 A Herbrand Theorem for Extensible Constraint Logic Programming

Our Herbrand theorem for constraint logic provides mathematical foundations for constraint solving. The approach is to instantiate the category-based Herbrand theorem (18) to constraint logic viewed as the CBEL determined by the forgetful functor $\mathcal{U}'_A$ of Definition 64. This version assumes that domains are many sorted sets; the more general version appears in [11].

Theorem 65. <u>*Herbrand Theorem*</u> *Let $\langle \mathcal{M}, \mathcal{D} \rangle : (\mathbb{A} \xrightarrow{\mathcal{U}} \mathbf{Set}^S) \to (\mathbb{A}' \xrightarrow{\mathcal{U}'} \mathbf{Set}^{S'})$ be a liberal morphism of category-based equational signatures, and let $A \in |\mathbb{A}|$ be a model. Assume the* **Deduction Framework** *for $\mathcal{U}'$, and that $\mathcal{U}'$ has a left adjoint $\mathcal{F}'$ and preserves filtered colimits. Let Γ be a set of conditional constraint equations with projective quantifiers, and let $(\exists B)q$ be a $\mathcal{U}'$-constraint query where B is projective. Then:*

1. *the initial Γ-constraint model 0_Γ exists;*
2. *$\Gamma \models (\exists B)q$ iff $0_\Gamma \models (\exists B)q$; and*
3. *if $\mathbb{A}'$ has non-empty sorts, then $\Gamma \models (\exists B)q$ iff $\Gamma \models (\forall y)q;[h, j_{A^{\mathcal{M}}}]$ for some S'-sorted set Y and some model morphism $h : B \to A[Y]$.*

□

In practice, the sentences in Γ don't usually involve elements of the built-in model A (i.e., Γ contains only Σ'-sentences, in the notation of the discussion opening this section), although queries may involve built-in constants. In this case, the initial constraint model 0_Γ has a simpler representation as a quotient of the free expansion of the built-in model. In our CB framework, $\mathcal{U}'$-sentences play the rôle of Σ'-sentences, and they can be canonically viewed as constraint sentences (i.e., $\mathcal{U}'_A$-sentences) via the translation along the morphism of CB equational signatures $\langle \mathcal{M}_A, 1_{\mathbf{Set}^{S'}} \rangle : \mathcal{U}' \to \mathcal{U}'_A$ (see Definition 51). The proof of the following is in [11].

Proposition 66. *Assuming the hypotheses of Theorem 65, suppose that Γ contains only $\mathcal{U}'$-equations. Then the initial constraint model 0_Γ is isomorphic to the canonical map $!_\Gamma = A \xrightarrow{A\eta} A^{\mathcal{M}}\mathcal{M} \xrightarrow{e\mathcal{M}} (A^{\mathcal{M}}/{\equiv_\Gamma})\mathcal{M}$, where $\equiv_\Gamma$ is the least congruence on $A^{\mathcal{M}}$ closed under Γ-substitutivity.* □

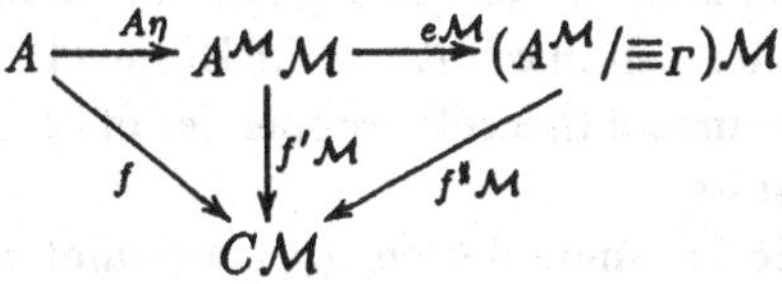

For order sorted Horn clause logic with equality, [21] proves the existence of initial constraint models when Γ contains only Σ'-sentences. This result is crucial for the semantics of extensible contraint logic programming in Eqlog, and follows from the above results.

As pointed out in [21], the notion of *protecting expansion* gives the right semantic condition for built-ins. This means that 0_Γ must *protect* the built-in model A, i.e., that 0_Γ is an isomorphism $A \cong (A^{\mathcal{M}}/{\equiv_\Gamma})\mathcal{M}$, where $\equiv_\Gamma$ is the least congruence on $A^{\mathcal{M}}$ closed under Γ-substitutivity. For order sorted Horn clause logic with equality, [21] gives conditions that guarantee protection but impose some restrictions on the sentences in Γ; these restrictions are almost always met in practice.

7 Conclusions and Open Problems

Category-based equational logic is a powerful new formalism that encompasses many applications beyond those that are usually considered equational, including Horn clause logic and constraint logic programming. This paper has developed both theoretical and computational aspects of CBEL, including CB equational deduction and its completeness, paramodulation modulo a morphism, semantic rewriting, confluence modulo a morphism, the CBEL institution, modularisation for CBEL, soundness and completeness for module imports (with respect to queries), semantics for extensible constraint logic programming (including generalised polynomials), and finally a Herbrand theorem for extensible constraint logic programming over any suitable CBEL, an entirely new result that uses nearly everything in the paper. This result provides mathematical foundations for powerful programming paradigms, e.g., that of Eqlog, which combines order sorted Horn clause logic with equality modulo axioms, plus constraints over arbitrary user-definable data types.

The following are some open problems in the theory of category-based equational logic:

1. Use the CB framework to develop equational logic programming over novel structures; e.g., let **X** be CPOs or graphs. This could integrate equational and constraint logic programming with paradigms like higher-order programming, object orientation, and concurrency.
2. Look for meaningful CBEL equations with non-free projective quantifiers.
3. Prove that closure under operations is equivalent to closure under context evaluation for reachable models, at the same level of abstraction as the CB treatment of semantic paramodulation. This involves the categorical treatment of finiteness for model operations.
4. Investigate novel notions of rewriting and paramodulation by suitably instantating context monoids in the CB treatment of semantic paramodulation.
5. Treat so-called "Lifing Lemmas" at the CBEL level.
6. Investigate the basic model theoretic properties of constraint logic, including axiomatisability results.
7. Investigate semantic paramodulation for constraint programming, and develop theory and technology for extensible modular constraint programming.

References

1. Hajnal Andréka and István Németi. A general axiomatizability theorem formulated in terms of cone-injective subcategories. In B. Csakany, E. Fried, and E.T. Schmidt, editors, *Universal Algebra*, pages 13–35. North-Holland, 1981. Colloquia Mathematics Societas János Bolyai, 29.
2. Jon Barwise. Axioms for abstract model theory. *Annals of Mathematical Logic*, 7:221–265, 1974.
3. Jon Barwise and Solomon Feferman. *Model-Theoretic Logics*. Springer, 1985.

4. M. Bauderon and Bruno Courcelle. Graph expressions and graph rewritings. *Math. Systems Theory*, 20, 1987.
5. Garrett Birkhoff. On the structure of abstract algebras. *Proceedings of the Cambridge Philosophical Society*, 31:433–454, 1935.
6. Virgil Căzănescu. Local equational logic. In Zoltan Esik, editor, *Proceedings, 9th International Conference on Fundamentals of Computation Theory FCT'93*, pages 162–170. Springer-Verlag, 1993. Lecture Notes in Computer Science, Volume 710.
7. Nachum Dershowitz. Computing with rewrite rules. Technical Report ATR-83(8478)-1, The Aerospace Corp., 1983.
8. Nachum Dershowitz and Jean-Pierre Jouannaud. Rewriting systems. In Jan van Leeuwen, editor, *Handbook of Theoretical Computer Science, Volume B: Formal Methods and Semantics*, pages 243–320. North-Holland, 1990.
9. Răzvan Diaconescu, Joseph Goguen, and Petros Stefaneas. Logical support for modularisation. In Gerard Huet and Gordon Plotkin, editors, *Logical Environments*, pages 83–130. Cambridge, 1993. Proceedings of a Workshop held in Edinburgh, Scotland, May 1991.
10. Răzvan Diaconescu. The logic of Horn clauses is equational. Technical Report PRG-TR-3-93, Programming Research Group, University of Oxford, 1990.
11. Răzvan Diaconescu. *Category-based Semantics for Equational and Constraint Logic Programming.* PhD thesis, Oxford University, 1994.
12. Răzvan Diaconescu. Category-based modularisation for equational logic programming. submitted for publication.
13. Răzvan Diaconescu. Completeness of category-based equational deduction. *Mathematical Structures in Computer Science*, to appear 1994.
14. M. Fay. First-order unification in an equational theory. In *Proceedings, 4th Workshop on Automated Deduction, Austin, Texas*, pages 161–167, 1979.
15. Laurent Fribourg. SLOG: A logic programming language interpreter based on clausal superposition and rewriting. In *Proceedings, SLP '85*, pages 172–185. 1985.
16. Joseph Goguen. Reusing and interconnecting software components. *Computer*, 19(2):16–28, February 1986. Reprinted in *Tutorial: Software Reusability*, Peter Freeman, editor, IEEE Computer Society, 1987, pages 251–263, and in *Domain Analysis and Software Systems Modelling*, Rubén Prieto-Díaz and Guillermo Arango, editors, IEEE Computer Society, 1991, pages 125–137.
17. Joseph Goguen. *Theorem Proving and Algebra.* MIT, to appear 1995.
18. Joseph Goguen and Rod Burstall. Institutions: Abstract model theory for specification and programming. *J. Assoc. Computing Machinery*, 39(1):95–146, January 1992. Also Report ECS-LFCS-90-106, Computer Science Department, University of Edinburgh, January 1990.
19. Joseph Goguen and Răzvan Diaconescu. An Oxford survey of order sorted algebra. *Mathematical Structures in Computer Science*, 4:363–392, 1994.
20. Joseph Goguen and José Meseguer. Completeness of many-sorted equational logic. *Houston Journal of Mathematics*, 11(3):307–334, 1985. Preliminary versions appeared in *SIGPLAN Notices*, July 1981, Volume 16, Number 7, pages 24–37, and SRI Computer Science Lab, Report CSL-135, May 1982.
21. Joseph Goguen and José Meseguer. Models and equality for logical programming. In Hartmut Ehrig, Giorgio Levi, Robert Kowalski, and Ugo Montanari, editors, *Proceedings, 1987 TAPSOFT*, pages 1–22. Springer, 1987. Lecture Notes in Computer Science, Volume 250.
22. Joseph Goguen and José Meseguer. Order-sorted algebra I: Equational deduction for multiple inheritance, overloading, exceptions and partial operations. *Theoreti-*

cal Computer Science, 105(2):217–273, 1992. Also Programming Research Group Technical Monograph PRG–80, Oxford University, December 1989.

23. Joseph Goguen, James Thatcher, and Eric Wagner. An initial algebra approach to the specification, correctness and implementation of abstract data types. Technical Report RC 6487, IBM T.J. Watson Research Center, October 1976. In *Current Trends in Programming Methodology, IV*, Raymond Yeh, editor, Prentice-Hall, 1978, pages 80–149.
24. Joseph Goguen, Timothy Winkler, José Meseguer, Kokichi Futatsugi, and Jean-Pierre Jouannaud. Introducing OBJ. In Joseph Goguen, editor, *Algebraic Specification with OBJ: An Introduction with Case Studies*. Cambridge, to appear 1995. Also Technical Report, SRI International.
25. George Gratzer. *Universal Algebra*. Springer, 1979.
26. William S. Hatcher. Quasiprimitive categories. *Math. Ann.*, (190):93–96, 1970.
27. Horst Herrlich and C.M.Ringel. Identities in categories. *Can. Math. Bull.*, (15):297–299, 1972.
28. Steffen Hölldobler. Foundations of equational logic programming. In *Lecture Notes in Artificial Intelligence*, number 353. Springer Verlag, 1988.
29. Gérard Huet. Confluent reductions: Abstract properties and applications to term rewriting systems. *J. Assoc. Computing Machinery*, 27(4):797–821, 1980. Preliminary version in *Proceedings*, 18th IEEE Symposium on Foundations of Computer Science, IEEE, 1977, pages 30–45.
30. Gérard Huet and Derek Oppen. Equations and rewrite rules: A survey. In Ron Book, editor, *Formal Language Theory: Perspectives and Open Problems*, pages 349–405. Academic, 1980.
31. Jean-Marie Hullot. Canonical forms and unification. In Wolfgang Bibel and Robert Kowalski, editors, *Proceedings, 5th Conference on Automated Deduction*, pages 318–334. Springer, 1980. Lecture Notes in Computer Science, Volume 87.
32. Horst Herrlich, Jiri Adamek and George Strecker. *Abstract and Concrete Categories*. John Wiley, 1990.
33. Joxan Jaffar and Jean-Louis Lassez. Constraint logic programming. In *14th ACM Symposium on the Principles of Programming languages*, pages 111–119. 1987.
34. Jean-Pierre Jouannaud and Hélène Kirchner. Completion of a set of rules modulo a set of equations. *Proceedings, 11th Symposium on Principles of Programming Languages*, 1984. In *SIAM Journal of Computing*.
35. Jan Willem Klop. Term rewriting systems: from Church-Rosser to Knuth-Bendix and beyond. In Samson Abramsky, Dov Gabbay, and Tom Maibaum, editors, *Handbook of Logic in Computer Science*. Oxford, 1992.
36. Saunders Mac Lane. *Categories for the Working Mathematician*. Springer, 1971.
37. Dallas Lankford and A.M. Ballantyne. Decision procedures for simple equational theories with permutative axioms: Complete sets of permutative reductions. Technical Report ATP-37, Dept. of Mathematics and Computer Science, Univ. of Texas, Austin, 1977.
38. Peter Mosses. Unified algebras and institutions. In *Proceedings, Fourth Annual Conference on Logic in Computer Science*, pages 304–312. IEEE, 1989.
39. Pieter-Hendrik Rodenburg. A simple algebraic proof of the equational interpolation theorem. *Algebra Universalis*, 28:48–51, 1991.
40. J.R. Slagle. Automatic theorem proving in theories with simplifiers, commutativity and associativity. *Journal of ACM*, 21:622–642, 1974.
41. Andrzej Tarlecki. Free constructions in algebraic institutions. In M.P. Chytil and V. Koubek, editors, *Proceedings, International Symposium on Mathematical Foun-*

dations of Computer Science, pages 526–534. Springer, 1984. Lecture Notes in Computer Science, Volume 176; extended version, University of Edinburgh, Computer Science Department, CSR-149-83.

42. Alfred Tarski. The semantic conception of truth. *Philos. Phenomenological Research*, 4:13–47, 1944.

Knowledge Based Computation* (Extended Abstract)

Rohit Parikh

Department of Computer Science
Brooklyn College of CUNY and CUNY Graduate Center
33 West 42nd St, New York, NY 10036
E-mail: ripbc@cunyvm.cuny.edu

1 Introduction

Since the mid-eighties, reasoning about knowledge has become an important part of distributed computing. Early papers include [HM], [PR], [CM] and others. In almost all the popular approaches, knowledge is thought of as a modal operator, looking syntactially rather like negation and semantically as a predicate transformer, monotonic and commuting with conjunction, but not with disjunction.

The case of many knowers has an interest far beyond the longer studied case of the knowledge of a single individual. If $\{1, 2, ...n\}$ are the individuals, then we have n knowledge operators $K_1, ..., K_n$. If L is the basic 'objective' language, speaking about the world, then the knowledge language $K(n, L)$ is obtained from L by repeatedly applying the operators K_i and closing under truth functions.

The semantics of this expanded language is obtained by taking a world W of models of L, i.e. each element s of W is an interpretation of L, plus n equivalence relations $\approx_i$ on W.

Here the intuitive meaning of $s \approx_i t$ is that s and t look the same to i. The notion $s \models A$ for a formula A of $K(n, L)$ is then defined by induction on the complexity of A; in the usual way if A belongs to L or is a truth functional combination of simpler formulas. If A is of the form $K_i(B)$, then we let $s \models A$ iff $(\forall t)(t \approx_i s \rightarrow t \models B)$. The operator K_i then has the semantic properties of S5. I.e. it is normal, $K_i(A)$ implies A and $K_i(K_i(A))$, and $\neg K_i(A)$ implies $K_i(\neg K_i(A))$.

In the actual context of a distributed system, we need concrete versions of L, W and $\approx_i$. Typically, the individuals i are various processes, communicating with each other by asynchronous or synchronous communication or via shared variables. Each process has a local state which may change as a result of a local action or because of a message received. The *global state* is simply the n-tuple of all local states, possibly augmented by an *environment state*. A *global history* is a sequence of global states (with time usually being the natural numbers, though sometimes there is no global time). Then W is simply the set of all possible global histories that can arise during some computation. The language L refers to properties of the global history and a particular formula A is true iff it is true

* Research supported by NSF grant CCR 92-08437

of the actual global history so far. If a formula has temporal operators which refer to the future, then it is evaluated by considering all possible extensions of the current global history.

Given two global histories h, h', they are equivalent to process i if they result in the same local state for i. If the processes remember all their past, then this amounts to saying that h, h' generate the same local history, i.e. the same sequence of local states of i. If the memory is shorter, e.g. if the processes are finite state, then they will retain less information, $\approx_i$ will be coarser, and the processes will have less knowledge.

2 Levels of Knowledge

[HM] point out, in their well known disussion of the co-ordinated attack problem, that synchronized actions require common knowledge. Here, to say that A is common knowledge means that everyone knows A, everyone knows that everyone knows A, and so on.

Defining

$$E(A) = E^1(A) = \bigwedge_i K_i(A)$$

and

$$E^{n+1}(A) = E(E^n(A))$$

we let $C(A) = \bigwedge_i E^i(A)$. As pointed out by [HM], common knowledge cannot be achieved through asynchronous communication. This result is generalized in [PK2] who characterize all possible levels of knowledge attainable by various means, and show that while broadcasts can achieve common knowledge, common knowledge among $n+1$ processes cannot be achieved by broadcasts limited to n processes (including the sender). This implies that certain kinds of knowledge depend on the kind of communication available and that may limit the sort of algorithms that can be carried out.

3 Knowledge Based Programs:

Usually, the processes operate according to certain protocols where the actions that an individual process may take will depend on certain tests or guards. In ordinary programs these guards depend only on the local variables of the process in question. In knowledge based programs, certain knowledge tests may also be used. I.e. instead of using formulas of L as tests (and then only those that are local to the process) we may augment them by certain formulas of the form $K_i(B)$ where i is the process in question.[2]

Two questions immediately arise. The first is why we should want to have such tests in our programs. And the second is how to interpret such tests. The

[2] Note that since $C(A)$ is equivalent to $K_i(C(A))$, $C(A)$ may also be used as a test.

first question is easier to answer. It turns out, [HZ] is a very good example, that actual protocols in many indiviudal cases can be best understood if we see them as implementations of knowledge programs. [DM] and [MT] are other examples of such knowledge based programs. Thus knowledge based programs become in a certain sense 'high-level' programs where certain kinds of inessential information are abstracted away.

The second question is subtler. While we can intuitively say that of course i knows what it knows and hence can act on it, in practice we need to reduce a knowledge formula to an ordinary one, dependent in the last analysis on i's local state only. Now $K_i(A)$ holds in a certain local state iff A is true of all global histories which yield this local state. Thus the only problem is the universal quantifier over all global histories. However, these global histories are themselves dependent on what the results of the knowledge tests are, so there is a circularity in the definition of the set W of all global histories. This set is used to *evaluate* the knowledge tests whose results are then used to *generate* the set of all possible global histories and we need that the two should be the same.

In practice, as [FHMV] point out, there may be a unique solution to finding W, or more than one solution, or no solution. Thus the instruction "set x to 1 if you know that x is ever going to be 1" has two possible interpretations. One where x is never set to 1 and hence the test is false, justifying not setting x to 1, or else x is set to 1 in the first round, justifying, via the knowledge test, the fact that it was set to 1. Thus there are two consistent interpretations.

On the other hand, "set x to 1 if you know that x is never going to be 1" has no consistent interpretation. If x *is* set to 1, then the test is falsified and so is the reason for setting x to 1. If it is never set to 1, then the test is true (and known to be true in W) and x *should* have been set to 1. Thus there is no consistent interpretation. Such problems can be avoided if a knowledge test only refers to that portion of global behaviour that has already taken place. Beverly Sanders [S] points out that another consequence of such facts is that knowledge programs are not monotonic in the set W and hence, paradoxically, a knowledge program that works correctly under a certain set of initial conditions may actually fail under more restricted initial conditions.[3] There is a similarity here with other non-monotonic situations, like that in default logic where a consistent set of formulas may fail to have an extension which is its own closure under the default rules. [Pa] contains a discussion of nonmonotonicity in logics of knowledge.

4 Global States and TIme

In previous discussion we assumed the existence of a global time which is linearly ordered. However, contexts in which this assumption fails or is unrealistic have

[3] It seems to me that this particular problem can be solved by *not* telling the individual processes that the initial conditions have been restricted so that they - incorrectly but successfully - evaluate their knowledge tests on the larger, unrestricted, W.

much interest. If the processes do not have synchronized clocks, then a process which is quiescent during a period may not be aware of time passing. Accepting this fact will lead to the conclusion that events are linearly ordered only at the local level (and only in the past) and any global ordering is partial and governed by causality.

In [KR] the issue of knowledge in such settings in discussed in detail and two classes of systems are studied. One is ACSA's, asynchronous communicating sequential agents and the other is KTS's, knowledge transition systems. In the first, each event has an 'owner' whose event it is and in the second case, each event has a 'knower' who knows that it happened. So in a certain sense, a computation of a KTS consists of an evolution of knowledge. The authors show the existence of maps going back and forth between these two classes of systems, which, in one direction, compose to isomorphism, and in the other to simulation. Thus these two ways of looking at computation are dual.

5 Identity and Locality

We assumed that the notion of an individual process and locality are well-defined and unproblematic. In particular, if some action is dependent on knowledge of a local variable, then the process has that knowledge and can act on it. However, these notions of individuality and locality themselves need investigation. A person may not know that she has cancer even though it is in some sense a fact that is local to her. We tend to treat the telephone company (at times) as an individual, but it is not unusual to send in a payment and receive a cancellation threat from another part of the same company which does not know of the payment. Since we do sometimes treat a group of processes as a single process on occasion (this is common in concurrency theory) both the notion of local variable and local action need refinement. If information is received by one of a group of processes regarded as an individual, and action is required of another part, then there are clear assumptions about communication and co-ordination which we need to make explicit before we can conclude that the proper action will indeed be forthcoming. Since even a Turing machine can be regarded as a distributed system, with the head and each square of the tape a subpart, notions like 'local fact' and 'local action' acquire a subtlety which lies behind computational complexity.

6 Probabilistic Considerations:

If there is a probability distribution on the initial conditions, on the possible actions for each process at each moment of time (and with asynchronous systems on the delivery profile of messages), then this will generate a probability distribution on the global histories.[4] Thus it will be possible to also write programs

[4] In existing studies, as for example in [PK], the processes are deterministic and messages are always delivered in unit time, so that only the first probability distribution on initial states is of relevance.

where guarded commands have the form: "if the probability of A is greater than x then do action a". This method can give programs which have a high probability of yielding correct results. Such commands can also have other advantages. In [Pa2] it is shown that under certain circumstances, an instruction of the form "do a if you know A" can take an infinite amount of time, but its cousin, "do a if the probability of A is $> .999$" is guaranteed to terminate in a finite amount of time. This fact opens up the possibility of looking for probabilistic knowledge algorithms where ordinary knowledge is, for whatever reason, unavailable.

References

[CM] Chandy, M. and Misra, J., "How processes learn", *Distributed Computing* **1** (1986) p.40-52.

[DM] Dwork, C. and Moses, Y., "Knowledge and common knowledge in a byzantine environment", *Information and Computation* **88** (1990) pp. 156–196.

[FHMV] Fagin, R., Halpern, J., Moses, Y. and Vardi, M., "Knowledge-based programs", to appear.

[HF] Halpern, J. and Fagin, R., "Modelling knowledge and action in distributed systems", *Distributed Computing* **3** (1989) pp. 159–179.

[HM] Halpern, J. and Moses, Y., "Knowledge and common knowledge in a distributed environment", *Jour. Assoc. Comp. Mach.* **37** (1990) pp. 549–587.

[HZ] Halpern, J. and Zuck, L., "A little knowledge goes a long way", *Jour. Assoc. Comp. Mach.* **39** (1992) pp. 449–478.

[KR] Krasucki, P. and Ramanujam, R., "Knowledge and the ordering of events in distributed systems", *TARK-5* (Ed. R. Fagin) Morgan Kaufmann (1994) pp. 267–283.

[MK] Moses, Y. and Kislev, O., "Knowledge-oriented programming", *12th ACM-PODC Symposium* (1993) pp. 261–270.

[MT] Moses, Y. and Tuttle, M., "Programming simultaneous actions using common knowledge", *Algorithmica* **3** (1988) pp. 121–169.

[KPN] Krasucki, P., Parikh, R. and Ndjatou, G., "Probabilistic knowledge and probabilistic common knowledge", *Int. Symp. Method. Int. Sys.*, (Ed. Ras, Zemankova and Emrich) North Holland 1990 pp. 1–8.

[Pa] Parikh, R., "Monotonic and Non-monotonic Logics of Knowledge", in *Fundamenta Informatica* special issue, *Logics for Artificial Intelligence* vol XV (1991) pp. 255–274.

[Pa2] Parikh, R., "Finite and infinite dialogues", in the *Proceedings of a Workshop on Logic from Computer Science*, Ed. Moschovakis, MSRI publications, Springer 1991, pp. 481–498.

[PK] Parikh, R. and Krasucki, P., "Communication, consensus and knowledge", *J. Economic Theory* **52** (1990) pp. 178–189.

[PK2] Parikh, R. and Krasucki, P., "Levels of knowledge in distributed computing", *Sadhana - Proc. Ind. Acad. Sci.* **17** (1992) pp. 167–191.

[PR] Parikh, R. and Ramanujam, R., "Distributed processing and the logic of knowledge", in *Logics of Programs*, Springer Lecture Notes in Computer Science #193, (1985) pp. 256–268.

[S] Sanders, B., "A Predicate transformer approach to knowledge and knowledge-based protocols", *10th ACM-PODC Symposium* (1991) pp. 217–230.

Order-sorted Algebraic Specifications with Higher-order Functions[1]

Anne Elisabeth Haxthausen

Department of Computer Science
Technical University of Denmark, bldg. 344,
DK-2800 Lyngby, Denmark

Abstract. This paper gives a proposal for how order-sorted algebraic specification languages can be extended with higher-order functions. The approach taken is a generalisation to the order-sorted case of an approach given by Möller, Tarlecki and Wirsing for the many-sorted case. The main idea in the proposal is to only consider reachable extensional algebras. This leads to a very simple theory, where it is possible to relate the higher-order specifications to first order specifications.

1 Introduction

The objective of this paper is to investigate how order-sorted algebraic specification languages can be extended with higher-order functions. The paper is a short version of a technical report [Hax94] in which more details and proofs can be found.

Below we describe our goals and their background and give a summary of the contents of the paper.

1.1 Background

During the last decades many languages for specifying data types and functions have been researched and developed. Two major approaches may be distinguished: the model-oriented and the algebraic. The algebraic specification languages have the advantage that they allow a high abstraction level where one can abstract away from implementation details like data type representations. However, in contrast to model-oriented specification languages and functional programming languages the algebraic specification languages (with a few exceptions like RSL [Rlg92]) do not allow higher-order functions. This is a pity, since higher order functions are useful, for instance for describing schematic algorithms

[1] The work described in this paper was carried out during a visit at Electrotechnical Laboratory in Tsukuba in Japan and was supported by Japan International Science and Technology Exchange Center. The production of the paper has been supported by the Danish Technical Research Council under the "Codesign" programme.

such as generic tree-walking operations or for describing programming language semantics. Goguen shows in [Gog88] how one can use parameterised specifications instead of using higher-order functions which take functions as parameters. However, a similar replacement is not possible for higher-order functions which return functions, and it may be more elegant to use higher-order functions.

We therefore wish to extend algebraic specification languages with higher-order functions. Such extended languages we will call *higher-order (algebraic) specification languages*. In particular, we wish to extend order-sorted algebraic specification languages like OBJ (see [FGJM85] and [GWM+92]) because they provide the possibility of specifying subtypes (subsorts) and thereby a clean treatment of partially defined functions. Below is an example of what we may like to write.

sorts nat, nznat .
subsorts nznat $\leq$ nat .
ops ..., twice : (nat $\rightarrow$ nznat) $\rightarrow$ (nat $\rightarrow$ nznat) .
vars f : nat $\rightarrow$ nznat, x : nat .
eqs twice(f)(x) = f(f(x)) .

Here 'twice' is a higher-order function, which takes a function, 'f', as argument and returns another function, which when applied to some argument, 'x', returns the same as if 'f' had been applied twice to 'x'.

Research has been done on how to give semantics to higher-order algebraic specification languages. One way is to use Cartesian closed categories as the mathematical foundations, as e.g. in [Poi86] (for the many-sorted case) and [MOM90] (for the order-sorted case). Another way is to try to extend the usual set-theoretical algebraic framework as e.g. in [Möl86] (for the many-sorted case) and [Qia90] (for the order-sorted case). In [MTW88b] and [MTW88a], Möller, Tarlecki and Wirsing do this implicitly for the many-sorted case by a transformational semantics: under the assumption that one is only interested in term-generated algebras, the semantics of a higher-order specification can be given by the semantics of a corresponding first order specification. This transformational approach is very attractive because, in contrast to the other approaches, many definitions and theorems can directly be derived from the first order case.

Therefore, one of the main goals of this paper is to investigate how Möller, Tarlecki and Wirsing's approach can be generalised to the order-sorted case.

1.2 Contents of Paper

This paper gives, in two steps, in sections 2 and 3, a proposal for higher-order order-sorted algebraic specification.

In section 2, we keep the subtype relation so simple, that it is possible to relate the higher-order order-sorted specifications to first order order-sorted specifications, where the subtype-as-inclusion principle is used. While, in section 3, we use a more general subtype relation for which the function type constructor is anti-monotonic in its first argument, such that for instance a function of type

Int $\rightarrow$ Int can be passed as an actual parameter to a (higher-order) function which requires a function of type Nat $\rightarrow$ Int. For this more general relation, it is not possible to relate the higher-order order-sorted specifications to usual first order order-sorted specifications — it is necessary to relate them to a notion of first order generalised order-sorted specifications, which includes two kinds of subtypes: subtypes-as-inclusion and subtypes-as-implicit-coercion. We have developed such a notion in appendix D. In [MOM90] Marti-Oliet and Meseguer also include and distinguish the two kinds of subtypes in higher-order specifications, however their semantics is not based on a transformation to first order specifications but on Cartesian closed categories. Also Qian, in [Qia90], distinguishes two kinds of subtypes, but the second of these is less general than ours and is defined in a framework which is quite different from the usual algebraic framework.

Finally, in section 4, a summary and discussion of the proposal is given.

For the convenience of the reader, appendices A and B contain some well-known definitions and results from (first order) many-sorted algebra (as defined in [EM85]) and (first order) order-sorted algebra (as defined in [GM92]) on which other definitions in this paper depend. Furthermore, these appendices also show which notation is used.

Appendix C contains (our own) definitions and results for the fundamentals of (first order) algebraic specification with subtypes as implicit coercions. This is used in our definition of generalised algebra in appendix D.

2 Higher-order Order-sorted Specification

When defining a notion of higher-order specifications, there are a number of tasks to be done: The notions of signatures, axioms, algebras, homomorphism and satisfaction relation should be decided. Furthermore, if initial algebra semantics should be used, the existence of initial algebras for specifications should be investigated. These tasks are done in the following.

The approach we take is a generalisation of Möller, Tarlecki and Wirsing's approach for the many-sorted case in [MTW88a]. The main idea in this approach is, that under the assumption that we are only interested in reachable (i.e. term-generated) extensional higher-order algebras, we can consider higher-order specifications as first order specifications with implicit higher-order sorts, apply-functions and extensionality axioms. In this way we easy get a number of definitions and theorems for the higher-order case more or less for free from the first order case.

2.1 Signatures

A higher-order order-sorted signature is like a first order order-sorted signature (see appendix B) but the sorts that may be used in the definitions of operation symbols are higher-order sorts, i.e. not just basic sorts, but also functional sorts as well. Functional sorts, s1 $\rightarrow$ s2, are built by applying a built-in sort constructor, $\rightarrow$, to sorts s1 and s2 which may be basic or functional. Operation symbols having functional sorts are intended to denote functions. Hence, in a higher-order specification one can specify functions that take functions as argument and/or return functions. Below we give the precise definitions.

Definition 2.1 A *higher-order order-sorted signature* HSIG is a triple (S, $\leq$, OP), where

1. (S, $\leq$) is a poset
2. OP is a family $(OP_s)_{s \in S^{\rightarrow}}$ of distinct $S^{\rightarrow}$-sorted constant symbols.

The set $S^{\rightarrow}$ of *higher-order sorts* generated from S is the least set for which:

1. $S \subseteq S^{\rightarrow}$
2. $w \rightarrow s \in S^{\rightarrow}$, if $w \in (S^{\rightarrow})^{+}$, $s \in S^{\rightarrow}$

Note that compared with first order order-sorted signatures we only allow constant symbols – we do not need operation symbols of the form $\sigma : w \mapsto s$, as we can now define constants having functional sorts, i.e. of the form $\sigma : w \rightarrow s$. This restriction is solely made in order to avoid confusion between the two forms – there would be no theoretical problems in allowing both. A consequence of the restriction is that function symbols cannot be overloaded as constant symbols cannot.

One of the choices we must make is how the subsort relation, $\leq$, on the basic sorts, S, should be extended to a relation on the higher-order sorts $S^{\rightarrow}$. In this section, we will use a relation, $\leq^{\rightarrow}$, for which $\rightarrow$ is constant in its first argument and monotonic in its second argument. With this relation we shall see that it is possible to relate our higher-order order-sorted specifications to first order order-sorted specifications. In section 3 we will consider a more general relation, $\leq^{\Rightarrow}$, for which $\rightarrow$ is anti-monotonic in its first argument and monotonic in its second argument.

Definition 2.2 Let (S,$\leq$) be a poset. Then $(S^{\rightarrow}, \leq^{\rightarrow})$ is a poset generated from (S,$\leq$) with the ordering relation $\leq^{\rightarrow}$ being the least relation satisfying:

1. $s \leq^{\rightarrow} s'$, if $s \leq s'$
2. $w \rightarrow s \leq^{\rightarrow} w \rightarrow s'$, if $s \leq^{\rightarrow} s'$

Definition 2.3 Given a higher-order order-sorted signature HSIG = (S,$\leq$,OP). Its *associated* first order order-sorted signature is: $HSIG^{\rightarrow} = (S^{\rightarrow}, \leq^{\rightarrow}, OP^{\rightarrow})$, where $OP^{\rightarrow} = OP \cup (\{apply\}_{\langle w \rightarrow s, w\rangle, s})_{w \in (S^{\rightarrow})^{+}, s \in S^{\rightarrow}}$

Notation: For terms (see next section) of the form apply(f,x) one could use a more appealing notation like f(x). This convention has been used in the example in the introduction.

Note, the requirement to HSIG that the constant symbols in OP must be distinct is a necessary and sufficient condition for ensuring that the monotonicity condition for $HSIG^{\rightarrow}$ is satisfied.

We are interested in when signatures have properties like being regular, as certain theorems only hold for signatures having these properties. Below we define these properties and give some facts about when they hold.

Definition 2.4 A higher-order order-sorted signature HSIG is *regular/ locally upwards filtered/ coherent* iff $HSIG^{\rightarrow}$ is regular/locally upwards filtered/coherent[2].

Fact 2.5 All higher-order order-sorted signatures are regular.

A proof of this is given in [Hax94].

Fact 2.6 HSIG is locally upwards filtered if the connected components of (S,$\leq$) are locally upwards filtered.

Proof: Follows from the fact that the sort constructor $\rightarrow$ is monotone wrt. $\leq^{\rightarrow}$.

Corollar 2.7 A higher-order signature HSIG = (S,$\leq$,OP) is coherent if the connected components of (S,$\leq$) are locally upwards filtered.

2.2 Terms and Axioms

Definition 2.8 Given a higher-order order-sorted signature HSIG.
A *higher-order order-sorted HSIG-term/axiom* is a first order order-sorted $HSIG^{\rightarrow}$-term/axiom[3].

Fact 2.9 Any HSIG-term has a least sort.

Proof: Follows from fact 2.5.

2.3 Specifications

Definition 2.10 A *higher-order order-sorted specification* HSPEC is a pair (HSIG, HE) consisting of a higher-order order-sorted signature HSIG and a set HE of higher-order order-sorted HSIG-axioms.

---

[2] For a definition, see appendix B.

[3] For a definition of the notion of first order order-sorted terms and axioms, see appendix B.

2.4 Algebras, Homomorphisms and Satisfaction Relation

As higher-order HSIG-algebras we will use extensional HSIG$^{\rightarrow}$-algebras. The extensionality ensures that any carrier, $A_{w \rightarrow s}$, of a functional sort, $w \rightarrow s$, is isomorphic to a subset of the function space, $A_w \rightarrow A_s$.

Definition 2.11 Given a higher-order order-sorted signature HSIG. A *higher-order order-sorted HSIG-algebra* A is an extensional first order order-sorted HSIG$^{\rightarrow}$-algebra[4].

Definition 2.12 A HSIG$^{\rightarrow}$-algebra A is *extensional* if, for all sorts $w \rightarrow s \in S^{\rightarrow}$ it holds that for all $f, g \in A_{w \rightarrow s}$ the following holds:

$$(\forall a \in A_w \bullet apply^A(f,a) = apply^A(g,a)) \Rightarrow f = g$$

As higher-order order-sorted HSIG-algebras, terms and axioms are first order order-sorted HSIG$^{\rightarrow}$-algebras, terms and axioms, the notions of HSIG-homomorphisms, HSIG-evaluation of terms, HSIG-satisfaction of HSIG-axioms by HSIG-algebras, term-generatedness (reachability) etc. carry directly over from the first order case and we will not bother the reader with repeating these definitions.

Notation: The HSIG-algebras and HSIG-homomorphisms form a category denoted Alg(HSIG). The reachable HSIG-algebras and the HSIG-homomorphisms between these form a category denoted RAlg(HSIG).

It is possible to give a first order order-sorted specification whose reachable algebras coincide with the reachable HSIG-algebras:

Theorem 2.13 Given a higher-order signature HSIG.
Then RAlg(HSIG) = RAlg(SPEC), where SPEC = (HSIG$^{\rightarrow}$, ext(HSIG)) and ext(HSIG) consists of exactly one ground infinitary conditional equation

$$t = t' \textbf{ if } \bigwedge_{t'' \in T_w} (apply(t,t'') = apply(t',t''))$$

for each $t, t' \in T_{w \rightarrow s}$, $w \in (S^{\rightarrow})^+$, $s \in S^{\rightarrow}$.

Fact 2.14 In general not all HSIG$^{\rightarrow}$-algebras satisfying ext(HSIG) are extensional and vice versa.

Definition 2.15
Given a higher-order order-sorted specification HSPEC = (HSIG,HE). A *higher-order order-sorted HSPEC-algebra* A is a HSIG-algebra satisfying each of the axioms in HE (in other words an extensional HSIG$^{\rightarrow}$-algebra satisfying each of the axioms in HE).

---

[4] For a definition of the notion of first order order-sorted algebras, see appendix B.

Notation: The HSPEC-algebras and HSPEC-homomorphisms form a category denoted Alg(HSPEC). The reachable HSPEC-algebras and the HSPEC-homomorphisms between these form a category denoted RAlg(HSPEC).

For each higher-order specification, HSPEC, it is possible to give a first order order-sorted specification, $\mathrm{HSPEC}^{\rightarrow}$, whose reachable algebras coincide with the reachable HSPEC-algebras:

Definition 2.16 The first order order-sorted specification *associated* with a higher-order order-sorted specification, HSPEC = (HSIG,HE), is:
$\mathrm{HSPEC}^{\rightarrow} = (\mathrm{HSIG}^{\rightarrow}, \mathrm{HE} \cup \mathrm{ext(HSIG)})$

Theorem 2.17 $\mathrm{RAlg(HSPEC)} = \mathrm{RAlg}(\mathrm{HSPEC}^{\rightarrow})$

2.5 Initiality Theorems

Theorem 2.18 Let HSIG be a coherent higher-order order-sorted signature. Then there exists an initial algebra in RAlg(HSIG). If furthermore HSIG is sensible (see definition 2.19), then the $\mathrm{HSIG}^{\rightarrow}$-term algebra, $\mathrm{T}(\mathrm{HSIG}^{\rightarrow})$, is one of the initial algebras in RAlg(HSIG).

Proof:
First part follows from the fact that $\mathrm{RAlg(HSIG)} = \mathrm{RAlg}(\mathrm{HSIG}^{\rightarrow},\mathrm{ext(HSIG)})$, and that there exists an initial algebra in $\mathrm{RAlg}(\mathrm{HSIG}^{\rightarrow}, \mathrm{ext(HSIG)})$ if $\mathrm{HSIG}^{\rightarrow}$ is coherent. The second part follows from the fact that $\mathrm{T}(\mathrm{HSIG}^{\rightarrow})$ satisfies ext(HSIG), if HSIG is sensible.

Definition 2.19 A higher-order signature, HSIG = (S,≤,OP), is *sensible*, if for all functional sorts $w \rightarrow s \in S^{\rightarrow}$ it holds that:
$\mathrm{T}_{w \rightarrow s} \neq \{\} \Rightarrow (\mathrm{T}_w \neq \{\} \vee \mathbf{card}\ \mathrm{T}_{w \rightarrow s} = 1)$.

Theorem 2.20 Let HSPEC = (HSIG,HE) be a higher-order order-sorted specification with coherent signature, HSIG. Then there exists an initial algebra in RAlg(HSPEC). If furthermore, HSIG is sensible, then the $\mathrm{HSPEC}^{\rightarrow}$ quotient ground term algebra, $\mathrm{T}(\mathrm{HSPEC}^{\rightarrow})$, is one of the initial algebras in RAlg(HSIG).

Proof: First part follows from the fact that $\mathrm{RAlg(HSPEC)} = \mathrm{RAlg}(\mathrm{HSPEC}^{\rightarrow})$, and that there exists an initial algebra in $\mathrm{RAlg}(\mathrm{HSPEC}^{\rightarrow})$ if $\mathrm{HSIG}^{\rightarrow}$ is coherent. The second part follows from the fact that $\mathrm{T}(\mathrm{HSPEC}^{\rightarrow})$ satisfies ext(HSIG), if HSIG is sensible.

3 Generalised Higher-order Order-sorted Specification

Assume given a higher-order order-sorted signature HSIG = (S, ≤, OP).

We now wish to use the same approach as in last section, but with a more general subtype relation $\leq^{\Rightarrow}$, than $\leq^{\rightarrow}$, as motivated by the following example.

Example 3.1 Consider the following specification:

> **sorts** nat, int .
> **subsorts** nat $\leq$ int .
> **ops**
> f : (nat $\to$ int) $\to$ int,
> g : int $\to$ int .

It would be reasonable if apply(f,g) was a term. However, this is only the case if int $\to$ int is a subtype of nat $\to$ int. Unfortunately, that is not the case for the subtype relation $\leq^{\to}$, but for the subtype relation $\leq^{\Rightarrow}$ we define below.

Definition 3.2 Let the relation $\leq^{\Rightarrow}$ on $S^{\to}$ be the least relation satisfying:

1. $s \leq^{\Rightarrow} s'$, if $s \leq s'$
2. $w' \to s \leq^{\Rightarrow} w \to s'$, if $w \leq^{\Rightarrow} w'$ and $s \leq^{\Rightarrow} s'$
3. $\ldots s \ldots \leq^{\Rightarrow} \ldots s' \ldots$, if $s \leq^{\Rightarrow} s'$

Fact 3.3 $\leq^{\to}$ is a sub-relation of $\leq^{\Rightarrow}$

The generalisered subtype relation ($\leq^{\Rightarrow}$) is the same as in [MOM90], but more general than the one in [Qia90], where $w \leq^{\Rightarrow} w'$ in the 2. rule above is replaced with $w \leq^{\to} w'$.

We could now try to proceed in the same way as before by defining an associated first order signature, $HSIG^{\Rightarrow} = (S^{\to}, \leq^{\Rightarrow}, OP^{\to})$, for each higher-order signature HSIG, such that RAlg(HSIG) = RAlg($HSIG^{\Rightarrow}$,ext(HSIG)). However, this time we encounter a number of problems:

1. the signatures are not regular
2. the signatures are typically not locally upwards filtered
3. subtypes as inclusion does not give all the desired models

In the following, we will show how the two first problems can be solved by modifying the notions of order-sorted algebra, and the last by also allowing subtypes as implicit coercions as well as subtypes as inclusion.

3.1 Problem: Signatures are not regular

We want $HSIG^{\Rightarrow}$ to be regular, otherwise there are terms which do not have a unique least parse.

Fact 3.4 For any signature HSIG, $HSIG^{\Rightarrow}$ is not regular because of the implicit apply functions, but pre-regular[5].

Example 3.5 Consider the following signature, HSIG:

[5] For a definition, see appendix B.

sorts s, s1, s2 .
subsorts s1 $\leq$ s2 .
ops
f : s2 $\rightarrow$ s,
a : s1 .

Consider the sort string, w0 = (s2 $\rightarrow$ s) s1. The apply operations in $OP^{\rightarrow}$ with arities w, for which w0 $\leq^{\Rightarrow}$ w, are:

1. apply : (s2 $\rightarrow$ s) s2 $\mapsto$ s
2. apply : (s1 $\rightarrow$ s) s1 $\mapsto$ s

None of these have a least rank, as s1 $\leq^{\Rightarrow}$s2 and s2 $\rightarrow$ s $\leq^{\Rightarrow}$ s1 $\rightarrow$ s. Therefore HSIG is not regular. This has the consequence that the term apply(f,a) has two possible parses — there is no unique "least parse":

1. apply : (s2 $\rightarrow$ s) s2 $\mapsto$ s, f : s2 $\rightarrow$ s, a : s2
2. apply : (s1 $\rightarrow$ s) s1 $\mapsto$ s, f : s1 $\rightarrow$ s, a : s1

However, HSIG is pre-regular, since there is always a least co-arity. Hence, the term apply(f,a) has a least sort: s

The problem, that there may be several possible parses is not serious if we add the extra requirement to the algebras, A, that whenever there are more than one possible parse of a term, t, the meaning of t in A, using either of them, is the same. For the example above it means that

$$\text{apply}^A_{(s2\rightarrow s)s2,s}(f^A_{s2\rightarrow s}, a^A_{s1}) = \text{apply}^A_{(s1\rightarrow s)s1,s}(f^A_{s2\rightarrow s}, a^A_{s1})$$

(Note, that the applications are well-formed, since $a^A_{s1} \in A_{s1} \subseteq A_{s2}$ and $f^A_{s2\rightarrow s} \in A_{s2\rightarrow s} \subseteq A_{s1\rightarrow s}$.)

In general it means that the problem can be solved if we change definition B.8 with the following definition.

Definition 3.6 Given a pre-regular first order order-sorted signature SIG = (S,$\leq$,OP). A *modified first order order-sorted SIG-algebra* is a first order many-sorted SIG-algebra satisfying the following condition:

If $\sigma \in OP_{w1,s1} \cap OP_{w2,s2}$, w0 $\leq$ w1 and w0 $\leq$ w2 then
$\forall$ a $\in A_{w0} \bullet \sigma^A_{w1,s1}(a) = \sigma^A_{w2,s2}(a)$

Note, this condition implies the usual monotonicity condition.

With this modified notion of algebras, it is sufficient in theorems about initiality etc. to require signatures to be pre-regular instead of regular.

3.2 Problem: Signatures are typically not locally upwards filtered

We are interested in when the connected components of $(S^{\rightarrow}, \leq^{\Rightarrow})$ are locally upwards filtered, as many of the theorems for order-sorted algebra only hold for locally upwards filtered signatures.

Fact 3.7 That the connected components of $(S, \leq)$ are locally upwards filtered does not imply that the connected components of $(S^{\rightarrow}, \leq^{\Rightarrow})$ are locally upwards filtered (it only implies that the connected components of $(S^{\rightarrow}, \leq^{\rightarrow})$ are locally upwards filtered). $(S^{\rightarrow}, \leq^{\Rightarrow})$ only becomes locally filtered, if $(S, \leq)$ is locally upwards and downwards filtered.

The problem is due to the fact that $\rightarrow$ is anti-monotone wrt. $\leq^{\Rightarrow}$ in its first argument. This is very unfortunate, since in practice it would be rare that the connected components of $(S, \leq)$ are locally downwards filtered.

Example 3.8 For $(S,\leq)$ defined by:

> **sorts** s1, s2, s3, s .
> **subsorts** s1 $\leq$ s3, s2 $\leq$ s3 .

we have: $s3 \rightarrow s \leq^{\Rightarrow} s1 \rightarrow s$ and $s3 \rightarrow s \leq^{\Rightarrow} s2 \rightarrow s$.
Therefore not all connected components of $(S^{\rightarrow}, \leq^{\Rightarrow})$ are locally upwards filtered.

This problem can be solved by changing the definition of the notion of axioms to the following:

Definition 3.9 Given a pre-regular first order order-sorted signature SIG = (S,$\leq$,OP). A *modified first order order-sorted SIG-axiom* is a conditional equation of the form

$$(\forall X)\ t = t' \textbf{ if } \wedge_{i \in I}\ (t_i = t'_i)$$

where $t, t', t_i, t'_i \in T(SIG,X)$ for $i \in I$, and there exists a least upper bound of $LS(t)$ and $LS(t')$ in $(S,\leq)$, and for each $i \in I$, there exists a least upper bound of $LS(t_i)$ and $LS(t'_i)$ in $(S,\leq)$.

With this modified notion of axioms, it is un-necessary in the theorems about initiality etc. to require the signatures to be locally upwards filtered. Actually it would have been enough to require that there exists upper bounds, however we find it more natural to require that there exists least upper bounds, and when we change to generalised algebra as proposed in section 3.3 we need that requirement.

3.3 Problem: Subtypes as inclusion not sufficient

If we use the subtype-as-inclusion principle, the algebras would not comprise all what we might expect.

Example 3.10 Assume s1 $\leq$ s2, and thereby s2 $\rightarrow$ s $\leq^{\Rightarrow}$ s1 $\rightarrow$ s. Using the subtype-as-inclusion principle for the algebras A, we would get:

$$A_{s2 \rightarrow s} \subseteq A_{s1 \rightarrow s}$$

which would rule out algebras, where $A_{s2 \rightarrow s} \subseteq A_{s2} \rightarrow A_s$ and $A_{s1 \rightarrow s} \subseteq A_{s1} \rightarrow A_s$, since $A_{s2} \rightarrow A_s$ and $A_{s1} \rightarrow A_s$ would in general be disjoint.

This problem can be solved by using the subtypes-as-implicit-coercion principle for subtypes. In appendix C we have developed a notion of coercion algebras. However, we wish, like Marti-Oliet and Meseguer in [MOM90], to include and distinguish between the two principles for subtypes. Therefore, it has been necessary to develop a generalised algebra, which includes both order-sorted algebra and coercion-algebra. This is presented in appendix D.

3.4 The Final Solution

So now we can do exactly as in section 2, however instead of using order-sorted algebra we use generalised algebra. First we define HSIG$^{\Rightarrow}$:

Definition 3.11 Given a higher-order order-sorted signature HSIG = (S,$\leq$,OP). Its *associated* first order generalised signature is: HSIG$^{\Rightarrow}$ = ($S^{\rightarrow}$, $\leq^{\rightarrow}$, $\leq^{\Rightarrow}$, $OP^{\rightarrow}$)

Then the remaining definitions and theorems can be given as in section 2, except that we use HSIG$^{\Rightarrow}$ instead of HSIG$^{\rightarrow}$ and "generalised" instead of "order-sorted".

4 Conclusions

4.1 Main Results

In sections 2 and 3, we gave proposals for the fundamentals of higher-order order-sorted algebraic specification. The approach we took was a generalisation of Möller, Tarlecki and Wirsing's approach in [MTW88a] from the many-sorted case to the order-sorted case.

The main idea in the approach is only to consider reachable extensional algebras. This leads to a very simple theory, where it is possible to relate the higher-order specifications to first order specifications. To be more precise, a notion of higher-order specifications is defined such that for each higher-order specification, HSPEC, a first order specification, HSPEC$^{\rightarrow}$, can be derived such that the class of reachable HSPEC$^{\rightarrow}$-algebras is equal to the class of reachable HSPEC-algebras.

One of the choices we had to make was how the subsort relation, $\leq$, on the basic sorts, S, should be extended to a relation on the higher-order sorts $S^{\rightarrow}$. First, in section 2, we tried the ideas out for a relation, $\leq^{\rightarrow}$, for which $\rightarrow$ was constant in its first argument and monotonic in its second argument. In this case everything turned out smoothly. The main results were:

1. definitions of syntactic and semantic notions
2. existence of an initial reachable extensional algebra

Then, in section 3, we tried to do the same, but for a more general relation $\leq^{\Rightarrow}$, for which $\rightarrow$ was anti-monotonic in its first argument and monotonic in its second argument. In this case it turned out that we could not relate our higher-order order-sorted specifications to first order order-sorted specifications. We therefore developed a new notion of first order generalised algebra, which includes and distinguishes between the two principles for subtypes: subtypes-as-inclusion and subtypes-as-implicit-coercion. (This notion of first order generalised algebra, may be seen as a result in itself). By relating the higher-order order-sorted specifications to first order generalised specifications instead of first order order-sorted specifications everything turned out smoothly.

4.2 Advantages and Disadvantages

Some of the advantages of this approach are:

1. the semantics is simple
2. properties may easily be derived from first order case
3. there is only one kind of function arrow

In order to avoid two kinds of function arrows, we choose only to allow constant symbols in the signatures. The price of this is that function symbols (which are then constant symbols) must be distinct. However, this decision can easily be changed, such that non-constant symbols are allowed. In this case all the definitions and results we have presented still hold (the only exception is that signatures are not necessarily always pre-regular/regular).

4.3 Acknowledgements

I would like to thank Professor Futatsugi for inviting me to Japan and for scientific discussions, all my Japanese colleagues, Professor Tarlecki and Professor Möller for scientific discussions, and the referees for comments to a draft version of this paper.

References

[EM85] H. Ehrig and B. Mahr. *Fundamentals of Algebraic Specification 1, Equations and Initial Semantics.* EATCS Monographs on Theoretical Computer Science, vol. 6. Springer-Verlag, 1985.

[FGJM85] K. Futatsugi, J. Goguen, J. Jouannaud, and J. Meseguer. Principles of OBJ2. In *12th Symposium on POPL*. Association for Computing Machinery, 1985.

[GM92] J. Goguen and J. Meseguer. Order-Sorted Algebra I: Equational Deduction for Multiple Inheritance, Overloading, Exceptions and Partial Operations. *Theoretical Computer Science*, 105(2), 1992.

[Gog88] J. Goguen. Higher-Order Functions Considered Unnecessary for Higher Order Programming. Technical Report SRI-CSL-88-1R, SRI Int., 1988.

[Rlg92] The RAISE Language Group. *The RAISE Specification Language*. The BCS Practitioners Series. Prentice Hall Int., 1992.

[GWM+92] J. Goguen, T. Winkler, J. Meseguer, K. Futatsugi, and J. Jouannaud. Introducing OBJ. Technical Report SRI-CSL-92-03, SRI Int., 1992. Draft.

[Hax94] A. E. Haxthausen. Algebraic Specification with Higher-Order Functions. Technical Report ETL TR-94-18, Electrotechnical Laboratory, 1994.

[Möl86] B. Möller. Algebraic Specification with Higher-Order Operations. In *IFIP TC2 Working Conference on Program Specification and Transformation*. North-Holland, 1986.

[MOM90] N. Marti-Oliet and J. Meseguer. Inclusions and Subtypes. Technical Report SRI-CSL-90-16, SRI Int., 1990.

[MTW88a] B. Möller, A. Tarlecki, and M. Wirsing. Algebraic Specification of Reachable Higher-Order Algebras. In *5th workshop on Specification of Abstract Datatypes*, number 332 in LNCS. Springer-Verlag, 1988.

[MTW88b] B. Möller, A. Tarlecki, and M. Wirsing. Algebraic Specifications with Built-in Domain Constructions. In *CAAP'88*, number 299 in LNCS. Springer-Verlag, 1988.

[Poi86] A. Poigné. On Specifications, Theories and Models with Higher Types. *Information and Control*, 68, 1986.

[Qia89] Z. Qian. Relation-sorted algebraic specifications with built-in coercers: Parameterization and parameter passing. In *Categorical Methods in Computer Science*, number 393 in LNCS. Springer-Verlag, 1989.

[Qia90] Z. Qian. Higher-Order Order-Sorted Algebras. In *Algebraic and Logic Programming*, number 463 in LNCS. Springer-Verlag, 1990.

[Rey80] J. Reynolds. Using category theory to design implicit conversions and generic operators. In *Semantics-Directed Compiler Generation*, number 94 in LNCS. Springer-Verlag, 1980.

A First-order Many-sorted Algebra

This appendix contains some well-known definitions from many-sorted algebra on which other definitions in this paper depend. Furthermore, it also shows which notation we use. For a full treatment of the topic, we refer to [EM85].

Definition A.1 A *many-sorted signature* is a pair (S, OP), where

1. S is a set of *sorts*
2. OP is a $S^* \times S$-sorted family $(OP_{w,s})_{w \in S^*, s \in S}$ of *operation symbols*

Notation: We write $\sigma : w \mapsto s \in OP$ for $\sigma \in OP_{w,s}$; $\sigma : s$ for $\sigma : \mapsto s$; and OP_s for $OP_{w,s}$, if w is the empty string λ. For an operation symbol $\sigma : w \mapsto s$, we call w $\mapsto$ s or $\langle$w,s$\rangle$ for its *rank*, w for its *arity* and s for its *co-arity*. Operation symbols whose arity is the empty string are called *constant symbols*.

Definition A.2 Given a many-sorted signature SIG = (S, OP). A *many-sorted SIG-algebra*, A, consists of

1. a *carrier* set A_s for each s $\in$ S
2. a constant $\sigma_s^A \in A_s$ for each $\sigma \in OP_s$
3. a function $\sigma_{w,s}^A : A_w \mapsto A_s$ for each $\sigma \in OP_{w,s}$, $w \neq \lambda$

Notation: We write A_w for $A_{s1} \times \ldots \times A_{sn}$, when w = s1...sn. We write σ^A for σ_s^A and $\sigma_{w,s}^A$, when this does not give confusion.

Definition A.3 Given a many-sorted signature SIG = (S, OP) and two SIG-algebras A and B. A *many-sorted SIG-homomorphism*, h : A $\mapsto$ B, from A to B, is an S-sorted family $(h_s)_{s \in S}$ of functions $h_s : A_s \mapsto B_s$ satisfying the homomorphism conditions

1. $h_s(\sigma_s^A) = \sigma_s^B$, for constant symbols $\sigma \in OP_s$
2. $\forall\, a \in A_w \bullet h_s(\sigma_{w,s}^A(a)) = \sigma_{w,s}^B(h_w(a))$, for non-constant symbols $\sigma \in OP_{w,s}$

Notation: We write $h_w(a)$ for $(h_{s1}(a1),\ldots,h_{sn}(an))$, when w = s1...sn and a = (a1,...,an).

B First-order Order-sorted Algebra

This appendix contains some well-known definitions and results from order-sorted algebra on which other definitions in this paper depend. For a full treatment of the topic, we refer to [GM92].

B.1 Signatures

Definition B.1 An *order-sorted signature* SIG is a triple (S,$\leq$,OP) such that:

1. (S,OP) is a many-sorted signature
2. (S,$\leq$) is a poset (i.e. reflexive, transitive and anti-symmetric)
3. the following *monotonicity condition* is satisfied:
   if $\sigma \in OP_{w1,s1} \cap OP_{w2,s2}$ and w1 $\leq$ w2 then s1 $\leq$ s2

Notation: We write $w \leq w'$ for $s1 \leq s1' \wedge \ldots \wedge sn \leq sn'$, when w = s1...sn and $w' = s1' \ldots sn'$.

Definition B.2 SIG = (S,$\leq$,OP) *regular* iff:
For any $\sigma \in OP_{w1,s1}$ and w0 $\leq$ w1, there exists a least $\langle$w,s$\rangle$ for which $\sigma \in OP_{w,s}$ and w0 $\leq$ w.

Definition B.3 SIG = (S,$\leq$,OP) *pre-regular* iff:
For any $\sigma \in OP_{w1,s1}$ and w0 $\leq$ w1, there exists a least s for which there exists a w such that $\sigma \in OP_{w,s}$ and w0 $\leq$ w.

Definition B.4 SIG = (S,≤,OP) is *locally upwards/downwards filtered* iff: each connected component of (S,≤) is locally upwards/downwards filtered.

Definition B.5 A poset (S,≤) is *locally upwards filtered* iff:
$\forall$ s,s′ $\in$ S • $\exists$ s″ $\in$ S • s,s′ $\leq$ s″

Definition B.6 A poset (S,≤) is *locally downwards filtered* iff:
$\forall$ s,s′ $\in$ S • $\exists$ s″ $\in$ S • s″ $\leq$ s,s′

Definition B.7 SIG = (S,≤,OP) is *coherent* iff: SIG is regular and locally upwards filtered.

B.2 Algebras and Homomorphisms

Definition B.8 Given an order-sorted signature SIG = (S,≤,OP). An *order-sorted SIG-algebra*, A, is a many-sorted (S,OP)-algebra satisfying the following *monotonicity conditions*:

1. $A_s \subseteq A_s{}'$, if $s \leq s'$
2. $\forall\, a \in A_{w1} \bullet \sigma^{A}_{w1,s1}(a) = \sigma^{A}_{w2,s2}(a)$, if $\sigma \in OP_{w1,s1} \cap OP_{w2,s2}$ and $w1 \leq w2$

Definition B.9 Given an order-sorted signature SIG = (S,≤,OP), and two SIG-algebras A and B. An *order-sorted SIG-homomorphism*, h : A $\mapsto$ B, from A to B, is a many-sorted (S,OP)-homomorphism from A to B satisfying the following *restriction condition*: $\forall \in A_s \bullet h_s(a) = h_{s'}(a)$, if $s \leq s'$

Notation: The category of order-sorted SIG-algebras and SIG-homomorphisms is denoted Alg(SIG).

B.3 Terms and Axioms

Definition B.10 Given an order-sorted signature SIG = (S,≤,OP).
Let $X = (X_s)_{s \in S}$ be a S-sorted set of *variables*. The sets, $T_s(SIG,X)$, $s \in S$, of *SIG-terms* of sort s with variables in X, are inductively defined by a number of rules:

1. $x \in T_s(SIG,X)$, if $x \in X_s$
2. $\sigma \in T_s(SIG,X)$, if $\sigma \in OP_s$
3. $\sigma(t1,...,tn) \in T_s(SIG,X)$, if $\sigma \in OP_{s1...sn,s}$, $ti \in T_{si}(SIG,X)$ for $1 \leq i \leq n$
4. $t \in T_s(SIG,X)$, if $t \in T_{s'}(SIG,X)$ and $s' \leq s$

Notation: We write $T_w(SIG,X)$ for $T_{s1}(SIG,X) \times ... \times T_{sn}(SIG,X)$, when w = s1...sn, and $\sigma(t)$ for $\sigma(t1,...,tn)$ when t = (t1,...,tn).

Fact B.11 A SIG-term may have several sorts, due to the last rule. For a regular signature SIG, any term, t, has a least sort, which we denote LS(t).

Definition B.12 Given an order-sorted signature SIG = (S,$\leq$,OP). The *SIG-term algebra* T(SIG,X) is defined as follows:

1. carriers: $T(SIG,X)_s = T_s(SIG,X)$
2. constants: $\sigma_s^{T(SIG,X)} = \sigma$
3. functions: $\forall\, t \in T(SIG,X)_w \bullet \sigma_{w,s}^{T(SIG,X)}(t) = \sigma(t)$

Notation: We write T(SIG) for T(SIG,X), if X is empty. We write T(X) for T(SIG,X), and T for T(SIG), when this does not give rise to confusion.

Theorem B.13 Given a regular order-sorted signature SIG. Then T(SIG) is initial in Alg(SIG) and T(SIG,X) is free over X in Alg(SIG).

Definition B.14 Given a regular order-sorted signature SIG = (S,$\leq$,OP). An *order-sorted SIG-axiom* is a conditional equation of the form

$$(\forall\, X)\ t = t' \textbf{ if } \wedge_{i \in I}\ (t_i = t'_i)$$

where $t, t', t_i, t'_i \in T(SIG,X)$ for $i \in I$, and LS(t) and LS(t') belong to the same connected component in (S,$\leq$), and for each $i \in I$, $LS(t_i)$ and $LS(t'_i)$ belong to the same connected component in (S,$\leq$).

Notation: If I is empty, we just write $(\forall\, X)\ t = t'$. When the variable set X of an axiom can be deduced from the context, we allow the quantification to be omitted.

Definition B.15 Given a regular order-sorted signature SIG = (S,$\leq$,OP) and a family X of SIG-variables. An ***assignment*** $a : X \mapsto A$ is an S-sorted family $(a_s)_{s \in S}$ of functions $a_s : X_s \mapsto A_s$. The natural extension of a is the unique SIG-homomorphism, $a^* : T(SIG,X) \mapsto A$, for which $a^*(x) = x$ for $x \in X_s$.

Definition B.16 Given a regular order-sorted signature SIG. Then a SIG-algebra, A, *satisfies* a SIG-axiom, $(\forall\, X)\ t = t' \textbf{ if } \wedge_{i \in I}\ (t_i = t'_i)$, if for every assignment, $a : X \mapsto A$, for which $a^*_{LS(t_i)}(t_i) = a^*_{LS(t'_i)}(t'_i)$ for every $i \in I$, it holds that $a^*_{LS(t)}(t) = a^*_{LS(t')}(t')$

Definition B.17 An order-sorted *specification*, SPEC, is a pair, (SIG,E), consisting of an regular order-sorted signature, SIG, and a set of order-sorted axioms, E. A *SPEC-algebra* is a SIG-algebra satisfying all the axioms in E, and a *SPEC-homomorphism* is a SIG-homomorphism between SPEC-algebras.

Notation: The category of order-sorted SPEC-algebras and SPEC-homomorphisms is denoted Alg(SPEC).

For a definition of the quotient SPEC-term algebras T(SPEC,X) and T(SPEC): see [GM92].

Theorem B.18 Given an order-sorted specification, SPEC, with coherent signature. Then T(SPEC) is initial in Alg(SPEC) and T(SPEC,X) is free over X in Alg(SPEC).

C First-Order Coercion Algebra

In [Rey80] and [Qia89] two notions of coercion algebras have been designed using category theory and set theory, respectively. In this section we develop our own notion of coercion algebra, which differs from [Qia89] and [Rey80] by certain details concerning requirements to signatures and algebras, and is more close to the exposition for order-sorted algebra given in [GM92].

Proofs of most theorems and facts can be found in [Hax94].

C.1 Signatures

Definition C.1 A *coercion signature* is a triple (S,≤,OP), such that

1. (S,OP) is a many-sorted signature
2. (S,≤) is a pre-order (i.e. reflexive and transitive)
3. the following *monotonicity condition* is satisfied:
   if $\sigma \in OP_{w1,s1} \cap OP_{w2,s2}$ and $w1 \leq w2$ then $s1 \leq s2$

If we compare coercion signatures with order-sorted signatures the only difference is that the ordering need not to be anti-symmetric.

Definition C.2 A coercion signature (S,≤,OP) is *pre-regular* iff for any $\sigma \in OP_{w1,s1}$ and $w0 \leq w1$, there exists a least sort s for which there exists a w such that $\sigma \in OP_{w,s}$ and $w0 \leq w$. Such a sort s is called *the least sort of σ over w0*, and is denoted LS(σ,w0).

C.2 Algebras and Homomorphisms

Definition C.3 Given a pre-regular coercion signature CSIG = (S,≤,OP). A *coercion CSIG-algebra*, A, consists of

1. a many-sorted (S,OP)-algebra A
2. coercion functions $c^A_{s,s'} : A_s \mapsto A_{s'}$, for each $s \leq s'$, satisfying the following conditions:
   (a) *reflexivity* for $s \in S$: $\forall\, a \in A_s \bullet c^A_{s,s}(a) = a$
   (b) *transitivity* for $s \leq s'$, $s' \leq s''$ in S: $\forall\, a \in A_s \bullet c^A_{s',s''}(c^A_{s,s'}(a)) = c^A_{s,s''}(a)$
   (c) The *monotonicity condition* for $\sigma \in OP_{w1,s1}$, w0, w2 for which $w0 \leq w1$ and $w0 \leq w2$ and $\sigma \in OP_{w2,s2}$ and $s2 = LS(\sigma, w0)$:
   $\forall\, a \in A_{w0} \bullet \sigma^A_{w1,s1}(c^A_{w0,w1}(a)) = c^A_{s2,s1}(\sigma^A_{w2,s2}(c^A_{w0,w2}(a)))$

Notation: We write $c_{w,w'}(a)$ for $(c_{s1,s1'}(a1),\ldots,c_{sn,sn'}(an))$, when $w = s1\ldots sn$, $w' = s1'\ldots sn'$ and $a = (a1,\ldots,an)$.

Definition C.4 Given a pre-regular coercion signature CSIG = (S,≤,OP), and two CSIG-algebras A and B. A *coercion CSIG-homomorphism*, $h : A \mapsto B$, from A to B, is a many-sorted (S,OP)-homomorphism from A to B satisfying the following *restriction condition*:

$\forall\, a \in A_s \bullet h_{s'}(c^A_{s,s'}(a)) = c^B_{s,s'}(h_s(a))$, if $s \leq s'$

Notation: The CSIG-algebras and CSIG-homomorphisms form a category we denote CAlg(CSIG).

C.3 Terms and Axioms

Coercion CSIG-terms are defined exactly as order-sorted terms:

Definition C.5 Given a coercion signature CSIG = (S,≤,OP). Let $X = (X_s)_{s \in S}$ be a S-sorted set of variables. The sets, T_s(CSIG,X), s ∈ S, of *CSIG-terms* of sort s with variables in X, are inductively defined by a number of rules:

1. $x \in T_s$(CSIG,X), if $x \in X_s$
2. $\sigma \in T_s$(CSIG,X), if $\sigma \in OP_s$
3. $\sigma(t) \in T_s$(CSIG,X), if $\sigma \in OP_{w,s}$ and $t \in T_w$(CSIG,X), n≥1
4. $t \in T_s$(CSIG,X), if $t \in T_{s'}$(CSIG,X) and $s' \leq s$

Fact C.6 A CSIG-term may have several sorts, due to the last rule. For a pre-regular signature CSIG, any term, t, has a least sort, which we denote LS(t).

Definition C.7 Given a coercion signature CSIG = (S,≤,OP). The coercion *CSIG-term algebra* T(CSIG,X) is defined as follows:

1. carriers: $T(CSIG,X)_s = T_s(CSIG,X)$
2. constants: $\sigma_s^{T(CSIG,X)} = \sigma$
3. functions: $\forall\, t \in T(CSIG,X)_w \bullet \sigma_{w,s}^{T(CSIG,X)}(t) = \sigma(t)$
4. coercion functions: $\forall\, t \in T(CSIG,X)_s \bullet c_{s,s'}^{T(CSIG,X)}(t) = t$

Notation: We write T(CSIG) for T(CSIG,X), if X is empty. We write T(X) for T(CSIG,X), and T for T(CSIG), when this does not give rise to confusion.

Theorem C.8 Given a pre-regular coercion signature CSIG. Then T(CSIG) is initial in CAlg(CSIG).

Theorem C.9 Given a pre-regular coercion signature CSIG. Then T(CSIG,X) is free over X in CAlg(CSIG).

The notions of CSIG-axioms and satisfaction of CSIG-axioms by CSIG-algebras are just as for modified order-sorted algebra:

Definition C.10 Given a pre-regular coercion signature CSIG = (S,≤,OP). A *coercion CSIG-axiom* is a conditional equation of the form

$(\forall\, X)\ t = t'$ **if** $\wedge_{i \in I}\ (t_i = t'_i)$

where $t, t', t_i, t'_i \in$ T(CSIG,X) for i ∈ I, and there exists a least upper bound of LS(t) and LS(t') in (S,≤), and for each i ∈ I, there exists a least upper bound of $LS(t_i)$ and $LS(t'_i)$ in (S,≤).

Definition C.11 Given a pre-regular coercion signature SIG = (S,$\leq$,OP) and a family X of SIG-variables. An *assignment* a : X $\mapsto$ A is an S-sorted family $(a_s)_{s \in S}$ of functions $a_s : X_s \mapsto A_s$. The natural extension of a is the unique SIG-homomorphism, a^* : T(SIG,X) $\mapsto$ A, for which $a^*(x) = x$ for $x \in X_s$.

Definition C.12 Given a pre-regular coercion signature CSIG. Then a CSIG-algebra, A, *satisfies* a CSIG-axiom, ($\forall$ X) t = t′ **if** $\wedge_{i \in I}$ ($t_i = t'_i$), if for every assignment, a : X $\mapsto$ A, for which $a^*_{si}(t_i) = a^*_{si}(t'_i)$ for every i $\in$ I, it holds that $a^*_s(t) = a^*_s(t')$, where s = LUB(LS(t),LS(t′)) and si = LUB(LS(ti),LS(ti′)), i $\in$ I.

Definition C.13 A coercion *specification*, CSPEC, is a pair, (CSIG,E), consisting of a pre-regular coercion signature, CSIG, and a set of coercion axioms, E. A *CSPEC-algebra* is a CSIG-algebra satisfying all the axioms in E, and a *CSPEC-homomorphism* is a CSIG-homomorphism between CSPEC-algebras.

Notation: The CSPEC-algebras and homomorphisms form a category we denote CAlg(CSPEC).

Theorem C.14 There exists an initial algebra in CAlg(CSPEC).

In [Hax94] it is explained how coercion algebra can be considered as a generalisation of modified order-sorted algebra and how it can be reduced to many-sorted algebra.

D First-order Generalised Algebra

Definition D.1 A *generalised signature* is a four-tuple (S, $\leq_1$,$\leq_2$, OP) for which

1. (S,$\leq_1$,OP) is a pre-regular order-sorted signature
2. (S,$\leq_2$,OP) is a pre-regular coercion signature
3. $\leq_1$ is a subrelation of $\leq_2$

Definition D.2 Given a generalised signature GSIG = (S,$\leq_1$,$\leq_2$, OP). Then GSIG-terms and -axioms are coercion (S,$\leq_2$,OP)-terms and axioms, as defined in appendix C.

Definition D.3 Given a generalised signature GSIG = (S, $\leq_1$,$\leq_2$, OP). A *generalised GSIG-algebra* is a coercion (S, $\leq_2$,OP)-algebra with the additional property that the following *injectivity condition* is satisfied:

$$\forall\, a1,a2 \in A_s \bullet c^A_{s,s'}(a1) = c^A_{s,s'}(a2) \Rightarrow a1 = a2, \text{ if } s \leq_1 s'$$

Since generalised algebras are just special coercion algebras, all the definitions of GSIG-homomorphisms etc. carry directly over to the generalised case.

In [Hax94] it is explained how generalised algebra can be considered as a generalisation of coercion algebra and modified order-sorted algebra, and how it can be reduced to many-sorted algebra.

Proving the Correctness of Behavioural Implementations

Michel Bidoit*, Rolf Hennicker**

*LIENS, CNRS & Ecole Normale Supérieure
45, Rue d'Ulm, 75230 Paris Cedex, FRANCE
**Institut für Informatik, Ludwig-Maximilians-Universität München
Leopoldstr. 11B, D-80802 München, GERMANY

Abstract. We introduce a concept of behavioural implementation for algebraic specifications which is based on an indistinguishability relation (called behavioural equality). The central objective of this work is the investigation of proof rules that first allow us to establish the correctness of behavioural implementations in a modular way and moreover are practicable enough to induce proof obligations that can be discharged with existing theorem provers. Our proof technique can also be applied for proving abstractor implementations in the sense of Sannella and Tarlecki.

1 Introduction

Algebraic specification techniques allow one to formalize correctness notions for program development steps. Thereby an important role is played by observability concepts since it is often essential to abstract from internal implementation details and to rely only on the observable behaviour of programs. Many approaches in the literature have considered behavioural concepts (cf. e.g. [GM 82], [R 87], [ST 88], [Sch 87], [NO 88]) but there is still a lack of appropriate and practically applicable proof methods for proving implementation correctness.

In this work we study proof techniques for behavioural implementations that are based on a behavioural equality. The underlying idea is that for any data structure A (formally given by a Σ-algebra) one may consider the behaviour of A by identifying all elements which are indistinguishable "from the outside", i.e. which are behaviourally equal. Then a specification SP-I is called a behavioural implementation of a specification SP w.r.t. a behavioural equality $\approx$ (given by a congruence relation on any Σ-algebra A) if the behaviour of any model of SP-I (restricted to the signature of SP) is a model of SP. Thus an "abstract" specification SP is regarded as a "specification of behaviours". The use of a behavioural equality for the description of behaviours is related to the use of representation equivalences that goes already back to the pioneering work of [H 72].

In our approach we consider structured specifications which are built on top of flat specifications (with arbitrary axioms) by enrichments and/or combinations of specifications. Since one of the essential advantages of modular system design is the ability to perform modular proofs (cf. [BB 91]) we investigate proof rules for behavioural implementations that allow us to perform correctness proofs according to the structure of the specification to be implemented. In particular we provide rules for proving behavioural implementations of flat specifications and we investigate conditions under which the proof of a behavioural implementation of the combination of specifications can be reduced to proving behavioural implementation relations for the constituent parts of the specification. Our proof rules will be demonstrated on a simple but representative example where we prove the correctness of a behavioural implementation of a specification of arithmetic expressions.

Although all results are presented here in the framework of behavioural implementations there is a close relationship to abstractor implementations in the sense of [ST 88] where an equivalence relation between algebras is used for abstracting from the model class of the specification to be implemented. Since, following a result of [BHW 94], behavioural semantics is in general more restrictive than abstractor semantics (for factorizable abstractors), all our proof rules are also correct for abstractor implementations. Moreover, since any specification to be implemented w.r.t. a behavioural equality $\approx$ should be compatible with $\approx$ (i.e. should be behaviourally consistent) and since for behaviourally consistent specifications there is no difference between behavioural and abstractor implementations, all proof rules exhibit their full power also for abstractor implementations.
Note that for lack of space there was no possibility to include proofs of theorems etc. in this paper. If no other reference is given the omitted proofs can be found in [BH 95b].

2 Basic Concepts

2.1 Algebraic Specifications

We assume the reader to be familiar with the basic notions of algebraic specifications (cf. e.g. [EM 85], [W 90]) like the notions of (many sorted) *signature* $\Sigma = (S, F)$, *total* Σ-*algebra* $A = ((A_s)_{s\in S}, (f^A)_{f\in F})$, Σ-*term algebra* $T_\Sigma(X)$, *valuation* $\alpha: X \to A$ and *interpretation* $I_\alpha: T_\Sigma(X) \to A$. Throughout this paper we assume that the carrier sets A_s of a Σ-algebra A are not empty and that $X = (X_s)_{s\in S}$ is a family of countably infinite sets X_s of variables of sort $s \in S$. A Σ-*congruence* on a Σ-algebra A is (as usual) a family $\approx_A = (\approx_{A,s})_{s\in S}$ of equivalence relations $\approx_{A,s}$ on A_s compatible with the operations of Σ. If $\Sigma 1$ is a subsignature of Σ then $A|_{\Sigma 1}$ denotes the restriction of A to $\Sigma 1$ and $(\approx_A)|_{\Sigma 1}$ denotes the restriction of $\approx_A$ to $\Sigma 1$ which is the $\Sigma 1$-congruence on $A|_{\Sigma 1}$ defined by $((\approx_A)|_{\Sigma 1})_s =_{def} \approx_{A,s}$ for all $s \in S1$.

The set of *first-order* Σ-*formulas* is defined as usual from equations $t = r$ (with $t, r \in T_\Sigma(X)$), the logical connectives $\neg, \wedge, \ldots$ and the quantifiers $\forall, \exists$. We will also use infinitary Σ-formulas of the form $\bigvee_{i\in I} \phi_i$ or $\bigwedge_{i\in I} \phi_i$ where ϕ_i is a countable set of Σ-formulas. A Σ-*sentence* is a Σ-*formula* which contains no free variables. The *(standard) satisfaction* of a Σ-formula ϕ by a Σ-algebra A, denoted by $A \models \phi$, is defined as usual in the first-order predicate calculus (with a straightforward extension to infinitary conjunctions and disjunctions).[1] The notation $A \models \phi$ is extended in the usual way to classes of algebras, specifications and sets of formulas.

A *basic (algebraic) specification* $SP = \langle\Sigma, Ax\rangle$ consists of a signature Σ and a set Ax of (possibly infinitary) Σ-sentences, called *axioms* of SP. The *model class* of SP is defined by $Mod(SP) =_{def} \{A \in Alg(\Sigma) \mid A \models Ax\}$. If the set Ax of axioms is empty then the specification $\langle\Sigma, \emptyset\rangle$ is denoted by the signature symbol Σ (if it is clear from the context).

If SP1 and SP2 are specifications with signatures $\Sigma 1$ and $\Sigma 2$ respectively then SP1+SP2 is a specification, called the *combination (sum)* of SP1 and SP2, with signature $\Sigma 1 \cup \Sigma 2$ and with model class $Mod(SP1+SP2) =_{def} \{A \in Alg(\Sigma 1 \cup \Sigma 2) \mid A|_{\Sigma 1} \in Mod(SP1)$ and $A|_{\Sigma 2} \in Mod(SP2)\}$.

As a derived operator we define the *enrichment* of a specification SP1 with signature $\Sigma = (S1, F1)$ by a set of sorts S, a set of function symbols F and a set of axioms Ax by

[1] Due to the requirement of non-empty carrier sets no pathodological situations can occur w.r.t. the satisfaction relation (cf. [KK 67]).

SP = **enrich** SP1 **by sorts** S **operations** F **axioms** Ax $=_{def}$ SP1+Δ-SP where Δ-SP is the basic specification ‹(S1∪S, F1∪F), Ax›.

Note that for any specification SP the model class Mod(SP) is closed under Σ-isomorphism. By convention, we will denote in the following the signatures of specifications SP, SP', SP1, etc. by Σ, Σ', Σ1, etc.

In the definition of basic specifications we have allowed infinitary sentences to be used as axioms. However in practice we consider basic specifications with finitary axioms together with a reachability constraint that allows us to express a generation principle for the elements of a Σ-algebra:

1. A *reachability constraint* over a signature Σ = (S, F) is a pair $\mathcal{R} = (S_{\mathcal{R}}, F_{\mathcal{R}})$ such that $S_{\mathcal{R}} \subseteq S$, $F_{\mathcal{R}} \subseteq F$ and for any $f \in F_{\mathcal{R}}$ with functionality $s_1, \ldots, s_n \to s$ the sort s belongs to $S_{\mathcal{R}}$. A sort $s \in S_{\mathcal{R}}$ is called *constrained sort* and a function symbol $f \in F_{\mathcal{R}}$ is called *constructor symbol* (or briefly *constructor*). We assume also that for each constrained sort $s \in S_{\mathcal{R}}$ there exists at least one constructor in $F_{\mathcal{R}}$ with range s.
2. A *constructor term* is a term $t \in T(\Sigma', X')_s$ of sort $s \in S_{\mathcal{R}}$ where $\Sigma' = (S, F_{\mathcal{R}})$, $X' = (X'_s)_{s \in S}$ with $X'_s = X_s$ if $s \in S \backslash S_{\mathcal{R}}$ and $X'_s = \varnothing$ if $s \in S_{\mathcal{R}}$. The set of constructor terms is denoted by $T_{\mathcal{R}}$.
3. A Σ-algebra A *satisfies a reachability constraint* $\mathcal{R} = (S_{\mathcal{R}}, F_{\mathcal{R}})$, denoted by $A \models \mathcal{R}$, if for any $s \in S_{\mathcal{R}}$ and $a \in A_s$ there exists a constructor term $t \in T_{\mathcal{R}}$ and a valuation $\alpha: X' \to A$ such that $I_\alpha(t) = a$. (Note that this definition is independent of the choice of X because X_s is assumed to be countably infinite for all $s \in S$.)

Fact 2.1 Let A be a Σ-algebra and $\mathcal{R} = (S_{\mathcal{R}}, F_{\mathcal{R}})$ be a reachability constraint over Σ. Then $A \models \mathcal{R}$ if and only if $A \models GEN_s$ for all $s \in S_{\mathcal{R}}$ where GEN_s is the following infinitary Σ-sentence: $GEN_s =_{def} \forall x{:}s. \bigvee_{t \in (T_{\mathcal{R}})_s} \exists Var(t).x = t$.

Thereby $\exists Var(t).x = t$ is an abbreviation for $\exists x_1{:}s_1. \ldots \exists x_n{:}s_n.\ x = t$ where $x_1, \ldots, x_n$ is the set of variables occurring in t of sort $s_1, \ldots, s_n$.

According to this fact specifications with finitary axioms and reachability constraints can be defined as a particular kind of basic specifications with infinitary axioms as follows: Let Σ be a signature, $\mathcal{R} = (S_{\mathcal{R}}, F_{\mathcal{R}})$ be a reachability constraint over Σ and Ax be a set of finitary Σ-sentences. Then the triple SP = ‹Σ, $\mathcal{R}$, Ax› is, by definition, the basic specification ‹Σ, Ax ∪ $\{GEN_s \mid s \in S_{\mathcal{R}}\}$›.

2.2 Behavioural Semantics

The underlying idea of behavioural semantics is to equip any Σ-algebra A with an indistinguishability relation $\approx_A$ (also called *behavioural equality*) such that for any two elements a, b of A, $a \approx_A b$ holds whenever a and b are considered to be behaviourally indistinguishable (cf. [BHW 94]). An important example is the observational equality where two elements are considered to be observationally equal if they cannot be distinguished by observable experiments (cf. e.g. [R 87]). Formally the behavioural equality is represented by a family $\approx = (\approx_A)_{A \in Alg(\Sigma)}$ of Σ-congruences on the algebras of Alg(Σ). In the following we summarize the basic definitions and results of behavioural semantics. Thereby we assume given an arbitrary behavioural equality ≈.

Definition 2.2 Let A be a Σ-algebra, $C \subseteq Alg(\Sigma)$ be a class of Σ-algebras, ϕ be a Σ-formula and let SP be a specification (of signature Σ).

(1) The *behaviour of* A *w.r.t.* $\approx$ is given by the quotient algebra $A/\approx_A$.

(2) The *behaviour class* of C w.r.t $\approx$ is the class $Beh_\approx(C) =_{def} \{A \in Alg(\Sigma) \mid A/\approx_A \in C\}$. In particular, the *behavioural semantics* of SP is given by $Beh_\approx(Mod(SP))$.

(3) C is called *behaviourally consistent w.r.t.* $\approx$ if $C \subseteq Beh_\approx(C)$ (or equivalently if for all $A \in C$, $A/\approx_A$ belongs to C). SP is called *behaviourally consistent* if $Mod(SP) \subseteq Beh_\approx(Mod(SP))$.

(4) We write $A \models_\approx \phi$ if and only if $A/\approx_A \models \phi$. If $A \models_\approx \phi$ for all $A \in C$ we write $C \models_\approx \phi$ (or similarly $SP \models_\approx \phi$ if C is the model class of a specification SP). Then ϕ is called *behavioural theorem* over C (over SP resp.).

Remark 2.3

(1) The behaviour $A/\approx_A$ of a Σ-algebra A identifies all elements of A which are indistinguishable "from the outside". Hence $A/\approx_A$ can be considered as the "black box view" of A.

(2) The behaviour class $Beh_\approx(C)$ consists of all algebras whose behaviour belongs to C. In particular, the behavioural semantics $Beh_\approx(Mod(SP))$ of a specification SP consists of all algebras whose behaviour fulfills the given specification SP. Thereby SP is interpreted as a specification of intended behaviours.

(3) A specification is behaviourally consistent if and only if any model of SP satisfies the intended behaviour, i.e. for any model $A \in Mod(SP)$ its behaviour $A/\approx_A$ is also a model of SP and hence fulfills the requirements of SP. If there are models of SP which do not satisfy the intended behaviour then the specification is *behaviourally inconsistent.* A specification SP is behaviourally inconsistent w.r.t. $\approx$ if the specification requires some property which is not compatible with the chosen behavioural equality $\approx$, i.e. the desired property cannot be behaviourally valid in the specification. In such a case, remembering our fundamental assumption that the essence of a behavioural equality is to provide a means for expressing intended behaviours, the specification SP and the chosen behavioural equality $\approx$ contradict each other, i.e. either the specification SP or the behavioural equality is chosen in the wrong way (in the same sense as in the standard case specifications with contradictory axioms are considered to be inconsistent, cf. Example 2.6).

(4) The notation $A \models_\approx \phi$ expresses that the behaviour of A satisfies the formula ϕ, i.e. ϕ is satisfied by A up to the identification of indistinguishable elements. This can be equivalently defined using an explicit notion of *behavioural satisfaction* of formulas where instead of the set theoretic equality the behavioural equality is used for the interpretation of the predicate symbol "=" (cf. e.g. [R 87], [BHW 94]).

The behavioural equality $\approx$ is called *strongly uniform* if it satisfies the following property which intuitively says that whenever an algebra B "lies between" an algebra A and its behaviour $A/\approx_A$ then B has the same behaviour as A (up to isomorphism). In particular, strong uniformity implies that isomorphic algebras have isomorphic behaviours.

Definition 2.4 The family $\approx = (\approx_A)_{A \in Alg(\Sigma)}$ of Σ-congruences is called *strongly uniform* if for all Σ-algebras A and B the following holds: If $g: A \to B$ and $h: B \to A/\approx_A$ are surjective homomorphisms such that their composition $h \bullet g: A \to A/\approx_A$ is the canonical epimorphism then $A/\approx_A$ and $B/\approx_B$ are isomorphic.

Proposition 2.5 Let $\approx = (\approx_A)_{A \in Alg(\Sigma)}$ be a strongly uniform family of Σ-congruences and let $C \subseteq Alg(\Sigma)$ be closed under isomorphism and behaviourally consistent w.r.t. $\approx$. Moreover, let $\approx' = (\approx'_A)_{A \in Alg(\Sigma)}$ be a family of Σ-congruences such that $\approx'_A \subseteq \approx_A$ for any Σ-algebra A. Then $Beh_{\approx'}(C) \subseteq Beh_{\approx}(C)$.

Example 2.6 *(Observational equality)* An important example of a behavioural equality is the *observational equality* of elements as defined e.g. in [R 87], [BH 94]. Formally, let $\Sigma = (S, F)$ be a signature and $Obs \subseteq S$ be a distinguished set of observable sorts. Then:

(1) An *observable Σ-context* is a Σ-term c of observable sort which contains besides variables in X exactly one distinguished variable z_s (called *context variable*) of some sort $s \in S$. By exception, Var(c) will denote the set of variables occurring in c apart from the context variable of c. The application of a context c with context variable z_s to a term t of sort s is denoted by c[t]. The set of the observable Σ-contexts is denoted by $C(\Sigma)^{Obs}$.

(2) If A is a Σ-algebra then two elements $a, b \in A_s$ are *observationally equal*, denoted by $a \approx_{Obs,A} b$ if for all observable contexts $c \in C(\Sigma)^{Obs}$ with context variable z_s, for all valuations $\alpha: X \to A$, we have $I_{\alpha 1}(c) = I_{\alpha 2}(c)$, where $\alpha 1, \alpha 2: X \cup \{z_s\} \to A$ are the unique extensions of α defined by $\alpha 1(z_s) = a$ and $\alpha 2(z_s) = b$.

(3) The family $(\approx_{Obs,A})_{A \in Alg(\Sigma)}$ of observational equalities is denoted by $\approx_{Obs}$. It is a strongly uniform family of Σ-congruences. (The proof is slightly technical but straightforward.)

For any specification SP the class $Beh_{\approx_{Obs}}(Mod(SP))$ is called *observational behaviour semantics of SP w.r.t. Obs*. According to a theorem of [R 87] a basic specification SP is behaviourally consistent w.r.t. Obs if the axioms of SP are conditional equations with observable premises (because under this syntactic restriction the model class of SP is closed under the observational quotient construction, cf. Remark 2.3 (3)). More general conditions for behavioural consistency of flat and structured specifications will be pointed out later. As a particular application of Proposition 2.5 in the observational framework we can consider two observational equivalences $\approx_{Obs}$ and $\approx_{Obs'}$ with $Obs \subseteq Obs'$. Then for any Σ-algebra A, $\approx_{Obs',A} \subseteq \approx_{Obs,A}$. Hence, if SP is a specification behaviourally consistent w.r.t. $\approx_{Obs}$ then $Beh_{\approx_{Obs'}}(Mod(SP)) \subseteq Beh_{\approx_{Obs}}(Mod(SP))$.

An explicit example of a behaviourally consistent specification will be considered in Example 3.7. Here we would like to point out by a very simple example how behavioural inconsistency can arise. Consider a specification SP which has one sort s, two constants a, b of sort s and the axiom $a \neq b$ requiring that a and b are different. Now choose the observational equality $\approx_{\emptyset}$ generated by the empty set of observable sorts. Then there are no observations that allow to distinguish elements hence all elements are observationally equal and in particular a and b are observationally equal. This contradicts the required inequation $a \neq b$ thus the specification SP is behaviourally inconsistent w.r.t. $\approx_{\emptyset}$. For obtaining a behaviourally consistent specification one has either to omit the axiom $a \neq b$ or one has to consider the sort s to be observable.

2.3 Proving Behavioural Theorems

For proving the correctness of behavioural implementations it will be essential to be able to prove behavioural theorems over specifications. According to Definition 2.2 (4) a

Σ-formula ϕ is a behavioural theorem over a class C of Σ-algebras, i.e. $C \models_{\approx} \phi$, if and only if $C/\approx \models \phi$ (where $C/\approx =_{def} \{A/\approx_A \mid A \in C\}$). The following results due to [BH 95a] show how to reduce proofs of behavioural theorems to proofs of standard theorems. To achieve this the first idea is to introduce an explicit denotation for the behavioural equality by corresponding predicate symbols equipped with an appropriate axiomatization.[2]

Definition 2.7 *(Axiomatization of the behavioural equality)* Given a signature Σ, $\mathcal{L}(\Sigma)$ is the signature Σ enriched by a binary predicate symbol $\sim_s$ for each sort $s \in S$, i.e. $\mathcal{L}(\Sigma) =_{def} \Sigma \cup \{\sim_s : s, s \mid s \in S\}$. Let $(Beh_s(x_s, y_s))_{s \in S}$ be a family of finitary first-order Σ-formulas with exactly two free variables x_s, y_s (both of sort s) and let BEH be the following $\mathcal{L}(\Sigma)$-sentence:

$$BEH =_{def} \bigwedge_{s \in S} \forall x_s, y_s : s.\ [Beh_s(x_s, y_s) \Leftrightarrow x_s \sim_s y_s].^3$$

$(Beh_s(x_s, y_s))_{s \in S}$ is called a (finitary) axiomatization of the behavioural equality $\approx$ with respect to a class C of Σ-algebras if for any Σ-algebra $A \in C$ and all elements $a, b \in A_s$ $(s \in S)$, $a \sim_s^{\mathcal{E}(A)} b$ if and only if $a \approx_A b$ where $\mathcal{E}(A)$ is the unique extension of A to a model of $\langle \mathcal{L}(\Sigma), BEH\rangle$.[4]

Example 2.8 Consider the observational equality $\approx_{Obs}$ w.r.t. a set Obs of observable sorts (cf. Example 2.6). Let $C_0 \subseteq C(\Sigma)^{Obs}$ be a finite subset of observable Σ-contexts such that, for any observable sort $s \in Obs$, $z_s \in C_0$. The set C_0 induces the following finitary Σ-formulas (for all $s \in S$):

$$Beh_s(x_s, y_s) =_{def} \bigwedge_{c \in C_0(s)} \forall Var(c).\ c[x_s] = [y_s]\ .$$

Thereby $C_0(s)$ denotes the subset of all the contexts in C_0 with context variable of sort s. According to Theorem 42 in [BH 95a], $(Beh_s(x_s, y_s))_{s \in S}$ is an axiomatization of the behavioural equality $\approx_{Obs}$ with respect to a class C of Σ-algebras if and only if $\mathcal{E}(C) \models CONG_\Sigma(\sim)$ where $CONG_\Sigma(\sim)$ is a finitary $\mathcal{L}(\Sigma)$-sentence requiring that the family $(\sim_s)_{s \in S}$ is a Σ-congruence and $\mathcal{E}(C) =_{def} \{\mathcal{E}(A) \mid A \in C\}$. A concrete example will be considered in Section 3.4. In particular, if SP-I is a specification with signature $\Sigma\text{-}I \supseteq \Sigma$ then $(Beh_s(x_s, y_s))_{s \in S}$ is an axiomatization of the behavioural equality $\approx_{Obs}$ with respect to $Mod(SP\text{-}I)|_\Sigma$ if and only if $SP\text{-}I + \langle \mathcal{L}(\Sigma), BEH\rangle \models CONG_\Sigma(\sim)$.[5]

The following theorem is a special case of Theorem 38 and Example 6 in [BH 95a]:

Theorem 2.9 *(Reduction of behavioural theorems to standard theorems)*
Let ϕ be a Σ-formula and let $\mathcal{L}(\phi) =_{def} \phi[t \sim_s r / t = r]$ be the $\mathcal{L}(\Sigma)$-formula obtained from ϕ by replacing any equation $t = r$ occurring in ϕ by $t \sim_s r$ (where s is the common sort of t and r).

(1) Let SP be a specification and let $(Beh_s(x_s, y_s))_{s \in S}$ be an axiomatization of the behavioural equality $\approx$ w.r.t. Mod(SP).

$SP \models_{\approx} \phi$ if and only if $SP + \langle \mathcal{L}(\Sigma), BEH\rangle \models \mathcal{L}(\phi)$.

---

[2] We assume the reader to be familiar with the usual notions of predicate symbols and their interpretations.

[3] We use an infix notation for the binary predicate symbols $\sim_s$.

[4] Such an extension (with the same carriers as A) exists since BEH defines unambiguously the interpretation of the predicate symbols $\sim_s$.

[5] As a particular case we have considered in [BH 94] a set C_0 of so-called crucial observable contexts where the corresponding formula $CONG_\Sigma(\sim)$ is called "observability kernel".

(2) Let SP-I be a specification with signature $\Sigma\text{-I} \supseteq \Sigma$ and let $(Beh_s(x_s, y_s))_{s\in S}$ be an axiomatization of the behavioural equality $\approx$ w.r.t. $Mod(SP\text{-}I)|_\Sigma$.

$Mod(SP\text{-}I)|_\Sigma \models_\approx \phi$ if and only if $SP\text{-}I + \langle \mathcal{L}(\Sigma), BEH\rangle \models \mathcal{L}(\phi)$.

As a first useful application of Theorem 2.9 we have:

Theorem 2.10 *(Characterization of behavioural consistency)*
Let $SP = \langle \Sigma, Ax\rangle$ be a basic specification and let $(Beh_s(x_s, y_s))_{s\in S})$ be an axiomatization of the behavioural equality $\approx$ w.r.t. Mod(SP). The following conditions are equivalent:
(1) SP is behaviourally consistent w.r.t. $\approx$.
(2) $SP \models_\approx Ax$.
(3) $\langle \mathcal{L}(\Sigma), Ax \cup BEH\rangle \models \mathcal{L}(Ax)$.

3 Proving Behavioural Implementations

3.1 Behavioural Implementations

The concept of behavioural implementations is motivated by the idea that a specification SP-I is a correct implementation of a specification SP if it fulfills the intended behaviour of SP with respect to some behavioural equality $\approx$. More precisely this means that the behaviour of any model of SP-I (after restriction to the signature of SP) is a model of SP. Formally this can be expressed as follows using the behaviour operator $Beh_\approx$:

Definition 3.1 Let SP, SP-I be specifications with $\Sigma \subseteq \Sigma\text{-I}$ and let $\approx$ be a family of Σ-congruences. SP-I is called *behavioural implementation of* SP *w.r.t.* $\approx$ (written $SP \overset{\approx}{\rightsquigarrow} SP\text{-}I$) if $Mod(SP\text{-}I)|_\Sigma \subseteq Beh_\approx(Mod(SP))$.

Remark 3.2 By definition of $Beh_\approx(Mod(SP))$, SP-I is a behavioural implementation of SP w.r.t. $\approx$ if and only if for any model $A\text{-}I \in Mod(SP\text{-}I)$ the following holds: If we first forget all sorts and operations of the "concrete" signature Σ-I which do not belong to the "abstract" signature Σ (i.e. if we construct the Σ-algebra $A\text{-}I|_\Sigma$) and if we then identify all elements of $A\text{-}I|_\Sigma$ w.r.t. the behavioural equality then we obtain a model of SP. Hence in principle behavioural implementations are related to the fundamental ideas of [H 72] whereby Hoare's representation equivalence is given by the behavioural equality. We do not consider in this paper the issue of restricting the representation domains to those concrete values which are used for the representation of abstract values. However this could be handled using partial congruences as studied in [BHW 94] and [BH 95a].

Fact 3.3 $SP \overset{\approx}{\rightsquigarrow} SP$ if and only if SP is behaviourally consistent w.r.t. $\approx$.

Theorem 3.4 Let SP, SP-I be specifications with $\Sigma \subseteq \Sigma\text{-I}$ and let $\approx$ and $\approx'$ be two families of Σ-congruences such that $\approx'_A \subseteq \approx_A$ for any Σ-algebra A. Assume $\approx$ is strongly uniform and SP is behaviourally consistent w.r.t. $\approx$. Then $SP \overset{\approx'}{\rightsquigarrow} SP\text{-}I$ implies $SP \overset{\approx}{\rightsquigarrow} SP\text{-}I$.

The following proposition shows that under simple conditions behavioural implementations can be vertically composed:

Theorem 3.5 *(Vertical composition)* Let SP, SP', SP'' be specifications with $\Sigma \subseteq \Sigma' \subseteq \Sigma''$, let $\approx$ be a strongly uniform family of Σ-congruences and let $\approx'$ be a family of Σ'-congruences such that for any $A' \in Alg(\Sigma')$, $(\approx'_{A'})|_{\Sigma} \subseteq \approx_{(A'|_\Sigma)}$ where $(\approx'_{A'})|_{\Sigma}$ is the restriction of $\approx'_{A'}$ to Σ (cf. Section 2.1). Then the following holds:

If SP $\overset{\approx}{\sim\sim\sim>}$ SP' and SP' $\overset{\approx'}{\sim\sim\sim>}$ SP'' then SP $\overset{\approx}{\sim\sim\sim>}$ SP''.

Example 3.6 Let SP, SP', SP'' be as in Theorem 3.5, let Obs be a set of observable sorts over Σ and Obs' be a set of observable sorts over Σ' such that Obs $\subseteq$ Obs'. Then it is easy to see that the conditions of Theorem 3.5 are satisfied for the observational equalities $\approx_{Obs}$ and $\approx_{Obs'}$. Hence in the observational case behavioural implementations can be vertically composed if the observable sorts on more abstract levels are also considered as observable on more concrete levels of a development. This is not a restriction because in practice one may even consider a fixed set of primitive data types, like Booleans, Integers etc. to be observable throughout the development.

Example 3.7 *(Optimizing the evaluation of expressions)*
In this example we construct a behavioural implementation of a specification of arithmetic expressions w.r.t. a set Obs of observable sorts. Since all axioms are conditional equations with observable premisses (positive or negative ones) the specification is behaviourally consistent.
The specification to be implemented is a usual specification of arithmetic expressions which contains a substitution function . [. /.] exp, exp, id → exp. The evaluation of expressions under an environment (also called state) is defined by the evaluation function eval: exp, state → nat. States and their usual operations like lookup and update are specified in the underlying specification STATE which is built over standard specifications NAT for the natural numbers and ID for identifiers.

spec EXP = **enrich** STATE **by**
  constrained sorts {exp}
  constructors {natexp: nat → exp, idexp: id → exp, plus: exp, exp → exp, mult: exp, exp → exp}
  operations { . [. /.]: exp, exp, id → exp, eval: exp, state → nat}
  axioms {∀ n:nat, i, j:id, s:state, e, e1, e2:exp.
    eval(natexp(n), s) = n ∧ eval(idexp(i), s) = lookup(i, s) ∧
    eval(plus(e1, e2), s) = eval(e1, s) + eval(e2, s) ∧
    eval(mult(e1, e2), s) = eval(e1, s) * eval(e2, s) ∧
    natexp(n)[e/i] = natexp(n) ∧ idexp(i)[e/i] = e ∧ [i ≠ j ⇒ idexp(j)[e/i] = idexp(j)] ∧
    plus(e1, e2)[e/i] = plus(e1[e/i], e2[e/i]) ∧ mult(e1, e2)[e/i] = mult(e1[e/i], e2[e/i]) }
endspec

spec STATE = **enrich** NAT+ID **by**
  constrained sorts {state}
  constructors {init: → state, update: id, nat, state → state}
  operations {lookup: id, state → nat}
  axioms {∀ n, m:nat, i, j:id, s:state.
    lookup(i, init) = 0 ∧ lookup(i, update(i, n, s)) = n ∧
    [i ≠ j ⇒ lookup(i, update(j, n, s)) = lookup(i, s)] ∧
    update(i, n, update(i, m, s)) = update(i, n, s) ∧
    [i ≠ j ⇒ update(i, n, update(j, m, s)) = update(j, m, update(i, n, s))] }
endspec

The implementation of the specification EXP is based on two ideas: First, we are interested in a (constructive) implementation of states which can directly be translated into a functional program. We will adopt here an ad hoc solution of this problem by simply removing the last two axioms of STATE which leads to the implementing specification STATE-I. Due to the observational approach to implementations we can check that this does not affect the observable behaviour of states if only natural numbers and identifiers but not states are considered to be observable.
The second, more interesting aspect of the implementation choice concerns the evaluation of expressions. In the abstract specification the evaluation eval(e1[e2/i], s) of a substitution expression e1[e2/i] is not efficient (assuming call by value) since first the substitution is performed recursively according to the structure of e1 and then the evaluation of the resulting expression is again computed recursively according to its structure. The idea of the implementation is to optimize the evaluation such that the value of eval(e1[e2/i], s) is directly computed by evaluating the expression e1 under the new state update(i, eval(e2, s), s) (cf. the last axiom of EXP-I). The implementation EXP-I is built on top of the specification STATE-I which is just the specification STATE where the last two axioms (concerning update) are omitted.

spec EXP-I = **enrich** STATE-I **by**
  constrained sorts {exp}
  constructors {natexp: nat → exp, idexp: id → exp, plus: exp, exp → exp, mult: exp, exp → exp,
    . [. / .]: exp, exp, id → exp}
  operations {eval: exp, state → nat}
  axioms {∀ n:nat, i:id, s:state, e, e1, e2:exp.
    eval(natexp(n), s) = n ∧ eval(idexp(i), s) = lookup(i, s) ∧
    eval(plus(e1, e2), s) = eval(e1, s) + eval(e2, s) ∧
    eval(mult(e1, e2), s) = eval(e1, s) * eval(e2, s) ∧
    eval(e1[e2/i], s) = eval(e1, update(i, eval(e2, s), s)) }
endspec

Note that the axioms of EXP defining the substitution function and the last two axioms of STATE are not satisfied (in the standard sense) by the implementation. Nevertheless EXP-I satisfies the required observable behaviour of EXP if only the sorts nat and id are considered to be observable, i.e. EXP-I is a behavioural implementation of EXP. How to prove the correctness of a behavioural implementation will be the topic of the next sections.

3.2 Correctness Proofs: The Basic Case

In this section we will consider how to prove behavioural implementations if the specification to be implemented is a basic (flat) specification. Due to the definition of the behavioural implementation relation we know that SP $\approx$~~~> SP-I holds if and only if, for any model A-I ∈ Mod(SP-I), the behaviour of $A\text{-}I|_\Sigma$ w.r.t. $\approx$ belongs to Mod(SP). Hence, if SP = ‹Σ, Ax› is a flat specification then SP $\approx$~~~> SP-I holds if and only if $\text{Mod(SP-I)}|_\Sigma$ $\models_\approx$ Ax, i.e. if all $\phi \in$ Ax are behavioural theorems over $\text{Mod(SP-I)}|_\Sigma$. Since we know from Theorem 2.9 how to reduce the proof of behavioural theorems to the proof of standard theorems, we obtain the following characterization of behavioural implementations of basic specifications:

Theorem 3.8 Let SP = ‹Σ, Ax› be a basic specification, let SP-I be a specification with signature Σ-I ⊇ Σ and let $(Beh_s(x_s, y_s))_{s\in S}$ be an axiomatization of the behavioural equality ≈ w.r.t. $Mod(SP\text{-}I)|_\Sigma$. Then the following inference rule (R0) for behavioural implementations is sound (in fact the condition above the line is even necessary for SP ≈~~~> SP-I):

$$\text{(R0)} \quad \frac{\text{SP-I} + \langle \mathcal{L}(\Sigma), \text{BEH}\rangle \models \mathcal{L}(\text{Ax})}{\text{SP} \overset{\approx}{\rightsquigarrow} \text{SP-I}}$$

Let us now consider the case where the specification to be implemented is a basic specification SP = ‹Σ, $\mathcal{R}$, Ax› with finitary axioms Ax and reachability constraint $\mathcal{R} = (S_\mathcal{R}, F_\mathcal{R})$. Since SP is, by definition (cf. Section 2.1), the basic specification ‹Σ, Ax ∪ {GEN_s | $s \in S_\mathcal{R}$}› Theorem 3.8 shows that SP ≈~~~> SP-I holds if and only if SP-I + ‹$\mathcal{L}(\Sigma)$, BEH› ⊨ $\mathcal{L}$(Ax) and SP-I + ‹$\mathcal{L}(\Sigma)$, BEH› ⊨ $\mathcal{L}(GEN_s)$ for all $s \in S_C$ (provided that an axiomatization of the behavioural equality ≈ w.r.t. $Mod(SP\text{-}I)|_\Sigma$ is given). For proving SP-I + ‹$\mathcal{L}(\Sigma)$, BEH› ⊨ $\mathcal{L}$(Ax) one can apply any standard proof technique since Ax consists only of finitary axioms. For proving the second condition one has to show (by definition of $\mathcal{L}(GEN_s)$):

(*) SP-I + ‹$\mathcal{L}(\Sigma)$, BEH› ⊨ $\forall x{:}s. \bigvee_{t\in(T_\mathcal{R})_s} \exists Var(t).x \sim_s t$ (for all $s \in S_\mathcal{R}$).

This means that for any model A-I of SP-I, any element a ∈ A-I of sort $s \in S_\mathcal{R}$ is behaviourally equal to an element b ∈ A-I that can be denoted by a constructor term $t \in (T_\mathcal{R})_s$. For proving this fact we introduce auxiliary predicate symbols $IsGen_s$ for all constrained sorts $s \in S_\mathcal{R}$ and a set of axioms IsGen-Ax such that all elements denoted by a constructor term belong to (the interpretations of) the predicate symbols $IsGen_s$. The set of axioms IsGen-Ax consists of the following sentences:
For any $f \in F_\mathcal{R}$ with functionality $s_1,\dots,s_n \to s$ the formula

$$\forall x_1{:}s_1. \dots x_n{:}s_n.\ IsGen_{s_{r1}}(x_{r1}) \wedge \dots \wedge IsGen_{s_{rk}}(x_{rk}) \Rightarrow IsGen_s(f(x_1, \dots, x_n))$$

belongs to IsGen-Ax where $s_{r1}, \dots, s_{rk}$ are the constrained sorts under the argument sorts of f and $x_{r1}, \dots, x_{rk}$ are the corresponding variables occurring in $x_1, \dots, x_n$. With this definition we have the following lemma:

Lemma 3.9 Let BEH-ISGEN $=_{def}$ ‹$\mathcal{L}(\Sigma)$ ∪ {$IsGen_s$: s | s ∈ S}, BEH ∪ IsGen-Ax›. Then Property (*) from above is satisfied if
SP-I + BEH-ISGEN ⊨ $\forall x{:}s. \exists y{:}s.\ IsGen_s(y) \wedge x \sim_s y$ for all $s \in S_\mathcal{R}$.

Remark 3.10 The proposed proof technique for proving "behavioural reachability" corresponds to the technique for proving reachability in the standard case as used in the Larch Prover (cf. [GH 93]).

The above results allow to specialize the rule (R0) of Theorem 3.8 to the following sound inference rule for behavioural implementations of specifications with constraints:

$$\text{(R0-Reach)} \quad \frac{\begin{array}{l}(1)\ \text{SP-I} + \langle \mathcal{L}(\Sigma), \text{BEH}\rangle \models \mathcal{L}(\text{Ax}) \\ (2)\ \text{SP-I} + \text{BEH-ISGEN} \models \forall x{:}s. \exists y{:}s.\ IsGen_s(y) \wedge x \sim_s y \quad \text{for all } s \in S_\mathcal{R}\end{array}}{\text{SP} \overset{\approx}{\rightsquigarrow} \text{SP-I}}$$

3.3 Correctness Proofs: The General Case

In the next step we study how to prove behavioural implementations of structured specifications. The general idea is to provide proof rules which allow us to reduce the proof of an implementation of a structured specification to the proof of implementations of parts of the specification. As a first useful fact we show that the behaviour operator $Beh_{\approx}$ is compatible with the combination of two specifications SP1 and SP2 if both specifications have the same signature.

Proposition 3.11 Let SP1 and SP2 be specifications with $\Sigma 1 = \Sigma 2$ and let $\approx$ be a family of $\Sigma 1$-congruences. Then $Beh_{\approx}(Mod(SP1+SP2)) = Beh_{\approx}(Mod(SP1)) \cap Beh_{\approx}(Mod(SP2))$.

This result is also useful for specifications with different signatures.

Corollary 3.12 Let SP1 and SP2 be specifications and $\approx$ be a family of $\Sigma 1 \cup \Sigma 2$-congruences. Then $Beh_{\approx}(Mod(SP1+SP2)) = Beh_{\approx}(Mod(SP1+\Sigma 2)) \cap Beh_{\approx}(Mod(SP2+\Sigma 1))$. (where Σi denotes the basic specification consisting only of the signature Σi for $i = 1, 2$).

Corollary 3.12 gives rise to a first inference rule for behavioural implementations:

Proposition 3.13 Let SP1, SP2 and SP-I be specifications such that $\Sigma 1 \cup \Sigma 2 \subseteq \Sigma\text{-}I$ and let $\approx$ be a family of $\Sigma 1 \cup \Sigma 2$-congruences. The following rule (R1) is sound (in fact the conditions above the line are even necessary for $SP1+SP2 \overset{\approx}{\rightsquigarrow} SP\text{-}I$):

$$\text{(R1)} \qquad \frac{SP1+\Sigma 2 \overset{\approx}{\rightsquigarrow} SP\text{-}I, \quad SP2+\Sigma 1 \overset{\approx}{\rightsquigarrow} SP\text{-}I}{SP1+SP2 \overset{\approx}{\rightsquigarrow} SP\text{-}I}$$

The above inference rule shows that we can always split the implementation proof of the combination of two specifications into implementation proofs of the single constituent parts if we add to each part the sorts and function symbols of the other part. By inductive reasoning over the structure of specifications one can show that in this way one can reduce the implementation proof of a structured specification to implementation proofs of basic specifications which however still have the full signature of the original structured specification. In the following we are interested in conditions which allow to split implementation proofs into implementation proofs for parts of a given specification. For this purpose we introduce the following notion of restrictability which roughly means that a Σ-congruence $\approx$ is restrictible to a $\Sigma 1$-congruence $\approx^1$ w.r.t. a class C of Σ-algebras if for all algebras $A \in C$ and for all carriers A_s of some sort s of $\Sigma 1 \subseteq \Sigma$ there is no difference between the behavioural equality $\approx$ and the behavioural equality $\approx^1$. Looking to practical examples this intuitively expresses that when specifications are built in a modular way then behavioural equalities defined for smaller specifications should be preserved when constructing large system specifications.

Definition 3.14 Let Σ and $\Sigma 1$ be signatures such that $\Sigma 1 \subseteq \Sigma$, let $\approx$ be a family of Σ-congruences, $\approx^1$ be a family of $\Sigma 1$-congruences and let C be a class of Σ-algebras.
$\approx$ is called *restrictible to* $\approx^1$ *on* C if for all $A \in C$, $(\approx_A)|_{\Sigma 1} = \approx^1_{(A|_{\Sigma 1})}$ where $(\approx_A)|_{\Sigma 1}$ is the restriction of $\approx_A$ to $\Sigma 1$ (cf. Section 2.1).

Proposition 3.15 Let SP1, SP-I be specifications and let $\Sigma2$ be a signature such that $\Sigma1 \cup \Sigma2 \subseteq \Sigma\text{-I}$. Moreover, let $\approx$ be a family of $\Sigma1\cup\Sigma2$-congruences and $\approx^1$ be a family of $\Sigma1$-congruences such that $\approx$ is restrictible to $\approx^1$ on $\mathrm{Mod}(\text{SP-I})|_{\Sigma1\cup\Sigma2}$. Then the following inference rule is sound:

$$(\mathrm{R2}) \qquad \frac{\mathrm{SP1} \overset{\approx^1}{\leadsto} \text{SP-I}}{\mathrm{SP1}+\Sigma2 \overset{\approx}{\leadsto} \text{SP-I}}$$

Theorem 3.16 *(Horizontal composition)* Let SP1, SP1-I, SP2, SP2-I be specifications such that $\Sigma1 \subseteq \Sigma1\text{-I}$, $\Sigma2 \subseteq \Sigma2\text{-I}$. Moreover, let $\approx^i$ be a family of Σi-congruences (for i = 1,2) and let $\approx$ be a family of $\Sigma1\cup\Sigma2$-congruences such that $\approx$ is restrictible to $\approx^1$ and to $\approx^2$ on $\mathrm{Mod}(\text{SP1-I}+\text{SP2-I})|_{\Sigma1\cup\Sigma2}$. Then the following inference rule is sound:

$$(\mathrm{R3}) \qquad \frac{\mathrm{SP1} \overset{\approx^1}{\leadsto} \text{SP1-I},\ \mathrm{SP2} \overset{\approx^2}{\leadsto} \text{SP2-I}}{\mathrm{SP1}+\mathrm{SP2} \overset{\approx}{\leadsto} \text{SP1-I} + \text{SP2-I}}$$

As a particular consequence of (R3) we obtain a derived rule for enrichments:

Corollary 3.17 Let SP1, SP1-I be specifications such that $\Sigma1 \subseteq \Sigma1\text{-I}$ and let $\approx^1$ be a family of $\Sigma1$-congruences. Moreover, let SP = **enrich** SP1 **by** Δ and SP-I = **enrich** SP1-I **by** Δ-I be enrichments of SP1 and SP1-I (resp.) such that $\Sigma \subseteq \Sigma\text{-I}$ and let $\approx$ be a family of Σ-congruences which is restrictible to $\approx^1$ on $\mathrm{Mod}(\textbf{enrich}\ \text{SP1-I}\ \textbf{by}\ \Delta\text{-I})|_{\Sigma}$. Then the following inference rule is sound:

$$(\mathrm{R4}) \qquad \frac{\mathrm{SP1} \overset{\approx^1}{\leadsto} \text{SP1-I},\ \Delta\text{-SP} \overset{\approx}{\leadsto} \textbf{enrich}\ \text{SP1-I}\ \textbf{by}\ \Delta\text{-I}}{\textbf{enrich}\ \mathrm{SP1}\ \textbf{by}\ \Delta \overset{\approx}{\leadsto} \textbf{enrich}\ \text{SP1-I}\ \textbf{by}\ \Delta\text{-I}}$$

Note that the specification Δ-SP in the second hypothesis of rule (R4) is a basic specification (cf. Section 2.1) and therefore for proving $\Delta\text{-SP} \overset{\approx}{\leadsto}$ **enrich** SP1-I **by** Δ-I one can directly apply the rule (R0) for the basic case (cf. Theorem 3.8). This shows that if we want to prove a behavioural implementation of an abstract specification that is constructed by a series of enrichment steps one can reduce the proof to successive proofs of implementations of basic specifications (one proof for each enrichment step provided that the congruences are restrictible).

Remark 3.18 As a particular application of rule (R4) we can consider the case where $\Delta = \Delta\text{-I}$. Then $\Delta\text{-SP} \overset{\approx}{\leadsto}$ **enrich** SP1-I **by** Δ always holds if Δ-SP is behaviourally consistent w.r.t. $\approx$. For instance in the observational case this is true whenever the axioms occurring in Δ satisfy the syntactic restrictions discussed in Example 2.6. Then we can derive under the restrictability assumption for $\approx$:

$$(\mathrm{R5}) \qquad \frac{\mathrm{SP1} \overset{\approx^1}{\leadsto} \text{SP1-I}}{\textbf{enrich}\ \mathrm{SP1}\ \textbf{by}\ \Delta \overset{\approx}{\leadsto} \textbf{enrich}\ \text{SP1-I}\ \textbf{by}\ \Delta}$$

For the application of the inference rules (R2) - (R5) it is essential to have practically applicable conditions which allow to check when a given Σ-congruence $\approx$ is restrictible to a $\Sigma 1$-congruence $\approx^1$ on a class C of Σ-algebras. The following proposition provides such a condition in terms of axiomatizations of $\approx$ and $\approx^1$.

Proposition 3.19 *(Restrictability criterion)* Let $\Sigma = (S, F)$ and $\Sigma 1 = (S1, F1)$ be signatures such that $\Sigma 1 \subseteq \Sigma$, let $\approx$ be a family of Σ-congruences, $\approx^1$ be a family of $\Sigma 1$-congruences and C be a class of Σ-algebras. Moreover, let $(Beh_s(x_s, y_s))_{s \in S}$ be an axiomatization of $\approx$ w.r.t. C and let $(Beh1_s(x_s, y_s))_{s \in S1}$ be an axiomatization of $\approx^1$ w.r.t. $C|_{\Sigma 1}$ such that $C \models (Beh_s(x_s, y_s) \Leftrightarrow Beh1_s(x_s, y_s))$ for all $s \in S1$. Then $\approx$ is restrictible to $\approx^1$ on C.

Example 3.20 *(Restrictability of observational equalities)* Let Σ, $\Sigma 1$ be as in Proposition 3.19, let $\approx_{Obs}$ be the observational Σ-equality induced by $Obs \subseteq S$ and let $\approx_{Obs1}$ be the observational $\Sigma 1$-equality induced by Obs1 such that $Obs1 = S1 \cap Obs$. Let $C_0 \subseteq C(\Sigma)^{Obs}$ be a finite set of observable Σ-contexts such that $z_s \in C_0$ for any observable sort $s \in S$ and let $(Beh_s(x_s, y_s))_{s \in S}$ be the family of Σ-formulas induced by C_0 (cf. Example 2.8). Now assume that any context $c \in C_0$ with context variable z_s of sort $s \in S1$ is indeed an observable $\Sigma 1$-context which means that c is built with operation symbols of $\Sigma 1$ only, i.e. $c \in C(\Sigma 1)^{Obs1}$. Then $\approx_{Obs}$ is restrictible to $\approx_{Obs1}$ on C for any class C of Σ-algebras with $\mathcal{E}(C) \models CONG_\Sigma(\sim)$. (For the definition of $\mathcal{E}(C)$ see Example 2.8.) In particular, if SP-I is a specification with signature $\Sigma\text{-I} \supseteq \Sigma$ then $\approx_{Obs}$ is restrictible to $\approx_{Obs1}$ on $Mod(SP\text{-}I)|_\Sigma$ if $SP\text{-}I + \langle \mathcal{L}(\Sigma), BEH \rangle \models CONG_\Sigma(\sim)$.

As a particular consequence of Theorem 3.16 we can derive a condition for the behavioural consistency of structured specifications:

Proposition 3.21 *(Behavioural consistency of structured specifications)*
Let SP1, SP2 be specifications, let $\approx^i$ be a family of Σi-congruences (for i = 1, 2) and let $\approx$ be a family of $\Sigma 1 \cup \Sigma 2$-congruences such that $\approx$ is restrictible to $\approx^1$ and to $\approx^2$ on Mod(SP1+SP2). If SP1 is behaviourally consistent w.r.t. $\approx^1$ and SP2 is behaviourally consistent w.r.t. $\approx^2$ then SP1+SP2 is behaviourally consistent w.r.t. $\approx$.

3.4 Proof of the Sample Implementation

The statement to prove

The specification EXP-I (cf. Example 3.7) is a behavioural implementation of EXP w.r.t. the observational Σ_{EXP}-congruence $\approx_{Obs}$ generated by the observable sorts Obs = {nat, id}, i.e. EXP $\overset{\approx_{Obs}}{\leadsto}$ EXP-I.

The implementation proof

In the following we will show how a "bottom up" implementation proof can be performed using the implementation rules given in the last sections. First, according to (R4), we have

$$\text{(R4)}\quad \frac{\text{STATE} \overset{\approx_{Obs1}}{\sim\sim\sim>} \text{STATE-I},\ \Delta\text{-EXP} \overset{\approx_{Obs}}{\sim\sim\sim>} \text{EXP-I}}{\text{EXP} \overset{\approx_{Obs}}{\sim\sim\sim>} \text{EXP-I}}$$

if $\approx_{Obs}$ is restrictible to $\approx_{Obs1}$ on Mod(EXP-I)$|_{\Sigma_{EXP}}$ where Obs1 = {nat, id} and $\approx_{Obs1}$ is the observational Σ_{STATE}-equality. Hence for proving EXP $\overset{\approx_{Obs}}{\sim\sim\sim>}$ EXP-I we have to show:

(1) $\approx_{Obs}$ is restrictible to $\approx_{Obs1}$ on Mod(EXP-I)$|_{\Sigma_{EXP}}$,
(2) Δ-EXP $\overset{\approx_{Obs}}{\sim\sim\sim>}$ EXP-I,
(3) STATE $\overset{\approx_{Obs1}}{\sim\sim\sim>}$ STATE-I.

For the proof of (1) and (2) we will first construct an axiomatization of $\approx_{Obs}$ w.r.t. Mod(EXP-I)$|_{\Sigma_{EXP}}$. Let C_0 be the following set of observable contexts:
$C_0 =_{def}$ {lookup(i, z_{state}), eval(z_{exp}, s), z_{nat}, z_{id}}.
The set C_0 induces the following formulas $(Beh_s(x_s, y_s))_{s\in S}$ (cf. Example 2.8):

Beh_{state}(sL, sR) $=_{def}$ ∀i:id. lookup(i, sL) = lookup(i, sR)
Beh_{exp}(eL, eR) $=_{def}$ ∀s:state. eval(eL, s) = eval(eR, s)
$Beh_s(x_s, y_s) =_{def} x_s = y_s$ (for s ∈ {nat, id})

We want to show that $(Beh_s(x_s, y_s))_{s\in S}$ is an axiomatization of $\approx_{Obs}$ w.r.t. Mod(EXP-I)$|_{\Sigma_{EXP}}$. But before we note that we can apply an obvious simplification. Since on observable sorts the observational equality coincides with the set-theoretic equality it is not necessary to introduce predicate symbols $\sim_{nat}$ and $\sim_{id}$. Hence it is sufficient to use the following definitions:

$\mathcal{L}(\Sigma_{EXP}) =_{def} \Sigma_{EXP} \cup$ {$\sim_{state}$: state, state, $\sim_{exp}$: exp, exp},
BEH-EXP $=_{def}$
∀sL, sR : state. [∀i:id. lookup(i, sL) = lookup(i, sR) ⇔ sL $\sim_{state}$ sR] ∧
∀eL, eR : exp. [∀s:state. eval(eL, s) = eval(eR, s) ⇔ eL $\sim_{exp}$ eR]

and CONG$_{\Sigma_{EXP}}$(~) is the conjunction of the following sentences:

∀sL, sR:state.
[(sL $\sim_{state}$ sR) ⇒ (∀j:id, n:nat, e:exp.
update(j, n, sL) $\sim_{state}$ update(j, n, sR) ∧ eval(e, sL) = eval(e, sR))] ∧
∀eL, eR:exp.
[(eL $\sim_{exp}$ eR) ⇒ (∀i:id, e:exp.
plus(eL, e) $\sim_{exp}$ plus(eR, e) ∧ plus(e, eL) $\sim_{exp}$ plus(e, eR) ∧
mult(eL, e) $\sim_{exp}$ mult(eR, e) ∧ mult(e, eL) $\sim_{exp}$ mult(e, eR) ∧
eL[e/i] $\sim_{exp}$ eR[e/i] ∧ e[eL/i] $\sim_{exp}$ e[eR/i])]

Proof Obligation 1 Prove that EXP-I + ‹$\mathcal{L}(\Sigma_{EXP})$, BEH-EXP› ⊨ CONG$_{\Sigma_{EXP}}$(~)

This proof obligation and the ones below are not trivial but nevertheless all of them can be discharged using for instance the Larch Prover (cf. [GH 93]).
The result of proof obligation 1 can be used for two purposes: First, since the only context of C_0 with context variable z_{state} is the context lookup(i, z_{state}) which is already an observable context over the signature Σ_{STATE} we know by Example 3.20 that $\approx_{Obs}$ is restrictible to $\approx_{Obs1}$ on Mod(EXP-I)$|_{\Sigma_{EXP}}$, i.e. (1) is already proved. Secondly, we know according to Example 2.8 that $(Beh_s(x_s, y_s))_{s\in S}$ is an axiomatization of $\approx_{Obs}$ w.r.t.

$Mod(EXP\text{-}I)|_{\Sigma_{EXP}}$. For proving (2), i.e. Δ-EXP $^{\approx Obs}$---> EXP-I, we can then apply rule (R0-Reach) given in Section 3.2.

Proof Obligation 2
Prove that EXP-I + ‹$\mathcal{L}(\Sigma_{EXP})$, BEH-EXP› $\models \mathcal{L}(\phi)$ for all axioms ϕ of Δ-EXP.

Proof Obligation 3 Let BEH-ISGEN-EXP $=_{def}$
‹$\mathcal{L}(\Sigma_{EXP}) \cup \{IsGen_{nat}: nat, IsGen_{id}: id, IsGen_{exp}: exp\}$, BEH-EXP ∪ IsGen-Ax› where
IsGen-Ax $=_{def}$

$$\forall i{:}id.\ IsGen_{exp}(idexp(i)) \wedge \forall n{:}nat.\ IsGen_{exp}(natexp(n)) \wedge$$
$$\forall e_1, e_2{:}exp.\ IsGen_{exp}(e_1) \wedge IsGen_{exp}(e_2) \Rightarrow IsGen_{exp}(plus(e_1, e_2)) \wedge$$
$$\forall e_1, e_2{:}exp.\ IsGen_{exp}(e_1) \wedge IsGen_{exp}(e_2) \Rightarrow IsGen_{exp}(mult(e_1, e_2))$$

Prove that EXP-I + BEH-ISGEN-EXP $\models \forall e{:}exp.\ \exists y{:}exp.\ IsGen_{exp}(y) \wedge e \sim_{exp} y$.

Proof Obligations 2 and 3 allow us to apply rule (R0-Reach) of Section 3.2. Hence we obtain Δ-EXP $^{\approx Obs}$---> EXP-I, i.e. (2) is proved. It remains to prove (3), i.e. STATE $^{\approx Obs1}$---> STATE-I, which can be done using again (R4):

$$\text{(R4)}\quad \frac{\text{NAT+ID}\ {}^{\approx Obs2}\text{--->}\ \text{STATE-I},\ \Delta\text{-STATE}\ {}^{\approx Obs1}\text{--->}\ \text{STATE-I}}{\text{STATE}\ {}^{\approx Obs1}\text{--->}\ \text{STATE-I}}$$

The implementation proof can easily be finished proving the preconditions of (R4). The first precondition is trivial. For the second precondition one uses again (R0-Reach).

4 Relating Behavioural and Abstractor Implementations

Another approach in the literature which allows to relax the standard semantics of algebraic specifications is abstractor semantics. The notion of "abstractor" was introduced in [ST 88] for abstracting from the model class of a specification with respect to an equivalence relation on the class of all Σ-algebras. According to [ST 88] abstractor implementations are defined as follows:

Definition 4.1 Let $\equiv \subseteq Alg(\Sigma) \times Alg(\Sigma)$ be an equivalence relation between Σ-algebras.
(1) For any class $C \subseteq Alg(\Sigma)$, $Abs_{\equiv}(C) =_{def} \{B \in Alg(\Sigma) \mid B \equiv A$ for some $A \in C\}$.
For any specification SP, the *abstractor semantics* of SP is given by $Abs_{\equiv}(Mod(SP))$.
(2) Let SP, SP-I be specifications with $\Sigma \subseteq \Sigma$-I. SP-I is called *abstractor implementation of* SP *w.r.t.* ≡ (written SP $^{\equiv}$---> SP-I) if $Mod(SP\text{-}I)|_{\Sigma} \subseteq Abs_{\equiv}(Mod(SP))$.

The key idea for relating behavioural and abstractor semantics is the notion of *factorizability* (cf. [BHW 94]). In the sequel of this section we always assume given a strongly uniform family $\approx\ = (\approx_A)_{A \in Alg(\Sigma)}$ of Σ-congruences and an equivalence relation ≡ on Alg(Σ) such that ≡ is factorizable by ≈, i.e.:
For all $A, B \in Alg(\Sigma)$, $A \equiv B$ if and only if $A/{\approx_A}$ and $B/{\approx_B}$ are isomorphic.

Example 4.2 For any set Obs of observable sorts the observational equivalence between algebras used in [R 87], denoted by $\equiv_{Obs}$, is factorizable by $\approx_{Obs}$ (cf. [BHW 94]).

From the strong uniformity of $\approx$ and the factorizability assumption one can easily derive that $A \equiv A/{\approx_A}$ for any Σ-algebra A. This fact is the only ingredient needed to prove the following theorem which provides important relationships between behavioural and abstractors semantics.[6]

Theorem 4.3 Let $C \subseteq Alg(\Sigma)$ be a class of Σ-algebras.
(1) $Beh_{\approx}(C) \subseteq Abs_{\equiv}(C)$.
(2) If C is closed under Σ-isomorphism, then
C is behaviourally consistent w.r.t. $\approx$ if and only if $Beh_{\approx}(C) = Abs_{\equiv}(C)$.

Theorem 4.3 implies the following relationships between behavioural and abstractor implementations:

Theorem 4.4
(1) If SP $^{\approx}$~~~> SP-I then SP $^{\equiv}$~~~> SP-I.
(2) If SP is behaviourally consistent w.r.t. $\approx$ then:
SP $^{\equiv}$~~~> SP-I if and only if SP $^{\approx}$~~~> SP-I.

By Theorem 4.4 (1) any behavioural implementation is also an abstractor implementation. Therefore the proof techniques for behavioural implementations investigated in Section 3 are also sound for abstractor implementations. Theorem 4.4 (2) shows that there is no difference between the concepts of behavioural and abstractor implementations whenever the specification to be implemented is behaviourally consistent. Since we claim that behavioural consistency is a fundamental property for any specification which is supposed to describe a behaviour of some data structures or programs (cf. Remark 2.3 (3)) the proof rules for behavioural implementations introduced in Section 3 exhibit their full power also for abstractor implementations. It is however interesting to note that it seems impossible to obtain directly such rules for abstractor implementations without the help of the behavioural framework.

5 Conclusion

Behavioural implementations provide a flexible implementation concept for algebraic specifications by capturing our intuition that an implementation is correct if it has the intended behaviour. In this paper we have considered behaviours determined by total congruence relations, but we believe that our results can be extended to partial congruences as studied in [BHW 94] and [BH 95a]. Moreover, for sake of clarity, we have always assumed that the signature of the specification to be implemented is included in the signature of the implementing specification. This concept can be easily generalized by embedding a derive construction in the definition of the implementation relation (cf. e.g. [ST 88]).

Acknowledgements We would like to thank Andrzej Tarlecki for pointing us out the usefulness of the notion of "strong uniformity" for behavioural equalities. Special thanks to Steve Garland for using LP and to anonymous referees for helpful comments. This work

---

[6] The same results were already presented in [BHW 94] for a slightly different definition of the behaviour operator (using fully abstract algebras) and using as a general precondition the *regularity* of $\approx$. Note that regularity also ensures $A \equiv A/{\approx_A}$ and hence Theorem 4.3.

is partially sponsored by the French-German cooperation programme PROCOPE, by the E. C. ESPRIT Working Group COMPASS and by the E. C. HCM project MeDiCis.

References

[BB 91] G. Bernot, M. Bidoit: Proving the correctness of algebraically specified software: modularity and observability issues. Proc. AMAST '91, 216-242, Springer-Verlag Workshops in Computing Series, 1992.

[BH 94] M. Bidoit, R. Hennicker: Proving behavioural theorems with standard first-order logic. In Proc. ALP '94, Fourth International Conference on Algebraic and Logic Programming, Springer Lecture Notes in Computer Science 850, 41-58, 1994.

[BH 95a] M. Bidoit, R. Hennicker: Behavioural theories and the proof of behavioural properties. Report LIENS-95-5, Ecole Normale Supérieure, 1995.

[BH 95b] M. Bidoit, R. Hennicker: Proving the correctness of behavioural implementations. Technical Report, Universität München, 1995.

[BHW 94] M. Bidoit, R. Hennicker, M. Wirsing: Behavioural and abstractor specifications. Report LIENS-94-10, Ecole Normale Supérieure, 1994. Revised version to appear in Science of Computer Programming.

[EM 85] H. Ehrig, B. Mahr: Fundamentals of algebraic specification 1, EATCS Monographs on Theoretical Computer Science 6, Springer, 1985.

[GH 93] J. Guttag, J. Horning: Larch: Languages and Tools for Formal Specification. Texts and Monographs in Computer Science, Springer, 1993.

[GM 82] J. A. Goguen, J. Meseguer: Universal realization, persistent interconnection and implementation of abstract modules. In Proc. ICALP '82, Springer Lecture Notes in Computer Science 140, 265-281, 1982.

[H 72] C. A. R. Hoare: Proofs of correctness of data representations. Acta Informatica 1, 271-281, 1972.

[KK 67] G. Kreisel, J. L. Krivine: Eléments de Logique Mathematique. Dunod (Paris), 1967.

[NO 88] P. Nivela, F. Orejas: Initial behaviour semantics for algebraic specifications. In: D. T. Sannella, A. Tarlecki (eds.): Proc. 5th Workshop on Algebraic Specifications of Abstract Data Types, Springer Lecture Notes in Computer Science 332, 184-207, 1988.

[R 87] H. Reichel: Initial computability, algebraic specifications, and partial algebras. International Series of Monographs in Computer Science No. 2, Oxford: Clarendon Press, 1987.

[Sch 87] O. Schoett: Data abstraction and correctness of modular programming. Ph. D. thesis, CST-42-87, University of Edinburgh, 1987.

[ST 88] D. T. Sannella, A. Tarlecki: Toward formal development of programs from algebraic specifications: implementation revisited. Acta Informatica 25, 233-281, 1988.

[W 90] M. Wirsing: Algebraic specification. In: J. van Leeuwen (ed.): Handbook of Theoretical Computer Science, 675-788, Elsevier Science Publishers B. V., 1990.

On the Decidability of Process Equivalences for the π-calculus

Mads Dam*

Swedish Institute of Computer Science

Abstract. We present general results for showing process equivalences applied to the finite control fragment of the π-calculus decidable. Firstly a Finite Reachability Theorem states that up to finite name spaces and up to a static normalisation procedure, the set of reachable agent expressions is finite. Secondly a Boundedness Lemma shows that no potential computations are missed when name spaces are chosen large enough, but finite. We show how these results lead to decidability for a number of π-calculus equivalences such as strong or weak, late or early bismulation equivalence. Furthermore, for strong late equivalence we show how our techniques can be used to adapt the well-known Paige-Tarjan algorithm. Strikingly this results in a single exponential running time not much worse than the running time for the case of for instance CCS. Our results considerably strengthens previous results on decidable equivalences for parameter-passing process calculi.

1 Introduction

The problem of obtaining a unified view of on the one hand sequential computation as embodied by the λ-calculus, and reactive systems such as CCS or CSP on the other has recently had considerable attention. The π-calculus [11] was proposed as a calculus for mobile processes, i.e. processes whose interconnection topology may be dynamically changed. It extends CCS by features for the transmission and generation of channel names. Considerable expressive power is gained by this. For instance, data types [9], lambda calculus [10], object-oriented programming languages [18], and higher-order processes [16] can all be captured, underlining the foundational importance of the calculus. Moreover the practical usefulness of the calculus have been demonstrated in application studies on mobile telecommunication networks and high speed networks [14, 13]. It is therefore important to investigate to what extent methods and tools developed for, say, CCS lift to the more expressive setting of the π-calculus.

One such set of tools of fundamental importance are process equivalence checking algorithms, as exemplified by the Paige-Tarjan algorithm [15, 8]. Algorithms like these apply in general only to finite-state processes, characterised, in the case of CCS, by disallowing occurrences of the parallel combinator as well

* Work partially supported by ESPRIT BRA project 8130 "LOMAPS". Authors address: SICS, Box 1263, S-164 28 Kista, Sweden. Email: mfd@sics.se.

as unguarded occurrences of process identifiers in recursive definitions. The corresponding fragment of the π-calculus is termed the *finite control* fragment. In this paper we show that for a range of equivalences the finite control conditions are in fact sufficient to lift algorithms to the π-calculus. This is far from a trivial result, since even very simple π-calculus agents exhibit infinite-state behaviour while satisfying these conditions. One example, using a CCS-like notation, is the memory cell

$$Mem(x) = in(y).Mem(y) + \overline{out}x.Mem(x)$$

that can either input a channel name y along channel in and then proceed as $Mem(y)$ or else output x along out and then proceed as $Mem(x)$. This is an example of a *data-independent* agent such as those considered previously by Jonsson and Parrow [7]. However, the finite-control fragment goes beyond this, since it allows synchronisation, or testing, on channel names passed as parameters. By adding a positive and negative conditional [2] **if** $x = y$ **then** P **else** Q to the π-calculus, as we do, we can for instance encode the memory cell

$$\begin{aligned} &KillableMem(x) = \\ &\quad in(y).(\textbf{if } y = KILL \textbf{ then } NIL \textbf{ else } Mem(y)) + \overline{out}x.Mem(x) \end{aligned}$$

that is killed in case the channel $KILL$ is passed to it. Other examples concern the facility of the π-calculus to declare new private names and to pass them on to other parallel components. Consider for instance the agent

$$(\nu x)(Gen(x)|Listen(x))$$

where

$$\begin{aligned} Gen(x) &= (\nu y)\overline{x}y.Gen(y) \\ Listen(x) &= x(y).Listen(y) \end{aligned}$$

In this system Gen repeatedly declares a new channel name y, transmits y to $Listen$ along x and then proceeds as $Gen(y)$. Since the y's are known to be fresh and thus different from any other name previously encountered during a computation, the state space generated by $(\nu x)(Gen(x)|Listen(x))$ is infinite.

In this paper we provide the basic tools to show decidability for the finite control fragment for a number of equivalences, including late or early, strong or weak bisimulation equivalence (c.f. [12]), and open (or uniform [1]) bisimulation equivalence [17] [3]. Though the technical details differ, the basic approach we use to show decidability is quite familiar from other recent decidability results in process algebra such as [2, 3, 6, 7], namely by showing that up to the equivalence concerned the state space can be represented in a finitary manner. The tools consist of two key Lemmas, proofs of which are given in the paper:

[2] In the paper we actually use the notation $[x = y]PQ$ instead of **if** $x = y$ **then** P **else** Q for the conditional.

[3] For open bisimulation equivalence decidability is already known [17]

1. A Finite Reachability Theorem showing that up to a finite name space and up to a deterministic static normalisation procedure only a finite number of distinct agents are reachable.
2. A proof that the number of distinct free names needed at any point during a computation can be bounded.

Put together these results imply that a bound can be put on the size of the name space, and decidability for a given process equivalence $\equiv$ then consists of showing for all agents A and B, that $A \equiv B$ iff $A\sigma \equiv_N B\sigma$ where $\equiv_N$ represents equivalence with respect to a large enough but finite name space N, and σ is a map representing names as names in N. We establish this result for all the equivalences mentioned above with particular focus on strong late bisimulation equivalence [9]. For this equivalence we show how the partition refinement algorithm of Paige and Tarjan [15] can be applied resulting, as for the case of e.g. CCS, in a single exponential worst case complexity.

Our results considerably strengthens previous results in the area of value-passing process calculi. Besides the decidability result of Jonsson and Parrow [7] for data-independent programs, Hennessy and Lin [5] showed decidability of bisimulation equivalence for a certain class of symbolic transition graphs. Both these results are subsumed by the work presented here. The notion of open bisimulation equivalence was specifically formulated with an eye on efficiency concerns. In the area of model checking a close relative to the present work is the decidability result with respect to an extended version of the modal μ-calculus of [4].

2 The Polyadic π-calculus, syntax

We use a slight extension of Milner's polyadic π-calculus, introduced in [4]. Letters $x, y, z, \ldots$ range over *names* of which there is a countably infinite supply, A, B range over *agents*, and D over *agent identifiers*. *Actions*, α, β, are either names, co-names of the form $\overline{x}$, or the distinguished constant τ. If α is a name x then $\mathrm{n}(\alpha)$ (the name of α) is x. and $\mathrm{p}(\alpha)$ (the polarity of α) is $-$. Otherwise if $\alpha = \overline{x}$ then $\mathrm{n}(\alpha) = x$ and $\mathrm{p}(x) = +$. The syntax of agents is the following:

$$A ::= 0 \,\Big|\, A + A \,\Big|\, \alpha.A \,\Big|\, A \mid A \,\Big|\, [x = y]AA \,\Big|$$
$$(\lambda x)A \,\Big|\, Ax \,\Big|\, (\nu x)A \,\Big|\, D \,\Big|\, \mathrm{fix}D.A \,\Big|\, [x]A$$

Name binders are the operators λ and ν. We use the notation $\mathrm{fn}(A)$ for the set of names occurring freely in A and $A\{x/y\}$ ($A\{A'/D\}$) for substitution of x (A') for y (D) in A. The intended meaning of connectives is familiar from CCS and the π-calculus. The present version is based on the polyadic π-calculus of [9]. There are three main differences:

Recursion. We use recursive definitions rather than replication. Just as for CCS we require restriction to those expressions that are well-guarded (in the sense of, for an expression fix$D.A$, only allowing free occurrences of D in A within the scope of a prefix operator, and for which uses of the parallel combinator $|$ within recursive definitions are disallowed. We refer to this fragment as the ***finite control*** fragment. In addition we require for technical reasons that recursions fix$D.A$ are ***fully parametrised*** in the sense that recursive agents fix$D.A$ have no free occurrences of names. The expressive power of the language is unaffected by this latter restriction since all equivalences considered here will respect the identification of $(\text{fix}D.A)(x_1, \ldots, x_n)$ with

$$(\text{fix}D.(\lambda x_1)\cdots(\lambda x_n)A\{Dx_1\cdots x_n/D\})x_1\cdots x_n.$$

Conditionals. We admit the conditional $[x = y]AB$, identified with A when $x = y$, and B when $x \neq y$. The admission of negative as well as positive matching has been an issue of some controversy in the π-calculus (c.f. [17]). It is accomodated (though not required) in our framework by a relativisation of the operational semantics to complete descriptions of name identities and inequalities.

Well-formedness. A well-formedness condition is imposed, reflecting the stratified syntax of [9]. Agents A that are to be considered ***well-formed*** are assigned an integer ***arity*** n, written $A : n$. ***Processes*** are agents of arity 0, ***abstractions*** are agents of negative arity, and ***concretions*** are agents of positive arity. Given an assignment $D : n$ of arities to agent identifiers, agent arities are computed as follows:

$$\frac{}{0:0} \qquad \frac{A:0 \quad B:0}{A+B:0} \qquad \frac{A:n \quad n\leq 0}{x.A:0} \qquad \frac{A:n \quad n\geq 0}{\overline{x}.A:0} \qquad \frac{A:0}{\tau.A:0}$$

$$\frac{A:0 \quad B:0}{A \mid B:0} \qquad \frac{A:n \quad B:n}{[x=y]AB:n} \qquad \frac{A:n \quad n\leq 0}{(\lambda x)A:n-1} \qquad \frac{A:n-1 \quad n\leq 0}{Ax:n}$$

$$\frac{A:n}{(\nu x)A:n} \qquad \frac{D:n \quad A:n}{\text{fix}D.A:n} \qquad \frac{A:n \quad n\geq 0}{[x]A:n+1}$$

For the remainder of the paper we restrict attention to well-formed agents.

3 Operational Semantics

In [9] the semantics of the π-calculus is given in terms of a structural congruence relation together with a relation of commitment. Here we choose a different, more operational approach, replacing the structural congruence relation with a normalisation procedure.

Name partitionings. Since the decision procedure handles names in a symbolic fashion, normalisation needs to know what identities and inequalities are assumed of names. This information is supplied by partitionings ε on the set of names. A partitioning ε identifies the names x and y (written $\varepsilon \models x = y$) if and only if x and y are members of the same element of ε. The operation $(\nu x)\varepsilon$ of ***name generation*** is defined by

$$(\nu x)\varepsilon = \{S - \{x\} \mid S \in \varepsilon\} \cup \{\{x\}\}.$$

Normal forms and normalisation. Processes in normal form, ranged over by P, are generated by the grammar

$$P ::= 0 \Big| P+P \Big| \alpha.A \Big| P \mid P \Big| (\nu x)P$$

Abstractions in normal form have the form $(\lambda x)A$, and concretions in normal form have one of the forms $[x]A$ or $(\nu x)[x]A$. The normalisation procedure is given by the pseudo-ML function **nf**:

```
fun nf(0,ε) = 0 |
    nf(A+B,ε) = nf(A,ε) + nf(B,ε) |
    nf(α.A,ε) = α.A |
    nf(A | B,ε) = (nf(A,ε) | nf(B,ε)) |
    nf([x = y]AB,ε) = if ε ⊨ x = y then nf(A,ε) else nf(B,ε) |
    nf((λx)A,ε) = (λx)A |
    nf(Ax,ε) = (case nf(A,ε) of (λy)A1 => nf(A1{x/y},ε)) |
    nf((νx)A,ε) =
      let A1 = nf(A,(νx)ε) in
      if x ∈ fn(A1)
      then if A1 : 0 then (νx)A1 else
        (case A1 of
           (λy)A2=> if x = y then (λy)A2 else (λy)(νx)A2 |
           [y]A2 => if x = y then (νx)[y]A2 else [y](νx)A2 |
           (νy)[y]A2 => if x = y then (νy)[y]A2 else (νy)[y](νx)A2)
      else A1
      end |
    nf(fixD.A,ε) = nf(A{fixD.A/D},ε) |
    nf([x]A,ε) = [x]A
```

The following points are easily verified:

1. **nf** preserves well-formedness.
2. **nf** is total (for agents without free occurrences of agent identifiers).
3. For all well-formed A and ε, **nf**(A,ε) is in normal form.

Restricting to well-formed agents in the conditional-free fragment it is possible to compare the normalisation procedure with the structural congruence relation $\equiv$ of [9]. It is quite easy to show, appealing to [9] for the definition of $\equiv$, that for all well-formed, conditional-free agents A, **nf**(A,ε) is independent of ε, and for all ε, $A \equiv$ **nf**(A,ε).

Commitment. The definition of the commitment relation needs the ancillary operations $\|$ and $\cdot$ on normal forms:

- $A \parallel B = A \mid B$ when $A : 0$ and $B : 0$. If $A : 0$ and $B : n \neq 0$ then $A \parallel [x]B' = [x](A \parallel B')$, $A \parallel (\nu x)[x]B' = (\nu y)[y](A \parallel B'\{y/x\})$, and $A \parallel (\lambda x)B' = (\lambda y)(A \parallel B'\{y/x\})$, where in the two last cases it is assumed that $y \notin \mathrm{fn}(A)$. The case for $A : n \neq 0$ and $B : 0$ is defined symmetrically.

- $A \cdot B$ is defined only when $A : -n$ and $B : n$ for some (positive or negative) n. For $n = 0$, $A \cdot B = A \mid B$. If $n > 0$, $(\lambda x)A' \cdot [y]B' = A\{y/x\} \cdot B'$ and $(\lambda x)A' \cdot (\nu y)[y]B' = A\{z/x\} \cdot B'\{z/y\}$ where $z \notin (\mathrm{fn}(A') - \{x\}) \cup (\mathrm{fn}(B') - \{y\})$. The case for $n < 0$ is defined symmetrically.

As the normalisation procedure the commitment relation is relativised to name partitions too. It is defined as follows:

$$\text{ACT:} \frac{}{\alpha.A \succ_\varepsilon \alpha.A} \qquad \text{SUM:} \frac{A_1 \succ_\varepsilon B}{A_1 + A_2 \succ_\varepsilon B}$$

$$\text{COMM:} \frac{A_1 \succ_\varepsilon x.B_1 \; A_2 \succ_\varepsilon \overline{y}.B_2 \quad \varepsilon \models x = y}{A_1 \mid A_2 \succ_\varepsilon \tau.(\mathtt{nf}(B_1,\varepsilon) \cdot \mathtt{nf}(B_2,\varepsilon))}$$

$$\text{PAR:} \frac{A_1 \succ_\varepsilon \alpha.B}{A_1 \mid A_2 \succ_\varepsilon \alpha.(\mathtt{nf}(B,\varepsilon) \parallel A_2\,)}$$

$$\text{RES-1:} \frac{A \succ_{(\nu x)\varepsilon} \tau.B}{(\nu x)A \succ_\varepsilon \tau.(\nu x)B} \qquad \text{RES-2:} \frac{A \succ_{(\nu x)\varepsilon} \alpha.B}{(\nu x)A \succ_\varepsilon \alpha.(\nu x)B} \; (x \neq \mathrm{n}(\alpha))$$

+ symmetrical versions of rules SUM, COMM and PAR

Relating to [9] let the *full* name partitioning ε_f be the one containing only singleton sets. The full partitioning identifies names only if they are literally the same. It can then be shown for the well-formed fragment without conditionals that $A \succ B$ according to [9] if and only if for some B', $\mathtt{nf}(A,\varepsilon_f) \succ_{\varepsilon_f} B'$, and $\mathtt{nf}(B,\varepsilon_f) = \mathtt{nf}(B',\varepsilon_f)$.

Definition 1. (Simulations, Bisimulations) A (strong, late) *partition-relativised simulation* (or pr-simulation) is an ε-indexed family of binary relations R_ε on well-formed agents satisfying the following conditions:

1. If $AR_\varepsilon B$ then $\mathtt{nf}(A,\varepsilon) R_\varepsilon \mathtt{nf}(B,\varepsilon)$.
2. If $AR_\varepsilon B$ and $A : n$ then $B : n$.
3. If $[x]A' R_\varepsilon [y]B'$ then $A' R_\varepsilon B'$ and $\varepsilon \models x = y$.
4. If $(\nu x)[x]A' R_\varepsilon (\nu y)[y]B'$ then $A'\{z/x\} R_{(\nu z)\varepsilon} B'\{z/y\}$ whenever $z \notin (\mathrm{fn}(A') - \{x\}) \cup (\mathrm{fn}(B') - \{y\})$.
5. If $(\lambda x)A' R_\varepsilon (\lambda y)B'$ then for all z, $A'\{z/x\} R_\varepsilon B'\{z/y\}$.
6. If $AR_\varepsilon B$ and $A \succ_\varepsilon \alpha.A'$ then $B \succ_\varepsilon \beta.B'$ for some B' such that $\varepsilon \models \alpha = \beta$ and $A' R_\varepsilon B'$.

Then R is a *partition-relativised bisimulation* (pr-bisimulation) if for each ε both R_ε and R_ε^{-1} are pr-simulations; A and B are ε-bisimilar ($A \sim_\varepsilon B$) if there is a pr-bisimulation R such that $AR_\varepsilon B$; and A and B are pr-bisimilar ($A \sim B$) if there is a pr-bisimulation R such that $AR_\varepsilon B$ for all ε.

The following is stated without proof:

Proposition 2. *For the fragment of well-formed, conditional-free agents, $\sim_{\varepsilon_f}$ is the (strong) bisimulation equivalence of [9], and $\sim$ is strong congruence.* □

4 A Finite Reachability Theorem

The main ingredient in the decidability proof is a finite reachability Theorem, showing that for finite control agents, if names are always chosen from a fixed finite number of candidates then only a finite number of distinct agent expressions are reachable.

Small enough names. Names are chosen at the following points:

- When computing $A \parallel B$ or $A \cdot B$.
- When instantiating names bound by λ or ν.

We restrict these choices by assuming an enumeration $x_0, x_1, \ldots$ of names and imposing a maximal index n_0 such that whenever a name is to be chosen then it is chosen *small enough*, i.e. with an index not exceeding n_0. If no such name exists (because otherwise confusion of names would ensue) then the result is left undefined. We show later that by picking n_0 large enough all choices can in fact be made.

Definition 3. (Reachability relation) Relative to a choice of n_0 the reachability relation $\rightsquigarrow$ on well-formed agents is defined as follows:

1. For all ε, $A \rightsquigarrow \mathtt{nf}(A,\varepsilon)$,
2. $[x]A \rightsquigarrow A$,
3. $(\nu x)[x]A \rightsquigarrow A\{y/x\}$ whenever y is small enough and $y \notin \mathrm{fn}(A) - \{x\}$,
4. $(\lambda x)A \rightsquigarrow A\{y/x\}$ whenever y is small enough,
5. If P is a process in normal form and for some ε, $P \succ_\varepsilon \alpha.A$ while choosing only names that are small enough, then $P \rightsquigarrow A$.

Theorem 4. (Finite reachability) *For all well-formed A and n_0, $\{B \mid A \rightsquigarrow^* B\}$ is finite.*

Proof. We define a reduction relation $\rightarrow$ such that $\rightarrow^*$ includes $\rightsquigarrow^*$, and such that we can prove $\{B \mid A \rightarrow^* B\}$ finite. The relation $\rightarrow$ is determined by the following closure properties:

0. $A \rightarrow B$ whenever A and B are alpha-congruent, and B results from A by replacing small enough bound names with small enough bound names
1. $A + B \rightarrow A$, $A + B \rightarrow B$
2. $\alpha.A \rightarrow A$
3. $[x = y]AB \rightarrow A$, $[x = y]AB \rightarrow B$
4. $(\lambda x)A \rightarrow A\{y/x\}$ whenever y is small enough
5. $Ax \rightarrow A$
6. $\mathrm{fix}D.A \rightarrow A\{\mathrm{fix}D.A/D\}$
7. $[x]A \rightarrow A$
8. If $A \rightarrow B$ and $x \in \mathrm{fn}(B)$ then $(\nu x)A \rightarrow (\nu x)B$
9. $(\nu x)A \rightarrow A$
10. $(\nu x)(\lambda y)A \rightarrow (\lambda y)(\nu x)A$

11. If $x \neq y$ then $(\nu x)[y]A \rightarrow [y](\nu x)A$
12. $(\nu x)(\nu y)[y]A \rightarrow (\nu y)[y](\nu x)A$
13. If $A \rightarrow A'$ then $A \mid B \rightarrow A' \mid B$ and $B \mid A \rightarrow B \mid A'$
14. $((\lambda x)A) \mid B \rightarrow (\lambda x)(A \mid B)$, $A \mid ((\lambda x)B) \rightarrow (\lambda x)(A \mid B)$
15. $([x]A) \mid B \rightarrow [x](A \mid B)$, $A \mid ([x]B) \rightarrow [x](A \mid B)$
16. $((\nu x)A) \mid B \rightarrow (\nu x)(A \mid B)$, $A \mid ((\nu x)B) \rightarrow (\nu x)(A \mid B)$

Proposition 5. $\leadsto^* \subseteq \rightarrow^*$.

Proof. We need to show

(i) For all $\varepsilon, A \rightarrow^* \mathtt{nf}(A,\varepsilon)$.
(ii) $[x]A \rightarrow^* A$.
(iii) $(\nu x)[x]A \rightarrow^* A\{y/x\}$ whenever y is small enough and $y \notin \mathrm{fn}(A) - \{x\}$.
(iv) $(\lambda x)A \rightarrow^* A\{y/x\}$ whenever y is small enough.
(v) $P \rightarrow^* A$ whenever P is a process in normal form and for some ε and α, $P \succ_\varepsilon \alpha.A$.

Of these (i) and (v) use structural induction, and (ii)–(iv) follow directly from the conditions given. □ (Proposition 5)

We then proceed to prove finiteness:

Lemma 6. (Finiteness) *For all A, $\{B \mid A \rightarrow^* B\}$ is finite.*

Proof. By well-guardedness it is sufficient to show that any infinite derivation

$$d = A_0 \rightarrow \cdots \rightarrow A_n \rightarrow \cdots$$

such that $A_0 = A$ visits only a finite number of distinct agents, i.e. $\mathcal{R}(d) = \{A_i \mid i \in \omega\}$ is finite. We define the *size*, $|A|$, of A in the following manner:

$$|0| = |D| = 2$$

$$|A + B| = |[x = y]AB| = |A| + |B| + 1$$

$$|\alpha.A| = |(\lambda x)A| = |Ax| = |[x]A| = |\mathrm{fix} D.A| = |A| + 1$$

$$|(\nu x)A| = 2 \cdot |A| + 1$$

$$|A \mid B| = |A| \cdot |B|$$

Lemma 7. *Axiom (0) does not increase size. All axioms among (1)–(16) except (6) decrease size. Rules (8), (11), (13) preserve size decrease* □ (Lemma 7)

We can assume that the unfolding axiom (6) is used infinitely often along d. By finite control each A_i will have the form $A_i = C_i(B_{i,1}, \ldots, B_{i,m})$ such that C_i is an m-ary context that does not contain occurrences of the fixed point operator, and for which each B_j has no occurrences of parallel composition operator. Moreover m will be independent of i. The number of occurrences in C_i of operators among $+$, prefixing, the conditional, or application will decrease with increasing i since the only reduction that can cause such occurrences to duplicate is axiom (6) which does not apply due to the finite control assumption. Moreover, for each occurrence of one of these operators, either it is never reduced, and then the subterm in question can be viewed as a constant, or else the number of occurrences of that particular operator in the C_i is reduced by 1. Thus there is no loss of generality in assuming that we can find some i_0 for which C_{i_0} is an m'ary context built using only operators of the form $[x]$, (λx), (νx), or $|$.

Now, for all $i \geq i_0$, A_i will have a similar form $C_i(B_{i,1}, \ldots, B_{i,m})$, and for each $j : 1 \leq j \leq m$, either $B_{i,j} = B_{i+1,j}$, or else $B_{i,j} \rightarrow B_{i+1,j}$. In addition we can assume that for infinitely many i, does $B_{i,j} \rightarrow B_{i+1,j}$, since otherwise it suffices to pick a larger i_0. Thus the proof has been reduced to showing

(i) only a finite number of distinct C_i are reachable
(ii) any derivation d that does not involve parallel composition visits a finite number of distinct agents only.

Contexts. To prove (i) we introduce a transition system on contexts, and prove it finite. *Contexts* are terms C generated by the abstract syntax

$$C ::= [\cdot] \,\big|\, (\nu x)C \,\big|\, (\lambda x)C \,\big|\, [x]C \,\big|\, C \mid C$$

Here $[\cdot]$ is the empty context. Say of a context C that x *is visible through* C if either there is some occurrence of $[\cdot]$ in C not within the scope of a binding occurrence of x, or else x occurs unbound in C. Rule (6) below shows where this notion is needed. The transition relation $\rightarrow$ is now determined in the following way where Ω ranges over operators among (νx), (λx), and $[x]$ with x small enough:

1. If C_1 and C_2 are alpha congruent then $C_1 \rightarrow C_2$
2. $[\cdot] \rightarrow \Omega[\cdot]$
3. $(\Omega C_1) \mid C_2 \rightarrow \Omega(C_1 \mid C_2)$, $C_1 \mid (\Omega C_2) \rightarrow \Omega(C_1 \mid C_2)$
4. $[x]C \rightarrow C$, $(\nu x)C \rightarrow C$, $(\lambda x)C \rightarrow C\{y/x\}$ whenever y is small enough
5. $(\nu x)\Omega C \rightarrow \Omega(\nu x)C$
6. if $C_1 \rightarrow C_1'$ and x is visible through C_1' then $(\nu x)C_1 \rightarrow (\nu x)C_1'$
7. if $C_1 \rightarrow C_1'$ then $C_1 \mid C_2 \rightarrow C_1' \mid C_2$ and $C_2 \mid C_1 \rightarrow C_2 \mid C_1'$

It is easy to verify that for $i \geq i_0$, if A_i has the form $C_i(B_{i,1}, \ldots, B_{i,m})$ and A_{i+1} similarly the form $C_{i+1}(B_{i+1,1}, \ldots, B_{i+1,m})$ and for each $j : 1 \leq j \leq m$, either $B_{i,j} = B_{i+1,j}$ or $B_{i,j} \rightarrow B_{i+1,j}$, then either $C_i = C_{i+1}$ or $C_i \rightarrow C_{i+1}$. To prove (i) it therefore suffices to show that only a finite number of contexts are reachable from any context C. We use the notion of *legitimate prefix* to establish the fact.

Definition 8. (Context prefix, Legitimate prefix)

1. A (context) *prefix* is a string $p = \Omega_1 \cdots \Omega_n$ where each Ω_i is either (νx), (λx), or $[x]$, where x is small enough. Write $p \triangleright C$ for the context obtained by prefixing C with p.
2. A prefix $\Omega_1 \cdots \Omega_n$ is *legitimate* if
   (a) at most one Ω_i has the form either (λx) or $[x]$, and
   (b) the total number of occurrences of operators of the form (νx) or (λx) for some small enough x is at most n_0.

Lemma 9. *For all C, $\{C' \mid C \rightarrow^* C'\}$ is finite.*

Proof. By induction in the size of C:

$C = [\cdot]$: If suffices to show that any context reachable from $[\cdot]$ has the form $p \triangleright [\cdot]$ where p is a legitimate prefix. To show this assume that p is legitimate and that $p \triangleright [\cdot] \rightarrow C'$. Then C' has the form $p' \triangleright [\cdot]$. Clearly condition (i) above is satisfied. To see that also (ii) is satisfied suppose for a contradiction that it is not, so that p' has $n_0 + 1$ occurrences of a binding operator. Then p' must have the form $p_1(\nu x)p_2\Omega p_3$ for some x where Ω binds x. But this cannot happen since the justification of $p \triangleright [\cdot] \rightarrow p' \triangleright [\cdot]$ must have appealed to rule (6) for justifying $(\nu x)p'' \triangleright [\cdot] \rightarrow (\nu x)p_2\Omega p_3 \triangleright [\cdot]$ for some p''. But x is not visible through $p_2\Omega p_3$—a contradiction.

$C = (\nu x)C'$: We show that any context reachable from C (in 1 step or more) has the form $p \triangleright C_1$ where p is a legitimate prefix and C_1 is reachable from C' (in 1 step or more). So assume that $p \triangleright C_1 \rightarrow C_2$. The only case that needs considering is when p has the form $p'(\nu x)$, C_1 the form $\Omega C_1'$, and C_2 the form $p\Omega(\nu x) \triangleright C_1'$. We then need to show that $p\Omega(\nu x)$ is legitimate, but this follows exactly as in the previous case.

$C = C_1 \mid C_2$: We show that any context reachable from C (in 1 step or more) has the form $p \triangleright (C_1' \mid C_2')$ where p is legitimate, C_1' is reachable from C_1, and C_2' reachable from C_2. The only cases that need considering are applications of rule (3), but these follow as in the case for restriction above.

The remaining cases are quite straightforward. □ (Lemma 9)

Non-parallel agents. We then proceed to the proof of (ii).

Lemma 10. *Suppose that A has no occurrences of $\mid$. For all derivations $d = A_0 \rightarrow \cdots \rightarrow A_n \rightarrow \cdots$ with $A_0 = A$, $\mathcal{R}(d)$ is finite.*

Proof. The proof proceeds by induction in the size of A. The cases for $\mathbf{0}$, $+$, prefixing, conditional, abstraction, application, and concretion follow directly from the induction hypothesis. This leaves two cases to be considered. For restriction the proof is a correlate of the corresponding case in the proof of Lemma 9. So assume that $A = \text{fix}D.A'$. We to show that any agent reachable from A has the form $p \triangleright (A''\{A/D\})$ where p is a legitimate prefix and A'' is reachable from A', thus completing the proof by the induction hypothesis. To each transition

$A_i \longrightarrow A_{i+1}$ is associated a unique justification, a proof using the axioms and rules among (1)–(16) together with alpha-conversion. Say that *step i refers to* A, if the justification of the transition $A_i \longrightarrow A_{i+1}$ involves an appeal to (6) with D instantiated to itself, and A to A'. Suppose now that A_i has the form $p \rhd (A''\{A/D\})$ such that A'' is reachable from A'. Handling the case where step i is an instance of one of the axioms (10)–(12) as in the proof of Lemma 9 only one potentially problematic case remains, namely where step i refers to A. This, however, can only be the case when A'' has the form $p' \rhd D$ for p' a prefix, and in this situation it must, as we have seen, be the case that the prefix pp' is legitimate. Thus A_{i+1} has been brought into the desired form. □
(Lemmas 10 and 6, Theorem 4)

5 Choosing names

The problem with Theorem 4 is that potential derivations might be lost because at some point it becomes impossible to choose a small enough name. In this section we show that we can avoid this problem by choosing n_0 sufficiently large. Let $\sharp_{fns}(A)$ be the maximal number of free names in any subterm of A, and $\sharp_{par}(A)$ be the number of occurrences of the parallel combinator $|$ in A.

Lemma 11. *For all A_n, if $A_0 \longrightarrow^* A_n$ then* $|\mathrm{fn}(A_n)| \leq \sharp_{fns}(A_0) \cdot (\sharp_{par}(A_0) + 1)$

Proof. Assume that

$$d = A_0 \longrightarrow \cdots \longrightarrow A_n \longrightarrow \cdots$$

We show $|\mathrm{fn}(A_n)| \leq \sharp_{fns}(A_0) \cdot (\sharp_{par}(A_0) + 1)$ by structural induction, using the notion of legitimate prefix introduced in the proof of 9. For all cases except λ, recursion, and parallel composition, the result follows directly from the induction hypothesis, so only these three are considered:

$A_0 = (\lambda x)A_0'$. If $n > 0$ it must be the case that (up to an initial sequence of alpha-conversions) $A_0 \longrightarrow A_1$ is an instance of (4), i.e. that $A_1 = A_0'$, so that $A_0' \longrightarrow^* A_n$. Then by the induction hypothesis,

$$\begin{aligned}|\mathrm{fn}(A_n)| &\leq \sharp_{fns}(A_0') \cdot (\sharp_{par}(A_0') + 1)\\ &= \sharp_{fns}(A_0) \cdot (\sharp_{par}(A_0) + 1)\end{aligned}$$

$A_0 = A_{0,1} \mid A_{0,2}$. In this case A_n has the form $p \rhd (A_{n,1} \mid A_{n,2})$ for some legitimate prefix p. Then for each $i \in \{1,2\}$ we find legitimate prefixes p_i such that $A_{0,i} \longrightarrow^* p_i \rhd A_{n,i}$, and p is the merge of p_1 and p_2 in a manner such that if $[x]$ occurs in p with x in a bound position then so it does in whichever p_i that contains $[x]$. By the induction hypothesis,

$$|\mathrm{fn}(p_i \rhd A_{n,i})| \leq \sharp_{fns}(A_{0,i}) \cdot (\sharp_{par}(A_{0,i}) + 1)$$

for $i = 1$ and $i = 2$. Now

$$\begin{aligned}|\mathrm{fn}(A_n)| &\le |\mathrm{fn}(p_1 \rhd A_{n,1})| + |\mathrm{fn}(p_2 \rhd A_{n,2})| \\ &\le \sharp_{fns}(A_{0,1}) \cdot (\sharp_{par}(A_{0,1}) + 1) \\ &\quad + \sharp_{fns}(A_{0,2}) \cdot (\sharp_{par}(A_{0,2}) + 1)\end{aligned}$$

Let B be whichever of $A_{0,1}/A_{0,2}$ such that $\sharp_{fns}(B)$ is maximal. Then

$$\begin{aligned}&\sharp_{fns}(A_{0,1}) \cdot (\sharp_{par}(A_{0,1}) + 1) \\ &+ \sharp_{fns}(A_{0,2}) \cdot (\sharp_{par}(A_{0,2}) + 1) \\ &\le \sharp_{fns}(B) \cdot (\sharp_{par}(A_{0,1}) + \sharp_{par}(A_{0,2}) + 2) \\ &= \sharp_{fns}(B) \cdot (\sharp_{par}(A_0) + 1) \\ &\le \sharp_{fns}(A_0) \cdot (\sharp_{par}(A_0) + 1)\end{aligned}$$

completing the case.

$A_0 = \mathrm{fix}D.A_0'$. Since $\sharp_{par}(A_0) = 0$ by the assumption of finite control it suffices to show that $|\mathrm{fn}(A_n)| \le \sharp_{fns}(A_0)$. We then find legitimate prefixes p and p' such that

$$\mathrm{fix}D.A_0' \rightarrow^* p \rhd \mathrm{fix}D.A \rightarrow^* A_n,$$

A_n has the form $p' \rhd A_n'\{\mathrm{fix}D.A/D\}$, and $A_0' \rightarrow^* A_n'$. Note that we can assume that p has no free occurrences of names since if it had, before $\mathrm{fix}D.A_0'$ would be subsequently unfolded, the free name (occurring in an output prefix) would be eliminated by an application of (7). By the induction hypothesis we know that

$$\begin{aligned}|\mathrm{fn}(A_n')| &\le \sharp_{fns}(A_0') \\ &= \sharp_{fns}(A_0)\end{aligned}$$

The only case in which $|\mathrm{fn}(A_n)|$ could be greater than $|\mathrm{fn}(A_n')|$ is when p' contains an occurrence of an output prefix $[y]$ such that the occurrence of y in $[y]$ is free in p'. This, however, can only happen if we can factorise the derivation $p \rhd \mathrm{fix}D.A_0' \rightarrow^* A_n$ as follows

$$p \rhd \mathrm{fix}D.A_0' \rightarrow^* p'' \rhd A_n''\{\mathrm{fix}D.A_0'/D\} \rightarrow^* A_n$$

such that p'' has no free occurrence of y (i.e. A_n results from $A_n''\{\mathrm{fix}D.A_0'/D\}$ by applications of 8, 9, and 11), and such that $A_0' \rightarrow^* A_n''$. Moreover $\mathrm{fn}(A_n'') = \mathrm{fn}(A_n') \cup \{y\}$. But, by the induction hypothesis, $|\mathrm{fn}(A_n'')| \le \sharp_{fns}(A_0)$, and the proof is complete. □ (Lemma 11)

Thus by choosing n_0 greater than $\sharp_{fns}(A_0) \cdot (\sharp_{par}(A_0) + 1)$ it will always be possible to choose a small enough name.

6 Decidability

Consider now a version of pr-bisimulation of Definition 1 modified such that z in 1.2 and 1.3 is required to be small enough, and such that commitment in 1.4 is conditional on only small enough names being chosen. Call the ensuing variant of pr-bisimulation for ***name-bounded pr-bisimulation***, or nbpr-bisimulation, for short. By König's Lemma, since the number of transitions that use only small enough names and that emanate from a given agent is finite, and by Lemma 11 any infinite path must visit the same agent expression infinitely often, the following decidability result obtains:

Theorem 12. *Name-bounded pr-bisimulation equivalence is decidable.* □

Using this result we can then easily establish our first main Theorem:

Theorem 13. *Strong late bisimulation equivalence is decidable.*

Proof. Both decidability results follow from decidability of ε-pr-bisimulation which we go on to demonstrate. Assume first if R_ε and R_ε^{-1} are both pr-simulations. Then they are also nbpr-simulations for any n_0. Suppose on the other hand that R_ε and R_ε^{-1} are both nbpr-bisimulations for some n_0 greater than

$$\sharp_{fns}(A) \cdot (\sharp_{par}(A) + 1) + \sharp_{fns}(B) \cdot (\sharp_{par}(B) + 1).$$

Let a ***name representation*** be any pair of maps (f_{free}, f_{bound}) such that f_{free} is an injection, and for each binding occurrence of a name in A or B f_{bound} maps that name into a small enough name, such that

1. if x and y are distinct, both occurs freely in some subterm of A or B, and both are occurrences of bound names, then $f_{bound}(x) \neq f_{bound}(y)$, and
2. if x and y are distinct, both occurs in a subterm of A or B and, say, x is an occurrence of a bound name and y an occurrence of a free name, then $f_{bound}(x) \neq f_{free}(y)$.

For a name partition ε let $f_{free}(\varepsilon) = \{\{f(x) \mid x \in U\} \mid U \in \varepsilon\}$. Because of the choice of n_0, a name representation exists. Let then $AS_\varepsilon B$ if and only if there is a name representation (f_{free}, f_{bound}) such that if A' and B' are the agents resulting from renaming free and bound names according to (f_{free}, f_{bound}), then A' and B' are $f_{free}(\varepsilon)$-nbpr-bisimulations. It is then easy to verify that, due to the choice of n_0, S_ε is an ε-pr-bisimulation. This completes the proof. □ (Theorem 13)

Other equivalences. In a similar manner we can prove decidability for other versions of bisimulation equivalence, notably early strong bisimulation equivalence, late and early weak bisimulation equivalence. Decidability of open bisimulation equivalence can also be shown in this manner. However, open bisimulation equivalence is already known to be decidable (indeed it was formulated with this as a central concern).

Early equivalence is characterised by permuting the quantifications over transitions and inputs which is implicit in clauses (3) and (4) of Def. 1 (c.f. [12] for a definition of early equivalence). The proofs of Theorems 12 and 13 are only minimally affected by this modification. For the weak late and early equivalences again only small modifications are needed (though alternative characterisations of these equivalences are likely to be mandatory for reasons of efficiency). We thus obtain:

Theorem 14. *1. Strong early bisimulation equivalence is decidable.*
2. Weak late and early bisimulation equivalence are decidable. □

7 Complexity and Discussion

The obvious backtracking-based algorithm for deciding name-bounded bisimulation equivalence is quite inefficient. As for standard bisimulation equivalence a better solution is obtained using the Paige-Tarjan algorithm [15, 8] with a worst-case running time of $\mathcal{O}(n_t \log n_s + n_s)$ where n_t is the number of transitions and n_s the number of states. With minor modifications to cater for bound output this algorithm is applicable once the full state spaces have been constructed, as pairs (A, ε). If the total number of reachable agents A is m and the sum of the length of the input agents is n then, since in the worst case n_0 is quadratic in n, n_s is $\mathcal{O}(m2^{n^2})$ and n_t is $\mathcal{O}(m^2n^22^{n^2})$. Thus the running time of the Paige-Tarjan algorithm is bounded by $\mathcal{O}(n^42^{n^2}m^2 \log m)$. To estimate m note first that up to the choice of names the number of agents reachable from one parallel component is $\mathcal{O}(n)$. Since each name can be instantiated in n_0 different ways the entire number of agents reachable from one parallel component is $\mathcal{O}(n2^{n^2 \log(n^2)})$. Thus m is bounded by $\mathcal{O}((n2^{n^2 \log(n^2)})^n) = \mathcal{O}(2^{n^3 \log(n^2) + n \log n})$, which is $2^{\mathcal{O}(n^3 \log(n^2))}$. This is strikingly close to the similarly approximated upper bound for CCS of $2^{\mathcal{O}(n \log n)})$. Both the parallel combinator already present in CCS and the π-calculus features of name generation and passing causes an exponential blow-up in the size of the state space. One might fear that these two causes of state space blow-up could interfere in a serious manner, resulting in double exponential running times or worse. However, even though some interference does take place because of scope extrusion, our results show that this fear is unfounded.

Lower bounds. Concerning lower bounds Jonsson and Parrow [7] shows that the bisimulation problem for data-independent programs (not including the parallel combinator |) is NP-hard. Since data-independent programs are subsumed by those considered here that lower bound applies here as well.

Efficiency. As for efficiency, based on the asymptotically quite similar worst-case bounds for CCS and for the π-calculus, since the Paige-Tarjan algorithm has been applied to quite realistically sized examples in CCS one might hope that this applies here too. Whether this in fact turns out to be the case remains to be seen. It may well be that alternative characterisations of the equivalences can be exploited to improve the efficiency of our algorithms, along the lines of for instance the efficient characterisation of strong open bisimulation equivalence [17], or the symbolic bisimulations of Hennessy and Lin [5]. For the weak equivalences in particular we expect such efficient characterisations to be indisposable.

References

1. R. Amadio. A uniform presentation of CHOCS and π-calculus. Rapport de Recherche 1726, INRIA-Lorraine, Nancy, 1992.
2. D. Caucal. Graphes canoniques des graphes algébriques. *Informatique Théorique et Applications (RAIRO)*, 24(4):339–352, 1990.
3. S. Christensen, H. Hüttel, and C. Stirling. Bisimulation equivalence is decidable for all context-free processes. In Proc. CONCUR'92, W. R. Cleaveland (ed.), *Lecture Notes in Computer Science*, 630:138–147, 1992.
4. M. Dam. Model checking mobile processes. In *Proc. CONCUR'93,* Lecture Notes in Computer Science, 715:22–36, 1993. Full version in SICS report RR94:1, 1994.
5. M. Hennessy and H. Lin. Symbolic bisimulations. Dept. of Computer Science, University of Sussex, Report 1/92, 1992.
6. Y. Hirschfeld and F. Moller. A fast algorithm for deciding bisimilarity of normed context-free processes. In Proc. CONCUR'94, B. jonsson, J. Parrow (eds.), *Lecture Notes in Computer Science*, 836:48–63, 1994.
7. B. Jonsson and J. Parrow. Deciding bisimulation equivalences for a class of non-finite-state programs. *Information and Computation*, 1992.
8. P. C. Kannellakis and S. A. Smolka. CCS expressions, finite state processes, and three problems of equivalence. *Information and Computation*, 86:43–68, 1990.
9. R. Milner. The polyadic π-calculus: A tutorial. Technical Report ECS-LFCS-91-180, Laboratory for the Foundations of Computer Science, Department of Computer Science, University of Edinburgh, 1991.
10. R. Milner. Functions as processes. *Mathematical Structures in Computer Science*, 2:119–141, 1992.
11. R. Milner, J. Parrow, and D. Walker. A calculus of mobile processes, I and II. *Information and Computation*, 100(1):1–40 and 41–77, 1992.
12. R. Milner, J. Parrow, and D. Walker. Modal logics for mobile processes. *Theoretical Computer Science*, 114:149–171, 1993.
13. F. Orava. *On the Formal Analysis of Telecommunication Protocols.* PhD thesis, Dept. of Computer Systems, Uppsala University and Swedish Institute of Computer Science, 1994. Forthcoming.
14. F. Orava and J. Parrow. An algebraic verification of a mobile network. *Formal Aspects of Computing*, pages 497–543, 1992.
15. R. Paige and R. E. Tarjan. Three partition refinement algorithms. *SIAM Journal of Computing*, 16(6):973–989, 1987.
16. D. Sangiorgi. From π-calculus to higher-order π-calculus—and back. To appear in Proc. TAPSOFT'93, 1993.
17. D. Sangiorgi. A theory of bisimulation for the π-calculus. in *Proc. CONCUR'93* Lecture Notes in Computer Science, 715:127–142, 1993.
18. D. Walker. Objects in the π-calculus. *Information and Computation*, 1994. (To appear).

Detecting Isomorphisms of Modular Specifications with Diagrams

Catherine Oriat

LGI – IMAG, BP 53, 38041 Grenoble Cedex 9, France
e-mail: Catherine.Oriat@imag.fr

Abstract. We propose to detect isomorphisms of algebraic modular specifications, by representing specifications as diagrams over a category $\mathcal{C}_0$ of base specifications and specification morphisms. We start with a formulation of modular specifications as terms, which are interpreted as diagrams. This representation has the advantage of being more abstract, i.e. less dependent of one specific construction than terms. For that, we define a category diagr $(\mathcal{C}_0)$ of diagrams, which is a completion of $\mathcal{C}_0$ with finite colimits. The category diagr $(\mathcal{C}_0)$ is *finitely cocomplete*, even if $\mathcal{C}_0$ is *not* finitely cocomplete. We define a functor $\mathcal{D}[\![\,]\!]$: Term $(\mathcal{C}_0) \rightarrow$ diagr $(\mathcal{C}_0)$ which maps specifications to diagrams, and specification morphisms to diagram morphisms. This interpretation is sound in that the colimit of a diagram representing a specification is isomorphic to this specification. The problem of isomorphisms of modular specifications is solved by detecting isomorphisms of diagrams.

1 Introduction

The specification of large systems requires the use of modular specifications. Small specifications are combined in a structured way to construct larger specifications. In this paper, we only consider specifications which are built with colimits over a category $\mathcal{C}_0$ of fixed base specifications. Formally, this means that we work in the category freely generated by finite colimits over $\mathcal{C}_0$. If we wish to add a new specification a posteriori in the category (for instance with an enrichment or hiding), we need to work in a new category which contains the already constructed specifications as well as the new defined specification.

The aim of this paper is to solve the problem of isomorphism of modular specifications in this setting of constructions with finite colimits. There are several kinds of isomorphisms we could be interested in:

- "Isomorphism of name". Two specifications are isomorphic if they have the same name. This simple definition provides a very weak form of isomorphism.
- "Isomorphism of structure". Two specifications are isomorphic if they have been constructed the same way, independently of aliases which may have been defined in the construction process. This isomorphism is slightly less weak than the previous one, but still not very interesting.

- "Semantic isomorphism". Two specifications are isomorphic if their associated classes of models (defined by the semantics) are the same. This isomorphism is very hard to treat, mainly because classes of algebras cannot be manipulated easily.
- "Isomorphism in SPEC". Two specifications are isomorphic if they are related by a bijective specification morphism. The difficulty here is first to exhibit this morphism, and secondly to check that it is indeed a *specification* morphism, i.e. that the equations of each specification hold in the other one.
- The isomorphisms we consider here are "construction isomorphisms". Two specifications are isomorphic if we can prove it using the general properties of colimits. We think this kind of isomorphism is interesting because it not too general in that it reflects the constructions of the modular specification. But it is more general than the isomorphism of structure because the specific steps chosen for the construction are abstracted. These isomorphisms do not depend on the actual definition of the base specifications, they only depend on how base specifications, related by base specification morphisms, are combined. Of course if two base specifications are isomorphic, and if this isomorphism is not part of $\mathcal{C}_0$, then we will fail to find it.

Most existing specification languages give more importance to the construction of a modular specification than to the result of the construction; for example CLEAR [3, 4], ACTONE [6], ASL [14], OBJ2 [7], PLUSS [8, 2], LPG [1]. This implies that the only tractable isomorphisms are isomorphisms of structure. We propose to adopt a less syntactic view of modular specification, by representing them as diagrams over the category of base specifications $\mathcal{C}_0$. This representation is more abstract than terms because irrelevant steps of the construction disappear. We need of course to work in a *category* of diagrams, and so we associate specification morphisms between modular specifications to diagram morphisms. This approach is similar to that adopted to describe the semantics of CLEAR [4]. Our diagrams correspond to based theories. We need a more general definition for arrows than that of based morphisms, because based morphisms only correspond to inclusions of modular specifications, whereas we want a diagram morphism to correspond to any specification morphism. So we define a category of diagrams diagr $(\mathcal{C}_0)$, and "construction isomorphisms" of modular specifications then correspond to isomorphisms in the category diagr $(\mathcal{C}_0)$.

This paper is organized as follows. In section 2, we present the category Term $(\mathcal{C}_0)$, which provides a syntax for modular specifications and specification morphisms. In section 3, we present examples of modular specifications to illustrate the syntax and motivate the definition of a category DIAGR $(\mathcal{C}_0)$. In section 4, we define the categories DIAGR $(\mathcal{C}_0)$ and diagr $(\mathcal{C}_0)$ and present some theoretical results. In section 5, we explain how terms denoting specifications or specification morphisms can be associated to diagrams and diagram morphisms. This mapping is described by a functor $\mathcal{D}[\![\,]\!]$: Term $(\mathcal{C}_0) \rightarrow$ diagr $(\mathcal{C}_0)$. In section 6, we present an algorithm to detect when two diagrams are isomorphic in the category diagr $(\mathcal{C}_0)$, when the base category $\mathcal{C}_0$ is finite and has no cycle.

2 Syntax for Modular Specifications

In this section, we present a syntax for modular algebraic specifications constructed with colimits. This syntax is formulated with the concept of *dependent types* as suggested by Cartmell [5]. Cartmell's generalized algebraic theories are a generalization of many-sorted algebras, which allow to define dependent types, i.e. types parameterized by terms. This approach has already been presented in [10, 11, 12]. We suppose we have a category $\mathcal{C}_0$ of base specifications and specification morphisms.

We have two types: the type of specifications Spec, and the type of specification morphisms Hom, which depends on two specifications. If A and B are specifications, $\mathsf{Hom}\,(A, B)$ is the type of specification morphisms from A to B.

$$\frac{}{\mathsf{Spec\ is\ a\ type}} \qquad \frac{A, B : \mathsf{Spec}}{\mathsf{Hom}\,(A, B)\ \mathsf{is\ a\ type}} \tag{1,2}$$

We now define *terms* of both types Spec and Hom, as well as axioms satisfied by these terms.

$$\frac{Sp \text{ specification of } \mathcal{C}_0}{Sp : \mathsf{Spec}} \qquad \frac{p : Sp_1 \rightarrow Sp_2 \text{ specification morphism of } \mathcal{C}_0}{p : \mathsf{Hom}\,(Sp_1, Sp_2)} \tag{3,4}$$

$$\frac{A, B, C : \mathsf{Spec}\ ;\ f : \mathsf{Hom}\,(A, B)\ ;\ g : \mathsf{Hom}\,(B, C)}{g \circ f : \mathsf{Hom}\,(A, C)} \tag{5}$$

$$\frac{A, B, C, D : \mathsf{Spec}\ ;\ f : \mathsf{Hom}\,(A, B)\ ;\ g : \mathsf{Hom}\,(B, C)\ ;\ h : \mathsf{Hom}\,(C, D)}{(h \circ g) \circ f = h \circ (g \circ f) : \mathsf{Hom}\,(A, D)} \tag{6}$$

$$\frac{A : \mathsf{Spec}}{\mathsf{id}_A : \mathsf{Hom}\,(A, A)} \tag{7}$$

$$\frac{A, B : \mathsf{Spec}\ ;\ f : \mathsf{Hom}\,(A, B)}{f \circ \mathsf{id}_A = f : \mathsf{Hom}\,(A, B)} \qquad \frac{A, B : \mathsf{Spec}\ ;\ f : \mathsf{Hom}\,(A, B)}{\mathsf{id}_B \circ f = f : \mathsf{Hom}\,(A, B)} \tag{8,9}$$

An algebra which satisfies the generalized algebraic theory specified by rules (1)–(2) and (5)–(9) is a category. We now give a syntax for colimit constructions. Here, for instance we give a syntax for an initial object and pushouts.

$$\frac{}{\emptyset : \mathsf{Spec}} \qquad \frac{A : \mathsf{Spec}}{\mathsf{j}_A : \mathsf{Hom}\,(\emptyset, A)} \qquad \frac{A : \mathsf{Spec}\ ;\ f, g : \mathsf{Hom}\,(\emptyset, A)}{f = g : \mathsf{Hom}\,(\emptyset, A)} \tag{10,11,12}$$

$$\frac{A, B, C : \mathsf{Spec}\ ;\ f : \mathsf{Hom}\,(A, B)\ ;\ g : \mathsf{Hom}\,(A, C)}{\mathsf{push}\,(A, B, C, f, g) : \mathsf{Spec}} \tag{13}$$

$$\frac{A, B, C : \mathsf{Spec}\ ;\ f : \mathsf{Hom}\,(A, B)\ ;\ g : \mathsf{Hom}\,(A, C)}{\&_1(A, B, C, f, g) : \mathsf{Hom}\,(B, \mathsf{push}\,(A, B, C, f, g))} \tag{14}$$

$$\frac{A,B,C:\mathsf{Spec}\ ;\ f:\mathsf{Hom}(A,B)\ ;\ g:\mathsf{Hom}(A,C)}{\&_2(A,B,C,f,g):\mathsf{Hom}(C,\mathsf{push}(A,B,C,f,g))} \tag{15}$$

$$\frac{A,B,C:\mathsf{Spec}\ ;\ f:\mathsf{Hom}(A,B)\ ;\ g:\mathsf{Hom}(A,C)}{\&_1(A,B,C,f,g)\circ f=\&_2(A,B,C,f,g)\circ g:\mathsf{Hom}(A,\mathsf{push}(A,B,C,f,g))} \tag{16}$$

$$\frac{\begin{array}{c}A,B,C,D:\mathsf{Spec}\ ;\ f:\mathsf{Hom}(A,B)\ ;\ g:\mathsf{Hom}(A,C)\\ f':\mathsf{Hom}(B,D)\ ;\ g':\mathsf{Hom}(C,D)\ ;\ f'\circ f=g'\circ g:\mathsf{Hom}(A,D)\end{array}}{\mathsf{up}(A,B,C,D,f,g,f',g'):\mathsf{Hom}(\mathsf{push}(A,B,C,f,g),D)} \tag{17}$$

$$\frac{\begin{array}{c}A,B,C,D:\mathsf{Spec}\ ;\ f:\mathsf{Hom}(A,B)\ ;\ g:\mathsf{Hom}(A,C)\\ f':\mathsf{Hom}(B,D)\ ;\ g':\mathsf{Hom}(C,D)\ ;\ f'\circ f=g'\circ g:\mathsf{Hom}(A,D)\end{array}}{\mathsf{up}(A,B,C,D,f,g,f',g')\circ\&_1(A,B,C,f,g)=f':\mathsf{Hom}(B,D)} \tag{18}$$

$$\frac{\begin{array}{c}A,B,C,D:\mathsf{Spec}\ ;\ f:\mathsf{Hom}(A,B)\ ;\ g:\mathsf{Hom}(A,C)\\ f':\mathsf{Hom}(B,D)\ ;\ g':\mathsf{Hom}(C,D)\ ;\ f'\circ f=g'\circ g:\mathsf{Hom}(A,D)\end{array}}{\mathsf{up}(A,B,C,D,f,g,f',g')\circ\&_2(A,B,C,f,g)=g':\mathsf{Hom}(C,D)} \tag{19}$$

$$\frac{\begin{array}{c}A,B,C,D:\mathsf{Spec}\ ;\ f:\mathsf{Hom}(A,B)\ ;\ g:\mathsf{Hom}(A,C)\\ u,v:\mathsf{Hom}(\mathsf{push}(A,B,C,f,g),D)\\ u\circ\&_k(A,B,C,f,g)=v\circ\&_k(A,B,C,f,g)\ ;\ (k=1,2)\end{array}}{u=v:\mathsf{Hom}(\mathsf{push}(A,B,C,f,g),D)} \tag{20}$$

An algebra which satisfies this specification is a finitely cocomplete category.

We must note that this specification is actually not a generalized algebraic theory, because the rules (17)–(20) contain equalities in their premises. To write a proper generalized algebraic theory, one has to axiomatize equality in the type system with a predicate eq [5]. These equalities raise another problem: it may not be decidable whether or not a term is well-formed. A rigorous construction of the category freely generated by a chosen initial object and chosen pushouts as a category of *terms* is under development.

Let Term $(\mathcal{C}_0)$ be the algebra freely generated on the specified colimit constructions. Let SPEC be the category of all specifications, with a *chosen* initial object, and *chosen* pushouts. SPEC is therefore an algebra which satisfies the equations. $\mathcal{C}_0$ is a subcategory of SPEC, and of Term $(\mathcal{C}_0)$. We note these inclusions

$$i:\mathcal{C}_0\rightarrow \text{SPEC},\quad e:\mathcal{C}_0\rightarrow \text{Term}(\mathcal{C}_0)$$

As Term $(\mathcal{C}_0)$ is a free algebra, there exists a unique homomorphism (which is also a functor)

$$\mathcal{S}[\![]\!]:\text{Term}(\mathcal{C}_0)\rightarrow \text{SPEC}$$

such that $\mathcal{S}[\![]\!]\circ e=i$. This functor associates to each term of type Spec the specification that it represents, and to each term of type Hom the specification morphism it represents. $\mathcal{S}[\![]\!]$ is a "standard semantics" for terms.

3 Example

The aim of this section is first to give some examples of modular specifications written with the syntax presented in the previous section, and secondly to motivate the definition of the category of diagrams presented in the next section.

We present different ways of specifying the theory of rings in the specification language LPG.

$$S = \boxed{\begin{array}{l}\textbf{property A-SORT} \\ \textbf{sorts} \quad \textbf{s}\end{array}} \qquad B = \boxed{\begin{array}{l}\textbf{property BIN-OP} \\ \textbf{sorts} \quad \textbf{s} \\ \textbf{operators op : s,s -> s} \\ \textbf{satisfies A-SORT[s]}\end{array}}$$

S specifies a single sort, B specifies a binary operator and a specification morphism $s : S \rightarrow B$, defined by the statement **satisfies A-SORT[s]** in B.

$$M = \boxed{\begin{array}{l}\textbf{property MONOID} \\ \textbf{sorts} \quad \textbf{s} \\ \textbf{operators * : s,s -> s} \\ \qquad\qquad \textbf{1 : -> s} \\ \textbf{equations 1 * x == x} \\ \qquad\qquad \textbf{x * 1 == x} \\ \textbf{(x * y) * z == x * (y * z)} \\ \textbf{satisfies BIN-OP[s,*]}\end{array}} \qquad G = \boxed{\begin{array}{l}\textbf{property ABEL-GROUP} \\ \textbf{sorts} \quad \textbf{s} \\ \textbf{operators + : s,s -> s} \\ \qquad\qquad \textbf{0 : -> s} \\ \qquad\qquad \textbf{i : s -> s} \\ \textbf{equations x + y == y + x} \\ \qquad\qquad \textbf{i(x) + x == 0} \\ \textbf{satisfies MONOID[s,+,0]}\end{array}}$$

M specifies a monoid, with a specification morphism $b : B \rightarrow M$. G specifies an Abelian group: we add to the specification of monoids an inverse function i and the commutativity of the binary operator. We also define a specification morphism $m : M \rightarrow G$.

$$D = \boxed{\begin{array}{l}\textbf{property DISTRIBUTIVE} \\ \textbf{sorts} \quad \textbf{s} \\ \textbf{operators +,* : s,s -> s} \\ \textbf{equations x * (y + z) == (x * y) + (x * z)} \\ \textbf{satisfies BIN-OP[s,+], BIN-OP[s,*]}\end{array}}$$

D specifies two operators related by the distributive law, and two specification morphisms $m_+, m_* : B \rightarrow D$. m_+ maps op to + and m_* maps op to *.

To summarize what we have defined so far, we work in the category C_0 (Fig. 1).

$$S \xrightarrow{\ s\ } B \xrightarrow{\ b\ } M \xrightarrow{\ m\ } G$$
$$m_+ \Big\downarrow\Big\downarrow m_*$$
$$D$$

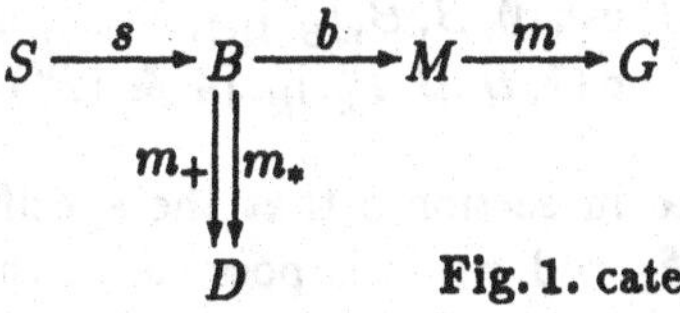

Fig. 1. category C_0

We can now define several modular specifications of rings with pushouts. In LPG, such specifications can be defined with the **combines** construction.

$R_1 = \mathsf{push}\,(B, M, \mathsf{push}\,(B, D, G, m_+, m\circ b), b, \&_1(B, D, G, m_+, m\circ b)\circ m_*)$
$R_1' = \mathsf{push}\,(B, \mathsf{push}\,(B, M, D, b, m_*), G, \&_2(B, M, D, b, m_*)\circ m_+, m\circ b)$

R_1 and R_1' are two specifications of rings. Here, the difference is somehow artificially introduced by the syntactic construction push . Indeed both constructions are a coding with pushouts of the colimit of the diagram δ_1 (Fig. 2).

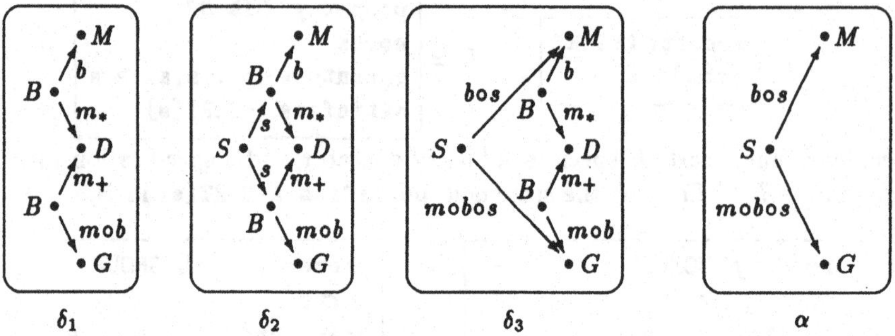

Fig. 2. diagrams δ_1, δ_2, δ_3 and α

In the following, if $p = \mathsf{push}\,(x, y, z, u, v)$, we will write $\&_i(p)$ for $\&_i(x, y, z, u, v)$. More complicated cases may arise. We can for instance define a "pseudo ring" i.e. a ring without distributivity either with the term P_3, or P_3' as follows.

$P_1 = \mathsf{push}\,(S, B, B, s, s)$
$P_2 = \mathsf{push}\,(B, M, P_1, b, \&_1(P_1))$
$P_3 = \mathsf{push}\,(B, P_2, G, \&_2(P_2)\circ\&_2(P_1), m\circ b)$
$P_3' = \mathsf{push}\,(S, M, G, b\circ s, m\circ b\circ s)$

(P_3' corresponds to the colimit of the diagram α.)

Now we can "add the distributivity" on two different ways and get two new specifications of rings R_2 and R_3.

$R_2 = \mathsf{push}\,(\mathsf{push}\,(\emptyset, B, B, \mathrm{j}_B, \mathrm{j}_B), D, P_3, \mathsf{up}\,(\emptyset, B, B, D, \mathrm{j}_B, \mathrm{j}_B, m_*, m_+),$
$\quad \mathsf{up}\,(\emptyset, B, B, P_3, \mathrm{j}_B, \mathrm{j}_B, \&_1(P_3)\circ\&_2(P_2)\circ\&_1(P_1),$
$\quad \&_1(P_3)\circ\&_2(P_2)\circ\&_2(P_1)))$
$R_3 = \mathsf{push}\,(\mathsf{push}\,(\emptyset, B, B, \mathrm{j}_B, \mathrm{j}_B), D, P_3', \mathsf{up}\,(\emptyset, B, B, D, \mathrm{j}_B, \mathrm{j}_B, m_*, m_+),$
$\quad \mathsf{up}\,(\emptyset, B, B, P_3', \mathrm{j}_B, \mathrm{j}_B, \&_1(P_3')\circ b, \&_2(P_3')\circ m\circ b))$

We will see in section 5 that the specifications R_2 and R_3 correspond to the diagrams δ_2 and δ_3. It is possible to check that the colimits of δ_1, δ_2 and δ_3 are all isomorphic. This comes from the equality $m_+ \circ s = m_* \circ s$ in C_0. In other words, the fact that both binary operations are defined on the same set is contained in the distributivity property D.

4 Categories of Diagrams

In the following, we assume the reader is familiar with basic notions of category theory. Vertices(α^Φ) and Edges(α^Φ) respectively denote the set of vertices and the set of edges of a graph α^Φ.

4.1 The Category DIAGR($\mathcal{C}_0$)

Definition 1 (Diagram). A diagram over a category $\mathcal{C}_0$ is a couple

$$\alpha = (\alpha^\Phi,\ \alpha : \alpha^\Phi \to \mathcal{C}_0),$$

and $\alpha : \alpha^\Phi \to \mathcal{C}_0$ is a graph morphism. A diagram is *finite* when its underlying graph is finite.

To get a *category* of diagrams, we need to define diagram morphisms. We could consider couples

$$\sigma : \alpha \to \beta = (\sigma^\Phi : \alpha^\Phi \to \beta^\Phi\ ;\ \sigma : \alpha \xrightarrow{\bullet} \beta \circ \sigma^\Phi)$$

where σ^Φ is a graph morphism, and σ a natural transformation. This definition appears in [13] (it is the "flatten" category **Funct** ($\mathcal{C}_0$), page 244, example 4), and in a dual form in [9] (it is the "super-comma category", page 111, exercise 5.b.) This definition is not general enough, because some specification morphisms have no corresponding diagram morphisms. For instance, there is a morphism

$$\mathsf{up}\,(S, M, G, R_1, b\circ s, m\circ b\circ s, \&_1(R_1), \&_2(R_1)\circ\&_2(B, D, G, m_+, m\circ b))$$

from P_3' to R_1, which corresponds to an arrow from Colim α to Colim δ_1, because $m_+ \circ s = m_* \circ s$. But there is no diagram morphism from α to δ_1 with the definition above (Fig. 3). We need a more general definition of arrows, and thus must consider ***generalized graph morphisms***, which associate a ***zigzag*** to each edge of a graph, and ***generalized natural transformations***. With this definition, we can define an arrow $\sigma : \alpha \to \delta_1$, which consists of a generalized graph morphism σ^Φ and a generalized natural transformation σ, defined for instance as follows: $\sigma^\Phi(1) = 1'$, $\sigma^\Phi(2) = 2'$, $\sigma^\Phi(3) = 3'$; $\sigma^\Phi(a_0) =$ zigzag from $1'$ to $2'$, $\sigma^\Phi(a_1) =$ zigzag from $1'$ to $3'$; $\sigma_1 = m_* \circ s$, $\sigma_2 = id_M$, $\sigma_3 = id_G$.

Definition 2 (Zigzag on a graph). Let α^Φ be a graph. A *zigzag* on α^Φ is a finite linear sequence of edges of α^Φ:

$$Z = n_0 \xrightarrow{a_0} n_1 \xleftarrow{a_1} n_2 \cdots \xrightarrow{a_{k-1}} n_k, \quad \text{noted } Z : n_0 \rightsquigarrow n_k,$$

each edge is oriented either from left to right or from right to left.

We get a graph Zigzag(α^Φ), with the same vertices as α^Φ, and with edges the zigzags of α^Φ.

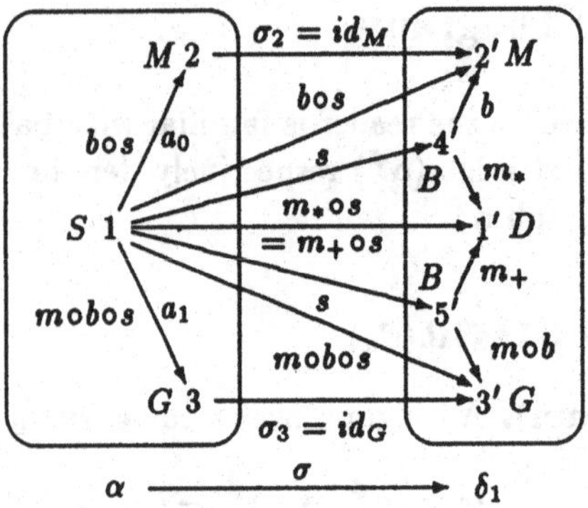

Fig. 3. diagram morphism $\sigma : \alpha \to \delta_1$

Definition 3 (Generalized graph morphism). Let us consider two graphs α^Φ and β^Φ. A *generalized graph morphism* $\sigma^\Phi : \alpha^\Phi \rightsquigarrow \beta^\Phi$ from α^Φ to β^Φ is a graph morphism from α^Φ to Zigzag(β^Φ). We can compose generalized graph morphisms by joining zigzags.

Definition 4 (Connection relation).
Let $\delta = (\delta^\Phi,\ \delta : \delta^\Phi \to \mathcal{C}_0)$ be a diagram. Two arrows $u : A \to B$ and $v : A \to C$ of $\mathcal{C}_0$ are said to be *connected* by the diagram δ if and only if there exist a zigzag on δ^Φ, $Z = n_0 \xrightarrow{a_0} n_1 \cdots n_{k-1} \xrightarrow{a_{k-1}} n_k$, and a set of arrows in $\mathcal{C}_0$, $\{c_i : A \to \delta(n_i);\ i \in \{0, \ldots, k\}\}$, such that:

- $u = c_0$ (and thus $\delta(n_0) = B$)
- $v = c_k$ (and thus $\delta(n_k) = C$)
- $\forall i \in \{0, \ldots, k-1\}$:
  $\delta(a_i) \circ c_i = c_{i+1}$, if a_i is oriented from n_i to n_{i+1};
  $\delta(a_i) \circ c_{i+1} = c_i$, if a_i is oriented from n_{i+1} to n_i.

We note $u \sim_\delta v$; or $u \sim_\delta v\ [Z]$, if we want to specify the zigzag Z.

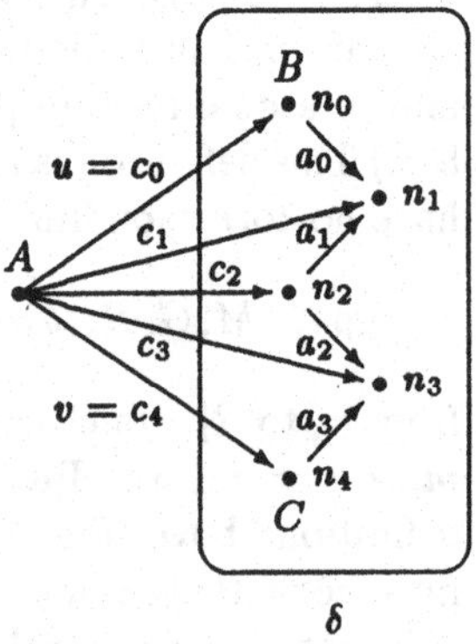

Definition 5 (Category of finite diagrams DIAGR ($\mathcal{C}_0$)).
Let $\mathcal{C}_0$ be a category.

- An *object* δ of DIAGR ($\mathcal{C}_0$) is a finite diagram.
- Let $\alpha = (\alpha^\Phi,\ \alpha : \alpha^\Phi \to \mathcal{C}_0)$ and $\beta = (\beta^\Phi,\ \beta : \beta^\Phi \to \mathcal{C}_0)$ be two diagrams. A *diagram morphism* from α to β is a couple

$$\tau : \alpha \to \beta = (\tau^\Phi : \alpha^\Phi \rightsquigarrow \beta^\Phi,\ \tau : \alpha \overset{\bullet}{\rightsquigarrow} \beta \circ \tau^\Phi),$$

  where
  - $\tau^\Phi : \alpha^\Phi \rightsquigarrow \beta^\Phi$ is a generalized graph morphism.
  - $\tau : \alpha \overset{\bullet}{\rightsquigarrow} \beta \circ \tau^\Phi$ is a "generalized natural transformation" i.e. a set of arrows $\tau_n : \alpha(n) \to \beta(\tau^\Phi(n)),\ \forall n \in \text{Vertices}(\alpha^\Phi)$ such that $\forall a : m \to n \in \text{Edges}(\alpha^\Phi),\ \ \tau_n \circ \alpha(a) \sim_\beta \tau_m\ \ [\tau^\Phi(a)]$

Note that if τ^{Φ} is a graph morphism, then τ is a natural transformation.

- Let us consider three diagrams and two diagram morphisms
$\alpha = (\alpha^{\Phi}, \alpha : \alpha^{\Phi} \to \mathcal{C}_0)$, $\beta = (\beta^{\Phi}, \beta : \beta^{\Phi} \to \mathcal{C}_0)$, $\gamma = (\gamma^{\Phi}, \gamma : \gamma^{\Phi} \to \mathcal{C}_0)$
$\sigma : \alpha \to \beta = (\sigma^{\Phi} : \alpha^{\Phi} \rightsquigarrow \beta^{\Phi}, \sigma : \alpha \overset{\bullet}{\rightsquigarrow} \beta \circ \sigma^{\Phi})$
$\tau : \beta \to \gamma = (\tau^{\Phi} : \beta^{\Phi} \rightsquigarrow \gamma^{\Phi}, \tau : \beta \overset{\bullet}{\rightsquigarrow} \gamma \circ \tau^{\Phi})$
The composition of σ and τ is the couple
$$\tau \circ \sigma : \alpha \to \gamma = (\tau^{\Phi} \circ \sigma^{\Phi} : \alpha^{\Phi} \rightsquigarrow \gamma^{\Phi}, \lambda : \alpha \overset{\bullet}{\rightsquigarrow} \gamma \circ \tau^{\Phi} \circ \sigma^{\Phi}),$$
where $\tau^{\Phi} \circ \sigma^{\Phi}$ is a composition of generalized graph morphisms, and λ is the "generalized natural transformation" defined by $\lambda_n = \tau_{\sigma^{\Phi}(n)} \circ \sigma_n$.

One can easily check that DIAGR $(\mathcal{C}_0)$ is indeed a category.

Colimits

The category $\mathcal{C}_0$ can be embedded in DIAGR $(\mathcal{C}_0)$ with a functor $I : \mathcal{C}_0 \to$ DIAGR $(\mathcal{C}_0)$. Colimits can be defined as usual [9]. In our setting, a cone from a diagram α is a couple $(C, \lambda : \alpha \to I(C))$ where C is an object of $\mathcal{C}_0$ and λ is an arrow of DIAGR $(\mathcal{C}_0)$.

The cone $(C, \lambda : \alpha \to I(C))$ is a *colimiting cone* from α if and only if for any cone $(D, \lambda' : \alpha \to I(D))$, there exists a unique arrow $\phi : C \to D$ such that $I(\phi) \circ \lambda = \lambda'$. C is called the *colimit* of α, and we note $C = \mathrm{Colim}\, \alpha$. One diagram may have several colimits, which are then isomorphic. Writing $C = \mathrm{Colim}\, \alpha$ means that we have *chosen* the object C for the colimit of α.

A category is finitely cocomplete when every finite diagram has a colimit.

Let α and β be two diagrams with colimiting cones $(\mathrm{Colim}\, \alpha, \eta_\alpha : \alpha \to I(\mathrm{Colim}\, \alpha))$ and $(\mathrm{Colim}\, \beta, \eta_\beta : \beta \to I(\mathrm{Colim}\, \beta))$.

Let $\sigma : \alpha \to \beta$ be a diagram morphism. Then there exists a unique arrow $\mathrm{Colim}\, \sigma : \mathrm{Colim}\, \alpha \to \mathrm{Colim}\, \beta$, such that $I(\mathrm{Colim}\, \sigma) \circ \eta_\alpha = \eta_\beta \circ \sigma$.

Theorem 6. *Let $\mathcal{C}_0$ be a finitely cocomplete category. Then*

1. Colim : DIAGR $(\mathcal{C}_0) \to \mathcal{C}_0$ *is a functor.*
2. *The mapping η, which associates to each diagram α the colimiting cone from α to $I(\mathrm{Colim}\, \alpha)$ is a natural transformation $\eta : Id_{\mathrm{DIAGR}(\mathcal{C}_0)} \overset{\bullet}{\to} I \circ \mathrm{Colim}$.*
3. *The functor* Colim : DIAGR $(\mathcal{C}_0) \to \mathcal{C}_0$ *is a left-adjoint for the functor I : $\mathcal{C}_0 \to$* DIAGR $(\mathcal{C}_0)$. *The unit of the adjunction* (Colim $\dashv I$) *is the natural transformation η.*

4.2 The Category diagr $(\mathcal{C}_0)$

In the category DIAGR $(\mathcal{C}_0)$, different arrows may have equal colimits. For instance, for defining the arrow σ from α to δ_1 (Fig. 3), we can associate the vertex 1 either to the vertex $2'$, $4'$, $1'$, $5'$ or $3'$. (Of course association of edges to zigzags must be done accordingly). Those different arrows have the same colimit. The same way, non isomorphic objects may have isomorphic colimits in DIAGR $(\mathcal{C}_0)$. For instance, $\delta_1 \not\cong \delta_3$, but $\mathrm{Colim}\, \delta_1 \cong \mathrm{Colim}\, \delta_3$ (Fig. 2). The aim of this paragraph is to define a category where equalities of colimiting arrows and isomorphisms of colimit objects will be reflected at the level of diagrams.

Definition 7. Let α and β be two diagrams. Let $\sigma, \tau : \alpha \to \beta$ be two arrows of DIAGR $(\mathcal{C}_0)$. We define the relation $\approx$ on arrows of DIAGR $(\mathcal{C}_0)$ as follows:

$$\sigma \approx \tau \Leftrightarrow \forall n \in \mathrm{Vertices}(\alpha^{\Phi}) : \sigma_n \sim_\beta \tau_n$$

Theorem 8. *The relation $\approx$ is a congruence*

Definition 9. As $\approx$ is a congruence, we can consider the quotient category

$$\mathrm{diagr}\,(\mathcal{C}_0) = \mathrm{DIAGR}\,(\mathcal{C}_0)/\approx$$

Let $[-] : \mathrm{DIAGR}\,(\mathcal{C}_0) \to \mathrm{diagr}\,(\mathcal{C}_0)$ be the associated projection functor.

Theorem 10. *Let $\mathcal{C}_0$ be a finitely cocomplete category. The functor* Colim *is compatible with $\approx$. In other words, let α, β be two diagrams, and $\sigma, \tau : \alpha \to \beta$ two arrows of* DIAGR $(\mathcal{C}_0)$. *We have: $\sigma \approx \tau \Rightarrow \mathrm{Colim}\,\sigma = \mathrm{Colim}\,\tau$.*

The category diagr $(\mathcal{C}_0)$ is finitely cocomplete. We have to show that every diagram over diagr $(\mathcal{C}_0)$ — i.e. every object of DIAGR (diagr $(\mathcal{C}_0)$) — has a colimit in diagr $(\mathcal{C}_0)$. We first define an operation of flattening

$$\mu : \mathrm{DIAGR}\,(\mathrm{DIAGR}\,(\mathcal{C}_0)) \to \mathrm{DIAGR}\,(\mathcal{C}_0)$$

Intuitively, flattening the diagram of diagrams $\boldsymbol{\Delta}$ consists in considering the union of all subdiagrams of $\boldsymbol{\Delta}$, and in transforming every arrow of DIAGR $(\mathcal{C}_0)$ into a set of arrows of $\mathcal{C}_0$.

Definition 11 (Flattening μ : DIAGR (DIAGR $(\mathcal{C}_0)$) $\to$ DIAGR $(\mathcal{C}_0)$).
Let $\boldsymbol{\Delta} = (\Delta^{\Phi}, \Delta : \Delta^{\Phi} \to \mathrm{DIAGR}\,(\mathcal{C}_0))$ be an object of DIAGR (DIAGR $(\mathcal{C}_0)$). We define the diagram $\mu(\boldsymbol{\Delta}) = \boldsymbol{\delta} = (\delta^{\Phi},\ \delta : \delta^{\Phi} \to \mathcal{C}_0)$ as follows:

- δ^{Φ} is a graph, given by Vertices(δ^{Φ}) and Edges(δ^{Φ}).

$$\mathrm{Vertices}(\delta^{\Phi}) = \{\, (N, n_N)\ ;\ N \in \mathrm{Vertices}(\Delta^{\Phi})\ ;\ n_N \in \mathrm{Vertices}(\Delta(N)^{\Phi}) \,\}$$

$$\begin{aligned}\mathrm{Edges}(\delta^{\Phi}) \;=\; &\{\, (N, a_N) : (N, n_N) \to (N, n'_N)\ ;\\ &\quad N \in \mathrm{Vertices}(\Delta^{\Phi})\ ;\ n_N, n'_N \in \mathrm{Vertices}(\Delta(N)^{\Phi})\ ;\\ &\quad a_N : n_N \to n'_N \in \mathrm{Edges}(\Delta(N)^{\Phi}) \,\}\\ \bar{\cup} &\{\, (A : N \to N', n_N) : (N, n_N) \to (N', \Delta(A)^{\Phi}(n_N))\ ;\\ &\quad N, N' \in \mathrm{Vertices}(\Delta^{\Phi})\ ;\ A : N \to N' \in \mathrm{Edges}(\Delta^{\Phi})\ ;\\ &\quad n_N \in \mathrm{Vertices}(\Delta(N)^{\Phi}) \,\}\end{aligned}$$

- $\delta : \delta^{\Phi} \to \mathcal{C}_0$ is a functor, given by its action on vertices and edges of δ^{Φ}.
  - Action on vertices: $\delta(N, n_N)$ is an object of $\mathcal{C}_0$, isomorphic to $\Delta(N)(n_N)$. We call this isomorphism $(J_N)_{n_N} : \Delta(N)(n_N) \to \delta(N, n_N)$.
  - Action on edges: $\delta(N, a_N) = (J_N)_{n'_N} \circ \Delta(N)(a) \circ (J_N)^{-1}_{n_N}$
    $\delta(A, n_N) = (J_{N'})_{\Delta(A)^{\Phi}(n_N)} \circ \Delta(A)_{n_N} \circ (J_N)^{-1}_{n_N}$

Theorem 12. *The category* diagr $(\mathcal{C}_0)$ *is finitely cocomplete.*

Proof sketch. Let Γ be an object of DIAGR(diagr($\mathcal{C}_0$)). We suppose we are able to choose a representative for each equivalence class of arrows, i.e. we have a graph morphism Rep : diagr($\mathcal{C}_0$) $\rightarrow$ DIAGR($\mathcal{C}_0$) which is the identity on objects. We define a diagram Δ in DIAGR(DIAGR($\mathcal{C}_0$)) as

$$\Delta = Rep \circ \Gamma = (\Gamma^{\Phi}, Rep \circ \Gamma : \Gamma^{\Phi} \rightarrow \mathrm{DIAGR}(\mathcal{C}_0))$$

Let $\delta = [\mu(\Delta)]$. Let $(\eta_\Gamma)_N = [J_N]$. Then $(\delta, \eta_\Gamma : \Gamma \rightarrow I(\delta))$ is a colimiting cone from Γ.

As usual, colimits are defined up to isomorphisms. To define δ, we made two choices: μ is defined up to isomorphisms on objects of $\mathcal{C}_0$, and Rep contains a choice of arrow to represent an equivalence class.

Theorem 13. *Let $\mathcal{C}$ be a finitely cocomplete category. For every functor F : $\mathcal{C}_0 \rightarrow \mathcal{C}$ there exists a functor H : diagr($\mathcal{C}_0$) $\rightarrow \mathcal{C}$, unique up to isomorphism, such that $H \circ [-] \circ I \cong F$ and for every diagram Γ of* DIAGR(diagr($\mathcal{C}_0$)), $\mathrm{Colim}(H \circ \Gamma) \cong H(\mathrm{Colim}\,\Gamma)$.

5 Representing Modular Specifications as Diagrams

We associate modular specifications to diagrams and specification morphisms to diagram morphisms in diagr($\mathcal{C}_0$). As diagr($\mathcal{C}_0$) is finitely cocomplete, with chosen colimits, there exists a unique homomorphism (which is also a functor)

$$\mathcal{D}[\![]\!] : \mathrm{Term}(\mathcal{C}_0) \rightarrow \mathrm{diagr}(\mathcal{C}_0)$$

such that $\mathcal{D}[\![]\!] \circ e = [-] \circ I$. $\mathcal{D}[\![]\!]$ associates to each categorical construction in Term($\mathcal{C}_0$) the corresponding construction in diagr($\mathcal{C}_0$):

$\mathcal{D}[\![Sp]\!]$	$= [I(Sp)]$,	if Sp is a specification of $\mathcal{C}_0$
$\mathcal{D}[\![p]\!]$	$= [I(p)]$,	if p is a specification morphism of $\mathcal{C}_0$
$\mathcal{D}[\![g \circ f]\!]$	$= \mathcal{D}[\![g]\!] \circ \mathcal{D}[\![f]\!]$	(composition of arrows in diagr($\mathcal{C}_0$))
$\mathcal{D}[\![\mathrm{id}_A]\!]$	$= id_{\mathcal{D}[\![A]\!]}$	(identity arrow of diagr($\mathcal{C}_0$))
$\mathcal{D}[\![\emptyset]\!]$	$= \bigcirc$	($\bigcirc$ is the empty diagram, initial object of diagr($\mathcal{C}_0$))
$\mathcal{D}[\![j_A]\!]$	$= j_{\mathcal{D}[\![A]\!]}$	(unique arrow from the empty diagram to $\mathcal{D}[\![A]\!]$)

$\mathcal{D}[\![\mathrm{push}(A,B,C,f,g)]\!] = push(\mathcal{D}[\![A]\!], \mathcal{D}[\![B]\!], \mathcal{D}[\![C]\!], \mathcal{D}[\![f]\!], \mathcal{D}[\![g]\!])$
$\mathcal{D}[\![\&_1(A,B,C,f,g)]\!] = \&_1(\mathcal{D}[\![A]\!], \mathcal{D}[\![B]\!], \mathcal{D}[\![C]\!], \mathcal{D}[\![f]\!], \mathcal{D}[\![g]\!])$
$\mathcal{D}[\![\&_2(A,B,C,f,g)]\!] = \&_2(\mathcal{D}[\![A]\!], \mathcal{D}[\![B]\!], \mathcal{D}[\![C]\!], \mathcal{D}[\![f]\!], \mathcal{D}[\![g]\!])$
$\mathcal{D}[\![\mathrm{up}(A,B,C,D,f,g,f',g')]\!] = up(\mathcal{D}[\![A]\!], \mathcal{D}[\![B]\!], \mathcal{D}[\![C]\!], \mathcal{D}[\![D]\!], \mathcal{D}[\![f]\!], \ldots, \mathcal{D}[\![g']\!])$

¿From Theorem 13, there exists a functor eval : diagr($\mathcal{C}_0$) $\rightarrow$ SPEC such that $\mathrm{eval} \circ [-] \circ I \cong i$. This functor maps each diagram to the specification it represents.

Theorem 14. *There is a natural isomorphism* $\mathrm{eval} \circ \mathcal{D}[\![]\!] \cong \mathcal{S}[\![]\!]$

This theorem states that the calculus of diagrams is *sound*. The specification associated to a diagram coincides with the specification given by the standard semantics. More precisely, the colimit of a diagram representing a specification is isomorphic to this specification. The calculus of diagrams is also *complete*, in the sense that two isomorphic specifications in Term $(\mathcal{C}_0)$ correspond to isomorphic diagrams in diagr $(\mathcal{C}_0)$. This comes from the fact that $\mathcal{D}[\![]\!]$ is well defined, because diagr $(\mathcal{C}_0)$ is finitely cocomplete.

Let us compute the diagram associated to the specification R_3 of section 3.

$$P_3' = \mathsf{push}\,(S, M, G, bos, mobos)$$
$$R_3 = \mathsf{push}\,(\mathsf{push}\,(\emptyset, B, B, \mathsf{j}_B, \mathsf{j}_B), D, P_3', \mathsf{up}\,(\emptyset, B, B, D, \mathsf{j}_B, \mathsf{j}_B, m_*, m_+),$$
$$\mathsf{up}\,(\emptyset, B, B, P_3', \mathsf{j}_B, \mathsf{j}_B, \&_1(P_3') \circ b, \&_2(P_3') \circ mob))$$

$$\mathcal{D}[\![P_3']\!] = push(\mathcal{D}[\![S]\!], \mathcal{D}[\![M]\!], \mathcal{D}[\![G]\!], \mathcal{D}[\![b]\!] \circ \mathcal{D}[\![s]\!], \mathcal{D}[\![m]\!] \circ \mathcal{D}[\![b]\!] \circ \mathcal{D}[\![s]\!])$$
$$\mathcal{D}[\![S]\!] = [I(S)] = \boxed{\bullet S} \qquad \mathcal{D}[\![M]\!] = \boxed{\bullet M} \qquad \mathcal{D}[\![G]\!] = \boxed{\bullet G}$$
$$\mathcal{D}[\![b]\!] \circ \mathcal{D}[\![s]\!] = \boxed{S\bullet} \xrightarrow{bos} \boxed{\bullet M} \qquad \mathcal{D}[\![m]\!] \circ \mathcal{D}[\![b]\!] \circ \mathcal{D}[\![s]\!] = \boxed{S\bullet} \xrightarrow{mobos} \boxed{\bullet G}$$

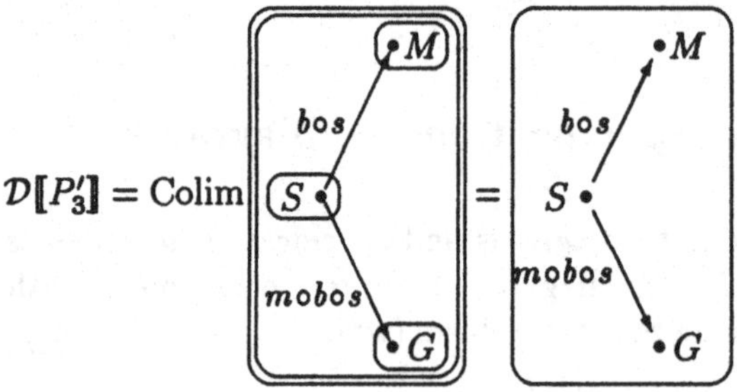

$$\mathcal{D}[\![\emptyset]\!] = \bigcirc$$
$$\mathcal{D}[\![B]\!] = \boxed{\bullet B}$$
$$\mathcal{D}[\![\mathsf{j}_B]\!] = \bigcirc \rightarrow \boxed{\bullet B}$$
$$\mathcal{D}[\![D]\!] = \boxed{\bullet D}$$
$$\mathcal{D}[\![m_*]\!] = \boxed{B\bullet} \xrightarrow{m_*} \boxed{\bullet D}$$
$$\mathcal{D}[\![m_+]\!] = \boxed{B\bullet} \xrightarrow{m_+} \boxed{\bullet D}$$

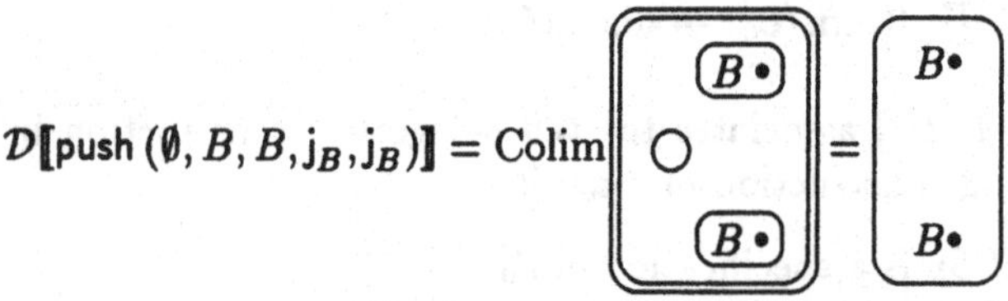

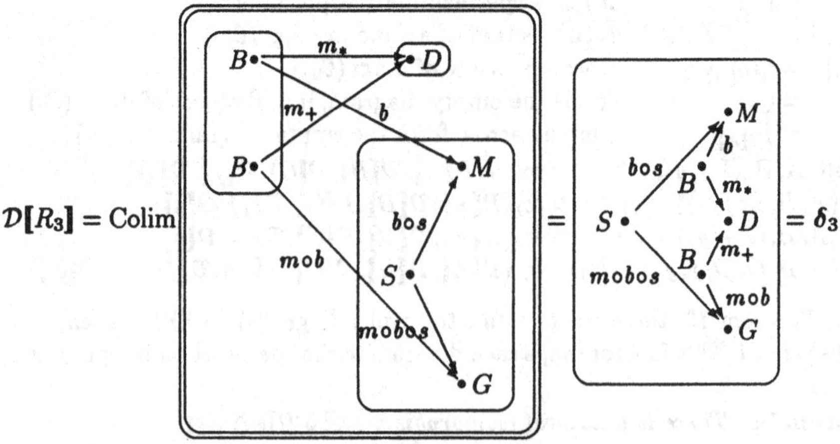

6 Isomorphisms of Diagrams

We have seen how to associate specifications to diagrams. Now the problem is to detect isomorphisms in diagr ($\mathcal{C}_0$). In this section, we briefly describe an algorithm allowing to detect isomorphisms of diagrams in the restricted case when the base category $\mathcal{C}_0$ is finite and has no cycle. By "having no cycle" we mean here that any arrow from an object A to itself is the identity.

We make this restriction because we have not solved the problem in the general case. This restriction is compatible with the LPG language, because the definition of cycling morphisms is syntactically forbidden.

In order to simplify the presentation, we also suppose that there is no isomorphism between objects in the category $\mathcal{C}_0$. So in $\mathcal{C}_0$, $A \cong B \Rightarrow A = B$. The algorithm is actually not much more complicated without this restriction.

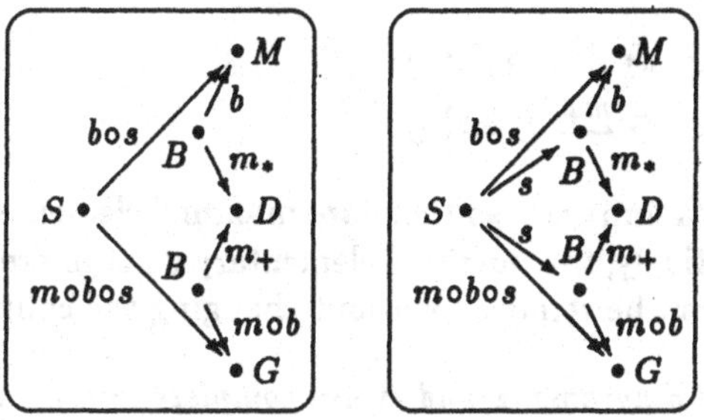

δ_3 δ_3'

Fig. 4. the completion of diagram δ_3 is δ_3'

The algorithm can be described in 3 steps:

1 Complete the diagram

For each pair of arrows $f : A \to B$ and $g : C \to B$, if there exists an arrow $h : A \to C$ such that $g \circ h = f$ in $\mathcal{C}_0$, we add the arrow h to the diagram. If h is the identity function id_A, then we merge both vertices labeled by A.
For instance completing the diagram δ_3 gives the diagram δ_3' (Fig. 4). The diagrams δ_1, δ_2 and α are already complete.

2 Match the "terminating vertices"

A "terminating vertex" is a vertex where no edge starts from.

Lemma 15. *If two diagrams α and β are complete and isomorphic, then for every terminating vertex m of α, there exists a terminating vertex n of β such that $\alpha(m) \cong \beta(n)$.*

Proof. use the fact that the diagrams are complete, and $\mathcal{C}_0$ has no cycle.

We can check that the diagrams δ_1, δ_2, and δ_3' have three terminating vertices, labeled by M, D and G.

3 Match the "elementary zigzags"
An elementary zigzag of a diagram α is a zigzag $n' \xleftarrow{f} n \xrightarrow{g} n''$ of α, where f and g may be compositions of arrows of $\mathcal{C}_0$, and $f \neq g$.

Definition 16 (Ordering on elementary zigzags). Let α be a complete diagram. We define an ordering on elementary zigzags of α as follows.
Let $Z = n'_0 \xleftarrow{u} n \xrightarrow{v} n'_k$ be an elementary zigzag of α. Let Z' be a zigzag of α, composed of the elementary zigzags $Z_1, ...Z_k$, with $Z'_i = n'_{i-1} \longleftarrow n_i \longrightarrow n'_i$.
- $Z < Z'_i$ iff ($u \sim_\alpha v$ $[Z']$ and $\forall i, n \neq n_i$).
- $Z \leq Z'_i$ iff ($Z = Z'$ or $Z < Z'$)

To prove that $\leq$ is indeed an ordering, one has to use the fact that $\mathcal{C}_0$ has no cycle, and that the vertices linked by an identity arrow have been merged in the diagram α.

For instance, in δ'_3, (and in δ_2 as well)

$$M \xleftarrow{bos} S \xrightarrow{mobos} G \leq M \xleftarrow{b} B \xrightarrow{m_*} D$$
$$\leq D \xleftarrow{m_+} B \xrightarrow{mob} G$$

$\leq$ is an ordering on a finite set, so there are maximal elements (maximal elementary zigzags). Intuitively, the maximal elementary zigzags are those which really count. The others can be removed without changing the colimit of the diagram.

Lemma 17. *If two diagrams α and β are complete and isomorphic, then there exists an isomorphism from α to β which associates every maximal elementary zigzag of α to a maximal elementary zigzag of β.*

Finally, two diagrams are isomorphic if they have the same terminating vertices, linked by the same maximal elementary zigzags. In particular, δ_1, δ_2 and δ'_3 are isomorphic.

7 Conclusion

We proposed to study "construction isomorphisms" of modular specifications. A "construction isomorphism" is an isomorphism which comes from general properties of colimits. We think this isomorphism is interesting, because it relies on the construction of a modular specification, without depending on the specific steps chosen for the construction. We showed in this paper that these isomorphisms of modular specifications correspond to isomorphisms of diagrams in the category diagr $(\mathcal{C}_0)$. The category diagr $(\mathcal{C}_0)$ is a completion with finite colimits of the category $\mathcal{C}_0$ of base specifications. In particular, diagr $(\mathcal{C}_0)$ is finitely cocomplete, even if $\mathcal{C}_0$ is not finitely cocomplete. We showed how specifications can be associated to diagrams. We gave an algorithm to detect isomorphisms in diagr $(\mathcal{C}_0)$, which strongly relies on the assumption that $\mathcal{C}_0$ has no cycle. We have not solved the problem in the general case, i.e. if there are cycles in $\mathcal{C}_0$

(which may introduce cycles in the diagram while completing it). We think this problem is much more difficult, because the number of arrows to consider may be infinite. However, in the case of algebraic specification, arrows are specification morphisms between finite signatures, so the number of arrows between two specifications remains finite, which suggests that isomorphisms should still be detectable.

References

1. D. Bert and R. Echahed. Design and implementation of a generic, logic and functional programming language. In *Proceedings of ESOP'86*, number 213 in LNCS, pages 119–132. Springer-Verlag, 1986.
2. M. Bidoit. The stratified loose approach: A generalization of initial and loose semantics. Technical Report 402, Université d'Orsay, France, 1988.
3. R.M. Burstall and J.A. Goguen. Putting theories together to make specifications. In *Int. Conf. Artificial Intelligence*, 1977.
4. R.M. Burstall and J.A. Goguen. The semantics of CLEAR, a specification language. In *Proc. Advanced Course on Abstract Software Specification*, number 86 in LNCS, pages 292–332. Springer-Verlag, 1980.
5. J. Cartmell. Generalized algebraic theories and contextual categories. *Annals of Pure and Applied Logic*, 32:209–243, 1986.
6. H. Ehrig and B. Mahr. *Fundamentals of algebraic specification 1. Equations and initial semantics*, volume 6 of *EATCS Monographs on Theoretical Computer Science*. Springer-Verlag, 1985.
7. K. Futatsugi, J.A. Goguen, J.-P. Jouannaud, and J. Meseguer. Principles of OBJ2. In *Proc. Principles of Programming Languages*, pages 52–66, 1985.
8. M.-C. Gaudel. A first introduction to PLUSS. Technical report, Université d'Orsay, France, 1984.
9. S. Mac Lane. *Categories for the Working Mathematician*. Springer-Verlag, 1971.
10. J.-C. Reynaud. Putting algebraic components together: A dependent type approach. Research Report 810 I IMAG, LIFIA, Apr. 1990.
11. J.-C. Reynaud. Putting algebraic components together: A dependent type approach. Number 429 in LNCS. Springer-Verlag, 1990.
12. J.-C. Reynaud. Isomorphism of typed algebraic specifications. Internal Report, LGI-IMAG, Feb. 1993.
13. A. Tarlecki, R.M. Burstall, and J.A. Goguen. Some fundamental algebraic tools for the semantics of computation: Part 3. indexed categories. *Theoretical Computer Science*, 91:239–264, 1991.
14. M. Wirsing. Structured Algebraic Specifications: A Kernel Language. *Theoretical Computer Science*, 42:123–249, 1986.

Higher-Order Narrowing with Convergent Systems

Christian Prehofer*

Technische Universität München**

Abstract. Higher-order narrowing is a general method for higher-order equational reasoning and serves for instance as the foundation for the integration of functional and logic programming. We present several refinements of higher-order lazy narrowing for convergent (terminating and confluent) term rewrite systems and their application to program transformation. The improvements of narrowing include a restriction of narrowing at variables, generalizing the first-order case. Furthermore, functional evaluation via normalization is shown to be complete and a partial answer to the eager variable elimination problem is presented.

1 Introduction and Overview

Higher-order narrowing is a method for solving higher-order equations modulo a set of rewrite rules. It forms the basis of functional-logic programming and has been extensively studied in the first-order case, for a survey see [10]. Motivated by functional programming, there exist several higher-order extensions for such languages [7, 18, 32]. Even more expressive than the latter is the language Escher, proposed in [17]. Higher-order narrowing [29] can be used as an operational semantics for such languages. The basis for narrowing are higher-order rewrite rules. Examples are the function *map* with

$$map(F, [X|R]) \rightarrow [F(X)|map(F, R)]$$

or a rule for pushing quantifiers inside:

$$\forall x.P \wedge Q(x) \rightarrow P \wedge \forall x.Q(x)$$

In the latter example the quantifier $\forall$ is a constant of type $(term \rightarrow bool) \rightarrow bool$, where $\forall(\lambda x.P)$ is written as $\forall x.P$ for brevity. For more examples on higher-order rewriting, we refer to [30], to [24] for formalizing logics and λ-calculi, and for Process Algebras to [27].

With higher-order narrowing we solve higher-order unification problems modulo such rewrite rules. Compared to the first-order case, also values for functional

* Research supported by the DFG under grant Br 887/4-2, *Deduktive Programmentwicklung* and by ESPRIT WG 6028, *CCL*.

** Full Address: Fakultät für Informatik, Technische Universität München, 80290 München, Germany. E-mail: prehofer@informatik.tu-muenchen.de

variables have to be computed via higher-order unification. To show the expressiveness of this method, we give an example for program transformation.

The framework for higher-order narrowing in [29] serves as a basis for the refinements of lazy narrowing we present here. For convergent higher-order rewrite systems, we show several techniques that use the determinism of convergent systems. The main contributions are as follows:

- We disallow narrowing at variable positions, generalizing the first-order case, as it is possible to restrict attention to R-normalized solutions. This is the gist of (first-order) narrowing, since narrowing into variables is undesirable.
- Simplification of equational goals via rewriting is shown to be complete. This is an important refinement as it performs deterministic evaluation without any search.
- Completeness of eager variable elimination (see below) is an open problem even for the first-order case [33]. By using oriented goals, this can be partially solved.
- Several deterministic operations for constructors, i.e. uninterpreted symbols, are shown.

Notice that the third item is also new for the first-order case. Another partial solution to this problem has been presented recently [20]. The significance of the other contributions has been argued in the first-order case in several papers, for references see [10].

Eager variable elimination means to solve a goal $X =^? t$ by binding X to t, without considering alternative rules. The result for eager variable elimination is based on a simpler notion of goals to be solved: we consider *oriented* equational goals of the form $s \rightarrow^? t$, where a substitution θ is a solution if $\theta s \xrightarrow{*} \theta t$. We show that for goals of the form $X \rightarrow^? t$, elimination is complete. We adopt this simpler operational model, which also eases technicalities, with no loss of expressiveness.

The higher-order case is more subtle in many respects. One of the typical technical problems is that higher-order substitutions and reducibility wrt a rewrite system R are harder to relate. For instance, if θt is R-normalized, then neither θ nor t must be R-normalized, which is the basis for first-order narrowing. The solution is to use patterns, a restricted class of λ-terms, for the left-hand sides of rules. This is no limitation in practice and allows to argue similar to the first-order case when needed.

The paper is organized as follows. Section 3 introduces a calculus for higher-order narrowing that utilizeses normalized solutions. This is followed by an analysis of deterministic operations for constructors in Section 4 and deterministic variable elimination in Section 5. Narrowing with simplification is the subject of Section 6. An application to program transformation is shown in Section 7.

2 Preliminaries

We briefly introduce simply typed λ-calculus (see e.g. [12]). We assume the following **variable conventions**:

- F, G, H, P, X, Y denote free variables,
- a, b, c, f, g (function) constants, and
- x, y, z bound variables.

Type judgments are written as $t : \tau$. Further, we often use s and t for terms and u, v, w for constants or bound variables. The set of types $\mathcal{T}$ for the simply typed λ-terms is generated by a set $\mathcal{T}_0$ of base types (e.g. int, bool) and the function type constructor $\rightarrow$. The syntax for **λ-terms** is given by

$$t \;=\; F \mid x \mid c \mid \lambda x.t \mid (t_1\, t_2)$$

A list of syntactic objects $s_1, \ldots, s_n$ where $n \geq 0$ is abbreviated by $\overline{s_n}$. For instance, n-fold abstraction and application are written as $\lambda\overline{x_n}.s = \lambda x_1 \ldots \lambda x_n.s$ and $a(\overline{s_n}) = ((\cdots(a\ s_1)\cdots)\ s_n)$, respectively.

Substitutions are finite mappings from variables to terms and are denoted by $\{\overline{X_n \mapsto t_n}\}$. Free and bound variables of a term t will be denoted as $\mathcal{FV}(t)$ and $\mathcal{BV}(t)$, respectively. The **conversions in λ-calculus** are defined as:

- **α-conversion:** $\lambda x.t =_\alpha \lambda y.(\{x \mapsto y\}t)$,
- **β-conversion:** $(\lambda x.s)t =_\beta \{x \mapsto t\}s$, and
- **η-conversion:** if $x \notin \mathcal{FV}(t)$, then $\lambda x.(tx) =_\eta t$.

For β-conversion (η-conversion), applying the rule from left to right is called β-reduction (η-reduction), and expansion in the other direction. A term is in $\beta\eta$-normal form if no β- or η-reductions apply, and η-expanded if no η-expansion applies. The **long $\beta\eta$-normal form** of a term t, denoted by $\updownarrow_\beta^\eta t$, is the η-expanded form of the $\beta\eta$-normal form of t. It is well known [12] that $s =_{\alpha\beta\eta} t$ iff $\updownarrow_\beta^\eta s =_\alpha \updownarrow_\beta^\eta t$. As long $\beta\eta$-normal forms exist for typed λ-terms, we will in general assume that terms are in long $\beta\eta$-normal form. For brevity, we may write variables in η-normal form, e.g. X instead of $\lambda\overline{x_n}.X(\overline{x_n})$. We assume that the transformation into long $\beta\eta$-normal form is an implicit operation, e.g. when applying a substitution to a term.

The convention that α-equivalent terms are identified and that free and bound variables are kept disjoint (see also [2]) is used in the following. Furthermore, we assume that bound variables with different binders have different names. Define $\mathcal{D}om(\theta) = \{X \mid \theta X \neq X\}$ and $\mathcal{R}ng(\theta) = \bigcup_{X \in \mathcal{D}om(\theta)} \mathcal{FV}(\theta X)$. Two **substitutions are equal on a set of variables W**, written as $\theta =_W \theta'$, if $\theta\alpha = \theta'\alpha$ for all $\alpha \in W$. A substitution θ **is idempotent** iff $\theta = \theta\theta$. We will in general assume that substitutions are idempotent. A substitution θ' is more general than θ, written as $\theta' \leq \theta$, if $\theta = \sigma\theta'$ for some substitution σ.

We describe positions in λ-terms by sequences over natural numbers. The subterm at a **position** p in a λ-term t is denoted by $t|_p$. A term t with the subterm at position p replaced by s is written as $t[s]_p$.

A term t in β-normal form is called a **(higher-order) pattern** if every free occurrence of a variable F is in a subterm $F(\overline{u_n})$ of t such that the $\overline{u_n}$ are η-equivalent to a list of distinct bound variables. Unification of patterns is decidable and a most general unifier exists if they are unifiable [21]. Also, the unification of a linear pattern with a second-order term is decidable and finitary, if they are variable-disjoint [28].

Examples of higher-order patterns are $\lambda x, y.F(x, y)$ and $\lambda x.f(G(\lambda z.x(z)))$, where the latter is at least third-order. Non-patterns are for instance $\lambda x, y.F(a, y)$ and $\lambda x.G(H(x))$.

2.1 Higher-Order Rewriting

The following definitions for higher-order rewriting are in the lines of [24, 19].

Definition 1. A **rewrite rule** is a pair $l \rightarrow r$ such that l is a pattern but not η-equivalent to a free variable, l and r are long $\beta\eta$-normal forms of the same base type, and $\mathcal{FV}(l) \supseteq \mathcal{FV}(r)$. A **Higher-Order Rewrite System (HRS)** is a set of rewrite rules. The letter R always denotes an HRS. Assuming a rule $(l \rightarrow r) \in R$ and a position p in a term s in long $\beta\eta$-normal form, a **rewrite step** from s to t is defined as

$$s \longrightarrow_{p,\theta}^{l \rightarrow r} t \;\Leftrightarrow\; s|_p = \theta l \wedge t = s[\theta r]_p.$$

For instance, with the quantifier rule of the first section, we have the following rewrite step:

$$\forall y.\forall x.p(y) \wedge q(x, y) \longrightarrow^{\forall x.P \wedge Q(x) \rightarrow P \wedge \forall x.Q(x)} \forall y.p(y) \wedge \forall x.q(x, y)$$

For a rewrite step we often omit some of the parameters $l \rightarrow r, p$ and θ. We assume that constants symbols are divided into free **constructor symbols** and defined symbols. A symbol f is called a **defined symbol**, if a rule $f(\ldots) \longrightarrow t$ exists. Constructor symbols are denoted by c and d. A term is in **R-normal form** if no rule from R applies and a substitution θ is **R-normalized** if if all terms in the image of θ are in R-normal form.

In contrast to the first-order notion of term rewriting, $\longrightarrow$ is not stable under substitution: reducibility of s does not imply reducibility of θs. Its transitive reflexive closure is however stable [19]:

Lemma 2. *Assume an GHRS R. If $s \stackrel{*}{\longrightarrow}^R t$, then $\theta s \stackrel{*}{\longrightarrow}^R \theta t$.*

A reduction is called **confluent**, if any two reductions from a term t are joinable, i.e. if $t \stackrel{*}{\longrightarrow} u$ and $t \stackrel{*}{\longrightarrow} v$ then there exists w with $u \stackrel{*}{\longrightarrow} w$ and $v \stackrel{*}{\longrightarrow} w$. For results on confluence of higher-order rewrite systems, we refer to [19]. A terminating and confluent reduction system is called **convergent**.

Termination orderings for higher-order rewriting can be found in [27, 16]. For our purpose, we need the following result, which can be shown similar to the first-order case [15]. A term $s = \lambda \overline{x_n}.s_0$ is a **subterm modulo binders** of $t = \lambda \overline{x_n}.t_0$, written as $s <_{sub} t$, if s_0 is a true subterm of t_0

Theorem 3. *The reduction $\longrightarrow_{sub}^{R} = \longrightarrow^{R} \cup >_{sub}$ is terminating for a GHRS R if $\longrightarrow^R$ is terminating.*[3]

[3] All missing proofs can be found in [30].

Notice that a subterm $s|_p$ may contain free variables which used to be bound in s. For rewriting it is possible to ignore this, as only matching of a left-hand side of a rewrite rule is needed. For narrowing, we need unification and hence we use the following construction to lift a rule into a binding context.

An $\overline{x_k}$**-lifter** of a term t **away from** W is a substitution $\sigma = \{F \mapsto (\rho F)(\overline{x_k}) \mid F \in \mathcal{FV}(t)\}$ where ρ is a renaming such that $\mathcal{D}om(\rho) = \mathcal{FV}(t)$, $\mathcal{R}ng(\rho) \cap W = \{\}$ and $\rho F : \tau_1 \to \cdots \to \tau_k \to \tau$ if $x_1 : \tau_1, \ldots, x_k : \tau_k$ and $F : \tau$. A term t (rewrite rule $l \to r$) is $\overline{x_k}$-lifted if an $\overline{x_k}$-lifter has been applied to t (l and r). For example, $\{G \mapsto G'(x)\}$ is an x-lifter of $g(G)$ away from any W not containing G'.

2.2 Higher-Order Unification

We introduce in the following the transformations for higher-order unification as in [34]. Although higher-order unification is undecidable in general, it performs remarkably well in systems such as λ-Prolog [22] and Isabelle [25]. For programming applications, there even exist decidable fragments [28, 30].

In contrast to first-order unification, we solve unification problems modulo the conversions of λ-calculus, i.e. θ is a unifier of $s =^? t$ if $\theta s =_{\alpha\beta\eta} \theta t$. We examine in the following the most involved case of higher-order unification: flex-rigid goals of the form $\lambda\overline{x_k}.F(\overline{t_n}) =^? \lambda\overline{x_k}.v(\overline{t'_m})$, where v is not a free variable. Clearly, for any solution θ to F the term $\theta F(\overline{t_n})$ must have (after β-reduction) the symbol v as its head. There are two possibilities:

- In the first case, v already occurs in (the solution to) some t_i. For instance, consider the equation $F(a) =^? a$, where $\{F \mapsto \lambda x.x\}$ is a solution based on a **projection**. In general, a projection binding for F is of the from $\{F \mapsto \lambda\overline{x_n}.x_i(\ldots)\}$. As some argument, here a, is carried to the head of the term, such a binding is called projection.
- In the second case, the head of the solution to F is just the desired symbol v. For instance, in the last example, an alternative solution is $\{F \mapsto \lambda x.a\}$. This is called **imitation.** Notice that imitation is not possible if v is a bound variable.

To solve a flex-rigid pair, the strategy is to guess an appropriate imitation or projection binding only for one rigid symbol, here a, and thus approximate the solution to F. Unification proceeds by iterating this process which focuses only on the outermost symbol. Roughly speaking, the rest of the solution for F is left open by introducing new variables.

Definition 4. Assume an equation $\lambda\overline{x_k}.F(\overline{t_n}) =^? \lambda\overline{x_k}.v(\overline{t'_m})$, where all terms are in long $\beta\eta$-normal form. An **imitation binding** for F is of the form

$$F \mapsto \lambda\overline{x_n}.f(\overline{H_m(\overline{x_n})})$$

where $\overline{H_m}$ are new variables of appropriate type. A **projection binding** for F is of the form

$$F \mapsto \lambda\overline{x_n}.x_i(\overline{H_p(\overline{x_n})})$$

where $\overline{H_p}$ are new variables with $\overline{H_p : \tau_p}$ and $x_i : \overline{\tau_p} \to \tau$. A **partial binding** is an imitation or a projection binding.

Notice that in the above definition, the bindings are not written in long $\beta\eta$-normal form. The long $\beta\eta$-normal form of an imitation or projection binding can be written as

$$F \mapsto \lambda\overline{x_n}.v(\overline{\lambda\overline{z_{j_p}}.H_p(\overline{x_n},\overline{z_{j_p}})}).$$

A full exhibition of the the types involved can be found in [34].

For lack of space, the transformation rules for higher-order unification are shown in Figure 1 together with the narrowing rules. The rules consist of the basic rules for unification, such as Deletion, Elimination and Decomposition plus the two rules explained above: Imitation and Projection. For the purpose of narrowing (to be detailed later), the rules work on oriented goals, which does not affect unification, and use subscripts (d), which only serve to improve narrowing.

It should be mentioned that the higher-order unification rules only perform so-called pre-unification. The idea of pre-unification is to handle **flex-flex pairs** as constraints and not to attempt to solve them explicitly. These are equations of the form $\lambda\overline{x_k}.P(\ldots) =^? \lambda\overline{x_k}.P'(\ldots)$. Huet [13] showed that for such pairs there may exist an infinite chain of unifiers, one more general than the other, without any most general one. Since flex-flex pairs are guaranteed to have at least one unifier, e.g. $\{P \mapsto \lambda\overline{x_m}.a, P' \mapsto \lambda\overline{x_n}.a\}$, pre-unification is sufficient.

3 Lazy Narrowing with Normalized Solutions

We introduce in this section higher-order lazy narrowing and refine it for R-normalized solutions. Consider a solution θ of an equational goal $s \rightarrow^? t$ with $\theta s \xrightarrow{*} \theta t$.[4] For any solution there exists an equivalent R-normalized one, assuming convergent rewrite systems. Hence it is a desirable restriction to consider only these. In the higher-order case, narrowing at (sub-)terms with variable heads such as $H(t)$ is needed [29]. The main improvement we discuss in this section is that narrowing is not needed at goals of the form $H(\overline{x_n}) \rightarrow^? t$ for normalized solutions, which covers many practical cases. The rules of System LNN for lazy higher-order narrowing, shown in Figure 1, consist of the rules for higher-order unification plus two narrowing rules; they are a refinement of System LN in [29].

Let $s \overset{?}{\leftrightarrow} t$ stand for one of $s \rightarrow^? t$ and $t \rightarrow^? s$. For a sequence $\Rightarrow^{\theta_1} \ldots \Rightarrow^{\theta_n}$ of LNN steps, we write $\overset{*}{\Rightarrow}{}^{\theta}$, where $\theta = \theta_n \ldots \theta_1$.

The subscripts (d) and d on goals only serve for a particular optimization and are not needed for soundness or completeness. The idea is to use **marked goals** $s \rightarrow^?_d t$. These are created only in the last rule, in order to avoid repeated application of Lazy Narrowing rules on these goals. The remaining rules work on both marked goals and unmarked goals, indicated by $\rightarrow^?_{(d)}$. For both $\overset{?}{\leftrightarrow}$ and $\rightarrow^?_{(d)}$ the rules are intended to preserve the orientation for $\overset{?}{\leftrightarrow}$ and marking for $\rightarrow^?_{(d)}$. Only the Decomposition rule and the Imitation rule, which includes decomposition,

[4] Although this corresponds only to equational matching, an equational unification problem $s =_R t$ can easily be encoded by adding a new rule $X =_R X \rightarrow true$ and solving the goal $s =_R t \rightarrow^? true$.

Deletion

$$\{t \rightarrow^?_{(d)} t\} \cup S \;\Rightarrow\; S$$

Decomposition

$$\{\lambda\overline{x_k}.f(\overline{t_n}) \rightarrow^?_{(d)} \lambda\overline{x_k}.f(\overline{t'_n})\} \cup S \;\Rightarrow\; \{\overline{\lambda\overline{x_k}.t_n \rightarrow^? \lambda\overline{x_k}.t'_n}\} \cup S$$

Elimination

$$\begin{array}{ll}\{F \overset{?}{\leftrightarrow}_{(d)} \lambda\overline{x_k}.t\} \cup S \Rightarrow^\theta & \theta S \text{ if } F \notin \mathcal{FV}(\lambda\overline{x_k}.t) \text{ and} \\ & \text{where } \theta = \{F \mapsto \lambda\overline{x_k}.t\}\end{array}$$

Imitation

$$\begin{array}{ll}\{\lambda\overline{x_k}.F(\overline{t_n}) \overset{?}{\leftrightarrow}_{(d)} \lambda\overline{x_k}.f(\overline{t'_m})\} \cup S \Rightarrow^\theta & \{\overline{\lambda\overline{x_k}.H_m(\overline{\theta t_n}) \overset{?}{\leftrightarrow} \lambda\overline{x_k}.\theta t'_m}\} \cup \theta S \\ & \text{where } \theta = \{F \mapsto \lambda\overline{x_n}.f(\overline{H_m(\overline{x_n})})\} \\ & \text{is an imitation binding with fresh variables}\end{array}$$

Projection

$$\begin{array}{ll}\{\lambda\overline{x_k}.F(\overline{t_n}) \overset{?}{\leftrightarrow}_{(d)} \lambda\overline{x_k}.v(\overline{t'_m})\} \cup S \Rightarrow^\theta & \{\lambda\overline{x_k}.\theta t_i(\overline{H_j(\overline{t_n})}) \overset{?}{\leftrightarrow}_{(d)} \lambda\overline{x_k}.v(\overline{\theta t'_m})\} \cup \theta S \\ & \text{where } \theta = \{F \mapsto \lambda\overline{x_n}.x_i(\overline{H_j(\overline{x_n})})\}, \\ & \text{is a projection binding with fresh variables}\end{array}$$

Lazy Narrowing with Decomposition

$$\begin{array}{ll}\{\lambda\overline{x_k}.f(\overline{t_n}) \rightarrow^? \lambda\overline{x_k}.t\} \cup S \;\Rightarrow & \{\overline{\lambda\overline{x_k}.t_n \rightarrow^? \lambda\overline{x_k}.l_n}\} \cup \\ & \{\lambda\overline{x_k}.r \rightarrow^? \lambda\overline{x_k}.t\} \cup S \\ & \text{where } f(\overline{l_n}) \rightarrow r \text{ is an } \overline{x_k}\text{-lifted rule}\end{array}$$

Lazy Narrowing at Variable

$$\begin{array}{ll}\{\lambda\overline{x_k}.H(\overline{t_n}) \rightarrow^? \lambda\overline{x_k}.t\} \cup S \;\Rightarrow & \{\lambda\overline{x_k}.H(\overline{t_n}) \rightarrow^?_d \lambda\overline{x_k}.l\} \cup \\ & \{\lambda\overline{x_k}.r \rightarrow^? \lambda\overline{x_k}.t\} \cup S \\ & \text{where } \overline{x_k}.H(\overline{t_n}) \text{ is not a pattern and} \\ & l \rightarrow r \text{ is an } \overline{x_k}\text{-lifted rule}\end{array}$$

Fig. 1. System LNN for Lazy Narrowing

transform marked goals to unmarked goals. In other words, on marked goals the Lazy Narrowing rules may only be applied after some decomposition took place.

Consider for instance the matching problem $\lambda x.H(f(x)) \rightarrow^? \lambda x.h(g(x), f(x))$, modulo the rule $f(f(X)) \rightarrow g(X)$. Here LNN yields

$$\{\lambda x.H_1(f(x)) \rightarrow^? \lambda x.g(x), \lambda x.H_2(f(x)) \rightarrow^? \lambda x.f(x)\}$$

by the imitation $\{H \mapsto \lambda y.h(H_1(y), H_2(y))\}$. Then the second goal can be solved by Projection, and the first by Lazy Narrowing to

$$\{\lambda x.H_1(f(x)) \longrightarrow^?_d \lambda x.f(f(X(x))), \lambda x.g(X(x)) \longrightarrow^? \lambda x.g(x)\}$$

Notice that the rewrite rule has been lifted over the variable x in the binding environment. As the first goal is marked, Lazy Narrowing does not re-apply. This is an important restriction and improves the similar system in [29], as otherwise infinite reductions occur, as in this case, very often. The two goals can be solved by several higher-order unification steps, which yield the solution

$$\{H_1 \mapsto \lambda y.f(y), X \mapsto \lambda x.x\}.$$

Observe that the last two rules of LNN can be integrated into one rule of the form

$$\{\lambda\overline{x_k}.s \longrightarrow^? \lambda\overline{x_k}.t\} \cup S \Rightarrow \{\lambda\overline{x_k}.s \longrightarrow^?_d \lambda\overline{x_k}.f(\overline{l_n}), \lambda\overline{x_k}.r \longrightarrow^? \lambda\overline{x_k}.t\} \cup S,$$

which is used in the completeness proofs and is called the **Lazy Narrowing** rule. From this rule, the narrowing rules of LNN can easily be derived, e.g. the first by decomposition on f.

Theorem 5. *If $s \longrightarrow^? t$ has solution θ, i.e. $\theta s \xrightarrow{*}{}^R \theta t$, and θ is R-normalized for a convergent HRS R, then $\{s \longrightarrow^? t\} \Rightarrow^{\delta}_{LNN} F$ such that δ is more general, modulo the newly added variables, than θ and F is a set of flex-flex goals.*

The proof proceeds as in the more general Theorem 9, which we show later. A key ingredient is the following lemma which generalizes the first-order case:

Lemma 6. *Assume an HRS R and a substitution θ. Then $\theta F(\overline{x_n})$ is R-reducible, iff θF is R-reducible.*

4 Deterministic Narrowing Rules for Constructors

In practice, rewrite systems often have a number of symbols, called constructors, that only serve as data structures. For constructor symbols, we give a few simple additional rules for Lazy Narrowing in Figure 2. Their main advantage is that their application is deterministic. The rules cover the cases where the root symbol of the left side of a goal is a constructor. Notice that the rules, except for the first, are only possible with oriented goals, where evaluation proceeds only from left to right. The correctness of the rules follows easily.

Deterministic Constructor Decomposition

$$\{\lambda\overline{x_k}.c(\overline{t_n}) \to^?_{(d)} \lambda\overline{x_k}.c(\overline{t'_n})\} \cup S \;\Rightarrow\; \{\overline{\lambda\overline{x_k}.t_n \to^? \lambda\overline{x_k}.t'_n}\} \cup S$$
if c is a constructor symbol

Deterministic Constructor Imitation

$$\{\lambda\overline{x_k}.c(\overline{t_n}) \to^?_{(d)} \lambda\overline{x_k}.F(\overline{x_m})\} \cup S \Rightarrow^{\theta} \{\overline{\lambda\overline{x_k}.t_n \to^? \lambda\overline{x_k}.H_n(\overline{x_m})}\} \cup \theta S$$
where $\theta = \{F \mapsto \lambda\overline{x_m}.f(\overline{H_n(\overline{x_m})})\}$
is an imitation binding with fresh variables

Constructor Clash

$$\{\lambda\overline{x_k}.c(\overline{t_n}) \to^?_{(d)} \lambda\overline{x_k}.v(\overline{t'_n})\} \cup S \;\Rightarrow\; \mathit{fail}$$
if $c \neq v$, where c is a constructor symbol
and v is not a free variable

Fig. 2. Deterministic Constructor Rules

5 Deterministic Variable Elimination

Eager variable elimination is a particular strategy of general E-unification systems. The idea is to apply the elimination rule as a deterministic operation whenever possible. It is an open problem of general (first-order) E-unification strategies if eager variable elimination is complete [33].

In our case, with oriented goals, we obtain more precise results by differentiating the orientation of the goal to be eliminated. We distinguish two cases of variable elimination, where in one case elimination is deterministic, i.e. no other rules have to be considered for completeness.

Theorem 7. *System LNN with eager variable elimination on goals of the form $X \to^? t$ with $X \notin \mathcal{FV}(t)$ is complete for a convergent HRS R.*

The main idea of the proof is that there can be no rewrite step in $\theta X \to^? \theta t$, thus we have $\theta X = \theta t$, assuming that θ is R-normalized.

In the general case, variable elimination may copy reducible terms with the result that the reductions have to be performed several times. Notice that this case of variable elimination does not affect the reductions in the solution considered, as only terms in normal form are copied: θt must be in normal form.

There are however a few important cases when elimination on goals of the form $t \to^? X$ is deterministic [30]: if t is either ground and in R-normal form or a pattern without defined symbols. Furthermore, for left-linear rewrite systems, elimination on goals of the form $t \to^? X$ is not needed, as shown in [30].

6 Lazy Narrowing with Simplification

Simplification by normalization of goals is one of the earliest [4] and one of the most important optimizations. Its motivation is to prefer deterministic reduction over search within narrowing. Notice that normalization coincides with deterministic evaluation in functional languages. For first-order systems, functional-logic programming with normalization has shown to be a more efficient control regime than pure logic programming [6, 9].

The main problem of normalization is that completeness of narrowing may be lost. For first-order (plain) narrowing, there exist several works dealing with completeness of normalization in combination with other strategies (for an overview see [10]). Recall from Section 4 that deterministic operations are possible as soon as the left-hand side of a goal has been simplified to a term with a constructor at its root. For instance, with the rule $f(1) \rightarrow 1$, we can simplify a goal $f(1) \rightarrow^? g(Y)$ by $\{f(1) \rightarrow^? g(Y), \ldots\} \Rightarrow \{1 \rightarrow^? g(Y), \ldots\}$ and deterministically detect a failure.

In the following, we show completeness of simplification for lazy narrowing. The result is similar to the corresponding result for the first-order case [11]. The technical treatment here is more involved in many respects due to the higher-order case. Using oriented goals, however, simplifies the completeness proof.

For oriented goals, normalization is only complete for goals $s \rightarrow^? t$, where θt is in R-normal form for a solution θ. For instance, it suffices if t is a ground term in R-normal form. For most applications, this is no real restriction and corresponds to the intuitive understanding of directed goals.

Definition 8. Normalizing Lazy Narrowing, called NLN, is defined as the rules of LNN plus arbitrary simplification steps on goals. A **simplification step** on a goal $s \rightarrow^? t$ is a rewrite step on s, written as $\{s \rightarrow^? t\} \Rightarrow_{NLN} \{s' \rightarrow^? t\}$ if $s \stackrel{R}{\longrightarrow} s'$.

We first need an auxiliary construct for the termination ordering in the completeness result. The **decomposition function** D is defined as

$$D(s \rightarrow^? t) = s \rightarrow^? t$$

$$D(\lambda \overline{x_k}.f(\overline{s_n}) \rightarrow^?_d \lambda \overline{x_k}.f(\overline{t_n})) = \overline{\lambda \overline{x_k}.s_n \rightarrow^? \lambda \overline{x_k}.t_n}$$

and is undefined otherwise. The function D extends component-wise to sets of goals. The idea of D is to view marked goals as goals with delayed decomposition.

Theorem 9 Completeness of NLN. *Assume a confluent HRS R that terminates with order $<^R$. If $s \rightarrow^? t$ has solution θ, i.e. $\theta s \stackrel{*}{\longrightarrow}{}^R \theta t$ where θt and θ are R-normalized, then $\{s \rightarrow^? t\} \stackrel{*}{\Rightarrow}{}^{\delta}_{NLN} F$ such that δ is more general modulo the newly added variables than θ and F is a set of flex-flex goals.*

Proof. Let $<^R_{sub} = <^R \cup <_{sub}$. Assume $\overline{G_n} = \overline{s_n \rightarrow^?_{(d)} t_n}$ is a system of goals with solution θ, i.e. $\overline{\theta s_n \stackrel{*}{\longrightarrow}{}^R \theta t_n}$. Let $\overline{s'_m \rightarrow^? t'_m} = D(\overline{G_n})$. The proof proceeds by induction on the following lexicographic termination order on $(\overline{G_n}, \theta)$:

- A: $<^R_{sub}$ extended to the multiset of $\{\overline{\theta s'_m}\}$,
- B: multiset of sizes of the bindings in θ,
- C: multiset of sizes of the goals $\overline{\theta G_n}$,
- D: $<^R$ extended to the multiset of $\{\overline{s_n}\}$.

By Theorem 3, item A is terminating. For the proof we need two invariants: first, all $\overline{t'_m}$ are R-normalized terms. Secondly, for marked goals $s \rightarrow^?_d t$, $Head(\theta s) = Head(\theta t)$ is not a free variable and furthermore, no rewrite step at root position occurs in $\theta s \xrightarrow{*}{}^R \theta t$. Except for the narrowing rule, it follows easily that the latter is invariant. E.g. Decomposition and Imitation on marked goals decompose the outermost symbol and yield unmarked goals.

In the following we show that normalization reduces this ordering and, furthermore, that for a non flex-flex goal some rule applies that reduces the ordering. In addition, we show in each of these cases that the above invariants are preserved. First, we select some non flex-flex goal $s \rightarrow^? t$ from $\overline{G_n}$; if none exists, the case is trivial.

We first consider the case where a simplification step is applied to an unmarked goal, i.e. $s \rightarrow^? t$ is transformed to $s' \rightarrow t$. We obtain $\theta s \xrightarrow{*} \theta s'$ from Lemma 2. As θt is in normal form, confluence of R yields $\theta s \xrightarrow{*} \theta s' \xrightarrow{*} \theta t$. Thus θ is a solution of $s' \rightarrow t$. For termination, we have two cases:

- If $\theta s = \theta s'$, measures A through C remain unchanged, whereas D decreases.
- If $\theta s \neq \theta s'$ measure A decreases.

Clearly, the invariants are preserved.

If no simplification is applied, we distinguish two cases: if $\theta s = \theta t$, then we proceed as in pure higher-order unification, as one of the rules of higher-order unification must apply. In case of the Deletion rule, measure A decreases. For Decomposition on marked goals, A and B remain unchanged, whereas C decreases. On unmarked goals, Decomposition reduces A. Imitation on marked goals does not change A, but reduces B; on unmarked goals, it reduces A. Projection only decreases B.

Normalization of the associated solution is preserved in these cases: In case of a Projection or Imitation, the partial binding computed maps a variable X to a higher-order pattern of the form $\lambda \overline{x_n}.v(\overline{H_m(\overline{x_n})})$. The new, intermediate solution constructed maps the newly introduced variables $\overline{H_m(\overline{x_n})}$ to subterms of θX, which are in R-normal form. Hence all $\overline{\theta H_m}$ must be in R-normal form. For the elimination rule, no new variables are introduced, thus the solution remains R-normalized.

Furthermore, the terms $\overline{\theta t'_m}$ do not change under Decomposition and Imitation on marked goals. On unmarked goals, Decomposition and Imitation yield new right hand sides t_n. These are subterms of θt and are thus R-normalized.

In the remaining case, there must be a rewrite step in $\theta s \xrightarrow{*} \theta t$. First, assume there is no rewrite step at the root position in $\theta s \xrightarrow{*} \theta t$. Hence all terms in this sequence have the same root symbol. Then similar to the last case, one of the unification rules must apply.

Now consider the case with rewrite steps in $\theta s \xrightarrow{*} \theta t$ at root position. Clearly, $s \rightarrow^? t$ cannot be marked. Further, s cannot be of the form $\lambda\overline{x_n}.X(\overline{y_m})$: with the invariant that θ' is R-normalized, it is clear that there can be no rewrite step in the solution θ of a goal $\lambda\overline{x_n}.X(\overline{y_m}) \rightarrow^? t$ as θX is in R-normal form.

Assume the first rewrite step in $\theta s \xrightarrow{*} \theta t$ is $\theta s \xrightarrow{*} \lambda\overline{y_k}.s_1 \longrightarrow_{\epsilon}^{l\rightarrow r} \lambda\overline{y_k}.t_1$, with the rule $l \rightarrow r$. Notice that $s_1 \rightarrow t_1$ must be an instance of $l \rightarrow r$ (modulo lifting).

We apply Lazy Narrowing (integrating the two lazy narrowing rules), yielding the subgoals:

$$s \rightarrow_d^? \lambda\overline{y_k}.l, \lambda\overline{y_k}.r \rightarrow^? t$$

As there exists δ such that $s_1 = \delta l$ and $t_1 = \delta r$, we can extend θ to the newly added variables: define $\theta' = \theta \cup \delta$. Let $\overline{s_m \rightarrow^? l_m} = D(\theta' s \rightarrow_d^? \lambda\overline{y_k}.\theta' l)$. Clearly, $s_i <_{sub}^R \theta' s$ holds, and $\theta' \lambda\overline{x_k}.r <_{sub}^R \theta' s$ follows from $\theta' s \xrightarrow{*} \theta' r$. Thus θ' is a solution of $\overline{s_m \rightarrow^? l_m}$ and $r \rightarrow^? t$ that coincides with θ on $\mathcal{FV}(\overline{G_n})$. It remains to show that θ' is in R-normal form. As the reduction is innermost, all $\theta\overline{l_m}$ are in R-normal form. As l is a pattern, this yields that θ' is R-normalized. Since we consider the first rewrite step a root position, the new marked goal $s \rightarrow_d^? \lambda\overline{x_k}.l$ fulfills the invariant, as $Head(\theta s) = Head(\theta l)$ and no rewrite step can occur at root position. □

The termination ordering in this proof is rather complex. For instance, the last item in the ordering is needed for the following example: assume a goal $\lambda x.c(F(x,t)) \rightarrow^? \lambda x.c(x)$ with solution $\theta = \{F \mapsto \lambda x, y.x\}$. Here, normalization of t does not change $\theta\lambda x.c(F(x,t))$.

7 An Example: Program Transformation

The utility of higher-order unification for program transformations has been shown nicely by Huet and Lang [14] and has been developed further in [26, 8]. The following models an example for unfold/fold program transformation in [5]. We assume the following rules for lists:

$$\begin{aligned} map(F,[X|R]) &\rightarrow [F(X)|map(F,R)] \\ foldl(G,[X|R]) &\rightarrow G(X, foldl(G,R)) \end{aligned}$$

Now assume writing a function $g(F,L)$ by

$$g(F,L) \rightarrow foldl(\lambda x, y.plus(x,y), map(F,L))$$

that first maps F onto a list and then adds the elements. This simple implementation for g is very inefficient, since the list must be traversed twice. The goal is now to find an equivalent function definition that is more efficient. We can specify this desired behavior in a syntactic fashion by one simple equation:

$$\lambda f, x, l.g(f,[x|l]) =^? B(f(x), g(f,l))$$

The variable B represents the body of the function to be computed. The schema on the right only allows recursing on l for g, indicated by the argument $g(f,l)$ to B, and similarly allows to use $f(x)$. Notice that the bound variables above can be viewed as $\forall$-quantified variables.

To solve this equation, we add a rule $X =^? X \rightarrow true$, where we view $=^?$ as a new (infix) constant and then apply narrowing, yielding the solution $\theta = \{B \mapsto \lambda fx, rec.plus(fx, rec)\}$ where

$$g(f, [x|l]) = \theta B(f(x), g(f,l)) = plus(f(x), g(f,l)).$$

This shows the more efficient definition of g. In this example, simplification can reduce the search space for narrowing drastically: it suffices to simplify the goal to

$$\lambda f, x, l.plus(f(x), foldl(plus, map(f,l))) =^? B(f(x), foldl(plus, map(f,l))),$$

where narrowing with the newly added rule $X =^? X \rightarrow true$ yields the two goals

$$\lambda f, x, l.plus(f(x), foldl(plus, map(f,l))) \rightarrow^? \lambda f, x, l.X(f,x,l),$$
$$\lambda f, x, l.B(f(x), foldl(plus, map(f,l))) \rightarrow^? \lambda f, x, l.X(f,x,l).$$

These can be solved by pure higher-order unification. Observe that simplification in this examples corresponds to (partial) evaluation.

8 Conclusions and Related Work

We have presented several refinements for narrowing, based on the determinism of reduction in convergent systems, in a highly expressive setting. The results apply to higher-order functional-logic programming, for which there exist several approaches and implementations [3, 7, 18, 32, 17] and to high-level reasoning, e.g. dealing with programs or mathematics [29]. Further development of higher-order narrowing towards functional-logic programming languages can be found in [30].

The work in [31] on higher-order narrowing considers only a restricted class of λ-terms, higher-order patterns with first-order equations, which does not suffice for modeling higher-order functional programs. The approach to higher-order narrowing in [1] aims at narrowing with higher-order functional programs, but restricts higher-order variables in the left-hand sides of rules and only permits first-order goals. These restrictions seem to be similar to the ones in [7].

Compared to higher-order logic programming [23], predicates and terms are not separated here. In the former, higher-order λ-terms are used for data structures and do not permit higher-order programming as in functional languages. For instance, the function *map* as in the last section cannot be written directly in higher-order logic programming.

References

1. J. Avenhaus and C. A. Loría-Sáenz. Higher-order conditional rewriting and narrowing. In Jean-Pierre Jouannaud, editor, *1st International Conference on Constraints in Computational Logics*, Lecture Notes in Computer Science, vol. 845, München, Germany, 7–9 September 1994. Springer-Verlag.
2. Hendrik Pieter Barendregt. *The Lambda Calculus, its Syntax and Semantics*. North Holland, 2nd edition, 1984.
3. P. G. Bosco and E. Giovannetti. IDEAL: An ideal deductive applicative language. In *Symposium on Logic Programming*, pages 89–95. IEEE Computer Society, The Computer Society Press, September 1986.
4. M. Fay. First order unification in equational theories. In *Proc. 4th Conf. on Automated Deduction*, pages 161–167. Academic Press, 1979.
5. Anthony J. Field and Peter G. Harrison. *Functional Programming*. Addison-Wesley, Wokingham, 1988.
6. L. Fribourg. SLOG: A logic programming language interpreter based on clausal superposition and rewriting. In *Symposium on Logic Programming*, pages 172–184. IEEE Computer Society, Technical Committee on Computer Languages, The Computer Society Press, July 1985.
7. J.C. González-Moreno, M.T. Hortalá-González, and M. Rodríguez-Artalejo. On the completeness of narrowing as the operational semantics of functional logic programming. In E. Börger, G. Jäger, H. Kleine Büning, S. Martini, and M.M. Richter, editors, *Computer Science Logic. Selected papers from CSL'92*, LNCS, pages 216–231, San Miniato, Italy, September 1992. Springer-Verlag.
8. John Hannan and Dale Miller. Uses of higher-order unification for implementing program transformers. In *Fifth International Logic Programming Conference*, pages 942–959, Seattle, Washington, August 1988. MIT Press.
9. M. Hanus. Improving control of logic programs by using functional logic languages. In *Proc. of the 4th International Symposium on Programming Language Implementation and Logic Programming*, pages 1–23. Springer LNCS 631, 1992.
10. M. Hanus. The integration of functions into logic programming: From theory to practice. *Journal of Logic Programming*, 19&20:583–628, 1994.
11. M. Hanus. Lazy unification with simplification. In *Proc. 5th European Symposium on Programming*, pages 272–286. Springer LNCS 788, 1994.
12. J.R. Hindley and Jonathan P. Seldin. *Introduction to Combinators and λ-Calculus*. Cambridge University Press, 1986.
13. Gérard Huet. *Résolution d'équations dans les languages d'ordre 1,2,...ω*. PhD thesis, University Paris-7, 1976.
14. Gérard Huet and Bernard Lang. Proving and applying program transformations expressed with second-order patterns. *Acta Informatica*, 11:31–55, 1978.
15. Jean-Pierre Jouannaud and Claude Kirchner. Completion of a set of rules modulo a set of equations. *SIAM Journal of Computing*, 15(4):1155–1194, 1986.
16. Stefan Kahrs. Towards a domain theory for terminatin proofs. In *International Conference on Rewriting Techniques and Applications, RTA*, 1995. To appear.
17. John Wylie Lloyd. Combining functional and logic programming languages. In *Proceedings of the 1994 International Logic Programming Symposium, ILPS'94*, 1994.
18. Hendrik C.R Lock. *The Implementation of Functional Logic Languages*. Oldenbourg Verlag, 1993.

19. Richard Mayr and Tobias Nipkow. Higher-order rewrite systems and their confluence. Technical report, Institut für Informatik, TU München, 1994.
20. Aart Middeldorp, Satoshi Okui, and Tetsuo Ida. Lazy narrowing: Strong completeness and eager variable elimination. In *Proceedings of the 20th Colloquium on Trees in Algebra and Programming*, Lecture Notes in Computer Science. Springer-Verlag, 1995. To appear.
21. Dale Miller. A logic programming language with lambda-abstraction, function variables, and simple unification. *J. Logic and Computation*, 1:497–536, 1991.
22. Gopalan Nadathur and Dale Miller. An overview of λ-Prolog. In Robert A. Kowalski and Kenneth A. Bowen, editors, *Proc. 5th Int. Logic Programming Conference*, pages 810–827. MIT Press, 1988.
23. Gopalan Nadathur and Dale Miller. Higher-order logic programming. Technical Report CS-1994-38, Department of Computer Science, Duke University, December 1994. To appear in *Volume 5* of *Handbook of Logic in Artificial Intelligence and Logic Programming*, D. Gabbay, C. Hogger and A. Robinson (eds.), Oxford University Press.
24. Tobias Nipkow. Higher-order critical pairs. In *Proceedings, Sixth Annual IEEE Symposium on Logic in Computer Science*, pages 342–349, Amsterdam, The Netherlands, 15–18 July 1991. IEEE Computer Society Press.
25. Lawrence C. Paulson. Isabelle: The next 700 theorem provers. In P. Odifreddi, editor, *Logic and Computer Science*, pages 361–385. Academic Press, 1990.
26. Frank Pfenning and Conal Elliott. Higher-order abstract syntax. In *Proc. SIGPLAN '88 Symp. Programming Language Design and Implementation*, pages 199–208. ACM Press, 1988.
27. Jaco van de Pol. Termination proofs for higher-order rewrite systems. In J. Heering, K. Meinke, B. Möller, and T. Nipkow, editors, *Higher-Order Algebra, Logic and Term Rewriting*, volume 816 of *Lect. Notes in Comp. Sci.*, pages 305–325. Springer-Verlag, 1994.
28. Christian Prehofer. Decidable higher-order unification problems. In *Automated Deduction — CADE-12*, LNAI 814. Springer-Verlag, 1994.
29. Christian Prehofer. Higher-order narrowing. In *Proceedings, Ninth Annual IEEE Symposium on Logic in Computer Science*, pages 507–516. IEEE Computer Society Press, 1994.
30. Christian Prehofer. *Solving Higher-order Equations: From Logic to Programming*. PhD thesis, TU München, 1995.
31. Zhenyu Qian. Higher-order equational logic programming. In *Proc. 21st ACM Symposium on Principles of Programming Languages*, Portland, 1994.
32. Yeh-Heng Sheng. HIFUNLOG: Logic programming with higher-order relational functions. In David H. D. Warren and Peter Szeredi, editors, *Proceedings of the Seventh International Conference on Logic Programming*, pages 529–545, Jerusalem, 1990. The MIT Press.
33. Wayne Snyder. *A Proof Theory for General Unification*. Birkäuser, Boston, 1991.
34. Wayne Snyder and Jean Gallier. Higher-order unification revisited: Complete sets of transformations. *J. Symbolic Computation*, 8:101–140, 1989.

Context-Free Event Domains are Recognizable *

Eric Badouel, Philippe Darondeau, and Jean-Claude Raoult

Irisa, Campus de Beaulieu, F-35042 Rennes Cedex, France
E-mail : {Eric.Badouel, Philippe.Darondeau, Jean-Claude.Raoult }@irisa.fr

Abstract. The possibly non distributive event domains which arise from Winskel's event structures with binary conflict are known to coincide with the domains of configurations of Stark's trace automata. We prove that whenever the transitive reduction of the order on finite elements in an event domain is a context-free graph in the sense of Müller and Schupp, that event domain may also be generated from a finite trace automaton, where both the set of states and the concurrent alphabet are finite. We show that the set of graph grammars which generate event domains is a recursive set. We obtain altogether an effective procedure which decides from an unlabelled graph grammar whether it generates an event domain and which constructs in that case a finite trace automaton recognizing that event domain. The advantage of trace automata over unlabelled graph grammars is to provide for a more concrete and therefore more tractable representation of event domains, well suited to an automated verification of their properties.

1 Introduction

This study arises from the junction of two related trends in concurrency theory, namely the comparison of semantic models and the concern for expressivity. The first trend, best represented in [NW93], aims at charting embeddings and equivalences between categories of models of concurrency, thus comparing their intrinsic power of representation. For instance, we showed in [BD93] that separated trace automata are equivalent with saturated trace nets and correspond at the level of domains of configurations with event structures with binary conflict. The resulting circuit between models preserves behaviours i.e. domains of configurations, and it also preserves finiteness. Such behaviour preserving correspondences between models help saving the effort of constructing explicitly several interpretations for a given programming language while remaining free to choose the most convenient model for analysing such and such semantic feature (e.g. nets for causality, automata for cycles and termination, etc.). The second trend, appeared in [deS84] and [BBK87a] and amplified in [Vaa92], aims at measuring the extension of one or several (fragments of) programming languages interpreted in a given model. Possible scales of measure are the polynomial and arithmetic hierarchies of sets or the Chomsky hierarchy of languages, applied

* This work was partly supported by the French P.R.C. *Modèles et Preuves*, by the H.C.M. Network *Express*, and by the Esprit B.R.A. *Asmics*.

to the objects of the models viewed as sets or languages. For example, it was shown in [deS84] that a combined use of unguarded recursion and of operational rules in de Simones's format yields enough power for assigning a finite expression to any recursively enumerable transition system considered up to strong bisimulation. We show in this paper that any event domain which is generated from a context-free graph grammar can also be generated from some finite trace automaton. In order to establish that result, we construct an effective (partial) mapping of unlabelled graph grammars into finite trace automata, preserving domains of configurations.

An event domain is the set of configurations of an event structure with binary conflict, ordered by set inclusion, and it is in particular a coherent and finitary Scott domain [Win80]. As such, an event domain is totally determined by the transitive reduction (i.e. Hasse diagram) of the order on its finite elements. An event domain is said to be context-free when that diagram may be generated from a context-free graph grammar or equivalently when a bounded number of connected components are left (up to isomorphism) after removing all nodes at depth less that some arbitrary constant [MS85]. We will show that the subfamily of unlabelled graph grammars generating event domains forms a recursive set. The mapping of that recursive set into the set of finite trace automata induces regular labellings on Hasse diagrams of context free event domains, where vertices are mapped to a finite set of states and arcs are mapped to a finite concurrent alphabet. The outcome is a more concrete and tractable presentation of event domains, well suited to an automated verification of their properties.

Let us add a few words about the background of this work. Context-free graphs, which may in general present cycles, have been studied at depth. For instance, they are known to coincide with the algebraic graphs of [Cou90] and with the pattern graphs of [Cau92]. On the other hand, recursively defined and effective domains have been thoroughly investigated [Smy77], but no attention has been paid so far to families of domains with a regular structure, such as the context-free event domains. The present work is a limited attempt in this direction. Context-free event domains correspond actually to a subclass of the context-free processes which are usually considered in the litterature on bisimulation [BBK87b]. In our opinion, the associated event structures provide a tractable alternative to the recursively defined event structures of [GL91].

The rest of the paper is organized as follows. Section 2 recalls the basic relationships between conflict event domains, event structures with binary conflict, and trace automata. Section 3 introduces a general definition of recognizable (but not necessarily context-free) event domains, and reports our early work towards their order theoretic characterization. We focus on context-free event domains in Section 4, which contains the main results. Section 5 concludes the paper. All the proofs may be found in the extended version of the paper, where we give in particular two alternative procedures deciding whether a graph grammar defines a conflict event domain. The first procedure relies on Müller and Schupp's theorem, which states the decidability of the monadic theory of context free graphs [MS85]. The second procedure is more explicit and relies solely on the computation of fixed points of monotone operators on finite lattices.

2 Trace Automata and Conflict Event Domains

This section recalls the definitions of event structures with binary conflict, conflict event domains, trace automata, and the basic theorems stating their tight relationships. All the material of the section is borrowed from the literature, especially from [Win80], [Win88], [Sta89a] and [Sta89b].

Definition 2.1 (Event Structures and their Configurations)
An event structure *is a triple* $(E, \#, \vdash)$ *where*

1. E *is a* countable *set of* events*;*
2. $\#$ *is a binary, symmetric and irreflexive relation on* E*, called the* conflict *relation;*
3. *let* Con *be the family of finite and conflict-free subsets of* E *then* $\vdash$ *is a subset of* $Con \times E$*, called the* enabling *relation, such that*
$$(X \vdash e \text{ and } X \subseteq Y \in Con) \Rightarrow Y \vdash e$$

A configuration *is a subset* X *of* E *which is*

1. conflict free*:* $e\#e' \Rightarrow (e \notin X \text{ or } e' \notin X)$ *and,*
2. secured*:* $\forall e \in X\ \exists e_0 \dots e_n \in X$ *such that* $e_n = e$ *and* $\{e_0, \dots e_{i-1}\} \vdash e_i$ *for* $i = 1, \dots, n$.

The set of configurations of an event structure ordered by inclusion form a CPO. It is ω-algebraic and satisfies the following axioms **F**, **C**, **R**, **V**. Here and in the sequel, we use the following notations:
$[x, y]$ or $x \prec y$ when $x < y$ and $(x \leq z \leq y \Rightarrow z = x \text{ or } z = y)$.
$[x, y] \prec [z, t]$ when $x = y \wedge z$ and $t = y \vee z$.
This relation generates the *projective order* $[x, y] \sqsubseteq [z, t]$ and the *projective equivalence* $[x, y] \equiv [z, t]$ Define a conflict event domain as a complete ω-algebraic partial order in which the finite elements satisfy the following four axioms:

(F) $\{y \mid y \leq x\}$ is finite
(C) $x \prec y,\ x \prec z,\ y \neq z,\ y \uparrow z \Rightarrow y \vee z$ exists & $y \prec (y \vee z)$ & $z \prec (y \vee z)$
(R) $[x, y] \equiv [x, y'] \Rightarrow y = y'$
(V) $[x, x'] \equiv [y, y']$ & $[x, x''] \equiv [y, y'']$ & $x' \uparrow x'' \Rightarrow y' \uparrow y''$

Then Winskel [Win80] proved the following representation theorem.

Theorem 2.2 (G. Winskel)
i) The set of configurations $D(E)$ *of an event structure* E*, ordered by inclusion, is a conflict event domain.*
ii) Conversely, let D *be a conflict event domain, then* $(E_D, \#_D, \vdash_D)$ *is an event structure with the following definitions.*

1. E_D *is the set of events (i.e. classes of projective prime intervals) of the event domain* D*,*
2. $a\#_D b$ *if and only if there exist* x*,* y *and* z *in* D *such that* $[x, y] \in a$*,* $[x, z] \in b$*, and* $\neg(y \uparrow z)$*,*

3. let $E(x)$ be the configuration representing the element x of D, i.e. $E(x) = \{[y, y']_\equiv \mid y' \le x\}$. Then $X \vdash_D a$ for a finite and conflict free set of events X if and only if there exist x and y in D such that $E(x) \subset X$ and $[x, y] \in a$

Moreover, E is an isomorphism from D onto $D(E_D)$.

In fact, one can prove (see for instance [Cur86]) that any two maximal chains between two finite elements are equivalent by permutations. Moreover, one can also prove (see [Cur86]) that axioms (F) and (C) imply that the order is consistently complete, thus a conflict event domain is a finitary Scott domain.

Let us now explain the relationship between conflict event domains and trace automata.

Definition 2.3 (Trace Automata) *An automaton $\mathcal{A} = (A, Q, q_0, T)$ consists of a* countable *set A of actions, a set Q of states with initial state $q_0 \in Q$, and a transition relation $T \subseteq Q \times A \times Q$. A* trace automaton *$\mathcal{A} = (A, \|, Q, q_0, T)$ is an automaton whose alphabet is equipped with a symmetric and irreflexive relation $\| \subset A \times A$, called the* independence relation, *such that the following conditions on transitions $p \xrightarrow{a} q \in T$ are satisfied:*

determinism: $q \xleftarrow{a} p \xrightarrow{a} r \Rightarrow q = r$

commutativity: $(r \xleftarrow{a} q \xrightarrow{b} s \text{ and } a \parallel b) \Rightarrow (r \xrightarrow{b} p \xleftarrow{a} s \text{ for some } p)$

Given a conflict event domain D, a labelled and acyclic transition system T_D may be derived from D by setting $x \xrightarrow{a} y$ in T_D when $[x, y] \in a$. Declare events a and b *independent*, written $a \parallel_D b$ when there exist x, y and z in D such that $[x, y] \in a$, $[x, z] \in b$, and $y \uparrow z$. It should be clear that T_D becomes in this way a trace automaton over the concurrent alphabet $(E_D, \|_D)$. Indeed, axiom (R) says exactly that T_D is deterministic, while axioms (C) and (V) entail commutativity.

Notation 2.4 *By an abuse of notation, we let $T_D = (T_D, \|_D)$ denote the trace automaton derived from the conflict event domain D.*

A trace automaton may be infinite (like T_D if D is infinite) and it may present cycles (unlike T_D). Two trace automata whether finite and cyclic, or infinite and acyclic, may nevertheless generate two isomorphic domains of configurations as defined below.

The *language* $\mathcal{L}(\mathcal{A})$ of an automaton $\mathcal{A}$ is the set $\{m \in A^* \mid (\exists q)\, q_0 \xrightarrow{m} q\}$ of words labelling paths from the initial state. The equivalence by permutations $\sim$ is the equivalence on $\mathcal{L}(\mathcal{A})$ generated by those pairs $(uabv, ubav)$ in $\mathcal{L}(\mathcal{A}) \times \mathcal{L}(\mathcal{A})$ such that $a \parallel b$ in $\mathcal{A}$. Let $\le$ be the prefix order, and $\lesssim\ = (\le \cup \sim)^*$ be the prefix preorder modulo permutations. The equivalence by permutations coincides with the equivalence induced by the preorder $m \sim n$ defined by $m \lesssim n$ and $n \lesssim m$; thus the following definition makes sense.

Definition 2.5 (Domain of Configurations of a Trace Automaton) *The set of* finite configurations *of a trace automaton $\mathcal{A}$ is $\mathcal{L}(\mathcal{A})/\sim$, and the* domain of configurations *$D(\mathcal{A})$ of $\mathcal{A}$ is the ideal completion of the ordered set $(\mathcal{L}(\mathcal{A})/\sim, \lesssim/\sim)$.*

Now $D \cong D(T_D)$ for any conflict event domain D. The explicit connection between conflict event domains and trace automata is stated by the following theorem.

Theorem 2.6 (E.W. Stark) *The domains of configurations of trace automata are the conflict event domains.*

Summing up theorems 2.2 and 2.6, we see that conflict event structures and trace automata are just alternative "syntactic" forms for event domains.

3 Recognizable Event Domains

This section presents the early steps of a general study aiming at graph-theoretic characterization of the recognizable event domains defined as follows.

Definition 3.1 (Recognizable Event Domains) *A conflict event domain is* recognizable *if it is isomorphic to the domain of configurations of some finite trace automaton.*

Figure 1 shows two recognizable event domains $D_1 \cong D(\mathcal{A}_1)$ and $D_2 \cong D(\mathcal{A}_2)$, where $\mathcal{A}_1$ and $\mathcal{A}_2$ are the two different trace automata with a single state q_0 and two transitions $q_0 \xrightarrow{a} q_0$ and $q_0 \xrightarrow{b} q_0$ which are obtained by setting $a \parallel b$ in $\mathcal{A}_1$ but not in $\mathcal{A}_2$. Given a (finite or infinite) trace automaton $\mathcal{A}$, and given an event

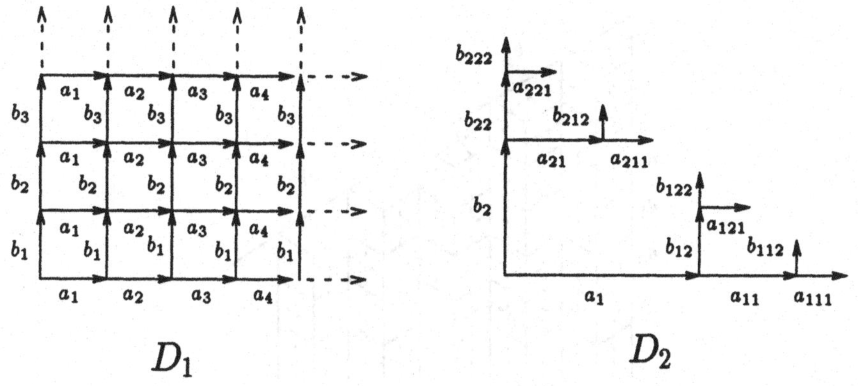

Fig. 1. two recognizable event domains

domain D inducing a trace automaton T_D as defined in section 2, the domain D is indeed isomorphic to the domain of configurations of the trace automaton $\mathcal{A}$ if and only if $\mathcal{A}$ is a *folding* of T_D according to the following definition, adapted from [BD93].

Definition 3.2 (Folding morphisms for Trace Automata) *A* folding morphism *$(\eta, \sigma) : \mathcal{A}_1 \rightarrow \mathcal{A}_2$ between trace automata is a pair of mappings $\eta : A_1 \rightarrow A_2$, and $\sigma : Q_1 \rightarrow Q_2$ between their respective sets of actions and states, satisfying conditions 1 to 4:*

1. *$\sigma(q_{0,1}) = q_{0,2}$,*
2. *$q \xrightarrow{a} q'$ in T_1 implies $\sigma(q) \xrightarrow{\eta(a)} \sigma(q')$ in T_2,*
3. *η restricts to a bijection between $\{a/(\exists r)\ q \xrightarrow{a} r\}$ and $\{b/(\exists r)\ \sigma(q) \xrightarrow{b} r\}$, at every $q \in Q_1$,*
4. *$q \xrightarrow{a} r$ and $q \xrightarrow{b} s$ in $\mathcal{A}_1$ imply $a \parallel b \Leftrightarrow \eta(a) \parallel \eta(b)$.*

$\mathcal{A}_2$ is said to be a folding *of $\mathcal{A}_1$ (notation: $\mathcal{A}_1 \geq \mathcal{A}_2$) if there exists some folding morphism (η, σ) from $\mathcal{A}_1$ to $\mathcal{A}_2$.*

Proposition 3.3 (E. Badouel & Ph. Darondeau) *Let $\mathcal{U}$ be the* unfolding *functor that maps a trace automaton $\mathcal{A}$ to the acyclic trace automaton $\mathcal{U}\mathcal{A} = T_{D(\mathcal{A})}$ derived from the domain of configurations of $\mathcal{A}$, then*

1. *$\mathcal{U}\mathcal{A} \geq \mathcal{A}$, and*
2. *$\mathcal{B} \geq \mathcal{A} \Rightarrow \mathcal{U}\mathcal{A} \cong \mathcal{U}\mathcal{B}$,*

where $\cong$ denotes isomorphism of trace automata.

Corollary 3.4 *A conflict event domain D is recognizable if and only if the derived trace automaton T_D admits a finite folding $T_D \geq \mathcal{A}$.*

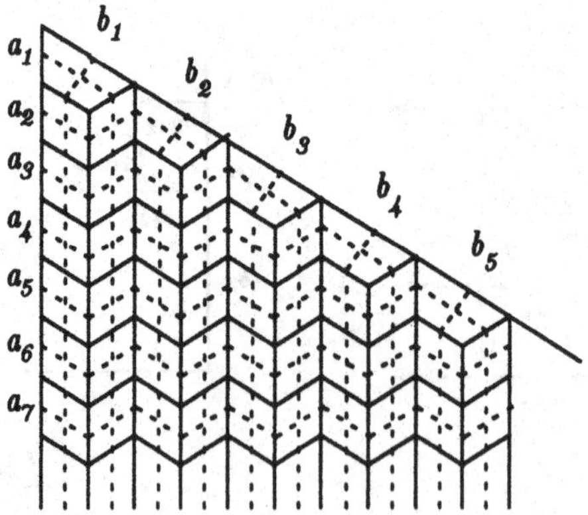

Fig. 2. a non recognizable event domain

This corollary entails the non-recognizability of the event domain D shown in Fig. 2. Indeed, any two events a_i and a_j $(i \neq j)$ are co-initial at some state and independent in T_D, hence from condition (4) in Def. 3.2, $\eta(a_i) \parallel \eta(a_j)$ and *a fortiori* $\eta(a_i) \neq \eta(a_j)$ for any folding morphism $(\eta, \sigma) : \mathcal{A} \rightarrow \mathcal{B}$, whereby

$\mathcal{B}$ must be infinite. Note however that the domain D has exactly five types of isomorphic types of ***residuals***, where the residual D_x is the upper restriction of D determined by x. We can now state two ***necessary*** conditions of recognizability for conflict event domains, inspired from the above, and propose a conjecture. First, a recognizable event domain D must have a finite number of types of isomorphic residuals, bounded by the number of states of any finite trace automaton $\mathcal{A}$ such that $D \cong D(\mathcal{A})$. Second, a recognizable event domain D must have bounded cliques for the least irreflexive and symmetric relation xRy on E_D containing $x \parallel y$ and $x\#y$ and $x \parallel z\#y$ for some event z co-initial with x and y at two different states in T_D. If $D \cong D(\mathcal{A})$, and since $x \parallel z\#y \Rightarrow \eta(x) \parallel \eta(z)$ but $\neg(\eta(z) \parallel \eta(y))$ by condition (4) in Def. 3.2, the size of the R-cliques must be smaller than the size of the alphabet of $\mathcal{A}$.

Conjecture 3.5 *A conflict event domain is recognizable if and only if it has simultaneously a finite number of non isomorphic residuals and bounded R-cliques.*

We give in Section 4 a complete answer to the question of recognizability for the sub-class of *context-free* conflict event domains.

4 Context-Free Event Domains

A context-free graph is a rooted graph of finite out-degree such that, by removing all vertices within some arbitrary distance from the root, one obtains a bounded number of connected components up to isomorphism. According to a result of Müller and Schupp [MS85], this characteristic property determines exactly the class of rooted graphs of finite out-degree which may be generated from deterministic graph grammars (see below for a precise definition). In this section, we focus our attention on context-free event domains, defined as follows.

Notation 4.1 *For any domain D, let H_D denote the Hasse diagram, or transitive reduction, of the order in D, restricted to the finite elements of D.*

Definition 4.2 (Context-Free Event Domains) *A conflict event domain D is* context-free *if H_D is a context-free graph.*

The purpose of this section is to provide a constructive proof for the following theorem.

Theorem 4.3 *Context-free event domains are recognizable.*

From a graph grammar for H_D, we shall deduce a finite trace automaton $\mathcal{A}$ such that $D(\mathcal{A}) \cong D$. The running example for the section is the context-free event domain sketched in Fig. 3 This dag may be seen as a thick binary tree in which branches are assembled in squares. Since zig-zags have length at most 6, the picture is "full of gaps", contrasting with the image of the event damain D_1 in Fig. 1, which is recognizable but not context-free.

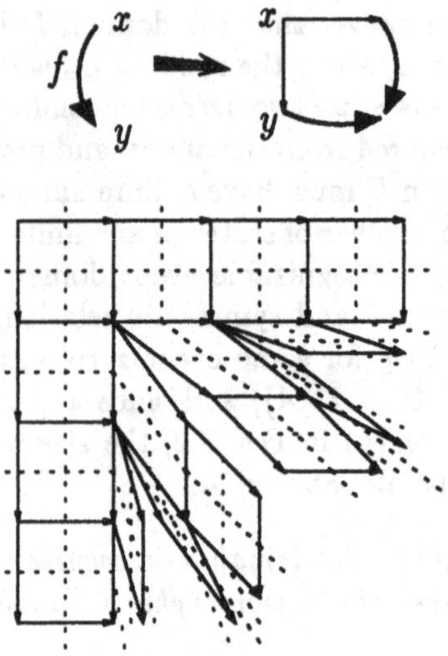

Fig. 3. a context-free event domain generated by a single replacement rule

4.1 Graph Grammars Generating Event Domains

In this subsection, we show that any graph grammar generating the transitive reduction of an event domain may be normalized to a *monotone* and *uniform* graph grammar, yielding a natural encoding of vertices and edges by words. The projective order on prime intervals is then encoded by a finite suffix-rewriting system on words. These various encodings are exploited in a second subsection, where trace automata are derived from monotone and uniform graph grammars.

Definition 4.4 (Graph Grammar) *A* hypergraph *on a graded alphabet* $F = \sum_{n>0} F_n$ *is a graph* (V, A)*, with set* V *of vertices and set* A *of arcs, equipped with a set of* hyperarcs $fx_1 \ldots x_n$ labelled *by* $f \in F_n$ *and connecting the vertices* $x_1, \ldots, x_n$ *in this order.* [2] *A* hyperarc replacement rule, *or simply a rule, is a production* $fx_1 \ldots x_n \to H_f$ *where* $f \in F_n$*, where* H_f *is a* finite *hypergraph on* F *and* $x_1, \ldots, x_n$ *are* distinct *vertices of* H_f*. A* graph grammar *on* F *is a finite set of rules.*

A graph grammar G is said to be *deterministic* if two rules never have the same label on the left-hand side. A deterministic graph grammar G defines a vector of graphs $(G^\omega(f))_{f \in F}$ obtained by infinitely unfolding the rules $fx_1 \ldots x_n \to H_f$ starting from the respective non terminal hyperarcs $fx_1 \ldots x_n$ See [Hab89] or [Cau92] for a formal definition.

[2] The arcs in A are hyperarcs as well — terminal hyperarcs for our grammars. We shall nevertheless call them arcs to tell them from the non-terminal hyperarcs.

Definition 4.5 (Context-Free Graph) *A* context-free *graph is a rooted graph of finite out-degree which is isomorphic to $G^\omega(f)$ for some deterministic graph grammar G and unary symbol f.*

Definition 4.6 (Monotone and Uniform Graph Grammar)
A graph grammar G is monotone *if the following four conditions on hypergraph H_f are satisfied for every production*
$fx_1 \dots x_n \rightarrow H_f$:

1. *the underlying graph $[H_f]$ of H_f is a directed acyclic graph (or dag),*
2. *$x_1, \dots, x_n$ are the minimal vertices of $[H_f]$, called the* input vertices *of H_f,*
3. *the vertices y_i connected by non terminal hyperarcs $gy_1 \dots y_m$ in H_f are maximal elements of $[H_f]$, called* ouput vertices *of H_f,*
4. *the input and ouput vertices of H_f are disjoint.*

A deterministic graph grammar G over F is uniform *if all graphs $G^\omega(f)$ for ($f \in F$) are connected and the following three conditions on the hypergraph H_f are satisfied for every rule $fx_1 \dots x_n \rightarrow H_f$, except for the possible case where H_f reduces to a single vertex:*

1. *all vertices of non terminal hyperarcs are shared with arcs, whereas one vertex can neither occur twice on a non terminal hyperarc nor be shared by two non terminal hyperarcs,*
2. *all vertices in the set $\{x_1 \dots x_n\}$ occur on arcs and they do not occur on non terminal hyperarcs,*
3. *at least one of the vertices $x_1 \dots x_n$ occurs on each arc.*

Proposition 4.7 (D. Caucal) *A rooted dag of finite degree containing no triangle is context-free if and only if it is isomorphic to $G^\omega(f)$ for some monotone and uniform grammar G and unary symbol f.*

Theorem 4.8 *Given a monotone and uniform graph grammar G over F and a unary non terminal symbol $f \in F$, one can decide whether the context-free graph $G^\omega(f)$ is the transitive reduction of the order on finite elements in some conflict free event domain.*

Let $G = \{fx_1 \dots x_n \rightarrow H_f;\ f \in F\}$ be a monotone and uniform graph grammar, generating the transitive reduction of the order on finite elements in an event domain D (i.e. $H_D \cong G^\omega(f)$). One may assume without loss of generality that every $g \in F$ occurs at most once on the right hand side of each production. Hence, unless H_f reduces to a single vertex, its underlying graph $[H_f]$ is the transitive reduction of a partial order of depth 1 with $a(f)$ minimal elements, and its set of maximal elements is partitioned by the non terminal hyperarcs. Because of this particular form, each production $fx_1 \dots x_n \rightarrow H_f$ may be encoded into a plain graph $G_f = (V_f, A_f)$, isomorphic to $[H_f]$. The set of vertices $V_f = I_f \cup O_f$ is divided into *input vertices* $I_f = \{(f,1), \dots, (f, a(f))\}$ and *output vertices* $O_f = \{(fg, i);\ g \in \mathrm{succ}(f)\ \&\ 1 \leq i \leq a(g)\}$, where $f \in F_{a(f)}$ and $g \in \mathrm{succ}(f)$ if g occurs in H_f. The set of arcs $A_f \subseteq I_f \times O_f$ is the set of pairs $((f,i),(fg,j))$ such that there exists an arc from x_i to the jth vertex of non terminal hyperarc

g in H_f. Thus, in particular, $V_f = \{(f,1)\}$ and $A_f = \emptyset$ for non terminal symbols f associated with productions of the special form $fx_1 \to x_1$.

Notation 4.9 *Let T be the set of F-words $f_1 \dots f_n$ such that $\forall i < n$ $f_{i+1} \in$ succ(f_i), and let T_f (respectively T^g) be the subset of words $f_1 \dots f_n$ in T such that $f_1 = f$ (resp. $f_n = g$). Finally, let $T_f^g = T_f \cap T^g$. The set T_f ordered by the prefix ordering may be seen as a deterministic tree, called the* parse-tree *of $G^\omega(f)$.*

Proposition 4.10 *$G^\omega(f) \cong \bigcup\{G_v \mid v \in T_f\}$ where $G_v = uG_g = (uV_g, uA_g)$ for every $v = ug \in T_f$, letting $u(w,i) = (uw,i)$ and $u((w,i),(w',i')) = ((uw,i),(uw',i'))$, and the union of graphs is defined componentwise:*

$$\bigcup_{i\in I}(V_i, A_i) = (\bigcup_{i\in I} V_i, \bigcup_{i\in I} A_i).$$

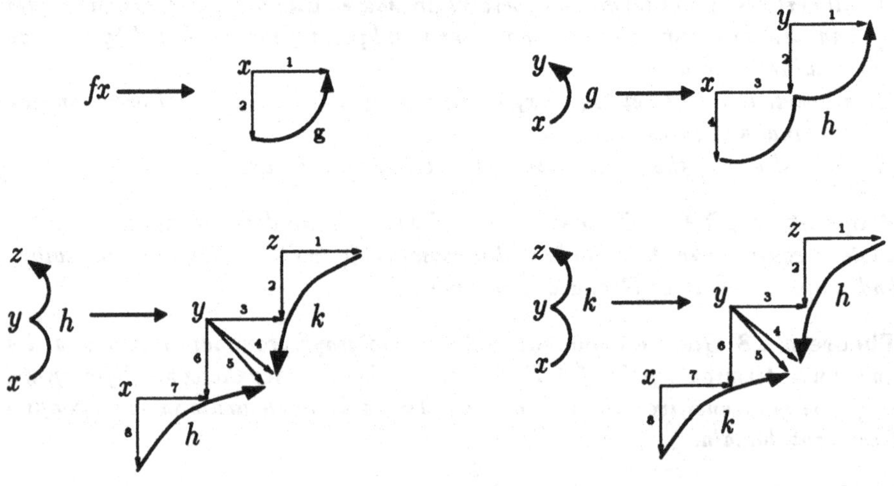

Fig. 4. a uniform graph grammar generating the event domain of Fig. 3

Henceforth, we adopt for $G^\omega(f)$ the definition $G^\omega(f) = \bigcup\{G_v \mid v \in T_f\}$. This variant definition, justified by Prop. 4.10, suggests simple notations for the vertices and arcs of $G^\omega(f)$. In fact, let $G^\omega(f) = (V_f^\omega, A_f^\omega)$. The vertices in V_f^ω are mapped bijectively to pairs (u,i) such that $u \in T_f^g$ and $1 \le i \le a(g)$. Fix for each $f \in F$ an enumeration of the arcs in H_f (as in Fig. 4); the arcs in A_f^ω are mapped bijectively to words ui such that $u \in T_f^g$ and $1 \le i \le n(g)$ where $n(g)$ is the number of arcs in A_g. Before we tackle the encoding of the projective order on prime intervals in H_D, let us introduce a few definitions which will be used in the second subsection.

Definition 4.11 (Sections) *For $v = ug \in T_f$, let $u : G^\omega(g) \hookrightarrow G^\omega(f)$ be the graph embedding $u(V_g^\omega, A_g^\omega) = (uV_g^\omega, uA_g^\omega)$. The set $\varphi(u) = \{(u,1), \ldots, (u,n)\}$ of minimal vertices of $uG^\omega(g)$ is called the section of $G^\omega(f)$ determined by $u \in T_f$ — thus, the set of sections of $G^\omega(f)$ defines a partition of V_f^ω.*

In section 2, events have been defined as equivalence classes of projective prime intervals in conflict event domains, and they have been equipped with two disjoint binary relations, called independence and conflict. All definitions and notations given in section 2 for partially ordered sets apply naturally to the transitive closure of $G^\omega(f)$. For instance, in Fig. 5, the prime intervals $fg1 \sqsubseteq fgh3 \sqsubseteq$

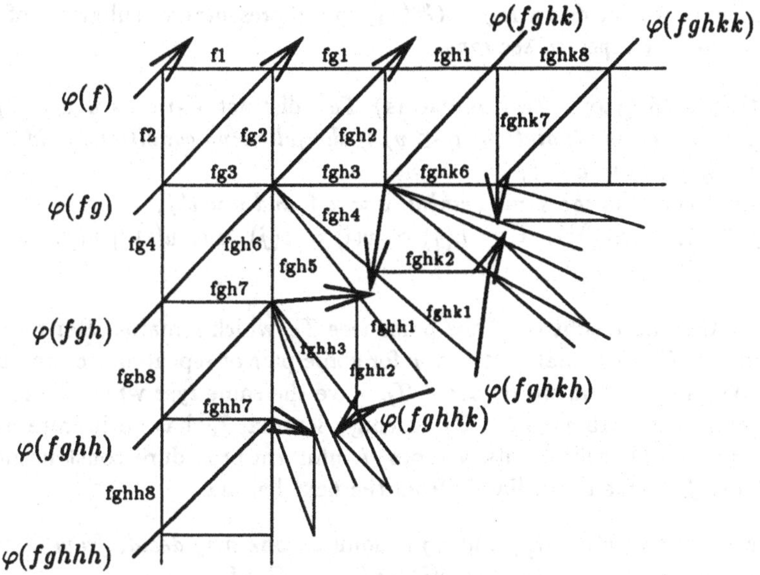

Fig. 5. sections in the domain of Fig. 3 induced by the graph grammar of Fig. 4

$fghk2$ are linearly ordered and belong to the same event. Now, with our encoding of prime intervals, the projective order may be represented by a finite suffix-rewriting system deduced from the graph grammar. To see it, observe that whenever $[x,y] \prec [x',y']$ in $G^\omega(f)$, x belongs to a section $\varphi(ug)$ such that, for some $h \in \mathbf{succ}(g)$, $[x,y] = ugi$, $[x,x'] = ugj$, $[y,y'] = ughj'$, and $[x',y'] = ughi'$ for some $i,j \in \{1,\ldots,n(g)\}$ and $i',j' \in \{1,\ldots,n(h)\}$.

Observation 4.12 *$wi \prec w'i'$ if and only if $w = ug$, $w' = ugh$ and $gi \prec ghi'$.*

Observation 4.13 *$ufi \sqsubseteq vgj$ if and only if $v = uw$ and $fi \xrightarrow{*} wgj$, where $\xrightarrow{*}$ is the derivation relation in the suffix rewriting system with rules $gi \rightarrow ghi'$ such that $g,h \in F$ and $gi \prec ghi'$ in $G^\omega(g)$ (or in its approximation at depth 2).*

Notation 4.14 *For $u \in T_f^g$ and $1 \leq i \leq n(g)$, let $u(i)$ denote the event of $G^\omega(f)$ which contains the prime interval ui.*

For instance, in Fig. 5, $fg(1) = fgh(3) = fghk(2)$. Thus each word $u \in T_f^g$ denotes a corresponding mapping from $\mathrm{dom}(u) = \{1, \ldots, n(g)\}$ to the set of events of $G^\omega(f)$; let $Im(u)$ denote the image of this mapping.

4.2 Extracting a Trace Automaton from a Graph Grammar

In this subsection we exploit the syntactic presentation of the context-free domain D induced by the uniform and monotone grammar G in order to derive a finite trace automaton $\mathcal{A}$ whose domain of configuration is isomorphic to D. In a first step, we cut down $H_D \cong G^\omega(f_0)$ to a representative subgraph of finite size, delimited by *repeated sections.*

Definition 4.15 (Repeated Sections) *Two* distinct *F-words $u, v \in T_{f_0}$ induce repeated sections (notation: $u \ll v$) if the following conditions hold :*
1. *u is a prefix of v: $u < v$,*
2. *u and v end with the same symbol: $u = u'f$ and $v = v'f$,*
3. *$\forall i, j \in \{1, \ldots, n(f)\}$ $u(i) = u(j) \Leftrightarrow v(i) = v(j)$, and $u(i) \parallel u(j) \Leftrightarrow v(i) \parallel v(j)$ in $G^\omega(f_0)$.*

Let $\mathcal{T}_\Delta$ be the initial subtree of the parse tree T_{f_0} which remains after removing all nodes $w \in F^*$ such that $u < v < w$ for some pair of repeated sections $u \ll v$ in H_D. We say that two words $u, v \in T_{f_0}$ have the same *type* when the last two conditions of Def. 4.15 are met. By Koenig's lemma, $\mathcal{T}_\Delta$ has no infinite branch since there are only finitely many types. A uniform procedure constructing $\mathcal{T}_\Delta$ from G and f_0 stems immediately from the next lemma.

Lemma 4.16 *Given $u \in T_{f_0}$ and $i, j \in \mathrm{dom}(u)$, one may decide from G and f_0 whether $u(i) = u(j)$, and whether $u(i) \parallel u(j)$, in $G^\omega(f_0)$.*

Let $\mathbf{Fr}(\mathcal{T}_\Delta)$ denote the *frontier* of $\mathcal{T}_\Delta$, i.e. the set of maximal words of $\mathcal{T}_\Delta$, and let $\mathcal{T}_\Delta^\circ = \mathcal{T}_\Delta \setminus \mathbf{Fr}(\mathcal{T}_\Delta)$ denote the *interior* of $\mathcal{T}_\Delta$. Let $\beta_1, \ldots, \beta_n$ be an enumeration of $\mathbf{Fr}(\mathcal{T}_\Delta)$, and for $1 \leq i \leq n$ let α_i be the unique word in $\mathcal{T}_\Delta^\circ$ such that $\alpha_i \ll \beta_i$. Clearly, the frontier of $\mathcal{T}_\Delta$ is a *cut* of T_{f_0} (i.e. a maximal subset of pairwise incomparable elements). Moreover, since α_i and β_i have the same ending symbol, the following holds by definition of T_{f_0}:

$$\forall w \in F^* \quad \alpha_i w \in T \Leftrightarrow \beta_i w \in T$$

Thus we can reduce any word $u \in T_{f_0}$ by the prefix rewriting system $\Delta = \{\beta_i \xrightarrow{i} \alpha_i \mid \alpha_i \ll \beta_i\}$: this rewriting system is deterministic (because the words β_i are incomparable) and noetherian (because the length of α_i is strictly less than the length of β_i), hence any word $u \in T_{f_0}$ reduces to a unique normal form $u' = u \bmod \Delta$ and clearly $u' \in \mathcal{T}_\Delta^\circ$. We can now clarify the role of the last condition stated for u and v in the definition of repeated sections. First of all, this condition entails that composite relations $u^{-1}v$ and $v^{-1}u$ (where u and v denote functions on events according to notation 4.14) are functional relations on E_D, inducing reciprocal bijections $\psi_u^v : Im(u) \rightarrow Im(v)$ and $\psi_v^u : Im(v) \rightarrow Im(u)$ such that:

$$\forall i \; 1 \leq i \leq n(f) \;\; \psi_u^v(u(i)) = v(i) \tag{1}$$

More importantly, the mapping ψ_u^v is a *partial isomorphism* of the graph $(E_D, \|_D)$, with domain $Im(u)$ and range $Im(v)$.

Definition 4.17 (Partial Isomorphism) *A partial isomorphism of a relational structure G is an isomorphism $\psi : D \rightarrow R$ between two substructures of G called respectively the domain and the range of ψ.*

We recall that H is a *substructure* of G or equivalently that G is a *conservative extension* of H if H has the structure induced by that of G:

$$R(x_1, \ldots, x_n) \in G \;\&\; (\forall i)\; x_i \in H \Rightarrow R(x_1, \ldots, x_n) \in H.$$

Thus, the repeated sections $\alpha_i \ll \beta_i$ $(1 \leq i \leq n)$ determine n partial isomorphisms $\psi_i = \psi_{\alpha_i}^{\beta_i}$ of the concurrent alphabet $(E_\Delta, \|_\Delta)$, where E_Δ is the set of events occurring in the representative part of H_D, i.e. $E_\Delta = \bigcup\{Im(u) \mid u \in T_\Delta\}$, and $\|_\Delta$ is the restriction of $\|_D$ to E_Δ. The following result, which is the crux of our construction, was first established by Hrushovski in [Hru92].

Theorem 4.18 (E. Hrushovski) *A finite graph equipped with a finite number of partial isomorphisms $\psi_1, \ldots, \psi_n$ can always be extended conservatively into a finite graph equipped with automorphisms $\Psi_1, \ldots, \Psi_n$ extending the former ones.*

> Hrushovski constructs in fact automorphisms Ψ_i which do not depend on the relation $\|_\Delta$ but only on the partial isomorphisms. His construction was extended to relational structures in [Her94]. A simpler proof of Hrushovski's theorem is due to Lascar, who constructs in [Las94] an extension $(E, \|)$ in which E does not depend upon the given partial isomorphisms but only on the given graph.

The above theorem indicates that one can always extend $(E_\Delta, \|_\Delta)$ into a finite alphabet $(E, \|)$, large enough to permit constructing from D a trace automaton with underlying graph H_D, by labelling the prime intervals of D according to a recursive procedure defined by the automorphisms Ψ_i and the reduction rules $\beta_i \xrightarrow{i} \alpha_i$ in Δ. Moreover, the possibly infinite trace automaton obtained in this way may always be folded to a finite and equivalent trace automaton.

Definition 4.19 (Labels of Prime Intervals) *Let $\Psi_1, \ldots, \Psi_n$ be automorphisms of $(E, \|)$ constructed according to Theo. 4.18 from $(E_\Delta, \|_\Delta)$ and the partial isomorphisms ψ_i induced by the repeated sections $\alpha_i \ll \beta_i$ $(1 \leq i \leq n)$ of H_D. Then each prime interval uj of domain D is given a* label $l_\Delta(uj) \in E$, *defined as follows:*

1. $l_\Delta(uj) = u(j) \in E_\Delta$ *if* $u \in T_\Delta^\circ$,
2. $l_\Delta(uj) = \Psi_i(l_\Delta(vj))$ *if* $u \xrightarrow{i} v$ *in* Δ.

We proceed below to a series of verifications, necessary for proving that the labelled version of H_D is a trace automaton.

Observation 4.20 *The label of a prime interval uj such that $u \in T_\Delta$ is given by $l_\Delta(uj) = u(j)$.*

Actually, this identity holds by definition of l_Δ for $u \in T_\Delta^\circ$; and for $u = \beta_i$ we have $l_\Delta(uj) = l_\Delta(\beta_i j) = \Psi_i(l_\Delta(\alpha_i j))$ [because $\beta_i \xrightarrow{i} \alpha_i$] $= \Psi_i(\alpha_i(j))$ [because $\alpha_i \in T_\Delta^\circ$] $= \beta_i(j)$ [because Ψ_i is the extension of $\psi_i = \psi_{\alpha_i}^{\beta_i}$] $= u(j)$.

Lemma 4.21 *Two prime intervals which belong to the same event have the same label.*

Definition 4.22 *Let $\eta_\Delta : E_D \to E$ be the labelling function on events induced from the mapping l_Δ.*

Lemma 4.23 *Two independent events are always labelled by independent letters in the concurrent alphabet $(E, \|)$.*

Lemma 4.24 *Two conflicting events are never labelled by independent letters in the concurrent alphabet $(E, \|)$.*

Lemmas 4.23 and 4.24 tell us that, by labelling the prime intervals of $H_D \cong G^\omega(f_0)$ according to the function l_Δ, one does obtain an unfolded trace automaton over the finite alphabet $(E, \|)$. The construction of this labelling relies chiefly on Hrushovski's theorem. Nevertheless, we have not yet completely exploited the finite group Ω of automorphisms of $(E, \|)$ generated by $\Psi_1 \ldots \Psi_n$. We finally construct from Ω and the representative part of

H_D a finite trace automaton $\mathcal{A}$ (over $(E, \|)$) that unfolds to the labelled version of H_D.

Define $\mathcal{A} = (E, \|, Q, q_0, T)$ with set of states Q, initial state q_0 and set of transitions T as follows. Let $Q = \Omega \times V_\Delta$ where $V_\Delta = \{(u,i) \in V_{f_0}^\omega \mid u \in T_\Delta^\circ\}$, and let $q_0 = (\Psi_\epsilon, (f_0, 1))$ where Ψ_ϵ is the identity on E. More generally, for $m \in \{1, \ldots n\}^*$, let Ψ_m be inductively defined by $\Psi_{jm} = \Psi_j \circ \Psi_m$. Then $T \subseteq Q \times E \times Q$ is the set of transitions $\sigma_\Delta(u,i) \xrightarrow{\eta_\Delta((u(k))} \sigma_\Delta(v,j)$ such that $u \in T_{f_0}$ and $uk : (u,i) \prec (v,j)$ in H_D, where $\sigma_\Delta : V_{f_0}^\omega \to Q$ is the mapping inductively defined by:

$$\sigma_\Delta(u,j) = \begin{cases} (\Psi_\epsilon, (u,j)) & \text{if } u \in T_\Delta^\circ \\ (\Psi_k \circ \Psi_m, (v,j)) & \text{if } u \xrightarrow{k} u' \text{ and } \sigma_\Delta(u',j) = (\Psi_m, (v,j)). \end{cases}$$

We state below a second series of lemmas, necessary for proving that $\mathcal{A}$ unfolds to H_D.

Lemma 4.25 *Let $(u,i) \in V_{f_0}^\omega$ and $uk : (u,i) \prec (v,j)$ in H_D. If $\sigma_\Delta(u,i) = (\Psi_m, (u',i))$ then $\eta_\Delta(u(k)) = \Psi_m(u'(k))$.*

Lemma 4.26 *For any vertex (u,i) in $V^{\omega}_{f_0}$, let $E_{(u,i)} = \{\eta_\Delta(u(k)) \mid (\exists v)(\exists j)\ uk : (u,i) \prec (v,j)\}$. If $\sigma_\Delta(u_1,i_1) = \sigma_\Delta(u_2,i_2)$ then $E_{(u_1,i_1)} = E_{(u_2,i_2)}$.*

Corollary 4.27 *The transition relation T depends only upon the pairs (u,i) in V_Δ, and therefore, the definition of $\mathcal{A}$ is constructive.*

Proposition 4.28 *$\mathcal{A} = (E, \|, Q, q_0, T)$ is a finite trace automaton; and the pair of mappings $(\sigma_\Delta, \eta_\Delta) : H_D \rightarrow \mathcal{A}$ is a folding morphism, where H_D is viewed as an unfolded trace automaton.*

This completes the proof of Theorem 4.3.

5 Conclusion

We have shown in this paper that every context-free event domain is recognizable, i.e. coincides with the domain of configurations of some finite trace automaton. This representation result has been established by a constructive proof, which extracts a finite trace automaton from any graph grammar defining the Hasse diagram of a conflict event domain. The correspondence between "context-free" event structures (i.e. event structures generating context-free event domains) and finite trace automata may hopefully serve to establish a tight relationship between logics of concurrency in the spirit of [MT92] or [Pen88], and Büchi's monadic second order logic S1S [Büc60]. A related question is to decide whether an event structure presented by rational sets and relations is context-free. Let us stress now some limitations of our work. The finite trace automata which generate context-free event domains form clearly a recursively enumerable and strict subset of the finite trace automata, but we do not know presently whether this set is recursive. For this reason, our representation result fails to establish a full correspondence between well identified classes of models. In another respect, we have failed to characterize the class of event domains which are recognized by finite trace automata. We have set the conjecture that such domains are recognizable if and only if they have finitely many non isomorphic residuals and bounded cliques for the relation R. A first step before proving or disproving this conjecture is to examine the particular case of the (possibly not context-free) distributive event domains, which are known to coincide with the domains of configurations of prime event structures.

References

[BBK87a] BAETEN, J.C.M., BERGSTRA, J.A., and KLOP, J.W., *On the consistency of Koomen's fair abstraction rule*, Theoretical Computer Science, 51 (1987) 129-176.

[BBK87b] BAETEN, J.C.M., BERGSTRA, J.A., and KLOP, J.W., *Decidability of bisimulation equivalence for processes generating context-free languages*, Parle 87, Lecture Notes in Computer Science 259 (1987) 94-111.

[BD93] BADOUEL, E., and DARONDEAU, PH., *Trace Nets.* REX Workshop "Semantics: Foundations and Applications", Lecture Notes in Computer Science 666 (1993) 21-50.

[Ber73] BERGE, C. *Graphs and Hypergraphs.* North-Holland, Amsterdam (1973).

[Büc60] BÜCHI, J.R., *On a decision method in restricted second order arithmetic*, in : E. Nagel et al., eds., proceedings of the International Congress on Logic, Methodology and Philosophy of Science, Stanford University Press (1960) 1-11.

[Cau92] CAUCAL, D., *On the Regular Structure of Prefix Rewriting*, Theoretical Computer Science 106 (1992) 61-86.

[Cou90] COURCELLE, B., *Graph rewriting : An algebraic and logic approach*, in : Handbook of Theoretical Computer Science, vol. B (J.v. Leeuwen, ed.), Elsevier, Amsterdam (1990) 193-242.

[Cur86] CURIEN, P.L., *Categorical combinators, Sequential algorithms and functional programming*, Research notes in Theoretical Computer Science (1986).

[deS84] DE SIMONE, R., *Calculabilité et expressivité dans l'algèbre de processus MEIJE*, Thèse de 3ème Cycle, Paris VII (1984).

[GL91] GOLTZ, U., and LOOGEN, R., *Modelling nondeterministic concurrent processes with event structures*, Fundamenta Informaticae XIV (1991) 39-74.

[Hab89] HABEL, A., *Hyperedge replacement: Grammars and languages.* Thesis, Bremen (1989).

[Her94] HERWIG, B., *Extending Partial Isomorphisms on Finite Structures*, to appear in Combinatorica (1994).

[Hru92] HRUSHOVSKI, E., *Extending Partial Isomorphisms of Graphs*, Combinatorica 12 (1992), 411–416.

[Las94] LASCAR, D., *A note on a theorem of Hrushovski*, Unpublished note, may be asked to the author : *lascar@logique.jussieu.fr* (1994).

[MS85] MULLER, D., and SCHUPP, P., *The Theory of Ends, Pushdown Automata, and Second Order Logic*, Theoretical Computer Science 37 (1985) 51-75.

[MT92] MUKUND, M., and THIAGARAJAN, P.S., *A logical characterization of well branching event structures*, Theoretical Computer Science 96 (1992) 35-72.

[NW93] NIELSEN, M., and WINSKEL, G., *Categories of Models for Concurrency*, Handbook of Logic in Computer Science, Oxford (1993).

[Pen88] PENCZEK, W., *A Temporal Logic for Event Structures*, Fundamenta Informaticae XI (1988).

[Smy77] SMYTH, M.B., *Effectively given domains*, Theoretical Computer Science 5 (1977) 257-274.

[Sta89a] STARK, E.W., *Connections between a Concrete and an Abstract Model of Concurrent Systems.* 5th Mathematical Foundations of Programming Semantics (1989) 53-79.

[Sta89b] STARK, E.W., *Compositional Relational Semantics for Indeterminate Dataflow Networks.* Summer Conference on Category Theory and Computer Science, Springer-Verlag Lecture Notes in Computer Science 389 (1989) 52-74.

[Vaa92] VAANDRAGER, F.W., *Expressiveness Results for Process Algebras*, REX Workshop "Semantics: Foundations and Applications", Lecture Notes in Computer Science 666 (1993) 609-638.

[Win80] WINSKEL, G., *Events in Computations.* Ph.D thesis, University of Edinburgh (1980).

[Win88] WINSKEL, G., *An Introduction to Event Structures.* in REX school "Linear Time, Branching Time and Partial Order in Logics and Models for Concurrency", Noordwijkerhout, Lecture Notes in Computer Science 354 (1988) 364-397.

Encoding Natural Semantics in Coq

Delphine Terrasse

INRIA - Sophia-Antipolis, 2004 Route des Lucioles, B.P. 93
F-06902 Sophia-Antipolis Cedex, France
Delphine.Terrasse@sophia.inria.fr

Abstract. We address here the problem of automatically translating the Natural Semantics of programming languages to Coq, in order to prove formally general properties of languages. Natural Semantics [18] is a formalism for specifying semantics of programming languages inspired by Plotkin's Structural Operational Semantics [22]. The Coq proof development system [12], based on the Calculus of Constructions extended with inductive types (CCind), provides mechanized support including tactics for building goal-directed proofs. Our representation of a language in Coq is influenced by the encoding of logics used by Church [6] and in the Edinburgh Logical Framework (ELF) [15, 3].

1 Introduction

The motivation for our work is the need for an environment to help develop proofs in Natural Semantics. The interactive programming environment generator Centaur [17] allows us to compile a Natural Semantics specification of a given language into executable code (type-checkers, evaluators, compilers, program transformers or optimizers), but does not provide us with theorem-proving facilities to verify the correctness of such tools in a mechanical way. The theorem prover developed under Centaur, Theo [9], was first-order and not equipped with induction. By linking the Centaur and Coq systems, we get a wide assistance in writing, executing and validating programs of any language under study.

In this paper, we formalize two possible translations from Natural Semantics into Coq and prove that each of these translations is correct. Consider a first-order programming language L given in Natural Semantics. The terms of L belong to an abstract syntax. The semantic relations on terms are defined inductively by a system of inference rules. A system in Natural Semantics is given in the Natural Deduction style, where the formulas derived are sequents, concerning some property on expressions E of the language. A sequent is indexed by a relation name: for example, the sequent $\rho \vdash_{type} E \ : \tau$ might state that the expression E is correctly typed of type τ in the typing context ρ. This style of definition is influenced by the theory of free algebras [5] and Logic Programming [19]. Our representation of L in Coq is influenced by Church [6] and the Edinburgh Logical Framework [15, 3]. We refer to the "LF encoding of logics" in the sense that formal symbols of L are implemented by typed constants of the CCind λ-calculus in our case, and that the construction of valid terms or proofs is under control of the CCind type system. Nevertheless, we can stress two main differences. First, we do not use the higher-order features of the target typed λ-calculus to treat binders, or to discharge hypotheses or side conditions of languages. This would involve a Natural Semantics extended with higher-order features like the one described in [14], as well as the possibility of describing higher-order abstract syntax in Coq (we could overcome this difficulty by using the higher-order representation developed in [8]). Encoding Natural Semantics using higher-order abstract syntax is not in the scope of this paper. Secondly, the syntactic categories also called *sorts* or *phyla* of L, are not pairwise distinct but belong to a partially ordered set. This raises the problem of representing subtypes in Coq, which does not currently provide this feature. The Coq system has the advantage of internalizing the induction principles and generating them automatically from inductive definitions, allowing structural induction and rule induction, often needed in the process of proving properties of languages.

We examine two possible representations of L, which differ in the way they represent sorts. The most natural choice is to translate each sort into an inductive type. Recognizing that a term is well-formed is then handled by the type system [16]. In this setting, for each pair of sorts $S_1 \subseteq S_2$, we have to represent explicitly the coercion of S_1 to S_2. For example, let natural numbers be a subset of expressions of L, n be a natural number and f be an expression denoting a function on naturals. The type *Exp* implements the set of expressions in Coq. Let the operation of applying one expression to another one be a constructor of the language. It is implemented in Coq by the following constant declaration denoting a function on *Exp*: '*apply* : *Exp* $\rightarrow$ *Exp* $\rightarrow$ *Exp*'. We encode the operation of applying f to n, with n of type *Nat*, by '(*apply* f (*coer_NatExp* n))', where *coer_NatExp* is a coercion operator that injects naturals into expressions. In contrast to the other operators of the language, the coercion operators do not have any semantic explanation. They are purely syntactical, like the conversion insertions mentioned in [21].

The alternative consists in *flattening* the sorts of L to only one inductive type $\mathcal{L}$ which contains all the constructors of the language. Then we define an auxiliary property to characterize the subsets that correspond to the sorts. Here the constant *apply* declared in Coq denotes a function on $\mathcal{L}$: '*apply* : $\mathcal{L} \rightarrow \mathcal{L} \rightarrow \mathcal{L}$'. We encode the operation of applying f to n by '(*apply* f n)', assuming that n satisfies the property of being a natural.

We have implemented both of these translations from Typol [11, 10], the language that implements Natural Semantics in Centaur, into Coq [24]. This tool already allows us to check mechanically the proofs of properties of programming languages described in Natural Semantics, using inductive reasoning over the resulting specifications in Coq.

The rest of the paper is organized as follows. In Sect. 2 we discuss related work. In Sect. 3 and 4, we define the input of the encoding consisting of an abstract syntax (specifying *terms*) and a semantics specification (specifying *derivations* of theorems). In Sect. 5, we study two encodings of terms and theorems in the Coq formalism. To prove that each representation is correct we show that the encoding of terms is a *compositional bijection* (as defined in [15]) and that the encoding of proofs of a given theorem is a bijection. Finally, in Sect. 6, we give some features of our implementation in Centaur.

2 Related Work

Many experiments have already been done in encoding particular programming languages, with semantics specified by inference systems, in a theorem prover. For example, Myra VanInwegen and Elsa Gunter have encoded a subset of SML in HOL [23], and Claire Jones has encoded a fragment of SPARK Ada in Lego. As far as we know, none of these experiments has been generalized to provide a formal translation from a definition of a language into a theorem prover, as we do here.

Different object logics have also been encoded in LF, such as FOL, HOL or Mini-ML. Each of these representations had to be validated by a corresponding adequacy theorem. In her thesis [13], Gardner introduces a new framework called ELF^+, based on ELF, in order to generalize the representations of logics and the adequacy theorems. She has given a method of defining adequate representations of logics, whereas we have provided an automatic translation which has been proved correct for once. As in ELF, she allows the treatment of higher-order terms or formulas, which we do not include in our setting yet, as Centaur does not provide a higher-order Natural Semantics implementation.

3 Abstract Syntax of a Programming Language

In the following, we formally define terms of an arbitrary language L in Natural Semantics. They will be interpreted as elements of an order-sorted algebra. A *sort* is a syntactic class, which defines a non-empty set of terms. The partially ordered set of sorts is denoted by $(\mathcal{S}, \subseteq)$, where $\mathcal{S}$ is a finite set of sort symbols and $\subseteq$ denotes the subsort relation. We suppose that $\mathcal{S}$ is closed under intersection. We use capital letters for sorts and small letters for operators. We write $\prod_{k=1}^{n} \mathrm{T}_k$ to denote the cartesian product $\mathrm{T}_1 \times \cdots \times \mathrm{T}_n$. Let $\mathrm{T}_j^k, \mathrm{T}_j \in \mathcal{S}$ for $j = 1 \ldots n$ and $k = 1 \ldots l_j$, and let Ω be the following set of operator declarations: $\Omega = \{\, \mathrm{c}_1 : \prod_{k=1}^{l_1} \mathrm{T}_1^k \rightarrow \mathrm{T}_1 \cdots \mathrm{c}_n : \prod_{k=1}^{l_n} \mathrm{T}_n^k \rightarrow \mathrm{T}_n \,\}$. The c_i's are the constructors of the language L. We fix the natural convention for $l_j = 0$ that c_j is a constant of sort T_j. The order-sorted algebra T_Σ, freely generated by the signature $\Sigma = (\mathcal{S}, \Omega)$, defines the abstract syntax of L, that is the set of well-formed terms. From now on, we call terms of L the well-formed terms. We denote by $\mathrm{T}_{\Sigma,\mathrm{S}}$ the set of terms of sort S. We can see terms as finite trees and sorts as classes of trees. If S is a subsort of S′ then a term of sort S is also of sort S′ ($\mathrm{T}_{\Sigma,\mathrm{S}} \subseteq \mathrm{T}_{\Sigma,\mathrm{S}'}$).

Example 1. To illustrate our discussion, we take a simple λ-calculus with constants that we call Li. In Sect. 4, we will consider the static semantics of this language which is given by a Church-style type system. The language Li contains natural numbers built from 'zero' and a successor operator, and expressions built from naturals, identifiers, lambda abstraction and application. The identifiers are encoded by naturals. We also give the specifications of types and contexts that will be needed to describe the type system associated to Li. The abstract syntax of Li is specified as follows:

Sorts : $\mathcal{S} =$ {Nat,Id,Exp,Type, Ctxt}
Subsorts : Nat, Id $\subseteq$ Exp
Constructors of Li :

$$\begin{array}{llll} \Omega = \{\text{zero} & : & \rightarrow \text{Nat}, & \text{int} : \rightarrow \text{Type}, \\ \quad \text{succ} & : \text{Nat} & \rightarrow \text{Nat}, & \text{arrow} : \text{Type} \times \text{Type} \rightarrow \text{Type}, \\ \quad \text{id} & : \text{Nat} & \rightarrow \text{Id}, & \\ \quad \text{lam} & : \text{Id} \times \text{Type} \times \text{Exp} & \rightarrow \text{Exp}, & \text{nil} : \rightarrow \text{Ctxt}, \\ \quad \text{app} & : \text{Exp} \times \text{Exp} & \rightarrow \text{Exp}, & \text{ctxt} : \text{Id} \times \text{Type} \times \text{Ctxt} \rightarrow \text{Ctxt}\} \end{array}$$

In the following, the operators zero, succ, id, lam, app, arrow, nil and ctxt are respectively pretty printed as '0', 'S_', 'id_', 'λ_ : _._', '(_ _)', '_ ⇒ _', '[]' and '[_ : _._]'. The term '$\lambda \mathsf{id}_0$: int ⇒ int. (id_0 S0)' belongs to the language, but 'Sid_0' does not because the expression 'id_0' is not a natural number.

We do not treat the case of mutual recursion, which can be naturally included by gathering mutually recursive operators in a single sort. Once the terms of the language are defined, we can express semantic properties on these terms using inductively defined relations.

4 Natural Semantics Specification

A semantics specification is a set of inference rules that inductively defines a relation (possibly several relations) on sorts of the language. Each rule explains a construct of the language in terms of its components, as in Structural Operational Semantics. Natural Semantics, also called *big step* ($\rightarrow^*$ in rewrite rules terminology), allows multiple premises and can be obtained by transitive closure from Structural Operational Semantics, called *small step* ($\rightarrow$). We write *Sem* for the Natural Semantics of the programming language L. In the following, we give formal definitions of inductively defined relations and representations of proof objects (*derivations*) in Natural Semantics. We will use these definitions in Sect. 5, where we state the correspondence between Natural Semantics and its representation in Coq. As in the definition of the syntax, we restrict our study to non mutually recursive definitions, without loss of generality.

4.1 Inductively Defined Relations in Natural Semantics

In the general setting, we represent sequents ($\_ \vdash_R \_$) of our object language by the usual relations on terms (R(_, _)), for the sake of clarity. We internalize the "holes" (classes of terms) appearing in schematic rules and axioms by the notion of *sorted variables*. Let $\mathcal{V}_S$ be mutually disjoint and countably infinite sets of variables distinct from the language constants for S $\in \mathcal{S}$, and $\mathcal{V} = \bigcup_{S \in \mathcal{S}} \mathcal{V}_S$. Small letters in italics will denote variables. Intuitively, a variable of sort S can be filled by any term of sort S. We write $T_\Sigma[\mathcal{V}]$ for the set of terms with variables in $\mathcal{V}$ and $T_{\Sigma,S}[\mathcal{V}]$ for the subset of sort S. Let t be a term in $T_\Sigma[\mathcal{V}]$. We define $\mathcal{V}ar(t)$ to be the set of variables appearing in t. We write $t[x \in \mathcal{V}_S]$ to indicate that the variable x of sort S occurs in t. We introduce next the substitution operation on terms.

Definition 1. A *substitution* is a set of bindings of the form $t/_{x\in\mathcal{V}_S}$, where $t \in T_{\Sigma,S}[\mathcal{V}]$. We write *Subst* for the set of all substitutions. We write $t[t'/_{x\in\mathcal{V}_S}]$ for the term of the language that can be obtained by substituting the term $t' \in T_{\Sigma,S}[\mathcal{V}]$ for all the occurrences of x in t. We also write $t[\sigma]$ for the result of applying the parallel substitution $\sigma \equiv \{t_1/_{x_1\in\mathcal{V}_{S_1}}, \ldots, t_n/_{x_n\in\mathcal{V}_{S_n}}\}$ to the term t. The new term $t[\sigma]$ is called an *instance* of t.

Example 2. Let $x \in \mathcal{V}_{Nat}$, we can interpret the term '$\lambda \mathsf{id}_0 : \mathsf{int} \Rightarrow \mathsf{int}.\ (\mathsf{id}_0\ \mathsf{x})$' as the class of terms '$(\lambda \mathsf{id}_0 : \mathsf{int} \Rightarrow \mathsf{int}.\ (\mathsf{id}_0\ \mathsf{x})\ [\sigma])$' where $\sigma \in \mathit{Subst}$.

Definition 2. We characterize the *Weakest sort constraint* $\mathrm{Wsc}(t[v])$ that can be associated with a variable v in t as the intersection of all the sorts associated to the occurrences of v in t (if the intersection is empty there exists no S such that $t[v \in \mathcal{V}_S]$ is a well-formed term). Note that the variable v in the term $t[v \in \mathcal{V}_S]$ has a sort S that must be included in $\mathrm{Wsc}(t[v])$.

Let P denote a relation in $U_1 \times \ldots \times U_n$ inductively defined by a set of inference rules of the following form: r $\dfrac{Q_1(t_1^1,\ldots,t_{n_1}^1),\ldots,Q_m(t_1^m,\ldots,t_{n_m}^m)}{P(t_1,\ldots,t_n)}$.

We use the convention that $t_k^0 \equiv t_k$ and $U_k^0 \equiv U_k$. The Q_i are defined respectively on the domains $U_1^i \times \cdots \times U_{n_i}^i$ for $i = 1 \ldots m$. The t_k^i denote terms in $T_{\Sigma,U_k^i}[\mathcal{V}]$ for $i = 0 \ldots m$ and $k = 1 \ldots n_i$. The relations above the line are called the *premises* and may involve the relation P being defined. The relation below the line is called the *conclusion*. The rule r denotes a basic deduction step of our system. If $m = 0$ then r is called an axiom. The relation P is inductively defined, meaning that the relation is the smallest set of sequents containing the axioms that is closed under the introduction rules [1].

Example 3. We consider now the example of the Church-style type system of Li in Natural Semantics. It is specified by the "type" property that holds for the typing context ρ, the expression e and the type τ whenever e has type τ in ρ. There is one rule for each constructor of expressions of the language. We only give here the rule concerning the lambda operator:

$$\mathsf{type}_{lam}\quad \frac{[\mathsf{id_X} : \tau.\rho] \vdash_{\mathsf{type}} \mathsf{e}\ :\ \tau'}{\rho \vdash_{\mathsf{type}} \lambda \mathsf{id_X} : \tau.\ \mathsf{e}\ :\ \tau \Rightarrow \tau'}$$

This system determines the valid typed expressions of Li. For example, we can derive the sequent '$[] \vdash_{type} \mathrm{S0}\ :\ \mathrm{int}$', that associates the object type "int" to the expression 'S0' in the empty context ('$[] \vdash_{type} \mathrm{S0}\ :\ \mathrm{int}$' stands by convention for 'type([], S0, int)').

We extend all previous definitions to sequents, axioms and rules in a natural way.

4.2 Proof Representation in Natural Semantics

We are interested in theorems that state general properties of programming languages. They are universally quantified relations on terms, that can be derived from axioms and inference rules in $\mathcal{S}em$. In this interpretation, we can see $\mathcal{S}em$ as an order-sorted logic program, and theorems as logical consequences. The derivation of a given theorem can be viewed as a tree: the root is labeled by the theorem statement, nodes by instances of rule conclusions (with sons labeled by the corresponding premises instances) and leaves by instances of axioms. A more compact form, called a "proof term" and defined from symbols annoted by substitutions, can be used to denote proof objects.

Definition 3. A *proof term* in $\mathcal{S}em$ is recursively defined by:

1. 'ax_σ', where '$\mathrm{ax} : \mathrm{Concl}$ ' $\in \mathcal{S}em$ and $\sigma \in \mathcal{S}ubst$, is a proof term for $\mathrm{Concl}[\sigma]$,
2. an application '$(\mathrm{r}_\sigma\ PT_1 \ldots PT_p)$', where '$\mathrm{r} : \frac{\mathrm{Prem}_1, \cdots, \mathrm{Prem}_m}{\mathrm{Concl}}$' $\in \mathcal{S}em$, $\sigma \in \mathcal{S}ubst$ and PT_i is a proof term for $\mathrm{Prem}_i[\sigma]$ for $i = 1 \ldots m$, is a proof term for $\mathrm{Concl}[\sigma]$.

We will keep this notation to define the correspondence between Natural Semantics and its representation in Coq. The equivalence relation induced by the different proof terms for a given theorem yields a set of equivalence classes $\mathcal{C}$, which corresponds precisely to the set of theorems that we are going to consider to show the correctness of the encoding. The theorems concerned (considered modulo α-conversion) are in a one-one correspondence with a certain set of theorems in Coq as shown in Sect. 5.

5 Encoding in Coq

An abstract syntax is represented in Coq by inductive types declared in the predefined type Set. For instance, '*Exp*' is an inductive type of type Set and the constructor 'app : Exp × Exp → Exp' is represented by the curried function '*app* : *Exp* → *Exp* → *Exp*'. Then, each inductively defined relation on terms has an associated, inductively defined dependent type in Coq that ranges over the predefined type Prop (for propositions). This type depends on terms of the language. Each rule is implemented by a constructor declaration in Coq. For example, the rule r of Sect. 4.1 is encoded in Coq by the constant r declared as a constructor of the inductive type P. As the semantic relations of the object language are implemented by propositions of the meta-logic Coq, the deduction is encoded by the primitive implication '→', and the variables appearing in rules are bound by the primitive universal quantifier '∀'. Let $\mathcal{X} = \{x_1, \ldots, x_n\}$ be the standard enumeration of $\mathcal{V}ar(\mathrm{r})$, and $\xi_\mathcal{X}$ be the encoding function from $\mathrm{T}_\Sigma[\mathcal{X}]$ to Coq terms. We can roughly define r by:

$$r : \forall x_1 \ldots x_n.((Q_1\ \xi_\mathcal{X}(t_1^1) \ldots \xi_\mathcal{X}(t_{n_1}^1)) \rightarrow \cdots \rightarrow (Q_m\ \xi_\mathcal{X}(t_1^m) \ldots \xi_\mathcal{X}(t_{n_m}^m)) \\ \rightarrow (P\ \xi_\mathcal{X}(t_1) \ldots \xi_\mathcal{X}(t_n)))$$

Proof validation in the resulting type theory is based on an extension of the Curry-Howard *Propositions as Types* principle to higher-order propositions.

According to this principle, the objects manipulated in Coq have both a logical and a functional interpretation. A complete study of inductive definitions in Coq can be found in [20].

We consider two different ways of encoding the abstract syntax of L given in Sect. 3. The most natural way is to encode each sort by a type. The other solution consists in *flattening* the sorts in one inductive type which contains all the constructors. In the latter case, we have to define an auxiliary property specifying that a term is well-formed. In both settings, we give a formal definition of the encoding of terms and proofs, following the approach described in [3, 15]. One difference between our approach and the LF encoding is due to our order-sorted setting, because we have to handle subtyping explicitly in the encoding map on terms. Another difference is that the terms and proofs in Coq remain first-order constructions, so our notion of context in Coq corresponds only to the variables *over* the object language and not *in* the object language. We write symbols in Coq in italics to distinguish them from the original ones.

5.1 Each Sort is encoded by a Type

Definition of Syntax in Coq. In this setting we have to deal with inclusion of types, since Coq does not have subtyping. Each sort of the language is implemented by one type in Coq. To represent subtyping we create new coercion operators. This representation may lead to inconsistency. For example, the following inclusion relation between the sorts S_1, S_2, S_3 and S_4: $\{S_4 \subseteq S_2 \subseteq S_1, S_4 \subseteq S_3 \subseteq S_1\}$ is inconsistent. A term t of sort S_4 may be coerced by two different ways to S_1, and may have two well typed representations in Coq. We cannot accept this kind of directed acyclic graph to model our subsort relation. We could overcome this difficulty by defining our own equality relation on terms, but the machinery needed to prove that this equality is preserved for all operations on terms would be a real burden at the mechanized proofs stage. To restrict the subsort relation, we could have defined each sort as a union of *basic* and *mutually disjoint* sorts as studied in [7]. To respect the initial specification, we have chosen instead to assume that the non-directed graphs induced by the subsort relation are acyclic. In other words, the graph connected to a given sort is a tree and the whole graph is a forest and so $\mathcal{S}$ satisfies the following well known property on trees.

Proposition 4. *Between two sorts in $\mathcal{S}$, the inclusion path is finite and unique.*

Notice that this restriction prevents us from defining a sort L containing the union of all sorts of the language. Instead of defining one single (overloaded) sequent over all terms of the language, we have to define one sequent for each sort. Let S be a sort and let Ant_S be the set of direct predecessors of S. The sorts and operators of the language are denoted in Coq by the same symbols written in italics. For each sort S in $\mathcal{S}$ we specify an inductive set S in Coq.

Definition 5. The constructors of S declared in Coq belong to the smallest set defined as follows:

1. $c : \prod_{k=1}^{p} S_k \rightarrow S$ is a constructor of S whenever $\mathrm{c} : \prod_{k=1}^{p} \mathrm{S}_k \rightarrow \mathrm{S} \in \Omega$,
2. the coercion operator $coer\_S'S : S' \rightarrow S$ is a constructor of S if $\mathrm{S}' \in \mathcal{A}nts_{\mathrm{S}}$.

Example 4. In our example, the terms of the language Li are encoded by elements of the inductive set *Exp*. They involve new coercion operators *Coer_NE* and *Coer_IE* to include elements of the inductive sets *Nat* and *Id*.

Inductive Set $Exp = coer\_NE : Nat \rightarrow Exp \mid coer\_IE : Id \rightarrow Exp$
$\mid lam : Id \rightarrow Type \rightarrow Exp \rightarrow Exp \mid app : Exp \rightarrow Exp \rightarrow Exp.$

Let us suppose that $\mathcal{S} = \{\mathrm{S}_1 \ldots \mathrm{S}_N\}$. The sorts $\mathrm{S}_1, \ldots, \mathrm{S}_N$ of our language L are implemented in Coq by $S_1, \ldots, S_N$ given in definition 5. Let $\mathcal{X} \subset \mathcal{V}$ be a finite set of variables and $\mathcal{X}_{\mathrm{S}} \subset \mathcal{V}_{\mathrm{S}}$

be the subset of sort $\mathrm{S} \in \mathcal{S}$. For each $\mathrm{S} \in \mathcal{S}$ we define a bijection ρ_{S} from $\mathcal{X}_{\mathrm{S}}$ to a set of variables of type S in Coq (denoted by new symbols). We write ρ_{S}^{-1} for the inverse of ρ_{S}. Let us call $\Gamma_{\mathcal{X}}$ the context that associates the variable $\rho_{\mathrm{S}}(x)$ for $x \in \mathcal{X}_{\mathrm{S}}$ to the type S in Coq, and $Coq_{\mathrm{L}}^{1}[\Gamma_{\mathcal{X}}]$

the set of terms in normal form of types $S_1, \ldots, S_N$ in Coq and $\Gamma_{\mathcal{X}}$. The encoding function is indexed by a sort symbol for coercion insertion.

Definition 6. The terms in $\mathrm{T}_{\Sigma}[\mathcal{X}]$ are encoded in Coq by a family of recursive functions $\xi_{\mathcal{X},\mathrm{S}}$, for $\mathrm{S} \in \mathcal{S}$, which maps a term t of a given sort S with variables in $\mathcal{X}$ to a term of type S in Coq and $\Gamma_{\mathcal{X}}$.

1. $\xi_{\mathcal{X},\mathrm{S}}(v) = \rho_{\mathrm{S}}(v)$ where $v \in \mathcal{X}_{\mathrm{S}}$,
2. $\xi_{\mathcal{X},\mathrm{S}}(v) = (coer\_S''S \;\; \xi_{\mathcal{X},\mathrm{S}''}(v))$ where $v \in \mathcal{X}_{\mathrm{S}'}$, $\mathrm{S}'' \in \mathcal{A}nts_{\mathrm{S}}$ and $\mathrm{S}' \subseteq \mathrm{S}''$,
3. $\xi_{\mathcal{X},\mathrm{S}}(\mathrm{c}(t_1, \ldots, t_p)) = (c \;\; \xi_{\mathcal{X},\mathrm{S}_1}(t_1) \;\; \ldots \;\; \xi_{\mathcal{X},\mathrm{S}_p}(t_p))$ where c ranges over S,
4. $\xi_{\mathcal{X},\mathrm{S}}(\mathrm{c}(t_1, \ldots, t_p)) = (coer\_S''S \;\; \xi_{\mathcal{X},\mathrm{S}''}(\mathrm{c}(t_1, \ldots, t_p)))$ where c ranges over S', $\mathrm{S}'' \in \mathcal{A}nts_{\mathrm{S}}$ and $\mathrm{S}' \subseteq \mathrm{S}''$.

The sort S" in (2) and (4) is unique by proposition 4. By construction, the term $\xi_{\mathcal{X},\mathrm{S}}(t)$ is well typed of type S in Coq and $\Gamma_{\mathcal{X}}$. Hence, the image of $\mathrm{T}_{\Sigma}[\mathcal{X}]$ by $\xi_{\mathcal{X},\mathrm{S}}$ is a subset of $Coq_{\mathrm{L}}^{1}[\Gamma_{\mathcal{X}}]$. Notice that the variables over the object language are identified with Coq variables. To prove the correctness of the correspondence, we prove the following theorem for each $\mathrm{S} \in \mathcal{S}$.

Theorem 7 (Adequacy for Syntax). *The function $\xi_{\mathcal{X},\mathrm{S}}$ is a compositional bijection from terms of sort* S *to terms of type S in Coq and $\Gamma_{\mathcal{X}}$.*

Proof. Bijectivity of the Encoding. The function $\xi_{\mathcal{X},\mathrm{S}}$ is injective by construction (by cases on the structure of terms) and by proposition 4. To demonstrate surjectivity, we define the following recursive function $\xi_{\mathcal{X},\mathrm{S}}^{-1}$ on terms of type S in Coq and $\Gamma_{\mathcal{X}}$:

1. $\xi_{\mathcal{X},\mathrm{S}}^{-1}((coer\_S''S \;\; t)) = \xi_{\mathcal{X},\mathrm{S}''}^{-1}(t)$,
2. $\xi_{\mathcal{X},\mathrm{S}}^{-1}(v) = \rho_{\mathrm{S}}^{-1}(v)$ where v is a Coq variable declared in $\Gamma_{\mathcal{X}}$,
3. $\xi_{\mathcal{X},\mathrm{S}}^{-1}((c \; t_1 \ldots t_p)) = \mathrm{c}(\xi_{\mathcal{X},\mathrm{S}_1}^{-1}(t_1), \ldots, \xi_{\mathcal{X},\mathrm{S}_p}^{-1}(t_p))$ where c is not a coercion.

By construction we have $\xi^{-1}_{\mathcal{X},\mathrm{S}}(t) \in \mathrm{T}_{\Sigma,\mathrm{S}}[\mathcal{X}]$. To show surjectivity we prove that $\xi_{\mathcal{X},\mathrm{S}}(\xi^{-1}_{\mathcal{X},\mathrm{S}}(t)) = t$ by induction on terms of type S in Coq and $\Gamma_{\mathcal{X}}$.

Compositionality of the Encoding. We suppose that $\mathcal{X}$ and $\mathcal{Y}$ are two disjoint and finite sets of variables. Let $\mathcal{X} = \{x_1, \ldots, x_n\}$, $t \in \mathrm{T}_{\Sigma,\mathrm{S}}[\mathcal{X}]$ and $t_i \in \mathrm{T}_{\Sigma,\mathrm{S_i}}[\mathcal{Y}]$ for $i = 1 \ldots n$. To show the compositionality we have to prove that $\xi_{\mathcal{Y},\mathrm{S}}(t[t_1/_{x_1 \in \mathcal{X}_{\mathrm{S}_1}}, \ldots, t_n/_{x_n \in \mathcal{X}_{\mathrm{S}_n}}]) = \xi_{\mathcal{X},\mathrm{S}}(t)[\xi_{\mathcal{Y},\mathrm{S}_1}(t_1)/\rho_{\mathrm{S}_1}(x_1), \ldots, \xi_{\mathcal{Y},\mathrm{S}_n}(t_n)/\rho_{\mathrm{S}_n}(x_n)]$. We can prove this by induction on the structure of terms. □

Notice that the notation for parallel substitution is overloaded. It is used both for variables in $\mathcal{V}$ and Coq variables.

Example 5. The representation of the expression '$\lambda \mathsf{id_0} : \mathsf{int} \Rightarrow \mathsf{int}.(\mathsf{id_0}\ \mathsf{S0})$' is the following Coq term of type *Exp* in the empty context:
(*lam* (*id zero*)(*arrow int int*) (*app* (*coer_IE* (*id zero*))(*coer_NE* (*succ zero*)))).

Specification of Semantics in Coq. Let us consider the inductive relation P on $\mathrm{U}_1 \times \cdots \times \mathrm{U}_n$ given in Sect. 4. The property P is encoded in Coq by an inductive definition P of type $U_1 \to \cdots \to U_n \to \mathsf{Prop}$. We suppose that the Q_i are either previously defined or equal to P. Let $\mathcal{X} = \{x_i \in \mathcal{V}_{\mathrm{S}_i} \mid i = 1 \ldots q\}$ be the standard enumeration of $\mathcal{V}ar(\mathrm{r})$. The rule r is then encoded by the following constructor declaration, using the encoding function on terms

$$r : \forall_{i=1}^{q} \rho_{\mathrm{S}_i}(x_i) : S_i\ .(Q_1\ \xi_{\mathcal{X},\mathrm{U}_1^1}(t_1^1)\ \ldots\ \xi_{\mathcal{X},\mathrm{U}^1_{n_1}}(t^1_{n_1})) \to \cdots \to (Q_m\ \xi_{\mathcal{X},\mathrm{U}_1{}^m}(t_1^m)\ \ldots\ \xi_{\mathcal{X},\mathrm{U}^m_{n_m}}(t^m_{n_m})) \to (P\ \xi_{\mathcal{X},\mathrm{U}_1}(t_1)\ \ldots\ \xi_{\mathcal{X},\mathrm{U}_n}(t_n)).$$

Example 6. Let us give next the representation of the rule $type_{lam}$ which is a constructor of the *type* relation in Coq. We write v instead of $\rho_{\mathrm{S}}(v)$ for the sake of clarity.

$$\begin{array}{l} \mathsf{Inductive\ Definition}\ type : Ctxt \to Exp \to Nat \to \mathsf{Prop} \\ = \cdots \mid type_{lam} : \forall \rho : Ctxt. \forall x : Nat. \forall \tau, \tau' : Type. \forall e : Exp. \\ \quad (type\ (ctxt\ (id\ x)\ \tau\ \rho)\ e\ \tau') \to (type\ \rho\ (lam\ (id\ x)\ \tau\ e)\ (arrow\ \tau\ \tau')) \cdots \end{array}$$

Let $\mathrm{P} \in \mathrm{U}_1 \times \cdots \times \mathrm{U}_n$ be a relation defined in *Sem* and let a_i denote a term in $\mathrm{T}_{\Sigma,\mathrm{U}_i}[\mathcal{Y}]$ for $i = 1 \ldots n$ such that the sequent $\mathrm{P}(a_1, \ldots, a_n)$ is a theorem in *Sem*. To encode Natural Semantics proofs in Coq we define a recursive map ξ from proof terms to Coq terms as follows.

Definition 8. We suppose that $\mathcal{X}$ and $\mathcal{Z}$ are two disjoint and finite sets of variables and that $\{x_i \in \mathcal{V}_{\mathrm{S}_i} \mid i = 1 \ldots q\}$ is the standard enumeration of $\mathcal{X}$. The encoding ξ is recursively defined over proof terms by the following:

1. $\xi((\mathrm{ax}\{t_1/_{x_1 \in \mathcal{V}_{\mathrm{S}_1}}, \ldots, t_q/_{x_q \in \mathcal{V}_{\mathrm{S}_q}}\})) = (ax\ \xi_{\mathcal{Z},\mathrm{S}_1}(t_1) \ldots \xi_{\mathcal{Z},\mathrm{S}_q}(t_q))$ where $\mathcal{X} = \mathcal{V}ar(\mathrm{ax})$ and $t_i \in \mathrm{T}_{\Sigma,\mathrm{S}_i}[\mathcal{Z}]$ for $i = 1 \ldots q$,
2. $\xi((\mathrm{r}\{t_1/_{x_1 \in \mathcal{V}_{\mathrm{S}_1}}, \ldots, t_q/_{x_q \in \mathcal{V}_{\mathrm{S}_q}}\}\ PT_1 \ldots PT_p)) = (r\ \xi_{\mathcal{Z},\mathrm{S}_1}(t_1) \ldots \xi_{\mathcal{Z},\mathrm{S}_q}(t_q)\ \xi(PT_1) \ldots \xi(PT_p))$ where $\mathcal{X} = \mathcal{V}ar(\mathrm{r})$ and $t_i \in \mathrm{T}_{\Sigma,\mathrm{S}_i}[\mathcal{Z}]$ for $i = 1 \ldots q$.

Theorem 9. *ξ defines a bijective map from proof terms for $P(a_1, \ldots, a_n)$ to normal terms of type $(P\ \xi_{\mathcal{Y},\mathrm{U}_1}(a_1)\ \ldots\ \xi_{\mathcal{Y},\mathrm{U}_n}(a_n))$ in Coq and $\Gamma_{\mathcal{Y}}$.*

Proof. We prove first that if PT is a proof term for $P(a_1, \ldots, a_n)$ then $\xi(PT)$ is a term in normal form of type $(P\ \xi_{\mathcal{Y},U_1}(a_1)\ \ldots\ \xi_{\mathcal{Y},U_n}(a_n))$ in Coq and $\Gamma_{\mathcal{Y}}$, by induction on the structure of proof terms, using compositionality of term encoding. The injectivity is straightforward by cases on the structure of proof terms, using the injectivity of term encoding and the injectivity property for constructors in Coq. To demonstrate surjectivity we define a right-inverse ξ^{-1} from terms in normal form of type $(P\ b_1 \ldots b_n)$ in Coq and $\Gamma_{\mathcal{Y}}$ to proof terms for $P(\xi^{-1}_{\mathcal{Y},U_1}(b_1), \ldots, \xi^{-1}_{\mathcal{Y},U_n}(b_n))$. The function ξ^{-1} is inductively defined in a simple way and is proven to be total by construction. Then we can prove by induction that $\xi(\xi^{-1}(Pr)) = Pr$, therefore ξ is surjective. □

5.2 Each Sort is encoded by a Property on Terms

Definition of Syntax in Coq. The idea is to specify subsets of a given set by a property on elements of this set (the "subset property"). This is used for example to implement set theory in Coq. The logical implication in Coq represents the usual set inclusion, and the logical connectives $\vee$ and $\wedge$ on Coq propositions represent the union and the intersection over sets. In this setting, the terms of L are represented by the union of the subsets that represent sorts. We specify the set $\mathcal{L}$ on which we will extract *valid* terms of L. We introduce the notation $\xrightarrow[k=1]{n} \mathcal{L}$ for $\mathcal{L} \rightarrow \overset{n}{\cdots} \rightarrow \mathcal{L}$. The inductive set $\mathcal{L}$ is defined in Coq as follows:

Inductive Set $\mathcal{L} = \quad c_1 : \xrightarrow[k=1]{l_1} \mathcal{L} \rightarrow \mathcal{L} \mid \cdots \mid \ c_n : \xrightarrow[k=1]{l_n} \mathcal{L} \rightarrow \mathcal{L}.$

This set $\mathcal{L}$, where all sorts have been merged, is obviously too big and contains terms which do not belong to the language.

Example 7. Let us come back to our example Li and its representation $\mathcal{L}i$:

Inductive Set $\mathcal{L}i = zero : \mathcal{L}i \mid succ : \mathcal{L}i \rightarrow \mathcal{L}i \mid id : \mathcal{L}i \rightarrow \mathcal{L}i$
$\mid lam : \mathcal{L}i \rightarrow \mathcal{L}i \rightarrow \mathcal{L}i \rightarrow \mathcal{L}i \mid app : \mathcal{L}i \rightarrow \mathcal{L}i \rightarrow \mathcal{L}i \mid int : \mathcal{L}i$
$\mid arrow : \mathcal{L}i \rightarrow \mathcal{L}i \rightarrow \mathcal{L}i \mid nil : \mathcal{L}i \mid ctxt : \mathcal{L}i \rightarrow \mathcal{L}i \rightarrow \mathcal{L}i \rightarrow \mathcal{L}i\,.$

We can see that the term '$(succ\ (id\ zero))$' belongs to $\mathcal{L}i$ although it does not represent a term of Li.

To specify the subset of valid terms in $\mathcal{L}$, we have to encode the sorts in $\mathcal{S}$ by subsets of $\mathcal{L}$. Instead of defining a property for each sort of $\mathcal{L}$ we have chosen to define a property 'L_is' that takes the sort symbol as parameter. The sort symbols are implemented by constants of the inductive set L_S:

Inductive Set L_S $=$ S_1 : L_S $\mid \cdots \mid$ S_N : L_S.

The property 'L_is' holds for a given term t in $\mathcal{L}$ and a given sort S in L_S whenever t corresponds to a term of the language of sort S. We can now define the property 'L_is' by primitive recursion over $\mathcal{L}$ and L_S, using the corresponding elimination schemes generated by Coq. We only give here the computation rules associated to 'L_is'. We recall that $c_i \in \Omega$ and $S_k \in \mathcal{S}$ for $i = 1 \ldots n$ and $k = 1 \ldots N$.

if $T_i \subseteq S_k$
then $(\text{L\_is}\ (c_i\ t^i_1\ \ldots\ t^i_{l_i})\ S_k) \triangleright (\text{L\_is}\ t^i_1\ T^1_i) \wedge \cdots \wedge (\text{L\_is}\ t^i_{l_i}\ T^{l_i}_i)$
else $(\text{L\_is}\ (c_i\ t^i_1\ \ldots\ t^i_{l_i})\ S_k) \triangleright$ False

The fact that a term belongs to L in this setting can be expressed by the following *well_formed* property: 'Definition *well_formed* $= \lambda t : \mathcal{L}. (\text{L\_is } t \ \text{S}_1) \vee \cdots \vee (\text{L\_is } t \ \text{S}_N)$'. In Coq, we can express the inclusion between two sorts S $\subseteq$ S' by the proposition '$\forall t : \mathcal{L}. (\text{L\_is } t \ \text{S}) \rightarrow (\text{L\_is } t \ \text{S'})$', which can be proved easily by structural induction on t. From now on, we call the proposition '(L_is t S)' a *sort constraint*. We have prefered a functional representation of the property 'L_is' because it seems more relevant at the stage of mechanized proofs.

Example 8. After having specified the set of sort symbols Li_S and the property 'Li_is' in Coq, we have for instance the proposition '(Li_is (*id zero*) Nat)' which is convertible with False as expected. We have also that '(Li_is t Nat) $\rightarrow$ (Li_is t Exp)' holds for every term t of type $\mathcal{L}i$, expressing the subsort relation.

Let $\mathcal{X} \subset \mathcal{V}$ be a finite set of variables. We define a bijection ρ from $\mathcal{X}$ to a finite set of variables of type $\mathcal{L}$ in Coq. We also define a set of hypotheses assuming the sort constraints on variables implementations in Coq. We denote by ρ^{-1} the inverse of ρ. Let $\{x_i \in \mathcal{V}_{\text{S}_i} | i = 1 \ldots q\}$ be the standard enumeration of $\mathcal{X}$, then the context $\Gamma_{\rho(\mathcal{X})}$ in Coq has the following shape: $[\rho(x_1) : \mathcal{L}, \ldots, \rho(x_q) : \mathcal{L}, H_1 : (\text{L\_is } \rho(x_1) \ \text{S}_1), \ldots, H_q : (\text{L\_is } \rho(x_q) \ \text{S}_q)]$. Let $Coq^2_{\text{L}}[\Gamma_{\rho(\mathcal{X})}]$ denote the set of terms in normal form of type $\mathcal{L}$ in Coq and $\Gamma_{\rho(\mathcal{X})}$, such that the *well_formed* property can be proved. As in the previous section, we can formalize the encoding of terms as follows.

Definition 10. $t \in \text{T}_\Sigma[\mathcal{X}]$ is encoded by a recursive function $\xi_\mathcal{X}$ mapping terms of L with variables in $\mathcal{X}$ to normal terms of type $\mathcal{L}$ in Coq and $\Gamma_{\rho(\mathcal{X})}$.

1. $\xi_\mathcal{X}(v) = \rho(v)$ where $v \in \mathcal{X}$
2. $\xi_\mathcal{X}(\text{c}(t_1, \ldots, t_p)) = (c \ \xi_\mathcal{X}(t_1) \ \ldots \ \xi_\mathcal{X}(t_p))$.

By construction, the resulting representation $\xi_\mathcal{X}(t)$ of the term t satisfies the proposition (*well_formed* $\xi_\mathcal{X}(t)$). More precisely, if the term $t \in \text{T}_{\Sigma,\text{S}}[\mathcal{X}]$ then the proposition '(L_is $\xi_\mathcal{X}(t)$ S)' can be proved in Coq and $\Gamma_{\rho(\mathcal{X})}$ (this can be shown by induction on the structure of terms). The image of the set $\text{T}_\Sigma[\mathcal{X}]$ by $\xi_\mathcal{X}$ is then a subset of $Coq^2_{\text{L}}[\Gamma_{\rho(\mathcal{X})}]$. To prove the correctness of the correspondence between $\text{T}_\Sigma[\mathcal{X}]$ and $Coq^2_{\text{L}}[\Gamma_{\rho(\mathcal{X})}]$ we prove the following theorem.

Theorem 11 (Adequacy for Syntax). *The function $\xi_\mathcal{X}$ is a compositional bijection.*

Proof. The proof is similar to that given for Th. 7, except that we have to deal here with the additional proofs of sort constraints. □

Example 9. The encoding of '$\text{id}_{\text{S}x}$' for $x \in \mathcal{X}_{\text{Nat}}$ is $(id \ (succ \ \rho(x)))$ in the Coq context $[\rho(x) : \mathcal{L}i, H_x : (\text{Li\_is } \rho(x) \ \text{Nat})]$.

Specification of Semantics in Coq. The property P that defines a subset of $U_1 \times \cdots \times U_n$ is encoded in Coq by an inductive definition P of type $\overset{n}{\underset{k=1}{\longrightarrow}} \mathcal{L} \rightarrow$ Prop. To handle the domains of relations we have to deal explicitly with the proofs that terms are well-formed. We can either express the sort constraints in each rule specification or on the arguments of P. We explain now these two possible representations of the relation P in Coq.

1. Specification of the Sort Constraints in the Inductive Definition. The first choice is to implement the constructor r of the inductive relation P in Coq with all the sort constraints. We suppose that the Q_i have either been previously implemented or denote the relation P. Let $\mathcal{X} = \{x_i \in \mathcal{V}_{S_i} | \, i = 1 \ldots q\}$ be the standard enumeration of $\mathcal{V}ar(\mathrm{r})$. The rule r is then encoded by the following constructor declaration of P:

$$r : \forall_{i=0}^{q} \rho(x_i) : \mathcal{L}. \overset{q}{\underset{i=1}{\longrightarrow}} (\text{L\_is } \rho(x_i) \text{ S}_i) \rightarrow \overset{m}{\underset{1=1}{\longrightarrow}} (Q_i \ \xi_{\mathcal{X}}(t_1^i) \ldots \xi_{\mathcal{X}}(t_{n_i}^i)) \rightarrow (P \ \xi_{\mathcal{X}}(t_1) \ldots \xi_{\mathcal{X}}(t_n)).$$

We can easily prove in Coq the following proposition that we call *P_dom*:

$$\forall_{k=1}^{n} v_k : \mathcal{L}. (P \ v_1 \ldots v_n) \rightarrow \wedge_{k=1}^{n} (\text{L\_is } v_k \ \text{U}_k).$$

Example 10. Let us give the rule $type_{lam}$ of the *type* relation of Li in Coq according to this choice. We write v instead of $\rho(v)$ for the sake of clarity.

Inductive Definition $type : \mathcal{L}i \rightarrow \mathcal{L}i \rightarrow \mathcal{L}i \rightarrow$ Prop
$= \cdots \ |type_{lam} : \forall \rho, x, \tau, \tau', e : \mathcal{L}i.$ (Li_is ρ Ctxt) $\rightarrow$ (Li_is x Nat) $\rightarrow$
(Li_is τ Type) $\rightarrow$ (Li_is τ' Type) $\rightarrow$ (Li_is e Exp) $\rightarrow$
$(type \ (ctxt \ (id \ x) \ \tau \ \rho) \ e \ \tau') \rightarrow (type \ \rho \ (lam \ (id \ x) \ \tau \ e) \ (arrow \ \tau \ \tau')) \cdots$

This representation is rather tedious but powerful tactics in Coq and Ctcoq [4] allow us to perform usual proofs quite automatically.

Let $\mathrm{P} \in \mathrm{U}_1 \times \cdots \times \mathrm{U}_n$ be a relation defined in $\mathcal{S}em$ and a_i denote terms in $\mathrm{T}_{\Sigma,\mathrm{U}_i}[\mathcal{Y}]$ for $i = 1 \ldots n$ such that the sequent $\mathrm{P}(a_1, \ldots, a_n)$ is a theorem. To encode Natural Semantics proofs in Coq we define a recursive map ξ from proof terms of $\mathrm{P}(a_1, \ldots, a_n)$ to terms of type $(P \ \xi_{\mathcal{Y}}(a_1) \ \ldots \ \xi_{\mathcal{Y}}(a_n))$ in Coq and $\Gamma_{\rho(\mathcal{Y})}$.

Definition 12. We suppose that $\mathcal{X}$ and $\mathcal{Z}$ are two disjoint and finite sets of variables and that $\{x_i \in \mathcal{V}_{S_i} | \, i = 1 \ldots q\}$ is the standard enumeration of $\mathcal{X}$. For $t \in \mathrm{T}_{\Sigma,\mathrm{S}}[\mathcal{Z}]$ let H_t denote the proof of the sort constraint '(L_is $\xi_{\mathcal{Z}}(t)$ S)' in Coq and $\Gamma_{\rho(\mathcal{Z})}$. The map ξ is recursively defined over proof terms by:

1. $\xi((\mathrm{ax}\{t_1/_{x_1 \in \mathcal{V}_{S_1}}, \ldots, t_q/_{x_q \in \mathcal{V}_{S_q}}\})) = (ax \ \xi_{\mathcal{Z}}(t_1) \ldots \xi_{\mathcal{Z}}(t_q) \ H_{t_1} \ldots H_{t_q})$ where $\mathcal{X} = \mathcal{V}ar(\mathrm{ax})$ and $t_i \in \mathrm{T}_{\Sigma,\mathrm{S}_i}[\mathcal{Z}]$ for $i = 1 \ldots q$,
2. $\xi((\mathrm{r}\{t_1/_{x_1 \in \mathcal{V}_{S_1}}, \ldots, t_q/_{x_q \in \mathcal{V}_{S_q}}\} \ PT_1 \ldots PT_p)) = (r \ \xi_{\mathcal{Z}}(t_1) \ldots \xi_{\mathcal{Z}}(t_q) \ H_{t_1} \ldots H_{t_q} \ \xi(PT_1) \ldots \xi(PT_p))$ where $\mathcal{X} = \mathcal{V}ar(\mathrm{r})$ and $t_i \in \mathrm{T}_{\Sigma,\mathrm{S}_i}[\mathcal{Z}]$ for $i = 1 \ldots q$.

Theorem 13. *The encoding ξ defines a bijective map.*

Proof. The proof is similar to that for theorem 9. □

2. Specification of the Sort Constraints outside *the Inductive Definition.* In this case, the sort constraints cannot be deduced as before from the inductive definition of P (P_dom can not be proved). The encoding of P is parametrized by the proofs of sort constraints on its arguments and has the form: $\lambda y_1, \ldots, y_n : \mathcal{L}. \xrightarrow[i=1]{n} (\text{L\_is } y_i \text{ U}_i) \rightarrow (P\ y_1 \ldots y_n)$. We denote by $Cond_r(v)$ the property that v is a variable of sort S_v appearing in the rule r and that either v does not appear in the conclusion $Concl$ of r or S_v is strictly included in the sort Wsc($Concl[v]$). Let $\mathcal{X} = \{x_i \in \mathcal{V}_{\text{S}_i} \mid i = 1 \ldots q\}$ be the standard enumeration of $\mathcal{V}ar(\text{r})$. The rule r is implemented in Coq by the following constructor declaration of p:

$$r : \forall_{i=1}^{q} \rho(x_i) : \mathcal{L}. \xrightarrow[\{x \in \mathcal{X}_\text{S} \mid Cond_r(x)\}]{} (\text{L\_is } \rho(x)\ \text{S}) \rightarrow \xrightarrow[k=1]{m} (Q_k\ \xi_\mathcal{X}(t_1^k) \mathinner{..} \xi_\mathcal{X}(t_{n_k}^k))$$
$$\rightarrow (P\ \xi_\mathcal{X}(t_1) \mathinner{..} \xi_\mathcal{X}(t_n))$$

Example 11. The representation of "type" in Coq according to this second choice is '$\lambda\rho, e, \tau : \mathcal{L}i.(\text{Li\_is } \rho \text{ Ctxt}) \rightarrow (\text{Li\_is } e \text{ Exp}) \rightarrow (\text{Li\_is } \tau \text{ Type}) \rightarrow (type\ \rho\ e\ \tau)$' where *type* is partially defined below. We write v instead of $\rho(v)$.

Inductive Definition $type : \mathcal{L}i \rightarrow \mathcal{L}i \rightarrow \mathcal{L}i \rightarrow$ Prop
$= \cdots \mid type_{lam} : \forall \rho, x, \tau, \tau', e : \mathcal{L}i.$
$(type\ (ctxt\ (id\ x)\ \tau\ \rho)\ e\ \tau') \rightarrow (type\ \rho\ (lam\ (id\ x)\ \tau\ e)\ (arrow\ \tau\ \tau')) \cdots$

To demonstrate the correctness of this encoding of the semantics, we define an encoding map ξ from proof terms of $\text{P}(a_1, \ldots, a_n)$ to terms of type $(P\ \xi_\mathcal{Y}(a_1)\ \ldots\ \xi_\mathcal{Y}(a_n))$ in Coq and $\Gamma_{\rho(\mathcal{Y})} \cup [A_1 : (\text{L\_is } \xi_\mathcal{Y}(a_1)\ \text{U}_1), \ldots, A_n : (\text{L\_is } \xi_\mathcal{Y}(a_n) \text{U}_n)]$.

Theorem 14. *The encoding ξ defines a bijective map.*

Proof. The proof is similar to previous one except for the demonstration that the right-inverse is total. There we use the additional hypotheses $A_1, \ldots, A_n$ and the following Lemma.

Lemma 15. *Let $t \in T_{\Sigma,\text{S}}[\mathcal{X}]$ and $x_i \in \mathcal{V}ar(t)$. If we can prove '(L_is $\xi_\mathcal{X}(t)$ S)' then we can prove '(L_is $\rho(x_i)$ Wsc($t[x_i]$))' in Coq and $\Gamma_{\rho(\mathcal{X})}$.*

Notice that we have mixed in this last statement the meta level and the object level: the lemma is parametrized by the language and the Wsc operator. There is an obvious proof object for each instantiation of the lemma. □

6 Implementation in Centaur

In Centaur, the concrete and abstract syntax of a language can be written in the formalism METAL, in order to produce parsers and abstract syntax tree generators. The abstract syntax is given by a list of sorts and operators on sorts. Each sort is assigned to a set of *root* operators that defines a class of trees. The semantics specification written in Typol, can be type-checked and then compiled into Lisp or Prolog-Sepia. The input of the automatic translation from Typol to Coq consists of the declarative part of the language implementation, that is the abstract syntax specification and a type-checked Typol program. The translation is implemented following the two alternatives we have presented. This tool has already been used to compile the specifications of Mini-ML and a small imperative language called little [2]. We have done several proofs in our team about the resulting Coq specifications, for example that the interpreter of Mini ML returns values, that the Subject Reduction Theorem holds for a subset of our Mini ML, and that some program transformations about little are correct.

7 Conclusion and Future Work

In spite of some difficulties in dealing with subtyping, we gain from our approach an automatic way of representing programming languages in CCind and the possibility of reasoning efficiently about these representations by induction. Hence we would like to extend our correspondence to theorems about the language specification, such as derived rules and inference rules. Then we could improve the translation by combining the power of type theory and the expressiveness of set theory. We could for example recognize the cycles in the graph induced by the subsort relation and flatten all the sorts belonging to a given cycle into one type (solution presented in Sect. 5.2). The remaining sorts that do not belong to any cycle would be implemented by one type (solution presented in Sect. 5.1). It would also be interesting to study the impact of the representation on proof development. We aim at providing a setting where we can automate proofs in programming languages as much as possible, writing tactics and using Ctcoq. Incrementality is another concern and we would like to study the consequences of the encoding when altering the syntax, for example by adding a constructor or by adding a sort. Finally, we would like to have a system where we can build proof trees in Natural Semantics independently from the implementation of the proof construction tool. This system would feature the display of proof trees and the capability to call external theorem provers like Coq transparently.

Acknowledgements

I am grateful to Joëlle Despeyroux, Yves Bertot, André Hirschowitz, Chet Murthy and Healf Goguen for many helpful discussions and to Gilles Kahn for many pointers. I also thank all the CROAP team for useful explanations on the Centaur system.

References

1. P. Aczel: An Introduction to Inductive Definitions. The Handbook of Mathematical Logic, J. Barwise ed., North-Holland, (1992) 739-782
2. Y. Bertot, R. Fraer: Reasoning with Executable Specifications. I. Joint Conference of Theory and Practice of Software Development, LNCS, Aarhus (1995)
3. Avron, Honsell, Mason: An Overview of the Edinburgh Logical Framework. Current Trends in Hardware Verification and Automated Theorem Proving (1988)

4. Y. Bertot, G. Kahn, L. Théry: Proof by Pointing. Proceedings of Theoretical Aspects Computer Science (TACS '94), Tohoku University, Sendai, Japan, LNCS (1994) 789
5. R. Burstall, J. Goguen: Algebras, theories and freeness: an introduction for computer scientists. Theoretical Foundations of Programming Methodology, (1982) 329-350
6. A. Church: A formulation of the simple theory of types. J. of Symbolic Logic, 5 (1940) 56-68
7. O. Dahl: Verifiable Programming. Prentice Hall International series in computer science (1992)
8. J. Despeyroux, A. Hirschowitz: Higher-Order Syntax and Induction in Coq. Pr. of the fifth Int. Conf. on Logic Programming and Automated Reasoning Kiev, (1994) 16–21
9. J. Despeyroux: Theo: an Interactive Proof Development System. Scandinavian J. on Computer Science and Numerical Analysis (BIT), 32 (1992) 15–29
10. T. Despeyroux: Typol and Natural Semantics. Notes de cours pour l'Ecole Jeunes Chercheurs du GRECO de Programmation (1991)
11. T. Despeyroux: Typol: a formalism to implement Natural Semantics. Technical Report 94, Inria, Sophia-Antipolis, France (1988)
12. G. Dowek, A. Felty, H. Herbelin, G. Huet, C. Murthy, C. Parent, C. Paulin, B. Werner: The Coq Proof Assistant User's guide, Version 5.8. Technical Report 1154, Inria, Rocquencourt, France (1991)
13. P. Gardner: Representing Logics in Type Theory. Phd Thesis, Department of Computer Science, The University of Edinburgh (1992)
14. J. Hannan: Extended Natural Semantics. J. of Functional Programming, Cambridge University Press, 2 (1993) 123-152
15. R. Harper, F. Honsell, G. Plotkin: A Framework for Defining Logics. J. of the ACM, 40(1) (1993) 143-184
16. G. Huet: A Uniform Approach to Type Theory. Research Report 795, Inria, Rocquencourt, France (1988)
17. I. Jacobs. The Centaur 1.2 Manual. Technical report, Inria, Sophia-Antipolis, France (1992)
18. G. Kahn: Natural Semantics. Proceedings of the Symp. on Theorical Aspects of Computer Science, TACS, Passau, Germany (1987)
19. J.W. Lloyd: Foundations of Logic Programming. Ed. by L.Bolc, A.Bundy, P.Hayes and J.Siekmann, Germany (1987)
20. C. Paulin-Mohring: Inductive Definitions in the System Coq. Rules and Properties. Pr. of the Int. Conf. on Typed Lambda Calculi and Applications, LNCS 664 (1993) 328–345
21. J.C. Mitchell: Type Inference with Simple Subtypes. J. of Functional Programming, 1(3) (1991) 245–286
22. G.D. Plotkin: A Structural Approach to Operational Semantics. Technical Report, Aarhus, (1981) DAIMI FN-19
23. M. VanInwegen, E. Gunter: HOL-ML. Pr. of the Tech. Work. BRA 'Types' on 'Proving Properties of Programming Languages', Ed. J. Despeyroux, INRIA,Sophia-Antipolis, France (1993)
24. D. Terrasse: Translation From Typol to Coq. Pr. of the Tech. Work. BRA on 'Proving Properties of Programming Languages', Ed. J. Despeyroux, (1993)

Mongruences and Cofree Coalgebras

Bart Jacobs

CWI, Kruislaan 413, 1098 SJ Amsterdam, The Netherlands.
Email: bjacobs@cwi.nl

Abstract. A coalgebra is introduced here as a model of a certain signature consisting of a type X with various "destructor" function symbols, satisfying certain equations. These destructor function symbols are like methods and attributes in object-oriented programming: they provide access to the type (or state) X. We show that the category of such coalgebras and structure preserving functions is comonadic over sets. Therefore we introduce the notion of a 'mongruence' (predicate) on a coalgebra. It plays the dual role of a congrence (relation) on an algebra.

An algebra is a set together with a number of operations on this set which tell how to form (derived) elements in this set, possibly satisfying some equations. A typical example is a monoid, given by a set M with operations $1 \to M$, $M \times M \to M$. Here $1 = \{\emptyset\}$ is a singleton set. In mathematics one usually considers only single-typed algebras, but in computer science one more naturally uses many-typed algebras like $1 \to \mathsf{list}(A)$, $A \times \mathsf{list}(A) \to \mathsf{list}(A)$.

Here we are not primarily interested in algebras, but in coalgebras. These consist of a set together with operations "going out" of the set; they tell how to "deconstruct" elements of the set. These kind of structures naturally arise within object-oriented programming—as explained in [16, 10], but see also [13], or [5] (where the "destructors" are called "observers"). A possible (set-theoretic) semantics for objects is to see them as pairs $\langle x \in X, c: X \to T(X)\rangle$ where c is a coalgebra of a functor T: **Sets** $\to$ **Sets** interpreting the methods, and x is an element of the carrier X of the coalgebra c. Usually c will be a tuple $c = \langle c_1, \ldots, c_n\rangle$ of maps. Message passing then means application of the appropriate component of c to $x \in X$.

This picture is very elementary but clear enough to serve as our motivation. Here we concentrate on equations in such coalgebras. The main conceptual novelty that we introduce is the notion of what we call a *mongruence* (predicate). A mongruence plays the same role in coalgebra that a congruence plays in algebra: it is a predicate which is closed under the coalgebra operations, just like a congruence is a relation which is closed under the algebra operations.

We use these mongruences in particular to construct cofree coalgebras satisfying certain equations. In Mac Lane's book on category theory [14] (but see also [15]) there is a section VI, 8 called "Algebras are T-algebras". It is shown there that algebras as sets with operations satisfying certain equations are monadic over sets. The free algebra is obtained by first forming the free term algebra built on the constructors, and then taking the quotient by the least congruence relation induced by the equations. Our main result Theorem 12 is the dual of this classical result: it shows that equationally defined coalgebras are comonadic over sets. The cofree construction also proceeds in two stages: first the cofree coalgebra on the destruvctors is formed, and then one takes the sub-coalgebra given by the greatest mongruence induced by the equations.

What we present here is a fundamental study. It may have ramifications into logic, algebra, category theory, datatype theory, automata theory, semantics of object-oriented programming, database theory and linguistics. A coalgebra is a fundamental notion since it gives us a formalization of a state as a "black box" to which we only have access via specified operations. No information is given as to what is in the black box.

An explanation of the notions of mongruence and congruence (and also of induction and coinduction) at the proper level of generality requires categorical logic (in terms of fibred categories). Since we only deal with interpretation in sets here, and since we don't wish to put off too many readers, we shall be describing these notions more concretely in the logic of sets. But the notation we use is a categorical one. This 'compromise' has the advantage that readers familiar with fibred category theory can see the underlying abstract level, and readers without this familiarity can compute in sets.

1 Constructor and destructor signatures

Let At be some non-empty set, the elements of which will be regarded as atomic types. Let X be an element, not in At. We write $\overline{\mathrm{At} \cup \{X\}}$ for the formal closure of the set $\mathrm{At} \cup \{X\}$ under finite products $(1, \times)$ and coproducts $(0, +)$. The elements of this set $\overline{\mathrm{At} \cup \{X\}}$ will be regarded as types. A **constructor signature** is then essentially just a type, say $\tau \in \overline{\mathrm{At} \cup \{X\}}$, sometimes written as $\tau(X)$ to express that X may occur in τ. We think of τ as the input type of a constructor constr: $\tau(X) \to X$. We usually assume that τ is in "disjunctive normal form", i.e. of the form $\tau \equiv \tau_1 + \cdots + \tau_n$ with τ_i consisting of a finite product $\alpha_1 \times \cdots \times \alpha_m \times X \times \cdots \times X$ of atoms α_i and X's. The single constructor constr: $\tau \to X$ may then be decomposed into n constructors $\mathrm{constr}_i \colon \tau_i \to X$. For example the constructor signature of monoids is given by the type $\tau(X) = 1 + X \times X$ (with constructors $1 \to X$ and $X \times X \to X$ for unit and multiplication) and the signature for binary α-labelled trees by the type $\tau(X) = 1 + X \times \alpha \times X$ (with constructors $1 \to X$ for the empty tree, and $X \times \alpha \times X \to X$ for forming a node).

(Such a signature is called **single-typed** if its set At of atomic types is empty. For example, the above signature of monoids is single-typed.)

Given such a signature, one can form typed terms $v_1 \colon \sigma_1, \ldots, v_n \colon \sigma_n \vdash M \colon X$ starting from the typed variables $v_i \colon \sigma_i$, the constructors from the signature and the term forming operation as associated with finite products and coproducts: tupling and projections and cotupling and coprojections (including empty ones), see e.g. [12, 6]. We use capital letters Γ to abbreviate contexts $v_1 \colon \sigma_1, \ldots, v_n \colon \sigma_n$ of variable declarations. An **equation** is then a pair of terms with type X in the same context, written as $\Gamma \vdash M =_X M'$. Given a set $\mathcal{E}$ of such equations, we can take these as axioms, and start equational logic. There one has as derivation rules that $=_\sigma$ is reflexive, symmetric and transitive, and also that it is compatible: if $\Gamma \vdash M =_\sigma M'$ is derivable, then for each term $\Gamma, v{:}\,\sigma \vdash N{:}\,\rho$, the equation $\Gamma \vdash N[M/v] =_\rho N[M'/v]$ is also derivable. Finally, one has the substitution rule: if $\Gamma \vdash M{:}\,\sigma$ and $\Gamma, v{:}\,\sigma \vdash N =_\rho N'$ are derivable, then so is $\Gamma \vdash N[M/v] =_\rho N'[M/v]$.

A **destructor signature** has "destructors" where a constructor signature has constructors. Instead of telling how to build up X's (as above), a destructor tells us how to act on X. Destructors have the form $X \to \tau(X)$. But we shall consider them with parameter type σ in $\sigma \times X \to \tau(X)$, with X not occuring in σ. Equivalently, we can write it as $X \to (\sigma \Rightarrow \tau(X))$ using the exponent type former $\Rightarrow$. Of course, σ may be the singleton type 1, in which case it can be ignored.

Formally a destructor signature is given by a set At of atomic types, by an n-tuple $\tau_1, \ldots, \tau_n \in \overline{\mathrm{At} \cup \{X\}}$ of "output" types in which X may occur, and by an n-tuple $\sigma_1, \ldots, \sigma_n \in \overline{\mathrm{At}}$ of "input" or "parameter" types, in which X does not occur. We may name the destructors explicitly as in

$$\begin{array}{lcl} \mathsf{destr}_1 : \sigma_1 \times X \longrightarrow \tau_1(X) & & \mathsf{destr}_1 : X \longrightarrow \sigma_1 \Rightarrow \tau_1(X) \\ \vdots & \text{or as in} & \vdots \\ \mathsf{destr}_n : \sigma_n \times X \longrightarrow \tau_n(X) & & \mathsf{destr}_n : X \longrightarrow \sigma_n \Rightarrow \tau_n(X) \end{array}$$

using the adjoint correspondence between $\sigma \times X \to Y$ and $X \to (\sigma \Rightarrow Y)$. This is also called 'Currying'. The type of such a destructor signature is

$$\tau(X) = (\sigma_1 \Rightarrow \tau_1(X)) \times \cdots \times (\sigma_n \Rightarrow \tau_n(X)).$$

Examples of destructor signatures are

$$\mathsf{head}{:}\, X \longrightarrow \alpha, \qquad \mathsf{tail}{:}\, X \longrightarrow X$$

where one may think of X as lists (of elements of type α). A very rudimentary bank account (of a single person) may be described as

$$\mathsf{bal}{:}\, X \longrightarrow \mathsf{N}, \qquad \mathsf{credit}{:}\, \mathsf{N} \times X \longrightarrow X.$$

And a deterministic automaton with alphabet type Σ and output type U has two destructors

$$\mathsf{out}{:}\, X \longrightarrow U, \qquad \mathsf{act}{:}\, \Sigma \times X \longrightarrow X.$$

(The restriction that X should not occur in parameter types σ is a severe one; it precludes a specification of points with a distance operation, as in

$$\mathsf{hor\_coord}{:}\, X \longrightarrow \mathsf{R}, \qquad \mathsf{vert\_coord}{:}\, X \longrightarrow \mathsf{R}, \qquad \mathsf{dist}{:}\, X \times X \longrightarrow \mathsf{R}.$$

Having such X's in parameter types introduces contravariance problems for exponents—which we wish to avoid at this stage. See also the discussion in [10] on this matter.)

A destructor $\text{destr}: \sigma \times X \to \tau(X)$ will be called an **attribute** if X does not occur in the output type τ. Otherwise, it will be called a **method.**

One can also do equational logic with such destructor signatures. One takes the destructors (of a specific signature) as atomic terms, and generate a term calculus, using the term forming operations for finite product & coproduct types and exponents. An equation is then a sequent of the form $\Gamma, x: X \vdash M =_\rho M'$ where M, M' are terms of type τ in context $\Gamma, x: X$, with the restriction that X does not occur in (the types in) Γ.

For example, for the above bank account there is an obvious equation

$$n: \mathsf{N}, x: X \vdash \mathsf{bal}(\mathsf{credit}(n, x)) =_{\mathsf{N}} n + \mathsf{bal}(x).$$

(Strictly speaking, we have to regard N not as an atomic type, but N with + as an imported signature; but this is not essential in what follows.)

2 Interpreting constructors and destructors

What follows applies to an arbitrary category $\mathbb{C}$ with finite product $(1, \times)$, coproducts $(0, +)$ and exponents $\Rightarrow$. Such a category is called bicartesian closed. The projections will be written as $\pi: X \times Y \to X$, $\pi': X \times Y \to Y$ and the coprojections as $\kappa: X \to X + Y$, $\kappa': Y \to X + Y$. Tupling of $f: Z \to X$ and $g: Z \to Y$ yields $\langle f, g\rangle: Z \to X \times Y$, and cotupling of $f: X \to Z$ and $g: Y \to Z$ gives $[f, g]: X + Y \to Z$.

Definition 1. A **polynomial functor** $T: \mathbb{C} \to \mathbb{C}$ is a functor which is built up in the following fashion.

(i) The identity functor $id: \mathbb{C} \to \mathbb{C}$ is polynomial.

(ii) For any object $A \in \mathbb{C}$, the constant functor $X \mapsto A$ is polynomial.

(iii) Given polynomial functors $T, S: \mathbb{C} \to \mathbb{C}$, the following functors are also polynomial,

$$X \mapsto T(X) \times S(X) \quad \text{and} \quad X \mapsto T(X) + S(X).$$

(iv) If $A \in \mathbb{C}$ and T is a polynomial functor, then

$$X \mapsto A \Rightarrow T(X)$$

is also polynomial. (This last clause will only be needed for destructor signatures.)

Let At be a set of atomic types, and assume that for each $\alpha \in \text{At}$, an interpretation $A = [\![\alpha]\!] \in \mathbb{C}$ is given. Then, associated with the type $\tau(X) = \tau_1(X) + \cdots + \tau_n(X)$ of a constructor signature, there is a polynomial functor $T: \mathbb{C} \to \mathbb{C}$, which follows the structure of τ. For example, the functor associated with $1 \to X, \alpha \times X \to X$ is $X \mapsto 1 + A \times X$, where $A = [\![\alpha]\!]$.

Similarly, associated with type type $\tau(X) = (\sigma_1 \Rightarrow \tau_1(X)) \times \cdots \times (\sigma_n \Rightarrow \tau_n(X))$ of a destructor signature there is a polynomial functor $T: \mathbb{C} \to \mathbb{C}$. Here one really needs clause (iv) about exponents. For instance, the functor associated with the above automaton $X \to U$, $\Sigma \times X \to X$ is $X \mapsto [\![U]\!] \times ([\![\Sigma]\!] \Rightarrow X)$.

For a functor $T: \mathbb{C} \to \mathbb{C}$, an **algebra** is a "carrier" object $I \in \mathbb{C}$ together with a map $a: T(I) \to I$. And a morphism of algebras (also called 'algebra map') from $(a: T(I) \to I)$ to $(b: T(J) \to J)$ is a morphism $u: I \to J$ in $\mathbb{C}$ between the carrier objects with $u \circ a = b \circ T(u)$. This yields a category Alg(T). A basic observation is that an algebra of the polynomial functor T associated with a constructor signature $\mathsf{constr}_1: \tau_1(X) \to X, \ldots, \mathsf{constr}_n: \tau_n(X) \to X$ gives an interpretation $I = [\![X]\!] \in \mathbb{C}$ of X and an interpretation

$$a = [\,[\![\mathsf{constr}_1]\!], \ldots, [\![\mathsf{constr}_n]\!]\,]: T(I) \longrightarrow I$$

of all the constructors in one n-cotuple. And algebra maps are morphisms in $\mathbb{C}$ between the interpretations of X, commuting with the interpretations of the constructors.

Dually, a **coalgebra** of a functor $T: \mathbb{C} \to \mathbb{C}$ consists of a carrier object I together with a map $c: I \to T(I)$. And a map of coalgebras from $(c: I \to T(I))$ to $(d: J \to T(J))$ is a map $u: I \to J$ in $\mathbb{C}$ with $d \circ u = T(u) \circ c$. We thus get a category CoAlg(T). And a coalgebra of the functor T associated with a destructor signature $\mathsf{destr}_1: X \to (\sigma_1 \Rightarrow \tau_1(X)), \ldots, \mathsf{destr}_n: X \to (\sigma_n \Rightarrow \tau_n(X))$ consists of a carrier object $I = [\![X]\!] \in \mathbb{C}$ together with a map

$$c = \langle [\![\mathsf{destr}_1]\!], \ldots, [\![\mathsf{destr}_n]\!] \rangle: I \longrightarrow T(I)$$

interpreting the destructors in a single n-tuple. Again, coalgebra maps commute with these interpretations of the destructors.

Example 1. Take $\mathbb{C} =$ **Sets**, the category of sets and functions. The empty set 0 (or $\emptyset$) is initial object and $1 = \{\emptyset\}$ is terminal object. The cartesian product $X \times Y$ of two sets X, Y is categorical product, and the disjoint union $X + Y$ is coproduct. We consider the bank account destructor signature $\mathsf{bal}: X \to \mathsf{N}$, credit: $\mathsf{N} \times X \to X$. Let's take the set of natural numbers $\mathbb{N}$ as interpretation of the atomic type N (and addition on $\mathbb{N}$ as interpretation of $+$ on N). The functor involved is $X \mapsto \mathbb{N} \times (\mathbb{N} \Rightarrow X)$. We mention the following examples of coalgebras of this functor.

(a) $\emptyset \mapsto \langle 0, \lambda n.\, \emptyset \rangle : 1 \longrightarrow \mathbb{N} \times (\mathbb{N} \Rightarrow 1)$
(b) $n \mapsto \langle n, \lambda m.\, n + m \rangle : \mathbb{N} \longrightarrow \mathbb{N} \times (\mathbb{N} \Rightarrow \mathbb{N})$
(c) $\alpha \mapsto \langle \mathsf{sum}(\alpha), \lambda n.\, n \cdot \alpha \rangle : \mathbb{N}^* \longrightarrow \mathbb{N} \times (\mathbb{N} \Rightarrow \mathbb{N}^*)$
(d) $\varphi \mapsto \langle \varphi([\,]), \lambda n.\, \lambda \alpha.\, \varphi(n \cdot \alpha) \rangle : (\mathbb{N}^* \Rightarrow \mathbb{N}) \longrightarrow \mathbb{N} \times (\mathbb{N} \Rightarrow (\mathbb{N}^* \Rightarrow \mathbb{N}))$.

One sees how the various interpretations of the 'balance' and 'credit' destructors are encoded in these coalgebras.

The next step is to see what validity of equations means in such (co)algebra models. Given the interpretations of the basic terms (the constructors and destructors), one can extend the interpretation to all terms, in such a way that a term in context, $\Gamma = x_1: \sigma_1, \ldots, x_n: \sigma_n \vdash M: \rho$ becomes a morphism, $[\![M]\!]: [\![\Gamma]\!]$

$\longrightarrow [\![\rho]\!]$, where $[\![\Gamma]\!] = [\![\sigma_1]\!] \times \cdots \times [\![\sigma_n]\!]$. As usual, one has that (co)algebra maps commute with these interpretations of terms (in the obvious sense).

An equation (or axiom) $\Gamma \vdash M =_\rho M'$ is then **valid** (or **holds**) if the two resulting maps $[\![M]\!], [\![M']\!]: [\![\Gamma]\!] \rightrightarrows [\![\rho]\!]$ are equal. There is the usual soundness result: all derivable equations are also valid.

In the bank account example above, the coalgebras (b) and (c) satisfy the equation mentioned at the end of the previous section.

For a suitable collection $\mathcal{E}$ of equations, we write $\mathrm{Alg}(T, \mathcal{E}) \hookrightarrow \mathrm{Alg}(T)$ and $\mathrm{CoAlg}(T, \mathcal{E}) \hookrightarrow \mathrm{CoAlg}(T)$ for the full subcategories of algebras and coalgebras satisfying the equations in $\mathcal{E}$.

In this paper we are interested in forcing the validity of equations. For algebras this is done in the familiar way via quotients: given a set of equations $\mathcal{E} = \{\Gamma_\ell \vdash M_\ell =_X M'_\ell \mid \ell \in L\}$, one takes the image of each tuple $\langle [\![M_\ell]\!], [\![M'_\ell]\!]\rangle$ as $[\![\Gamma_\ell]\!] \twoheadrightarrow R_\ell \hookrightarrow [\![X]\!] \times [\![X]\!]$, and one forms the union $R = \bigcup_\ell R_\ell \hookrightarrow [\![X]\!] \times [\![X]\!]$ of these. Quotienting $[\![X]\!]$ by the least congruence relations containing R (that is also an equivalence relation) yields the free algebra (on the given one) which satisfies the equations in $\mathcal{E}$.

In the coalgebra case, one does not quotient by a congruence, but one takes the subcoalgebra induced by what we call a **mongruence**. To explain properly what the latter is, we need to say more about the logic of predicates on sets.

3 Predicates and relations

First we define the categories **Pred** and **Rel** of predicates and relations on sets, and mention their basic properties. Then we describe how polynomial functors T: **Sets** $\rightarrow$ **Sets** can be lifted to polynomial functors $\mathrm{Pred}(T)$: **Pred** $\rightarrow$ **Pred** and $\mathrm{Rel}(T)$: **Rel** $\rightarrow$ **Rel** by induction on the structure of T. This is as in [7, 9, 8]. Algebras and coalgebras of these lifted functors $\mathrm{Pred}(T)$ and $\mathrm{Rel}(T)$ give important (logical) information about T.

Definition 2. (i) The category **Pred** has as objects pairs (I, P) where $P \subseteq I$ is a subset of I—or a predicate on I. Morphisms $(P \subseteq I) \rightarrow (Q \subseteq J)$ are functions $u: I \rightarrow J$ for which one has an implication $\forall i \in I.\, P(i) \supset Q(u(i))$.

(ii) Similarly, the category **Rel** has pairs (I, R) where $R \subseteq I \times I$ as objects. And morphisms $(R \subseteq I \times I) \rightarrow (S \subseteq J \times J)$ are functions $u: I \rightarrow J$ satisfying $\forall i, i' \in I.\, R(i, i') \supset S(u(i), u(i'))$.

There are obvious forgetful functors **Pred** $\rightarrow$ **Sets** and **Rel** $\rightarrow$ **Sets**, namely $(I, P) \mapsto I$ and $(I, R) \mapsto I$. These are (split) fibrations: for each function $u: I \rightarrow J$ there are substitution functors:

$$(Q \subseteq J) \mapsto (\{i \mid u(i) \in Q\} \subseteq I)$$
$$(S \subseteq J \times J) \mapsto (\{(i, i') \mid (u(i), u(i')) \in S\} \subseteq I \times I).$$

Both these functors are denoted by u^*; they are functors indeed, namely from the "fibre" poset of predicates/relations on J to the "fibre" poset of predicates/relations on I (with inclusion as ordering). The assignment $u \mapsto u^*$ is functorial: one has $id^* = id$ and $(v \circ u)^* = u^* \circ v^*$.

There are special functors $1\colon \mathbf{Sets} \to \mathbf{Pred}$ and $\mathrm{Eq}\colon \mathbf{Sets} \to \mathbf{Rel}$, which map a set $I \in \mathbf{Sets}$ to the truth predicate $1(I) = (I \subseteq I)$ on I, and to the equality relation $\mathrm{Eq}(I) = (\{(i,i) \mid i \in I\} \subseteq I \times I)$ on I.

We mention some basics in the logic of sets.

Proposition 3. (i) *Quantification: for a function $u\colon I \to J$ the substitution functor u^* has (both for predicates and for relations) a left adjoint $\coprod_u$ and a right adjoint $\prod_u$. These are given on $P \subseteq I$ and $R \subseteq I \times I$ by*

$$\begin{aligned}
\textstyle\coprod_u(P) &= \{j \in J \mid \exists i \in I.\, u(i) = j \wedge P(i)\} \\
\textstyle\prod_u(P) &= \{j \in J \mid \forall i \in I.\, u(i) = j \supset P(i)\} \\
\textstyle\coprod_u(R) &= \{(j,j') \in J \times J \mid \exists i, i' \in I.\, u(i) = j \wedge u(i') = j' \wedge R(i,i')\} \\
\textstyle\prod_u(R) &= \{(j,j') \in J \times J \mid \forall i, i' \in I.\, u(i) = j \wedge u(i') = j' \supset R(i,i')\}.
\end{aligned}$$

(The "Beck-Chevalley" conditions holds for these $\prod$'s and $\coprod$'s.)

(ii) *Subsets (or comprehension) and quotients: the truth functor $1\colon \mathbf{Sets} \to \mathbf{Pred}$ has a right adjoint $\{-\}$; it maps a predicate $(P \subseteq I)$ to its extent $\{i \in I \mid P(i)\} \in \mathbf{Sets}$. And the equality functor $\mathrm{Eq}\colon \mathbf{Rel} \to \mathbf{Sets}$ has a left adjoint Q; it maps a relation $(R \subseteq I \times I)$ to the quotient $I/\widehat{R} \in \mathbf{Sets}$, where $\widehat{R} \subseteq I \times I$ is the least equivalence relation containing R.*

(iii) *Logical predicates and logical relations: both $\mathbf{Pred}$ and $\mathbf{Rel}$ are bicartesian closed, and the functors $\mathbf{Pred} \to \mathbf{Sets}$, $\mathbf{Rel} \to \mathbf{Sets}$ strictly preserve this structure. These BiCCC-structures are given by the so-called "logical" formulas; e.g.*

$$\begin{aligned}
(R \subseteq I \times I) \times (S \subseteq J \times J) &= (\pi \times \pi)^*(R) \wedge (\pi' \times \pi')^*(S) \\
&= \{(\langle i,j\rangle, \langle i',j'\rangle) \mid R(i,i') \wedge S(j,j')\} \\
(P \subseteq I) + (Q \subseteq J) &= \textstyle\coprod_\kappa(P) \vee \coprod_{\kappa'}(Q) \\
&= \{z \in I + J \mid (\exists i \in I.\, z = \kappa i \wedge P(i)) \\
&\qquad \vee (\exists j \in J.\, z = \kappa' j \wedge Q(j))\}.
\end{aligned}$$

See [7, 8] for a more abstract description. □

A polynomial functor $T\colon \mathbf{Sets} \to \mathbf{Sets}$ acting on sets can be lifted to two new polynomial functors $\mathrm{Pred}(T)\colon \mathbf{Pred} \to \mathbf{Pred}$ acting on predicates and to $\mathrm{Rel}(T)\colon \mathbf{Rel} \to \mathbf{Rel}$ acting on relations, in commuting squares

$$\begin{array}{ccc}
\mathbf{Pred} & \xrightarrow{\mathrm{Pred}(T)} & \mathbf{Pred} \\
\downarrow & & \downarrow \\
\mathbf{Sets} & \xrightarrow[T]{} & \mathbf{Sets}
\end{array}
\qquad \text{and} \qquad
\begin{array}{ccc}
\mathbf{Rel} & \xrightarrow{\mathrm{Rel}(T)} & \mathbf{Rel} \\
\downarrow & & \downarrow \\
\mathbf{Sets} & \xrightarrow[T]{} & \mathbf{Sets}
\end{array}$$

One defines these functors by induction on the structure of T. One builds $\mathrm{Pred}(T)$ and $\mathrm{Rel}(T)$ with the same structure as T, except that the constant functor $X \mapsto A$ in T—for $A \in \mathbf{Sets}$—is replaced by the constant functors

$$X \mapsto 1(A) \text{ in } \mathrm{Pred}(T) \qquad \text{and} \qquad X \mapsto \mathrm{Eq}(A) \text{ in } \mathrm{Rel}(T).$$

The following results are proved by induction on the structure of the functor T.

Lemma 4. (i) *These* Pred(T) *and* Rel(T) *are fibred functors: they commute with substitution functors* u^*.

(ii) *They are also cofibred: they commute with the* $\coprod_u$*'s. (In order to handle clause (iv) on exponents in the definition of polynomial functors, one needs the Axiom of Choice.)*

(iii) Pred(T) *commutes with truth, and* Rel(T) *with equality:* $\text{Pred}(T) \circ 1 = 1 \circ T$ *and* $\text{Rel}(T) \circ \text{Eq} = \text{Eq} \circ T$. □

4 Mongruences

In this section we turn to algebras and coalgebras of the lifted functors Pred(T) and Rel(T). The following table gives an overview.

	algebra	coalgebra
Pred(T)	induction assumption	**mongruence**
Rel(T)	congruence	coinduction assumption
	(on the underlying T-algebra)	(on the underlying T-coalgebra)

This table defines a mongruence (predicate) $P \subseteq I$ as a coalgebra in **Pred** of the form $P \to \text{Pred}(T)(P)$. If the underlying map is $c: I \to T(I)$, then we say that P is a mongruence on the (T-)coalgebra c.

(From [9, 8] it follows that in the logic of sets one always has induction and coinduction: for an initial T-algebra $a: T(I) \to I$, the truth predicate $1(I)$ on I carries $\text{Pred}(T)(1(I)) = 1(T(I)) \overset{1(a)}{\to} 1(I)$ as initial Pred(T)-algebra. And for a terminal coalgebra $c: I \to T(I)$ the equality relation Eq(I) on I has $\text{Eq}(I) \overset{\text{Eq}(c)}{\to} \text{Eq}(T(I)) = \text{Rel}(T)(\text{Eq}(I))$ as terminal Rel(T)-coalgebra. The reader unfamiliar with this (co)algebraic description of (co)induction may wish to see that for example finite-list-induction amounts to initiality of the truth predicate on the initial algebra $\text{list}(A) = A^*$ of $X \mapsto 1 + A \times X$. And coinduction for infinite lists amounts to terminality of the equality relation on the terminal coalgebra A^ω of $X \mapsto A \times X$.)

We turn to congruences and mongruences. First we notice that a relation $R \subseteq I \times I$ is a congruence on an algebra $a: T(I) \to I$ if and only if $\text{Rel}(T)(R) \subseteq a^*(R)$, if and only if $\coprod_a(\text{Rel}(T)(R)) \subseteq R$. Similarly a predicate $P \subseteq I$ is a mongruence on a coalgebra $c: I \to T(I)$ if and only if $P \subseteq c^*\text{Pred}(T)(P)$, if and only if $\coprod_c(P) \subseteq \text{Pred}(T)(P)$.

Example 2. Consider the functor $T(X) = A \times X$ on **Sets**, and a T-coalgebra $\langle h, t\rangle: I \to A \times I$. One thinks of I as a set of lists of A's, with $h: I \to A$ as head and $t: I \to I$ as tail operation. A predicate $P \subseteq I$ is then a mongruence on $\langle h, t\rangle$ if and only if for each $i \in I$ one has $i \in P$ implies $t(i) \in P$, since

$\mathrm{Pred}(T)(P) = \{\langle a, i\rangle \mid a \in A \wedge t(i) \in P\} \subseteq A \times I$. Thus a mongruence is closed under application of the tail method.

Notice that for an arbitrary predicate $P \subseteq I$ there is a greatest mongruence $\underline{P} \subseteq P$, namely the predicate $\underline{P} = \{i \in I \mid \forall n \in \mathbb{N}.\, t^{(n)}(i) \in P\}$. This will be formulated more generally in the next section.

The following are basic examples of congruences and mongruences.

Proposition 5. (i) *Equality is a congruence and truth is a mongruence.*

(ii) *Kernel relations of algebra maps are congruences, and images of coalgebra maps are mongruences.*

Proof. (i) Because $\mathrm{Rel}(T)(\mathrm{Eq}(I)) = \mathrm{Eq}(T(I))$ and $\mathrm{Pred}(T)(1(I)) = 1(T(I))$.

(ii) The kernel relation of an algebra map $u: I \to J$ from a to b is the relation $\mathrm{Ker}(u) = u^*(\mathrm{Eq}(J))$. It is a congruence since

$$\begin{aligned} a^*\mathrm{Ker}(u) &= a^*u^*\mathrm{Eq}(J) = T(u)^*b^*\mathrm{Eq}(J) \\ &\supseteq T(u)^*\mathrm{Eq}(T(J)) \\ &= T(u)^*\mathrm{Rel}(T)(\mathrm{Eq}(J)) \\ &= \mathrm{Rel}(T)(u^*\mathrm{Eq}(J)) = \mathrm{Rel}(T)(\mathrm{Ker}(u)). \end{aligned}$$

The image of a coalgebra map $u: I \to J$ from c to d is $\mathrm{Im}(u) = \coprod_u 1(I)$. It is a mongruence since

$$\begin{aligned} \textstyle\coprod_d \mathrm{Im}(u) &= \textstyle\coprod_d \coprod_u 1(I) = \coprod_{T(u)} \coprod_c 1(I) \\ &\subseteq \textstyle\coprod_{T(u)} 1(T(I)) \\ &= \textstyle\coprod_{T(u)} \mathrm{Pred}(1(I)) \\ &= \mathrm{Pred}(T)(\textstyle\coprod_u 1(I)) = \mathrm{Pred}(T)(\mathrm{Im}(u)). \end{aligned}$$ □

We have already seen that the predicate lifting of T commutes with the fibration and with truth; but also with comprehension, as we shall see next.

Lemma 6. *The canonical map $T \circ \{-\} \to \{-\} \circ \mathrm{Pred}(T)$ is an isomorphism.*

Proof. By induction on the structure of T, using that the comprehension functor $\{-\}$: **Pred** → **Sets** preserves finite products and coproducts. For exponents we explicitly check: for $T(-) = A \Rightarrow S(-)$ and $P \subseteq I$,

$$\begin{aligned} \{\mathrm{Pred}(T)(P)\} &= \{f \in A \Rightarrow S(I) \mid \forall a.\, f(a) \in \mathrm{Pred}(S)(P)\} \\ &= A \Rightarrow \{\mathrm{Pred}(S)(P)\} \\ &\cong A \Rightarrow S(\{P\}) \qquad \text{by induction hypothesis} \\ &= T(\{P\}). \end{aligned}$$ □

Corollary 7. *A predicate $P \subseteq I$ is a mongruence on a coalgebra $c: I \to T(I)$ if and only if the extent $\{P\} \hookrightarrow I$ carries a (necessarily unique) sub-coalgebra structure c_P in*

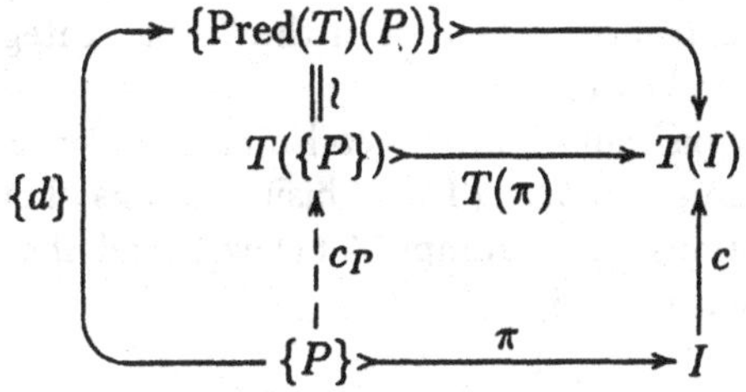

Proof. Because there are bijective correspondences between

(a) coalgebras $d: P \to \mathrm{Pred}(T)(P)$ over $c: I \to T(I)$.

(b) functions $\{d\}: \{P\} \to \{\mathrm{Pred}(T)(P)\}$ making the outer rectangle commute.

(c) functions (or subcoalgebras) $c_P: \{P\} \to T(\{P\})$ making the inner rectangle commute. □

Remark. This result is like: quotients by congruences inherit algebra structures. It is important to stress that the essence of mongruences and congruences (and also of induction and coinduction) is "logical": these notions live in **Pred** and **Rel**—and in other fibred categories where one need not have subsets $\{-\}$ or quotients—and the induced coalgebras and algebras in **Sets** are secondary.

One may be tempted to define a mongruence on a coalgebra $c: I \to T(I)$ in a seemingly equivalent (and much simpler) way as a predicate $P \hookrightarrow I$ for which there is a map c' in

$$\begin{array}{ccc} P & \xrightarrow{c'} & T(P) \\ {\scriptstyle m}\downarrow & & \downarrow{\scriptstyle T(m)} \\ I & \xrightarrow{c} & T(I) \end{array} \qquad (*)$$

The content of the above corollary is that the definition (*) is the same as the earlier definition of mongruence as a Pred(T)-coalgebra—as long as we restrict ourselves to polynomial functors T on **Sets**.

But this "subcoalgebra" definition (*) has two serious drawbacks.

(i) It makes being a mongruence a structure instead of a property. One has to know that T preserves injections (or monos in an arbitrary category) so that there can be at most one such c'.

(ii) But a more serious drawback is that this definition is not "logical" in the sense that it is on the wrong level: it is not about (abstract) predicates, but about types. To explain this criticism properly one needs some fibred category theory. There may be a "base" category that one wishes to reason about in different ways (i.e. with different logics). In fibred category theory one can do this by putting different fibrations on this "base" category. By using the liftings Pred(T) and Rel(T) one also gets these different ways of reasoning about T.

For example, one may wish to reason about a base category of metric spaces M with closed subsets $X \subseteq M$, or with morphisms $M \to [0,1]$ as predicates on M. This involves different logics, which have their own "logical liftings". The formulation (∗) cannot take such differences into account. It is on the wrong level: in the "base" category instead of in the "total" category of the fibration, which is where the predicates live.

The definition of bisimulation by Aczel and Mendler [2] is not on the right level for the very same reasons. (These bisimulations should be what we call coinduction assumptions, see Example 4 below.) And the same applies to the definition of congruence in [15].

5 Greatest mongruences

In algebra one always finds phrases like "let $\overline{R}$ be the least congruence containing a relation R". In coalgebra we shall be using "the greatest mongruence $\underline{P}$ contained in a predicate P". In this section we investigate how to obtain such $\underline{P}$ via fixed points.

Let $T\colon \mathbf{Sets} \to \mathbf{Sets}$ be a polynomial functor and $c\colon I \to T(I)$ a coalgebra. For a predicate $P \subseteq I$ we seek the greatest mongruence $\underline{P} \subseteq I$ on c with $\underline{P} \subseteq P$. Recall that $Q \subseteq I$ is a mongruence on c if and only if $Q \subseteq c^*\mathrm{Pred}(T)(Q)$.

Lemma 8. *The function $\mathcal{P}(I) \to \mathcal{P}(I)$ given by $Q \mapsto c^*\mathrm{Pred}(T)(Q)$ preserves filtered infima $\bigwedge^{\downarrow}$ (with respect to inclusion).*

Proof. By induction on T. The most difficult case is $T(-) = T_1(-) + T_2(-)$. Then, for $Q = \bigwedge^{\downarrow}_{\ell} Q_\ell$ in $\mathcal{P}(I)$, we have $c^*\mathrm{Pred}(T)(Q) = c^* \coprod_\kappa \bigwedge^{\downarrow}_{\ell} \mathrm{Pred}(T_1)(Q_\ell) \cup c^* \coprod_{\kappa'} \bigwedge^{\downarrow}_{\ell} \mathrm{Pred}(T_2)(Q_\ell)$, where

$$\begin{aligned} c^* \textstyle\coprod_\kappa \bigwedge^{\downarrow}_{\ell} \mathrm{Pred}(T_1)(Q_\ell) &= \{i \mid \exists x \in T_1(I).\, c(i) = \kappa x \wedge x \in \textstyle\bigwedge^{\downarrow}_{\ell} \mathrm{Pred}(T_1)(Q_\ell)\} \\ &\overset{(*)}{=} \textstyle\bigwedge^{\downarrow}_{\ell} \{i \mid \exists x \in T_1(I).\, c(i) = \kappa x \wedge x \in \mathrm{Pred}(T_1)(Q_\ell)\} \\ &= \textstyle\bigwedge^{\downarrow}_{\ell} c^* \coprod_\kappa \mathrm{Pred}(T_1)(Q_\ell). \end{aligned}$$

The equation (∗) holds because we have a filtered meet (so that there is at least one ℓ), and because the coprojections κ, κ' are injective. □

Proposition 9. *For an arbitrary predicate $P \subseteq I$, put*

$$P_0 = P \qquad \text{and} \qquad P_{n+1} = P_n \cap c^*\mathrm{Pred}(T)(P_n).$$

We get a descending chain $(P_n)_{n \in \mathbb{N}}$, so that we can take $\underline{P} = \bigwedge^{\downarrow}_n P_n$.

This $\underline{P} \subseteq I$ is then the greatest mongruence contained in P.

Proof. The predicate $\underline{P}$ is the greatest fixed point $\underline{P} = \bigwedge^{\downarrow}_n \Phi^{(n)}(P)$ of the $\bigwedge^{\downarrow}$-preserving function $\Phi\colon {\downarrow} P \to {\downarrow} P$ given by $Q \mapsto Q \cap c^*\mathrm{Pred}(T)(Q)$. From $\underline{P} = \Phi(\underline{P})$ we get $\underline{P} \subseteq P$ and $\underline{P} \subseteq c^*\mathrm{Pred}(T)(\underline{P})$, so that $\underline{P}$ is a mongruence contained in P. And if $Q \subseteq P$ is also a mongruence, then $Q \subseteq c^*\mathrm{Pred}(T)(Q)$, so that $Q = Q \cap c^*\mathrm{Pred}(T)(Q) = \Phi(Q)$. But since $\underline{P}$ is the greatest fixed point of Φ, we get $Q \subseteq \underline{P}$. □

6 Cofree coalgebras without equations

The standard way to obtain the terminal coalgebra of a functor T: **Sets** $\to$ **Sets** is to take the limit of the ω-chain $1 \leftarrow T(1) \leftarrow T^2(1) \leftarrow \cdots$, under suitable cocontinuity assumptions on T. These are always satisfied by the polynomial functors T that come from destructor signatures. The (carriers of the) terminal coalgebras that one gets in this way are sets of (generally infinite) trees, which can be described in terms of observations, see [11]. We mention a particular case (which is essentially as used in [13, 16]).

Proposition 10. *The terminal coalgebra of the functor* $X \mapsto A \times (B \Rightarrow X)$ *associated with the signature* $X \longrightarrow A$, $B \times X \longrightarrow X$ *is the set of functions* $B^\star \Rightarrow A$, *with operations* $(B^\star \Rightarrow A) \longrightarrow A$ *by* $\varphi \mapsto \varphi([\,])$, *and* $B \times (B^\star \Rightarrow A) \longrightarrow (B^\star \Rightarrow A)$ *by* $(b, \varphi) \mapsto \lambda\beta.\, \varphi(b \cdot \beta)$. □

With this result, we know that the terminal coalgebra for the bank account signature bal: $X \to \mathsf{N}$, credit: $\mathsf{N} \times X \to X$ is $\mathbb{N}^\star \to \mathbb{N}$. By suitably rearranging destructors, we can also use the proposition for an extended bank account, with additional destructors name: $X \to \mathsf{M}$, debit: $\mathsf{N} \times X \to X$. The name and balance attributes may be combined into a single attribute $X \to \mathsf{N} \times \mathsf{M}$. Similarly, the credit and debit methods may be combined into a single method $(\mathsf{N} + \mathsf{N}) \times X \to X$. As terminal coalgebra we thus get $(\mathbb{N} + \mathbb{N})^\star \Rightarrow M \times \mathbb{N}$, where $M = [\![\mathsf{M}]\!]$ is a set of 'names' and $\mathbb{N} = [\![\mathsf{N}]\!]$.

Theorem 11. *For a polynomial functor* T: **Sets** $\to$ **Sets** *we get a right adjoint* G_0 *to the forgetful functor* $U: \mathrm{CoAlg}(T) \to$ **Sets** *by*

$$C \mapsto [\pi' \circ c \text{ where } c \text{ is the terminal coalgebra of } X \mapsto C \times T(X)].$$

Moreover, the forgetful functor U *is comonadic: the induced comonad* $H_0 = UG_0$ *on* **Sets** *has an isomorphism* $K: \mathrm{CoAlg}(T) \to \mathbf{Sets}^{H_0}$ *as comparison functor—where* $\mathbf{Sets}^{H_0}$ *is the category of coalgebras of the comonad* H_0.

Proof. By definition we have $c: H_0C \xrightarrow{\cong} C \times T(H_0C)$ and $G_0C = (\pi' \circ c: H_0C \to T(H_0C))$. For a T-coalgebra $d: W \to TW$ there is a bijective correspondence between functions $f: W \to C$ and coalgebra maps $g: W \to H_0C$. Indeed, given f, one obtains $\overline{f}: W \to H_0C$ with $c \circ \overline{f} = id \times T(\overline{f}) \circ \langle f, d\rangle$ by terminality of c. This $\overline{f}$ is then also a coalgebra map $d \to G_0(C)$. And conversely, given a coalgebra map $g: W \to H_0C$, one takes $\overline{g} = \pi \circ c \circ g: W \to C$.

Comonadicity follows from "co-Beck". But it is not hard to see that the inverse of the comparison functor K maps a (comonad-)coalgebra $\psi: C \to H_0C$ to the T-coalgebra

$$C \xrightarrow{\psi} H_0C \xrightarrow{\pi' \circ c} T(H_0C) \xrightarrow{T(\pi \circ c)} T(C).$$ □

The algebra version of this result is quite familiar: for a polynomial functor T, the *left* adjoint to the forgetful functor $\mathrm{Alg}(T) \to \mathbf{Sets}$ maps a "set of variables" V to the algebra of terms, with variables from V; this algebra is obtained as $v \circ \kappa'$ from the initial algebra v of $X \mapsto V + T(X)$. The resulting adjunction occurs as Proposition 5 in [4, Section 9.4], with this initial algebra computed explicitly as a colimit. In the coalgebra case the set C on which one constructs the cofree coalgebra may be seen as a set of "colours", which are used as extra labels in the trees in the terminal coalgebra of T, see [11].

7 Cofree coalgebras with equations

Assume a destructor signature with $T\colon \mathbf{Sets} \to \mathbf{Sets}$ as associated functor, and with a set of equations, $\mathcal{E} = \{\Gamma_\ell, x\colon X \vdash M_\ell =_{\rho_\ell} M'_\ell \mid \ell \in K\}$. If we also assume a coalgebra $c\colon I \to T(I)$, then we can interprete these terms as maps $[\![M_\ell]\!], [\![M'_\ell]\!]\colon [\![\Gamma_\ell]\!] \times [\![X]\!] \rightrightarrows [\![\rho_\ell]\!]$, where $[\![X]\!] = I$. By abstraction, we get two maps $[\![X]\!] \rightrightarrows ([\![\Gamma_\ell]\!] \Rightarrow [\![\rho_\ell]\!])$, the equalizer of which will be written as $P_\ell \hookrightarrow [\![X]\!]$. Hence $P_\ell = \{x \in [\![X]\!] \mid \forall y \in [\![\Gamma_\ell]\!]. [\![M_\ell]\!](y, x) = [\![M'_\ell]\!](y, x)\}$. Put $P = \bigcap_\ell P_\ell \hookrightarrow [\![X]\!]$, and let $\underline{P}$ be the greatest mongruence contained in this intersection P. Then $\underline{P} \hookrightarrow I$ carries a sub-coalgebra structure $c_{\mathcal{E}}\colon \underline{P} \to T(\underline{P})$, which satisfies the equations in $\mathcal{E}$. The latter follows from the inclusion $\underline{P} \subseteq P$.

Example 3. Let us reconsider the bank account signature $\mathsf{bal}\colon X \to \mathsf{N}$, $\mathsf{credit}\colon \mathsf{N} \times X \to X$ and its (terminal) coalgebra with carrier $\mathsf{N}^* \Rightarrow \mathsf{N}$ as in Example 1. Its operations are $\mathsf{bal}(\varphi) = \varphi([\,])$ and $\mathsf{credit}(n, \varphi) = \lambda\alpha.\, \varphi(n \cdot \alpha)$. We consider the singleton set of equations,

$$\mathcal{E} = \{n\colon \mathsf{N}, x\colon X \vdash \mathsf{bal}(\mathsf{credit}(n, x)) =_{\mathsf{N}} n + \mathsf{bal}(x)\}.$$

We get $P \subseteq \mathsf{N}^* \Rightarrow \mathsf{N}$, as above, consisting of those $\varphi \in \mathsf{N}^* \Rightarrow \mathsf{N}$ with $\forall n \in \mathsf{N}.\, \varphi(n \cdot [\,]) = n + \varphi([\,])$. If we construct the greatest mongruence $\underline{P}$ contained in P as in Proposition 9, then $P_0 = P$ and

$$P_1 = P_0 \cap \{\varphi \mid \forall n_1, n_2.\, \varphi(n_1 \cdot n_2 \cdot [\,]) = n_2 + \varphi(n_1 \cdot [\,])\}.$$

Carrying on like this, we get

$$\underline{P} = \{\varphi \mid \forall \alpha \in \mathsf{N}^*. \forall n \in \mathsf{N}.\, \varphi(\alpha \cdot n \cdot [\,]) = n + \varphi(\alpha \cdot [\,])\}.$$

This means that $\varphi \in \underline{P}$ is completely determined by its value $\varphi([\,]) \in \mathsf{N}$ on the empty list. Hence $\underline{P}$ can be identified with N, and the inclusion $\mathsf{N} \hookrightarrow (\mathsf{N}^* \Rightarrow \mathsf{N})$ is $n \mapsto \lambda\alpha.\, n + \mathsf{sum}(\alpha)$. The induced coalgebra structure $\mathsf{N} \to \mathsf{N} \times (\mathsf{N} \Rightarrow \mathsf{N})$ on N is $n \mapsto \langle n, \lambda m.\, n + m\rangle$ as in Example 1 (b). It satisfies the equations.

(Notice, by the way, that the value $n = \varphi([\,]) \in \mathsf{N}$ can be seen as the amount of money put in the bank account at initialization. The resulting account $\lambda\alpha.\, n + \mathsf{sum}(\alpha)$ may be read as $\mathsf{new}(n)$ in OO-terminology.)

The above procedure of obtaining coalgebras satisfying certain equations yields an outcome with a universal property. This is our main result.

Theorem 12. *For a polynomial functor* T: **Sets** $\to$ **Sets** *we get a right adjoint* G_1 *to the forgetful functor* U: CoAlg$(T, \mathcal{E}) \to$ **Sets**, *where* $\mathcal{E}$ *is a suitable set of equations. Moreover, this functor* U *is comonadic.*

Proof. For $C \in$ **Sets**, we write $c: H_0C \xrightarrow{\cong} C \times T(H_0C)$ for the terminal coalgebra as in Theorem 11. Let $H_1C \hookrightarrow H_0C$ be the greatest mongruence induced by the equations in $\mathcal{E}$, as above, and write $G_1C = (d: H_1C \to T(H_1C))$ for the induced subcoalgebra (which satisfies the equations). For a coalgebra $e: W \to TW$ satisfying the equations in $\mathcal{E}$, and a function $f: W \to C$, we get a coalgebra map $\overline{f}: W \to H_0C$ as in the proof of Theorem 11. Since $\overline{f}$ is a coalgebra map, it commutes with the interpretations of terms (in e and in $\pi' \circ c$). And since the equations hold in e, $\overline{f}$ factors through the intersection of the equalizers induced by the equations in $\mathcal{E}$. But the image of $\overline{f}$ is a mongruence, so it must be contained in the greatest mongruence H_1C induced by this intersection. Hence $\overline{f}$ factors through $H_1C \hookrightarrow H_0C$.

Comonadicity follows from "co-Beck": U creates equalizers for coalgebra maps $u, v: I \rightrightarrows J$ which have an absolute equalizer $m: I' \rightarrowtail I$ in **Sets**. By absoluteness, one gets a sub-coalgebra structure $I' \dashrightarrow T(I')$ on I'. It satisfies the equations since it is a subcoalgebra. It is easy to see that $m: I' \rightarrowtail I$ is also equalizer of u, v in CoAlg$(T, \mathcal{E})$. □

We have constructed the cofree coalgebra (satisfying the equations) on an arbitrary set C, with $H_1(C)$ as carrier. In [6] only the cofree coalgebra on the singleton (terminal) set 1 is described, namely as

$$H_1(1) = \bigcup\{\mathrm{Im}(!_c) \subseteq H_0(1) \mid c \in \mathrm{CoAlg}(T, \mathcal{E})\}$$

where $!_c: c \to H_0(1)$ is the unique coalgebra map from c to the terminal coalgebra $H_0(1)$.

8 Adding powerset

One possible extension of the above-used notion of polynomial functor is to add the powerset functor $\mathcal{P}$. This may be finite subsets only, so that more terminal coalgebras exist, see [3]. One adds a fifth formation rule in Definition 1 (for $\mathbb{C} =$ **Sets**),

(v) if T: **Sets** $\to$ **Sets** is a polynomial functor, then so is $X \mapsto \mathcal{P}(T(X))$.

Then we can consider functors like $X \mapsto \mathcal{P}(A \times X)$, coalgebras of which are A-labelled transition systems, see [1, 17]. In our approach we first lift this powerset functor to predicates and relations.

Definition 13. Define the lifted powerset functor Pred$(\mathcal{P})$: **Pred** $\to$ **Pred** by

$$(Q \subseteq I) \mapsto (\{a \mid \forall i \in a. Q(i)\} \subseteq \mathcal{P}(I)).$$

And Rel($\mathcal{P}$): **Rel** → **Rel** by

$$(R \subseteq I \times I) \mapsto (\{(a,b) \mid \forall i \in a.\, \exists j \in b.\, R(i,j) \\ \wedge \forall j \in b.\, \exists i \in a.\, R(i,j)\} \subseteq \mathcal{P}(I) \times \mathcal{P}(I)).$$

This lifting is chosen because it satisfies the same properties of lifted functors that we have been using before, so that the same results apply for the extended notion of polynomial functor.

Proposition 14. (i) *The functor* Pred($\mathcal{P}$) *is fibred and cofibred:*

$$\mathrm{Pred}(\mathcal{P})(u^*(Q)) = \mathcal{P}(u)^*\mathrm{Pred}(\mathcal{P})(Q)$$
$$\mathrm{Pred}(\mathcal{P})(\coprod_u(Q)) = \coprod_{\mathcal{P}(u)} \mathrm{Pred}(\mathcal{P})(Q).$$

It also commutes with truth and comprehension:

$$\mathrm{Pred}(\mathcal{P})(1(I)) = 1(\mathcal{P}(I)) \quad \textit{and} \quad \{\mathrm{Pred}(\mathcal{P})(Q)\} \cong \mathcal{P}(\{Q\}).$$

(ii) *Similarly,* Rel($\mathcal{P}$) *is a fibred and cofibred functor, which commutes with equality.*

(It also commutes with quotients in the sense that $\mathcal{P}(I/R) \cong \mathcal{P}(I)/\mathrm{Rel}(\mathcal{P})(R)$ *for* $R \subseteq I \times I$ *an equivalence relation.)* □

Example 4. Consider a coalgebra $c: I \to \mathcal{P}(I \times A)$; it can be understood as an A-labelled transition system where $i \xrightarrow{a} j \Leftrightarrow (a,j) \in c(i)$. A mongruence on c is then a predicate $P \subseteq I$ satisfying for $i \in I$,

$$\text{if } P(i) \text{ then } \forall j \in I.\, \forall a \in A.\, i \xrightarrow{a} j \supset P(j).$$

And a coinduction assumption on c is a relation $R \subseteq I \times I$ such that for $i, i' \in I$,

$$\text{if } R(i,i') \text{ then } \forall j.\, \forall a.\, i \xrightarrow{a} j \supset \exists k.\, i' \xrightarrow{a} k \wedge R(j,k) \\ \wedge \forall k.\, \forall a.\, i' \xrightarrow{a} k \supset \exists j.\, i \xrightarrow{a} j \wedge R(j,k)$$

Such a relation is usually called a bisimulation.

9 Further work

We have described some of the basics of coalgebra. Much remains to be investigated, for example whether or not one has: completeness of a suitable logic in coalgebra, a "Birkhoff variety" theorem, or tensors of coalgebras.

References

1. P. Aczel. *Non-well-founded sets.* Number 14 in Lecture Notes. CSLI, Stanford, 1988.
2. P. Aczel and N. Mendler. A final coalgebra theorem. In D.H. Pitt, A. Poigné, and D.E. Rydeheard, editors, *Category Theory and Computer Science*, number 389 in Lect. Notes Comp. Sci., pages 357–365. Springer, Berlin, 1989.
3. M. Barr. Terminal coalgebras in well-founded set theory. *Theor. Comp. Sci.*, 114(2):299–315, 1993. Corrigendum in *Theor. Comp. Sci.* 124:189–192, 1994.
4. M. Barr and Ch. Wells. *Toposes, Triples and Theories.* Springer, Berlin, 1985.
5. W.R. Cook. Object-oriented programming versus abstract data types. In J.W. de Bakker, W.P. de Roever, and G. Rozenberg, editors, *Foundations of Object-Oriented Languages*, number 489 in Lect. Notes Comp. Sci., pages 151–178. Springer, Berlin, 1990.
6. U. Hensel and H. Reichel. Defining equations in terminal coalgebras. Manuscript, Univ. of Dresden. To appear in E. Astesiano et al. (eds.), *Recent trends in Data Type Specification* Springer LNCS, 1995.
7. C. Hermida. *Fibrations, Logical Predicates and Indeterminates.* PhD thesis, Univ. Edinburgh, 1993. Techn. rep. LFCS-93-277. Also available as Aarhus Univ. DAIMI Techn. rep. PB-462.
8. C. Hermida and B. Jacobs. Induction and coinduction via subset types and quotient types. Manuscript available from `ftp.cwi.nl` in `/pub/bjacobs`, December 1994.
9. C. Hermida and B. Jacobs. An algebraic view of structural induction. To appear in proceedings of *Computer Science Logic 1994.* Springer LNCS, 1995.
10. M. Hofmann and B. Pierce. An abstract view of objects and subtyping. *Journ. Funct. Progr.*, 1995.
11. B. Jacobs. Bisimulation and apartness in coalgebraic specification. Manuscript available from `ftp.cwi.nl` in `/pub/bjacobs`, January 1995.
12. B. Jacobs. Parameters and parametrization in specification using distributive categories. *Fund. Informaticae*, 1995.
13. S. Kamin. Final data types and their specification. *ACM Trans. on Progr. Lang. and Systems*, 5(1):97–123, 1983.
14. S. Mac Lane. *Categories for the Working Mathematician.* Springer, Berlin, 1971.
15. E.G. Manes. *Algebraic Theories.* Springer, Berlin, 1974.
16. H. Reichel. An approach to object semantics based on terminal co-algebras. *Math. Struct. Comp. Sci.*, 1995.
17. J. Rutten and D. Turi. On the foundations of final semantics: non-standard sets, metric spaces and partial orders. In J.W. de Bakker, W.P. de Roever, and G. Rozenberg, editors, *Semantics: Foundations and Applications*, number 666 in Lect. Notes Comp. Sci., pages 477–530. Springer, Berlin, 1993.

Semantic Typing for Parametric Algebraic Specifications

María Victoria Cengarle

Institut für Informatik
Ludwig-Maximilians-Universität München
Leopoldstr. 11b, 80802 Munich, Germany
cengarle@informatik.uni-muenchen.de

Abstract. The implementation relation of refinement of specifications is studied in the framework of the calculus of higher-order parameterization of specifications. An existing system of derivation of the relation among non-parametric specifications is enlarged so as to comprise parametric specifications. The new system is correct and complete under certain assumptions. By means of this system the calculus of parametric specifications can be enhanced with semantic types, and in this way a specification is a valid argument of a parametric specification as long as it shows the particular behavior demanded by the semantic parameter restriction. This typing can be derived, and so function application is conditional to the derivability of the parameter restrictions instantiated with the actual argument.
Keywords: algebraic specification, parametric specification, proof, implementation relation.

1 Introduction

A specification language is usually equipped with a notion of proof and a notion of implementation (see [6, 10]). The first one assists the inference of further relationships, other than the ones provided when the particular specification was defined. The second is a reflexive and transitive binary relation among specifications, which supports stepwise derivation: starting from an abstract description, a concrete specification or program is reached.

In [3] we have presented a specification language supporting programming in the large by two different means: specification-building operators and parameterization over higher-order variables. Specification-building operators allow the construction or combination of specifications, whose semantics is loosely defined as the collection of all models satisfying the inherited axioms. Parameterization abstracts just in the same way function definition does in traditional programming languages and is based on the simply typed λ-calculus. In this context, function application maps specification expressions into specification expressions, this means, parametric specifications are neither interpreted as specification morphisms nor define parametric algebras or structures (cf. [8]). Because of the parameter restrictions associated with the specification-building operators, requirements can be calculated that express the (syntactic) conditions to be fulfilled by a specification in order to be a valid argument to a parametric specification. These requirements define themselves a parametric calculus and their validity can be derived. A contextual proof system was also provided that allows the derivation of formulas valid in a specification term under certain assumptions for its free variables. What is missing in our earlier work is any definition of an implementation relation, and thus of semantic typing to be explained below.

In the present paper we revisit the refinement implementation relation. We extend this partial order relation defined by signature equality and model class inclusion to parametric specifications in the standard way. We present a system of derivation of this extended relation based on the contextual proof system, that is, derivations are relative to a context of value assumptions. It extends the one presented in [10], and in the same way as there it generates proof obligations. The system is correct and complete w.r.t. denotable assumptions.

Let us motivate with an example. Suppose we have the specification NAT of natural numbers and the specification INT of integers. By examination of the corresponding axioms, we may conclude that the natural numbers are implemented by the integers, usually denoted by NAT $\rightsquigarrow$ INT (where signature adjustment may be necessary).[1] Working with structured specifications, we can also prove that SET + NAT $\rightsquigarrow$ LIST + INT. There are two ways to test this latter relation. We can decompose the structure of both specifications and calculate the set of inherited axioms of both expressions (the specification arising by such a process is called a normal form in [1]). The drawback of this procedure is the loss of structure, and normalization itself can be a task of importance in the case of bigger specifications. The alternative is to take advantage of the structure, as for example in the system in [10], thus supporting the reuse of derivations. There we have an inference rule for each specification-building operator. Given that parameterization adds even more structure to the definition of specifications, we work exclusively with this structured inference system. This system extended allows the derivation of $\mathtt{SET}_{par}(\mathtt{NAT}) \rightsquigarrow \mathtt{LIST}_{par}(\mathtt{INT})$, where $\mathtt{SET}_{par}$ and $\mathtt{LIST}_{par}$ are *parameterized* specifications of sets and lists, respectively.

These results enable the introduction of semantic conditions to be fulfilled by the argument passed to a parametric specification. That is, an actual parameter may need not only to have some syntactic characteristics but also to exhibit a particular behavior in order to be a valid argument of a parametric specification. In a simple case, the satisfaction of formulas may be expressed using the implementation relation, that is, an actual parameter is a valid argument if it proves a set of formulas. In general, different parameters may interact, for instance we can demand that a parameter implements another argument. We define further a mechanism for typing the semantics of the parameters of a specification term. The additional type of a parameter variable is a set of conditions. A condition is defined as a pair of specification terms (which may contain free occurrences of

[1] This inference depends on the logic chosen: it is not always the case that
$\mathtt{INT} \vdash (\forall n)(0 \neq \mathtt{succ}(n))$.
Such conclusions are valid in the so-called *ultra-loose* approach to semantics of specifications, which allows the reasoning with constructors; see [10, 11].

variables, among them possibly the formal parameter one) in the implementation relation. This typing mechanism generalizes the one presented in [4]. In the same way as in that work, a step of β-reduction is possible only if the conditions can be derived, otherwise the application is undefined.

An illustrative example is a parameterized specification `LISTwithSORT` of lists equipped with a sorting algorithm. An application of `LISTwithSORT` makes sense only if the actual parameter provides an ordering relation among the elements to be listed. That means, any argument `ELEM` has to define a sort `Elem` and a binary relation `R` (up to now, just syntactic requirements), and moreover, `R` has to be reflexive, transitive and antisymmetric.

We begin in section 2 by giving a brief summary of [3], on which this work is based (for further details, consult the report cited). Afterwards in section 3 we revisit the refinement implementation relation, and give an extended definition of it, which is essentially extensionality since refinement is a partial order. This extended relation can be derived using a system of axioms and inference rules. By means of this system, we can soundly derive pairs in the refinement relation, where the specifications involved may be parameterized. Under certain conditions, we have also completeness. Finally in section 4 we add semantic types to the signature and specification terms, that can be tested for validity by virtue of a system defined in terms of the one for refinement. In this way, we achieve a powerful typing mechanism, in which the syntactic parametric restrictions are relaxed, and the language for stating the semantic ones is given more expressiveness.

2 Calculus of Parameterization

We assume that the reader is familiar with the basic notions of: many-sorted signature, usually denoted by Σ, Σ', Σ_1, etc.; signature morphisms $\sigma : \Sigma \to \Sigma'$; Σ-structure. For any signature Σ, the class of all Σ-structures is denoted by $\mathbf{CStrs}(\Sigma)$. $\mathbf{CStrs}(\sigma) : \mathbf{CStrs}(\Sigma') \to \mathbf{CStrs}(\Sigma)$ is the reduct defined in terms of a signature morphism $\sigma : \Sigma \to \Sigma'$ in the usual way (in the literature also denoted by $\cdot|_\sigma$).

A signature Σ induces a language of terms and of atomic formulas. Wrt. well-formed formulas, no particular approach is chosen, we only require an institution-independent semantics; see [5]. A simple specification SP is a signature Σ together with a set of Σ-sentences E. The semantics $Mod(SP)$ of SP is defined as the class of all Σ-structures that satisfy every sentence in E.

For attacking problems of a larger nature, signature- and specification-building operators are defined, which support programming in the large. Only three operators are provided, namely amalgamated union, bijective renaming and restriction (also called export). The first one has three operators: the two specifications to be unified and a subset of the common signature containing the symbols to be identified. If SP_1 and SP_2 are specifications and Σ is a subsignature both of $sig(SP_1)$ and of $sig(SP_2)$ (written $\Sigma \subseteq sig(SP_i)$), then $SP_1 +_\Sigma SP_2$ denotes the amalgamated union of SP_1 and SP_2 with shared signature Σ. In this case, there are injective morphisms inl and inr from $sig(SP_1)$ resp. $sig(SP_2)$ to $sig(SP_1 +_\Sigma SP_2)$, which simply label the symbols in order to discriminate different copies of homonymous names that are not to be identified.

If $r : sig(SP) \rightarrow \Sigma$ is a bijective signature morphism, then $r \cdot SP$ denotes the specification whose symbols are those of SP renamed using r. If $\Sigma \subseteq sig(SP)$, then $SP|_\Sigma$ denotes the specification whose *visible* symbols are the ones contained in Σ; the symbols in $sig(SP) \setminus \Sigma$ are called *hidden* symbols.

In contrast to the simple ones, signatures and specifications obtained by use of these operators are called structured. Simple and structured signatures (specifications) define the collection **CSign** (**CSpec**) of signature (specification) expressions. Every structured signature denotes a simple one, and thus the structures of the former are defined to be the structures of the latter. The morphisms between structured signatures are defined similarly.

Signature $sig(SP)$ and semantics $Mod(SP)$ of a structured specification SP are given in terms of the specification-building operator used and its arguments:

$$\begin{aligned} sig(SP_1 +_\Sigma SP_2) &= sig(SP_1) +_\Sigma sig(SP_2), \text{ and} \\ Mod(SP_1 +_\Sigma SP_2) &= \{A \in \mathbf{CStrs}(sig(SP_1 +_\Sigma SP_2)) : \\ &\qquad \mathbf{CStrs}(inl)(A) \in Mod(SP_1) \\ &\qquad \text{and } \mathbf{CStrs}(inr)(A) \in Mod(SP_2)\}, \text{ etc.}^2 \end{aligned}$$

Higher-order parameterization is based on the simply typed λ-calculus. There are two basic types 0 and 1, representing signatures and specifications, respectively. Other types are obtained, as usual, by closing the set of types under arrow application. A type is called a signature type if it is 0 or an arrow type that represents a signature returning function (e.g. $((\ldots \rightarrow (\rho \rightarrow 0))))$. A specification type is either 1 or a type representing a specification returning function (e.g. $((\ldots \rightarrow (\rho' \rightarrow 1))))$. Signature types and variables are denoted by τ, τ', τ_1 and y, y', y_1. Specification types and variables are denoted by π, π', π_1 and z, z', z_1. We also use p, p_1, p_2 and ρ, ρ_1, ρ_2 to denote variables and types which may be either signature or specification variables and types.

In the calculus of parameterization, signature terms are the *simple* signatures (the so-called constants) and signature variables; this set is closed under application of the signature-building operators (over terms of type 0) and under abstraction and application as usual. Specification terms are the *simple* specifications over signature terms of basic type (that is, not only over signature constants but over any signature term of type 0) and specification variables; this set is likewise closed under application of the specification-building operators (over terms of type 1) and under abstraction and application. As a consequence of this definition, signature terms have signature types and specification terms have specification types. There is an additional term constructor, denoted by sig_λ, that applied to a specification term yields a signature term. This means, signature terms and specification terms are defined by mutual recursion.

[2] In the literature one can find many approaches in which **CSign** and **CSpec** are characterized as categories, whose arrows are the signature morphisms resp. the specifications morphisms (which we leave undefined in this work). In our framework, we just need the definitions we have presented.

The notions of β-redex and β-equality are defined as usual. The domains of interpretation are $[\![0]\!] = \mathbf{CSign} \cup \{\mathsf{E}\}$ and $[\![1]\!] = \mathbf{CSpec} \cup \{\mathsf{F}\}$ for the basic signature and specification types, respectively. As there is no recursion in this framework (there are no terms e.g. of the form $M = z +_y M$), the semantics of an arrow type $(\rho_1 \to \rho_2)$ is the set of all mappings between the interpretation of ρ_1 and the interpretation of ρ_2, i.e. $[\![(\rho_1 \to \rho_2)]\!] = ([\![\rho_1]\!] \to [\![\rho_2]\!])$. Abstractions and applications are interpreted as usual. Notice that we have two layers of semantics. On the one hand and given an environment e assigning values in $[\![\rho]\!]$ to each variable p of type ρ, a specification term of type 1 (e.g. $N = ((\lambda z\!:\!1.z +_\Sigma z')\ SP))$ has a specification expression as semantics (in our example, $[\![N]\!]_e = SP +_\Sigma SP'$ if $e(z') = SP'$). On the other hand, a specification expression has a collection of models as associated semantics. Therefore in the case of our term N, it makes sense to write $Mod([\![N]\!]_e)$.

The distinguished values E and F denote "error" objects. A well-typed term may denote an error, that is, simple types are not enough to ensure the well-definedness of a term. This depends also on the parameter restrictions of the specification-building operators (for our N is $[\![N]\!]_e = \mathsf{F}$ if $\Sigma \not\subseteq sig(SP')$). In order to calculate and test such restrictions a calculus of parameter requirements is defined as well as a function *Req* from signature and specification terms to requirement terms. Because of abstractions (i.e. signature and specification terms of arrow type) we have to keep track of the abstracted variable. The set of requirements is thus closed under simply typed abstraction and application. Semantically, a signature or specification term M is well-defined in an environment e ($[\![M]\!]_e \notin \{\mathsf{E}, \mathsf{F}\}$ if M is of basic type) iff its associated requirement term $Req(M)$ is satisfied (i.e. $[\![Req(M)]\!]_e = \mathsf{true}$).[3] A term of arrow type $N\!:\!(\rho_1 \to \rho_2)$ is undefined in e if there is *no* argument making the application meaningful, i.e. if there is no $N_1 : \rho_1$ such that $(N\ N_1)$ is defined in e. Syntactically, requirements associated with terms of basic type can be derived by means of an inference system. Derivations are denoted by $\Gamma \rhd R$, where R is a requirement term and Γ a context assuming values for variables, among them possibly the ones occurring free in R. Derivations are correct: if $\Gamma \rhd R$, then R holds in any environment e induced by Γ. Consequently, if R is the requirement term associated with a signature or specification term M (which necessarily is of basic type), then $[\![M]\!]_e \notin \{\mathsf{E}, \mathsf{F}\}$ for any such e. This system is complete w.r.t. denotable environments (environments that assign to any variable a value denotable by a ground term).

In [10] a structured proof system is presented that allows the derivation of formulas that hold in any model of a specification expression. The proof system extends any correct proof system Π for the underlying logic to a proof system Π_s for structured specifications. Π_s is called structured because it is not based on normalization but on the structure of the target specification, that is, there is an inference rule for every specification-building operator. This system is correct, and moreover complete if Π is so and in the presence of the

---

[3] Nevertheless, M may denote a contradictory specification expression. That is, $[\![Req(M)]\!]_e = \mathsf{true} \not\Rightarrow Mod([\![M]\!]_e) \neq \emptyset$.

interpolation property; see also [1]. In [3], the structured proof system Π_s is extended to a contextual Π_{ps} for parameterized specifications that allows the derivation of formulas starting from specification terms of basic type. Given a specification term M of type 1 and a context Γ that assumes values for variables, if a formula φ can be derived by means of the contextual proof system (written $\Gamma \triangleright M \vdash_{\Pi_{ps}} \varphi$), then φ holds in any model associated with the semantics of M in the environments induced by Γ (i.e. $Mod([\![M]\!]_e) \models \varphi$ if e does not contradict Γ). Completeness is achieved w.r.t. denotable environments in the presence of the interpolation property and assuming Π complete.

A proof system is the key to the derivation of the refinement relation between specifications. Two specifications are in the refinement relation if they have the same signature and the class of models of the concrete specification is contained in the class of models of the abstract one. In the case of two simple specifications $SP_1 = \langle \Sigma, E_1 \rangle$ and $SP_2 = \langle \Sigma, E_2 \rangle$, SP_1 is refined by SP_2 (written $SP_1 \rightsquigarrow SP_2$) if $E_2 \models \varphi$ for every $\varphi \in E_1$. Given the functions *sig* and *Mod*, this relation obviously extends to arbitrary specification expressions. In [10] an inference system is presented that allows the derivation of pairs of specification expressions in the refinement relation. That means, we can derive **SETNAT** $\rightsquigarrow$ **LISTNAT** and **NAT** $\rightsquigarrow$ **INT**, for instance. The system is based on the proof system Π_s and is moreover structured, in the sense that it is not necessary to calculate the set of inherited axioms of a particular specification expression in order to test implementation.

In the next section, we extend refinement to parametric specifications. In this way, derivations of the kind $((\lambda z : 1.\mathtt{SET})\ \mathtt{NAT}) \rightsquigarrow ((\lambda z : 1.\mathtt{LIST})\ \mathtt{INT})$ can be accomplished, i.e. refinement among terms possibly involving parameterized specifications and/or signatures can be inferred.

3 Implementation

Normally a system is specified in a rather "abstract" way, and any implementation chooses one particular structure in the class of models. This selection is done by successively performing refinements. The process consists in the introduction of design decisions in the specification until a program that implements the specification is reached. Here we introduce the refinement approach to implementation, and study how its properties are preserved and/or modified in the framework of parameterization.

Refinement is a partial order among specifications. Two specifications are in this relation if they have the same syntax, i.e. the same signature, and their model classes are included one in the other. This relation is extended to parametric specifications. On the one hand, lambda abstraction may be present as well as application. With this respect, refinement extends to functions in the standard way (i.e. $f \leq g$ if $f(x) \leq g(x)$ for every x). On the other hand, there is an error symbol F interpreting undefined specification terms. In order to preserve reflexivity, F implements F.

Definition 1 (Refinement). Let π be a specification type, let S_1, S_2 be values in $[\![\pi]\!]$. S_2 *implements* or *refines* S_1, written $S_1 \rightsquigarrow S_2$, if:

1. (a) $\pi = 1$ and $S_1 = \mathsf{F} = S_2$, or
   (b) $\pi = 1$, $S_1 \neq \mathsf{F} \neq S_2$, $sig(S_2) = sig(S_1)$ and $Mod(S_2) \subseteq Mod(S_1)$.
2. $\pi = (\rho \rightarrow \pi')$ and $S_1(S) \rightsquigarrow S_2(S)$ for all $S \in [\![\rho]\!]$.

Lemma 2. *Specification-building operators are monotonic w.r.t.* $\rightsquigarrow$.

An environment e is called denotable if it assigns to every variable a denotable value, that is, if for any variable p there is a ground term M such that $e(p) = [\![M]\!]$.

Lemma 3. *For any denotable environment e, for any specification terms M, N_1 and N_2, if $[\![N_1]\!]_e \rightsquigarrow [\![N_2]\!]_e$ and both N_1 and N_2 are well-defined in e, then $[\![M[z := N_1]]\!]_e \rightsquigarrow [\![M[z := N_2]]\!]_e$.*

This means, denotable functions are also monotonic w.r.t. $\rightsquigarrow$. Moreover, given denotable F_1, F_2, S_1 and S_2, if $F_1 \rightsquigarrow F_2$ and $S_1 \rightsquigarrow S_2$, then $F_1(S_1) \rightsquigarrow F_2(S_2)$. On the one hand, by definition, $F_1 \rightsquigarrow F_2$ implies

$$F_1(S_1) \rightsquigarrow F_2(S_1),$$

and on the other, by monotonicity,

$$F_2(S_1) \rightsquigarrow F_2(S_2)$$

Thus, by transitivity, $F_1(S_1) \rightsquigarrow F_2(S_2)$.

Notation. Recall contexts are denoted by $\Gamma = \{p_1 = M_1, \ldots, p_n = M_n\}$, where p_i are different variables ($p_i \neq p_j$ if $i \neq j$) and M_i terms of the corresponding type. We let $dom(\Gamma)$ denote the set of variables $\{p_1, \ldots, p_n\}$, $\Gamma(p_i)$ denote the term M_i ($1 \leq i \leq n$), and $FV(\Gamma)$ denote $\cup\{FV(\Gamma(p)) : p \in dom(\Gamma)\}$. We call $Env(\Gamma)$ the set of environments e which verify $e(p) = [\![\Gamma(p)]\!]_e$ for all $p \in dom(\Gamma)$, where $[\![\cdot]\!]_e$ denotes the extension of the environment e to arbitrary terms in the standard way.

In the following we introduce a derivation system for $\rightsquigarrow$, that will be proved to be correct, i.e. if $M \rightsquigarrow N$ is derivable then $[\![M]\!]_e \rightsquigarrow [\![N]\!]_e$ in every environment e compatible with the context of derivation.

Remember we had the operators of amalgamated union and the embeddings $inl : sig(SP_1) \rightarrow sig(SP_1 +_\Sigma SP_2)$ and $inr : sig(SP_2) \rightarrow sig(SP_1 +_\Sigma SP_2)$. In the calculus of parameterization, and in order to derive requirements and perform proofs, these morphisms become term constructors.

Definition 4 (Refinement Derivation). The relation $\rightsquigarrow$ is axiomatized by the following axiom schemes and derivation rules.

[GROUND RULES]

$$\text{(ref-basic)}\ \frac{\Gamma \rhd Req(\langle K, E\rangle), K \equiv sig_\lambda(M), M \vdash_{\Pi_\mu} \varphi \text{ for all } \varphi \in E}{\Gamma \rhd \langle K, E\rangle \rightsquigarrow M}\ {}^4$$

[4] If M and N are terms of type 0, "$M \equiv N$" abbreviates "$M \subseteq N$" and "$N \subseteq M$," which are terms of the requirement language and thus if valid derivable.

$$(\text{ref-refl}_l)\ \frac{\Gamma \rhd M_1 \rightsquigarrow N}{\Gamma \rhd M_2 \rightsquigarrow N} \quad \text{if } M_1 =_\beta M_2$$

$$(\text{ref-refl}_r)\ \frac{\Gamma \rhd M \rightsquigarrow N_1}{\Gamma \rhd M \rightsquigarrow N_2} \quad \text{if } N_1 =_\beta N_2$$

[CONTEXTUAL RULES]

$$(\text{ref-ctxt}_l)\ \frac{\Gamma, z = M \rhd M \rightsquigarrow N}{\Gamma, z = M \rhd z \rightsquigarrow N}$$

$$(\text{ref-ctxt}_r)\ \frac{\Gamma, z = N \rhd M \rightsquigarrow N}{\Gamma, z = N \rhd M \rightsquigarrow z}$$

$$(\text{ref-hyp-add})\ \frac{\Gamma \rhd M_1 \rightsquigarrow M_2}{\Gamma, p = N \rhd M_1 \rightsquigarrow M_2} \quad \text{if } p \notin dom(\Gamma)$$

[OPERATOR RULES]

$$(\text{ref-monot}_l)\ \frac{\Gamma \rhd M \rightsquigarrow N}{\Gamma \rhd inl(M) \rightsquigarrow inl(N)}$$

$$(\text{ref-monot}_r)\ \frac{\Gamma \rhd M \rightsquigarrow N}{\Gamma \rhd inr(M) \rightsquigarrow inr(N)}$$

$$(\text{ref-sum})\ \frac{\begin{array}{c}\Gamma \rhd Req(M_1 +_K M_2), sig_\lambda(N) \equiv sig_\lambda(M_1 +_K M_2),\\ inl(M_1) \rightsquigarrow N|_{sig_\lambda(inl(M_1))}, inr(M_2) \rightsquigarrow N|_{sig_\lambda(inr(M_2))}\end{array}}{\Gamma \rhd M_1 +_K M_2 \rightsquigarrow N}$$

$$(\text{ref-restr})\ \frac{\Gamma \rhd K \equiv sig_\lambda(M_2), Req(M_1 +_K M_2)}{\Gamma \rhd M_1|_K \rightsquigarrow M_2}$$

if, under the assumptions of Γ, the extension $M_1 +_K M_2$ of $inr(M_2)$ is persistent[5]

$$(\text{ref-ren})\ \frac{\Gamma \rhd M \rightsquigarrow r^{-1} \cdot N}{\Gamma \rhd r \cdot M \rightsquigarrow N}$$

[ABSTRACTION RULES]

$$(\text{ref-app}_l)\ \frac{\Gamma \rhd Req(N_1), Req(N_2), N_1 \rightsquigarrow N_2}{\Gamma \rhd ((\lambda z{:}\pi.M)\ N_1) \rightsquigarrow ((\lambda z{:}\pi.M)\ N_2)}$$

$$(\text{ref-app}_r)\ \frac{\Gamma \rhd N_1 \rightsquigarrow N_2}{\Gamma \rhd (N_1\ M) \rightsquigarrow (N_2\ M)}$$

$$(\text{ref-abstr})\ \frac{\Gamma, p = N \rhd M_1[p := N] \rightsquigarrow M_2[p := N]}{\Gamma, p = N \rhd M_1 \rightsquigarrow M_2}$$

$$(\text{ref-apply})\ \frac{\Gamma, p = N \rhd M_1 \rightsquigarrow M_2}{\Gamma \rhd M_1[p := N] \rightsquigarrow M_2[p := N]} \quad \text{if } p \notin FV(\Gamma \cup \{p = N\})$$

[5] An extension SP' of a specification SP is called *persistent* if

$$Mod(SP) = \{A|_{in} : A \in Mod(SP')\}$$

where in is the embedding morphism from $sig(SP)$ to $sig(SP')$; see [10].

In practice, the rule (ref-restr) is not used: one usually proves the "hidden" axioms of M_1 by adding symbols and recursive definitions to M_2. The rule given is in fact an instance of a more powerful rule

$$\frac{\Gamma \rhd M_1 \rightsquigarrow (M_2 +_K \Delta)^{-1}}{\Gamma \rhd M_1|_K \rightsquigarrow M_2} \quad \text{if, under the assumptions of } \Gamma, \text{ the extension } M_2 +_K \Delta \text{ of } inr(M_2) \text{ is persistent}$$

where M^{-1} denotes the result of removing every label in the signature of M. (The premise additionally needs some well-definedness hypotheses like for example $Req(M_1|_K)$.)

Theorem 5 (Correctness). *$\Gamma \rhd M \rightsquigarrow N$ implies $[\![M]\!]_e \rightsquigarrow [\![N]\!]_e$ for any $e \in Env(\Gamma)$.*

This theorem can easily be demonstrated by induction on the length of the derivation of $\Gamma \rhd M \rightsquigarrow N$; see [2]. Monotonicity rules are not enough for completeness even in the non-parametric case (cf. [10]). The system introduced in definition 4 is complete at the cost of proof obligations (rule (ref-restr)). As refinement can be derived under the assumptions of a context, completeness is only achieved w.r.t. value assumptions for the free variables of the terms involved.

Notation. Given a denotable e and terms $M_1, \ldots, M_k$, we let $\Gamma(e, M_1, \ldots, M_k)$ denote any context verifying:

$\Gamma(e, y{:}0) = \{y = \Sigma\}$ where $\Sigma = e(y)$[6]

$\Gamma(e, z{:}1) = \{z = SP\}$ where $SP = e(z)$[7]

$\Gamma(e, p{:}(\rho_1 \to \rho_2)) = \{p = N\}$ where N is ground and s.t. $e(p) = [\![N]\!]$ (such an N exists given that e is denotable)

$\Gamma(e, M) = \cup_{p \in FV(M)} \Gamma(e, p)$

$\Gamma(e, M_1, \ldots, M_k) = \cup_{1 \leq i \leq k} \Gamma(e, M_i)$ where the choice for every variable is the same if it occurs free in more than one term M_i.

Theorem 6 (Relative Completeness). *Let M and N be specification terms of type 1, let e be a denotable environment. If the interpolation property holds and Π is complete, then $[\![M]\!]_e \neq \mathsf{F} \neq [\![N]\!]_e$ and $[\![M]\!]_e \rightsquigarrow [\![N]\!]_e$ imply $\Gamma(e, M, N) \rhd M \rightsquigarrow N$ for any $\Gamma(e, M, N)$.*

This theorem is proved by induction on M, where the ordering used –and considering rules e.g. (ref-monot$_l$) or (ref-restr)– is not the structure of M but another Noetherian ordering defined ad-hoc. The interpolation property also plays a role in the completeness of the system Π_s and therefore in the completeness of Π_{ps} when considering combined specifications $SP_1 +_\Sigma SP_2$; see [1, 2, 10].

The implementation relation meets the following property: if $SP_1 \rightsquigarrow SP_2$ and $Mod(SP_1) \models \varphi$, then $Mod(SP_2) \models \varphi$. We have developed different systems

[6] If $e(y) = \mathsf{E}$, we assume any erroneous signature expression for y, for instance $\Sigma_1|_{\Sigma_2}$ with $\Sigma_2 \not\subseteq \Sigma_1$.

[7] Analogously, if $e(z) = \mathsf{F}$ we assume any erroneous specification expression for z.

that allow the derivation of the well-definedness of specification terms, the proof of formulas, and the derivation of the implementation relation. All these systems work in relation to a context assuming values for variables. Suppose we can derive $\Gamma \rhd M \rightsquigarrow N, M \vdash_{\Pi_{\mathbf{m}}} \varphi$, and suppose $Env(\Gamma)$ is non-empty. We therefore know that there is a possible interpretation of the terms above such that $[\![M]\!]_e \rightsquigarrow [\![N]\!]_e$ and $Mod([\![M]\!]_e) \models \varphi$. By relative completeness and under the assumption of interpolation, $\Gamma(e, N) \rhd N \vdash_{\Pi_{\mathbf{m}}} \varphi$. One wonders if there is a relation between contexts Γ and $\Gamma(e, N)$. They can indeed be related by the notion of compatibility which is defined as follows.

Definition 7. A context Γ_2 is *compatible* with a context Γ_1 if for all $env_2 \in Env(\Gamma_2)$ there exists $env_1 \in Env(\Gamma_1)$, such that $env_1(p) = env_2(p)$ for every $p \in dom(\Gamma_1) \cap dom(\Gamma_2)$.

Proposition 8. *Let M, N be terms of type 1, let Γ be any environment and φ any sentence. Suppose that $\Gamma \rhd M \rightsquigarrow N$, and $\Gamma \rhd M \vdash_{\Pi_{\mathbf{m}}} \varphi$. If the interpolation property holds and $Env(\Gamma) \neq \emptyset$, then there is a context Γ' compatible with Γ such that $\Gamma' \rhd N \vdash_{\Pi_{\mathbf{m}}} \varphi$.*

The proof is straightforward using the tools defined and the lemmas proved so far. It relies on correctness and relative completeness, which in its turn requires interpolation.

Let us now consider the specifications

$\mathtt{NAT} = \langle \Sigma_{\mathrm{NAT}}, E_{\mathrm{NAT}} \rangle$, $\qquad \mathtt{SET}_{par} = (\lambda z{:}1.\mathtt{SET})$,

$\mathtt{INT} = \langle \Sigma_{\mathrm{INT}}, E_{\mathrm{INT}} \rangle$, $\qquad \mathtt{LIST}_{par} = (\lambda z{:}1.\mathtt{LIST})$.

Using a proof system Π for the ultra-loose semantics (see [2]), we can derive $E_{\mathrm{INT}} \vdash_{\Pi} \varphi$ for every $\varphi \in E_{\mathrm{NAT}}$. Therefore by rule (ref-basic),

$\rhd \mathtt{NAT} \rightsquigarrow \mathtt{INT}|_{\Sigma_{\mathrm{NAT}}}$.

On the one hand, by (ref-app$_l$),

$\rhd \mathtt{SET}_{par}(\mathtt{NAT}) \rightsquigarrow \mathtt{SET}_{par}(\mathtt{INT}|_{\Sigma_{\mathrm{NAT}}})$, and

$\rhd \mathtt{LIST}_{par}(\mathtt{NAT}) \rightsquigarrow \mathtt{LIST}_{par}(\mathtt{INT}|_{\Sigma_{\mathrm{NAT}}})$.

On the other hand, as it is classically proved in the literature,

$\rhd \mathtt{SET}[z := \mathtt{NAT}] \rightsquigarrow \mathtt{LIST}[z := \mathtt{NAT}]$.

Hence, by transitivity and (ref-refl$_r$),

$\rhd \mathtt{SET}[z := \mathtt{NAT}] \rightsquigarrow \mathtt{LIST}_{par}(\mathtt{INT}|_{\Sigma_{\mathrm{NAT}}})$

and by (ref-refl$_l$),

$\rhd \mathtt{SET}_{par}(\mathtt{NAT}) \rightsquigarrow \mathtt{LIST}_{par}(\mathtt{INT}|_{\Sigma_{\mathrm{NAT}}})$.

If SET and LIST are of the form $z +_{\{\mathrm{Elem}\}}$ SETBODY and $z +_{\{\mathrm{Elem}\}}$ LISTBODY, respectively, thanks to amalgamation we know that any value for z cannot interfere with the theorems we can derive both from SETBODY and from LISTBODY. Therefore it is also possible to derive a reusable result

$\{z = \langle \{\mathtt{Elem}\}, \emptyset \rangle\} \rhd \mathtt{SET}_{par}(z) \rightsquigarrow \mathtt{LIST}_{par}(z)$.

Once the refinement relation is extended to parametric specifications and can be derived by means of the system of definition 4, we are able to give to the designer of a parameterized specification more control over the type of arguments his/her specification will accept. This additional type information can be coded using refinement as defined in the next section.

4 Semantic Typing

In the present section we enhance the calculus of parametric specifications and signatures with semantic typing. The new characteristics we add allow the parametric system designer to demand not only syntactic requirements for an argument term, but also semantic ones, in order for an application over that argument to be defined. As an example, a parameter may need not only to have a particular signature but also to satisfy a number of formulas.

This additional type information is collected in a set of conditions. A condition is a relation $M \preceq N$ between terms of type 1, which in a given environment e is interpreted as $[\![M \preceq N]\!]_e = \mathsf{true}$ iff $sig([\![N]\!]_e) \subseteq sig([\![M]\!]_e)$, and $Mod([\![M]\!]_e|_{sig([\![N]\!]_e)}) \subseteq Mod([\![N]\!]_e)$.

For the parametric specification **LISTwithSORT** mentioned in the introduction, a parameter restriction demanding that a parameter z be an implementation of a partial order specification **P-ORD** is written $z \preceq$ **P-ORD**, and in this way an argument may have more symbols than those necessary to specify a partial order relation. By abuse of language a set of conditions is also called a condition.

Definition 9 (Semantic Conditions). The set of *conditions* and their semantics in an environment e is inductively defined by

1. $\emptyset$ is a condition,
   $[\![\emptyset]\!]_e = \mathsf{true}$;
2. if M and N are specification terms of type 1, then $\{M \preceq N\}$ is a condition,
   $[\![\{M \preceq N\}]\!]_e = \mathsf{true}$ iff $[\![N]\!]_e \rightsquigarrow [\![M]\!]_e|_{sig([\![N]\!]_e)}$;
3. if C_1 and C_2 are conditions, then $C_1 \cup C_2$ is a condition,
   $[\![C_1 \cup C_2]\!]_e = \mathsf{true}$ iff $[\![C_1]\!]_e = [\![C_2]\!]_e = \mathsf{true}$.

In contrast to the calculus of requirements, conditions are just sets of such expressions, there are neither types nor notions of abstraction and application. As they are based on the implementation relation, we can use the system deriving refinement in order to derive also conditions. The β-reduction, therefore, cannot be a syntactic relation anymore. It must be *derived* provided the condition typing the parameter can be derived for the formal parameter variable replaced by the actual argument. In this way, the enhanced calculus supports the specification with parameterization over higher-order variables and with semantic conditions.

A parametric signature or specification term has therefore a twofold type information, which is denoted by $(\lambda p^{\rho_1} : C.M) : (\rho_1 \rightarrow \rho_2)$ if ρ_i are simple types, p is a variable of type ρ_1, C a condition, and M a term of type ρ_2. Similarly, the requirement terms and the function Req have to be modified. The only case to be revised is that of abstraction: $Req((\lambda p^{\rho_1} : C.M)) = (\lambda p^{\rho_1} : C.Req(M))$.

Note that the terms in C are not tested for well-definedness. Moreover, p may occur in C, and thus some syntactic requirements for p may arise from C. These are ignored by $Req((\lambda p^{\rho} : C.M))$. However, if C is derivable then those terms and requirements are necessarily fulfilled.

The definition of semantics we had both for signature and specification terms and for requirements has to be changed. Now we have to take the conditions into account.

Definition 10 (Semantics). Given an environment e, its extension $[\![\cdot]\!]_e$ to signature and specification terms as well as to requirements is redefined[8] for abstractions as follows.

- If M has type ρ_2, $[\![(\lambda p^{\rho_1}:C.M)]\!]_e : [\![\rho_1]\!] \rightarrow [\![\rho_2]\!]$ is the map defined by
  $[\![(\lambda p^{\rho_1}:C.M)]\!]_e(S) = \begin{cases} [\![M]\!]_{e[p:=S]} & \text{if } [\![C]\!]_{e[p:=S]} = \mathsf{true} \\ \mathsf{fault}_{\rho_2} & \text{otherwise,} \end{cases}$
  where fault_{ρ_2} is defined on the structure of type ρ_2 by:
  - if $\rho_2 = 0$ then $\mathsf{fault}_{\rho_2} = \mathsf{E}$,
  - if $\rho_2 = 1$ then $\mathsf{fault}_{\rho_2} = \mathsf{F}$,
  - if $\rho_2 = (\rho_3 \rightarrow \rho_4)$ then fault_{ρ_2} is the constant function over $[\![\rho_3]\!]$ defined by $\mathsf{fault}_{\rho_2}(S) = \mathsf{fault}_{\rho_4}$ for all $S \in [\![\rho_3]\!]$.
- If R has type B, $[\![(\lambda p^{\rho}:C.R)]\!]_e : [\![\rho]\!] \rightarrow [\![B]\!]$ is the map defined by
  $[\![(\lambda p^{\rho}:C.R)]\!]_e(S) = \begin{cases} [\![R]\!]_{e[p:=S]} & \text{if } [\![C]\!]_{e[p:=S]} = \mathsf{true} \\ \bot_B & \text{otherwise,} \end{cases}$
  where $\bot_B$ is defined on the structure of type B by:
  - if B is basic, then $\bot_B = \mathsf{false}$,
  - if B is $(\rho' \rightarrow B')$, then $\bot_B$ is the constant function over $[\![\rho']\!]$ defined by $\bot_B(S) = \bot_{B'}$ for all $S \in [\![\rho']\!]$.

The requirement function *Req* preserves its nice properties:

M is well-defined in e iff $Req(M)$ is not contradictory in e,

since the requirement associated with an abstraction does not ignore the conditions that semantically type the parameter. The conditions themselves, however, are not tested for well-definedness. The system deriving requirements is modified in such a way that $Req(M)$ can be derived only if the conditions in the term M can be proved, which assures the well-definedness of M. Moreover the conditions can be proved only if the terms they involve are well-defined.

Since the abstraction terms were modified, the definition of substitution has to be changed:

$$(\lambda p^{\rho}:C.M)[p' := N] = (\lambda p^{\rho}:C[p' := N].M[p' := N]),$$

where $C[p' := N]$ is defined by

1. $\emptyset[p' := N] = \emptyset$,
2. $\{M_1 \preceq M_2\}[p' := N] = \{M_1[p' := N] \preceq M_2[p' := N]\}$,
3. $(C_1 \cup C_2)[p' := N] = C_1[p' := N] \cup C_2[p' := N]$.

The notion of β-reduction also needs to be altered. Not every abstraction applied to an arbitrary term constitutes a β-redex. For example consider the term

$$M = ((\lambda p^{\rho_2}:C.N_1)\ N_2) \text{ of type } \rho_1$$

where $C \neq \emptyset$. Given an environment e, suppose $N_1[p := N_2]$ is defined in e and C is contradictory in e (i.e. $[\![C]\!]_e = \mathsf{false}$, and thus M is not well-defined in e).

[8] The complete definition of semantics of the calculus of parameterized signatures and specifications and of requirements can be found in [3].

That means, $[\![M]\!]_e \neq [\![N_1[p := N_2]]\!]_e$. Therefore, we cannot ignore C and state $M \longrightarrow_\beta N_1[p := N_2]$, since it would be contrary to the philosophy of β-reduction, the property $[\![M]\!]_e = [\![M']\!]_e$ if $M =_\beta M'$ would be lost. The equation

$$((\lambda p^{\rho_2} : C.N_1)\ N_2) =_\beta N_1[p := N_2],$$

is *derived* provided C is derivable.

Definition 11 (Condition Derivation). Conditions can be derived by the following system of axiom and derivation rules.

[CONDITION RULES]

(cond-empty) $\rhd \emptyset$

$$\text{(cond-basic)}\ \frac{\Gamma \rhd N \rightsquigarrow M|_{sig_\lambda(N)},\ Req(M|_{sig_\lambda(N)})}{\Gamma \rhd M \preceq N}$$

$$\text{(cond-union)}\ \frac{\Gamma \rhd C_1, C_2}{\Gamma \rhd C_1 \cup C_2}$$

$$\text{(cond-hyp-add)}\ \frac{\Gamma \rhd C}{\Gamma, p = N \rhd C} \quad \text{if } p \notin dom(\Gamma)$$

$$\text{(cond-abstr)}\ \frac{\Gamma, p = N \rhd C[p := N]}{\Gamma, p = N \rhd C}$$

Given the system of derivation of conditions, we can also derive the β-relation as introduced in the following definition.

Definition 12 (Conditional β-Equality). The system of derivation of requirements (see [3]) is enriched with the following *conditional* β-equality rules:

[CONDITIONAL BETA EQUALITY]

$$\text{(cond-beta}_t\text{)}\ \frac{\Gamma, p = M \rhd C,\ Req(N)}{\Gamma \rhd ((\lambda p^\rho : C.N)\ M) =_\beta N[p := M]} \quad \text{if } p \notin FV(\Gamma \cup \{p = M\})$$

$$\text{(cond-beta}_r\text{)}\ \frac{\Gamma \rhd M_1 =_\beta M_2}{\Gamma \rhd Req(M_1) =_\beta Req(M_2)}$$

The β-equality commutes with the operators of sum, restriction and renaming, and with abstraction and application.

Lemma 13. *The condition rules as well as the conditional β-equality rules are correct. That is, if $\Gamma \rhd C$ then $[\![C]\!]_e =$ true for all $e \in Env(\Gamma)$, and if $\Gamma \rhd M =_\beta N$, then $[\![M]\!]_e = [\![N]\!]_e$ for all $e \in Env(\Gamma)$.*

In the rule (cond-beta$_t$), the hypothesis $Req(N)$ could be eliminated since the β-equality in fact does not depend on this requirement. We write this premise because we want to stress the character of the system in which we apply a function (i.e. we perform a step of β-reduction) only if requirements and conditions are fulfilled.

Let us reconsider the rules (ref-app$_l$) and (ref-app$_r$) presented in definition 4 for the derivation of the implementation relation. The first one, rephrased using the double typing of the present section, is no longer correct as it ignores the conditions. It has to be reformulated as follows:

$$(\text{ref-app}'_l)\quad \frac{\begin{array}{c}\Gamma, z = N_1 \ \triangleright C \\ \Gamma, z = N_2 \ \triangleright C \\ \Gamma \ \triangleright \mathit{Req}(N_1), \mathit{Req}(N_2), N_1 \rightsquigarrow N_2\end{array}}{\Gamma \ \triangleright ((\lambda z^{\pi} : C.M)\ N_1) \rightsquigarrow ((\lambda z^{\pi} : C.M)\ N_2)}$$

In the case of rule (ref-app$_r$) and by definition, the application of N_1 to *any* actual parameter is implemented by the application of N_2 to the same parameter, even if the argument does not comply with the conditions that the functions may possibly impose on their parameter. This rule is therefore correct.

Definition 14 (Derivation of Requirements with Conditions). The system of derivation of requirements associated with parameterized terms with semantic conditions is defined as the union of the system of derivation of requirements (see [3]), the contextual proof system (idem), the system of derivation of the implementation relation (definition 4) with (ref-app$_l$) modified as above, the system of derivation of conditions (definition 11), and the conditional β-equality rules (definition 12).

Unfortunately due to lack of space, we cannot process the example of a specification `LISTwithSORT` as mentioned in the introduction. It is indeed possible to instantiate such a specification using a specification of `WORDS`, where the ordering relation is the `prefix` one.

The correctness of the system presented is guaranteed by the correctness of the individual systems, since the derivations of one system cannot be achieved or refused by any other. Relative completeness is also inherited.

Theorem 15 (Correctness and Completeness). *The system of derivation of requirement terms, proofs, refinement, conditions, and conditional β-equality is correct. Assuming the underlying proof system Π complete, and in the presence of interpolation, the system is also complete in relation to denotable environments.*

Thus, we have reached our goal: a step of β-reduction is only performed if the requirements and conditions associated with a traditional β-redex are derivable. That is, a parametric specification term is applied to an actual parameter only if the argument fulfills the necessary requirements and conditions.

5 Conclusions and Future Work

We have studied the refinement approach to implementation relation among parameterized specifications. We can soundly derive pairs in this relation by virtue of a system of axioms and inference rules. Completeness is achieved under certain conditions and at the cost of proof obligations. The refinement derivation system of [10] was enlarged and combined with the contextual proof system of [3]. This new system allows derivation of the refinement relation among parameterized specifications, thus providing means –namely the conditions– for the designer to type the semantics of the arguments of a parameterized specification. By the use of these tools, the correct application of function terms is guaranteed: ***a parametric signature or specification term is applied to an actual parameter only if the parameter fulfills the associated requirements and conditions (i.e. both syntactic and semantic parameter restrictions).***

This parameterization mechanism generalizes the one presented in [8], since the syntactic conditions are relaxed and the semantic ones are given more expressiveness. Firstly, the constraints on the signature of a given parameter are deduced automatically in our system, the designer has only to care about the definition of the parametric specification itself. The enormous disadvantage of [8] is that the argument has to have *exactly* the signature associated with the parameter restriction. This restriction is very strong; even if we allow a supersignature instead, the idea is unsatisfactory since there are situations in which one wants precisely the opposite. Consider a parametric specification $(\lambda y : 0.SP|_y)$ that defines views of a database SP: in this case, we want the actual parameter to be a ***subsignature*** of that of SP. The semantic conditions, secondly, are more powerful than the typing proposed in the work cited. There, as well as in the λ-calculus approach defined in [9], the designer can only demand of the argument to be an implementation of a formal parameter restriction, whereas in our proposal a language for the typing mechanism is provided that allows e.g. the combination of different arguments in the refinement relation, and more important, syntactic and semantic typing can be *derived* using the mechanisms presented.

The proposal of [4] very much resembles the parameterization mechanism of the papers mentioned above. Here the parameter to an abstraction has to be less than a so-called parameter restriction, where the ordering relation may be any partial order, typically the implementation relation. In the same way as in the present proposal, β-equality is derivable, provided the parameter restriction has been met.

The other classical approach to parameterization, namely pushout parameterization, as for example in CLEAR and in ACT ONE, and if we consider only injective specification morphisms, can be simulated here using renaming and amalgamation, that is, using the specification-building operators provided.

We have to experiment with more examples in order to test how easy to use is the inference system proposed. Maybe the insertion of derived rules would make it smoother, and certainly a strategy is needed in order not to run into a loop especially if we want to implement a theorem prover based on it. A disadvantage of adding so much structure to the specification mechanism is that a parametric specifications for which its designer wants e.g. just a sort supplied becomes quite abstruse. We could define a complementary parameterization mechanism, as proposed in [7].

Following [8], parameterization over specifications is useful in the software design phase. On the contrary in the software implementation phase, it is better to have parameterization over structures, that is, specifications of parametric structures, as in e.g. Extended ML or OBSCURE. In the article mentioned, both forms of parameterization are supplied. It may be interesting to see what are the consequences of such specification instead of (or in addition to) the parameterization mechanism we propose. What is the impact on the calculus of requirements, how is the contextual proof system affected by this apparently orthogonal concept, these are issues that may be worth studying.

References

1. J. A. Bergstra, J. Heering, and P. Klint. Module Algebra. *Journal of the Association for Computing Machinery*, 37(2):335–372, Apr. 1990.
2. M. V. Cengarle. *Formal Specifications with Higher-Order Parameterization.* PhD thesis, Institut für Informatik, Ludwig-Maximilians-Universität München, 1994.
3. M. V. Cengarle and M. Wirsing. A Calculus of Higher-Order Parameterization for Algebraic Specifications. Technical Report 9417, Institut für Informatik, Ludwig-Maximilians-Universität München, 1994. Available at URL `http://www.pst.informatik.uni-muenchen.de/~cengarle/part-I.ps`.
4. L. M. G. Feijs. The calculus $\lambda\pi$. In M. Wirsing and J. A. Bergstra, editors, *Algebraic Methods: Theory, Tools and Applications*, volume 394 of *Lecture Notes in Computer Science*, pages 307–328. Springer Verlag, 1989.
5. J. A. Goguen and R. M. Burstall. Introducing Institutions. In E. Clarke and D. Kozen, editors, *Logic of Programming (Proceedings)*, volume 164 of *Lecture Notes in Computer Science*, pages 221–256. Springer Verlag, 1984.
6. J. A. Goguen, J. W. Thatcher, and E. G. Wagner. An Initial Algebra Approach to the Specification, Correctness, and Implementation of Abstract Data Types. In R. T. Yeh, editor, *Current Trends in Programming Methodology*, volume Four: Data Structuring, pages 80–149. Englewood Cliffs/Prentice Hall, 1978.
7. R. Grosu and D. Nazareth. Towards a new way of parameterization. In *Proceedings of the Third Maghrebian Conference on Software Engineering and Artificial Intelligence*, pages 383–392, 1994.
8. D. Sannella, S. Sokołowski, and A. Tarlecki. Toward Formal Development of Programs from Algebraic Specifications: Parameterisation Revisited. *Acta Informatica*, 29:689–736, 1992.
9. M. Wirsing. Algebraic Specification. In J. van Leeuwen, editor, *Handbook of Theoretical Computer Science*, chapter 13, pages 677–788. Elsevier Science Publishers B. V., 1990.
10. M. Wirsing. Structured Specifications: Syntax, Semantics and Proof Calculus. In F. L. Bauer, W. Brauer, and H. Schwichtenberg, editors, *Logic and Algebra of Specification*, volume 94 of *NATO ASI Series F: Computer and Systems Sciences*, pages 411–442. Springer Verlag, 1991.
11. M. Wirsing and M. Broy. A Modular Framework for Specification and Information. In J. Díaz and F. Orejas, editors, *International Joint Conference on Theory and Practice of Software Development (TapSoft '89, Proceedings, volume 1)*, volume 351 of *Lecture Notes in Computer Science*, pages 42–73. Springer Verlag, 1989.

Causality and True Concurrency: A Data-flow Analysis of the Pi-Calculus (Extended Abstract)

Lalita Jategaonkar Jagadeesan
Software Production Research Dept.
AT&T Bell Laboratories
Naperville, IL 60566 (USA)
lalita@research.att.com

Radha Jagadeesan*
Math. Sciences
Loyola University
Chicago, IL 60626 (USA)
radha@math.luc.edu

1 Introduction

The PI–CALCULUS [18, 17] is a process algebra for describing networks of processes with dynamically evolving communication structure. The key idea underlying the PI–CALCULUS is the notion of *naming*: names are used to refer to channels — the links between processes, and can be dynamically created or hidden. Names together with a rich algebra of process combinators that includes parallel composition, allow the PI–CALCULUS to encode asynchronous networks of processes that evolve dynamically. In turn, mobility — this ability to change the network configuration during execution — gives the calculus enormous expressive power; for example, it supports a compositional translation of the lambda calculus [17], supports a compositional translation of the higher order PI–CALCULUS into the first order fragment [23], and has allowed it to be exploited as the target language of semantics of object oriented programming [27].

Most of the reasoning principles proposed for the PI–CALCULUS have been based on *interleaving semantics*, in which only temporal behavior of processes is observed. In particular, all interleaving-based theories satisfy the equation

$$a \parallel b = ab + ba$$

However, a significant difference between these processes is that in the process $a \parallel b$, the actions a and b are *causally independent* and may occur simultaneously. In contrast, the process $ab + ba$ forces either the a to occur first, causing b, or vice-versa. Thus, as is well-known, interleaving-based semantics cannot capture the notion of causality.

Our approach and results

In this paper, we use Kahn semantics of static determinate data-flow as our starting point for the examination of causality in the PI–CALCULUS. The choice of this starting point is influenced by the elegant encoding of causality in Kahn semantics: an event a is a necessary cause for event b if the minimal input required

* This research was supported by grants from NSF and ONR.

to produce b includes a. However, there are two complications that arise when one attempts to use such ideas in analyzing the PI–CALCULUS.

- Kahn semantics applies to static networks, whereas the PI–CALCULUS can encode dynamic networks. The solution to this problem is motivated by tools used in the study of computation in Linear Logic [8], in particular the Geometry of Interaction program [9, 1]. For the purposes of this paper, the key relevant idea is to mimic dynamic networks (of the lambda calculus) by flow of *structured* tokens in a static network.
- The second complication that arises is the non-determinism in PI–CALCULUS processes. The approach we adopt here is to use a variant of the generalization of the determinate data-flow semantics to a semantics for indeterminate dataflow networks [22, 2].

By changing the structure of the tokens, the resulting semantics applies uniformly to the Calculus of Communicating Systems [19] and the PI–CALCULUS. This change of structure of tokens is suggested by the nature of computation in the PI–CALCULUS. The definitions of the process combinators essentially stand unchanged. In particular, the definition of parallel composition is unchanged. Thus our treatment of interaction is "generic", *an essential criterion for any good description of parallel composition.*

Furthermore, for all of these calculi, our semantics distinguishes the processes $a \parallel b$ and $ab + ba$. More generally, for CCS, we show that our semantics induces the same process equivalence as a pomset-based semantics.

Related work

Many "true concurrency" semantics [3, 5, 25, 11, 13, 21, 26, 28], which capture causality to varying degrees, have been proposed for other models of concurrency, including process algebras such as CSP and CCS, Petri Nets, and event structures. These semantics have typically generalized interleaving semantics to encode some degree of concurrency, such as "steps" of concurrent actions rather than single actions, and "pomsets" of partially-ordered multisets of actions rather than linear temporal sequences. However, the investigation of true concurrency semantics for the PI–CALCULUS has not achieved this level of maturity. As far as the authors are aware, the semantics developed in this paper is one of the first true concurrency semantics to generalize smoothly from CCS to the PI–CALCULUS.

The only exceptions of which we are aware is the work of Boreale and Sangiorgi on location bisimulation [7, 16] in the PI–CALCULUS [24] and causality bisimulation [5] in the PI–CALCULUS [4]. Since location bisimulation is incomparable with causal equivalences [4, 5], the former does not address the problem that we are aiming to solve.

On the other hand, [4] is quite relevant to our aims. The primary contribution of [4] is to show that their definition of causal bisimulation can be reduced to observational equivalence in the PI–CALCULUS. As a special case this shows that causal bisimulation on CCS can be reduced to observation equivalence on the monadic PI–CALCULUS. This *beautiful* theorem justifies the observing of some forms of "true concurrency" in CCS.

However, Boreale and Sangiorgi's extension of causal bisimulation to the PI–CALCULUS does not have a uniform notion of causality. Hence, as they point out, it distinguishes the processes

$$P = \nu b(\bar{a} < b > .0|\bar{b} < y > .0), \quad Q = \nu b(\bar{a} < b > .\bar{b} < y > .0)$$

for purely technical reasons. Indeed, the problem of describing a homogeneous causality-based semantics, that would perforce equate these processes, is left as an open problem there. We note that our semantics does equate these processes.

Organization

We start off with a description of the model for CCS and its full abstraction properties. This section brings out many of the key intuitions underlying our interpretation in a simpler setting. We follow with a description of the model for the PI–CALCULUS.

Notation and background

Domain Theory We assume some familiarity with domain theory and stable domain theory. A Scott domain is a bounded complete algebraic poset with a countable basis. We will use $\sqsubseteq$ for the ordering, $\sqsubset$ for the strict ordering, $\sqcup$ for least upper bounds, and $\sqcap$ for greatest lower bounds. We will use $B(D)$ to stand for the basis elements of a Scott domain D. Two elements of a Scott domain are consistent if they have an upper bound. A dI-domain is a Scott domain that is distributive, in which there are no infinite descending chains of compact elements. Given a set of labels $\mathcal{A} = \{ai\}$, we will use $\Pi_{\mathcal{A}}D$ for the A–indexed product of copies of the domain D. We will use $\mathbf{x}, \mathbf{y}$ etc for elements of $\Pi_{\mathcal{A}}D$. The ai'th projection will be denoted by π_{ai}. We will sometimes write x_{aj} for $\pi_{aj}(\mathbf{x})$.

A function between Scott domains is continuous if it preserves limits of omega chains. A function is *stable* if it is continuous and in addition preserves meets of consistent elements. If f is stable, and $y \sqsubseteq f(x)$, the least element $z \sqsubseteq x$ such that $y \sqsubseteq f(z)$ exists and is denoted $\mathtt{Min}(f, x, y)$. We will consider stable functions between dI-domains ordered by the extensional ordering, $f \sqsubseteq g$ if $(\forall x)$ $[f(x) \sqsubseteq g(x)]$.

We will also use the relational powerdomain construction on Scott domains. The relational powerdomain of a Scott domain D will be denoted as $P_R(D)$. $B(P_R(D))$ is the non-empty finite subsets of $B(D)$, ordered by $(\forall d_i)$ $(\exists e_j) d_i \sqsubseteq d_j \Rightarrow$ $\{d_1, \ldots, d_m\} \sqsubseteq$ $\{e_1, \ldots, e_n\}$. $B(D)$ is a lattice, and hence $P_R(D)$ is a complete algebraic lattice. The top element of $P_R(D)$ is denoted $\top$. For any domain D, $P_R(D)$ supports a continuous operation $\uplus : P_R(D) \times P_R(D) \rightarrow P_R(D)$ that is idempotent, commutative and associative and satisfies $x \sqsubseteq x \uplus y$. There is a natural embedding of D into $P_R(D)$, denoted $\{\cdot\}$; and any continuous function f from a Scott domain E into $P_R(D)$ extends uniquely to a $\uplus$ preserving function $f^\dagger$ from $P_R(E)$ to $P_R(D)$.

We will also use the following constructions on dI–domains. Let D be a dI-domain. Then, the domain $\mathbf{Seq}(D)$ ("sequences on D") is the dI-domain with basis elements $\{\langle d_1, \ldots, d_n\rangle \mid d_i \in B(D) - \{\bot\}\}$, with ordering $\langle d_1, \ldots, d_m\rangle \sqsubseteq \langle e_1, \ldots, e_n\rangle$ if $m \leq n \wedge (\forall 1 \leq i \leq m)\ [d_i \sqsubseteq e_i]$. Let D be a dI-domain. Then, the domain $D_\bot$ ("lift of D") is the domain with basis elements $\{d_\bot\rangle \mid d \in B(D)\} \cup \{\bot\}$, with ordering $\bot \sqsubseteq d_\bot, d \sqsubseteq e \Rightarrow d_\bot \sqsubseteq e_\bot$. We will use the following functions on $\mathbf{Seq}(D_\bot)$. We define them as monotone functions on $B(\mathbf{Seq}(D_\bot))$.

$$\mathtt{cons}(d, x) = \begin{cases} \langle (d)_\bot\rangle, & \text{if } x = \bot \\ \langle (d)_\bot, x_1, \ldots, x_n\rangle, & \text{if } x = \langle x_1, \ldots, x_n\rangle \end{cases}$$

$$\mathtt{tail}(\langle d_1 \ldots d_n\rangle) = \left\{ \begin{array}{l} \bot, \text{ if } n \leq 1 \\ \langle d_2, \ldots d_n\rangle \text{ otherwise} \end{array} \right\} \quad \mathtt{head}(x) = \left\{ \begin{array}{l} (d), \ x = \mathtt{cons}((d), \_) \\ \bot, \text{ otherwise} \end{array} \right\}$$

Pomsets A *pomset* is a labeled partial order. Formally, a pomset, p, consists of a set Events_p whose elements are called *events*, a set Labels_p whose elements are called *labels*, a function $\text{label}_p : \text{Events}_p \to \text{Labels}_p$, and a partial order relation $\leq_p$ on Events_p.

We say that event e *causes* event e' in a pomset p iff $e <_p e'$. We write initial(p) to denote the set of events in p that are minimal with respect to $<_p$, *i.e.*, events that do not have any causes in p. A *cut* of p is any subset C of Events_p such that no two distinct events in C are causally related by $<_p$.

A pomset p is a *prefix* of a pomset q iff p is a restriction of q to a downward-closed subset of Events_q. A pomset p is an *augmentation* of a pomset q simply adds more ordering to q; namely, $\text{Events}_p = \text{Events}_q$, $\text{Labels}_p = \text{Labels}_q$, $\text{label}_p = \text{label}_q$, and for all events $e, e' \in \text{Events}_q$, $e <_q e'$ implies $e <_p e'$.

2 CCS

The Calculus of Communicating Systems [19] is a calculus for describing static networks of processes that synchronize via channels. Let $\mathcal{A}$ be a set of actions, let $\bar{\mathcal{A}} = \{\bar{a} : a \in \mathcal{A}\}$ such that $\mathcal{A}$ and $\bar{\mathcal{A}}$ are disjoint, and let $\Sigma = \mathcal{A} \cup \bar{\mathcal{A}}$. For all actions $a \in \mathcal{A}$, $\bar{\bar{a}} = a$. We will use the names ai to refer to elements of Σ. The syntax of CCS processes is as follows, where a ranges over $\mathcal{A}$, α ranges over Σ, and X ranges over some set of process variables V:

$$p ::= \text{NIL} \mid \alpha.p \mid \nu a.p \mid p + q \mid (p|p) \mid X \mid recX.p$$

where . is the prefixing operator, ν is the restriction operator, + is the non-deterministic choice operator, | is the parallel composition operator, and rec is the recursion operator.

2.1 The semantics

The domain D_{CCS} N^v is the dI-domain of vertical natural numbers. Formally, $|N^v| = \{\infty\} \cup N$ with the ordering relation $n \leq m \Rightarrow n \sqsubseteq m, (\forall n)\ [n \sqsubseteq \infty]$. Note that $N^v \simeq \mathtt{Seq}((\cdot)_\perp)$ where $(\cdot)$ stands for the one point domain. In the dataflow metaphor that we will be using to motivate the definitions, the elements of N^v correspond to the number of tokens in a channel; the `tail` function defined earlier returns the "cdr" of the sequence of tokens in a channel, and the `cons` function defined earlier adds a token to the sequence of tokens in a channel. The domain D_{CCS} is defined as the indexed product $\Pi_{\mathcal{A}} N^v$, *i.e* a product with a copy of N^v for each action name. D_{CCS} is a dI-domain, that is also a lattice.

Basic intuitions We will exploit the intuitions gleaned from the analysis of computation in Linear Logic, in particular, as in the geometry of interaction program. A process will be a set of operators on D_{CCS}— thus, indeterminacy gets modeled by describing a process as a set of functions. The functions in a process can be represented as in the picture. Informally, the outputs on a channel

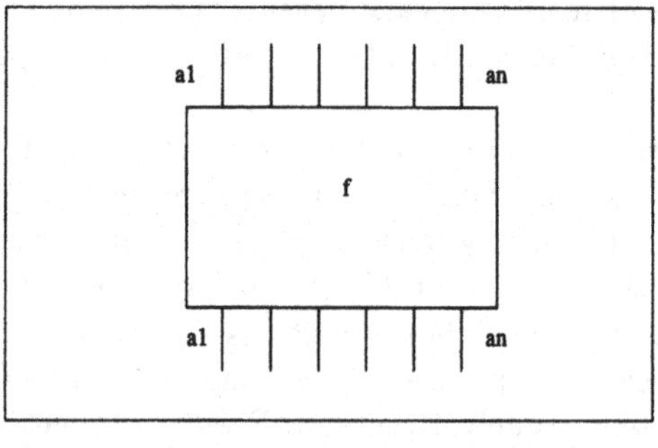

"a" can be thought of as "willingness" of the process to synchronize on 'a". Inputs on a channel "a" are thought of as the "willingness" of the environment to synchronize on "a".

Pomset operators We first identify the class of functions on D_{CCS} that are of interest. Our definition is motivated by two key properties of pomsets. Firstly, every event has a unique cause; stability is the domain theoretic representation of this property. Secondly, there is an underlying partial order and every event is reachable finitely from the initial events. The following definition identifies a reachability property of stable functions. In the context of the dataflow metaphor, we can read the following strict causality condition intuitively as follows. Consider the minimum input required to cause the output of a finite number n of tokens on channel ai. Then the number of tokens on channel ai in this input is strictly lesser than n. The full abstraction theorem brings out the precise content of this connection between pomset traces and the pomset operators.

Definition 1 *A* pomset operator f *on* D_{CCS} *is a stable function from* D_{CCS} *to* D_{CCS} *that satisfies* strict causality*:*

$$(\forall \mathbf{x} \in D_{\text{CCS}})\ (\forall x^f_{ai} \in B(N^v))\ [x^f_{ai} \sqsubseteq \pi_{ai} f(\mathbf{x}) \Rightarrow \pi_{ai}(\mathtt{Min}(\pi_{ai} \circ f, \mathbf{x}, x^f_{ai})) \sqsubset x^f_{ai}]$$

P_{CCS} is the set of pomset operators on D_{CCS}, ordered extensionally. The denotation of a process will be an element of $P_R(P_{\text{CCS}})$; *i.e* the denotation of a process is a set of pomset operators on D_{CCS}.

Nil $[\![\mathtt{nil}]\!]\mathbf{v} = \bot$. The **nil** process produces no information.

Prefixing We define a prefixing operation $[\![ai.[\cdot]]\!]$ on pomset operators as follows. The prefixing operation on processes is defined as $[\![ai.[\cdot]]\!]^{\dagger}$; *i.e.* by lifting the operation to sets of functions pointwise.

Pictorially, prefixing on a pomset operator is done as follows: in this picture, the box is used to indicate synchronization: the function inside the box is only activated if the indicated channel has non-bottom information.

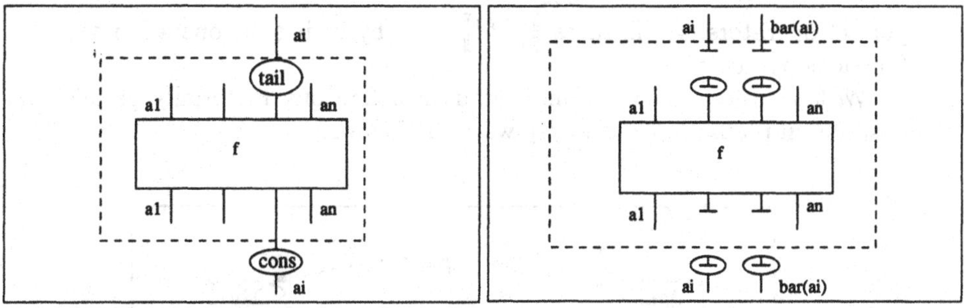

Fig. 1. Prefixing (left) and Restriction (right) in CCS

Formally, we proceed as follows. Define the functions $\mathtt{tail}_i, \mathtt{cons}_i : D_{\text{CCS}} \to D_{\text{CCS}}$ as follows. $\mathtt{tail}_i(\mathbf{x}) = \mathbf{y}$ if $j \neq i$ implies $\pi_{aj}(\mathbf{x}) = \pi_{aj}(\mathbf{y})$ and $j = i$ implies that $\pi_{ai}(\mathbf{y}) = \mathtt{tail}(\pi_{ai}(\mathbf{x}))$. Similarly, $\mathtt{cons}_i(\mathbf{x}) = \mathbf{y}$ if $j \neq i$ implies $\pi_{aj}(\mathbf{x}) = \pi_{aj}(\mathbf{y})$ and $j = i$ implies that $\pi_{ai}(\mathbf{y}) = \mathtt{cons}(\bot, \pi_{ai}(\mathbf{x}))$.

$$[\![\mathtt{ai.f}]\!]\mathbf{v} = \begin{cases} \mathtt{cons}_i(\bot), \text{ if } \pi_{ai}(\mathbf{v}) = \bot \\ \mathtt{cons}_i(f(\mathtt{tail}_i(\mathbf{v}))), \text{ otherwise} \end{cases}$$

We draw the attention of the reader to the fact that the above definition is merely a textual representation of the transformations indicated in the figure.

Restriction We define a restriction operation $[\![(\nu ai)\ [\cdot]]\!]$ on pomset operators as follows. The restriction operation on sets of pomset operators is defined as $[\![(\nu ai)\ [\cdot]]\!]^{\dagger}$; *i.e.* by lifting the operation to sets of functions pointwise.

Pictorially, restriction on a pomset operator f is done in figure 1. Ignore all input on channel $ai, \bar{ai}$ and produce no output on channel $ai, \bar{ai}$; otherwise behave like f.

Formally, we proceed as follows. Define a function $\mathtt{rest}_{ai} : D_{\mathrm{CCS}} \rightarrow D_{\mathrm{CCS}}$ as follows: $\mathtt{rest}_{ai}(\mathbf{x}) = \mathbf{y}$ if $\pi_{ai}(\mathbf{y}) = \pi_{\bar{ai}}(\mathbf{y}) = \perp$ and $\pi_{aj}(\mathbf{x}) = \pi_{aj}(\mathbf{y})$ otherwise. Define

$$[\![(\nu ai)[f]]\!]\mathbf{v} = \mathtt{rest}_{ai}(f(\mathtt{rest}_{ai}(\mathbf{v})))$$

Plus Plus is interpreted by the union operation of $P_R(P_{\mathrm{CCS}})$. This is the primary reason for the correspondence of the semantics with a variant of pomset traces (as opposed to a richer notion of pomset failures, for example).

$$[\![P_1 + P_2]\!] = [\![P_1]\!] \uplus [\![P_2]\!]$$

Parallel Composition We define a parallel composition operation $[\![[\cdot]|[\cdot]]\!]$ on pomset operators as follows; this composition will yield a set of pomset operators, *i.e.* an element of $P_R(P_{\mathrm{CCS}})$. The parallel composition operation on sets of pomset operators is defined as $[\![[\cdot]|[\cdot]]\!]^{\dagger}$; *i.e.* by lifting the operation to sets of functions pointwise.

We first expose the intuitions behind the definitions. Pictorially, parallel composition on pomset operators f, g works as follows.

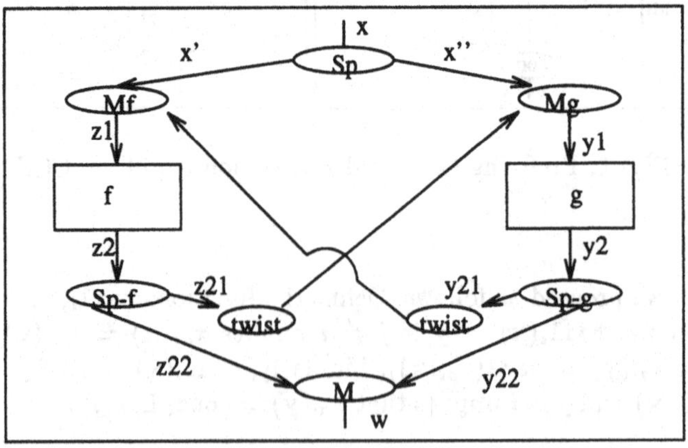

Fig. 2. Parallel composition in CCS

Note that the feedback loop in the above picture meshes with the intuitions that were stated earlier about inputs and outputs. The "willingness" of f to synchronize on channel a gets transmitted to g as the "willingness" of the environment to synchronize on $\bar{a}$, and vice versa. Inputs from the environment are split (Sp in the above figure) amongst f, g. Outputs created by f, g are split (Spf and Spg in the figure) between outputs to the environment and outputs to the dual channel name in the other process. The outputs on any channel name are obtained by merging the outputs to the environment from the individual processes(M in the above figure). The inputs to f, g are created by merging inputs from the environment and outputs from the dual channel name in the other process (Mf and Mg in the above figure). The above construction is repeated for all possible split and merge processes. The formal development follows.

Let $\mathtt{Sp}, \mathtt{Spf}, \mathtt{Spg}$ be determinate splits and $\mathtt{M}, \mathtt{Mf}, \mathtt{Mg}$ be determinate merges such that $(\mathtt{Sp}, \mathtt{M})$, $(\mathtt{Spf}, \mathtt{Mf})$ and $(\mathtt{Spg}, \mathtt{Mg})$ are pairs of functions that are inverses to each other. Let the function $\mathtt{twist}$ that interchanges the data on dual names be defined as follows: $\mathtt{twist}(\mathbf{x}) = \mathbf{y}$, if $\pi_{ai}(\mathbf{x}) = \pi_{\bar{a}i}(\mathbf{y})$.

The parallel composition of f, g for a given choice of determinate split and merge processes is described as as the least solution of the following equations; these equations are solved using the fixpoint iteration in the style of the semantics for determinate dataflow networks [14, 15]. More precisely, the following equations describe a pomset operator $h_{\mathtt{M},\mathtt{Mf},\mathtt{Mg}}$, where $h_{\mathtt{M},\mathtt{Mf},\mathtt{Mg}}(\mathbf{x}) = \mathbf{w}$, if $\mathbf{w}$ arises in the least solution of the equations:

$$\begin{array}{llll}
\mathtt{Sp}(\mathbf{x}) & = \langle \mathbf{x}', \mathbf{x}'' \rangle & \mathtt{Mf}\langle \mathbf{x}', \mathtt{twist}(\mathbf{y}_{21}) \rangle & = \mathbf{z}_1 \\
\mathtt{Mg}\langle \mathbf{x}'', \mathtt{twist}(\mathbf{z}_{21}) \rangle & = \mathbf{y}_1 & f(\mathbf{z}_1) & = \mathbf{z}_2 \\
g(\mathbf{y}_1) & = \mathbf{y}_2 & \mathtt{Spf}(\mathbf{z}_2) & = \langle \mathbf{z}_{21}, \mathbf{z}_{22} \rangle \\
\mathtt{Spg}(\mathbf{y}_2) & = \langle \mathbf{y}_{21}, \mathbf{y}_{22} \rangle & \mathtt{M}\langle \mathbf{z}_{22}, \mathbf{y}_{22} \rangle & = \mathbf{w}
\end{array}$$

Note that each merge (and hence the corresponding split) can be encoded as a (possibly infinite) sequence over $0, 1$. Let $\mathtt{Z}$ be the flat domain over the set $\{0, 1\}$. Thus, $\mathtt{Seq}(Z)$ is the domain of finite and infinite binary sequences under the prefix ordering. In this light, the fixpoint equation above describes a continuous function H from $\mathtt{Seq}(\mathtt{Z}) \times \mathtt{Seq}(\mathtt{Z}) \times \mathtt{Seq}(\mathtt{Z})$ (one copy to encode each merge) to pomset operators. Thus, $H^\dagger$ is a continuous function preserving $\uplus$ from $P_R(Seq(\mathtt{Z}) \times Seq(\mathtt{Z}) \times Seq(\mathtt{Z}))$ to P_{CCS}.

$$[\![f|g]\!] = H^\dagger(\top)$$

Recursion All the process combinators mentioned earlier are monotone and continuous in their process arguments. Recursion is interpreted via the usual fixpoint principles (valid since $P_R(P_{\mathrm{CCS}})$ is a cpo).

Examples Let $\mathtt{onea}$ denote the element of D_{CCS} that satisfies $\pi_i(\mathtt{onea}) = \perp$ for $i \neq a$ and $\pi_i(\mathbf{y}) = 1$ for $i = a$. Symmetrically define $\mathtt{oneb}$. Then, the denotation of the process $a||b$ has the following maximum element f: f is the constant function returning $\mathtt{onea} \sqcup \mathtt{oneb}$. The denotation of the process $a.b$ has the following maximum element g: $g(\mathbf{x}) = \mathtt{onea}$ when $\pi_a(\mathbf{x}) = \perp$, and $g(\mathbf{x}) = \mathtt{onea} \sqcup \mathtt{oneb}$ when

$\pi_a(\mathbf{x}) \neq \perp$. The denotation of the process $b.a$ has a maximum element h with the roles of a, b reversed in the definition of g. Thus, we have:

$$[\![a||b]\!] \neq [\![ab + ba]\!]$$

2.2 Results

We introduce the following variation on pomsets. Let p be a pomset, let a be a label in p, let C be a cut of p, and let C_a be the restriction of C to events labeled a. Then C has *temporally-ordered autoconcurrency over a* iff there is some total ordering $e_1 \ldots e_n$ of all the events in C_a such that for all $1 \leq i < j \leq n$; for all events $e \in Events_p$, $e <_p e_i$ implies that $e <_p e_j$, and for all events $e' \in Events_p$, $e_j <_p e'$ implies $e_i <_p e'$. Hence, if e_j is enabled, then e_i must be enabled as well; conversely, if e_j is a cause of some event e, then e_i is a cause of event e as well.

Definition 2 *A pomset p is a* TP *(for temporally ordered autoconcurrency) iff for all labels* $a \in Labels_p$ *and all cuts C of p, C has temporally-ordered autoconcurrency over a.*

We first give a pomset-trace based semantics to CCS by translating CCS terms to 1-safe Petri Nets as in [10, 12], and taking the augmentation-closure of the pomset-traces of these Petri Nets as in [13]. The details are standard and are omitted here due to space restrictions. We then restrict these sets of pomset-traces to TPs, to get the semantics $[\![(\cdot)]\!]_{\mathtt{POM}}$.

Correspondence theorems The space of TPs corresponds precisely to the space of pomset operators. This result in essentially "predicted" in the work on (stable) event structures [20].

Let f be a pomset operator; then, f induces the following (possibly infinite) TP. The nodes of the pomset are $\{ai^n \mid (\exists \mathbf{x})\ [n \sqsubseteq \pi_{ai}(f\mathbf{x})]$. The label on a node ai^n is ai. The ordering relation is induced by the canonical trace of the stable function f, $\mathtt{Tr}(f)$ that associates with each finite output the minimal input required to cause the output; aj^m is a cause for ai^n if $(\exists(\mathbf{y}, \mathbf{cons}_i(n)) \in \mathtt{Tr}(f))\ [m \sqsubseteq \pi_{aj}(\mathbf{y})]$. Stability ensures the unique causal prefix property required for pomsets. Strict causality ensures that we have a partial order and that all nodes of the induced pomset are reachable from the initial nodes. By construction, the pomsets are TPs.

Let C be a TP. Then, C induces the following pomset operator. $(\mathbf{x}, \langle \perp, \ldots, \perp, n_{ai}, \perp, \ldots, \rangle)$ is in the trace of the pomset operator if $\mathbf{x}$ is the minimum causal prefix in the pomset for n occurrences of ai: the TP condition ensures the existence of such a minimum causal prefix. Stability is guaranteed by the unique causal prefix properties of pomsets. The partial order underlying the pomset and the reachability of the pomset ensures the strict causality property.

This close correspondence between the space of pomset operators and the space of TPs is the key tool in proving full abstraction.

Theorem 1 *(FULL ABSTRACTION)* $[\![P]\!] = [\![Q]\!] \Leftrightarrow [\![P]\!]_{\mathtt{POM}} = [\![Q]\!]_{\mathtt{POM}}$.

3 PI–CALCULUS

This section describes the extension of the earlier treatment to the pi-calculus. Most of the theory developed for CCS goes through unchanged. The only changes will be the replacement of the base domain N^v by a domain T.

The pi-calculus [18, 17] extends CCS by allowing the passage of names along channels. Let $\mathcal{A}$ be a set of actions, let $\bar{\mathcal{A}} = \{\bar{a} : a \in \mathcal{A}\}$ such that $\mathcal{A}$ and $\bar{\mathcal{A}}$ are disjoint, and let $\Sigma = \mathcal{A} \cup \bar{\mathcal{A}}$. For all actions $a \in \mathcal{A}$, $\bar{\bar{a}} = a$. The syntax of PI–CALCULUS processes is as follows, where a, x range over $\mathcal{A}$:

$$p ::= \mathrm{NIL} |\ a(x).p|\ \bar{a}(x).p|\ \nu(x).p|\ p + q|\ (p|p)|\ !p$$

where $a(x).(.)$ is the input prefixing operator, and $\bar{a}(x).(.)$ is the output prefixing operator. The input prefixing operator binds the occurrence of x. ν is the restriction operator, $+$ is the non-deterministic choice operator, $|$ is the parallel composition operator, and $!p$ is the copying operator.

3.1 The semantics

The domain $D_{\mathtt{PiC}}$ The domain T (for "tree–like") is given by the solution of the recursive domain equation

$$T \simeq \mathsf{Seq}((T \times T)_\perp)$$

This solution is obtained in the category of dI–domains and rigid embeddings [6]. The domain $D_{\mathtt{PiC}}$ is defined as the indexed product $\Pi_{\mathcal{A}} T$. Thus, T is the domain associated with each channel name, just as N^v was the domain associated with each channel name in CCS. The structure of T is suggested by the information carried along names in the PI–CALCULUS. Intuitively, in the course of the evolution of a process, a name carries many (possibly different) names. The Seq construction above models this. Also, note that when a name is sent along another name, the co-name is also carried along implicitly. This gets modeled by the product in the above construction.

Pomset operators are defined exactly as before, *i.e.* strictly causal and stable operators on $D_{\mathtt{PiC}}$. $P_{\mathtt{PiC}}$ is the set of pomset operators on $D_{\mathtt{PiC}}$, ordered extensionally. Following the earlier recipe, processes are interpreted as elements of $P_R(P_{\mathtt{PiC}})$.

The definitions of restriction, nil and parallel composition are exactly as before. (apart from the changes in the domain and range of definition, of course!!). Furthermore, plus is again interpreted by the $\uplus$ operation on $P_R(P_{\mathtt{PiC}})$. As before, all combinators yield monotone and continuous functions on their process arguments, yielding an immediate interpretation of full recursion (hence including the pi-calculus operation $!P$). We treat prefixing below.

Input prefixing We define an input prefixing operation $\llbracket ai(x).[\cdot] \rrbracket$ on pomset operators as follows. The input prefixing operation on processes is defined as $\llbracket ai.[\cdot] \rrbracket^{\dagger}$; *i.e.* by lifting the operation to sets of functions pointwise.

Pictorially, input prefixing on a pomset operator is done as follows. The box indicates synchronization as before. The modeling of the fact that input prefixing $ai(x)$ binds the name x (and hence $\bar{x}$) is based on the treatment of multiplicatives in the geometry of interaction. In particular, the input information for x is extracted from the input on a, and the output information on x is included in the output on a. Similarly for $\bar{x}$. Also, note the absence of any connection between the external inputs/outputs on x and the internal inputs/outputs on x. Similarly for $\bar{x}$.

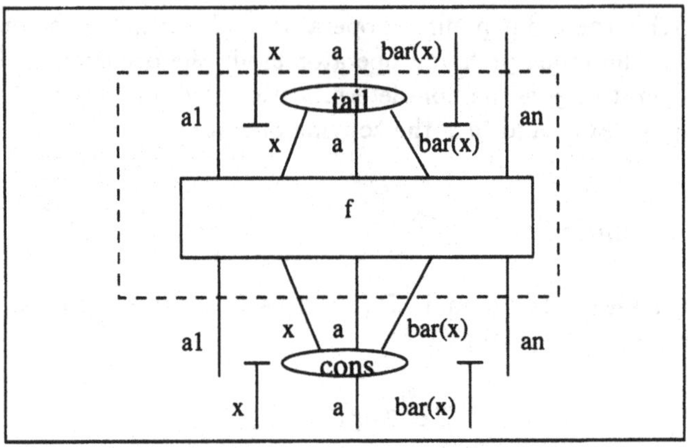

Fig. 3. Input prefixing in the pi-calculus

Formally, we have the following definition. $\llbracket \mathbf{ai(x).f} \rrbracket \mathbf{v} = \mathbf{w}$ where

$$v'_{aj} = \left\{ \begin{array}{l} \pi_{aj}(\mathbf{v}),\ \text{if } aj \notin \{ai, x, \bar{x}\} \\ \mathtt{tail}(v_{aj}),\ \text{if } aj = ai \\ \pi_1(\mathbf{head})(v_{ai}),\ \text{if } aj = x \\ \pi_2(\mathbf{head})(v_{ai}),\ \text{if } aj = \bar{x} \end{array} \right\} \quad \mathbf{w}' = \left\{ \begin{array}{l} f(\mathbf{v}'),\ \text{if } v_{ai} \neq \perp \\ \perp,\ \text{otherwise} \end{array} \right\}$$

$$w_{aj} = \left\{ \begin{array}{l} w'_{aj},\ \text{if } aj \notin \{ai, x, \bar{x}\} \\ \mathbf{cons}((w'_x, w'_{\bar{x}}), v_{aj}),\ \text{if } aj = ai \\ \perp,\ \text{if } aj \in \{x, \bar{x}\} \end{array} \right\}$$

Output prefixing We define an output prefixing operation $\llbracket \bar{ai}(x).[\cdot] \rrbracket$ on pomset operators; this operation yields an element of $P_R(P_{\mathrm{Pic}})$. The output prefixing operation on processes is defined as $\llbracket \bar{ai}(x).[\cdot] \rrbracket^{\dagger}$; *i.e.* by lifting the operation to sets of functions pointwise.

Pictorially, output prefixing on a pomset operator is done as follows. The synchronization box has been omitted to avoid cluttering up the figure. The important difference from the input prefixing is that the name x is not bound. This leads to two issues:

- The environment can interact independently with the name x. The nodes titled **M1x** and **S1x** in the figure mix the information on x from the environment and from the channel $\bar{a}$. Simlarly for $\bar{x}$.
- There is one further level of complication arising from *scope extrusion*. Output prefixing $\bar{a}(x)$ can export the environment's interaction on x to the output on $\bar{a}$ pertaining to $\bar{x}$. Similarly for $\bar{x}$. The required interaction is modeled by the dotted arrows in the figure.

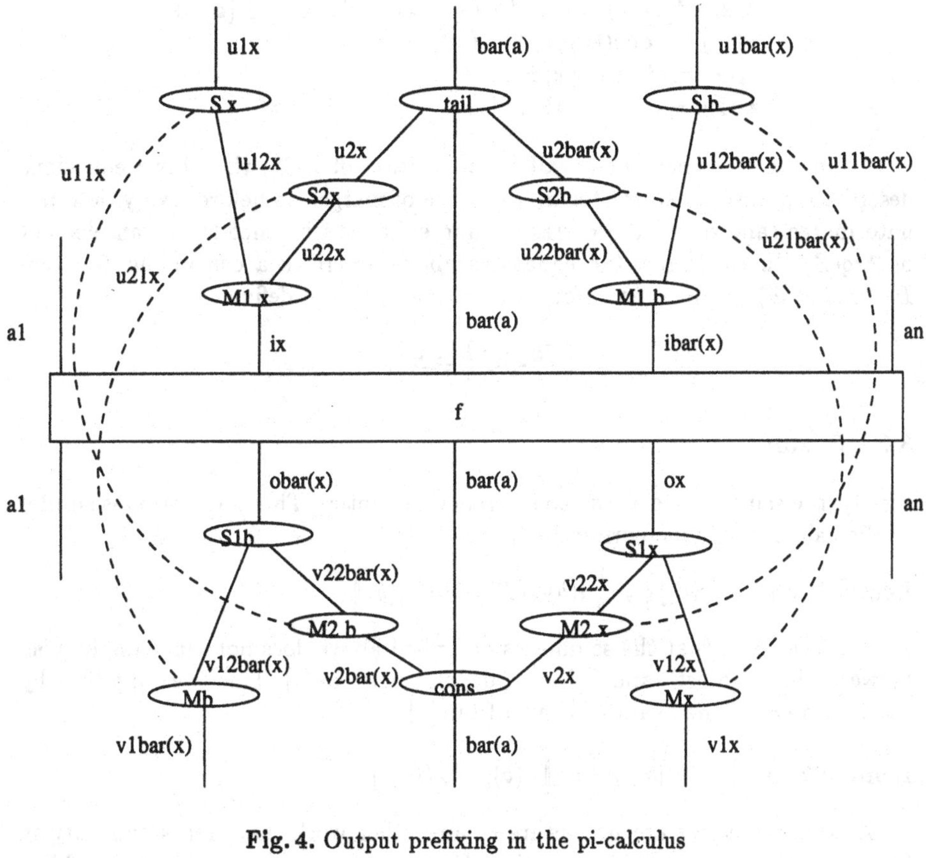

Fig. 4. Output prefixing in the pi-calculus

Formally, we proceed by writing down the equations implicit in the above figure. Let $S_s, S1_s, S2_s$ where $s \in \{x, \bar{x}\}$, be determinate splits. Let $M_s, M1_s, M2_s$, where $s \in \{x, \bar{x}\}$ be determinate merges such that (S_s, M_s), $(S1_s, M1_s)$ and $(S2_s, M2_s)$, where $s \in \{x, \bar{x}\}$ are pairs of functions that are inverses to each other. We are going to describe $\mathbf{z} = [\![\bar{a}i(x).f]\!](\mathbf{r})$. There are two cases. If $\pi_{\bar{a}}(\mathbf{r}) = \perp$, $\mathbf{z} = \mathbf{cons}_{\bar{a}}((\perp, \perp), \perp)$.

Otherwise, we proceed as follows. $\mathbf{w}$, the input into f, is defined by the following equations:

$$
\begin{array}{l}
u_2^x = \pi_2(\mathbf{head}(\pi_{\bar{a}}(\mathbf{r}))),\ u_2^{\bar{x}} = \pi_1(\mathbf{head}(\pi_{\bar{a}}(\mathbf{r}))) \\
\pi_s(\mathbf{r}) = u_1^s, S_s(u_1^s) = (u_{11}^s, u_{12}^s),\ \text{for } s \in \{x, \bar{x}\} \\
S2_s(u_2^s) = (u_{21}^s, u_{22}^s), M1_s(u_{12}^s, u_{22}^s) = i^s,\ \text{for } s \in \{x, \bar{x}\} \\
\pi_{\bar{a}}(\mathbf{w}) = \mathtt{tail}(\pi_{\bar{a}}(\mathbf{r})), \pi_s(\mathbf{w}) = i^s,\ \text{for } s \in \{x, \bar{x}\} \\
\pi_t(\mathbf{w}) = \pi_t(\mathbf{r}),\ \text{otherwise}
\end{array}
$$

Let $f(\mathbf{w}) = \mathbf{w}'$. The result $\mathbf{z}$, is defined by the following equations:

$$
\begin{array}{l}
\pi_s(\mathbf{w}') = o^s, S1_s(o^s) = (v_{12}^s, v_{22}^s),\ \text{for } s \in \{x, \bar{x}\} \\
M2_s(u_{11}^{\bar{s}}, v_{22}^s) = v_s^2, M_s(u_{21}^{\bar{s}}, v_{12}^s) = v_1^s,\ \text{for } s \in \{x, \bar{x}\} \\
\pi_{\bar{a}}(\mathbf{z}) = \mathbf{cons}((v_2^{\bar{x}}, v_2^x), \pi_{\bar{a}}(\mathbf{w}')) \\
\pi_s(\mathbf{z}) = v_1^s,\ s \in \{x, \bar{x}\} \\
\pi_t(\mathbf{x}) = \pi_t(\mathbf{w}'),\ \text{otherwise}
\end{array}
$$

As in the treatment of parallel composition of CCS, the above equations describe a pomset operator for every choice of merge. As before, every determinate merge (and hence the corresponding split) can be encoded as an element of $\mathtt{Seq}(Z)$. In this light, the equations above describes a continuous function $H : \Pi_6\mathtt{Seq}(Z) \to P_{\mathrm{CCS}}$; as before, we consider $H^\dagger$ and define

$$[\![\bar{a}i(x).f]\!] = H^\dagger(\top)$$

3.2 Results

Firstly, the semantics is a true concurrency semantics. This calculation is similar to the calculation done earlier for CCS.

Lemma 1. $[\![\bar{a}(b)|\bar{c}(d)]\!] \neq [\![\bar{a}(b).\bar{c}(d) + \bar{c}(d).\bar{a}(b)]\!]$.

Next, we note that the semantics described above does not make distinguish between the "subject" and "object" dependencies of [4]. This is exemplified by the treatment of the example drawn from [4].

Lemma 2. $[\![\nu(b)[\bar{a}(b)|\bar{b}(y)]]\!] = [\![\nu(b)[\bar{a}(b).\bar{b}(y)]]\!]$.

At this point, we are not aware of any other work that treats causality in the PI–CALCULUS as a uniform notion. Consequently, we are unable to establish a point of reference for comparison. (This was possible for CCS because of the rich literature that investigates true concurrency.) Here, we restrict ourselves to a basic adequacy theorem for interleaving traces on the standard transition system [18] for the PI–CALCULUS. We follow the notation of [18]; $\equiv$ stands for structural congruence and $\to$ stands for the one-step reduction relation.

Theorem 2

- $P \equiv Q \Rightarrow [\![P]\!] = [\![Q]\!]$
- $P \rightarrow Q \Rightarrow [\![Q]\!] \sqsubseteq [\![P]\!]$
- $[\![P]\!] = [\![Q]\!]$ *implies that for all process contexts* $C[\cdot]$, $\mathtt{Traces}(C[P]) = \mathtt{Traces}(C[Q])$.

The first item above states that the structural congruence is reflected as equality in the semantics. The proof proceeds by a straightforward inductive analysis of rules used in [18] to deduce that $P \equiv Q$. The second item is the analogue of the "one step reduction preserves meaning" property. We note that the inequality (as opposed to the more familiar equality in semantics of deterministic transition systems) is characteristic of non-deterministic transition systems. Intuitively, when P reduces one-step to Q, some potential reductions of P may not be available for Q. The key case in this proof is the reduction $a(x).P||\bar{a}(y).Q \rightarrow P[x/y]||Q$. The essence of the proof — namely that the model is correct with respect to substitution — is handled by an adaptation of the soundness proof of geometry of interaction models (see for example [1]).

Acknowledgments The authors thank the anonymous referees for valuable comments. This paper owes a great deal to the work and ideas of Samson Abramsky — indeed this work is a flushing out (in the setting of the PI–CALCULUS) of the analysis of computation in the geometry of interaction. The authors also wish to acknowledge the inspiration afforded by the work of Simon Gay on the relationship between interaction nets and determinate CCS. The authors are of course responsible for any errors and omissions.

References

1. S. Abramsky and R. Jagadeesan. New foundations for the geometry of interaction. *Information and Computation*, 111(1):53–119, 1994.
2. Samson Abramsky. A generalized Kahn principle for abstract asynchronous networks. In *Mathematical Foundations of Programming Semantics*, pages 1–21. Springer-Verlag, 1989. Lecture Notes in Computer Science Vol. 442.
3. L. Aceto and M. Hennessy. Towards action-refinement in process algebras. In *Proceedings, Fourth Annual Symposium on Logic in Computer Science*, pages 138–145. IEEE Computer Society Press, 1989.
4. M. Boreale and S. Sangiorgi. A fully abstract semantics for causality in the π-calculus. Technical report, University of Edinburgh, 1994. Tec. Rep.ECS–LFCS–94–297.
5. G. Boudol, I. Castellani, Matthew C. Hennessy, and A. Kiehn. A theory of processes with localities. In *Proceedings of International Conference on Concurrency Theory, Volume 630 of Lecture Notes in Computer Science*, pages 108–122, 1992.
6. T. Coquand, C. Gunter, and G. Winskel. dI-domains as a model of polymorphism. In *Third Workshop on the Mathematical Foundations of Programming Language Semantics*, pages 344–363. Springer-Verlag, 1987.

7. D. Degano and D. Darondeau. Causal trees. In *Proceedings of ICALP '89, Volume 372 of Lecture Notes in Computer Science*, pages 234–248, 1989.
8. J.-Y. Girard. Linear Logic. *Theoretical Computer Science*, 50(1):1–102, 1987.
9. J.-Y. Girard. Geometry of interaction 1: Interpretation of System F. In R. Ferro et al., editor, *Logic Colloquium 88*. North Holland, 1989.
10. Ursula Goltz. CCS and Petri nets. Technical report, GMD, July 1990.
11. Roberto Gorrieri. *Refinement, Atomicity, and Transactions for Process Description Languages*. PhD thesis, University of Pisa, 1991.
12. Lalita Jategaonkar. *Observing "True" Concurrency*. PhD thesis, Massachusetts institute of Technology, September 1993. Supervised by Albert R. Meyer.
13. Lalita Jategaonkar and Albert R. Meyer. Testing equivalence for Petri nets with action refinement. In *Proceedings of International Conference on Concurrency Theory, Volume 630 of the Lecture Notes in Computer Science*, pages 17–31, 1992.
14. G. Kahn. The semantics of a simple language for parallel programming. In *Information Processing 74*, pages 993–998. North-Holland, 1977.
15. G. Kahn and D. MacQueen. Coroutines and networks of parallel proceses. In *Proceedings of IFIP Congress*. North-Holland, 1977.
16. A. Kiehn. Local and global causes. Technical report, Technische Universitat Munchen, 1991.
17. R. Milner. Functions as processes. In *Proceedings of ICALP 90*, volume 443 of *Lecture Notes in Computer Science*, pages 167–180. Springer-Verlag, 1990.
18. R. Milner. The polyadic pi-calculus: a tutorial. Technical Report ECS-LFCS-91-180, University of Endinburgh, 1991.
19. Robin Milner. *Communication and Concurrency*. Series in Computer Science. Prentice Hall, 1989.
20. M. Nielsen, G. D. Plotkin, and G. Winskel. Petri nets, event structures and domains i. *Theoretical Computer Science*, 13(1):85–108, 1981.
21. Vaughan Pratt. Modeling concurrency with partial orders. *International Journal Of Parallel Programming*, 15(1):33–71, 1986.
22. J. R. Russell. On oraclizable networks and kahn's principle. In *Proceedings of ACM Symposium on Principles of Programming Languages*, 1990.
23. D. Sangiorgi. *Expressing Mobility in Process Algebras: First-Order and Higher-Order Paradigms*. PhD thesis, University of Edinburgh, 1992.
24. D. Sangiorgi. Locality and non-interleaving semantics in calculi for mobile processes. In *Proceedings of ICALP '94, Lecture Notes in Computer Science*, 1994.
25. Rob van Glabbeek and Frits Vaandrager. Petri net models for algebraic theories of concurrency. In *Proceedings of PARLE, Volume 259 of the Lecture Notes in Computer Science*, pages 224–242, 1987.
26. Walter Vogler. *Modular Construction and Partial Order Semantics of Petri Nets*, volume 625 of *Lecture Notes in Computer Science*. Springer-Verlag, 1992. 252 pp.
27. David Walker. π-calculus semantics of object-oriented programming languages. In *Theoretical Aspects of Computer Software*, pages 532–547, 1991.
28. Glynn Winskel. Event structures. In *Petri Nets: Applications and Relationships to Other Models of Concurrency, Volume 255 of Lecture Notes in Computer Science*, pages 325–392, 1987.

Verification in Continuous Time by Discrete Reasoning*

Luca de Alfaro and Zohar Manna

Computer Science Department
Stanford University
Stanford, CA 94305, USA
{luca,zm}@cs.stanford.edu

1 Introduction

There are two common choices for the semantics of real-time and hybrid systems. The first is a *discrete semantics*, in which the temporal evolution of the system is represented as an enumerable sequence of snapshots, each describing the state of the system at a certain time. The second is a *continuous semantics*, in which the system evolution is represented by a sequence of intervals of time, together with a description of the system state during each interval.

The weakness of the discrete semantics is that the sequence of snapshots does not correspond directly to the physical behavior of the system [6]. Moreover, the truth of temporal logic formulas depends not only on the behavior of the system, but also on the way in which the behavior has been sampled to produce the sequence of snapshots. In other words, the discrete semantics is not *sample invariant* [12]. What has made the discrete semantics popular, however, is the simplicity and power of the verification rules that can be formulated for temporal logic [7, 17]. The continuous semantics overcomes the weaknesses of the discrete one, but no comparable verification rules have been formulated for it.

This paper shows how the advantages of both semantics can be combined by adapting the simple verification rules of the discrete semantics to the continuous one. Specifically, we show that if a temporal logic formula has the property of *finite variability* (FV), its validity in the discrete semantics implies its validity in the continuous one. Thus, verification rules that have as conclusion a FV formula are sound also in the continuous semantics. Most formulas that arise in practice are FV, and we give criteria that help to characterize them. The ability to transfer results between the semantics, together with a deductive system for the continuous semantics, provides a powerful methodology for verifying temporal logic properties of real-time and hybrid systems in the continuous semantics.

We first present the case for real-time systems, and then we extend the results to hybrid systems.

---

* This research was supported in part by the National Science Foundation under grant CCR-92-23226, by the Defense Advanced Research Projects Agency under contract NAG2-892, and by the United States Air Force Office of Scientific Research under contract F49620-93-1-0139.

2 Real-Time Systems

Real-time systems will be modeled by ***timed transition systems*** [7, 15]. A timed transition system $S = \langle \mathcal{V}, \Sigma, \Theta, \mathcal{T}, L, U \rangle$ consists of the following components.

1. A set $\mathcal{V}$ of variables called ***state variables***, where each variable has its own type.
2. A set Σ of states: each state $s \in \Sigma$ is a type-consistent interpretation of all the variables in $\mathcal{V}$. We denote by $s(x)$ the value at state s of variable x, for every $x \in \mathcal{V}$.
3. A set $\Theta \subseteq \Sigma$ of initial states. Θ has an associated assertion $\Theta_f(\mathcal{V})$, such that $\Theta = \{s \in \Sigma \mid s \models \Theta_f\}$.
4. A set $\mathcal{T}$ of transitions, where $\tau \subseteq \Sigma \times \Sigma$ for all $\tau \in \mathcal{T}$. Each $\tau \in \mathcal{T}$ has an associated assertion $\rho_\tau(\mathcal{V}, \mathcal{V}')$ such that $\tau = \{(s, s') \mid (s, s') \models \rho_\tau\}$, where (s, s') interprets $x \in \mathcal{V}$ as $s(x)$ and x' as $s'(x)$.
5. Two sets L, U of minimum and maximum delays of transitions, s.t. $0 \leq l_\tau \leq u_\tau \leq \infty$ for all $\tau \in \mathcal{T}$.

We denote by c_τ the enabling condition of transition τ, defined by $c_\tau = \{s \mid \exists s'.\, (s, s') \in \tau\}$. If $s \in c_\tau$, we say that transition τ is enabled in state s. We will assume that transitions are self-disabling: $(s, s') \in \tau \rightarrow s' \notin c_\tau$.

The temporal behavior of a real-time system will be represented by ***discrete traces*** or by ***continuous traces***, thus giving rise to the discrete and continuous semantics [6, 15, 17].

Definition 1 (discrete trace). A *discrete trace* σ_d is an enumerable sequence of observations $\langle s_0, t_0 \rangle, \langle s_1, t_1 \rangle, \langle s_2, t_2 \rangle, \ldots$, with $s_n \in \Sigma$, $t_n \in \mathbb{R}^+$ for $n \in \mathbb{N}$, such that $t_0 = 0$, $\lim_{n \to \infty} t_n = \infty$, and $t_n \leq t_{n+1}$, $(t_n = t_{n+1}) \vee (s_n = s_{n+1})$ for all $n \in \mathbb{N}$. A *moment* of σ_d is simply an integer $n \in \mathbb{N}$. □

In the continuous semantics, the behavior of the system is represented by a mapping from intervals of non-negative real numbers (representing intervals of time) to states of the system. The intervals can overlap at most at the endpoints [6, 8, 2].

Definition 2 (continuous trace). A *continuous trace* σ_c is a sequence of pairs σ_c: $\langle s_0, I_0 \rangle$, $\langle s_1, I_1 \rangle$, $\langle s_2, I_2 \rangle$, $\ldots$, where I_n is an interval of $\mathbb{R}$ (open or closed at the endpoints) and $s_n \in \Sigma$ for all $n \in \mathbb{N}$, such that $\bigcup_{n \in \mathbb{N}} I_n = \mathbb{R}^+$ and $\sup I_n = \inf I_{n+1}$ for all $n \in \mathbb{N}$. For every $n \in \mathbb{N}$, either $s_n = s_{n+1}$, or I_n must be right-closed and I_{n+1} left-closed. A *moment* of a continuous trace σ_c is a pair (n, t) such that $t \in I_n$. The moments are ordered in lexicographic order: $(n, t) \leq (n', t')$ iff $n < n' \vee (n = n' \wedge t \leq t')$. □

Definition 3 (admission). A timed transition system S ***admits*** a trace σ, written $S \triangleright \sigma$, if the following conditions are satisfied.

1. For all $n \in \mathbb{N}$, either $s_n = s_{n+1}$ or $(s_n, s_{n+1}) \in \tau$ for a transition $\tau \in \mathcal{T}$ that has been continuously enabled at least for its minimum delay l_τ.
2. Transitions are never continuously enabled for a time longer than their maximum delay. □

3 Logic and Verification

Temporal logic. It is possible to define linear-time temporal logics for both the discrete and the continuous semantics. In both cases, we will use the same syntax, but the logics will have two different sets of models. We denote by $\mathrm{TL_D}$ the logic defined for the discrete semantics, and by $\mathrm{TL_C}$ the logic defined for the continuous semantics.

Syntax. The logic we consider includes the future temporal operators $\Box$, $\Diamond$, $\mathcal{U}$, and the corresponding past operators $\boxminus$, $\diamondminus$, $\mathcal{S}$ [7, 16]. However, the logic does not include the next and previous time operators $\bigcirc$, $\ominus$. The logic includes the *age function* Γ: for a formula ϕ, at any point in time, the term $\Gamma(\phi)$ denotes for how long in the past ϕ has been continuously true [17]. The language also includes the special symbol T, whose value represents the global time. Thus, for any point in time, the equality $T = \Gamma(\mathit{true})$ holds.

The logic allows quantification over *rigid* variables, whose value is constant in time [5]. We use Greek letters, like ξ, ζ, for the rigid variables and Latin letters, like x, y, for the state variables of the system.

Semantics. The truth of temporal logic formulas is evaluated with respect to a model and an interpretation. The models of $\mathrm{TL_D}$ are the discrete traces, the models of $\mathrm{TL_C}$ are the continuous traces. The interpretation $\mathcal{I}$ assigns the value $\mathcal{I}(\xi)$ to each rigid variable ξ. In both cases, the truth of a formula is evaluated at each moment of a trace. We write $\mathcal{I}, \sigma_\mathrm{d} \models^{\mathrm{D}}_{n} \phi$ (resp. $\mathcal{I}, \sigma_\mathrm{c} \models^{\mathrm{C}}_{(n,t)} \phi$) to denote that, under the interpretation $\mathcal{I}$, ϕ is true at moment n of σ_d (resp. (n, t) of σ_c). If the interpretation $\mathcal{I}$ is omitted, the above statement holds under all possible interpretations.

In our notation, $\sigma_\mathrm{d} \models^{\mathrm{D}} \phi$ (resp. $\sigma_\mathrm{c} \models^{\mathrm{C}} \phi$) means that ϕ is true in every moment of σ_d (resp. σ_c): this is called *floating* semantics, and differs from the *anchored* semantics of [16], where $\sigma \models \phi$ means that ϕ is true in the *first* moment of σ. This leads to a simpler axiomatization of $\mathrm{TL_C}$.

Validity of the formula ϕ in $\mathrm{TL_D}$ (resp. $\mathrm{TL_C}$) is denoted by $\models^{\mathrm{D}} \phi$ (resp. $\models^{\mathrm{C}} \phi$), and the provability relation is denoted by $\vdash^{\mathrm{D}}$ (resp. $\vdash^{\mathrm{C}}$). We say that a system S satisfies a formula ϕ of $\mathrm{TL_D}$ (resp. $\mathrm{TL_C}$) if all the traces of S satisfy ϕ, and we denote this by $S \models^{\mathrm{D}} \phi$ (resp. $S \models^{\mathrm{C}} \phi$).

An Overview of Verification

The logics $\mathrm{TL_D}$ and $\mathrm{TL_C}$ have different sets of valid formulas [18]. From the point of view of system specification, $\mathrm{TL_C}$ is preferable, as the continuous semantics corresponds more closely to the physical behavior of the system. The following is an example of a formula expressing a $\mathrm{TL_C}$-valid property of time that is intuitively obvious, yet is not valid in $\mathrm{TL_D}$ — in fact, its negation is valid in $\mathrm{TL_D}$!

Example 1. The formula

$$\phi_1 \; : \; \forall\xi\forall\zeta \left[\Diamond(T=\xi) \wedge \Diamond(T=\zeta) \rightarrow \Diamond\left(T = \frac{\xi+\zeta}{2}\right)\right]$$

expressing the density of time is such that $\models^{\mathrm{C}} \phi_1$, $\models^{\mathrm{D}} \neg\phi_1$. □

The methods proposed in [7, 15, 17] to verify properties written in $\mathrm{TL_D}$ rely on two concepts: *verification conditions* and *verification rules*. The proof of the soundness of the verification conditions and of the verification rules makes an essential use of the discreteness of the semantics. Therefore, this approach cannot be immediately transferred to $\mathrm{TL_C}$.

In this paper, we propose a methodology for verifying system properties expressed in the continuous semantics based on two components.

The first component consists of the SFV-theorem, relating the discrete and continuous semantics. The SFV-theorem states that if a temporal formula ϕ has the property of *syntactic finite variability*, or SFV, then $S \models^{\mathrm{D}} \phi$ implies $S \models^{\mathrm{C}} \phi$. This will allow us to use existing verification rules of $\mathrm{TL_D}$, and then transfer their conclusions to $\mathrm{TL_C}$.

The second component is a deductive system for $\mathrm{TL_C}$ that includes axioms about time. This deductive system can be used to construct a proof of the system specification from the time axioms and from the conclusions of the verification rules. The reasoning in $\mathrm{TL_C}$ can often be kept to a minimum, if desired.

Related work. A related approach to proving $S \models^{\mathrm{C}} \phi$ has been proposed in [6] for similar semantics and logics. It suggests *rephrasing* the property ϕ into a form ϕ' which is suited for the discrete semantics. If the rephrasing is perfect, then $S \models^{\mathrm{D}} \phi' \leftrightarrow S \models^{\mathrm{C}} \phi$; otherwise, it is sometimes possible to find a stronger property ϕ' such that $S \models^{\mathrm{D}} \phi' \rightarrow S \models^{\mathrm{C}} \phi$. In [6] it is explained how to rephrase some formulas, and how to approximate others with stronger conditions. However, no methodology is given for verifying properties that have no useful rephrasing or strengthening, and the case of hybrid systems is not considered.

4 Relationship Between the Semantics and SFV-Theorem

We say that a trace is a *refinement* of another if it has been obtained by sampling the state of the system more frequently in time [12, 13, 2].

Definition 4 (partitioning function). A *partitioning function* μ is a sequence $\mu_0, \mu_1, \mu_2, \ldots$ of adjacent and disjoint intervals of $\mathbb{N}$ covering $\mathbb{N}$. Formally, $\bigcup_{n\in\mathbb{N}} \mu_n = \mathbb{N}$, and $\max \mu_n = \min \mu_{n+1} - 1$ for all $n \in \mathbb{N}$. □

Definition 5 (refinement). A discrete trace σ'_{d}: $\langle s'_0, t'_0\rangle, \langle s'_1, t'_1\rangle, \langle s'_2, t'_2\rangle, \ldots$ is a *refinement* of σ_{d}: $\langle s_0, t_0\rangle, \langle s_1, t_1\rangle, \langle s_2, t_2\rangle, \ldots$ by the partitioning function μ, written $\sigma'_{\mathrm{d}} \succeq^{\mu} \sigma_{\mathrm{d}}$, if $t'_{\min \mu_n} = t_n$ and $s'_j = s_n$ for all $n \in \mathbb{N}$ and $j \in \mu_n$.

A continuous trace σ'_c: $\langle r'_0, I'_0\rangle$, $\langle r'_1, I'_1\rangle$, $\langle r'_2, I'_2\rangle$, ... is a *refinement* of σ_c: $\langle r_0, I_0\rangle$, $\langle r_1, I_1\rangle$, $\langle r_2, I_2\rangle$, ... by the partitioning function μ, written $\sigma'_c \succeq^\mu \sigma_c$, if $I_n = \bigcup_{j\in\mu_n} I'_j$ and $r'_j = r_n$ for all $n \in \mathbb{N}$ and $j \in \mu_n$.

We write $\sigma \succeq \sigma'$ if there is some μ such that $\sigma \succeq^\mu \sigma'$. □

If two traces have a common refinement, they are said to be *sample equivalent* [12, 10]. Sample equivalent traces correspond to different samplings of the same physical behavior of the system.

Definition 6 (sample equivalence). Two discrete (resp. continuous) traces σ, σ' are *sample equivalent*, written $\sigma \approx \sigma'$, if there is a discrete (resp. continuous) trace σ'' such that $\sigma'' \succeq \sigma$, $\sigma'' \succeq \sigma'$. □

Theorem 7. *If $\sigma_d \approx \sigma'_d$, then $S \rhd \sigma_d$ iff $S \rhd \sigma'_d$. If $\sigma_c \approx \sigma'_c$, then $S \rhd \sigma_c$ iff $S \rhd \sigma'_c$.*

We say that a temporal logic is *sample invariant* if $\sigma \approx \sigma'$ implies $\mathcal{I}, \sigma \models \phi \leftrightarrow \mathcal{I}, \sigma' \models \phi$. This means that the logic does not distinguish between sample equivalent traces, which is a desirable property. While it is known that $\mathrm{TL_D}$ is not sample invariant, it can be proved that $\mathrm{TL_C}$ is sample invariant.

Theorem 8 (sample invariance of $\mathrm{TL_C}$). *If $\sigma'_c \succeq^\mu \sigma_c$ and $j \in \mu_i$, then*

$$\mathcal{I}, \sigma'_c \overset{\mathrm{c}}{\models}_{(j,t)} \phi \leftrightarrow \mathcal{I}, \sigma_c \overset{\mathrm{c}}{\models}_{(i,t)} \phi \ .$$

for all $t \in I_i \cap I'_j$. If $\sigma'_c \approx \sigma_c$, then $\sigma'_c \models \phi \leftrightarrow \sigma_c \models \phi$.

Next, we define a translation from continuous to discrete traces that preserves admission and refinement.

Definition 9 ($\Omega : \sigma_c \mapsto \sigma_d$). The translation function Ω associates with σ_c: $\langle r_0, I_0\rangle$, $\langle r_1, I_1\rangle$, $\langle r_2, I_2\rangle$, ... the discrete trace σ_d: $\langle s_0, t_0\rangle$, $\langle s_1, t_1\rangle$, $\langle s_2, t_2\rangle$, ... such that, for all $n \in \mathbb{N}$, $s_{2n} = s_{2n+1} = r_n$, and:

1. if I_n is closed, then $t_{2n} = \inf I_n$, $t_{2n+1} = \sup I_n$;
2. if I_n is left open, then $t_{2n} = t_{2n+1} = \sup I_n$;
3. if I_n is right open, then $t_{2n} = t_{2n+1} = \inf I_n$;
4. if I_n is open, then $t_{2n} = t_{2n+1} = (\inf I_n + \sup I_n)/2$. □

Lemma 10. *If $S \rhd \sigma_c$, then $S \rhd \Omega(\sigma_c)$. If $S \rhd \sigma_c$ and $\sigma'_c \succeq \sigma_c$, then $S \rhd \Omega(\sigma'_c)$ and $\Omega(\sigma'_c) \succeq \Omega(\sigma_c)$.*

We now define finite variability, the main property that will be used in establishing the connection between the discrete and the continuous semantics.

Definition 11 (finite variability). A formula ϕ has the property of *finite variability* (FV) if for every σ_c and every $\mathcal{I}$ there exists a $\sigma'_c \succeq \sigma_c$ such that

$$\mathcal{I}, \sigma'_c \overset{\mathrm{c}}{\models}_{(i,t)} \psi \leftrightarrow \mathcal{I}, \sigma'_c \overset{\mathrm{c}}{\models}_{(i,t')} \psi$$

for all subformulas ψ of ϕ, and for all pairs (i,t), (i,t') of moments belonging to the same interval I'_i of σ'_c, for all $i \geq 0$. The trace σ'_c is called a ground trace for ϕ, σ_c, $\mathcal{I}$. □

Example 2. The formula $\phi_2 : T < 4 \rightarrow \Diamond(\cos 1/(T-4) > 0)$ is not FV, as it is not possible to subdivide $\mathbb{R}^+$ into a finite number of intervals in which the subformula $\cos(1/(T-4)) > 0$ has constant value.

Also the formula ϕ_1 of Example 1 is not FV. In fact, for each value of ξ and ζ it is possible to find a σ_c' such that the subformulas $\xi = T$, $\zeta = T$ and $T = (\xi + \zeta)/2$ have constant value in the intervals. However, it is not possible to find a σ_c' that has this property for all possible values of ξ and ζ. □

The interest of finite variability lies in the following theorems, that relate the satisfaction relation of the two semantics. Note that the converse of Theorem 13 is not valid.

Theorem 12. *If σ_c' is a ground trace for ϕ, σ_c, $\mathcal{I}$, then for all $n \in \mathbb{N}$ and for all moments (n,t) of σ_c',*

$$\mathcal{I}, \Omega(\sigma_c') \overset{\text{D}}{\models}_{2n} \phi \;\leftrightarrow\; \mathcal{I}, \sigma_c' \overset{\text{C}}{\models}_{(n,t)} \phi \ .$$

Theorem 13. *If $S \overset{\text{D}}{\models} \phi$ and ϕ is FV, then $S \overset{\text{C}}{\models} \phi$. If $\overset{\text{D}}{\models} \phi$ and ϕ is FV, then $\overset{\text{C}}{\models} \phi$.*

Proof. We prove only the first statement, as the proof of the second is similar. We prove the counterpositive: assume $S \overset{\text{C}}{\not\models} \phi$. Then there are $\mathcal{I}$, σ_c and a moment (n,t) of σ_c such that $S \triangleright \sigma_c$, and $\mathcal{I}, \sigma_c \not\models_{(n,t)} \phi$. As ϕ is FV, there is a trace $\sigma_c' \succeq^{\mu} \sigma_c$ that is ground for ϕ, σ_c, $\mathcal{I}$. There is a $k \in \mu_n$ such that (k,t) is a moment of σ_c', and from Theorem 8 we have that $\mathcal{I}, \sigma_c' \not\models_{(k,t)} \phi$. As σ_c' is ground for $\mathcal{I}$, ϕ, by Theorem 12 we have $\mathcal{I}, \Omega(\sigma_c') \not\models_{2k} \phi$. Lemma 10 ensures that $S \triangleright \Omega(\sigma_c')$, and we finally get $S \overset{\text{D}}{\not\models} \phi$, which concludes the proof. □

Syntactic finite variability. According to Definition 11, to show that a formula is FV it is necessary to give an argument for the existence of a ground trace for each trace and variable interpretation. As this might be complex, we propose instead a syntactic criterion, called ***syntactic finite variability*** (SFV), sufficient to establish that a formula is FV.

To be able to define SFV, we first define *well-behaved* functions as functions that are analytical (in the calculus sense) along the real axis in some of their variables. If a function $f(z)$ is analytical at the point b, there exists a region of positive radius centered on b in which $f(z)$ can be expanded as a series $f(z) = \sum_{n=0}^{\infty} c_n z^n$, where the coefficients c_n do not depend on z. This implies that around each zero of $f(z)$ there is a region of positive radius in which $f(z)$ has no other zero. This, in turn, leads to the well-known theorem of calculus stating that a function that is analytical along the real axis has at most a finite number of zeros in any finite interval of the real axis.

Definition 14 (well-behaved function). We say that a function

$$f(z_0, \ldots, z_n, v_1, \ldots, v_k) : \mathbb{R}^{n+k+1} \mapsto \mathbb{R}$$

is *well-behaved* if, for all $0 \le i \le n$, when considered as a function of z_i only, f is analytical along the real axis. □

Definition 15 (syntactic finite variability). The set of SFV formulas is defined inductively as follows.

1. If P is a predicate symbol and $u_1, \ldots, u_n$ are terms not containing T or Γ, then $P(u_1 \ldots u_n)$ is SFV.
2. If $f(z_0, \ldots, z_n, v_1, \ldots, v_k)$ is a well-behaved function, then

$$f(T, \Gamma(\phi_1), \ldots, \Gamma(\phi_n), c_1, \ldots, c_k) = 0 \ ,$$
$$f(T, \Gamma(\phi_1), \ldots, \Gamma(\phi_n), c_1, \ldots, c_k) > 0$$

   are SFV formulas, where $c_1, \ldots, c_k$ are constants or rigid variables of the logic (thus different from T), and $\phi_1, \ldots \phi_n$ are formulas not containing T or Γ. Such SFV formulas are called *T-atoms.*
3. A formula constructed from SFV formulas using propositional connectives or temporal operators is a SFV formula.
4. If ϕ is a SFV formula, and ξ does not occur in any T-atom of ϕ, then $\forall \xi . \phi$ is a SFV formula. □

If ϕ is SFV, the requirement that $f(z_0, \ldots, z_n, v_1, \ldots, v_k)$ is well-behaved insures that all the inequalities in ϕ, and thus all subformulas of ϕ, change truth value at most finitely often within each interval of a continuous trace. We will say that a formula is SFV even if it is not in a form described by the above definition, but can be easily transformed and put in such a form. SFV is a sufficient, but not necessary criterion for FV, but it is general enough to encompass many formulas used in the specification and verification of systems.

Example 3. The formula ϕ_2 of Example 2 is not SFV, as the function $\cos(1/(x - 4))$ is not analytical at $x = 4$, a point of the real axis. The formula $T < 4 \rightarrow \Diamond[T = 4]$ is SFV. The formula ϕ_1 of Example 1 is not SFV, as it quantifies over ξ and ζ that appear in the T-atoms $T = \xi$, $T = \zeta$ and $T = (\xi + \zeta)/2$. □

Theorem 16 (SFV-theorem). *If ϕ is SFV, it is also FV. Hence, the following inference rules are sound, with the proviso that ϕ is SFV:*

$$\frac{S \overset{D}{\vdash} \phi}{S \overset{C}{\vdash} \phi} \ , \qquad \frac{\overset{D}{\vdash} \phi}{\overset{C}{\vdash} \phi} \ , \qquad \frac{S \overset{D}{\vdash}_0 \phi}{S \overset{C}{\vdash}_{(0,0)} \phi} \ , \qquad \frac{\overset{D}{\vdash}_0 \phi}{\overset{C}{\vdash}_{(0,0)} \phi} \ .$$

Using syntactic finite variability, we can also establish a connection with propositional temporal logic. Let PTL be the propositional temporal logic of discrete linear time, with temporal operators $\mathcal{U}$, $\mathcal{S}$, $\Box$, $\boxminus$, $\Diamond$, $\diamondminus$, and based on the floating semantics. This logic is the same as the one presented in [14], except for the absence of the next and previous time operators $\bigcirc$, $\ominus$.

Theorem 17 (from PTL to TL$_C$). *Let α be a formula of PTL, containing occurrences of the propositional letters $p_1, \ldots, p_n$. If $\overset{PTL}{\models} \alpha[p_1, \ldots p_n]$, then*

$\models^{C} \alpha[\phi_1, \ldots, \phi_n]$ *provided* $\phi_1, \ldots, \phi_n$ *are FV. Similarly for initial validity. Therefore, the following inference rules*

$$\frac{\vdash^{PTL} \alpha[p_1, \ldots p_n]}{\vdash^{C} \alpha[\phi_1, \ldots, \phi_n]}, \qquad \frac{\vdash_0^{PTL} \alpha[p_1, \ldots p_n]}{\vdash_{(0,0)}^{C} \alpha[\phi_1, \ldots, \phi_n]},$$

with the proviso that $\phi_1, \ldots, \phi_n$ *are SFV, are sound.*

This result is of interest, as deductive systems for PTL are well-studied [14], and efficient decision algorithms for the problem of initial validity exist [9].

5 Verification in the Continuous Semantic

We propose the following methodology for verifying properties expressed in TL_C. Suppose that the property to be verified is $S \models^{C} \phi$, or $S \models_{(0,0)}^{C} \phi$.

First, we use the verification rules for the discrete semantics, like the ones proposed in [15, 7, 17], to establish temporal properties of the system that hold in TL_D. Let $S \models_0^{D} \psi_1, \ldots, S \models_0^{D} \psi_k$ be the conclusions reached. Then, we can use a deductive system for TL_D to reason about $S \models_0^{D} \psi_1, \ldots, S \models_0^{D} \psi_k$. Although TL_D does not have a complete axiomatization [5, 1, 2], a deductive system for initial validity in TL_D is given in [16]. Results can be transferred from initial validity to global validity using the rules

$$\frac{\vdash_0^{D} \Box\phi}{\vdash^{D} \phi}, \qquad \frac{S \vdash_0^{D} \Box\phi}{S \vdash^{D} \phi}.$$

Let $S \models_0^{D} \gamma_1, \ldots, S \models_0^{D} \gamma_m$ be the conclusions reached by reasoning in TL_D.

If the formulas $\gamma_1, \ldots, \gamma_m$ are SFV, the SFV-theorem enables us to conclude $S \models_{(0,0)}^{C} \gamma_1, \ldots, S \models_{(0,0)}^{C} \gamma_m$. We can then use temporal reasoning in TL_C, if necessary, to reach the desired conclusion $S \models^{C} \phi$ or $S \models_{(0,0)}^{C} \phi$. TL_C, like TL_D, also lacks a complete axiomatization. An (incomplete) deductive system for TL_C is presented below.

Temporal reasoning in TL_C is usually limited to the use of axioms about the completeness and divergence of time. To prove $S \models^{C} \phi$, it is often possible to find a formula $\alpha_1 \wedge \ldots \wedge \alpha_n \rightarrow \phi$, where $\alpha_1, \ldots, \alpha_n$ are TL_C-valid formulas stating the progress of time. Then, if it is possible to prove $S \models^{D} \alpha_1 \wedge \ldots \wedge \alpha_n \rightarrow \phi$, the only reasoning necessary in TL_C is to use axioms about time to obtain $S \models^{C} \phi$.

5.1 Deductive System for TL_C

Propositional and first-order axioms. A list of axioms for propositional linear-time temporal logic with $\mathcal{U}$, $\mathcal{S}$ has been presented in [3]. The axiomatization given there refers to an irreflexive temporal logic, i.e. to a logic in which the future and the past do not include the present; but it can be adapted to a reflexive logic, such as TL_C. A list of the adapted propositional axioms is given in [4], along with an entirely classical list of first-order axioms.

Table 1. Time axiom schemas for TLc.

$T \geq 0$	$\vdash^{C}_{(0,0)} T = 0$	$T = \xi \rightarrow \Box(T \geq \xi)$
$\neg\phi \rightarrow \Gamma(\phi) = 0$	$0 \leq \Gamma(\phi) \leq T$	$\xi \geq T \rightarrow \Diamond(T = \xi)$
$T = \xi \wedge \boxminus\neg[\phi\, \mathcal{U}\,(T = \xi)] \rightarrow \Gamma(\phi) = 0$		
$(T = \xi + \zeta) \wedge \phi\, \mathcal{S}\, (T = \xi \wedge \Gamma(\phi) = v) \rightarrow \Gamma(\phi) = v + \zeta$		

Table 2. Inference rules. The rules denoted by (†) have the proviso that ξ must not occur free in ϕ. The rules denoted by (§) have the proviso that $\phi, \phi_1, \ldots, \phi_n$ are SFV. In all of them, if the premise(s) is (are) S-valid, the conclusion is S-valid.

$\dfrac{\vdash^{C} \phi \rightarrow \psi, \quad \vdash^{C} \phi}{\vdash^{C} \psi}$	$\dfrac{\vdash^{C}_{(0,0)} \phi \rightarrow \psi, \quad \vdash^{C}_{(0,0)} \phi}{\vdash^{C}_{(0,0)} \psi}$	$\dfrac{\vdash^{C} \phi}{\vdash^{C} \Box\phi}$	$\dfrac{\vdash^{C} \phi}{\vdash^{C} \boxminus\phi}$
$\dfrac{\vdash^{C} \phi \rightarrow \psi}{\vdash^{C} \phi \rightarrow \forall\xi\, \psi}$ (†)	$\dfrac{\vdash^{C}_{(0,0)} \phi \rightarrow \psi}{\vdash^{C}_{(0,0)} \phi \rightarrow \forall\xi\, \psi}$ (†)	$\dfrac{\vdash^{C}_{(0,0)} \Box\phi}{\vdash^{C} \Box\phi}$	$\dfrac{\vdash^{C} \phi}{\vdash^{C}_{(0,0)} \phi}$
$\dfrac{\vdash^{PTL} \alpha[p_1, \ldots p_n]}{\vdash^{C} \alpha[\phi_1, \ldots, \phi_n]}$ (§)	$\dfrac{\vdash^{PTL}_{0} \alpha[p_1, \ldots p_n]}{\vdash^{C}_{(0,0)} \alpha[\phi_1, \ldots, \phi_n]}$ (§)	$\dfrac{\vdash^{D} \phi}{\vdash^{C} \phi}$ (§)	$\dfrac{\vdash^{D}_{0} \phi}{\vdash^{C}_{(0,0)} \phi}$ (§)

Time axioms. The set of axioms listed in Table 1 is used to reason about time. As usual, we list an axiom ϕ to mean $\vdash^{C} \phi$. In the case where we claim only the initial validity of the axiom, as in the case of the second one, we write it explicitly.

Inference rules. The inference rules we propose are listed in Table 2.

5.2 An Example of Verification

Imprecise oscillator. Consider the system OSC, consisting of an oscillator whose state is represented by the variable x. The oscillator can be in any of two states, $x = 0$ and $x = 1$, and it can stay in each of them for 3 to 5 seconds before switching to the other one. The oscillator starts in the state $x = 0$. The system can be described by:

$$\Theta_f : x = 0 \qquad \mathcal{T} : \{\tau_0, \tau_1\}$$

$$\begin{array}{lll} \rho_{\tau_0} : x = 0 \wedge x' = 1 & l_{\tau_0} : 3 & u_{\tau_0} : 5 \\ \rho_{\tau_1} : x = 1 \wedge x' = 0 & l_{\tau_1} : 3 & u_{\tau_1} : 5 \,. \end{array}$$

We want to verify that OSC satisfies the following property:

> *The oscillator is in the state $x = 1$ some time between 5 and 6 seconds after it is started.*

This property can be written as

$$\text{OSC} \models^{C}_{(0,0)} \Diamond(x = 1 \land 5 < T < 6) \ . \tag{1}$$

Note that the corresponding property in TL_D, $\text{OSC} \models^{D}_{0} \Diamond(x = 1 \land 5 < T < 6)$, does not hold. To prove (1), define the abbreviations

$$\psi: \ x = 0 \land [\Gamma(x = 0) = T] \land T \leq 3 \ ,$$
$$\phi: \ T \leq 5.5 \rightarrow [x = 1 \land \Gamma(x = 1) \leq T - 3] \ .$$

The following implications hold: $\phi \land T = 5.5 \rightarrow x = 1$, $\psi \land T = 5.5 \rightarrow x = 1$. The proof of (1) proceeds as follows:

$\text{OSC} \vdash^{D}_{0} \psi \mathcal{W} \phi$	from wait-for rule for TL_D	(2)
$\text{OSC} \vdash^{D} \phi \rightarrow \Box\phi$	from invariance rule for TL_D	(3)
$\text{OSC} \vdash^{D}_{0} \psi \mathcal{W} \Box\phi$	from (3) by temporal reasoning in TL_D	(4)
$\text{OSC} \vdash^{D}_{0} \Box(T = 5.5 \rightarrow x = 1)$	from (4) by temporal reasoning in TL_D	(5)
$\text{OSC} \vdash^{C}_{(0,0)} \Box(T = 5.5 \rightarrow x = 1)$	from (5), as it is SFV	(6)
$\text{OSC} \vdash^{C}_{(0,0)} \Diamond(T = 5.5)$	from the time axioms of TL_C	(7)
$\text{OSC} \vdash^{C}_{(0,0)} \Diamond(T = 5.5 \land x = 1)$	from (6), (7) by temp. reas. in TL_C	(8)
$\text{OSC} \vdash^{C}_{(0,0)} \Diamond(x = 1 \land 5 < T < 6)$	from (8).	(9)

6 Hybrid Systems

The results obtained for real-time systems can be transferred to hybrid systems, by giving a new definition of SFV to account for the fact that the state of the system can change continuously in time.

We will model hybrid systems by *phase transition systems* derived from those of [15, 17]. A phase transition system (PTS) $S = \langle \mathcal{V}, \Sigma, \mathcal{P}, \mathcal{T}, L, U, \Theta \rangle$ consists of the following components.

1. A set $\mathcal{V}$ of state variables. $\mathcal{V}$ is partitioned into two disjoint subsets: $\mathcal{V}_d$ and $\mathcal{V}_c$. The variables in $\mathcal{V}_d$ are called *discrete* variables, they can be of any type and they can change only in an instantaneous way. The variables in $\mathcal{V}_c$ are called *continuous* variables, have the real numbers as domain, and can change both in an instantaneous and in a continuous way.

2. A set Σ of states: each state is a type consistent interpretation of the variables. Again, we write $s(x)$ to denote the interpretation of $x \in \mathcal{V}$ at state s. We write $s|_{\mathcal{V}_d}$, $s|_{\mathcal{V}_c}$ to denote the restrictions of the interpretation s to only discrete and only continuous variables, respectively.
3. A set $\Theta \subseteq \Sigma$ of initial states.
4. A set $\mathcal{P}$ of *phases.* $\mathcal{P}$ is partitioned into disjoint subsets, one for each variable in $\mathcal{V}_c$. The subset corresponding to $x \in \mathcal{V}_c$ will be denoted by $\mathcal{P}_x$.
5. A set $\mathcal{T}$ of transitions, where $\tau \subseteq \Sigma \times \Sigma$ for each $\tau \in \Sigma$. $\mathcal{T}$ is partitioned into two disjoint subsets $\mathcal{T}_i$ and $\mathcal{T}_d$. The set $\mathcal{T}_i$ is the set of *immediate* transitions, that must be executed no later than the time at which they become enabled. The set $\mathcal{T}_d$ is the set of *delayed* transitions, whose enabling does not depend on the continuous variables.
6. Two sets L, U of minimum and maximum delays for the transitions in $\mathcal{T}_d$.

Phases. For each $x \in \mathcal{V}_c$, every phase $\wp \in \mathcal{P}_x$ consists of an *enabling condition* $c_\wp \subseteq \Sigma$ and a *phase function* $f_\wp : \Sigma \mapsto \mathbb{R}$. The phase $\wp$ is used to represent a differential equation governing x: the intended meaning is that if $c_\wp$ holds, then $\dot{x} = f_\wp(s)$ whenever no transition is taken. The enabling condition $c_\wp$ can depend on the discrete variables only: formally, for all $s, s' \in \Sigma$, $s|_{\mathcal{V}_d} = s'|_{\mathcal{V}_d} \rightarrow (s \in c_\wp \leftrightarrow s' \in c_\wp)$.

We say that a phase $\wp$ is *linear* if the function $f_\wp$ is a linear function of the continuous variables. It is not required that $f_\wp$ is linear in the discrete variables.

Transitions. We define the *enabling condition* c_τ of a transition $\tau \in \mathcal{T}$ as the set of states that have a successor according to the transition, or $c_\tau = \{s \mid \exists s'[(s, s') \in \tau]\}$. Transitions must be self-disabling, that is, $(s, s') \in \tau \rightarrow s' \notin c_\tau$.

If an immediate transition becomes enabled at time t, it has to be taken or disabled by some other transition before time advances past t. There are no restrictions on the enabling conditions of immediate transitions: they can depend on both the continuous and the discrete part of the state.

Each delayed transition $\tau \in \mathcal{T}_d$ has an associated minimum delay $l_\tau \in L$ and maximum delay $u_\tau \in U$, with $0 \leq l_\tau \leq u_\tau \leq \infty$. After τ is enabled, it can wait for a time t_d: $l_\tau \leq t_d \leq u_\tau$ before being taken. The enabling condition of delayed transitions can depend only on the discrete component of the state: for all $s, s' \in \Sigma$, $\;s|_{\mathcal{V}_d} = s'|_{\mathcal{V}_d} \rightarrow (s \in c_\tau \leftrightarrow s' \in c_\tau)$.

Continuous semantics. The continuous semantics of hybrid systems is defined in terms of *hybrid traces.* They differ from the continuous traces used for real-time systems, as the value of the continuous variables can change in the intervals composing the trace.

Definition 18 (hybrid trace). A *hybrid trace* σ_h is a sequence of pairs σ_h: $\langle g_0, I_0\rangle, \langle g_1, I_1\rangle, \langle g_2, I_2\rangle, \ldots$, where I_n is an interval of $\mathbb{R}$ and $g_n : I_n \mapsto \Sigma$, for all $n \in \mathbb{N}$. The intervals can overlap at most at the endpoints, and they cover all $\mathbb{R}^+$: $\bigcup_{n \in \mathbb{N}} I_n = \mathbb{R}^+$, and $\sup I_n = \inf I_{n+1}$ for all $n \in \mathbb{N}$.

Each function g_n gives the state of the system $g_n(t) \in \Sigma$ corresponding to time $t \in I_n$. The value of variable x at time t of interval I_n is thus $g_n(t)(x)$. The discrete variables must have constant value in an interval: for all $n \in \mathbb{R}$, all $t_1, t_2 \in I_n$, and all $x \in \mathcal{V}_d$, $g_n(t_1)(x) = g_n(t_2)(x)$. Moreover, for all $n \in \mathbb{N}$, let $t_n = \sup I_n = \inf I_{n+1}$. We require that either $\lim_{t \to t_n^-} g_n(t) = \lim_{t \to t_n^+} g_{n+1}(t)$, or that I_n is right-closed and I_{n+1} is left-closed. □

Definition 19 (admission, hybrid traces). A PTS S *admits* a trace σ_h: $\langle g_0, I_0 \rangle, \langle g_1, I_1 \rangle, \langle g_2, I_2 \rangle, \ldots$, written $S \triangleright \sigma_h$, if the following conditions are satisfied:

1. All phases are respected: for each $x \in \mathcal{V}_c$ and $n \in \mathbb{N}$, if $\inf I_n < \sup I_n$, there is a $\wp \in \mathcal{P}_x$ such that, for all $t \in I_n$:
$$g_n(t) \in c_\wp \ , \qquad f_\wp(g_n(t)) = \left. \frac{d\, g_n(u)(x)}{du} \right|_{u=t} \ , \tag{10}$$
where it is assumed that for $\inf I_n < t < \sup I_n$ the derivative $dg_n(u)(x)/du$ is defined in $u = t$, and for $u = \inf I_n$, $u = \sup I_n$, the left-hand and right-hand derivatives, respectively, exist.
2. No immediate transition is skipped: for all $n \in \mathbb{N}$, $\tau \in \mathcal{T}_i$, and $\inf I_n \leq t < \sup I_n$, it is $g_n(t) \notin c_\tau$.
3. All discrete state changes are due to a transition: for all $n \in \mathbb{N}$, either $g_n(\sup I_n) = g_{n+1}(\inf I_{n+1})$ or there is $\tau \in \mathcal{T}$ such that
$$(g_n(\sup I_n), g_{n+1}(\inf I_{n+1})) \in \tau \ .$$
If such a τ is a delayed transition, we also require that it has been continuously enabled for at least l_τ.
4. Delayed transitions are never continuously enabled for a time longer than their maximum delay. □

Discrete semantics. The discrete semantics of hybrid systems is defined in terms of discrete traces, which are defined as in Definition 1 except that the last clause is replaced by $(t_n = t_{n+1}) \vee (s_n|_{\mathcal{V}_d} = s_{n+1}|_{\mathcal{V}_d})$ for all $n \in \mathbb{N}$. The definition of admission of discrete traces is given in the next section.

Temporal logic. The temporal logic corresponding to discrete semantics is denoted by $\mathrm{TL_D}$, as before. The logic corresponding to the hybrid semantics will be denoted by $\mathrm{TL_H}$, its satisfaction relation will be denoted with $\overset{\mathrm{H}}{\models}$, and its provability relation with $\overset{\mathrm{H}}{\vdash}$.

7 From Discrete to Continuous Validity

Refinement of discrete traces was defined in Definition 5. Refinement of hybrid traces is defined as follows.

Definition 20 (refinement, hybrid traces). A hybrid trace σ_h: $\langle g_0, I_0\rangle$, $\langle g_1, I_1\rangle$, $\langle g_2, I_2\rangle$, ... is a *refinement* of σ'_h: $\langle g'_0, I'_0\rangle$, $\langle g'_1, I'_1\rangle$, $\langle g'_2, I'_2\rangle$, ... by the partitioning function μ, denoted $\sigma'_h \succeq^{\mu} \sigma_h$, if $I_i = \bigcup_{j\in\mu_i} I'_j$, and for every $i, j \in \mathbb{N}$ such that $j \in \mu_i$, it is $\forall t \in I'_j\left[g_j(t) = g_i(t)\right]$. □

Sampling equivalence is then defined as before. We then define a translation function Υ from discrete traces to hybrid traces. A discrete trace does not encode all the information required to reconstruct a hybrid trace: it contains the information about the state at the beginning and at the end of each closed interval, but it does not represent the evolution of the state in the interior of the interval. Therefore, to a single discrete trace will correspond all the hybrid traces that agree with the discrete one at the endpoints of the intervals.

Definition 21 ($\Upsilon : \sigma_d \mapsto \sigma_h$). The translation function Υ associates to σ_d: $\langle s_0, t_0\rangle$, $\langle s_1, t_1\rangle$, $\langle s_2, t_2\rangle$, ... a set of hybrid traces $\Upsilon(\sigma_d)$, such that, for every σ_h: $\langle g_0, I_0\rangle$, $\langle g_1, I_1\rangle$, $\langle g_2, I_2\rangle$, ... $\in \Upsilon(\sigma_d)$, and for every $n \in \mathbb{N}$, it is $\min I_n = t_n$, $\max I_n = t_{n+1}$, $g_n(\min I_n) = s_n$, $g_n(\max I_n) = s_{n+1}$. □

It is not possible to check directly whether a discrete trace respects the phases of a system, as in (10) for hybrid traces. Therefore, the definition of admission is indirect: a PTS S admits a discrete trace σ_d if there is a hybrid trace corresponding to σ_d admitted by S. This is the implicit meaning of the definition given in [17].

Definition 22 (admission, discrete traces). A PTS S *admits* a discrete trace σ_d, written $S \triangleright \sigma_d$, if there is a $\sigma_h \in \Upsilon(\sigma_d)$ such that $S \triangleright \sigma_h$. □

In defining finite variability for hybrid systems, it is essential to define it with respect to a given PTS, so that the behavior of the continuous variables is constrained by the phases of the PTS. It is then possible to prove the corresponding of Theorem 16.

Definition 23 (hybrid finite variability). A formula ϕ is *hybrid finite variability* (HFV) with respect to a PTS S if for every σ_h admitted by S and every $\mathcal{I}$, there exists a $\sigma'_h \succeq \sigma_h$ such that: $\mathcal{I}, \sigma'_c \overset{\mathrm{C}}{\models}_{(i,t)} \psi \leftrightarrow \mathcal{I}, \sigma'_c \overset{\mathrm{C}}{\models}_{(i,t')} \psi$ for all subformulas ψ of ϕ, and for all pairs (i,t), (i,t') of moments belonging to the same interval I'_i of σ'_h, for all $i \geq 0$. □

Theorem 24. *If $S \overset{\mathrm{D}}{\models} \phi$ and ϕ is HFV with respect to S, then $S \overset{\mathrm{H}}{\models} \phi$. If $\overset{\mathrm{D}}{\models} \phi$ and ϕ is HFV with respect to S, then $S \overset{\mathrm{H}}{\models} \phi$. Similarly for initial validity.*

The property of *hybrid syntactic finite variability* (HSFV) provides a sufficient criterion for HFV, and is also defined with respect to a PTS S.

Definition 25 (simple age function). We say that an age function $\Gamma(\phi)$ is *simple* with respect to a system S if its argument ϕ contains neither T, nor Γ, nor occurrences of continuous state variables of S. □

Definition 26 (hybrid syntactic finite variability). A formula is *hybrid syntactic finite variability* (HSFV) with respect to a PTS S if the phases of S are linear, and if the formula is constructed in the following inductive way.

1. If P is a predicate symbol and $u_1, \ldots, u_n$ are terms not containing T, Γ, nor continuous state variables, then $P(u_1, \ldots, u_n)$ is HSFV.
2. If $f(z_0, \ldots, z_n, v_1, \ldots, v_k)$ is a well-behaved function, then

$$f(b_0, \ldots, b_n, c_1, \ldots, c_k) = 0 , \qquad f(b_0, \ldots, b_n, c_1, \ldots, c_k) > 0$$

   are HSFV formulas, provided $b_0, \ldots, b_k$ are either constants, rigid variables, state variables (continuous or discrete), T or simple age functions, and c_1, $\ldots$, c_k are rigid variables or constants. We call these types of HSFV formulas *HT-atoms*.
3. A formula constructed from HSFV formulas using propositional connectives or temporal operators is a HSFV formula.
4. If ϕ is a HSFV formula, and ξ does not occur in any *HT*-atom of ϕ, then $\forall\xi.\phi$ is a HSFV formula. □

The restriction requiring the linearity of the phases is important, and cannot be lifted without being substituted by some other kind of condition insuring that the solutions of the differential equations are well-behaved in the sense of Definition 14. Again, HSFV is a sufficient condition for finite variability of a formula with respect to a PTS, providing thus the connection between the discrete and continuous semantics.

Theorem 27 (from TL_D to TL_H). *If ϕ is HSFV with respect to a PTS S, then it is SFV with respect to S. Therefore, the inference rules*

$$\frac{S \overset{\text{D}}{\vdash} \phi}{S \overset{\text{H}}{\vdash} \phi}, \qquad \frac{S \overset{\text{D}}{\vdash}_0 \phi}{S \overset{\text{H}}{\vdash}_{(0,0)} \phi},$$

with the proviso that ϕ is HSFV with respect to S, are sound.

A deductive system for TL_H. Since the definition of syntactic finite variability is now relative to a PTS, we need to modify slightly the deductive system proposed for TL_C. We take the same set of axioms, and all the inference rules listed in Table 2 apart from the last four, denoted by (§). Those four are replaced by the following rules:

$$\frac{\overset{\text{PTL}}{\vdash} \alpha[p_1, \ldots p_n]}{S \overset{\text{H}}{\vdash} \alpha[\phi_1, \ldots, \phi_n]}, \qquad \frac{\overset{\text{PTL}}{\vdash}_0 \alpha[p_1, \ldots p_n]}{S \overset{\text{H}}{\vdash}_{(0,0)} \alpha[\phi_1, \ldots, \phi_n]}, \qquad \frac{S \overset{\text{D}}{\vdash} \phi}{S \overset{\text{H}}{\vdash} \phi}, \qquad \frac{S \overset{\text{D}}{\vdash}_0 \phi}{S \overset{\text{H}}{\vdash}_{(0,0)} \phi},$$

with the proviso that ϕ, ϕ_1, $\ldots$, ϕ_n are HSFV with respect to S.

Acknowledgments. We wish to thank Anuchit Anuchitanukul, Nikølaj Bjorner, Arjun Kapur, Amir Pnueli and Henny Sipma for their valuable comments and suggestions.

References

1. R. Alur, T. Feder, and T.A. Henzinger. The benefits of relaxing punctuality. In *Proc. 10th ACM Symp. Princ. of Dist. Comp.*, pages 139–152, 1991.
2. R. Alur and T.A. Henzinger. Logics and models of real time: a survey. In J.W. de Bakker, K. Huizing, W.-P. de Roever, and G. Rozenberg, editors, *Real Time: Theory in Practice*, Lecture Notes in Computer Science 600, pages 74–106. Springer-Verlag, 1992.
3. J.P. Burgess. Axioms for tense logic I. "since" and "until". *Notre Dame journal of Formal Logic*, 23(4):367–374, October 1982.
4. L. de Alfaro and Z. Manna. Continuous verification by discrete reasoning. Technical Report STAN-CS-TR-94-1524, Stanford University, September 1994.
5. J.W. Garson. Quantification in modal logic. In D. Gabbay and F. Guenthner, editors, *Handbook of Philosophical Logic*, volume 2, chapter 5. D. Reidel Publishing Company, 1984.
6. T. Henzinger, Z. Manna, and A. Pnueli. What good are digital clocks? In W. Kuich, editor, *Proc. 19th Int. Colloq. Aut. Lang. Prog.*, volume 623 of *Lect. Notes in Comp. Sci.*, pages 545–558. Springer-Verlag, 1992.
7. T. Henzinger, Z. Manna, and A. Pnueli. Temporal proof methodologies for timed transition systems. *Inf. and Comp.*, 112(2):273–337, August 1994.
8. T. Henzinger, X. Nicollin, J. Sifakis, and S. Yovine. Symbolic model checking for real-time systems. *Information and Computation*, (111):193–244, 1994.
9. Y. Kesten, Z. Manna, H. McGuire, and A. Pnueli. A decision algorithm for full propositional temporal logic. In *Computer Aided Verification, 5th International Workshop*, Lect. Notes in Comp. Sci. Springer-Verlag, 1993.
10. Y. Kesten, Z. Manna, and A. Pnueli. Temporal verification of simulation and refinement. In *Proc. of the REX Workshop "A Decade of Concurrency"*, volume 803 of *Lect. Notes in Comp. Sci.*, pages 273–346. Springer-Verlag, 1994.
11. L. Lamport. What good is temporal logic? In R.E.A. Mason, editor, *Proc. IFIP 9th World Congress*, pages 657–668. Elsevier Science Publishers (North-Holland), 1983.
12. L. Lamport. The temporal logic of actions. Technical Report 79, DEC SRC, Palo Alto, CA, December 1991.
13. L. Lamport. Verification and specification of concurrent programs. Technical report, DEC SRC, Palo Alto, CA, 1993.
14. O. Lichtenstein, A. Pnueli, and L. Zuck. The glory of the past. In *Proc. Conf. Logics of Programs*, volume 193 of *Lect. Notes in Comp. Sci.*, pages 196–218. Springer-Verlag, 1985.
15. O. Maler, Z. Manna, and A. Pnueli. From timed to hybrid systems. In J.W. de Bakker, C. Huizing, W.P. de Roever, and G. Rozenberg, editors, *Proc. of the REX Workshop "Real-Time: Theory in Practice"*, volume 600 of *Lect. Notes in Comp. Sci.*, pages 447–484. Springer-Verlag, 1992.
16. Z. Manna and A. Pnueli. *The Temporal Logic of Reactive and Concurrent Systems: Specification.* Springer-Verlag, New York, 1991.
17. Z. Manna and A. Pnueli. Models for reactivity. *Acta Informatica*, 30:609–678, 1993.
18. A. Pnueli. Development of hybrid systems. In *Proceedings of the Third International Symposium on Formal Techniques in Real-time and Fault-tolerant Systems*, volume 863 of *Lect. Notes in Comp. Sci.*, 1994. Extended abstract.

Dynamic Matrices and the Cost Analysis of Concurrent Programs*

GianLuigi Ferrari   Ugo Montanari

Dipartimento di Informatica, Università di Pisa
{giangi,ugo}@di.unipi.it

Abstract. The problem of the cost analysis of concurrent programs can be formulated and studied by dynamic methods based on matrix calculi. However, standard matrix calculi can handle only the case of programs whose dimensions are rigidly fixed. In this paper, the notion of dynamic matrix is presented. Dynamic matrices are special matrices having extensible dimensions (rows and columns) which allow matrix product to be always defined. We put forward the theory of dynamic matrices as the correct framework to study the problems of cost analysis of concurrent programs which can change dynamically their dimensions, i.e. their amount of parallelism.

1 Introduction

There have been considerable controversies in concurrency theory about which kind of criteria should be adopted to distinguish or identify concurrent programs, e.g. *linear time* vs. *branching time*, *interleaving* vs. *true concurrency*, to cite a few. Not surprisingly, several theories of equivalences have been proposed. They have been used to prove that a particular implementation is correct with respect to a given specification but it has not been easy to use them to calculate quantities which give the measure of the use of computational resources: the 'cost' of the behaviours of a system. The standard example of cost is *time* a program takes to be executed. Other examples of costs of practical interests are the cycle time of a system, the number of times certain operations are performed.

A semantical approach to cost analysis of concurrent behaviours would be of clear practical relevance as it puts on formal grounds engineering issues that would be otherwise considered as implementation details.

In this paper, we put forward a mathematical framework to express cost measures of computations of concurrent programs. The framework relies on the use of dynamic methods. To explain the basic idea of the approach we focus on timing analysis. Assume that behaviours are described by labelled partial orders[2]: nodes of the partial order represent the events, while labels are natural

* Work partially supported by ESPRIT BRA Project 6454 CONFER, and by CNR Progetto "Specifica ad alto livello e verifica formale di sistemi digitali" 94.01874.CT7.

[2] Labelled partial orders are a widely accepted truly concurrent model: the ordering relation reflects causal dependencies among events, concurrency is represented by the absence of ordering.

numbers which give the duration of the event. Suppose, further, that the events are $e_1, \ldots, e_n$. We then write $e_i <_d e_j$ to indicate that event e_i, whose label is the natural number d, strictly precedes event e_j in the partial order (meaning that the occurrence of e_j happens at least d-unit of time later than the occurrence of e_i). Our goal is to compute the time of the occurrence of events. The time of the occurrence of the event e is computed by taking all the paths in the partial order (causal chains) leading to the event e, and then by picking up the path having maximal weight, where the weight of a path is given by the the sum of the durations of the events in the path. Formally, the time of occurrence of the event e, $t(e)$, is given by the formula

$$t(e) = max\{t(e') + d \mid e' <_d e\}$$

where we assume that the time of the occurrence of the minimal events of the partial order is 0. Clearly, the problem of computing $t(e)$ is the problem of finding the path of maximal weight in a weighted acyclic directed graph. This algorithmic problem can be efficiently implemented by resorting to dynamic methods (e.g. see Chapter 16 of [3]).

Several cost measures of practical importance are profitably analyzed by employing dynamic methods. The idea is to specify the behaviour of a system by a matrix which describes the state transformations. Then, properties of the behaviours are derived by studying the properties of the associated matrix. For instance, [2] showed that the eigenvalue of the matrix gives the average time-cycle of timed systems (see also the recent work of Gunawardena [11, 12] on the timing analysis of digital circuits). An important point, however, is that these techniques can only deal with systems whose dimensions are rigidly fixed: *states are vectors of a fixed dimension.*

Let us turn our attention to process calculi. In the case of process calculi, state transformations are usually defined by structural operational semantics (SOS) following Plotkin's style [16]. Our idea is to view the SOS rules as defining certain matrices. Namely, any labelled transition $P \xrightarrow{a} P'$ is a square matrix whose dimension expresses the amount of parallelism of the transition. The main consequence of this choice is that the matrices associated to transitions have a canonical form which allows us to naturally define an operation of composition of matrices in presence of some interaction conditions (we are able to interpret synchronization as an operation on matrices).

In this setting the state transformation determined by a computation (i.e. a sequence of labelled transition), via the operation of matrix multiplication, is again a matrix. However, this does work only when the amount of parallelism of processes is fixed: it does not handle the case of processes which can fork and increase their amount of parallelism. For instance, let us consider the process $P = a.((b.nil \mid c.nil) + d.nil)$, and, for simplicity, assume that actions have unitary duration. Below, two computations of P are shown.

(i) $P \xrightarrow{a} (b.nil \mid c.nil) + d.nil \xrightarrow{b} nil \mid c.nil;$

(ii) $P \xrightarrow{a} (b.nil \mid c.nil) + d.nil \xrightarrow{d} nil.$

It is immediate to see that the transition $\xrightarrow{b}$ resolves the choice, and makes explicit the parallelism of the process $(b.nil \mid c.nil) + d.nil$. Hence, the first computation yields the sequence of matrices

$$\begin{vmatrix}1\end{vmatrix} \; \begin{vmatrix}1 & 0\\ 0 & 0\end{vmatrix}$$

where there is a mismatch between the number of columns of matrix associated to the transition $\xrightarrow{a}$, and the number of rows of the matrix associated to the transition $\xrightarrow{b}$. Instead, the second computation yields the sequence of matrices

$$|1| \; |1|.$$

To deal with processes which can change dynamically their amount of parallelism we need a more flexible operation of matrix multiplication which is defined also when there is a mismatch between columns and rows of two matrices.

This paper attempts to lay the foundations of methods for the cost analysis of concurrent programs of "dynamic" dimensions. First, we address the problem of finding out the right mathematical objects to represent states which can reconfigure dynamically their structure. Instead of considering natural numbers, we introduce the notion of *locality* as underlying mechanism to index the components of the state vector. A first contribution of this paper is the introduction of the notion of *locality types* as a suitable mechanism to specify the structure of state vectors. Intuitively, a locality type specifies the "degree" of concurrency of a program, where with degree of concurrency we mean the addresses of the components which can be considered as autonomous entities.

Next, a particular matrix calculus based on what we call dynamic matrices is presented. The innovative feature of dynamic matrices is that rows and columns are locality types. The use of structured entities as rows and columns has the main advantage that operations, called *coercions*, are naturally defined. Coercions are special dynamic matrices which coerce one locality type to another. Therefore, dynamic matrices can expand and reduce the structure of their rows and columns, thus allowing matrix product also in presence of state transformations which change the inner structure of the state.

In this paper, the basic theory of dynamic matrices is developed. It is proved that dynamic matrices of different dimensions which are obtained by coercion can be grouped into equivalence classes, and each equivalence class has a *unique* canonical representative dynamic matrix which is *minimal* w.r.t. the dimensions of the equivalent matrices.

Matrix calculi are completely characterized in terms of categories with biproducts [7], where morphisms are matrices and morphism composition is matrix product. Dynamic matrices are exactly the morphisms of a category with biproducts where the objects are locality types. We refer to the full paper for the details of the categorical characterization.

The first part of the paper is devoted to the introduction of locality types, dynamic matrices and the study of their properties. The second part of the paper demonstrates the usefulness of the approach by presenting some concrete examples of cost analysis based on dynamic matrices.

2 Preliminaries

We begin by reviewing the basic properties of semirings. Then, we discuss the special case of matrix calculi over idempotent semirings.

Let S be a set with two distinguished elements 0 and 1 ($0 \neq 1$), and two binary operations, denoted by $+$ and $\times$. The operation $+$ is called the *additive operation*, and $\times$ is called the *multiplicative operation*. The structure $(S, +, \times, 0, 1)$ is a *semiring* provided that $(S, +, 0)$ is a commutative monoid; $(S, \times, 1)$ is a monoid; the multiplicative operation distributes over the additive operation and $s \times 0 = 0 \times s = 0$, for all $s \in S$. A semiring is called *idempotent* provided that $s + s = s$. Idempotent semirings are also called *diods* (see [11] and the references therein). In this paper, we will always consider idempotent semirings.

Let $\mathbf{N}$ be the set of natural numbers, and let $\mathbf{N}_\perp$ be the natural numbers with an additional element $\perp$ adjoined. We equip $\mathbf{N}_\perp$ with operations max and $+$, where $max(\perp, n) = n = max(n, \perp)$ and $\perp + n = n + \perp = \perp$, for all natural numbers n. The structure max-plus $= (\mathbf{N}_\perp, max, +, \perp, 0)$ forms an idempotent semiring: max is the additive operation whose neutral element is the adjoined element $\perp$ (a sort of minus infinity); the sum $+$ is the multiplicative operation whose neutral element is 0. It is idempotent as $max(n, n) = n$.

There is close connection between matrices over the max-plus algebra (transformations which can be considered as linear in the max-plus algebra) and properties of graphs (as the problem of finding the path of maximal weight discussed in the Introduction). Much of the results which can be proved in the max-plus algebra can be generalized to arbitrary idempotent semirings. For instance, many problems which arise in performance modelling are solved by considering matrices over idempotent semirings [2].

3 Locality Types

When we think about a distributed system what we have in mind is a set of processing sites where processes run. The notion of *locality types* is inspired by this idea.

Definition 1 A locality ℓ is a finite word over the alphabet $\{l, r\}$. ■

A locality ℓ gives the position of a processing site within a parallel system. The locality ε (the empty word) can be interpreted as the address of the current system; instead l (r) is used to indicate that the process is located in the left (right) component of a parallel composition. For instance, the address of process P_2 in $(P_1 \mid P_2) \mid P_3$ is the locality lr. Similarly, the address of $P_1 \mid P_2$ is l. The notion of locality has been introduced in [8], and it has been often used in the literature to describe addresses of processes (e.g. see [4]).

In the following, we adopt the standard operations over languages without formally introducing them, for instance, the symbol $+$ will denote the operation of union of languages. Furthermore, localities ℓ will be coerced to denote singletons $\{\ell\}$ of localities.

Definition 2 (Locality Types)
Locality types Λ, whose typical elements are $\lambda, \lambda', \lambda_1, \lambda_2, \ldots$, are languages of localities inductively defined as follows:

$$\varepsilon \in \Lambda \qquad \{l, r\} \in \Lambda \qquad \frac{\lambda \in \Lambda \quad \ell \in \lambda}{(\lambda - \ell) + \ell\lambda' \in \Lambda}$$

■

A locality type specifies the degree of concurrency of a system by giving the addresses of the components which can be considered as autonomous entities. Locality types are pictorially represented in terms of certain binary trees where the set of paths from the root of the tree to the leaves give the set of localities of the locality type. For instance, the locality type $\{ll, lr, r\}$ is represented by the binary tree of Figure 1. Notice that the tree representing the locality type ε consists of a single node.

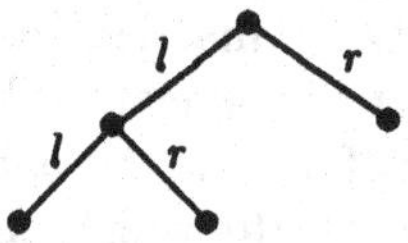

Fig. 1. Tree-like representation of Locality Types

Locality types are naturally equipped with a partial ordering relation. Let $\preceq$ be the partial ordering relation generated by the following clause.

$$\frac{\ell \in \lambda}{\lambda \preceq (\lambda - \ell) + \ell\lambda'}$$

It is immediate to see that $\varepsilon \preceq \lambda$ for any locality type λ. The partial order $\preceq$ has a natural interpretation in terms of the tree-like representation of locality types: it corresponds to the standard subtree ordering.

Given two locality types λ_1, λ_2 there always exists their least upper bound $\lambda_1 \sqcup \lambda_2$: the locality type $\lambda_1 \sqcup \lambda_2$ is built by adding the minimal details which make λ_1 and λ_2 in full agreement. If we view the tree as a term where the leaves are distinct variables, then the least upper bound is computed by the unification procedure which determines the most general unifier of terms. Similarly, there always exists also the greatest lower bound $\lambda_1 \sqcap \lambda_2$ (it is Plotkin's anti-unification [15]).

Proposition 3 *The partial order* $(\Lambda, \preceq)$ *is a lattice.* ■

The lattice of locality types is not a complete lattice as the least upper bound of subsets of Λ does not always exist. For instance, consider the chain visualized in Figure 2. Clearly, this chain has no least upper bound in Λ. However, the lattice $(\Lambda, \preceq)$ is ***down-complete*** as any non-empty subset X of Λ has greatest lower bound $\sqcap X$ in Λ.

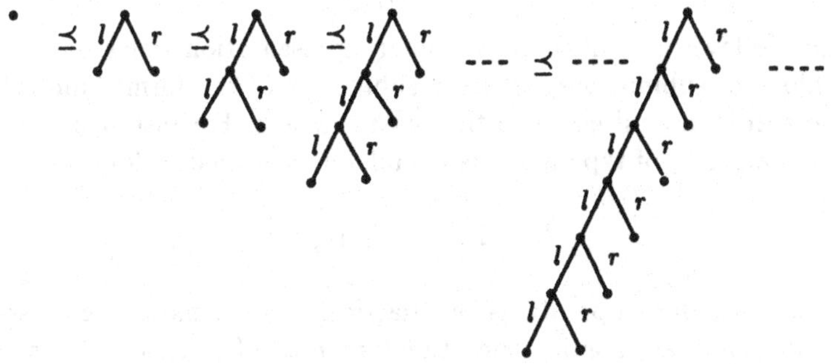

Fig. 2. A Chain of Locality Types

We now introduce an alternative view of locality types.

Definition 4 Let λ_1 and λ_2 be locality types. Then $\lambda_1 \mid \lambda_2$ is the locality type $\lambda_1 \mid \lambda_2 = l\lambda_1 + r\lambda_2$. ∎

For instance, $\varepsilon \mid \varepsilon$ is the locality type $\{l, r\}$ and $(\varepsilon \mid \varepsilon) \mid \varepsilon$ is the locality type $\{ll, lr, r\}$.

Proposition 5 *Let λ be a locality type. Then either $\lambda = \varepsilon$, or $\lambda = \lambda_1 \mid \lambda_2$, for unique locality types λ_1 and λ_2.* ∎

4 Dynamic Matrices

In this section, we study the basic properties of a matrix calculus where columns and rows of matrices are locality types.

Definition 6 Let S be an idempotent semiring, and let λ_1 and λ_2 be locality types. A dynamic matrix μ, of type $\lambda_1 \times \lambda_2$, is a function $\mu : \lambda_1 \times \lambda_2 \longrightarrow S$. ∎

A dynamic matrix μ is a rectangular array of elements of the idempotent semiring S whose entries are indexed by pairs of localities. The scalar $\mu(\ell_1, \ell_2)$ is the value of μ at row ℓ_1 and column ℓ_2. Dynamic matrices can be described by a tabular representation; for instance, the dynamic matrix μ (over the max-plus semiring) specified by

$$\mu(l,ll)=1 \quad \mu(l,lr)=2 \quad \mu(l,r)=1$$
$$\mu(r,ll)=3 \quad \mu(r,lr)=3 \quad \mu(r,r)=1$$

is represented by the following table of submatrices, where the internal subdivision of matrices (single and multiple lines) reflects the structure of locality types of the rows and columns of the matrix.

$$\begin{array}{|c|c||c|} \hline 1 & 2 & 1 \\ \hline 3 & 3 & 1 \\ \hline \end{array}$$

In what follows, we will adopt a tabular representation which partitions matrices into block of submatrices, where a submatrix of the form s indicates a matrix whose entries are all equal to the scalar value s. For instance, the *identity* dynamic matrix I_λ of type $\lambda \times \lambda$ is completely specified as follows.

$$I_\varepsilon = \begin{array}{|c|} 1 \end{array} \qquad I_{\lambda_1|\lambda_2} = \begin{array}{|c|c|} I_{\lambda_1} & 0 \\ \hline 0 & I_{\lambda_2} \end{array}$$

Many standard operations on matrices have a natural correspondence on dynamic matrices. For instance the *transpose* of $\mu : \lambda_1 \times \lambda_2$ is the dynamic matrix $(\mu)^\mathsf{T} : \lambda_2 \times \lambda_1$ defined as $(\mu)^\mathsf{T}(\ell_2, \ell_1) = \mu(\ell_1, \ell_2)$. It is immediate to see that $((\mu)^\mathsf{T})^\mathsf{T} = \mu$.

Definition 7 Let $\mu_1 : \lambda_1 \times \lambda_1'$ and $\mu_2 : \lambda_2 \times \lambda_2'$ be dynamic matrices. The product of μ_1 and μ_2 is defined provided that $\lambda_1' = \lambda_2$ and it is the dynamic matrix $\mu = \mu_1 \times \mu_2$ given by

$$\mu(\ell_1, \ell_2) = \sum_{\ell \in \lambda_2} \mu_1(\ell_1, \ell) \times \mu_2(\ell, \ell_2)$$

■

Proposition 8 *The product is associative.* ■

Proposition 9 *Let μ be a dynamic matrix of type $\lambda_1 \times \lambda_2$. Then $I_{\lambda_1} \times \mu = \mu = \mu \times I_{\lambda_2}$.* ■

After having shown that dynamic matrices form a matrix calculus it remains to be justified the reason of the adjective "dynamic". We show that dynamic matrices can be equipped with a more flexible operation of matrix product which allows matrix multiplication to be defined even if the columns of a matrix do not match the rows of the other matrix.

Definition 10 (Coercions)
Let λ_1 and λ_2 be locality types. Then the *coercion* $\gamma_{\lambda_1,\lambda_2}$ is the dynamic matrix of type $\lambda_1 \times \lambda_2$ inductively defined in Table 1. ■

Below, an example of coercion (in the max-plus semiring) is presented.

$$\gamma_{\varepsilon|\varepsilon,(\varepsilon|\varepsilon)|\varepsilon} = \begin{array}{|c|c||c|} \hline 0 & 0 & \bot \\ \hline \bot & \bot & 0 \\ \hline \end{array}$$

Coercions have nice properties as it is shown by the following propositions.

$$\gamma_{\epsilon,\lambda} = \begin{vmatrix}1\end{vmatrix}\ldots\begin{vmatrix}1\end{vmatrix} \qquad\qquad \gamma_{\lambda,\epsilon} = \begin{vmatrix}1\\ \vdots \\ 1\end{vmatrix}$$

$$\gamma_{\lambda_1|\lambda_2,\lambda_1'|\lambda_2'} = \begin{vmatrix}\gamma_{\lambda_1,\lambda_1'} & 0 \\ 0 & \gamma_{\lambda_2,\lambda_2'}\end{vmatrix}$$

Table 1. **The inductive definition of coercions**

Proposition 11 $\gamma_{\lambda,\lambda} = I_\lambda$. ■

Proposition 12 $\gamma_{\lambda_1,\lambda_2} = (\gamma_{\lambda_2,\lambda_1})^\mathsf{T}$. ■

Proposition 13 $\lambda_1 \preceq \lambda_2$ *if and only if* $\gamma_{\lambda_1,\lambda_2} \times \gamma_{\lambda_2,\lambda_1} = I_{\lambda_1}$. ■

Proposition 14 $\lambda_1 \preceq \lambda_2$ *implies* $\gamma_{\lambda_2,\lambda_1} \times \gamma_{\lambda_1,\lambda_2} \times \gamma_{\lambda_2,\lambda_1} \times \gamma_{\lambda_1,\lambda_2} = \gamma_{\lambda_2,\lambda_1} \times \gamma_{\lambda_1,\lambda_2}$. ■

Definition 15 (Generalized Product)
Let $\mu_1 : \lambda_1 \times \lambda_1'$ and $\mu_2 : \lambda_2 \times \lambda_2'$ be dynamic matrices. The *generalized product* $\mu_1 \cdot \mu_2$ is defined to be $\mu_1 \times \gamma_{\lambda_1',\lambda_2} \times \mu_2$. ■

The operation of generalized product is a total function: it is always defined. Clearly, when the columns of the first matrix coincide with the rows of the second matrix, generalized product reduces to the *standard* product. In the following examples (and in the rest of the paper) we will always consider dynamic matrices over the max-plus idempotent semiring. Calculations of generalized product between dynamic matrices are presented below.

$$\begin{vmatrix}4\end{vmatrix} \cdot \begin{vmatrix}0 & \bot \\ \bot & 0\end{vmatrix} = \begin{vmatrix}4\end{vmatrix} \times \begin{vmatrix}0 & 0\end{vmatrix} \times \begin{vmatrix}0 & \bot \\ \bot & 0\end{vmatrix} = \begin{vmatrix}4 & 4\end{vmatrix}$$

$$\begin{vmatrix}5 & 2 \\ \bot & 1\end{vmatrix} \cdot \begin{vmatrix}0 \\ \bot \\ 0\end{vmatrix} = \begin{vmatrix}5 & 2 \\ \bot & 1\end{vmatrix} \times \begin{vmatrix}0 & 0 & \bot \\ \bot & \bot & 0\end{vmatrix} \times \begin{vmatrix}0 \\ \bot \\ 0\end{vmatrix} = \begin{vmatrix}5 & 5 & 2 \\ \bot & \bot & 1\end{vmatrix} \times \begin{vmatrix}0 \\ \bot \\ 0\end{vmatrix} = \begin{vmatrix}5 & 1\end{vmatrix}$$

4.1 Principal Type and Canonical Representative

Because of coercions, dynamic matrices are always able to expand and reduce their dimensions. However, we can associate to each dynamic matrix a unique canonical dynamic matrix which has principal (minimal) dimension.

Definition 16 Let $\mu_1 : \lambda_1 \times \lambda_1'$ and $\mu_2 : \lambda_2 \times \lambda_2'$ be dynamic matrices. Define $\mu_1 \sqsubseteq \mu_2$ if and only if $\mu_1 = \gamma_{\lambda_1,\lambda_2} \times \mu_2 \times \gamma_{\lambda_2',\lambda_1'}$. ■

It is immediate to see that relation $\sqsubseteq$ is a preorder. Let $\equiv$ be the kernel of the preorder $\sqsubseteq$.

Definition 17 The *cone* $[\mu]_\equiv$ of a dynamic matrix $\mu : \lambda_1 \times \lambda_2$ is defined to be $\{\mu' \,:\, \mu \equiv \mu'\}$. ■

The cone of a dynamic matrix μ is an equivalence class of dynamic matrices: the matrices which behave almost as the dynamic matrix μ.

For instance, the dynamic matrix $\mu' : \varepsilon \mid \varepsilon \times \varepsilon \mid \varepsilon$ belongs to $[\mu]_\equiv$, where $\mu : (\varepsilon \mid \varepsilon) \mid (\varepsilon \mid \varepsilon) \times (\varepsilon \mid \varepsilon) \mid (\varepsilon \mid \varepsilon)$.

$$\mu' = \begin{array}{|c|c|}\hline 4 & 1 \\ \hline 1 & 0 \\ \hline\end{array} \qquad \mu = \begin{array}{|c|c||c|c|}\hline 4 & 4 & 1 & 1 \\ \hline 4 & 4 & 1 & 1 \\ \hline\hline 1 & 1 & 0 & 0 \\ \hline 1 & 1 & 0 & 0 \\ \hline\end{array}$$

In fact, $\mu \sqsubseteq \mu'$ as

$$\begin{array}{|c|c||c|c|}\hline 4 & 4 & 1 & 1 \\ \hline 4 & 4 & 1 & 1 \\ \hline\hline 1 & 1 & 0 & 0 \\ \hline 1 & 1 & 0 & 0 \\ \hline\end{array} = \begin{array}{|c|c|}\hline 0 & \bot \\ \hline 0 & \bot \\ \hline\hline \bot & 0 \\ \hline \bot & 0 \\ \hline\end{array} \times \begin{array}{|c|c|}\hline 4 & 1 \\ \hline 1 & 0 \\ \hline\end{array} \times \begin{array}{|c|c||c|c|}\hline 0 & 0 & \bot & \bot \\ \hline \bot & \bot & 0 & 0 \\ \hline\end{array}$$

Similarly, it can be showed that $\mu' \sqsubseteq \mu$. Notice that the dynamic matrix $\mu_1 = \mid 4 \mid$ is not equivalent to the dynamic matrix $\mu_1' : \varepsilon \mid \varepsilon \times \varepsilon \mid \varepsilon$

$$\mu_1' = \begin{array}{|c|c|}\hline 4 & 3 \\ \hline 2 & 1 \\ \hline\end{array}$$

In fact, $\mu_1 \sqsubseteq \mu_1'$ since

$$\mid 4 \mid = \begin{array}{|c|c|}\hline 0 & 0 \\ \hline\end{array} \times \begin{array}{|c|c|}\hline 4 & 3 \\ \hline 2 & 1 \\ \hline\end{array} \times \begin{array}{|c|}\hline 0 \\ \hline 0 \\ \hline\end{array}$$

while it is not true that $\mu_1' \sqsubseteq \mu_1$ (see below).

$$\begin{array}{|c|c|}\hline 4 & 3 \\ \hline 2 & 1 \\ \hline\end{array} \neq \begin{array}{|c|}\hline 0 \\ \hline 0 \\ \hline\end{array} \times \mid 4 \mid \times \begin{array}{|c|c|}\hline 0 & 0 \\ \hline\end{array} = \begin{array}{|c|c|}\hline 4 & 4 \\ \hline 4 & 4 \\ \hline\end{array}$$

Proposition 18 *Let $\mu : \lambda_1 \times \lambda_2$ be a dynamic matrix. Assume that λ_1' and λ_2' are locality types such that $\lambda_1 \preceq \lambda_1'$ and $\lambda_2 \preceq \lambda_2'$. Then, there is exactly one dynamic matrix $\mu' : \lambda_1' \times \lambda_2'$ in $[\mu]_\equiv$.* ■

Proposition 19 *Let $\mu : \lambda_1 \times \lambda_2$ be a dynamic matrix. Assume that λ_1' and λ_2' are locality types such that $\lambda_1' \preceq \lambda_1$ and $\lambda_2' \preceq \lambda_2$. Then, there is at most one dynamic matrix $\mu' : \lambda_1' \times \lambda_2'$ in $[\mu]_\equiv$.* ■

Proposition 20 *Let $\mu_1 : \lambda_1 \times \lambda_1'$ and $\mu_2 : \lambda_2 \times \lambda_2'$ be dynamic matrices in $[\mu]_\equiv$. Then, $[\mu]_\equiv$ contains also a dynamic matrix μ' of type $(\lambda_1 \sqcap \lambda_1') \times (\lambda_2 \sqcap \lambda_2')$.* ■

Proposition 21 *The cone of a dynamic matrix μ contains a dynamic matrix μ' whose type is minimal in the obvious ordering of $\Lambda \times \Lambda$.* ■

The results above show that any dynamic matrix μ can be always put into a canonical form, namely the dynamic matrix μ' having minimal type, and this canonical form is unique for each cone. Moreover, the canonical dynamic matrix is *minimal* with respect to the types of the matrices in the cone. In other words, the canonical dynamic matrix is the *optimal* representation of the set of equivalent matrices.

5 Cost Analysis of a Simple Process Calculus

In this section, we provide some examples of cost analysis of the behaviours of concurrent programs based on dynamic matrices. To this purpose, we consider a simple process calculus (a subset of Milner's CCS [13]).

Our idea is to associate a dynamic matrix to each SOS rule [16] of the operational semantics of the calculus. The mapping from labelled transitions to dynamic matrices makes use of a system of type assignment. The typing judgement $\vdash E : \lambda$ indicates that the locality type λ is assigned to agent E. Table 2 illustrates the rules of type assignment. Intuitively, the typing rules associate to each agent as many processing sites as the number of sequential components of the agent which can be considered autonomous entities (an action prefix is a fully sequential entity). As an example take the type assignment $\vdash a.E_1 \mid b.E_2 : \varepsilon \mid \varepsilon$. Notice that the locality type of a non deterministic agent is not a single processing site (e.g. $\vdash (a.E_1 \mid b.E_2) + c.E_3 : \varepsilon \mid \varepsilon$). This assumption corresponds to having a distributed mechanism to deal with non deterministic choices (see [5, 14]).

$$\vdash \mathit{nil} : \varepsilon \qquad\qquad \vdash a.E : \varepsilon$$

$$\frac{\vdash E_1 : \lambda_1, \vdash E_2 : \lambda_2}{\vdash E_1 + E_2 : \lambda_1 \sqcup \lambda_2} \qquad\qquad \frac{\vdash E_1 : \lambda_1, \vdash E_2 : \lambda_2}{\vdash E_1 \mid E_2 : \lambda_1 \mid \lambda_2}$$

Table 2. Typing Rules

We shall now introduce other operations on dynamic matrices. Assume that $\mu_1 : \lambda_1 \times \lambda_1'$ and $\mu_2 : \lambda_2 \times \lambda_2'$ are dynamic matrices. The *tensor product* $\mu_1 \otimes \mu_2$ is the dynamic matrix of type $(\lambda_1 \mid \lambda_2) \times (\lambda_1' \mid \lambda_2')$ defined as

$$\mu_1 \otimes \mu_2 = \begin{vmatrix} \mu_1 & 0 \\ 0 & \mu_2 \end{vmatrix}.$$

The operation of tensor product models parallel composition of behaviours. What it is missing is an operation which permits to model a parallel composition of matrices in presence of some interaction conditions.

A dynamic matrix $\mu : \lambda \times \lambda$ is called *simple* if it is obtained from an identity dynamic matrix by replacing a multiplicative unit down on the main diagonal with a scalar (different from the multiplicative unit). For instance, the dynamic matrices below are simple.

$$\begin{array}{|c|c||c|} 0 & \bot & \bot \\ \bot & 5 & \bot \\ \hline \bot & \bot & 0 \end{array} \qquad \begin{array}{|c|c|} 0 & \bot \\ \bot & 2 \end{array}$$

A simple dynamic matrix can be written as μ_{ℓ}^{s}. This indicates that the value of the entry on the main diagonal at row ℓ and column ℓ is the scalar s. Intuitively, a simple matrix represents an asynchronous move of the component whose address is given by the locality ℓ: all the other components stay idle.

Let $\mu_1 = \mu_{\ell_1}^{s_1}$ and $\mu_2 = \mu_{\ell_2}^{s_2}$ be simple dynamic matrices of type, respectively, $\lambda_1 \times \lambda_1$, $\lambda_2 \times \lambda_2$. The interaction of μ_1 and μ_2 is the dynamic matrix $\mu_1 \parallel \mu_2$ of type $\lambda_1 \mid \lambda_2 \times \lambda_1 \mid \lambda_2$ defined as

$$\mu_1 \parallel \mu_2 = \begin{array}{|c|c|} \mu_1 & \mu_1' \\ \hline \mu_2' & \mu_2 \end{array}$$

where $\mu_1' : \lambda_1 \times \lambda_2$, $\mu_2' : \lambda_2 \times \lambda_1$ are the dynamic matrices defined below.

$$\mu_1'(\ell, \ell') = \begin{cases} s_1 & \text{if } \ell = \ell_1, \ell' = \ell_2 \\ 0 & \text{otherwise} \end{cases} \qquad \mu_2'(\ell, \ell') = \begin{cases} s_2 & \text{if } \ell = \ell_2, \ell' = \ell_1 \\ 0 & \text{otherwise} \end{cases}$$

An example of interaction (in the max-plus semiring) is reported below.

$$\begin{array}{|c|c||c|} 0 & \bot & \bot \\ \bot & 5 & \bot \\ \hline \bot & \bot & 0 \end{array} \parallel \begin{array}{|c|c|} 0 & \bot \\ \bot & 2 \end{array} = \begin{array}{|c|c||c|||c|c|} 0 & \bot & \bot & \bot & \bot \\ \bot & 5 & \bot & \bot & 5 \\ \hline \bot & \bot & 0 & \bot & \bot \\ \hline\hline \bot & \bot & \bot & 0 & \bot \\ \bot & 2 & \bot & \bot & 2 \end{array}$$

Now, multiply the row vector

$$\begin{array}{|c|c||c|||c|c|} n_1 & 1 & n_3 & n_4 & 20 \end{array}$$

with the matrix obtained above. The result of this multiplication is the row vector

$$\begin{array}{|c|c||c|||c|c|} n_1 & max(6,22) & n_3 & n_4 & max(6,22) \end{array} = \begin{array}{|c|c||c|||c|c|} n_1 & 22 & n_3 & n_4 & 22 \end{array}$$

Because of the synchronization the second (faster) process located at llr must wait for the slower process (the fifth process) located at rr. This corresponds to the busy-waiting synchronization mechanism of [10]. Of course, interaction policies other than binary synchronization would require a more refined treatment. This is left to reader intuition.

Table 3 presents the mapping which associates a dynamic matrix to each (proof of a) labelled transition. We use $[\![E_1 \xrightarrow{\alpha} E_2]\!]$ for denoting the dynamic matrix associated to the labelled transition $E_1 \xrightarrow{\alpha} E_2$, where α is either an action a or the distinguished action τ. The mapping makes use of an auxiliary function $[\![a]\!]$ which yields the dynamic matrix of type $\varepsilon \times \varepsilon$ whose unique entry is the scalar value of the idempotent semiring S associated with the action a.

$$[\![a.E \xrightarrow{a} E]\!] = [\![a]\!]$$

$$\frac{[\![E_1 \xrightarrow{\alpha} E_1']\!] = \mu_1}{[\![E_1 + E \xrightarrow{\alpha} E_1']\!] = \mu_1} \qquad \frac{[\![E_1 \xrightarrow{\alpha} E_1']\!] = \mu_1}{[\![E + E_1 \xrightarrow{\alpha} E_1']\!] = \mu_1}$$

$$\frac{[\![E_1 \xrightarrow{\alpha} E_1']\!] = \mu_1, \vdash E : \lambda}{[\![E_1 \mid E \xrightarrow{\alpha} E_1' \mid E]\!] = \mu_1 \otimes I_\lambda} \quad \frac{[\![E_1 \xrightarrow{\alpha} E_1']\!] = \mu_1, \vdash E : \lambda}{[\![E \mid E_1 \xrightarrow{\alpha} E \mid E_1']\!] = I_\lambda \otimes \mu_1}$$

$$\frac{[\![E_1 \xrightarrow{a} E_1']\!] = \mu_1, [\![E_2 \xrightarrow{\bar{a}} E_2']\!] = \mu_2}{[\![E_1 \mid E_2 \xrightarrow{\tau} E_1' \mid E_2']\!] = \mu_1 \parallel \mu_2}$$

Table 3. From SOS Inference Rules to Dynamic Matrices

It is not too difficult to see that $[\![E \xrightarrow{\alpha} E']\!]$ for $\alpha \neq \tau$ is a simple dynamic matrix. This ensures that the synchronization rule is well defined. Notice that $[\![a.E \xrightarrow{a} E]\!]$ is a matrix of type $\varepsilon \times \varepsilon$ meaning that the state transformation of action prefixing does not contain any parallelism.

In the following examples, for simplicity, we will use the shorthand $a(s)$ to indicate that s is the scalar value of the underlying semiring associated with the action a. Moreover, we will feel free to omit writing the *nil* process.

To perform cost analysis we will study computations of agents. As a first example we will consider the case of timing analysis (the scalar values associated to actions give the *duration* of the action [1]). In the timing analysis we are interested in estimating the speed of computations: states are vectors whose entries indicate the delays caused by the computation on the components of the state. For instance, consider the following computation:

$$\begin{aligned} a(2).(b(1) \mid c(3)) \mid (\bar{b}(1) + \bar{c}(2).nil) &\xrightarrow{a} (b(1) \mid c(3)) \mid (\bar{b}(1) + \bar{c}(2)) \\ &\xrightarrow{\tau} (b(1) \mid nil) \mid nil \\ &\xrightarrow{b} nil \mid nil \mid nil \end{aligned}$$

which yields the matricial expression:

$$\mu = \left|\begin{array}{c|c} 2 & \bot \\ \hline \bot & 0 \end{array}\right| \cdot \left|\begin{array}{c|c||c} 0 & \bot & \bot \\ \hline \bot & 3 & 3 \\ \hline \bot & 2 & 2 \end{array}\right| \cdot \left|\begin{array}{c|c||c} 1 & \bot & \bot \\ \hline \bot & 0 & \bot \\ \hline \bot & \bot & 0 \end{array}\right| = \left|\begin{array}{c|c||c} 3 & 5 & 5 \\ \hline \bot & 2 & 2 \end{array}\right|$$

Notice the use of the generalized product to handle the case of a forking agent: the number of rows of the second matrix are larger than the number of columns of the first matrix. The suitable coercion makes columns and rows in full agreement.

To get the measure of the speed of the computation it suffices to apply the state transformation given by μ to the initial state of the computation. Here, we assume that the initial state consists of the vector $|\,0\,|$. This is not a restriction since a zero vector having as many components as the localities of the initial agent of the computation would have worked as well. Formally, we have:

$$|\,0\,| \cdot \left|\begin{array}{c|c||c} 3 & 5 & 5 \\ \hline \bot & 2 & 2 \end{array}\right| = |0|0| \times \left|\begin{array}{c|c||c} 3 & 5 & 5 \\ \hline \bot & 2 & 2 \end{array}\right| = |3|5||5|$$

where the resulting vector gives the delays of the components of the system. Notice that because of the synchronization the third component (located at r) got the delay of the second slower process.

Another example of timing analysis concerns the so called "ill timed but well caused phenomenon" [1]. Consider the computations of the agent $a(1) \mid b(2).c(1)$:

$$a(1) \mid b(2).c(1) \xrightarrow{b} a(1) \mid c(1) \xrightarrow{c} a(1) \mid nil \xrightarrow{a} nil \mid nil$$
$$a(1) \mid b(2).c(1) \xrightarrow{b} a(1) \mid c(1) \xrightarrow{a} nil \mid c(1) \xrightarrow{c} nil \mid nil$$
$$a(1) \mid b(2).c(1) \xrightarrow{a} nil \mid b(2).c(1) \xrightarrow{b} nil \mid c(1) \xrightarrow{c} nil \mid nil$$

These give rise to the three timed traces below.

$$\langle b@0, c@2, a@0\rangle \qquad \langle b@0, a@0, c@2\rangle \qquad \langle a@0, b@0, c@2\rangle$$

The authors [1] argue that the first trace even if ill timed (it does not reflect the passage of time: the action a cannot be observed at time 0 after having observed the execution of action b which takes two time unit) it is a legal trace since it is well caused: there is no causal connections between the occurrence of the action c and the occurrence of the action a.

In our approach, the matrix which corresponds to the first timed trace is:

$$\left|\begin{array}{c|c} 0 & \bot \\ \hline \bot & 2 \end{array}\right| \cdot \left|\begin{array}{c|c} 0 & \bot \\ \hline \bot & 1 \end{array}\right| \cdot \left|\begin{array}{c|c} 1 & \bot \\ \hline \bot & 0 \end{array}\right| = \left|\begin{array}{c|c} 1 & \bot \\ \hline \bot & 3 \end{array}\right|$$

Applying the resulting matrix to the vector $|\,0\,|$ we obtain the vector $|\,1\,|\,3\,|$ which makes clear that the delay of the first component is exactly the delay of the action a, while the delay of the second component is given by the sum of the durations of the actions b and c. In other words, the matrix calculus reflects the causal relationships among occurrences of actions. Indeed, the other two computations will yield the same matrix: this is the well known property that concurrent transitions can be executed in any order.

This example clearly shows that the notion of cost captured by our framework considers explicitly the causal dependencies among the events of a computation: the cost of an event e in a certain computation depends only on the costs of events which have caused the occurrence of e.

The previous examples show that timing analysis of concurrent programs falls naturally inside our framework. However, other notions of cost of behaviours which could be expressed as *maximal constraints* are interesting as well. The general idea is to associate to each computation a suitable dynamic matrix which gives the measure of the costs of the steps of the computation. When this matrix is applied to the initial state of the computation we obtain a state vector whose components are numerical quantities estimating the resource consumed from the beginning of the computation by each component. We refer to the full paper for other examples of cost analysis.

6 Concluding Remarks

We presented a mathematical framework to perform cost analysis of concurrent programs which change dynamically their amount of parallelism. We developed the calculus of dynamic matrices and showed that it is useful in the cost analysis of process calculi. The results of this paper provide the basis for a series of investigations.

Relating processes with respect to their costs
There is a general agreement in the theory of process calculi that the semantical objects which really matter are equivalence classes of processes under some observational equivalence. In this respect, the mapping $[\![E_1 \xrightarrow{a} E_2]\!]$ can be also viewed as an *observation function* which associates an "observation" to each labelled transition. On this ground, bisimulation semantics can be naturally introduced. Then E_1 and E_2 are bisimilar if

$$E_1 \xrightarrow{a} E_1' \text{ implies } E_2 \xrightarrow{a} E_2' \text{ and } [\![E_1 \xrightarrow{a} E_1']\!] \equiv [\![E_2 \xrightarrow{a} E_2']\!]$$

for some E_1' bisimilar to E_2'. This bisimulation semantics is able to distinguish between concurrency and non determinism but is very concrete. For instance, consider the agents $a(1) \mid nil$ and $a(1)$. These two agents are not bisimilar since

$$[\![a(1) \mid nil \xrightarrow{a} nil \mid nil]\!] = \begin{vmatrix} 1 & \perp \\ \perp & 0 \end{vmatrix} \not\equiv [\![a(1) \xrightarrow{a} nil]\!] = \mid 1 \mid$$

More abstract equivalences can be defined by adopting the well known technique of *history preserving bisimulation* [17, 9, 6].

Coercions
Another interesting theme is the introduction of the notion of coercions inside concurrency models other than matrix calculi. Here, coercions are needed to provide a proper treatment of the operation of generalized product which turns out to be the operation of composition of morphisms in the categorical characterization of dynamic matrices. A question is, thus, whether there are natural and

interesting definitions of coercions (and correspondingly of general operations for morphism composition) in other categorical paradigms for concurrency. Currently, we are investigating the notion of coercions inside monoidal categories: the categories where Petri Nets live.

References

1. Aceto, L., Murphy, D., On the Ill-timed but Well-caused. *CONCUR'93, LNCS 715*, 97-111, 1993.
2. Baccelli, F., Cohen, G., Olsder, G., Quadrat, J-P., *Synchronization and Linearity, Wiley Series in Probability and Mathematical Statistics*, Wiley and Sons, 1992.
3. Cormen, T., Leiserson, C., Rivest, R., *Introduction to Algorithms*, *MIT-Press*, 1989.
4. Cleaveland, R., Yankelevich, An Operational Framework for Value-Passing *POPL'94*, 326-338, 1994.
5. Degano, P. De Nicola, R., Montanari, U., A Distributed Operational Semantics for CCS Based on Condition/Event Systems, *Acta Informatica, 26*, 59-91, 1988.
6. Degano, P. De Nicola, R., Montanari, U., Partial Ordering Descriptions and Observations of Nondeterministic Concurrent Processes, *LNCS 354*, 438-466, 1989.
7. Elgot, C., Matricial Theory, *Journal of Algebra 42*, 391-422, 1976.
8. Ferrari, G., Gorrieri, R., Montanari, U., An Extended Expansion Theorem, *TAPSOFT 91, LNCS 494*, 29-48, 1991.
9. van Glabbeek, R., Goltz, U., Equivalence Notions for Concurrent Systems and Refinement of Actions, *MFCS'89, LNCS 397*, 237-248, 1989.
10. Gorrieri, R., Roccetti, Towards Performance Evaluation in Process Algebras (Extended Abstract) Proc. *AMAST 93*, 1993.
11. Gunawardena, J., Min-Max Functions, To appear in *Discrete Event Dynamic Systems*, 1994.
12. Gunawardena, J., A Dynamic Approach to Timed Behaviours, *CONCUR'94, LNCS 836*, 178-193, 1994.
13. Milner, R., *Communication and Concurrency, Prentice Hall*, 1989.
14. Olderog, E-R., Operational Petri Nets Semantics for CCSP, *Advances in Petri Nets, LNCS 266*, 1987.
15. Plotkin, G. Lattice-Theoretic Properties of Subsumption. *Memo MI-R-77*, University of Edinburgh, 1970.
16. Plotkin, G. A Structured Approach to Operational Semantics, *DAIMI FN-19*, Computer Science Dept., University of Aarhus, 1981.
17. Rabinovich, A., Trakhtenbrot, B., Behaviour Structures and Nets, *Fundamenta Informaticae XI*, 357-404, 1988.

Petri Nets, Traces, and Local Model Checking*

Allan Cheng**

Computer Science Department
Cornell University
Ithaca, New York 14853, USA
e-mail:acheng@cs.cornell.edu

Abstract. It has been observed that the behavioural view of concurrent systems that all possible sequences of actions are relevant is too generous; not all sequences should be considered as likely behaviours. By taking progress fairness assumptions into account one obtains a more realistic behavioural view of the systems. In this paper we consider the problem of performing model checking relative to this behavioural view. We present a CTL-like logic which is interpreted over the model of concurrent systems labelled 1-safe nets. It turns out that Mazurkiewicz trace theory provides a useful setting in which the progress fairness assumptions can be formalized in a natural way. We provide the first, to our knowledge, set of sound and complete tableau rules for a CTL-like logic interpreted under progress fairness assumptions.
keywords: fair progress, labelled 1-safe nets, local model checking, maximal traces, partial orders, inevitability

1 Introduction

Recently attention has focused on behavioural views of concurrent systems in which concurrency or parallelism is represented explicitly [Rei85, Maz86, Win86, Sta89, WN94]. This is is done by imposing more structure on models for concurrent systems, in our case an independence relation on the transitions.

Our main objective is to explore the use of the extra structure of independence in the context of *specification logics*. This paper introduces and studies a CTL-like branching time temporal logic, P-CTL, interpreted over the reachability graph of labelled 1-safe nets.

Labelled 1-safe nets are Petri nets whose transitions are labelled by actions from a set *Act* and whose reachable markings have at most one token on any place. Labelled 1-safe nets are for example obtained by translating agents from various process algebras or constructed as the synchronization of finite automata.

* This work has been supported by The Danish Research Councils and the Danish Research Academy. Part of this work was done at **BRCIS**.

** Visiting from Aarhus, **BRICS**, Basic Research in Computer Science, Centre of the Danish National Research Foundation. e-mail:acheng@daimi.aau.dk

As an example let us consider the process agent *fix* $(X = a.X)|(\tau.b.0)$. Its transition graph is given below to the left. The initial state is i and s_1 and s_2 are the only other reachable states. The agent can also be represented by the labelled 1-safe net to the right, containing three transitions labelled a, τ, and b, respectively [MN92, WN94].

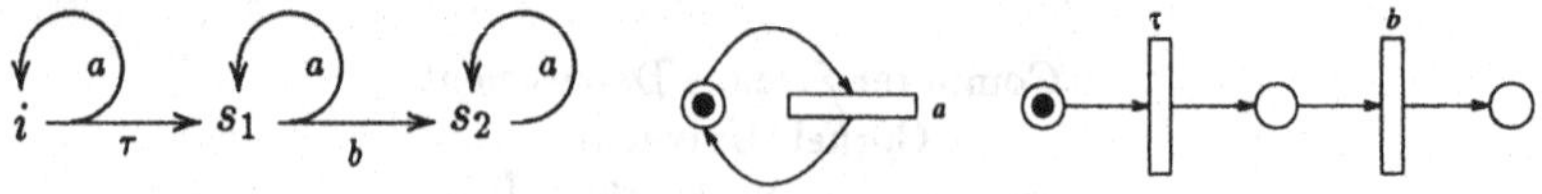

The net gives us a more concrete model of the process agent. It shows that the transition labelled a is independent of those labelled τ and b. We can therefore add *more structure* to the above transition system by providing a relation which explicitly states this *independence*. The new transition system is an example of a labelled asynchronous transition system (*l-ATS*) [Shi85, Bed88, WN94, Old91]. In general, we can obtain such a labelled asynchronous transition system as the case graph, extended with implicit information about independence, of a labelled 1-safe net. In this paper, we will concentrate on labelled 1-safe nets.

The logic P-CTL contains one important feature which is the model-theoretic incorporation of progress. What corresponds to quantified "until" path formulas in CTL is in our setting interpreted over firing sequences of labelled 1-safe nets respecting certain *progress assumptions*. This is formalized using maximal traces in the framework of Mazurkiewicz trace-theory, where we make explicit use of the notion of independence between transitions. As an example the formula Ev(<*b*> *tt*), "eventually b is enabled" (*tt* means "True"), is true of the process agent example under the assumption of progress (our interpretation), but not without (standard CTL interpretation). Our interpretation is conservative in the sense that if we interpret P-CTL over standard labelled transition systems (*lts*) we get the standard CTL interpretation. Intuitively, in process algebraic terms our notion of fair progress corresponds to a progress fair "parallel operator" (progress of independent events).

Work on expressing *fairness assumptions* can be found in for example Manna and Pnueli's book on temporal logic [MP92]. Often it involves "coding" these assumptions using linear time temporal logic formulas of the form $\phi_{fair} \Rightarrow \psi$, which require a more detailed knowledge of the particular system. When handling progress fairness we are able to avoid this obstacle and treat progress assumptions *uniformly* by using Mazurkiewicz trace-theory.

In the standard setting of CTL-like logics interpreted over *lts*, model checking has been described in [CES86] using a state based algorithm, and in [Lar88, SW89] using tableaux rules. Model checking in the framework of partial order semantics has been described in [Pen93, PP90].

In this paper we present the first, to our knowledge, set of sound and complete tableau rules in the style of [Lar88, SW89] for a CTL-like logic interpreted in the trace theoretic framework. The rules are a generalization of those in [Lar88, SW89] in the sense that if we restrict model checking to labelled 1-safe nets without independent transitions, our tableau rules work in the same way. Using the distinction between "local" and "global" model checking as advocated by

Stirling and Walker in [SW89] our method must be classified as "local" model checking. Local model checking has the advantage that it isn't necessary to have an explicit representation of all the states of the system being investigated. This is however necessary for the global model checking algorithm of [CES86]. Labelled 1-safe P/T nets are examples of models which can be "locally" model checked without necessarily generating the entire reachability graph/state space.

The rest of the paper is organized as follows. In Sect. 2 we provide the necessary definitions. In Sect. 3 we present the logic and its interpretation. Section 4 contains a motivating example followed by the tableau rules and the definition of tableaux. In Sect. 5 we state the main result, soundness and completeness of the proposed tableau rules, and state the complexity of our model checking problem. Finally Sect. 6 contains the conclusion and suggestions for future work.

2 Basic Definitions

In this section we recall some basic definitions. Furthermore we state some facts and lemmas. First we define *concurrent alphabets*, the fundamental structure in Mazurkiewicz trace theory [Maz86].

Definition 1. Concurrent alphabet and traces

- A *concurrent alphabet* (A, I) consists of a set A and a relation $I \subseteq A \times A$, the independence relation, which is symmetric and irreflexive.

  In the following assume a fixed concurrent alphabet (A, I).

- Given a set A, we define $A^\infty = A^* \cup A^\omega$, i.e. A^∞ is the set of all finite and infinite sequences of elements from A.
- Define concatenation $\circ$ of elements in A^∞ as:

$$u \circ v = \begin{cases} u & if\ |u| = \infty \\ uv & else \end{cases}$$

  For notational convenience we will write uv instead of $u \circ v$.
- Let $\leq_{pref}$ be the usual prefix ordering on sequences and $\pi_{(a,b)}$ the projection on $\{a, b\}^\infty$. We define a preorder $\preceq$ on A^∞ which demands that the relative order of arbitrary elements a and b, which are in conflict, i.e. $(a, b) \notin I$, must be the same when ignoring other elements of the sequences. Formally:

$$u \preceq v \text{ iff } (\forall (a, b) \notin I.\ \pi_{(a,b)}(u) \leq_{pref} \pi_{(a,b)}(v))$$

- Define an equivalence relation $\equiv$ on A^∞ as $u \equiv v$ if $u \preceq v$ and $v \preceq u$. Let $[u]$ denote the equivalence class containing u.
- Fact: $\equiv$ is a congruence with respect to $\circ$.
- The elements of $A^\infty/\equiv$ will be called traces.
- For $[u], [v] \in A^\infty/\equiv$ we define $[u] \preceq [v]$ if $u \preceq v$. It can be shown that this relation is a partial order. We write $[u] \prec [v]$ if and only if $u \preceq v$ and $u \not\equiv v$.

- Fact: for $u, v \in A^*$:
  - $[u] \preceq [v]$ iff $(\exists u' \in A^*.\, [uu'] = [v])$
  - $u \equiv v$ iff $u \equiv_M v$, where $\equiv_M$ is the well known equivalence used by Mazurkiewicz when defining finite traces, see [Maz86].

We have chosen to present traces using projections $\pi_{(a,b)}$. In this way finite as well as infinite traces are handled in a uniform way. Similar definitions can be found in for example [Kwi89]. We now define *labelled 1-safe nets*, the labelled version of 1-safe nets[3].

Definition 2. 1-safe nets

A *1-safe net*, or just a *net*, is a fourtuple $N = (P, T, F, M_0)$ such that

- P and T are finite nonempty disjoint sets; their elements are called *places* and *transitions*, respectively.
- $F \subseteq (P \times T) \cup (T \times P)$; F is called the *flow relation.*
- $M_0 \subseteq P$; M_0 is called the *initial marking* of N; in general, a set $M \subseteq P$ is called a *marking* or a *state* of N.

Given $a \in P \cup T$, the *preset* of a, denoted by ${}^\bullet a$, is defined as $\{a' \mid a'Fa\}$; the *postset* of a, denoted by $a^\bullet$, is defined as $\{a' \mid aFa'\}$. The union of ${}^\bullet a$ and $a^\bullet$ will be denoted ${}^\bullet a^\bullet$. We define the *independence relation* I to be the irreflexive symmetric relation over T defined by $t_1 I t_2$ *iff* ${}^\bullet t_1^\bullet \cap {}^\bullet t_2^\bullet = \emptyset$. t_1 and t_2 will be said to be independent if $t_1 I t_2$ and in conflict otherwise. For $D \subseteq T$, $t \in T$, we define $tID = DIt = \{t' \in D \mid t'It\}$.

Definition 3. Firing sequences

Let $N = (P, T, F, M_0)$ be a net.

- A transition $t \in T$ is *enabled* at a marking M of N if ${}^\bullet t \subseteq M$ and $t^\bullet \cap (M - {}^\bullet t) = \emptyset$. We denote the set of enabled transitions at a marking M by $next(M)$.
- Given a transition t, we define a relation $\xrightarrow{t}$ between markings as follows: $M \xrightarrow{t} M'$ if t is enabled at M and $M' = (M - {}^\bullet t) \cup t^\bullet$. The transition t is said to *occur* (or *fire*) at M. If $M_0 \xrightarrow{t_1} M_1 \xrightarrow{t_2} \cdots \xrightarrow{t_n} M_n$ for some markings $M_1, M_2, \ldots M_n$, then the sequence $\sigma = t_1 \ldots t_n$ is called an *occurrence sequence.* M_n is the marking *reached* by σ, and this is denoted $M_0 \xrightarrow{\sigma} M_n$. A marking M is *reachable* if it is the marking reached by some occurrence sequence. $M \not\rightarrow$ denotes that there are no enabled transitions at M, i.e. $next(M) = \emptyset$.
- Given a marking M of N, the set of reachable markings of (P, T, F, M) (i.e., the net obtained replacing the initial marking M_0 by M) is denoted by $[M\rangle$.
- A labelled 1-safe net $N = (P, T, F, M_0, l)$ is just a 1-safe net together with a map $l : T \rightarrow Act$, mapping each transition to an action in Act.

The behaviour of a net is captured by the reachability graph.

[3] An equivalent definition can be given in terms of Place/Transition nets, see [CEP93].

Definition 4. Reachability graph

- The reachability graph of a net N is the edge-labelled graph, $(V, E)_N$, whose vertices, V, are $[M_0\rangle$, the reachable markings of N; if $M \xrightarrow{t} M'$ for a reachable marking M, then there is an edge in E from M to M' labelled with t.

In the following we assume a fixed labelled 1-safe net N and consider its reachability graph $(V, E)_N$. We will use the symbols $p, q, \ldots$ to denote vertices in $(V, E)_N$ and $p \xrightarrow{t} q$ to denote that there is an edge between p and q labelled with t. Notice that (T, I) is a concurrent alphabet. If nothing else is mentioned it is implicitly assumed that (T, I) is used to generate the congruence $\equiv$.

Definition 5. Paths

- Define a *path from* $p \in V$ as a sequence, finite or infinite, of transitions $t_1, t_2, \ldots$, for which there exists states $p_1, p_2, \ldots$ such that $p \xrightarrow{t_1} p_1 \xrightarrow{t_2} p_2 \cdots$. Notice that the firing rules of the net ensure the uniqueness of the p_i's if they exist. We therefore also refer to $p \xrightarrow{t_1} p_1 \xrightarrow{t_2} p_2 \cdots$ as a path from p and use the notation $p \xrightarrow{\sigma}$ where $\sigma = t_1 t_2 \cdots$. Define $path(p) \subseteq T^\infty$ to be all paths from p. The notation $p \xrightarrow{\sigma}$ is used to indicate that $\sigma \in path(p)$.
- A path from p is *maximal* if it is either infinite or ends in a deadlocked state (or just a deadlock) p_n, i.e. a state p_n such that $p_n \not\rightarrow$.
- Due to the firing rules of the nets we have that $\equiv$ respect the path property, formally: $(\forall \sigma \in path(p).\ (\forall \sigma' \in [\sigma].\ p \xrightarrow{\sigma'}))$. Hence, $path(p)$ can be partitioned into elements of $T^\infty / \equiv$. Moreover, if σ is finite then $p \xrightarrow{\sigma} q$ implies $(\forall \sigma' \in [\sigma].\ p \xrightarrow{\sigma'} q)$.
- Given $\sigma \in path(p), |\sigma| = \infty, \sigma = t_1 t_2 \cdots$. A transition t is said to be *continuously concurrently enabled along* $p \xrightarrow{\sigma} = p \xrightarrow{t_1} p_1 \xrightarrow{t_2} p_2 \cdots$ if and only if t is enabled from a certain point and independent of the rest of the transitions taken along $p \xrightarrow{\sigma}$, formally: $(\exists n \in \mathbb{N}.\ (\forall j \geq n.\ p_j \xrightarrow{t} \ \wedge\ t I t_{j+1}))$. Notice that the irreflexivity of I implies that from a certain point t is never taken along the path $p \xrightarrow{\sigma}$. Whenever p is understood we simply say that t is continuously concurrently enabled along σ. In the process agent example from the introduction, τ is continuously concurrently enabled along $i \xrightarrow{a^\infty}$, when we use a, b, and τ to refer to the corresponding transitions.
- Define $comp(p)$ as the maximal elements with respect to $\preceq$ of $path(p)/\equiv$. For $\sigma \in [\sigma'] \in comp(p)$ we refer to $p \xrightarrow{\sigma}$ as a computation from p. In the process agent example, $\tau b a^\infty$ is a computation from i while a^∞ is not.

Lemma 6. *If t is continuously concurrently enabled along $\sigma \in path(p)$, then for any $\sigma' \in [\sigma]$ t is continuously concurrently enabled along σ', i.e. $\equiv$ respects* continuously concurrently enabled. *Hence, for* $\sigma \in path(p)$ *we say that* $t \in T$ *is* continuously concurrently enabled along $[\sigma]$ *if* t *is* continuously concurrently enabled along σ.

Lemma 7. *Given* $\sigma \in path(p), |\sigma| = \infty$. *Then there exists a* $t \in T$ *which is continuously concurrently enabled along* σ *if and only if there exists a* $\sigma' \in path(p)$ *such that* $[\sigma] \prec [\sigma']$.

Above, we have identified the maximal traces as maximal elements in a partial order. Lemma 7 explains why we concentrate on these traces. They represent executions (of a concurrent system) which are fair with respect to progress of independent processes. In [MOP89] the term "concurrency fairness" is used for such behaviours. Compared to other notions of "fairness" in the context of concurrent systems, "progress fairness" is a very weak assumption, see [MP92] for a comparison.

3 The Logic P-CTL and its Interpretation

In this section, we assume a fixed labelled 1-safe net $N = (P, T, F, M_0, l)$. Our logic has the following syntax, where $\alpha \in Act$.

$$A \ ::= \ tt \mid \neg A \mid A_1 \wedge A_2 \mid \bigcirc_\alpha A \mid A_1 \ U_\exists \ A_2 \mid A_1 \ U_\forall \ A_2$$

tt is an abbreviation for *TRUE*. In Hennessy-Milner logic [Mil89], $<a> A$ expresses the fact that one can perform an action a from a state and, in doing so, reach another state at which A holds. Similarly, the $\bigcirc_\alpha A$ expresses that a transition labelled α can be performed reaching a state where A holds. The "until" operators $U_\exists$ and $U_\forall$ are introduced as generalisations of their counterparts in [CES86], here interpreted over maximal traces, following Mazurkiewicz [Maz86].

The logic is interpreted over the reachability graph $(V, E)_N$ of N as follows, where $p \in V$, $\alpha \in Act$, and we have written $\models$ instead of $\models_N$ since N was fixed. Only the non-trivial cases are presented.

- $p \models \bigcirc_\alpha A$ iff $(\exists t \in T, q \in V.\ l(t) = \alpha \ \wedge \ p \xrightarrow{t} q \ \wedge \ q \models A)$
- $p \models A_1 \ U_\exists \ A_2$ iff $(\exists [\sigma] \in comp(p),\ p \xrightarrow{\sigma}= p_0 \xrightarrow{t_1} p_1 \xrightarrow{t_2} p_2 \cdots.$ $(\exists 0 \leq n \leq |\sigma|.\ (p_n \models A_2) \ \wedge \ (\forall 0 \leq i < n.\ p_i \models A_1)))$
- $p \models A_1 \ U_\forall \ A_2$ iff $(\forall [\sigma'] \in comp(p).\ (\forall \sigma \in [\sigma'],\ p \xrightarrow{\sigma} = p \xrightarrow{t_1} p_1 \xrightarrow{t_2} p_2 \cdots.$ $(\exists 0 \leq n \leq |\sigma|.\ (p_n \models A_2) \ \wedge \ (\forall 0 \leq i < n.\ p_i \models A_1))))$

Furthermore, we define $ff \equiv \neg tt$, $<\alpha> A \equiv \bigcirc_\alpha A$, $\mathsf{F}(A) \equiv tt\, U_\exists\, A$, $\mathsf{G}(A) \equiv \neg \mathsf{F}(\neg A)$, $\mathsf{Ev}(A) \equiv tt\, U_\forall\, A$, and $\mathsf{Al}(A) \equiv \neg \mathsf{Ev}(\neg A)$. The meaning of $\mathsf{Ev}(A)$ is that eventually/inevitably A will hold along any computation, while $\mathsf{Al}(A)$ means that along some computation A always holds. In the process agent example from the introduction we have $i \models \mathsf{Ev}(<b> tt)$.

Definition 8. Given a labelled 1-safe net $N = (P, T, F, M_0, l)$ and a formula A. The *model checking problem of N and A* is the problem of deciding whether or not $M_0 \models A$.

4 A Tableau Method for Model Checking

In this section we present a local model checker based on a tableau system for model checking formulas from our logic.

Local model checking based on tableau systems has been presented in [SW89]. As opposed to a global model checker [CES86], which checks if all states of the system satisfies a formula, a local model checker only checks if a specific state satisfies a given formula. For local model checkers based on tableau systems this is done by only visiting (other) states if the tableau rules require it. Hence, the local model checker may well be able to show that a state satisfies a formula without visiting all states of the system. For systems with a compact representation, such as 1-safe nets (where a state of the system/net is considered to be a marking), a local model checker only has to generate new parts of the reachability graph when the tableau rules require it. Since the size of the reachability graph can be exponentially bigger than the size of the net, a local model checker sometimes has an advantage over a global model checker, since it can perform model checking using less memory.

We begin by considering an example to give some intuition about the problems we are faced with when looking for a tableau system. Since our interpretation of the logical operators in P-CTL coincides with the usual interpretation when there is no concurrency in the nets, we would also like the tableau system to be a conservative extension of those presented in [Lar88, SW89]. The main difficulty is how to generalize the unfolding of formulas in P-CTL which correspond to minimal fixed-point assertions.

4.1 Unfolding Minimal Fixed-Point Assertions

Below we consider a very simple reachability graph, g_1, which is generated by the 1-safe net N_1 to the right.

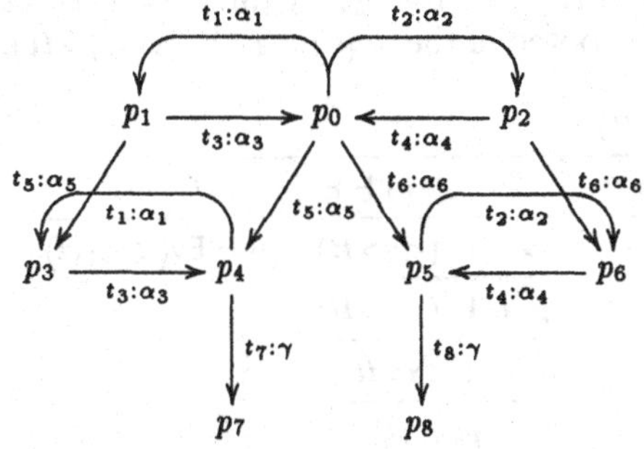

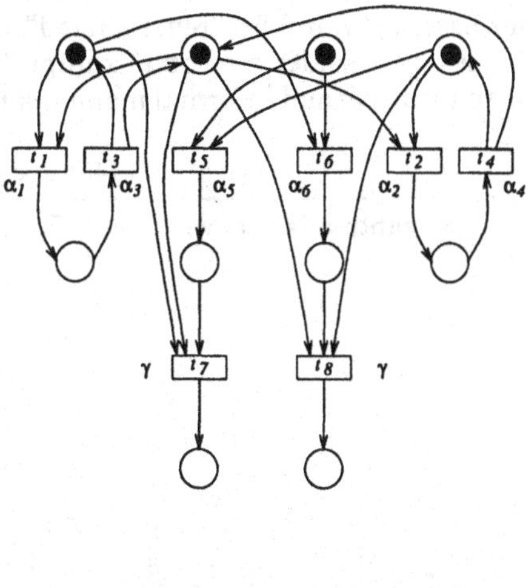

The t_i's are the transitions, the Greek letters the labels, and p_0 the initial marking. The independence relation is the smallest such containing (t_1, t_5), (t_3, t_5), (t_2, t_6), and (t_4, t_6). Now $p_0 \models_{N_1} \neg\mathsf{Ev}(<\gamma> tt)$ because $[(t_1 t_3 t_2 t_4)^\infty] \in comp(p_0)$ and no state along the computation $(t_1 t_3 t_2 t_4)^\infty$ satisfies $<\gamma> tt$. However, if we drop the transitions t_2, t_4, t_6, and t_8 and call this reduced net N_2, we do indeed have $p_0 \models_{N_2} \mathsf{Ev}(<\gamma> tt)$, since every computation from p_0 must eventually reach p_4 — t_5 cannot be continuously ignored while repeatedly firing t_1 and t_3; they are both *independent* of t_5.

Let us consider what a tableau (proof tree) for $p_0 \models_{g_2} \mathsf{Ev}(<\gamma> tt)$ might look like:

$$\frac{p_0 \vdash \mathsf{Ev}(<\gamma> tt)}{\dfrac{p_1 \vdash \mathsf{Ev}(<\gamma> tt)}{p_0 \vdash \mathsf{Ev}(<\gamma> tt) \quad \dfrac{p_3 \vdash \mathsf{Ev}(<\gamma> tt)}{\dfrac{p_4 \vdash \mathsf{Ev}(<\gamma> tt)}{\dfrac{p_4 \vdash <\gamma> tt}{p_7 \vdash tt}}}} \qquad \dfrac{p_4 \vdash \mathsf{Ev}(<\gamma> tt)}{\dfrac{p_4 \vdash <\gamma> tt}{p_7 \vdash tt}}}$$

The above tree is constructed according to some intuitive tableau rules. Although informal, the example provides the first important observation. The leftmost branch begins and ends with the sequent $p_0 \vdash \mathsf{Ev}(<\gamma> tt)$. In the μ-calculus $\mathsf{Ev}(<\gamma> tt)$ is expressed by the formula $\mu X.\ <\gamma> tt \vee ([Act]X \wedge\ <A> tt)$. Hence, based on the tableau methods from [Lar88, SW89], one might expect that the above tree should be discarded as a tableau since the unfolding of the minimal fixed-point assertion reaches itself. However, in the current framework, we interpret the logic over maximal traces, and the detected loop, $(p_0 \xrightarrow{t_1} p_1 \xrightarrow{t_3} p_0)^\infty$, is not a computation from p_0 since the transition t_5 is continuously concurrently enabled. This example suggests that we might allow the unfolding of a minimal fixed-point assertion to reach itself. The cases in which this will be allowed should include the existence of an transition that is continuously concurrently enabled along the loop represented by such a branch. Our solution to this problem is to annotate the logic used in the tableau rules. The idea is to keep track of the transitions which are continuously concurrently enabled and update this information as one unfolds the reachability graph via the tableau rules. So in our case, t_5 would be "remembered" along the $p_0 \to p_1 \to p_0$ branch.

Let us consider a second example. This time we use g_1. Again, we construct in an intuitive and informal manner a tree rooted in the sequent $p_0 \vdash \mathsf{Ev}(<\gamma> tt)$:

$$\frac{p_0 \vdash \mathsf{Ev}(<\gamma> tt)}{\text{Two subtrees as above} \qquad \dfrac{p_5 \vdash \mathsf{Ev}(<\gamma> tt)}{\dfrac{p_5 \vdash <\gamma> tt}{p_8 \vdash tt}} \qquad \dfrac{p_2 \vdash \mathsf{Ev}(<\gamma> tt)}{\dfrac{p_6 \vdash \mathsf{Ev}(<\gamma> tt)}{\dfrac{p_5 \vdash \mathsf{Ev}(<\gamma> tt)}{\dfrac{p_5 \vdash <\gamma> tt}{p_8 \vdash tt}}} \quad p_0 \vdash \mathsf{Ev}(<\gamma> tt)}}$$

Again, the interesting parts are the branches that unfold a minimal fixed-point assertion into itself. There are two such branches, the leftmost and the rightmost. However, along both of these there are transitions which are continuously concurrently enabled — t_5 for the left branch and t_6 for the right branch. So according to the previous remarks, these branches shouldn't discard the tree from being a tableau. But we do wish to discard the tree as a tableau since $p_0 \models_{g_1} \neg \mathsf{Ev}(<\gamma> tt)$. The problem is that by composing the two loops $(p_0 \xrightarrow{t_1} p_1 \xrightarrow{t_3} p_0)$ and $(p_0 \xrightarrow{t_2} p_2 \xrightarrow{t_4} p_0)$ we can obtain an infinite path $(p_0 \xrightarrow{t_1} p_1 \xrightarrow{t_3} p_0 \xrightarrow{t_2} p_2 \xrightarrow{t_4} p_0)^\infty$. Along this path there is no transition which is continuously concurrently enabled, that is, it is a computation from p_0. Moreover no state along the loop satisfies $<\gamma> tt$. This fact should discard the tree from being a tableau.

One solution to the problem of detecting such "combined" loops is to continue to unfold the minimal fixed-point assertions $p_0 \vdash \mathsf{Ev}(<\gamma> tt)$. If we unfold the fixed point assertion once more in the above example, still updating and propagating the information kept in the annotation, we will obtain a leaf with the information that we have found a looping path along which no transition is continuously concurrently enabled. This could discard the tree from being a tableau.

It turns out that the remaining problem is to find some general bound on the number of times we allow the unfolding of a minimal fixed-point assertion. In the next section we provide the necessary definitions. The bound we use is at most $|T|$, the number of transitions in the labelled 1-safe net.

4.2 Tableau Rules

In this section we consider a fixed labelled 1-safe net N and its reachability graph $(V, E)_N$.

We want to perform "local" model checking by unfolding parts of the reachability graph into a tree structure. The tableau rules are supposed to guide this unfolding by imposing constraints which restrict the size and shape of the tree structure. The main difficulty is handling the $U_\forall$ operator.

Consider a state q such that $q \not\models A_1 \; U_\forall \; A_2$. Then either there exists 1) a computation σ such that $A_1 \wedge \neg A_2$ holds at all states along $q \xrightarrow{\sigma}$

and either a deadlock is reached or a state such that $\neg A_1 \wedge \neg A_2$ holds reached; or 2) an infinite computation σ such that $A_1 \wedge \neg A_2$ holds at all states along $q \xrightarrow{\sigma}$, referred to as an invalid computation.

Since the formulas are interpreted at states and the state space is finite, case 2) reduces (simply by removing a finite number of loops from σ) to the existence of an infinite computation $\sigma_1\sigma_2$ from q where σ_1 is finite and all states along $q \xrightarrow{\sigma_1}$ occur only once along $q \xrightarrow{\sigma_1} p \xrightarrow{\sigma_2}$ while all states along $p \xrightarrow{\sigma_2}$ occur infinitely often. Also, $A_1 \wedge \neg A_2$ holds at all states, as will be the case for the following computations. Using Lemma 7 it is possible to obtain from σ_2 an infinite path σ_3 from p of the form $(\gamma_{p,p_1}\gamma_{p_1 loop}\gamma_{p_1,p} \cdots \gamma_{p,p_k}\gamma_{p_k loop}\gamma_{p_k,p})^\infty$, where all γ's are finite and made up from subsequences of σ_2 and $1 \leq k \leq |T|$. The indices are intended to illustrate the structure of the loops as follows.

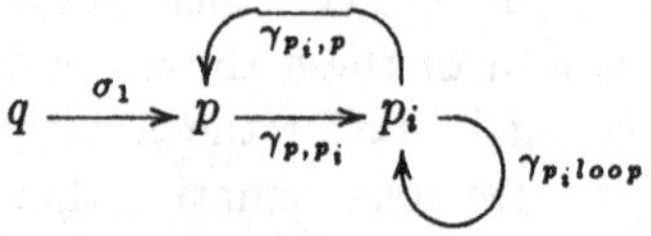

Also, since σ_2 was a computation from p, the γ's can be chosen such that for any $t \in next(p)$ one of the $\gamma_{p_i loop}$'s will contain a transition in conflict with t. Hence, σ_3 is a computation from p. We refer to the illustrated loops as *critical loops*. To conclude $\sigma_1\sigma_3$ is a computation from q along which all states satisfy $A_1 \wedge \neg A_2$. In the example from Sect. 4.1, if we chose p_0 as p, $p_0 \xrightarrow{t_1\,t_3} p_0$ and $p_0 \xrightarrow{t_2\,t_4} p_0$ would constitute critical loops. Actually, we can bound the sizes of the γ's since the state space was finite. The important observation is that together with $|T|$ we have a bound on the length and number of γ's we have to consider. This is what we will encode in the the tableau system. For that purpose we define an annotated logic.

The Annotated Logic. The syntax of the annotated logic which is used in the tableau rules differs only from the previous in that the $U_\exists$ and $U_\forall$ operators are replaced by labelled counterparts. The $U_\exists$ operator is replaced by $U_\exists^C$, where $C \subseteq V$. The intuition is that C keeps track of which states have been visited and prevents unnecessary unfolding. For the $U_\forall$ operator we use a more elaborate annotation, $U_\forall^C$, $U_\forall^{(p,n,T')}$, $U_\forall^{(p,n,T',V',\rightarrow)}$, and $U_\forall^{(p,n,T',V',\leftarrow)}$, where $p \in V$, $T' \subseteq T$, $V' \subseteq V$, and $0 \leq n \leq |T|$. V' plays a role similar to C, n bounds the number of critical loops the tableau rules allow to explore, and T' keeps track of which transitions have been concurrently enabled but ignored so far along a path. The emptyness of T' will indicate that an invalidating computation has been found.

The Tableau Rules. The tableau rules will consist of rules for sequents of the form $p \vdash B$. The rules can be read from top to bottom as: "the top sequent holds (B holds at p) if the bottom sequents and side conditions hold".

Tableau Rules

1) $$\frac{p \vdash B_1 \wedge B_2}{p \vdash B_1 \quad p \vdash B_2}$$

2) $$\frac{p \vdash \bigcirc_\alpha B}{q \vdash B}$$
- $t \in T, q \in V, p \xrightarrow{t} q$,
- $l(t) = \alpha$.

3) $$\frac{p \vdash B_1\, U_\exists^C\, B_2}{p \vdash B_2}$$
- $p \notin C$.

4) $$\frac{p \vdash B_1\, U_\exists^C\, B_2}{p \vdash B_1 \quad q \vdash B_1\, U_\exists^{C \cup \{p\}}\, B_2}$$
- $p \notin C, t \in T, q \in V, p \xrightarrow{t} q$.

5) $$\frac{p \vdash B_1\, U_\forall^C\, B_2}{p \vdash B_2}$$
- $p \notin C$.

6) $$\frac{p \vdash B_1\ U_\forall^C\ B_2}{p \vdash B_1 \quad q_1 \vdash B_1\ U_\forall^{C\cup\{p\}}\ B_2 \cdots q_m \vdash B_1\ U_\forall^{C\cup\{p\}}\ B_2}$$
- $next(p) = \{t_1, \ldots, t_m\}$, $0 < m \in I\!N, p \notin C$,
- $(\forall 1 \le i \le m.\ p \xrightarrow{t_i} q_i)$.

7) $$\frac{p \vdash B_1\ U_\forall^C\ B_2}{p \vdash B_1\ U_\forall^{(p,|next(p)|,next(p))}\ B_2}$$
- $p \in C$.

8) $$\frac{p \vdash B_1\ U_\forall^{(p,n,T')}\ B_2}{p \vdash B_1\ U_\forall^{(p,n-1,T',\emptyset,\rightarrow)}\ B_2}$$
- $0 < n \in I\!N, T' \neq \emptyset$.

9) $$\frac{q \vdash B_1\ U_\forall^{(p,n,T',V',\rightarrow)}\ B_2}{q \vdash B_1 \quad q_i \vdash B_1\ U_\forall^{(p,n,t_i!T',V'\cup\{q\},\rightarrow)}\ B_2}$$
- $q \notin V', next(q) = \{t_1, \ldots, t_m\}$, $0 < m \in I\!N$,
- $(\forall 1 \le i \le m.\ q \xrightarrow{t_i} q_i)$,

10) $$\frac{q \vdash B_1\ U_\forall^{(p,n,T',V',\rightarrow)}\ B_2}{q \vdash B_2}$$
- $q \notin V'$.

11) $$\frac{q \vdash B_1\ U_\forall^{(p,n,T',V',\rightarrow)}\ B_2}{q \vdash B_1\ U_\forall^{(p,n,T',\emptyset,\leftarrow)}\ B_2}$$
- $q \in V'$.

12) $$\frac{q \vdash B_1\ U_\forall^{(p,n,T',V',\leftarrow)}\ B_2}{q \vdash B_1 \quad q_i \vdash B_1\ U_\forall^{(p,n,t_i!T',V'\cup\{q\},\leftarrow)}\ B_2}$$
- $q \notin V', next(q) = \{t_1, \ldots, t_m\}$, $0 < m \in I\!N, q \neq p$,
- $(\forall 1 \le i \le m.\ q \xrightarrow{t_i} q_i)$.

13) $$\frac{q \vdash B_1\ U_\forall^{(p,n,T',V',\leftarrow)}\ B_2}{q \vdash B_2}$$
- $q \notin V'$.

14) $$\frac{p \vdash B_1\ U_\forall^{(p,n,T',V',\leftarrow)}\ B_2}{p \vdash B_1\ U_\forall^{(p,n,T')}\ B_2}$$

Rule 1 to 4 need no further explanation. Referring to the notation from Sect. 4.2, Rule 5 and 6 should detect σ_1, Rule 7 detect the "switch" to σ_3, Rule 8 to 10 detect $\gamma_{p,p_i}\gamma_{p_i loop}$, Rule 11 detect the "switch" to $\gamma_{p_i,p}$, and Rule 12 to 14 detect $\gamma_{p_i,p}$.

The next step is to define derivation trees which are build up according to the tableau rules.

The Derivation Trees and Tableaux. In this section we define the tableaux. This is done by first defining a larger class of trees, derivation trees, which are generated according to the tableau rules. The next step is to restrict the class of derivation trees, using the annotation of the formulas, to a subclass of derivation trees which will be defined to be the tableaux.

Derivation trees are defined inductively in the usual manner, except perhaps for negation. That is, if $\mathcal{T}_1, \ldots, \mathcal{T}_n$ are derivation trees with roots matching the sequents under the bar of a rule and the side conditions are fulfilled, then one obtains a new derivation tree by "pasting the derivation trees together" according to the rule. The root of the new derivation tree is labelled by the sequent above the bar. A tree consisting of a single node labelled with one of the following sequents is a derivation tree.

- $p \vdash tt$
- $p \vdash \neg B$
- $p \vdash B_1 \; U_\forall^{(p,n,T')} \; B_2$, where $n = 0$ or $T' = \emptyset$
- $q \vdash B_1 \; U_\forall^{(p,n,T',V',\leftarrow)} B_2$, where $q \in V'$

By applying the rules we can obtain new derivation trees, for example:

- If T_1 is a derivation tree with root $p \vdash B_1$, T_2 is a derivation tree with root $q \vdash B_1 \; U_\exists^{C \cup \{p\}} \; B_2$, where $p \notin C$, and there exists a $t \in T$ such that $p \xrightarrow{t} q$ then $\dfrac{p \vdash B_1 \; U_\exists^C \; B_2}{T_1 \quad T_2}$ is a derivation tree with root $p \vdash B_1 \; U_\exists^C \; B_2$.
- If T is a derivation tree with root $p \vdash B_2$ and $p \notin C$ then $\dfrac{p \vdash B_1 \; U_\forall^C \; B_2}{T}$ is a derivation tree with root $p \vdash B_1 \; U_\forall^C \; B_2$.

Nothing else is a derivation tree. Next, using the annotation of the formulas, we obtain two useful definitions:

- Let *Ann* be the obvious homomorphism which annotates a formula A (generated by the grammar in Sect. 3) by transforming every $U_\exists$ and $U_\forall$ into $U_\exists^\emptyset$ and $U_\forall^\emptyset$, respectively. An annotated formula B is said to be *clean* if there exists a formula A such that B equals $Ann(A)$.
- Sequents of the form $q \vdash B_1 \; U_\forall^{(q,n,\emptyset)} \; B_2$, where $n \in \mathbb{N}$ and $q \in V$, are called terminal sequents.

We continue by defining the tableaux. In this step we get rid of derivation trees as for example $p \vdash \neg tt$. A tableau is a derivation tree T with root $p \vdash Ann(A)$ such that either

- $A = tt$ or
- $A = \neg A'$ and there exists no tableau with root $p \vdash Ann(A')$ or
- A is not of the above form and
  1. every proper subtree T' of T whose root is labelled with a clean formula is itself a tableau and
  2. T has no leaves labelled with terminal sequents.

5 Soundness and Completeness

Having given the necessary definitions we are now ready to state the main result.

Theorem 9. *Given a finite labelled net N. Then, for any $p \in V$ we have:*

Soundness: *If T is a tableau with root $p \vdash Ann(A)$, then $p \models A$.*

Completeness: *If $p \models A$, then there exists a tableau with root $p \vdash Ann(A)$.*

Proof. SKETCH: The proof proceeds by structural induction, showing soundness and completeness simultaneously. The main difficulty is the $U_\forall$ operator. For the soundness, our observations from Sect. 4.2 provide the basis for a proof by contradiction. For the completeness, using the induction hypothesis one can give a direct construction of a proof tree. Intuitively, if $p \models A_1\ U_\forall\ A_2$, then a proof tree will be constructed (top-down from p) by always proving $q \vdash Ann(A_2)$ if $q \models A_2$ for any reached state q. Else, if $q \not\models A_2$, one proves $q \vdash Ann(A_1)$, starts unfolding the graph from q, and continues by trying to prove $Ann(A_1\ U_\forall\ A_2)$ at the states that are reached. This procedure can be shown to produce a proof tree for $p \vdash Ann(A_1\ U_\forall\ A_2)$. A complete proof can be found in [Che95b]. □

As an example, we show that the process agent from the introduction will eventually be able to fire a transition labelled by a b action (assume the transitions are t_1, t_2, and t_3, and are labelled a, τ, and b, respectively). By the previous theorem, to show $i \models \mathsf{Ev}(<b>tt)$ it is sufficient to construct a tableau with root $i \vdash tt\ U_\forall^{\emptyset}\ (<b>tt)$.

$$\frac{i \vdash tt\, U_\forall^{\emptyset}(<b>tt)}{i \vdash tt \qquad \dfrac{\dfrac{i \vdash tt\, U_\forall^{\{i\}}(<b>tt)}{i \vdash tt\, U_\forall^{(i,2,\{t_1,t_2\})}(<b>tt)}}{\mathcal{T}_1} \qquad \dfrac{\dfrac{s_1 \vdash tt\, U_\forall^{\{i\}}(<b>tt)}{s_1 \vdash <b>tt}}{s_2 \vdash tt}}$$

where $\mathcal{T}_1$ is

$$\frac{i \vdash tt\, U_\forall^{(i,1,\{t_1,t_2\},\emptyset,\rightarrow)}(<b>tt)}{\dfrac{\dfrac{\dfrac{i \vdash tt\, U_\forall^{(i,1,\{t_2\},\{i\},\rightarrow)}(<b>tt)}{i \vdash tt\, U_\forall^{(i,1,\{t_2\},\emptyset,\leftarrow)}(<b>tt)}}{i \vdash tt\, U_\forall^{(i,1,\{t_2\})}(<b>tt)}}{\mathcal{T}_2} \qquad \dfrac{\dfrac{s_1 \vdash tt\, U_\forall^{(i,1,\{t_1\},\{i\},\rightarrow)}(<b>tt)}{s_1 \vdash <b>tt}}{s_2 \vdash tt} \qquad i \vdash tt}$$

where $\mathcal{T}_2$ is

$$\frac{i \vdash tt\, U_\forall^{(i,0,\{t_2\},\emptyset,\rightarrow)}(<b>tt)}{i \vdash tt \qquad \dfrac{\dfrac{i \vdash tt\, U_\forall^{(i,0,\{t_2\},\{i\},\rightarrow)}(<b>tt)}{i \vdash tt\, U_\forall^{(i,0,\{t_2\},\emptyset,\leftarrow)}(<b>tt)}}{i \vdash tt\, U_\forall^{(i,0,\{t_2\})}(<b>tt)} \qquad \dfrac{\dfrac{s_1 \vdash tt\, U_\forall^{(i,0,\emptyset,\{i\},\rightarrow)}(<b>tt)}{s_1 \vdash <b>tt}}{s_2 \vdash tt}}$$

Notice that if we restrict ourselves to labelled 1-safe nets where the independence relation is empty and translate $A_1\ U_\exists\ A_2$ into $\mu X.\, A_2 \vee (A_1 \wedge <Act>X)$ and $A_1\ U_\forall\ A_2$ into $\mu X.\, A_2 \vee (A_1 \wedge <Act>\, tt \wedge [Act]X)$ (actually applying this translation recursively on the subformulas A_1 and A_2), our proof rules will work in essentially the same manner as those presented in [Lar88, SW89].

Choosing an instance of the model checking problem to be a pair (N, A) consisting of a labelled 1-safe net and a formula A and defining its size to be the sum of the size of the net and the length of the formula (see e.g. [CEP93, Che95a]), we obtain the following complexity result.

Theorem 10. *The model checking problem is PSPACE-complete.*

Proof. The hardness result follows from easy modifications of the results in [CEP93], while obtaining the PSPACE upper bound follows from a modification of the results in [Che95a] based on the observations in Sect. 4.2 (the bound on the number and length of the γ's). □

6 Conclusion and Future Work

Partial order semantics for concurrent systems have gained interest because interleaving models of concurrency have failed to provide an acceptable interpretation of what it means for events of a concurrent system to be independent. Much work has been devoted to transfer obtained results and notions from the interleaving models to the "true concurrency" models [JNW93, JM93, WN94, LPRT93]. Trying to contribute to the "transferring of results" we have provided proof rules for a CTL-like logic interpreted over maximal traces. The work which we have tried to transfer can be found in [Lar88, SW89]. Our work supports automatic verification of distributed systems whose liveness properties are only provable under the assumption of progress.

There is a trade off between the rules and the definition of tableaux. One can obtain simple rules at the cost of a complicated definition of tableaux[4]. At the cost of presenting less simple tableau rules we keep the definition of tableaux simple.

Future research might consider how to handle a more expressive logic (perhaps one containing a recursion operator) in a similar way, that is, define the interpretation of the formulas over maximal traces and proving soundness and completeness of some tableau proof rules.

Finally, the general satisfiability problem the logic is still an open problem (restricted versions of the satisfiability problem are known to be undecidable, see e.g. [Che94]). Consider the following example

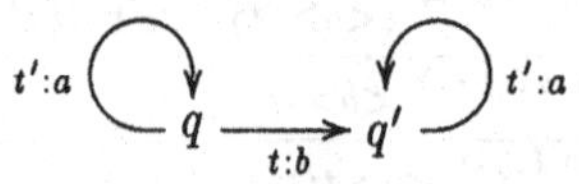

where t is independent of t'. Let p be an atomic proposition that holds at q and but not at q' (p can be simulated by having a p labelled enabled transition at all states where p should hold). Then, the formula $(\neg \mathsf{AI}(p)) \wedge (p \wedge \mathsf{Inv}(p \Rightarrow \langle a \rangle p))$ is satisfied at q. But under the usual CTL-interpretation the formula (where $\langle a \rangle$ is read as the "next" operator) is unsatisfiable. The important observation

[4] The set of simple rules we have identified requires a global side condition in the definition of tableaux.

is that our interpretation of AI does not quantify over the path $p_1 \xrightarrow{\sigma}$, where $\sigma = (t't'')^\infty$.

Acknowledgments: I thank Mogens Nielsen and Nils Klarlund for inspiring discussions and the anonymous referees for their helpful comments.

References

[Bed88] M. A. Bednarczyk. *Categories of asynchronous systems.* PhD thesis, University of Sussex, 1988. PhD in computer science, report no.1/88.

[CEP93] Allan Cheng, Javier Esparza, and Jens Palsberg. Complexity results for 1-safe nets. In *Proc. FST&TCS 13, Thirteenth Conference on the Foundations of Software Technology & Theoretical Computer Science*, pages 326–337. Springer-Verlag (*LNCS* 761), Bombay, India, December 1993. To appear in TCS, volume 148.

[CES86] Edmund M. Clarke, E. A. Emerson, and A. P. Sistla. Automatic verification of finite state concurrent system using temporal logic. *ACM Transactions on Programming Languages and Systems*, 8(2):244–263, 1986.

[Che94] Allan Cheng. Local Model Checking and Traces. Technical report, Daimi, Computer Science Department, Aarhus University, May 1994. BRICS Report Series RS-94-17.

[Che95a] Allan Cheng. Complexity Results for Model Checking. Research Series RS-95-18, BRICS, Department of Computer Science, University of Aarhus, 1995.

[Che95b] Allan Cheng. Petri Nets, Traces, and Local Model Checking. Research series, BRICS, 1995. Full version of paper to appear in proceedings of AMAST'95. In preparation.

[JM93] Lalita Jategaonkar and Albert Meyer. Deciding true concurrency equivalences on finite safe nets. In *Proc. ICALP'93*, pages 519–531, 1993.

[JNW93] André Joyal, Mogens Nielsen, and Glynn Winskel. Bisimulation and open maps. In *Proc. LICS'93, Eighth Annual Symposium on Logic in Computer Science*, pages 418–427, 1993.

[Kwi89] Marta Z. Kwiatkowska. Event fairness and non-interleaving concurrency. *Formal Aspects of Computing*, 1:213–228, 1989.

[Lar88] Kim G. Larsen. Proof systems for Hennessy-Milner logic with recursion. In *Proceedings of CAAP, Nancy France*, pages 215–230. Springer-Verlag (*LNCS* 299), March 1988.

[LPRT93] Kamal Lodaya, Rohit Parikh, R. Ramanujam, and P. S. Thiagarajan. A logical study of distributed transition systems. Technical report, School of Mathematics, SPIC Science Foundation, Madras, 1993. To appear in Information and Computation, a preliminary version appears as Report TCS–93–8.

[Maz86] Antoni Mazurkiewicz. Trace theory. In *Petri Nets: Applications and Relationships to Other Models of Concurrency*, pages 279–324. Springer-Verlag (*LNCS* 255), 1986.

[Mil89] Robin Milner. *Communication and Concurrency.* Prentice Hall International Series In Computer Science, C. A. R. Hoare series editor, 1989.

[MN92] Madhavan Mukund and Mogens Nielsen. CCS, Locations and Asynchronous Transition Systems. *Proc. Foundations of Software Technology and Theoretical Computer Science 12*, pages 328–341, 1992.

[MOP89] Antoni Mazurkiewicz, Edward Ochmański, and Wojciech Penczek. Concurrent systems and inevitability. *Theoretical Computer Science*, 64():281–304, 1989.

[MP92] Zohar Manna and Amir Pnueli. *The Temporal Logic of Reactive and Concurrent Systems*. Springer Verlag, 1992.

[Old91] Ernst R. Olderog. *Nets, Terms and Formulas*. Cambridge University Press, 1991. *Number 23 Tracts in Theoretical Computer Science*.

[Pen93] Wojciech Penzcek. Temporal logics for trace systems: On automated verification. *International Journal of Foundations of Computer Science*, 4 (1):31–67, 1993.

[PP90] Doron Peled and Amir Pnueli. Proving partial order liveness properties. In *Proc. ICALP'90*, pages 553–571. Springer-Verlag (*LNCS* 443), 1990.

[Rei85] Wolfgang Reisig. *Petri Nets - An Introduction*. EATCS Monographs in Computer Science Vol.4, 1985.

[Shi85] M. W. Shields. Concurrent machines. *Computer Journal*, 28:449–465, 1985.

[Sta89] Eugene W. Stark. Concurrent transition systems. *Theoretical Computer Science*, 64():221–269, 1989.

[SW89] Colin P. Stirling and David Walker. Local model checking in the modal mu-calculus. Technical Report ECS–LFCS–89–78, Laboratory for Foundations of Computer Science, Department of Computer Science - University of Edinburgh, May 1989.

[Win86] Glynn Winskel. Event structures. In *Petri Nets: Applications and Relationships to Other Models of Concurrency*, pages 325–390. Springer-Verlag (*LNCS* 255), 1986.

[WN94] Glynn Winskel and Mogens Nielsen. Models for concurrency. Research Series RS-94-12, BRICS, Department of Computer Science, University of Aarhus, May 1994. 144 pp. To appear as a chapter in the *Handbook of Logic and the Foundations of Computer Science*, Oxford University Press.

An Algebraic Framework for Developing and Maintaining Real-Time Systems*

Elizabeth I. Leonard[1] and Amy E. Zwarico[2]**

[1] Department of Computer Science, Johns Hopkins University
Baltimore, MD 21218, USA
leonard@cs.jhu.edu
[2] Bell South Telecommunications, S/E8H1 3535 Colonnade Parkway
Birmingham, AL 35213
amy@cs.jhu.edu, amy.zwarico@bst.bls.com

Abstract. In this paper we address the problem of safely replacing components of a real-time system, especially with faster ones. We isolate a class of real-time processes we call the *nonpre-emptive processes.* These processes can be related by their speed (relative efficiency) as well as their relative degrees of nondeterminism. A process algebra of nonpre-emptive processes, N-CCS, is presented that includes a language that expresses exactly the nonpre-emptive processes, testing preorders, and sound and complete axiomatizations of the preorders for finite N-CCS. The utility of this framework is demonstrated by an example.

1 Introduction

The development of real-time systems encompasses all of the difficulties encountered in building concurrent systems and introduces further complexity by requiring the processes to satisfy explicit timing constraints. Most of the formal techniques developed for real-time focus on ensuring that processes satisfy timing constraints. Applying formal methods to *system evolution* in real-time systems has received little attention. The importance of evolution is evident when we consider a module in a flight controller that samples the airspeed every two seconds, compares the sampled rate to the desired rate, and sends the results to a central controller which alters the plane's speed if necessary. When the system was built, the technology was such that a sampling rate of two seconds was the fastest rate possible. Two years later, the engineers wish to replace the old module with a newly developed airspeed sampler with a rate of one second.

The following two questions arise: (1) is the new, faster airspeed sampler *functionally equivalent* to the old one; and (2) will substitution of the new module for the old one result in overall system malfunction, due for example to timing errors? Provided the old and new systems are functionally equivalent, the second question can be answered affirmatively provided that no part of the system, including the original controller, depends on the slowness of that controller.

* This work was supported by AFOSR grants F49620-93-1-0169 and F49620-93-1-0616
** Currently on leave from Johns Hopkins

In this paper we develop a process algebra that allows us to answer both of these questions. It extends the class of nonpre-emptive processes reported in [CZ91]. Intuitively, a process is nonpre-emptive if its internal computation does not "eliminate" possibilities for external behavior. That is, it preserves "slow" behavior. As we prove here, nonpre-emptive processes can be replaced by faster (nonpre-emptive) processes without compromising overall system performance. Along with the general real-time model developed in [CZ91], this provides a unified framework for reasoning about both timing constraints (system correctness) and relative speeds (system evolution).

The nonpre-emptive process algebra consists of a set of operators, three testing preorders, and axiomatizations of the preorders. We use a method developed by Aceto, Bloom and Vaandrager [ABV92a, ABV92b] to develop the axiomatizations by simplifying the language to CCS with only +, action, and *nil*.

The remainder of this paper is structured as follows. We begin with a presentation of our framework for study in Section 2. This includes the simple process algebra FINTREE. Section 3 provides the motivation and a formal definition of nonpre-emptive processes. It also presents the nonpre-emptive preorders: a means of relating processes on the basis of their relative speeds as well as by their degrees of nondeterminism. These preorders were first presented in [CZ91]. Section 4 gives sound and complete axiomatizations of the nonpre-emptive preorders with respect to FINTREE. In Sections 5 and 6 we develop the nonpre-emptive process algebra N-CCS and provide a sound and complete axiomatization of the nonpre-emptive preorders with respect to N-CCS. The axiomatization is derived using the results of [ABV92a, ABV92b] and the axiomatization of Section 4. Section 7 shows a simple example of using this algebra to analyze a nonpre-emptive system. We end with discussions of future work and some related work.

2 Testing

In the introduction we described a nonpre-emptive process as a real-time process in which internal computation does not eliminate possibilities for external behavior. Before developing an algebra of such a class of processes, we present the framework for study which includes *labeled transition systems*, a general process algebra FINTREE, and timed testing.

Definition 1 *A* labeled transition system *is a quadruple* $\langle P, A, \rightarrow, p_0 \rangle$ *where*

- P *is a countable set of* states;
- A *is a countable set of* actions, *with* $\tau \in A$ *a distinguished* silent, *or* internal, *action;*
- $\rightarrow \subseteq P \times A \times P$ *is the* transition relation; *and*
- $p_0 \in P$ *is the* start state.

Intuitively, P consists of the set of states that the process may enter, and A contains the set of actions that it may perform. The distinguished state p_0 is the *start state*. The relation $\rightarrow$ describes the actions available in a particular state and the state that may result when the action is executed. In what follows we write $p \xrightarrow{a} p'$ in lieu of $\langle p, a, p' \rangle \in \rightarrow$, and we also use $p \xrightarrow{a}$ to indicate that there is a p' such that $p \xrightarrow{a} p'$. We restrict our attention to *image-finite* transition systems, namely, systems $\langle P, A, \rightarrow, p_0 \rangle$ such that for all $p \in P$ and $a \in A$, the set $\{ p' \mid p \xrightarrow{a} p' \}$ is finite. To simplify the presentation we use the following constructions on transition systems.

Definition 2

1. *The transition system* $nil = \langle \{p_0\}, \{\tau\}, \emptyset, p_0 \rangle$.
2. *Let* $\mathcal{P} = \langle P, A, \rightarrow, p_0 \rangle$ *be a transition system. The* prefixing operation *is defined as follows. Choose* $p' \notin P$. *Then* $a.\mathcal{P} = \langle P \cup \{p'\}, A \cup \{a\}, \rightarrow \cup \{p' \xrightarrow{a} p_0\}, p' \rangle$
3. *Let* $\mathcal{P} = \langle P, A, \rightarrow, p_0 \rangle$ *and* $\mathcal{P}' = \langle P', A', \rightarrow', p'_0 \rangle$ *be transition systems. The* nondeterministic composition *of* $\mathcal{P}$ *and* $\mathcal{P}'$ *is defined as follows. Choose* $p' \notin P \cup P'$.

$$\mathcal{P} + \mathcal{P}' = \langle P \cup P' \cup \{p'\}, A \cup A', \rightarrow \cup \rightarrow' \cup \{ p' \xrightarrow{a} p'' \mid p_0 \xrightarrow{a} p'' \vee p'_0 \xrightarrow{a} p'' \}, p' \rangle$$

So *a.nil*, *a.nil* + *b.c.nil*, and *a.*(*b.nil* + *c.nil*) all represent transition systems. In what follows we sometimes abuse notation and leave off trailing *nil*'s; so *b.c* stands for the process *b.c.nil*. Readers should note that these operators comprise the simple process algebra FINTREE [ABV92a, ABV92b], which is a subset of CCS [Mil89]. We will return to FINTREE later in the paper.

A standard semantic model of concurrency is testing, as developed in [DeN87, NH83, Hen88], in which tests are applied to a process to ascertain the amount of nondeterminism exhibitted by a process. That is, one can distinguish between the *possibility* of success ($\mathcal{P}$ **may** $\mathcal{T}$) and the *inevitability* of success ($\mathcal{P}$ **must** $\mathcal{T}$). However, these relations are insensitive to all differences in finite internal computation, and for any process $\mathcal{P}$ and action a, $a.\tau.\mathcal{P}$ and $a.\mathcal{P}$ are *testing equivalent*. However, from a timing perspective, one may argue that $a.\tau.\mathcal{P}$ and $a.\mathcal{P}$ should *not* be considered equivalent, since after engaging in an a action $a.\tau.\mathcal{P}$ must wait for one "tick" to begin acting like $\mathcal{P}$, while $a.\mathcal{P}$ need not. The timed testing framework provides a mechanism for capturing this intuition.

A timed test is a process in which the set of actions A contains the distinguished success actions w and w_s. The availability of the action w signifies that the test has reached a successful state. w_s, on the other hand, must first be executed before the test can be determined to have succeeded; it plays a role similar to the role played by the test $\tau.w$ in untimed testing. A process' behavior in response to a test is given in terms of the *computations* that result when the test is "applied" to the process. These intuitions may be formalized as follows.

Definition 3 *Let* $\mathcal{P} = \langle P, A, \rightarrow, q_0 \rangle$ *be a process and* $\mathcal{T} = \langle T, A' \cup \{w, w_s\}, \rightarrow, t_0 \rangle$ *a test with fail* $\notin T$ *a special "failure state" and succeed* $\notin T$ *a special "success state". Let* $p \in P$ *and* $t \in T$, *with* $a \in A$. *(Note that* a *may be* τ.*)*

1. *A* configuration *is a pair* $\langle p, t \rangle \in P \times T$. *Following [NH83, Hen88] we write* $p \| t$ *in lieu of* $\langle p, t \rangle$.

2. *Configuration* $p||t$ *is* successful *if* $t \xrightarrow{w}$ *or* $t \equiv$ *succeed.*
3. *The* timed configuration transition relation, $p||t \xrightarrow{a}_{\chi} p'||t'$, *is defined by the following rules.*
   - $p \xrightarrow{a} p', t \xrightarrow{a} t' \Rightarrow p||t \xrightarrow{a}_{\chi} p'||t'$.
   - $t \xrightarrow{\tau} t', p \not\xrightarrow{\tau} \Rightarrow p||t \xrightarrow{\tau}_{\chi} p||t'$.
   - $p \xrightarrow{\tau} p', t \not\xrightarrow{\tau} \Rightarrow p||t \xrightarrow{\tau}_{\chi} p'||fail$.
   - $t \xrightarrow{w_s} \Rightarrow p||t \xrightarrow{w_s}_{\chi} p||succeed$.
4. *A* computation *of* $p||t$ *is a maximal (hence potentially infinite) sequence* $p_1||t_1 \xrightarrow{a_1}_{\chi} p_2||t_2 \xrightarrow{a_2}_{\chi} \cdots$, *where* p_1 *is* p *and* t_1 *is* t. *A computation is* successful *if for some* $i \geq 1$, $p_i||t_i$ *is successful.*
5. p **may**$_{\chi}$ t *if* $p||t$ *has a successful computation.* $\mathcal{P}$ **may** $\mathcal{T}$ *if* p_0 **may**$_{\chi}$ t_0.
6. p **must**$_{\chi}$ t *if every computation of* $p||t$ *is successful.* $\mathcal{P}$ **must** $\mathcal{T}$ *if* p_0 **must**$_{\chi}$ t_0.

In a timed testing scenario, the test may be thought of as a "client" and the process as a "server". The client wishes to engage in certain actions with the server and is willing to wait a certain amount of time between these interactions. If the server "finishes" more quickly than the client, then this is of no consequence; however, if the server is not ready when the client wishes to engage in an interaction, then the client deems the server to have failed. More specifically, we assume that τ actions performed by a process have a duration of one "clock tick", and we wish to interpret the τ actions that a test can perform as the delay that the test will allow. So, when the test is capable of engaging in a τ, it is willing to permit a delay of one "tick", and the process being tested may therefore engage in a τ. A process and a test may also "synchronize" on any non-τ action that both are capable of. These observations motivate the first part of the above definition of $\longrightarrow_{\chi}$. The second part arises from the following property of reactive systems: if a system cannot execute internally, it "idles" (*i.e.*, remains in its current state) until its environment offers an interaction. Accordingly, if a test is willing to delay (by executing internally), and the process is incapable of internal computation, then the test may perform its delay and continue with the rest of the test.

The third and fourth parts of the definition depart somewhat from the classical untimed transition configuration relation, as they mention "states"—*fail* and *succeed*—that are not states in the test. Intuitively, the third part models the fact that the test does not have control over the internal computation that the process engages in. It may be the case that the test is only capable of engaging in a visible action, while the process may perform an internal computation step. When this is the case, the process may elect to execute its τ and thereby miss the "deadline" imposed by the test and fail the test. To model this, we introduce a "state", *fail*, that is not in the state space of the test. In the fourth part, our intention is to require a test actually to "execute" a w_s action before it succeeds. Thus we introduce the *succeed* "state". Note that by the definition of $\longrightarrow_{\chi}$, a configuration containing either *fail* or *succeed* is incapable of further transitions.

It is worthwhile to note the asymmetry of the definition of $\longrightarrow_\chi$, which reflects the asymmetry of the relationship between clients and servers. Although a process may wait for a test, the test cannot wait unless it can do a τ. Note also, that visible actions in the process are not given priority over invisible actions.

As shown in [CZ91] delays, time-outs, upper and lower bounds on timing constraints, and stability (no internal computation possible) can all be expressed in this framework.

3 Nonpre-emptive Processes

In order to replace a slower process by a faster one, it is necessary to ensure that none of the other components of the system fail because the new system is "too fast". We illustrate this by considering the airspeed sampler mentioned in the introduction. Suppose the original speed controller Error_Slow_Controller was designed to continuously sample the airspeed (*get_speed*), compare it to the desired speed for that segment of the flight (*too_slow*, *too_fast*, *ok*), and adjust the speed accordingly. Furthermore, if the airspeed has not been sampled within two seconds, it signals an error and tries sampling again. The original controller required one unit of processing time before it could do the comparison.

$$\begin{aligned}\text{Error\_Slow\_Controller} &= \mathit{get\_speed}.\tau.\text{Error\_Rest}\\ \text{Error\_Rest} &= (\mathit{too\_slow}.\mathit{increase}.\text{Error\_Slow\_Controller}\\ &\quad +\mathit{too\_fast}.\mathit{decrease}.\text{Error\_Slow\_Controller}\\ &\quad +\mathit{ok}.\tau.\text{Error\_Slow\_Controller}\\ &\quad +\tau.\mathit{error}.\text{Error\_Slow\_Controller})\end{aligned}$$

The "faster" version of the sampler, Error_Speed_Controller = *get_speed*.Error_-Rest′, removes the delay between sampling and comparison. Error_Rest′ is identical to Error_Rest except that all occurrences of Error_Slow_Controller are replaced by Error_Speed_Controller. To decide whether Error_Slow_Controller can be replaced by Error_Speed_Controller in the flight controller without compromising system performance, we consider two tests representing potential behaviors of the environment, including other system components. The first test, $t_1 = \mathit{get\_speed}.\tau.\mathit{too\_slow}.w$, requires *too_slow* to be available one time unit after *get_speed*,. Error_Slow_Controller $\mathbf{may}_\chi$ pass this test, but Error_Speed_-Controller does not $\mathbf{may}_\chi$ pass this test since it can do a *too_slow* only immediately after doing a *get_speed*. A second test, $t_2 = \mathit{get\_speed}.(\mathit{too\_slow} + \mathit{error}.w + \tau.(\mathit{error}.w + \tau.\mathit{error}.w))$, considers the availability of *too_slow* immediately after *get_speed* to be an error, and allows *error* to occur between 0 and 2 time units after *get_speed*. Error_Slow_Controller $\mathbf{must}_\chi$ passes this test, whereas Error_Speed_Controller does not $\mathbf{must}_\chi$ pass it. The discrepancy between the behavior of Error_Slow_Controller and Error_Speed_Controller is that there is no overlap between the times at which *too_slow* is available in the two processes. Thus, Error_Slow_Controller cannot be replaced by Error_Speed_Controller without compromising system performance. The difference in availability of *too_slow* between the two processes is a result of what we call *pre-emptive* behavior.

We now consider a pair of speed controllers Slow_Controller and Controller that differ from Error_Slow_Controller and Error_Speed_Controller in that the error condition in Error_Rest is replaced by retesting the speed.

$$\begin{aligned}\text{Slow\_Controller} &= \textit{get\_speed}.\tau.\text{Rest}\\ \text{Rest} &= (\textit{too\_slow}.\textit{increase}.\text{Slow\_Controller}\\ &\quad +\textit{too\_fast}.\textit{decrease}.\text{Slow\_Controller}\\ &\quad +\textit{ok}.\tau.\text{Slow\_Controller} + \tau.\text{Test})\\ \text{Test} &= (\textit{too\_slow}.\textit{increase}.\text{Slow\_Controller}\\ &\quad +\textit{too\_fast}.\textit{decrease}.\text{Slow\_Controller} + \textit{ok}.\tau.\text{Slow\_Controller})\end{aligned}$$

Now Controller = *get_speed*.Rest′, where Rest′ differs from Rest in that Test is replaced by Test′ and all occurrences of Slow_Controller are replaced by Controller in both Rest′ and Test′.

Now both Slow_Controller and Controller $\mathbf{must}_\chi$ pass t_1 because *too_slow* is still available after a delay in both processes. However if we consider a test $t_3 = \textit{get\_speed}.(\textit{too\_slow} + \textit{ok}.w + \tau.(\textit{ok}.w + \tau.\textit{ok}.w))$ we see that Slow_Controller $\mathbf{must}_\chi$ pass this test, but Controller does not. Here the problem is that the test expects *too_slow* NOT to be available immediately after *get_speed*. Note that its availability at any other time will not cause a problem.

As these examples illustrate, some processes do appear to speed up performance, but do so by eliminating previously existing behaviors (replacing Error_Slow_Controller by Error_Speed_Controller). However sometimes the problems are caused by the expectations of the environment, represented by tests. In both cases though, the problem is processes (or tests) in which the passage of time eliminates behaviors, what we call pre-emptive behavior. This observation, in fact, is adequate for describing the class of processes (and the environments in which they run) that can be "speeded up" without causing systems failure. We call these processes *nonpre-emptive* because they have the property that behaviors remain available in spite of delays. Formally we define them as follows.

Definition 4 *A labeled transition system $\langle P, A, \rightarrow, p_0\rangle$ is* nonpre-emptive *if the following holds for every $p \in P$.*

1. *If p has distinct transitions $p \xrightarrow{\tau} p_1$ and $p \xrightarrow{a} p_2$ and $a \neq \tau$, then either $p_1 \xrightarrow{a} p_2$; or $p_1 \xrightarrow{a} p_3$ for all $p_3 \in P'$, where $P' = \{p_3 | p_2 \xrightarrow{\tau} p_3\}$ and $P' \neq \emptyset$; and*
2. *If $p \xrightarrow{w}$ then for all $a \neq w$, $p \not\xrightarrow{a}$*

Let $\mathcal{N}$ be the class of (image-finite) nonpre-emptive transition systems.

Error_Slow_Controller, Error_Speed_Controller, t_2 and t_3 are not nonpre-emptive, whereas Slow_Controller, Controller and t_1 are. Additionally, $\tau.a + \tau.b$ and $a + w_s$ are both nonpre-emptive, whereas $a + w$ is not.

With the formalization of nonpre-emption, we are now able to provide a set of preorders that allow timed testing to provide a means of determining substitutivity with respect to process speed.

3.1 Nonpre-emptive Testing Preorders

As noted in the examples in the previous section, sometimes the problem of substitutivity of "faster" for "slower" arises not from the process, but from the environment in which it runs. This is captured by the tests. As we saw the nonpre-emptive processes Slow_Controller and Controller both failed to pass the pre-emptive t_3. As documented in [CZ91], this lead us to postulate that the system interaction with nonpre-emptive processes must be representable by linear tests. Subsequent work, including the broadened definition of nonpre-emption presented above, has lead us to expand the set of tests to include all nonpre-emptive tests. Thus, a process P_f may be substituted for a slower process P_s, provided that the environment in which P_s runs interacts with P_s in a nonpre-emptive manner.

The notion of processes being related based on their behavior with respect to a set of tests, in our case timed tests, is captured by temporal extensions of the classical De Nicola–Hennessy testing preorders $\sqsubseteq_{may}, \sqsubseteq_{must}, \sqsubseteq_{test}$. In our framework, there are two parameters: a class of processes (transition systems), and a class of tests.

Definition 5 *Let $\mathcal{C}$ be a class of transition systems, with $\mathcal{P}, \mathcal{P}' \in \mathcal{C}$. Also let $\mathcal{E}$ be a class of tests. Then the associated* timed testing preorders *are defined by:*

1. $\mathcal{P} \sqsubseteq^{\mathcal{E}}_{may_\chi} \mathcal{P}'$ *if for all* $\mathcal{T} \in \mathcal{E}, \mathcal{P}$ $\mathbf{may}_\chi$ $\mathcal{T}$ *implies* $\mathcal{P}'$ $\mathbf{may}_\chi$ $\mathcal{T}$.
2. $\mathcal{P} \sqsubseteq^{\mathcal{E}}_{must_\chi} \mathcal{P}'$ *if for all* $\mathcal{T} \in \mathcal{E}, \mathcal{P}$ $\mathbf{must}_\chi$ $\mathcal{T}$ *implies* $\mathcal{P}'$ $\mathbf{must}_\chi$ $\mathcal{T}$.
3. $\mathcal{P} \sqsubseteq^{\mathcal{E}}_{test_\chi} \mathcal{P}'$ *if* $\mathcal{P} \sqsubseteq^{\mathcal{E}}_{may_\chi} \mathcal{P}'$ *and* $\mathcal{P} \sqsubseteq^{\mathcal{E}}_{must_\chi} \mathcal{P}'$.

Thus to obtain nonpre-emptive preorders, we restrict both $\mathcal{C}$ and $\mathcal{E}$ to the nonpre-emptive processes, $\mathcal{N}$.

If we consider the framework in which nonpre-emptive processes are only tested by nonpre-emptive tests, we obtain preorders that capture an intuitive notion of relative efficiency.

Theorem 1. *Let $\mathcal{P} = \langle P, A, \rightarrow, p_0 \rangle$ be a nonpre-emptive process. Then $\tau.\mathcal{P} \sqsubseteq^{\mathcal{N}}_{may_\chi} \mathcal{P}$, and $\tau.\mathcal{P} \sqsubseteq^{\mathcal{N}}_{must_\chi} \mathcal{P}$.*

This theorem is proven by reasoning about the longest possible path of τ's leading to a stable state in the process. The interested reader can see [CZ93] for the full proof of this.

4 Complete Axiom Systems for the Preorders

Having established a semantic framework, we now develop tools for representing and reasoning about nonpre-emptive processes. We begin by axiomatizing the nonpre-emptive preorders in FINTREE, shown in Figure 1. Provability in this system is assumed to be closed under reflexivity, equality, transitivity, substitutivity and instantiation. Both of the axiom systems incorporate the usual

(A1) $p+q =^{\mathcal{N}}_{may_x} q+p$	(U1) $p+q =^{\mathcal{N}}_{must_x} q+p$
(A2) $(p+q)+r =^{\mathcal{N}}_{may_x} p+(q+r)$	(U2) $(p+q)+r =^{\mathcal{N}}_{must_x} p+(q+r)$
(A3) $p+p =^{\mathcal{N}}_{may_x} p$	(U3) $p+p =^{\mathcal{N}}_{must_x} p$
(A4) $p+nil =^{\mathcal{N}}_{may_x} p$	(U4) $p+nil =^{\mathcal{N}}_{must_x} p$
(A5) $a.p+a.q =^{\mathcal{N}}_{may_x} a.(p+q)$	(U5) $a.p+a.q \sqsubseteq^{\mathcal{N}}_{must_x} a.p$
(A6) $\tau.p \sqsubseteq^{\mathcal{N}}_{may_x} p$	(U6) $a.p+a.q \sqsubseteq^{\mathcal{N}}_{must_x} a.(p+q)$
(A7) $p \sqsubseteq^{\mathcal{N}}_{may_x} p+q$	(U7) $\dfrac{p \sqsubseteq^{\mathcal{N}}_{must_x} q}{a.p \sqsubseteq^{\mathcal{N}}_{must_x} a.q}$
(A8) $\dfrac{p \sqsubseteq^{\mathcal{N}}_{may_x} q}{a.p \sqsubseteq^{\mathcal{N}}_{may_x} a.q}$	(U8) $\dfrac{q \sqsubseteq^{\mathcal{N}}_{must_x} q'; p \xrightarrow{\tau} \text{ or } q \xrightarrow{\tau}, q' \xrightarrow{\tau} \text{ or } q \not\xrightarrow{\tau}, q' \not\xrightarrow{\tau}}{p+q \sqsubseteq^{\mathcal{N}}_{must_x} p+q'}$
(A9) $\dfrac{q \sqsubseteq^{\mathcal{N}}_{may_x} q'}{p+q \sqsubseteq^{\mathcal{N}}_{may_x} p+q'}$	(U9) $\dfrac{q \xrightarrow{\tau}}{p+q \sqsubseteq^{\mathcal{N}}_{must_x} q}$
(A10) $nil \sqsubseteq^{\mathcal{N}}_{may_x} p$	(U10) $\tau.p \sqsubseteq^{\mathcal{N}}_{must_x} p$

Fig. 1. The axiom systems A and U.

commutativity, associativity, idempotency, and nil laws for +. It is assumed that these axioms hold only for nonpre-emptive FINTREE terms.

The statement of Theorem 1 is captured in A and U by the axioms (A6) and (U10). Both preorders are preserved by prefixing (A8, U7). Distributivity yields an equality (A5) for $\sqsubseteq^{\mathcal{N}}_{may_x}$ but an inequality (U6) for $\sqsubseteq^{\mathcal{N}}_{must_x}$. A and U both contain axioms governing when $p+q$ is less than $p+q'$ (A9, U8). In A this is true simply when q is less than q', while in U there are additional restrictions on the stability of the processes. Additionally, the axiom system for $\sqsubseteq^{\mathcal{N}}_{may_x}$, A, also contains axioms stating that p is less than $p+q$ in the preorder (A7) and that nil is less than any other process (A10). In the axiom system for $\sqsubseteq^{\mathcal{N}}_{must_x}$, U, there is an axiom (U5) stating that the sum of two processes $a.p$ and $a.q$ is less than either one of the summands. Another axiom (U9) expresses the fact that when a sum includes a process of the form $\tau.p$, then other processes in that sum may be removed and all tests originally passed must still be passed.

Theorem 2.

- *A is a sound and complete axiom system for $\sqsubseteq^{\mathcal{N}}_{may_x}$ in FINTREE.*
- *U is a sound and complete axiom system for $\sqsubseteq^{\mathcal{N}}_{must_x}$ in FINTREE.*

The soundness proofs rely on the development of an alternative characterization of the nonpre-emptive preorders. The proofs of completeness proceed by structural induction.

5 Nonpre-emptive Process Algebra

In order for the class $\mathcal{N}$ of processes to be useful in the design and maintenance of real-time systems, it is necessary to have a language to express exactly those processes. In this section we present an algebra, N-CCS, for nonpre-emptive processes. This algebra is an extension of the one presented in [CZ91].

Let A be a set of actions (containing τ) ranged over by a, let $\mathcal{V}$ be a set of *process variables* ranged over by X, and let $S \subseteq (A - \{\tau\})$. Also let $f : A \to A$ be such that $f(a) = \tau$ if and only if $a = \tau$. As in the standard CCS framework, τ represents internal action. The formal syntax of our language is given as follows.

$$p ::= nil \mid X \mid a.p \mid p[\!]p \mid p||_S p \mid p[f] \mid p;p \mid p\langle L\rangle \mid rec(X.p)$$

Occurrences of X in the term $rec(X.p)$ are bound in the usual sense. We write $p[q/X]$ to represent the simultaneous substitution of q for all free occurrences of X in p.

Informally, the operators have the following description. Process $a.p$ may engage in action a and then behave like p. Process $p[\!]q$ may engage in any of the external actions that p or q may, and thereafter behave like the process whose initial external action was chosen; notice that *internal actions* may *not* resolve the choice. Similar choice operators are used in [Hen88] and MEIJE [AB84]. The process $p||_S q$ behaves like the *interleaving* of p and q, with the added stipulation that p and q must synchronize on actions in S (note that syntax restrictions prohibit τ being an element of S); this operator appears in TCSP [BHR84, Hoa85]. We use this form of concurrent composition because standard CCS composition is not a congruence with respect to the timed preorders. The process $p[f]$ behaves like p with actions "relabeled" by f. The process $p\langle L\rangle$ can be thought of as exporting the actions in L and hiding all other actions. The process $p;q$ first does the actions in p and, once p has terminated (is a member of $\surd$), executes the action in q. Finally, the process $rec(X.p)$ can be thought of as a distinguished solution to the equation $X = p$.

As is usual in a process algebra, our intention is to define a *global* transition system, $\langle P, A, \rightarrow\rangle$, where P is the set of closed terms generated by the above grammar and $\rightarrow$ is given by the *operational semantics* of terms. Then processes are defined by giving the term that corresponds to their start states in the global transition system. The formal semantics of the language appear in Figure 2.

The next lemma establishes that the global transition system induced by N-CCS is nonpre-emptive.

Lemma 3. *Let p be a closed process term. Then p is nonpre-emptive.*

We state without proof that the nonpre-emptive preorders are precongruences for finite N-CCS.

Theorem 4. *The relations $\sqsubseteq^{\mathcal{N}}_{may_x}$ and $\sqsubseteq^{\mathcal{N}}_{must_x}$ are precongruences for finite N-CCS.*

$$a.p \xrightarrow{a} p$$
$$p \xrightarrow{a} p', a \neq \tau \Rightarrow p \| q \xrightarrow{a} p' \qquad q \xrightarrow{a} q', a \neq \tau \Rightarrow p \| q \xrightarrow{a} q'$$
$$p \xrightarrow{\tau} p' \Rightarrow p \| q \xrightarrow{\tau} p' \| q \qquad q \xrightarrow{\tau} q' \Rightarrow p \| q \xrightarrow{\tau} p \| q'$$
$$p \xrightarrow{a} p', a \notin S \Rightarrow p ||_S q \xrightarrow{a} p' ||_S q \quad q \xrightarrow{a} q', a \notin S \Rightarrow p ||_S q \xrightarrow{a} p ||_S q'$$
$$p \xrightarrow{a} p', q \xrightarrow{a} q', a \in S \Rightarrow p ||_S q \xrightarrow{a} p' ||_S q'$$
$$p \xrightarrow{a} p' \Rightarrow p[f] \xrightarrow{f(a)} p'[f]$$
$$p \xrightarrow{a} p', a \in L \cup \{\tau\} \Rightarrow p\langle L\rangle \xrightarrow{a} p'\langle L\rangle$$
$$p \xrightarrow{a} p', a \notin L \cup \{\tau\}, p \overset{b}{\not\rightarrow}, b \in L \Rightarrow p\langle L\rangle \xrightarrow{\tau} p'\langle L\rangle$$
$$p \xrightarrow{a} p' \Rightarrow p;q \xrightarrow{a} p';q \quad p \in \surd, q \xrightarrow{a} q' \Rightarrow p;q \xrightarrow{a} q'$$
$$p[rec(X.p)/X] \xrightarrow{a} p' \Rightarrow rec(X.p) \xrightarrow{a} p'$$

where

$$nil \in \surd \qquad p[rec(X.p)/X] \in \surd \Rightarrow rec(X.p) \in \surd$$
$$p \in \surd \Rightarrow p[f] \in \surd, p\langle L\rangle \in \surd \quad p \in \surd, q \in \surd \Rightarrow p \| q \in \surd, p;q \in \surd, p ||_X q \in \surd$$

Fig. 2. The operational semantics of T-CCS.

6 Axiomatizing N-CCS

We now present complete axiomatizations of the may and must preorders for finite N-CCS (that is, N-CCS with no occurrences of *rec*). Our axiomatizations consist of two parts. First, we use the method developed in [ABV92a, ABV92b] to reduce N-CCS to FINTREE. We can then use the complete axiomatizations developed in Section 4.

6.1 A Method to Reduce Languages to a Subset of CCS

In [ABV92a, ABV92b] an algorithm is presented for generating a finite complete axiom system, with respect to a given axiomatized relation on FINTREE, for languages specified by rules in structural operational semantics with guarded recursion (GSOS rules). The resulting axiom systems consist of the rules for FINTREE and rules for reducing all terms to equivalent FINTREE terms.

The transformation of the non-FINTREE operators is done in three steps. First corresponding smooth operators are defined for all of the nonsmooth operators. Second, each smooth but nondistinctive operator has corresponding distinctive operators defined for it. Finally, axioms are defined that eliminate all occurrences of nonsmooth and nondistinctive operators from terms. Intuitively, if a rule is smooth, then if the ability of a process p to do an action and evolve to p' (p tests *positively*) is a necessary antecedent, the resulting process of the consequent must contain that evolution. Further a rule cannot use a process in more than one clause of the antecedent. An operator is distinctive if for each possible component process, either that component is tested positively in every rule or in none of them. Additionally, for every possible pair of different rules for the operator, both rules require some process p to be able to do distinct actions. Formally, smoothness and distinctiveness are defined as follows.

Definition 6

- *A GSOS rule for an operator f is* smooth *if it has the form*

$$\frac{\{p_i \xrightarrow{a_i} p_i' \mid i \in I\} \cup \{p_i \stackrel{b_{ij}}{\not\rightarrow} \mid i \in K, 1 \leq j \leq n_i\}}{f(p_1, \ldots, p_l) \xrightarrow{c} C[\mathbf{p}, \mathbf{p}']}$$

where I, K indexing sets with $I \cap K = \emptyset$, $I \cup K = \{1, \ldots, l\}$, and no p_i with $i \in I$ appears in $C[\mathbf{p}, \mathbf{p}']$, where $C[\mathbf{p}, \mathbf{p}']$ is a term in which at most the variables $\mathbf{p}$ and $\mathbf{p}'$ appear. A GSOS operation f is smooth if all the rules for the operation are smooth.

- *A smooth operation f is* distinctive *if both of the following conditions hold: (1) for each argument i, either all the rules for f test i positively or none of them do; and (2) for each pair of different rules for f, there is an argument for which both rules have a positive antecedent but with a different action.*

In general the smooth operator corresponding to the nonsmooth operator contains extra arguments to account for each distinct use of the component processes. The axioms relating a smooth and nonsmooth operator copy the multiply used process the appropriate number of times in the smooth operator. A set of distinctive operators corresponding to a nondistinctive one can be obtained by partitioning the rules of the nondistinctive ones by their use of a process, and creating a new operator for each of these partitions. There are three types of axioms relating the distinct set of operators to the nondistinctive operators: distributivity over +, action laws describing when a process using a distinctive operator can do an action, and inaction laws describing when it cannot. The smoothing and distinguishing axioms provide a means of reducing any term to an equivalent FINTREE term.

6.2 Reducing N-CCS to FINTREE

To reduce N-CCS to FINTREE we must eliminate all occurrences of relabelling, choice, parallel composition, hiding, and sequential composition from terms. To reduce the relabelling operator we use the standard simplification rules from [Mil89]. The remaining rules are smooth with the exception of hiding. Hiding is not smooth because its second rule has both a positive and a negative test on p. We create a smooth hiding operator $(p_0, p_1)\langle L\rangle'$, with the following operational semantics, by introducing separate processes to be tested positively and negatively.

$$\frac{p \xrightarrow{a} p', a \in \{L \cup \tau\}}{(p,p)\langle L\rangle' \xrightarrow{a} p'\langle L\rangle} \quad \frac{p_0 \xrightarrow{a} p', a \notin \{L \cup \tau\}, p_1 \stackrel{b}{\not\rightarrow}, b \in L}{(p_0, p_1)\langle L\rangle' \xrightarrow{\tau} p'\langle L\rangle}$$

So now $p\langle L\rangle = (p,p)\langle L\rangle'$. The remaining operators, as well as the smooth hiding operator are not distinctive. Thus we partition each of these operators as follows.

$$\frac{p \xrightarrow{a} p', a \notin S}{p \|_{1_S} q \xrightarrow{a} p' \|_S q} \qquad \frac{p \xrightarrow{a} p', q \xrightarrow{a} q', a \in S}{p \|_{2_S} q \xrightarrow{a} p' \|_S q'}$$

$$\frac{p \xrightarrow{a} p', a \neq \tau}{p \mathbin{[\!]}_1 q \xrightarrow{a} p'} \qquad \frac{p \xrightarrow{\tau} p'}{p \mathbin{[\!]}_2 q \xrightarrow{\tau} p' \mathbin{[\!]} q}$$

$$\frac{p \xrightarrow{a} p'}{p ;_1 q \xrightarrow{a} p' ; q} \qquad \frac{p \in \surd, q \xrightarrow{a} q'}{p ;_2 q \xrightarrow{a} q'}$$

$$\frac{p \xrightarrow{a} p', a \in \{L \cup \tau\}}{(p,p)\langle L\rangle'_1 \xrightarrow{a} p'\langle L\rangle} \qquad \frac{p_0 \xrightarrow{a} p', a \notin \{L \cup \tau\}, p_1 \not\xrightarrow{b}, b \in L}{(p_0,p_1)\langle L\rangle'_2 \xrightarrow{\tau} p'\langle L\rangle}$$

In this partitioning, $\|_{1_S}$ is a left composition operator and $\|_{2_S}$ plays the role of a synchronizing composition operator. $[\!]_1$ is a left choice operator for non-τ actions and $[\!]_2$ is a left choice operator for τ. $;_1$ is a sequential composition operator for when $p \notin \surd$ and $;_2$ is the sequential composition operator for when $p \in \surd$. $\langle L\rangle'_1$ is the hiding operator for actions in $\{L \cup \tau\}$, while $\langle L\rangle'_2$ does the actual hiding of actions not in $\{L \cup \tau\}$. These new operators are smooth and distinctive, and thus can be fully characterized by distributivity, action, and inaction rules. The reduction laws are in Figure 3. The correctness of these laws is a direct result from [ABV92a, ABV92b].

Finally, we combine (TF1)-(TF32) with A and U to generate sound and complete axiom systems for the nonpre-emptive preorders.

Theorem 5.
- (TF1)-(TF32) *together with A form a sound and complete axiom system for* $\sqsubseteq^{\mathcal{N}}_{may_x}$ *in finite N-CCS.*
- (TF1)-(TF32) *together with U form a sound and complete axiom system for* $\sqsubseteq^{\mathcal{N}}_{must_x}$ *in finite N-CCS.*

7 Example

To demonstrate the utility of the axiom systems in determining whether a faster process can be substituted for a slower one, we will consider finite versions of the speed controllers introduced in Section 3 without error conditions. We use N-CCS to specify both a slow controller and a faster one, and use the axiom systems to show that the faster controller may be substituted for the slower one.

$$\text{F\_Slow\_Controller} = \mathit{get\_speed}.\tau.\text{F\_Rest}$$
$$\text{F\_Rest} = (\mathit{too\_slow.increase} \mathbin{[\!]} \mathit{too\_fast.decrease} \mathbin{[\!]} \mathit{ok}.\tau)$$

The faster controller is specified by F_Controller = *get_speed*.F_Rest. In order to compare these two speed controllers using the axiomatizations for the nonpre-

(TF1) $p\|_S q = (p\|_{1S} q) + (q\|_{1S} p) + (p\|_{2S} q)$

(TF2) $(p_1 + p_2) * q = (p_1 * q) + (p_2 * q),\ * \in \{\|_{1S}, \|_{2S}, \boldsymbol{\|}_1, \boldsymbol{\|}_2, ;_1\}$

(TF3) $(p + q)[f] = p[f] + q[f]$

(TF4) $p * (q_1 + q_2) = (p * q_1) + (p * q_2),\ * \in \{\|_{2S}, ;_2\}$

(TF5) $p * nil = nil,\ * \in \{\|_{2S}, ;_2\}$ (TF6) $nil * q = nil,\ * \in \{\|_{1S}, \|_{2S}, \boldsymbol{\|}_1, \boldsymbol{\|}_2,\}$

(TF7) $a.p\|_{1S} q = a.(p\|_S q), a \notin S$ (TF8) $a.p\|_{1S} q = nil, a \in S$

(TF9) $a.p\|_{2S} a.q = a.(p\|_S q), a \in S$ (TF10) $a.p\|_{2S} b.q = nil, a = b \notin S$ or $a \neq b$

(TF11) $(a.p)[f] = f(a).p[f]$ (TF12) $nil[f] = nil$

(TF13) $p\boldsymbol{\|}q = (p\boldsymbol{\|}_1 q) + (q\boldsymbol{\|}_1 p) + (p\boldsymbol{\|}_2 q) + (q\boldsymbol{\|}_2 p)$

(TF14) $a.p\boldsymbol{\|}_1 q = a.p, a \neq \tau$ (TF15) $\tau.p\boldsymbol{\|}_1 q = nil$

(TF16) $a.p\boldsymbol{\|}_2 q = nil, a \neq \tau$ (TF17) $\tau.p\boldsymbol{\|}_2 q = \tau.(p\boldsymbol{\|}q)$

(TF18) $p\langle L\rangle = (p,p)\langle L\rangle'$ (TF19) $(p,p)\langle L\rangle' = (p,p)\langle L\rangle'_1 + (p,p)\langle L\rangle'_2$

(TF20) $(nil, nil)\langle L\rangle'_1 = nil$ (TF21) $(a.p, a.p)\langle L\rangle'_1 = nil, a \notin \{L \cup \tau\}$

(TF22) $(a.p, a.p)\langle L\rangle'_1 = a.(p\langle L\rangle), a \in \{L \cup \tau\}$

(TF23) $(p + q, p + q)\langle L\rangle'_1 = (p,p)\langle L\rangle'_1 + (q,q)\langle L\rangle'_1$

(TF24) $(a.p, b.q)\langle L\rangle'_2 = \tau.(p\langle L\rangle), a \notin \{L \cup \tau\}, b \notin L$

(TF25) $(a.p, q)\langle L\rangle'_2 = nil, a \in \{L \cup \tau\}$

(TF26) $(p, b.q)\langle L\rangle'_2 = nil, b \in L$ (TF27) $(nil, q)\langle L\rangle'_2 = nil$

(TF28) $(p_1 + p_2, q)\langle L\rangle'_2 = (p_1, q)\langle L\rangle'_2 + (p_2, q)\langle L\rangle'_2$

(TF29) $p; q = p;_1 q + p;_2 q$

(TF30) $a.p;_1 q = a.(p; q)$

(TF31) $p;_2 a.q = a.q, p \in \surd$ (TF32) $p;_2 a.q = nil, p \notin \surd$

Fig. 3. The set of axioms to reduce N-CCS to FINTREE.

emptive preorders we first eliminate $\boldsymbol{\|}$ using (TF13), (TF14) and (TF16).

$$\begin{aligned}\text{F\_Rest} = (&too\_slow.increase\\ &+[(too\_fast.decrease + ok.\tau + nil + nil)\boldsymbol{\|}_1 too\_slow.increase]\\ &+nil + [(too\_fast.decrease + ok.\tau + nil + nil)\boldsymbol{\|}_2 too\_slow.increase])\end{aligned}$$

We obtain a simple FINTREE term by applications of (TF2), (TF14), and (TF16).

$$\begin{aligned}\text{F\_Rest} = (&too\_slow.increase + (too\_fast.decrease + ok.\tau + nil + nil)\\ &+nil + (too\_fast.decrease + ok.\tau + nil + nil))\end{aligned}$$

We now have an equivalent FINTREE term for F_Rest and thus we have FINTREE expressions for F_Slow_Controller and F_Controller. We can now compare the two controllers using the axiomatizations. $get_speed.\tau.$F_Rest $\sqsubseteq^{\mathcal{N}}_{may_x}$ $get_speed.$F_Rest By A6 and U10 $\tau.$F_Rest $\sqsubseteq^{\mathcal{N}}_{may_x}$ F_Rest and $\tau.$F_Rest $\sqsubseteq^{\mathcal{N}}_{must_x}$ F_Rest.Then by A8 and U7 $get_speed.\tau.$F_Rest $\sqsubseteq^{\mathcal{N}}_{may_x}$ $get_speed.$F_Rest and $get_speed.\tau.$F_Rest $\sqsubseteq^{\mathcal{N}}_{must_x}$ $get_speed.$F_Rest. So we can substitute F_Controller for F_Slow_Controller without compromising the performance of the system.

8 Conclusions and Related Work

We have presented a process algebra for nonpre-emptive processes and complete axiomatizations for the nonpre-emtive testing preorders of the algebra. These axiomatizations are somewhat more complex than untimed axiomatizations because we must account for temporal differences as well as nondeterministic ones. Much of the complexity in the axiomatization occurs in the 32 rules for reducing a N-CCS term to a FINTREE term. The testing preorders presented can also be applied to the general real-time case and axiomatizations for these general preorders exist. We are currently undertaking the automation of these preorders.

In contrast with our work, most of the real-time process algebra work [Jon82, Mil83, KSdR+85, HGdR87, GL90] [GB87, RR87, Car82, HR90, NRSV90] [BB89, Zwa88, AKH90, Mil85] treats actions as *durationless.* The passage of time is then modeled by assigning times to the occurrence of actions (as in [BB89, GB87]) or by introducing special actions that denote "ticks" or "idles", as in [HR90, NRSV90]. The equivalences and preorders that result typically do not allow one to reason about differences in relative delays. One exception may be found in [AKH90], where a preorder is developed which discriminates between two processes in CCS which are bisimulation equivalent but one is faster than the other in the sense that it does fewer internal actions. The resulting preorder more closely models actual delays in processes than the usual asynchronous treatment of internal actions. A sound and complete axiomatization for this preorder is also presented. The axiom system is similar to that for observational congruence, modifying the τ laws to take into account the number of delays done by each process and adding a rule to allow removal of excess terms in summands. Another exception can be found in [MT91] where a preorder that relates processes based on relative speed is defined for a class of processes similar to our nonpre-emptive processes and a sound and complete equational theory is presented for a subset of terms which is similar to FINTREE.

References

[AB84] D. Austry and G. Boudol. Algèbre de processus et synchronisations. *Theoretical Computer Science*, 30:91–131, 1984.

[ABV92a] L. Aceto, B. Bloom, and F. Vaandrager. Turning SOS Rules into Equations. In *Proc. of Seventh Annual IEEE Symposium on Logic in Computer Science*, pages 113–124, 1992.

[ABV92b] L. Aceto, B. Bloom, and Frits Vaandrager. Turning SOS rules into Equations. Technical Report CS-R9218, Center for Mathematics and Computer Science, June 1992.

[AKH90] S. Arun-Kumar and M. Hennessy. An Efficiency Preorder for Processes. Technical Report 5/90, Department of Computing Science, University of Sussex, July 1990.

[BB89] J.C.M. Baeten and J.A. Bergstra. Real Time Process Algebra. Technical report, CWI, 1989.

[BHR84] S.D. Brookes, C.A.R. Hoare, and A.W. Roscoe. A Theory of Communicating Sequential Processes. *Journal of the ACM*, 31(3):560–599, July 1984.

[Car82] L. Cardelli. Real Time Agents. In *Proc. of Ninth Colloquium in Automata, Languages and Programming*, pages 94–106, 1982.

[CZ91] R. Cleaveland and A. Zwarico. A Theory of Testing for Real-Time. In *Proceedings of the Sixth Symposium on Logic in Computer Science*, 1991.

[CZ93] R. Cleaveland and A. Zwarico. A Theory of Testing for Real-Time. Technical report, Department of Computer Science, Johns Hopkins University, 1993.

[DeN87] R. DeNicola. Extensional Equivalences for Transition Systems. *Acta Informatica*, 24:211–237, 1987.

[GB87] R. Gerth and A. Boucher. A Timed Failure Semantics for Extended Communicating Processes. In *Proceedings of ICALP '87, LNCS 267*, 1987.

[GL90] R. Gerber and I. Lee. CCSR: A Calculus for Communicating Shared Resources. In *Proceedings, CONCUR '90, Lecture Notes in Computer Science 458*. Springer-Verlag, 1990.

[Hen88] M.C.B. Hennessy. *Algebraic Theory of Processes*. MIT Press, Boston, 1988.

[HGdR87] C. Huizing, R. Gerth, and W.P. de Roever. Full Abstraction of a Denotational Semantics for Real-time Concurrency. In *POPL87*, pages 223–237, 1987.

[Hoa85] C.A.R. Hoare. *Communicating Sequential Processes*. Prentice-Hall, London, 1985.

[HR90] M. Hennessy and T. Regan. A Temporal Process Algebra. Technical Report 2/90, Department of Computing Science, University of Sussex, April 1990.

[Jon82] G. Jones. *A Timed Model for Communicating Processes*. PhD thesis, Oxford University, 1982.

[KSdR$^+$85] R. Koymans, R.K. Shyamasundar, W.P. de Roever, R. Gerth, and S. Arun-Kumar. Compositional Semantics for Real-Time Distributed Computing. In *Logic of Programs Workshop '85, LNCS 193*, 1985.

[Mil83] R. Milner. Calculi for synchrony and asynchrony. *Theoretical Computer Science*, 25:267–310, 1983.

[Mil85] G.J. Milne. CIRCAL and the Representation of Communication, Concurrency, and Time. *ACM Transactions on Programming Languages and Systems*, 7(2):270–298, April 1985.

[Mil89] R. Milner. *Communication and Concurrency*. Prentice Hall, 1989.

[MT91] Faron Moller and Chris Tofts. Relating Processes with Respect to Speed. In *Proceedings, CONCUR'91, Lecture Notes in Computer Science 527*. Springer-Verlag, 1991.

[NH83] R. De Nicola and M.C.B. Hennessy. Testing Equivalences for Processes. *Theoretical Computer Science*, 34:83–133, 1983.

[NRSV90] Xavier Nicollin, Jean-Luc Richier, Joseph Sifakis, and Jacques Voiron. ATP: An Algebra for Timed Processes. In *Proceedings of the IFIP TC 2 Working Conference on Programming Concepts and Methods Principles of Programming Languages*, pages 402–429, April 1990.

[RR87] G.M. Reed and A.W. Roscoe. Metric Spaces as Models for Real-Time Concurrency. In *Proceedings of Math. Found. of Computer Science, LNCS 298*, 1987.

[Zwa88] A. Zwarico. *Timed Acceptance: An Algebra of Time Dependent Computing*. PhD thesis, Department of Computer and Information Science, University of Pennsylvania, 1988.

Logical Foundations for Compositional Verification and Development of Concurrent Programs in UNITY *

Pierre Collette[1] and Edgar Knapp[2]

[1] University of Manchester, Department of Computer Science, Manchester M13 9PL, United Kingdom
[2] Purdue University, Department of Computer Science, West Lafayette IN 47907-1398, U.S.A.

Abstract. To achieve modularity, we view UNITY specifications as describing open (rather than closed) systems. These may be composed in parallel or through hiding of global variables. Adopting the assumption-commitment paradigm, conventional properties of UNITY programs are extended with an explicit rely condition on interference; previous variants of the logic can be retrieved by specialising or omitting this rely condition. The outcome is a complete compositional proof system for both safety and progress properties.

1 Introduction

Compositional methods for software development, e.g. [1, 7, 11, 20], require proof rules for both verifying the correctness of a module from the correctness of its components (bottom-up design) and conversely for validating the decomposition of a module specification into specifications of components that are easier to implement (top-down design). *Compositional proof systems* thus provide a foundation for modularity in both software verification and software development. Examples are given in [3, 12] (temporal logic specifications), [11, 14, 20] (trace specifications in message-based concurrency), and [17, 19] (extended pre/post specifications for state-based concurrency). Following the same line, we present logical foundations for modular verification and development in UNITY. Designed by Chandy and Misra [4], the UNITY formalism has proved useful in the formal specification, development, and verification of concurrent programs; examples are [9, 16]. Our proposal overcomes a number of limitations that have become apparent in the course of this promising work:

1. Chandy and Misra propose compositional proof rules, but their axiomatic semantics of specifications does not match the operational one. Aware of the problem, Chandy and Misra introduce the Substitution Axiom, but the

* Research supported in part by the Belgian National Funds for Scientific Research, in part by the UK Science and Engineering Research Council, and in part by the Midwest University Consortium for International Activities, U.S.A.

logic is then inconsistent. The problem is fixed by Sanders [15] who also proposes a complete proof system. However, compositionality is lost: in order to infer properties about the parallel composition $F \| G$ of two programs F and G from the properties of F and G, additional invariants of $F \| G$ have to be taken into account. In this paper, we introduce specifications whose axiomatic semantics match the operational one and present a proof system which is both ***compositional and complete.***
2. Unlike the proof systems of [4, 15] in which composition rules are restricted to safety and basic (**ensures**) progress properties, ours includes rules for the ***composition of*** more general (**leadsto**) ***progress properties*** as well. Udink, Herman, and Kok [18] prove a composition result for a new operator, lying between **ensures** and **leadsto** but that operator is not intended to be used in specifications; it forms a basis for the definition of compositional refinement of programs.
3. In contrast with the non-compositional computational model of [4, 15], our operational semantics is geared towards the description of ***open*** systems in that the execution of a program may also include environment (other program) steps. As in [18], we thus introduce a distinction between local variables that are visible only within the current program and shared variables that can be accessed and modified by the environment. The ability to hide information is imperative for meeting the main challenge in programming, the management of complexity. Yet, the UNITY programs of [4] lack a hiding operator and it is only in [18] that it is first introduced. With respect to hiding, our approach is very similar to [18]; formulating adequate ***proof rules for hiding*** is straightforward.

To achieve compositionality, interference between concurrent components must be taken into account at the specification level [7]. The key idea is the extension of conventional properties of the UNITY logic with a ***rely condition*** on environment steps. Adopting the assumption/commitment paradigm of e.g. [1, 7, 11], a program is then required to satisfy the specified property only when it interacts with an environment that respects the rely condition. UNITY-like properties with rely conditions have been first introduced in [5]; the main contributions w.r.t that previous work are emphasised below.

1. This paper presents an ***axiomatic*** semantics for assumption-commitment specifications. The behaviours of [1, 5] are abstracted away in favour of an equivalent semantics based on appropriate extensions to the predicate transformers **sp** (strongest postcondition) and **wlt** (weakest leads-to). Then, the ***strongest invariant*** of a program subject to interference is defined as the strongest fixed point of an equation involving those new predicate transformers. This new predicate plays a central role. In particular, the relation between strongest invariants of concurrent programs lies at the heart of the composition rules.
2. This paper presents a complete proof system for UNITY ***programs.*** This contrasts with the focus on the logic in [5]. Essentially, specifications are ***conjunctions*** of requirements whereas programs are ***disjunctions*** of actions.

Because the computational model of UNITY programs presented in this paper is compositional, the union and hiding rules are easily adapted from [5]. Programs can be decomposed into elementary ones and thus rules for these are also provided.

Organisation of the Paper. Section 2 contains a compositional definition of UNITY programs together with a computational model for open systems. In Sect. 3, we introduce rely conditions, predicate transformers, and operational and axiomatic semantics of UNITY specifications. Section 4 presents compositional proof rules, illustrates them with an example, and discusses completeness issues. We draw some conclusions in Sect. 5.

Notations. A state predicate over V is a predicate whose free variables are included in V; a binary predicate over V is a predicate whose free variables are included in $V \cup V'$ where V' is the 'primed' version of V. Binary predicates specify relations, with primed variables referring to the next state. The binary predicate $\mathrm{cst}(V)$ stands for $\mathbf{v}' = \mathbf{v}$ where $\mathbf{v}$ is the vector of the variables in V. States map program variables to values; $s \in [P]$ indicates that predicate P holds at state s; $(r, s) \in [R]$ indicates that R holds for the pair of states (r, s), i.e. R evaluates to *true* if v and v' are replaced with $r(v)$ and $s(v)$ respectively. State predicates in **unless** and **leadsto** properties may also refer to logical (freeze) variables from a set LV that we assume to be disjoint from the set of all programming variables. Assuming ξ maps logical variables to values, $s \in [P]_\xi$ indicates that P holds at state s under the valuation ξ; in the sequel, ξ is always within the scope of a universal quantification.

2 Programs

2.1 Compositional Definition of Programs

Since we focus on the specification and development of open systems, we extend the definition of [4] by making a distinction between *local* and *shared* variables. A program F thus consists of:

- Two disjoint finite sets $F.S$ of shared variables and $F.L$ of local variables. The set of all variables is $F.V = F.S \cup F.L$.
- A state predicate $F.I$ over $F.V$ that gives initial conditions. We assume that $F.I$ is not equivalent to *false*.
- A finite set $F.A$ of deterministic assignment statements over the variables in $F.V$.

Such programs can be defined in a compositional way, from elementary programs, by union and hiding. An elementary program has a unique statement and does not have local variables. The union (parallel composition) of F and G is denoted by $F \| G$; the assign section of $F \| G$ is obtained by appending the assign sections of F and G; local variables are renamed first if necessary. Hiding transforms a shared variable into a local one; program $F \lfloor v$ is obtained from program F by making v local.

Table 1. Example Programs

Program F	Program G
Shared x, y : *integer*	**Shared** x : *integer*
Initially $x = 1 \wedge y = 1$	**Initially** $x = 1$
Assign	**Assign**
$x, y := 2x - y + 1, y + 1$	$x := 2x$
Program $F \| G$	**Program** $H = (F \| G)\lfloor y$
Shared x, y : *integer*	**Shared** x : *integer*
Initially $x = 1 \wedge y = 1$	**Local** y : *integer*
Assign	**Initially** $x = 1 \wedge y = 1$
$x, y := 2x - y + 1, y + 1$	**Assign**
$\| \; x := 2x$	$x, y := 2x - y + 1, y + 1$
	$\| \; x := 2x$

Definition 1. Program F is elementary iff $F.L = \emptyset$ and $F.A$ consists of a unique assignment statement over the variables in $F.S$. Let F, G be such that no local variable of one program is a variable of the other and such that $F.I \wedge G.I$ is not equivalent to *false*; let $v \in F.S$. Then, the programs $F \| G$ and $F\lfloor v$ are defined by:

$$\begin{array}{ll} (F\|G).S = F.S \cup G.S & (F\lfloor v).S = F.S \setminus \{v\} \\ (F\|G).L = F.L \cup G.L & (F\lfloor v).L = F.L \cup \{v\} \\ (F\|G).I = F.I \wedge G.I & (F\lfloor v).I = F.I \\ (F\|G).A = F.A \cup G.A & (F\lfloor v).A = F.A \end{array}$$

Examples are given in Table 1: programs F and G are elementary and program H is obtained from the union of F and G by hiding the variable y. The combination of union and hiding is especially useful. Indeed, when program components have been combined to run concurrently, their mutual interactions (through shared variables) are usually regarded as internal to the composite program; they are intended to occur without the knowledge or intervention of the overall environment. An example is given in [5] where the interaction between arbiters is invisible to the controlled processes. Another typical example is the composition of two concurrent queues of size N into a single one of size $2N+1$ by concealment of the intermediate variable. How hiding is handled in proofs will be made clear in Sect. 4.

2.2 Computations

Computations of UNITY programs start in some initial state and proceed by repeatedly executing one of the assignment statements ad infinitum. The choice of the next statement to be executed is ***nondeterministic*** but each statement must be selected infinitely often. This ***fairness*** assumption is the main difference with conventional guarded command programs; all progress properties of UNITY programs are consequences of it.

Such computations correspond to the execution of a program in *isolation*. When viewing a UNITY program as an *open system*, the interference of the environment with its execution must be taken into account. A so-called *potential computation* of a program with shared variables is a computation of the program interleaved with arbitrary environment steps; these may change the values of all variables except for the local ones. Omitting a precise definition, we let $[\![F]\!]$ be the set of potential computations of F; program steps are labeled with **p** and environment steps are labeled with **e**. For instance, a computation of the program H in Table 1 might start as follows:

$$(x, y \mapsto 1, 1) \xrightarrow{\mathbf{p}} (x, y \mapsto 2, 2) \xrightarrow{\mathbf{e}} (x, y \mapsto 0, 2) \xrightarrow{\mathbf{p}} (x, y \mapsto -1, 3) \xrightarrow{\mathbf{p}} (x, y \mapsto -2, 3)$$

Potential computations define a compositional model of programs: $[\![F [\!] G]\!]$ can be constructed from $[\![F]\!]$ and $[\![G]\!]$, and $[\![F \lfloor v]\!]$ can be constructed from $[\![F]\!]$; similar compositional models of programs can be found in e.g. [3, 17].

3 Specifications with Interference Predicates

We first present a variant of the UNITY logic where specifications explicitly record facts about interference. Specifications are then given both an operational and an axiomatic semantics. The latter is based on fixed points and predicate transformers. This axiomatic semantics eases the comparison with previous work on the UNITY logic and shows how definitions of predicate transformers can be extended to cope with interference in concurrency. We finally illustrate how previous variants of the UNITY logic can be retrieved from ours as special cases.

3.1 Specifications

To illustrate the use of interference predicates in specifications, consider the elementary program F that starts with $x = 0$ and repeats the statement $x := x + 1$. Next, suppose $x = 3$ at some state in a potential computation of F. Then, we cannot conclude that $x \geq 10$ holds eventually because x can be decreased by the environment between two steps of F. But, *under the assumption* that x is not decreased by the environment, we can conclude that $x \geq 10$ holds eventually. This assumption is represented by a binary predicate $x' \geq x$, called the *rely condition*. In this case, we write:

$$F \;\; \text{SAT} \;\; x = 3 \;\; \textbf{leadsto} \;\; x \geq 10 \;\; \textbf{wrt} \;\; x' \geq x$$

UNITY logic includes the operator **unless** for specifying next-state properties of a program. For instance, $x = n$ **unless** $x > n$ asserts that x is not decreased by the program. Equivalently, we say that the binary predicate $x' \geq x$ holds for all program steps and write **guarantee** $x' \geq x$. Using binary predicates such as $x' \geq x$ for specifying next-state properties of a program yields simpler composition rules, because the rely condition is a binary predicate, too. No expressive power is lost: all **unless** properties can be encoded as **guarantee** properties (and conversely for reflexive binary predicates).

The syntax of specifications with interference predicates is summarised in Definition 2. We require specifications to be over shared variables only. Indeed, in a compositional approach, the specification of a program should not refer to its internal features.

Definition 2. Let I be a state predicate over $F.S$, P and Q be state predicates over $F.S \cup LV$, R and T be binary predicates over $F.S$. Then, the following formulas are specifications of F:

$$\begin{array}{ll} \textbf{initially } I & \textbf{guarantee } T \textbf{ wrt } R \\ \textbf{invariant } I \textbf{ wrt } R & P \textbf{ leadsto } Q \textbf{ wrt } R \end{array}$$

Preconditions in Specifications. In this paper, we take the view that a component is responsible for the initial value of its shared variables. Alternatively, initial conditions could be dropped from the definition of programs, and the environment be considered as responsible for them. Which view is better is debatable. We choose the former view because the UNITY programs of [4] include initial conditions and the comparison with other variants of the logic is thus easier. Should the latter view be adopted, specifications would include a precondition J in addition to the rely condition R [6]; the operational and axiomatic semantics presented below can be adapted accordingly.

3.2 Operational Semantics

Recall that a potential computation is a sequence of states whose p-labeled and e-labeled transitions indicate program steps and environment steps respectively; the only constraint on the latter is that local variables are kept unchanged. The rely condition R further restricts potential computations by retaining only those where R holds for all e-labeled transitions.

Definition 3. Let σ be the computation $s_0 \xrightarrow{l_1} s_1 \xrightarrow{l_2} s_2 \xrightarrow{l_3} \ldots$. Then:

$$\begin{aligned} F \text{ SAT } \textbf{initially } I &\equiv \langle \forall \sigma : \sigma \in [\![F]\!] : s_0 \in [\![I]\!] \rangle \\ F \text{ SAT } \textbf{invariant } I \textbf{ wrt } R &\equiv \langle \forall \sigma : \sigma \in [\![F]\!]^R : \langle \forall k : k \geq 0 : s_k \in [\![I]\!] \rangle\rangle \\ F \text{ SAT } \textbf{guarantee } T \textbf{ wrt } R &\equiv \langle \forall \sigma : \sigma \in [\![F]\!]^R : \langle \forall k : k > 0 : \\ &\qquad l_k = \mathrm{p} \Rightarrow (s_{k-1}, s_k) \in [\![T]\!] \rangle\rangle \\ F \text{ SAT } P \textbf{ leadsto } Q \textbf{ wrt } R &\equiv \langle \forall \sigma : \sigma \in [\![F]\!]^R : \langle \forall \xi :: \langle \forall k : k \geq 0 : \\ &\qquad s_k \in [\![P]\!]_\xi \Rightarrow \langle \exists j : j \geq k : s_j \in [\![Q]\!]_\xi \rangle\rangle\rangle\rangle \end{aligned}$$

where $\sigma \in [\![F]\!]^R \equiv \sigma \in [\![F]\!] \wedge \langle \forall k : k > 0 : l_k = \mathrm{e} \Rightarrow (s_{k-1}, s_k) \in [\![R]\!] \rangle$

3.3 Predicate Transformers

Strongest Postcondition. Let S be a deterministic statement. Following [15], we first lift the conventional predicate transformer $\mathbf{sp}.S$ to $\mathbf{sp}.F$. The predicate $\mathbf{sp}.F.P$ is the strongest predicate that holds after the execution of *any* statement of F from a P state. In operational terms, $s \in [\mathbf{sp}.F.P]$ if and only if there exists $r \in [P]$ such that state r is transformed into state s by executing a statement of F.

Definition 4. $\mathbf{sp}.F.P = \langle \vee S : S \in F.A : \mathbf{sp}.S.P \rangle$

We next define the transformer $\mathbf{sp}.R$ in case R is a binary predicate; $\mathbf{sp}.R.P$ is the strongest predicate that holds after any transition that satisfies R and starts in a P state. In operational terms, $s \in [\mathbf{sp}.R.P]$ if and only if there exists $r \in [P]$ such that $(r, s) \in [R]$. For instance, $[\,\mathbf{sp}.(x' \geq x + y).(x \geq 3 \wedge y \geq 1) \equiv x \geq 4\,]$.

Definition 5. Let R be a binary predicate over V, P and Q be state predicates over V, $\mathbf{v}$ be the vector of the variables in V, $\mathbf{v}'$ be the corresponding vector of primed variables, and $\mathbf{w}$ be a vector of fresh variables. Then:

$$\mathbf{sp}.R.P = \langle \exists \mathbf{w} :: (\mathbf{v}, \mathbf{v}' := \mathbf{w}, \mathbf{v}).(P \wedge R) \rangle$$

where $(\mathbf{v}, \mathbf{v}' := \mathbf{w}, \mathbf{v}).(P \wedge R)$ denotes the substitution of $\mathbf{w}$ and $\mathbf{v}$ for respectively $\mathbf{v}$ and $\mathbf{v}'$ in the predicate $P \wedge R$.

Strongest Invariant. Sanders [15] reconciles the axiomatic semantics of UNITY properties with the operational interpretation sketched in [4]. Her semantics is based on a predicate $\mathbf{Sin}_{\mathbf{Sa}}.F$, the strongest invariant of F, that characterises the set of reachable states. It is the strongest solution X of $[\,X \equiv F.I \vee \mathbf{sp}.F.X\,]$. Indeed, when the program is executed in isolation, a state is reachable if and only if it is either an initial state or can be reached from another reachable state by the execution of some statement. When the program is executed as an open system, more states become reachable because of environment steps. In case the latter are constrained by R, this is captured by $[\,\mathbf{sp}.(R \wedge \mathrm{cst}(F.L)).X \Rightarrow X\,]$. Definition 6 below of $\mathbf{Sin}.R.F$ corresponds to Lamport's definition of $sin(\Delta, Q)$ [10]: it is the strongest invariant of a set Δ of actions (state transitions) that is weaker than Q. In this case, Q is $F.I$ and Δ is the set of all transitions of F plus all transitions of the environment that respect R and do not modify the local variables of F. In operational terms, $s \in [\mathbf{Sin}.R.F]$ if and only if there exists a computation $\sigma \in [F]^R$ (Definition 3) such that s occurs in σ.

Definition 6. $\mathbf{Sin}.R.F$ is the strongest solution X of

$$[\,X \equiv F.I \vee \mathbf{sp}.F.X \vee \mathbf{sp}.(R \wedge \mathrm{cst}(F.L)).X\,]$$

Weakest leads-to. To cope with interference, we generalise Jutla, Knapp, and Rao's predicate transformer **wlt**.F [8] to **wlt**.$R.F$. The predicate **wlt**.$R.F.Q$ is the weakest predicate leading to Q by a fair execution of the statements of F under the assumption R on environment steps. It is defined from **via**.$R.F.S.Q$ which is the weakest predicate leading to Q via statement S of F under the same assumption R.

Definition 7. **wlt**.$R.F.Q$ is the strongest solution X of

$$[Q \Rightarrow X] \wedge \langle \wedge S : S \in F.A : [\mathbf{via}.R.F.S.X \Rightarrow X] \rangle$$

where **via**.$R.F.S.Q$ is the weakest solution X of

$$[\mathbf{sp}.S.(X \wedge \neg Q) \Rightarrow Q] \wedge [\mathbf{sp}.F.(X \wedge \neg Q) \Rightarrow X] \wedge [\mathbf{sp}.R.(X \wedge \neg Q) \Rightarrow X]$$

Binary Predicates for Statements. The transitions of a program F can be characterised by a binary predicate over $F.V$. In the simplest case where F is the elementary program with unique statement $\mathbf{v} := f(\mathbf{v})$, this binary predicate is $\mathbf{v}' = f(\mathbf{v})$. More generally, **Rel**.F is obtained by priming the variables on the left-hand side of assignments, replacing ':=' with '=', and taking the disjunction over all statements. However, when programs are composed by union, one must take care of the variables of one program that are not modified by the other.

Definition 8. By induction on the structure of a program, assuming $\mathbf{v}$ is the vector of all the variables of an elementary program:

$$\begin{aligned} \mathbf{Rel}.(\mathbf{v} := f(\mathbf{v})) &\equiv \mathbf{v}' = f(\mathbf{v}) \\ \mathbf{Rel}.(F\lfloor v) &\equiv \mathbf{Rel}.F \\ \mathbf{Rel}.(F \| G) &\equiv (\mathbf{Rel}.F \wedge \mathbf{cst}(G.V \setminus F.V)) \vee (\mathbf{Rel}.G \wedge \mathbf{cst}(F.V \setminus G.V)) \end{aligned}$$

3.4 Axiomatic Semantics

Based on predicate transformers, the axiomatic semantics of properties with interference predicates is given in Proposition 9; a proof can be found in [6].

Proposition 9.

$$\begin{aligned} F \text{ SAT } \mathbf{initially}\ I &\equiv [F.I \Rightarrow I] \\ F \text{ SAT } \mathbf{invariant}\ I\ \mathbf{wrt}\ R &\equiv [\mathbf{Sin}.R.F \Rightarrow I] \\ F \text{ SAT } \mathbf{guarantee}\ T\ \mathbf{wrt}\ R &\equiv [\mathbf{Sin}.R.F \wedge \mathbf{Rel}.F \Rightarrow T] \\ F \text{ SAT } P\ \mathbf{leadsto}\ Q\ \mathbf{wrt}\ R &\equiv [\mathbf{Sin}.R.F \wedge P \Rightarrow \mathbf{wlt}.R.F.Q] \end{aligned}$$

The semantics of **invariant** properties follows from the observation that **Sin**.$R.F$ characterises reachable states when F is executed under the assumptions R. Next, compare [**Rel**.$F \Rightarrow T$] with the operational semantics of **guarantee** properties in Definition 3. The former means that the binary predicate T holds for steps of F. But this is too strong because the definition of **Rel**.F only considers the statements in $F.A$ without discarding unreachable states. To take them into account, the left-hand side of the implication is strengthened with **Sin**.$R.F$.

Finally, compare $[P \Rightarrow \textbf{wlt}.R.F.Q]$ with the operational semantics of **leadsto** properties in Definition 3. The former asserts that any fair execution of the statements of F from a P state eventually reaches a Q state. The predicate $\textbf{Sin}.R.F$ has been introduced to discard unreachable P states.

3.5 Comparison with Other Variants of the UNITY Logic

Translating **unless** into **guarantee** properties, the axiomatic semantics of [4, 15] can be given as:

$$F \text{ SAT}_{\text{CM}} \textbf{ guarantee } T \equiv [\textbf{Rel}.F \Rightarrow T]$$
$$F \text{ SAT}_{\text{Sa}} \textbf{ guarantee } T \equiv [\textbf{Sin}_{\text{Sa}}.F \wedge \textbf{Rel}.F \Rightarrow T]$$

where $\textbf{Sin}_{\text{Sa}}.F$ is the strongest solution X of $[X \equiv F.I \vee \textbf{sp}.F.X]$. This difference between [4] and [15] corresponds to a choice between $R = \textit{true}$ (arbitrary interference) and $R = \textit{false}$ (no interference). Indeed, in case programs do not have local variables, $[\textbf{Sin}.\textit{true}.F \equiv \textit{true}]$ and $[\textbf{Sin}.\textit{false}.F \equiv \textbf{Sin}_{\text{Sa}}.F]$. Therefore:

$$F \text{ SAT}_{\text{CM}} \textbf{ guarantee } T \equiv F \text{ SAT } \textbf{guarantee } T \textbf{ wrt } \textit{true}$$
$$F \text{ SAT}_{\text{Sa}} \textbf{ guarantee } T \equiv F \text{ SAT } \textbf{guarantee } T \textbf{ wrt } \textit{false}$$

In their work on the refinement of UNITY programs with local variables [18],Udink, Herman and Kok introduce properties subscripted by a set V of variables. In that approach, the axiomatic semantics of **guarantee** properties would be:

$$F \text{ SAT}_{\text{UHK}} \textbf{ guarantee}_V\ T \equiv [\textbf{Sin}_{\text{UHK}}.V.F \wedge \textbf{Rel}.F \Rightarrow T]$$

where $\textbf{Sin}_{\text{UHK}}.V.F$ is the strongest predicate that holds initially, is preserved by all statements of F and depends on the variable in V only; it characterises reachable states when variables in V are not modified by the environment. $\textbf{Sin}_{\text{UHK}}.V.F$ is the strongest solution X of $[X \equiv F.I \vee \textbf{sp}.F.X \vee \textbf{sp}.(\text{cst}(V) \wedge \text{cst}(F.L)).X]$. Clearly, $[\textbf{Sin}_{\text{UHK}}.V.F \equiv \textbf{Sin}.\text{cst}(V).F]$ and thus:

$$F \text{ SAT}_{\text{UHK}} \textbf{ guarantee}_V\ T \equiv F \text{ SAT } \textbf{guarantee } T \textbf{ wrt } \text{cst}(V)$$

4 Compositional Proof System

A compositional proof system is syntax-directed. Therefore, it comprises rules for elementary programs, hiding rules, and union rules. In addition, it includes adaptation rules such as weakening rules or invariant-dependent rules. Only the most typical rules are presented; a complete exposition with soundness proofs can be found in [6].

4.1 Syntax-directed Rules

Elementary programs. The rules for proving properties of elementary programs highlight the role of interference. Suppose that the unique statement of a program transforms $P \wedge \neg Q$ states into Q states. Since this statement is executed infinitely often, Q states are eventually reached from P states *provided that* environment steps transform $P \wedge \neg Q$ states into $P \vee Q$ states; this is captured by the second premise of Rule (1). In Rule (2), the third premise explicitly deals with environment steps.

Proposition 10. *Let S be the unique statement of the elementary program F.*

$$\frac{[\mathbf{sp}.S.(P \wedge \neg Q) \Rightarrow Q]\ [\mathbf{sp}.R.(P \wedge \neg Q) \Rightarrow P \vee Q]}{F \ \textsc{sat}\ P \ \mathbf{leadsto}\ Q \ \mathbf{wrt}\ R} \tag{1}$$

$$\frac{[F.I \Rightarrow I]\ [\mathbf{sp}.S.I \Rightarrow I]\ [\mathbf{sp}.R.I \Rightarrow I]}{F \ \textsc{sat}\ \mathbf{invariant}\ I \ \mathbf{wrt}\ R} \tag{2}$$

$$\frac{[\mathbf{Rel}.F \Rightarrow T]}{F \ \textsc{sat}\ \mathbf{guarantee}\ T \ \mathbf{wrt}\ R} \tag{3}$$

Hiding. Coping with hiding is straightforward: a concealed variable v amounts to a shared variable that is not subject to interference ($v' = v$). Consider the typical composition $H = (F \| G)\lfloor \mathbf{v}$ (hiding can be easily extended to a vector of variables). Then, properties of H can be proved as properties of $F \| G$ simply by adding $\mathbf{v}' = \mathbf{v}$ to the rely condition.

Proposition 11. *Let prop* $\in$ {**invariant** I, **guarantee** T, P **leadsto** Q}.

$$\frac{F \ \textsc{sat}\ prop \ \mathbf{wrt}\ R \wedge v = v'}{F\lfloor v \ \textsc{sat}\ prop \ \mathbf{wrt}\ R} \tag{4}$$

Union. When programs are composed by union, each program is part of the environment of the other. Therefore, all assumptions made by one program about its environment must be satisfied by the other. Discharging mutual assumptions is at the core of all composition rules for assumption-commitment specifications.

Definition 12. The programs F_1 and F_2 cooperate w.r.t. the predicates R_1 and R_2 if and only if there exists binary predicates T_1 and T_2 such that:

- $[T_1 \wedge \mathsf{cst}(F_2.S \setminus F_1.S) \Rightarrow R_2]$
- $[T_2 \wedge \mathsf{cst}(F_1.S \setminus F_2.S) \Rightarrow R_1]$
- F_1 SAT **guarantee** T_1 **wrt** R_1
- F_2 SAT **guarantee** T_2 **wrt** R_2

This cooperation condition lists premises that are common to all the union rules. To some extent, it generalises Owicki-Gries' test for interference-freedom [13]: no step of F_1 invalidates the assumption made by F_2, and vice versa. However, Owicki-Gries' test requires the *full text* of programs F_1 and F_2 to be available. In contrast, our cooperation condition is validated from the *specifications* of F_1 and F_2 *only*; there are no references to the statements of the programs nor to their local variables. Compositionality is thus preserved and design decisions can be justified before the development process goes on.

Rule (5) below reads as follows: invariants are preserved by parallel composition. Next, Rule (6) asserts that progress properties of a single component can be lifted to progress properties of the composite program. Finally, Rule (7) captures the interleaving between steps of programs composed by union. The conjunction $R_1 \wedge R_2$ appears in the conclusion of all the union rules: the overall environment of $F_1 \| F_2$ is part of the environment of F_1 *and* part of the environment of F_2.

Proposition 13. *If F_1 and F_2 cooperate w.r.t. the predicates R_1 and R_2, then*

$$\frac{\begin{array}{ll} F_1 \text{ SAT } \textbf{invariant } I_1 & \textbf{wrt } R_1 \\ F_2 \text{ SAT } \textbf{invariant } I_2 & \textbf{wrt } R_2 \end{array}}{F_1 \| F_2 \text{ SAT } \textbf{invariant } I_1 \wedge I_2 \ \ \textbf{wrt } R_1 \wedge R_2} \tag{5}$$

$$\frac{F_i \text{ SAT } P \textbf{ leadsto } Q \ \ \textbf{wrt } R_i}{F_1 \| F_2 \text{ SAT } P \textbf{ leadsto } Q \ \ \textbf{wrt } R_1 \wedge R_2} \tag{6}$$

$$\frac{\begin{array}{ll} F_1 \text{ SAT } \textbf{guarantee } T_1 & \textbf{wrt } R_1 \\ F_2 \text{ SAT } \textbf{guarantee } T_2 & \textbf{wrt } R_2 \end{array}}{F_1 \| F_2 \text{ SAT } \textbf{guarantee } T \ \ \textbf{wrt } R_1 \wedge R_2} \tag{7}$$

where $i \in \{1,2\}$ *and* $T = (T_1 \wedge \mathsf{cst}(F_2.S \setminus F_1.S)) \vee (T_2 \wedge \mathsf{cst}(F_1.S \setminus F_2.S))$.

Computations are not even mentioned in the soundness proof of Proposition 13; the proof is entirely carried out by fixed point reasoning and manipulation of predicate transformers. The key step is to establish a relation between strongest invariants of concurrent components. In fact, if the cooperation condition holds, one may conclude:

$$[\,\mathbf{Sin}.(R_1 \wedge R_2).(F_1 \| F_2) \Rightarrow \mathbf{Sin}.R_1.F_1 \wedge \mathbf{Sin}.R_2.F_2\,]$$

4.2 Invariant-dependent Rules

The substitution axiom of UNITY asserts that invariants can be replaced with *true* and conversely. As illustrated below, this eases the proof of other properties.

Proposition 14.

$$\frac{\begin{array}{ll} F \text{ SAT } \textbf{invariant } I & \textbf{wrt } R \\ F \text{ SAT } \textbf{guarantee } I \Rightarrow T & \textbf{wrt } R \end{array}}{F \text{ SAT } \textbf{guarantee } T \ \ \textbf{wrt } R} \tag{8}$$

$$\frac{\begin{array}{lll} F \text{ SAT} & \textbf{invariant } I & \textbf{wrt } R \\ F \text{ SAT } P \wedge I & \textbf{leadsto } Q & \textbf{wrt } R \end{array}}{\begin{array}{lll} F \text{ SAT} & P \textbf{ leadsto } Q & \textbf{wrt } R \end{array}} \tag{9}$$

4.3 A Small Example

Compositional proof systems are best used in support for program *development*. In particular, specifications must be decomposed so that the cooperation condition holds; an example in a UNITY-like formalism is given in [5]. In this paper, we only illustrate the rules for the compositional *verification* of

$$\begin{array}{ll} \vdash H \text{ SAT } \textbf{guarantee } x' \geq x & \textbf{wrt } x' \geq x \\ \vdash H \text{ SAT } x = k \textbf{ leadsto } x > k & \textbf{wrt } x' \geq x \end{array}$$

where H is the toy program in Table. 1 (Sect. 2). From the definition of the predicate transformers:

$$\begin{array}{l} [\textbf{Rel}.F \equiv x' = 2x - y + 1 \wedge y' = y + 1] \\ [\textbf{Rel}.G \equiv x' = 2x] \\ [\textbf{sp}.(x, y := 2x - y + 1, y + 1).(x \geq y) \equiv even(x - y) \wedge x \geq y] \\ [\textbf{sp}.(x' \geq x \wedge y' = y).(x \geq y) \equiv x \geq y] \\ [\textbf{sp}.(x := 2x).(x > 0) \equiv x > 0 \wedge even(x)] \\ [\textbf{sp}.(x' \geq x).(x > 0) \equiv x > 0] \\ [\textbf{sp}.(x := 2x).(x = k \wedge x > 0) \equiv x = 2k \wedge x > 0] \\ [\textbf{sp}.(x' \geq x).(x = k \wedge x > 0) \equiv x \geq k \wedge x > 0] \end{array}$$

Then, by Rules (1), (2), and (3):

$$\begin{array}{ll} \vdash F \text{ SAT } \textbf{invariant } x \geq y & \textbf{wrt } x' \geq x \wedge y' = y \\ \vdash G \text{ SAT } \textbf{invariant } x > 0 & \textbf{wrt } x' \geq x \\ \vdash F \text{ SAT } \textbf{guarantee } x \geq y \Rightarrow x' \geq x & \textbf{wrt } x' \geq x \wedge y' = y \\ \vdash G \text{ SAT } \textbf{guarantee } x > 0 \Rightarrow x' \geq x & \textbf{wrt } x' \geq x \\ \vdash G \text{ SAT } x = k \wedge x > 0 \textbf{ leadsto } x > k & \textbf{wrt } x' \geq x \end{array}$$

Next, by Rules (8) and (9):

$$\begin{array}{ll} \vdash F \text{ SAT } \textbf{guarantee } x' \geq x & \textbf{wrt } x' \geq x \wedge y' = y \\ \vdash G \text{ SAT } \textbf{guarantee } x' \geq x & \textbf{wrt } x' \geq x \\ \vdash G \text{ SAT } x = k \textbf{ leadsto } x > k & \textbf{wrt } x' \geq x \end{array}$$

By Rules (6), (7), and $(F.S \setminus G.S) = \{y\}$:

$$\begin{array}{ll} \vdash F[\!]G \text{ SAT } \textbf{guarantee } x' \geq x & \textbf{wrt } x' \geq x \wedge y' = y \\ \vdash F[\!]G \text{ SAT } x = k \textbf{ leadsto } x > k & \textbf{wrt } x' \geq x \wedge y' = y \end{array}$$

Finally, by Rule (4) and $H = (F[\!]G)\lfloor y$:

$$\begin{array}{ll} \vdash H \text{ SAT } \textbf{guarantee } x' \geq x & \textbf{wrt } x' \geq x \\ \vdash H \text{ SAT } x = k \textbf{ leadsto } x > k & \textbf{wrt } x' \geq x \end{array}$$

4.4 Completeness

Without extensions, the completeness of the proof system is restricted to programs without local variables. In the general case [6], completeness has been established for specifications with auxiliary variables. These are interpreted as in [17]: a program F is said to satisfy a specification w.r.t. a set X of auxiliary variables if and only if its statements can be augmented with assignments to X so that the augmented program satisfies the specification. Let $X \cdot spec$ indicates that *spec* (conjunction of properties) is w.r.t a set X of auxiliary variables. Then,

$$F \ \text{SAT} \ \ X \cdot spec \equiv \langle \exists G : G \rhd_X F : G \ \text{SAT} \ \ spec \rangle$$

where $G \rhd_X F$ means that program G is an augmentation of program F with assignments to the variables in X (in particular, $G.S = F.S \cup X$) and G SAT *spec* is given by Definition 3. For instance, let

$$\textbf{guarantee} \ \langle out' \rangle \circ q' \circ \langle in' \rangle = \langle out \rangle \circ q \circ \langle in \rangle \ \ \textbf{wrt} \ q = q'$$

be a safety specification of a concurrent queue program with auxiliary variable q; the symbol $\circ$ denotes concatenation. This specification reads: there exists assignments to a variable q, which is no modified by the environment, so that $\langle out \rangle \circ q \circ \langle in \rangle$ is not modified by any program step. Obviously, q models the current contents of the queue. This specification is satisfied by program F with local variables seq_1, seq_2, mid and statements:

$$\begin{array}{lll}
 & in, seq_1 := \bot, \ seq_1 \circ \langle in \rangle & \text{if } in \neq \bot \\
[] & mid, seq_1 := hd(seq_1), \ tail(seq_1) & \text{if } mid = \bot \wedge seq_1 \neq \bot \\
[] & mid, seq_2 := \bot, \ seq_2 \circ \langle mid \rangle & \text{if } mid \neq \bot \\
[] & out, seq_2 := hd(seq_2), \ tail(seq_2) & \text{if } out = \bot \wedge seq_2 \neq \bot
\end{array}$$

where $\bot$ denotes the empty value. Intuitively, q can be chosen as $seq_2 \circ \langle mid \rangle \circ seq_1$. The statements of the augmented program are:

$$\begin{array}{lll}
 & in, seq_1, q := \bot, \ seq_1 \circ \langle in \rangle, \ q \circ \langle in \rangle & \text{if } \ in \neq \bot \\
[] & mid, seq_1 := hd(seq_1), \ tail(seq_1) & \text{if } \ mid = \bot \wedge seq_1 \neq \bot \\
[] & mid, seq_2 := \bot, \ seq_2 \circ \langle mid \rangle & \text{if } \ mid \neq \bot \\
[] & out, seq_2, q := hd(seq_2), \ tail(seq_2), tail(q) & \text{if } \ out = \bot \wedge seq_2 \neq \bot
\end{array}$$

5 Conclusion

With respect to [4, 15], we have proposed a proof system that is both compositional and complete, that includes composition rules for **leadsto**, and that copes with hiding. As highlighted in Sect. 3.5, the key idea is to let the rely condition vary between the extremes *true* and *false*. The specifications of [18] use rely conditions of the form cst(V) which are sufficient to define hiding. This establishes a clear connection with this work, but [18] addresses a different problem. Indeed, the major concern of [18] is compositional *program* refinement whereas our proof system is intended to support the compositional verification/development of programs against/from temporal specifications.

Although the proof system is compositional and thus in principle eases the decomposition of a task into subtasks, there is no evidence yet that UNITY specifications with rely conditions can be kept simple in the development of sophisticated systems. In particular, a bad choice of the interface between components will probably result in complicate rely conditions. Moreover, this extension of the UNITY logic does not eliminate one of its drawbacks in large specifications of a component: a specification is a conjunction of properties. With increase in size comes the danger that one property affects another one in an unexpected way, especially when **leadsto** is involved. If a component is only split when the separate development of its subcomponents is easier than the development of the component as a whole (e.g. subcomponents do not depend too much on each other) and if individual specifications remain relatively small, we think that the logic presented in this paper can prove useful in practice.

From a theoretical point of view, this paper on UNITY shows that the assumption-commitment paradigm spreads across formalisms. We indeed inherit from previous work on the subject. Via common *semantic* rules [1, 6], our rules for parallel composition can be related to rules for assumption-commitment specifications in other frameworks, e.g. the ones proposed in [2, 11, 14, 17, 19].

References

1. M. Abadi and L. Lamport, Composing specifications, *ACM Transactions on Programming Languages and Systems*, 15:73-132, 1993.
2. M. Abadi and L. Lamport, Decomposing specifications of concurrent systems, in E.R. Olderog, ed., *Proc. IFIP Conference on Programming Concepts, Methods and Calculi*, 1994, pp. 323-336.
3. H. Barringer, R. Kuiper, and A. Pnueli, Now you may compose temporal logic specifications, in *Proc. 16th ACM Symposium on Theory of Computing*, 1984, pp. 51-63.
4. K.M. Chandy and J. Misra, *Parallel Program Design: a Foundation*, Addison-Wesley, 1988.
5. P. Collette, Composition of assumption-commitment specifications in a UNITY style, *Science of Computer Programming*, 23:107-125, 1994.
6. P. Collette, *Design of Compositional Proof Systems Based on Assumption-Commitment Specifications - Application to* UNITY, Ph.D. Thesis, 1994, Université Catholique de Louvain.
7. C.B. Jones, *Development Methods for Computer Programs Including a Notion of Interference*, Ph.D. Thesis, 1981, Oxford University.
8. C.S. Jutla, E. Knapp, and J.R. Rao, A predicate transformer approach to the semantics of parallel programs, *Proc. 8th ACM Symposium on Principles of Distributed Computing*, 1989, pp. 249-263.
9. E. Knapp, Derivation of concurrent programs: two examples, *Science of Computer Programming*, 19:1-23, 1992.
10. L. Lamport, *win* and *sin*: predicate transformers for concurrency, *ACM Transactions on Programming Languages and Systems*, 1990, 12:396-428, 1990.

11. J. Misra and K.M. Chandy, Proofs of networks of processes, *IEEE Transactions on Software Engineering*, 7:417-426, 1981.
12. A. Mokkedem and D. Méry, On using temporal logic for refinement and compositional verification of concurrent systems, *Theoretical Computer Science*, 140: 95-138,1995.
13. S. Owicki and D. Gries, An axiomatic proof technique for parallel programs, *Acta Informatica*, 6:319-340, 1976.
14. P.K. Pandya and M. Joseph, P–A logic - a compositional proof system for distributed programs, *Distributed Computing*, 5:37-54, 1991.
15. B. Sanders, Eliminating the substitution axiom from UNITY logic, *Formal Aspects of Computing*, 3:189-205, 1991.
16. M. Staskauskas, Formal derivation of concurrent programs: an example from industry, *IEEE Transactions on Software Engineering*, 19:503-528, 1993.
17. K. Stølen, A method for the development of totally correct shared-state parallel programs, in J.C.M. Baeten and J.F. Groote, eds., *Concurrency Theory*, Springer-Verlag, 1991, LNCS 527, pp. 510-525.
18. R.T. Udink, T. Herman, and J.N. Kok, Progress for local variables in UNITY, in E.R. Olderog, ed., *Proc. IFIP Conference on Programming Concepts, Methods and Calculi*, 1994, pp. 124-143.
19. Q. Xu and J. He, A theory of state-based parallel programming: part I, in J. Morris and R.C. Shaw, eds., *Proc. 4th Refinement Workshop*, Springer-Verlag, 1991, pp. 326-359.
20. J. Zwiers, *Compositionality, Concurrency, and Partial Correctness*, Springer-Verlag, 1989, LNCS 321.

CPO models for infinite term rewriting *

Andrea Corradini and Fabio Gadducci

Università di Pisa, Dipartimento di Informatica, Corso Italia 40, 56125 Pisa, Italy
({andrea,gadducci}@di.unipi.it)

Abstract. Infinite terms in universal algebras are a well-known topic since the seminal work of the ADJ group [1]. The recent interest in the field of *term rewriting* (TR) for infinite terms is due to the use of *term graph rewriting* to implement TR, where terms are represented by graphs: so, a cyclic graph is a finitary description of a possibly infinite term. In this paper we introduce *infinite rewriting logic*, working on the framework of *rewriting logic* proposed by José Meseguer [13, 14]. We provide a simple algebraic presentation of infinite computations, recovering the *infinite parallel term rewriting*, originally presented by one of the authors ([6]) to extend the classical, set-theoretical approach to TR with infinite terms. Moreover, we put all the formalism on firm theoretical bases, providing (for the first time, to the best of our knowledge, for infinitary rewriting systems) a clean algebraic semantics by means of (internal) 2-categories.

1 Introduction

Term rewriting systems (briefly, TRS's; see [7]) are a simple yet powerful formalism, based on the notion of *(sequence of) rewrites*: a binary, transitive relation over terms, usually generated from a finite set of rules, where each element $\langle t, s\rangle$ states the transformation from term t to term s. Despite their simplicity, TRS's can be considered a basic paradigm for computational devices: terms are states of an abstract machine, while rewriting rules are state-transforming functions: in this framework, computations simply are sequences of rewrites. Usually, however, TRS's deal with finite terms and finite computations: despite the seminal work on continuous algebras by Goguen et al. dates back to the mid-Seventies ([1]), the extension of term rewriting to infinite terms is a subject raised to a certain interest only in recent years, mainly due to the use of graphs to model rewrites. In fact, in *term graph rewriting*, a finite, cyclic graph may represent an infinite term, and a single rewriting step can be equivalent to an infinite sequence of rewrites (see e.g. [6, 8, 10]).

In this paper we introduce a new approach to infinitary term rewriting. Our starting point has been the seminal work of José Meseguer about *rewriting logic* (RL). The idea underlying RL is to take a logical viewpoint, regarding a TRS $\mathcal{R}$ as a logical theory, and any rewrite as a sequent entailed by that theory. The entailment relation is defined inductively by suitable deduction rules, showing

* Research partially supported by the Fixed Contribution Contract n. SC1*-CT92-0776 "Mathematical Structures in Semantics for Concurrency" (MASK).

how sequents can be derived from other sequents. Sequents themselves are triples $\langle \alpha, t, s \rangle$, where α is an element of an *algebra of proof terms*, encoding a justification of the rewriting of t into s. In [13, 14], the finitary case of rewriting logic (i.e., such that sequents represent *finite* sequences of rewrites over *finite* terms) is studied. Instead, in this paper we consider proof terms as elements of a continuous algebra. Since the elements of a continuous algebra form a complete partial order, fully exploiting this structure over proof terms we are able to introduce *infinitary rules*: whenever there exists a suitable chain of sequents, then we add the derivation corresponding to its supremum. Therefore, continuous algebras allow us to define a natural extension of RL we call *infinite rewriting logic*. The formalism has a nice and clean presentation and, despite its straightforwardness, it is quite powerful: indeed, it consistently includes the *infinite parallel term rewriting* previously proposed by one of the author ([6]).

In the standard approach, an operational model for TRS's is represented by their *derivation spaces*: a class of (structured) elements representing all the possible computations the system can perform, subject to a suitable equivalence. Alternatively, more abstract models for TRS's have been recently proposed, relying on the use of *2-categories* ([4, 15, 16, 17]) and *functor categories* ([13]). A functor category has functors as objects and natural transformations as arrows. A cat-enriched structure as a 2-category is given by a category such that also each hom-set forms a category: the class of arrows (called *cells*) of these hom-categories are closed under certain composition operators, and are subject to suitable *coherence* axioms. Given a TRS $\mathcal{R}$, we show how to freely generate a particular 2-category $\underline{\mathbf{2\text{-}Th}(\mathcal{R})}$, called the *Lawvere 2-theory* of $\mathcal{R}$, such that its arrows are in a one-to-one correspondence with terms, while its cells represent the derivation space of $\mathcal{R}$. Then, models for a TRS $\mathcal{R}$ are given by the functor category $[\underline{\mathbf{2\text{-}Th}(\mathcal{R})} \rightarrow \underline{\mathbf{C}}]$ of functors from $\underline{\mathbf{2\text{-}Th}(\mathcal{R})}$ to a given 2-category $\underline{\mathbf{C}}$. This notion of model has been shown ([13]) to be adequate for finitary RL (in the sense that a soundness and completeness result holds with respect to the set of finite sequents entailed by a given rewriting system) when considering the case $\underline{\mathbf{C}} = \underline{\mathbf{Cat}}$, where $\underline{\mathbf{Cat}}$ is the 2-category of categories, functors and natural transformations. In the paper we show that this method lifts smoothly to infinite RL, when considering the 2-category $\underline{\mathbf{Cat(CPO)}}$ of categories internal to **CPO** (i.e., to categories such that their components are not sets and functions, but cpo's and strict continuous functions) proving (as a main result of the paper) a soundness and completeness theorem for the infinitary case.

The paper has the following structure: in Section 2 we recall some basic notions of category theory and the CPO-structure of continuous algebras of infinite terms. In Section 3 we present *rewriting logic*, showing its tight relation with the classical theory of term rewriting, and we introduce *infinite rewriting logic*, showing the consistency result with the *infinite parallel term rewriting* proposal. Finally, in Section 4 we introduce the categorical models for our notion of infinite rewriting, showing a soundness and completeness result with regard to these models. The paper ends with a short discussion about the relevance of the proposed formalism, suggesting further directions for future works.

2 Background

We present here most of the technical notions we use along the paper. For some basic categorical definitions (as functor, product, etc.) we refer the reader to [3].

2.1 Algebras and continuous algebras

In this section we recall the definitions of the categories of algebras and of continuous algebras, introducing them in two equivalent ways: first in the traditional set-theoretical style, and then in a more categorical way using Lawvere theories.

Definition 1 (the category of Σ-algebras). Let Σ be a (one-sorted) *signature*, i.e., a ranked alphabet of operator symbols $\Sigma = \bigcup_{n\in\mathbf{N}}\Sigma_n$ (saying that f is of *arity* n for $f \in \Sigma_n$). A *Σ-algebra* is a pair $A = \langle |A|, \rho_A\rangle$ such that $|A|$ is a set (the *carrier*), and $\rho_A = \{f_A \mid f \in \Sigma\}$ is a family of functions such that for each $f \in \Sigma_n$, $f_A : |A|^n \to |A|$. Let A, B be two Σ-algebras: a *Σ-homomorphism* τ: $A \to B$ is a function τ: $|A| \to |B|$ preserving operators, i.e., such that for every $f \in \Sigma_n$, $\tau \circ f_A = f_B \circ \tau^n$. Since Σ-homomorphism are closed under (functional) composition and the identity function of the carrier is a Σ-homomorphism, Σ-algebras and Σ-homomorphisms form a category we denote by Σ-**Alg**. □

Continuous algebras were introduced in [1]: they are algebras where the carrier is not just a set, but rather a complete partial order. Correspondingly, since homomorphisms must preserve the algebraic structure, they are required to be continuous functions. The importance of these algebras in computer science has been stressed by a great deal of work during the Seventies, for example in providing denotational semantics for functional languages, semantical models for flow diagrams and, in general, for formalisms dealing with unbounded computations.

Definition 2 (complete partial orders). A partial order $\langle D, \leq\rangle$ is *complete* (is a *cpo*) if it has an element $\perp$ (called *bottom*) such that $\perp \leq d$ for all $d \in D$, has *greatest lower bounds* (glb's) for every pair of elements, and *least upper bounds* (lub's) for all ω-chains of elements. If $\{d_i\}_{i<\omega}$ is an ω-chain (i.e., $d_i \leq d_{i+1}$ for all $i < \omega$), we denote its lub by $\bigsqcup_{i<\omega}\{d_i\}$. A *continuous function* $f : \langle D, \leq_D\rangle \to \langle D', \leq_{D'}\rangle$ between cpo's is a function $f : D \to D'$ which preserves lub's of ω-chains, i.e., $f(\bigsqcup_{i<\omega})\{d_i\} = \bigsqcup_{i<\omega}\{f(d_i)\}$; it is *strict* if $f(\perp_D) = \perp_{D'}$. **CPO** denotes the category of cpo's and continuous functions. □

Definition 3 (the category of continuous Σ-algebras). A *continuous Σ-algebra* is a pair $A = \langle\langle |A|, \leq_A\rangle, \rho_A\rangle$ such that $\langle |A|, \leq_A\rangle$ is a cpo, and $\rho_A = \{f_A \mid f \in \Sigma\}$ is a family of *continuous* functions such that for every $f \in \Sigma_n$, $f_A : \langle |A|, \leq_A\rangle^n \to \langle |A|, \leq_A\rangle$. If A, B are continuous Σ-algebras, a *continuous Σ-homomorphism* τ: $A \to B$ is a continuous function τ: $\langle |A|, \leq_A\rangle \to \langle |B|, \leq_B\rangle$ which preserves the operators. A continuous Σ-homomorphism τ is called *strict* if $\tau(\perp_A) = \perp_B$. Σ-**CAlg** denotes the category having continuous Σ-algebras as objects and continuous Σ-homomorphism as arrows, and Σ-**SCAlg** its subcategory having the same objects but *strict* homomorphisms as arrows. □

2.2 Categories of algebras as functor categories

The categories of algebras just introduced can be presented in an equivalent way by using simple categorical techniques. Although such definitions are slightly more involved than the ones above, they have the advantage of separating in a better way the "Σ-structure" from the additional algebraic structure that the carrier can enjoy. We start defining the *Lawvere theory* $\mathbf{Th}(\Sigma)$ associated with a signature Σ [12]. This is a cartesian category having natural numbers as objects, generated in a free way from the operators of Σ. The relevant property (on which we will come back later) is that arrows from m to 1 are in one-to-one correspondence with terms of the free Σ-algebra with at most m variables.

Definition 4 (Lawvere theories). Given a signature Σ, the associated *Lawvere theory* is the cartesian category $\mathbf{Th}(\Sigma)$ such that

- its objects are underlined natural numbers: $\underline{0}$ is the terminal object and the product is defined as $\underline{n} \times \underline{m} = \underline{n+m}$;
- the arrows are those of a free cartesian category, such that for every operator $f \in \Sigma_n$, there is a basic arrow $f_\Sigma : \underline{n} \to \underline{1}$. □

Each arrow $t_\Sigma : \underline{n} \to \underline{1}$ identifies a Σ-term t with variables among $x_1,\ldots,x_n$; an arrow $\underline{n} \to \underline{m}$ is a m-uple of Σ-terms with n variables, and arrow composition is term substitution. The Lawvere theory can be regarded as an alternative presentation of a signature. Indeed, the additional structure it contains (besides the operators of the signature) is generated in a completely free way, so, in a sense, it does not add "information" to the original signature. The advantage of this presentation is that, using categorical techniques, we can easily define a very general notion of *model* for a Lawvere theory, which subsumes both algebras and continuous algebras. Let us recall the definition of functor category.

Definition 5 (functor category). Let $\mathbf{C}$ and $\mathbf{D}$ be two categories.[2] The *functor category* $[\mathbf{C} \to \mathbf{D}]$ is defined as follows:

- objects of $[\mathbf{C} \to \mathbf{D}]$ are functors $F : \mathbf{C} \to \mathbf{D}$;
- an arrow $\alpha : F \Rightarrow G$ between two parallel functors $F, G : \mathbf{C} \to \mathbf{D}$ is a *natural transformation*, i.e., a family of arrows of $\mathbf{D}$ indexed by the objects of $\mathbf{C}$, $\{\alpha_a : F(a) \to G(a) \mid a \in |\mathbf{C}|\}$, satisfying the *naturality* requirement, i.e., such that for each arrow $f : a \to b$ in $\mathbf{C}$, $\alpha_b \circ F(f) = G(f) \circ \alpha_a$. □

Definition 6 (models of Lawvere theories). Let $\mathbf{C}$ be a cartesian category. A $\mathbf{C}$-*model* of the Lawvere theory associated to a signature Σ is a cartesian functor $\mathcal{M}$: $\mathbf{Th}(\Sigma) \to \mathbf{C}$, while a *model morphism* is a natural transformation between models. The category $\mathbf{C}\text{-}\mathbf{Mod}_\Sigma$ of $\mathbf{C}$-models is the functor category $[\mathbf{Th}(\Sigma) \to \mathbf{C}]$: its objects are $\mathbf{C}$-models of $\mathbf{Th}(\Sigma)$, while its arrows are model morphisms. □

[2] Actually, in order for the functor category $[\mathbf{C} \to \mathbf{D}]$ to be defined, $\mathbf{C}$ and $\mathbf{D}$ should be *locally small*, i.e., the collection of morphisms between any two objects should form a set. We will not deal with foundational issues, assuming that our categories are locally small whenever this requirement is necessary.

By replacing the generic cartesian category **C** with specific categories, we obtain models which are equivalent (in a strong sense) to the various kinds of Σ-algebras introduced above. In particular, the **Set**-models of $\mathbf{Th}(\Sigma)$ turn out to be essentially the Σ-algebras, while its **CPO**-models are nothing else than the continuous Σ-algebras (indeed, it is easy to show that **CPO** is cartesian).

Theorem 7 (categories of algebras as functor categories). *For any signature Σ, the categorical equivalences Σ-***Alg** $\cong$ **Set-Mod**$_\Sigma$ *and Σ-***CAlg** $\cong$ **CPO-Mod**$_\Sigma$ *hold. Moreover, let* **S-CPO-Mod**$_\Sigma$ *be the subcategory of* **CPO-Mod**$_\Sigma$ *having the same objects and as arrows natural transformations such that all components are* strict *continuous functions. Then Σ-***SCAlg** $\cong$ **S-CPO-Mod**$_\Sigma$. □

The result for **Set**-models was proved by William Lawvere [12] in his doctoral dissertation, and is well-known: here we only sketch the underlying ideas. Given a **Set**-model $\mathcal{M}: \mathbf{Th}(\Sigma) \to \mathbf{Set}$, consider the pair $A_\mathcal{M} = \langle \mathcal{M}(\underline{1}), \rho = \{\mathcal{M}(f_\Sigma) \mid f \in \Sigma\}\rangle$. It is easy to check that $A_\mathcal{M}$ is a Σ-algebra: indeed, $\mathcal{M}(\underline{1})$ is a set, and for any $f \in \Sigma_n$, since $f_\Sigma : \underline{n} \to \underline{1}$ in $\mathbf{Th}(\Sigma)$, we have $f_{A_\mathcal{M}} \stackrel{def}{=} \mathcal{M}(f_\Sigma) : \mathcal{M}(\underline{n}) \to \mathcal{M}(\underline{1})$, and thus $\mathcal{M}(f_\Sigma) : \mathcal{M}(\underline{1})^n \to \mathcal{M}(\underline{1})$ (because $\mathcal{M}$ is product preserving and $\underline{n} = \underline{1}^n$ in $\mathbf{Th}(\Sigma)$), showing that $f_{A_\mathcal{M}}$ has the correct type. Besides, each natural transformation $\alpha : \mathcal{M} \Rightarrow \mathcal{N}$ between **Set**-models characterizes a homomorphism between $A_\mathcal{M}$ and $A_\mathcal{N}$: in fact, the naturality requirement implies that for any operator $f \in \Sigma_n$, we have $\alpha_{\underline{1}} \circ \mathcal{M}(f) = \mathcal{N}(f) \circ \alpha_n$, i.e., $\alpha_{\underline{1}} \circ f_{A_\mathcal{M}} = f_{A_\mathcal{N}} \circ \alpha_{\underline{1}}^n$.

These ideas apply also to **CPO**-models: here $\mathcal{M}$ is by definition a CPO, all operators of the signature are mapped by $\mathcal{M}$ to continuous functions of the right type, and natural transformations correspond to continuous homomorphisms.

2.3 Initial algebras

It is well-known that, for each signature Σ, the category Σ-**Alg** has an initial object, often called the *word algebra* and denoted by T_Σ. Its elements are all the terms freely generated from the constants and the operators of Σ, and can be regarded as finite trees whose nodes are labelled by operator symbols. As shown in [1], also the category Σ-**SCAlg** has an initial object, denoted CT_Σ. Its elements are possibly infinite, possibly partial terms freely generated from Σ, and they form a cpo where the ordering relation is given by $t \leq t'$ iff t' is "more defined" than t. We introduce directly CT_Σ, since T_Σ can be recovered as a suitable sub-algebra: definitions are borrowed from [1], with minor changes.

Let ω^* be the set of all finite strings of positive natural numbers: its elements are called *occurrences*. The empty string is denoted by λ, and $u \leq w$ means that u is a prefix of w. Occurrences u, w are *disjoint* (written $u|w$) if neither $u \leq w$ nor $w \leq u$. Let Σ be a signature and X a set of variables. A *term* over (Σ, X) is a partial function $t : \omega^* \to \Sigma \cup X$, such that the domain of definition of t, $\mathcal{O}(t)$, satisfies (for $w \in \omega^*$ and $i \in \omega$):

- $wi \in \mathcal{O}(t) \Rightarrow w \in \mathcal{O}(t)$;
- $wi \in \mathcal{O}(t) \Rightarrow t(w) \in \Sigma_n$ for some $n \geq i$.

$\mathcal{O}(t)$ is called the *set of occurrences* of t. Given an occurrence $w \in \omega^*$ of t, the *subterm* of t at (occurrence) w is the term t/w defined as $t/w(u) = t(wu)$ for all $u \in \omega^*$. A term t is *finite* if $\mathcal{O}(t)$ is finite; it is *total* if $t(w) \in \Sigma_n \Rightarrow wi \in \mathcal{O}(t)$ for all $0 < i \leq n$; it is *linear* if no variable occurs more than once in it; Given terms t, s and an occurrence $w \in \omega^*$, the *replacement* of s in t at (occurrence) w, denoted $t[w \leftarrow s]$, is the term defined as $t[w \leftarrow s](u) = t(u)$ if $w \not\leq u$ or $t/w = \bot$, and $t[w \leftarrow s](wu) = s(u)$ otherwise.

The set of terms over (Σ, X) is denoted by $CT_\Sigma(X)$ (CT_Σ stays for $CT_\Sigma(\emptyset)$). In the paper we will often use (for finite terms) the equivalent and more usual representation of terms as operators applied to other terms. Partial terms are made total in this representation by introducing the undefined term $\bot$ (called *bottom*), which represents the empty function $\bot : \emptyset \to \Sigma \cup X$. Thus, for example, if $x \in X$, $t = f(\bot, g(x))$ is the term such that $\mathcal{O}(t) = \{\lambda, 2, 2 \cdot 1\}, t(\lambda) = f \in \Sigma_2$, $t(2) = g \in \Sigma_1$, and $t(2 \cdot 1) = x \in X$. It is well known that $CT_\Sigma(X)$ forms a cpo with respect to the "approximation" relation. We say that t *approximates* t' (written $t \leq t'$) iff t is less defined than t' as partial function. The least element of $CT_\Sigma(X)$ with respect to $\leq$ is clearly $\bot$. An *ω-chain* $\{t_i\}_{i<\omega}$ is an infinite sequence of terms $t_0 \leq t_1 \leq \ldots$. Every ω-chain $\{t_i\}_{i<\omega}$ in $CT_\Sigma(X)$ has a lub $\bigcup_{i<\omega}\{t_i\}$ characterized as:

$$t = \bigcup_{i<\omega}\{t_i\} \quad \Leftrightarrow \quad \forall w \in \omega^* . \exists i < \omega . \forall j \geq i . t_j(w) = t(w).$$

Moreover, each pair of terms has a glb. All this amounts to saying that $CT_\Sigma(X)$ is an *ω-complete lower semilattice*, hence a CPO. CT_Σ is initial in **Σ-SCAlg**, and it is called the *continuous word algebra* of Σ.

3 Term Rewriting

We introduce now term rewriting over $CT_\Sigma(X)$. A term rewriting system is a (labelled) set of rules, i.e., of pairs of finite, total terms. In the classical viewpoint, a rule can be applied to a term t if its left-hand side matches a subterm of t, and the result is the term t where the matched subterm is replaced by a suitable instantiation of the right-hand side of the rule. After presenting the standard approach to rewriting, we show an equivalent definition using inference rules.

Definition 8 (term rewriting systems). Let X be a set of variables. A *term rewriting system* (briefly, TRS) $\mathcal{R}$ is a tuple (Σ, L, R), where Σ is a signature, L is a set of labels, and R is a function $R : L \to T_\Sigma(X) \times T_\Sigma(X)$, such that for all $d \in L$, if $R(d) = \langle l, r\rangle$ then $var(r) \subseteq var(l) \subseteq X$ and l is not a variable. We write $d : l \to r \in R$ if $d \in L$ and $R(d) = \langle l, r\rangle$; sometimes, to make explicit the variables contained in a rule, we will write $d(x_1, \ldots, x_n) : l(x_1, \ldots, x_n) \to r(x_1, \ldots, x_n) \in R$ where $\{x_1, \ldots, x_n\} = var(l)$. A rule $\langle l, r\rangle$ is *left-linear* if l is linear; a TRS $\mathcal{R}$ is *orthogonal* if all its rules are left-linear and it is *non-overlapping*: that is, the left-hand side of each rule does not unify with a non-variable subterm of any other rule in $\mathcal{R}$, or with a proper, non-variable subterm of itself. □

Definition 9 (Substitutions). Let X and Y be two sets of variables. A *substitution* (from X to Y) is a function $\sigma : X \to CT_\Sigma(Y)$ (used in postfix notation). A substitution σ from X to Y uniquely determines a continuous Σ-homomorphism (also denoted by σ) from $CT_\Sigma(X)$ to $CT_\Sigma(Y)$, which extends σ as follows

- $\bot\sigma = \bot$;
- $f(t_1, \ldots, t_n)\sigma = f(t_1\sigma, \ldots, t_n\sigma)$;
- $\left(\bigsqcup_{i<\omega}\{t_i\}\right)\sigma = \bigsqcup_{i<\omega}\{t_i\sigma\}$.

If X is finite, a substitution σ from X to Y can be represented as a finite set $\{x_1/t_1, \ldots, x_n/t_n\}$ with $t_i = x_i\sigma$ for all $1 \le i \le n$ and, given a term t such that $var(t) \subseteq \{x_1, \ldots, x_n\}$, we usually write $t(t_1, \ldots, t_n)$ for $t\sigma$. □

We need now to define the notion of *redex* (for *reducible expression*). In the literature it is usually defined as a subterm of a given term, matching the left-hand side of a rule. We use a slightly different definition, identifying a redex as a pair $\Delta = (w, d)$ where w is the occurence of the root of the matched subterm, and d is a rewrite rule. For the sake of simplicity (and without loss of generality) we consider just the rewriting of ground terms (i.e., elements of CT_Σ).

Definition 10 (redex, rewriting and derivations). Given a TRS $\mathcal{R}$, a *redex* Δ of t is a pair $\Delta = (w, d)$ where $w \in \omega^*$ is an occurrence, $d : l \to r \in R$ is a rule, and there exists a substitution $\sigma : var(l) \to CT_\Sigma$ such that $t/w = l\sigma$. The result of its *application* is $s = t[w \leftarrow r\sigma]$: we also write $t \to_\Delta s$, and we say that t *rewrites* to s (via Δ). We say that there is a *derivation* from t to t' if there are redexes $\Delta_1, \ldots, \Delta_n$ such that $t \to_{\Delta_1} t_1 \to_{\Delta_2} \ldots \to_{\Delta_n} t_n = t'$. □

We introduce now an alternative definition of rewriting, borrowed from the seminal work by José Meseguer [13, 14]. The idea is to take a logical viewpoint, regarding a term rewriting system $\mathcal{R}$ as a theory, and any rewrite like $t \to_\Delta s$ as a sequent entailed by the theory. The entailment relation is defined inductively by the deduction rules of *rewriting logic*: these rules will be introduced stepwise in this paper, then extended with an infinitary rule, carefully relating (unlike [14]) the "logical" presentation of rewriting to the more traditional one.

Definition 11 (proof terms and rewriting sequents). Let $\mathcal{R} = (\Sigma, L, R)$ be a TRS, and Λ the signature satisfying $\Lambda_n = \{d \mid d(x_1, \ldots, x_n) : l(x_1, \ldots, x_n) \to r(x_1, \ldots, x_n) \in R\}$ for each $n \in \mathbb{N}$. A *proof term* α is a term of $CT_{(\Sigma \cup \Lambda \cup \{\cdot\})}$, where "$\cdot$" is a binary operator (we assume that no name-clashing among the various sets of operators occur). A proof term α is *one-step* if it does not contain the operator "$\cdot$"; it is *linear* if it is one-step and contains exactly one operator in Λ; it is *many steps* if $\alpha = \alpha_1 \cdot \ldots \cdot \alpha_n$ with $1 \le n < \omega$ and α_i is one-step for each $i \in \{1, \ldots, n\}$;[3] finally, it is *sequential* if it is many steps and all the component one-step proof terms are linear. A *(rewriting) sequent* is a triple $\langle \alpha, t, s \rangle$ (usually written as $\alpha : t \to s$) where α is a proof term and $t, s \in CT_\Sigma$: it is *one-step* (*linear*, *many steps*, *sequential*) if so is α. □

[3] We assume that "$\cdot$" is associative, in order to get rid of a clumsy notation.

The rules of deduction we introduce now are the fragment of rewriting logic necessary to describe just the application of a redex to a term or a derivation from a term: they allow to derive only sequential sequents.

Definition 12 (sequential rewriting logic). Let $\mathcal{R} = (\Sigma, L, R)$ be a TRS. We say that $\mathcal{R}$ *entails* the sequential sequent α: $s \rightarrow t$ if it can be obtained by a finite number of applications of the following rules of deduction:

$$(\textit{reflexivity}) \quad \frac{t \in CT_\Sigma}{t : t \rightarrow t}; \qquad (\textit{transitivity}) \quad \frac{\alpha : s \rightarrow t, \beta : t \rightarrow u}{\alpha \cdot \beta : s \rightarrow u};$$

$$(\textit{instantiation}) \quad \frac{d : l \rightarrow r \in R, d \in \Lambda_n, t_i \in CT_\Sigma \text{ for } i = 1, \ldots, n}{d(t_1, \ldots, t_n) : l(t_1, \ldots, t_n) \rightarrow r(t_1, \ldots, t_n)};$$

$$(\textit{linear congruence}) \quad \frac{f \in \Sigma_n, \alpha : s \rightarrow s' \text{ linear}, t_j \in CT_\Sigma \text{ for } j \in \{1, \ldots, n\} \backslash i}{f(t_1^{i-1}, \alpha, t_{i+1}^n) : f(t_1^{i-1}, s, t_{i+1}^n) \rightarrow f(t_1^{i-1}, s', t_{i+1}^n)}$$

where t_p^q, $p \leq q$, stands for the tuple $t_p, \ldots, t_q$. □

Next fact states the precise relationship between the classical presentation of rewriting and the one using entailment of sequents (in the sequential case).

Proposition 13 (sequential sequents and derivations). *Let $\mathcal{R}$ be a* TRS. (1) *If Δ is a redex of t and $t \rightarrow_\Delta s$, then there is a linear proof term α_Δ such that $\mathcal{R}$ entails the sequent $\alpha_\Delta : t \rightarrow s$ (using the rules of Definition 12). Viceversa,* (2) *if $\mathcal{R}$ entails a linear sequent $\alpha : t \rightarrow s$, then there is a redex Δ_α of t such that $t \rightarrow_{\Delta_\alpha} s$. Hence, there is a derivation from t to t' iff $\mathcal{R}$ entails a sequential sequent $\alpha : t \rightarrow t'$.* □

3.1 Parallel rewriting

Sequential term rewriting can be generalized to parallel term rewriting by allowing for the simultaneous application of two or more redexes to a term. Clearly, the result of such a parallel rewriting must be well defined, and should be related in some way to the result obtained by applying the redexes in any order. This is easily achieved by allowing only for the parallel application of non-overlapping, left-linear redexes: the definitions below summarize those in [6, 5] (see also [2]). Intuitively, *finite* parallel rewriting can be defined easily by exploiting the confluence of orthogonal term rewriting. In fact, the parallel reduction of a finite number of redexes is defined simply as any *complete development* of them: any such development ends with the same term, so the result is well defined. Note however that, given two redexes of a term, the reduction of one of them can transform the other in various ways: the second redex can be destroyed, can be left intact, or can be copied a number of times. The situation is captured by the definition of *residual* (we assume here that the rules belong to an orthogonal TRS, thus two redexes of a term are either the same or do not overlap).

Definition 14 (residuals). Let $\Delta = (w, d)$ and $\Delta' = (w', d' : l' \to r')$ be two redexes in a term t. The *set of residuals of Δ by Δ'*, denoted by $\Delta \backslash \Delta'$, is defined

$$\Delta \backslash \Delta' = \begin{cases} \emptyset & \text{if } \Delta = \Delta'; \\ \{\Delta\} & \text{if } w \not> w'; \\ \{(w'w_x u, d) \mid r'/w_x = l'/v_x\} & \text{if } w = w'v_x u \text{ and } l'/v_x \text{ is a variable.} \end{cases}$$

If $\Phi \cup \{\Delta\}$ is a finite set of redexes of t, then the set $\Phi \backslash \Delta$ of residuals of Φ by Δ is defined as the union of $\Delta' \backslash \Delta$ for all $\Delta' \in \Phi$. If Φ is a set of redexes of t and $\rho = (t \to_{\Delta_1} t_1 \ldots \to_{\Delta_n} t_n)$ is a reduction sequence, then $\Phi \backslash \rho$ is defined as Φ if $n = 0$, and as $(\Phi \backslash \Delta_1) \backslash \rho'$, where $\rho' = (t_1 \to_{\Delta_2} t_2 \ldots \to_{\Delta_n} t_n)$, otherwise. □

In the last definition, if $t \to_\Delta t'$ and Δ' is a redex of t, the orthogonality of the system ensures that every element of $\Delta' \backslash \Delta$ is a redex of t'.

Definition 15 (complete development). Let Φ be a finite set of redexes of t. A *development of Φ* is a reduction sequence such that after each initial segment ρ, the next reduced redex is an element of $\Phi \backslash \rho$. A *complete development of Φ* is a development ρ such that $\Phi \backslash \rho = \emptyset$. □

Next fact derives from the *parallel moves lemma* (see e.g. Lemma 2.2 in [9]).

Proposition 16. *All complete developments ρ and ρ' of a finite set of redexes Φ in a term t are finite, and end with the same term.* □

Exploiting this result, we define the parallel reduction of a finite set of redexes as any complete development of them.

Definition 17 (finite parallel redex reduction). A *parallel redex Φ* of a term t is a (possibly infinite, necessarily countable) set of distinct redexes in t. If Φ is a *finite* parallel redex of t, we write $t \to_\Phi t'$ and say that there is a *(finite) parallel reduction* from t to t' if there exists a complete development $t \to_{\Delta_1} t_1 \ldots \to_{\Delta_n} t'$ of Φ. □

Let us now consider again rewriting logic: small changes to the deduction rules of Definition 12 are sufficient to generate many steps sequents.

Definition 18 (parallel rewriting logic). Let $\mathcal{R} = (\Sigma, L, R)$ be a term rewriting system. We say that $\mathcal{R}$ *entails* the many steps sequent $\alpha : s \to t$ if it can be obtained by a finite number of applications of the following rules of deduction:

(*reflexivity*) and (*transitivity*) as in Definition 12;

$$(\mathit{replacement}) \quad \frac{d : l \to r \in R, d \in \Lambda_n, \alpha_i : t_i \to s_i \text{ one-step for } i = 1, \ldots, n}{d(\alpha_1, \ldots, \alpha_n) : l(t_1, \ldots, t_n) \to r(s_1, \ldots, s_n)};$$

$$(\mathit{congruence}) \quad \frac{f \in \Sigma_n, \alpha_i : t_i \to s_i \text{ one-step for } i = 1, \ldots, n}{f(\alpha_1, \ldots, \alpha_n) : f(t_1, \ldots, t_n) \to f(s_1, \ldots, s_n)}.$$

□

Next proposition generalizes the relationship between the two presentations of rewriting, stated in Proposition 13, to the parallel case. We restrict our attention to orthogonal rules (because the version of parallel rewriting we considered is defined only for such TRS's), although the result holds in a more general context.

Proposition 19 (many steps sequents and parallel derivations). *Let $\mathcal{R}$ be an orthogonal* TRS. (1) *If Φ is a parallel redex of t and $t \rightarrow_{\Phi} s$, then there is a one-step proof term α_{Φ} such that $\mathcal{R}$ entails the sequent $\alpha_{\Phi} : t \rightarrow s$ (using the rules of Definition 18). Viceversa,* (2) *if $\mathcal{R}$ entails a one-step sequent $\alpha : t \rightarrow s$, then there is a parallel redex Φ_{α} of t such that $t \rightarrow_{\Phi_{\alpha}} s$. Hence, there is a parallel derivation from t to t' iff $\mathcal{R}$ entails a many steps sequent $\alpha : t \rightarrow t'$.* □

It is worth noting here that although the parallel rewriting logic of Definition 18 is slightly different from that of Meseguer (in the unconditional case), they are equivalent. Actually, in [14] the sequents appearing in the premises of the congruence and replacement rules are not restricted to be one-step, but they can be arbitrary. As a consequence, unlike in our definition, the operator "$\cdot$" can appear inside other operators in the proof term of a sequent entailed by $\mathcal{R}$.

Definition 20 (full rewriting logic). Let $\mathcal{R} = (\Sigma, L, R)$ be a TRS. We say that $\mathcal{R}$ *entails* the sequent $\alpha : s \rightarrow t$ if it can be obtained by a finite number of applications of the following rules of deduction:

(*reflexivity*) and (*transitivity*) as in Definition 12;

$$(\textit{replacement}) \quad \frac{d : l \rightarrow r \in R, d \in \Lambda_n, \alpha_i : t_i \rightarrow s_i \text{ for } i = 1, \ldots, n}{d(\alpha_1, \ldots, \alpha_n) : l(t_1, \ldots, t_n) \rightarrow r(s_1, \ldots, s_n)};$$

$$(\textit{congruence}) \quad \frac{f \in \Sigma_n, \alpha_i : t_i \rightarrow s_i \text{ for } i = 1, \ldots, n}{f(\alpha_1, \ldots, \alpha_n) : f(t_1, \ldots, t_n) \rightarrow f(s_1, \ldots, s_n)}.$$

□

Full rewriting logic coincides with Meseguer's logic. Next proposition shows that the more general format of rules does not change the relation among terms induced by the entailed sequents.

Proposition 21 (equivalence of rewriting logics). *Let $\mathcal{R}$ be a* TRS*: it entails a sequent $\alpha : t \rightarrow s$ in full rewriting logic iff it entails a many steps sequent $\alpha' : t \rightarrow s$ in parallel rewriting logic.* □

3.2 Infinite parallel rewriting

Parallel rewriting allows to reduce a finite set of redexes of a term in a single, parallel step. If we consider an infinite term, there might be infinitely many distinct redexes in it: since the simultaneous rewriting of any finite subset of those redexes is well-defined, by a continuity argument one would expect that also the simultaneous rewriting of infinitely many redexes in an infinite term can be properly defined. Note however that, since in the finite case the parallel application of a redex Φ is defined as the sequential application of all the contained redexes, a naïve extension to infinity could not work. We present here a definition which makes use of a suitable limit construction: for details we refer again to [6, 5].

Definition 22 (parallel redex reduction). Given an infinite parallel redex Φ of a term t, let $t_0 \leq t_1 \leq \ldots t_n \leq \ldots$ be any chain approximating t (i.e., such that $\bigsqcup_{i<\omega}\{t_i\} = t$) and satisfying:

- for each $i < \omega$, every redex $(w, d) \in \Phi$ is either a redex of t_i or $t_i(w) = \bot$ (that is, the image of the lhs of every redex in Φ is either all in t_i, or it is outside, but does not "cross the boundary");
- for each $i < \omega$, the subset Φ_i of all redexes in Φ which are also redexes of t_i is finite.

For each $i < \omega$, let s_i be the result of the (finite) parallel reduction of t_i via Φ_i (i.e., $t_i \rightarrow_{\Phi_i} s_i$). Then we say that there is an *(infinite) parallel reduction* from t to $s \stackrel{def}{=} \bigsqcup_{i<\omega}\{s_i\}$ via Φ, and we write $t \rightarrow_\Phi s$. □

Theorem 23 (parallel redex reduction is well-defined). *In the hypotheses of Definition 22:*

1. *for each $i < \omega$, $s_i \leq s_{i+1}$, thus $\{s_i\}_{i<\omega}$ is a chain.*
2. *Definition 22 is well-given i.e., the result of the infinite parallel reduction of t via Φ does not depend on the choice of the chain approximating t, provided that it satisfies the required conditions.* □

Proposition 24 (strong confluence of parallel reduction). *Let $\mathcal{R}$ be an orthogonal* TRS. *Then parallel reduction is strongly confluent, i.e., if $t' \;{}_{\Phi'}\!\!\leftarrow t \rightarrow_\Phi t''$, then there exist t''', Ψ, Ψ' such that $t' \rightarrow_{\Psi'} t''' \;{}_{\Psi}\!\!\leftarrow t''$. Hence, parallel reduction is confluent.* □

Now we introduce a natural extension of rewriting logic which allows for the generation of sequents corresponding to infinite parallel derivations. Since each sequent is composed of three terms which belong (by definition) to continuous algebras, the "infinitary" sequents we are looking for are obtained by taking the lub's of suitable chains.

Definition 25 (infinite parallel rewriting logic). Let $\mathcal{R} = (\Sigma, L, R)$ be a TRS. We say that $\mathcal{R}$ *entails* the *infinite, parallel* sequent $\alpha : s \rightarrow t$ if it can be obtained by a finite number of applications of the rules of deduction of parallel rewriting logic (see Definition 18), extended with the following one:

$$(\textit{infinite parallel rewr.}) \quad \frac{\alpha_i : t_i \rightarrow s_i \text{ one-step for } i \in \mathbb{N}, \alpha_i \leq \alpha_{i+1} \text{ for } i \in \mathbb{N}}{\bigsqcup_i \alpha_i : \bigsqcup_i t_i \rightarrow \bigsqcup_i s_i}.$$

□

The rule is well-defined and, accordingly to Definition 11, the sequent $\bigsqcup_i \alpha_i : \bigsqcup_i t_i \rightarrow \bigsqcup_i s_i$ is also one-step.

Proposition 26 (many steps sequents and infinite parallel derivations). *Let $\mathcal{R}$ be an orthogonal* TRS *and Φ be a (possibly infinite) parallel redex.* (1) *If $t \rightarrow_{\Phi} s$, then there is a one-step proof term α_{Φ} such that $\mathcal{R}$ entails the sequent $\alpha_{\Phi} : t \rightarrow s$ (using the rules of Definition 25). Viceversa,* (2) *if $\mathcal{R}$ entails a one-step sequent $\alpha : t \rightarrow s$, then there is a (possibly infinite) parallel redex Φ_α such that $t \rightarrow_{\Phi_\alpha} s$. As a consequence, there is a parallel derivation from t to t' iff $\mathcal{R}$ entails a many step sequent $\alpha : t \rightarrow t'$.* □

Finally, we introduce the full version of infinite parallel rewriting logic.

Definition 27 (full infinite rewriting logic). Let $\mathcal{R} = (\Sigma, L, R)$ be a term rewriting system. We say that $\mathcal{R}$ *entails* the *infinite* sequent $\alpha : s \rightarrow t$ if it can be obtained by a finite number of applications of the rules of deduction of full rewriting logic (see Definition 20), extended with the following one:

$$(\textit{infinite rewriting}) \quad \frac{\alpha_i : t_i \rightarrow s_i \text{ for } i \in \mathbb{N}, \alpha_i \leq \alpha_{i+1} \text{ for } i \in \mathbb{N}}{\bigsqcup_i \alpha_i : \bigcup_i t_i \rightarrow \bigcup_i s_i}.$$

□

Note that the full version of infinitary rewriting logic is stronger than its parallel counterpart, in the sense that there exists rewriting systems such that some sequents entailed in the full case have no equivalent in the parallel one.

4 Models of Term Rewriting

Along the presentation in Section 2.1, an algebra can be considered as a "model" of a signature. The presentation of categories of algebras as functor categories makes this interpretation explicit, showing that categories of models in different universes (like **Set** and **CPO**) can be taken into account. Essentially the same ideas can be applied to TRS's as well: such systems can be considered as syntactical specifications, and models for them are algebraic structures where all the "possible rewritings" have a suitable interpretation. According to [14], a reasonable model for a TRS is defined as follows.

Definition 28 ($\mathcal{R}$-models). Let $\mathcal{R} = (\Sigma, L, R)$ be a TRS. A *$\mathcal{R}$-model* $\mathcal{S}$ is a category **S** together with

- a Σ-algebraic structure, i.e., for each $f \in \Sigma_n$ a functor $f_{\mathcal{S}} : \mathbf{S}^n \rightarrow \mathbf{S}$;
- for each rewrite rule $d : s \rightarrow t \in R$, a natural transformation $\alpha_{\mathcal{S}} : s_{\mathcal{S}} \Rightarrow t_{\mathcal{S}}$, where the functors $s_{\mathcal{S}}, t_{\mathcal{S}}$ are defined inductively from the basic functors $f_{\mathcal{S}}$.

An *$\mathcal{R}$-homomorphism* $F : \mathcal{S} \rightarrow \mathcal{S}'$ is a functor $F : \mathbf{S} \rightarrow \mathbf{S}'$ preserving the algebraic structure (i.e., $f_{\mathcal{S}'} \circ F^n = F \circ f_{\mathcal{S}}$ for each $f \in \Sigma_n$) and the rewriting rules (i.e., given $id_F : F \rightarrow F$ the identity natural transformation, $id_F \circ \alpha_{\mathcal{S}} = \alpha_{\mathcal{S}'} \circ id_{F^n}$ holds for every rule $\alpha \in R$). $\mathcal{R}$-**Mod** denotes the category of $\mathcal{R}$-models and $\mathcal{R}$-homomorphisms. □

The reasonableness of this notion of model is confirmed by the characterization of the initial object $\mathcal{I}_{\mathcal{R}}$ of $\mathcal{R}$-**Mod**: it is the category having as objects terms of the algebra T_Σ, and as arrows the elements of the algebra $T_{(\Sigma \cup \Delta \cup \{\cdot\})}/E$, where E is the following set of axioms

$$(associativity)\ \frac{\alpha, \beta, \gamma \text{ proof terms}}{\alpha \cdot (\beta \cdot \gamma) = (\alpha \cdot \beta) \cdot \gamma}; \qquad (identity)\ \frac{\alpha : s \rightarrow t}{s \cdot \alpha = \alpha = \alpha \cdot t};$$

$$(distributivity)\ \frac{f \in \Sigma_n, \alpha_i, \beta_i \text{ proof terms for } i = 1, \ldots, n}{f(\alpha_1 \cdot \beta_1, \ldots, \alpha_n \cdot \beta_n) = f(\alpha_1, \ldots, \alpha_n) \cdot f(\beta_1, \ldots, \beta_n)};$$

$$(interchange)\ \frac{d : l \rightarrow r \in R, d \in \Lambda_n, \alpha_i : t_i \rightarrow s_i \text{ for } i = 1, \ldots, n}{\begin{array}{l} d(\alpha_1, \ldots, \alpha_n) = l(\alpha_1, \ldots, \alpha_n) \cdot d(s_1, \ldots, s_n) \\ \qquad = d(t_1, \ldots, t_n) \cdot r(\alpha_1, \ldots, \alpha_n) \end{array}}.$$

This characterization allows for an intuitive soundness and completness result: namely, that a theory $\mathcal{R}$ entails a sequent $\alpha : t \rightarrow s$ iff there exists a natural transformation $\alpha_{\mathcal{I}_{\mathcal{R}}} : t_{\mathcal{I}_{\mathcal{R}}} \Rightarrow s_{\mathcal{I}_{\mathcal{R}}}$ (then, iff there exists a natural transformation $\alpha_S : t_S \Rightarrow s_S$ for each $\mathcal{R}$-model S). Anyway, a more abstract notion of model can be defined, as already done for the category of (continuous) algebras: the main concern of the next section will be for functorial models.

4.1 Functorial models of term rewriting systems

As for the case of signatures, also for TRS's the models can be presented as suitable functors. Once again, this approach has the advantage of separating in a clear way what we can call the "$\mathcal{R}$-structure" (i.e., the algebraic structure defined by the signature and by the rewriting rules of the TRS) from the additional algebraic structure that can be enjoyed by the model. As for signatures, two kinds of categorical models for a system are considered, namely the **Set**-based and the **CPO**-based ones. The **Set**-based models have been studied by Meseguer [13], showing that they are equivalent to $\mathcal{R}$-models (exactly in the same way **Set**-models of Σ turned out to be Σ-algebras) when finite terms are considered, providing this way a soundness and completeness result for the finitary version of full rewriting logic. On the other hand, **CPO**-based models of a TRS $\mathcal{R}$ have not been studied before. The main result of this section is that soundness and completeness lift smoothly to this category of models, if we add to full rewriting logic the infinitary rule of Definition 27.

The categorical presentation of TRS's and of their models is built over the categorical presentation of signatures and algebras of Section 2.1. For a given TRS $\mathcal{R} = (\Sigma, L, R)$, the leading idea is to use the arrows of the Lawvere theory $\mathbf{Th}(\Sigma)$ as the "states" of the system (because those arrows represent terms of the word algebra T_Σ), and to regard rewrites as *2-cells*, i.e., as arrows between arrows. The resulting structure, obtained via a suitable closure operation, turns out to be a 2-category called the *Lawvere 2-theory* of $\mathcal{R}$. First of all, let us introduce the notion of 2-category: for further details we refer to [11].

Definition 29 (2-categories). A 2-category $\underline{C}$ consists of a collection $\{a, b, \ldots\}$ of *objects*, or 0-cells, a collection $\{f, g, \ldots\}$ of *arrows*, or 1-cells, and a collection $\{\alpha, \beta, \ldots\}$ of *transformations*, or 2-cells. 1-cells are assigned a source and a target 0-cell, written as $f : a \to b$, and 2-cells are assigned a source and a target 1-cell, say f and g, in such a way that $f, g : a \to b$, and this is indicated as $\alpha : f \Rightarrow g : a \to b$, or simply $\alpha : f \Rightarrow g$. The following operations are given:

- a partial operation $\_;\_$ of *horizontal composition* of 1-cells, which assignes to each pair $(f : a \to b, g : b \to c)$ a 1-cell $f; g : a \to c$;
- a partial operation $\_ * \_$ of *horizontal composition* of 2-cells, which assigns to each pair $(\alpha : f \Rightarrow g : a \to b, \beta : h \Rightarrow k : b \to c)$ a 2-cell $\alpha * \beta : f; h \Rightarrow g; k$;
- a partial operation $\_ \cdot \_$ of *vertical composition* of 2-cells, which assignes to each pair $(\alpha : f \Rightarrow g : a \to b, \beta : g \Rightarrow h : a \to b)$ a 2-cell $\alpha \cdot \beta : f \Rightarrow h$.

To each object a there is an associated identity 1-cell id_a and to each arrow f there is an associated 2-cell identity id_f. These data satisfy the following axioms:

- the objects and the arrows with the horizontal composition of 1-cells and the identities id_a form a category **C**, the *underlying* category of $\underline{C}$;
- for any pair of objects a and b, the morphisms of the kind $f : a \to b$ and their 2-cells form a category under the given operations of vertical composition of 2-cells with identities id_f;
- the objects and the 2-cells form a category under the operation of horizontal composition of 2-cells with identities id_{id_a};
- for all $f : a \to b$ and $g : b \to c$, $id_f * id_g = id_{(f;g)}$;
- whenever the two sides are defined, then $(\gamma * \alpha) \cdot (\delta * \beta) = (\gamma \cdot \delta) * (\alpha \cdot \beta)$.[4]

A 2-category $\underline{C}$ has *(finite) 2-products* if its underlying category **C** has (finite) products, and if for each pair of objects $\langle a, b\rangle$ with product diagram $\langle a \times b, \pi_0 : a \times b \to a, \pi_1 : a \times b \to b\rangle$, it holds that for every pair of 2-cells $\alpha : f \Rightarrow g : c \to a$ and $\beta : h \Rightarrow i : c \to b$, there exists a unique 2-cell $\langle \alpha, \beta\rangle : \langle f, h\rangle \Rightarrow \langle g, i\rangle : c \to a \times b$ satisfying $\langle \alpha, \beta\rangle * \pi_0 = \alpha$ and $\langle \alpha, \beta\rangle * \pi_1 = \beta$. A *2-terminal objects* for $\underline{C}$ is an object 0_C such that for every object a there exists a unique arrow $!_a : a \to 0_C$ satisfying $\alpha * !_b = !_a$ for every 2-cell $\alpha : f \Rightarrow g : a \to b$. Finally, $\underline{C}$ is *2-cartesian* if it has both 2-products and a 2-terminal object. □

Definition 30 (Lawvere 2-theories). Let $\mathcal{R} = (\Sigma, L, R)$ be a TRS. The associated *Lawvere 2-theory* $\underline{\mathbf{2\text{-}Th}(\mathcal{R})}$ is the 2-cartesian 2-category having as underlying category $\mathbf{Th}(\Sigma)$, and whose 2-cells are freely generated from the set $R_c = \{\alpha_d : s_\Sigma \Rightarrow t_\Sigma : \underline{n} \to \underline{1} \mid d : s \to t \in R \text{ and } d \in \Lambda_n\}$.

We defined the models of the Lawvere theory of a signature as cartesian functors to a suitable cartesian category. Similarly, we can define the models of the Lawvere 2-theory associated to a TRS $\mathcal{R}$ as functors to a suitable universe: however, those functors (as well as the corresponding natural transformations) have to preserve the relevant structure, which is now much richer.

[4] Note that identity 2-cells are almost always denoted by the correspondent 1-cell: therefore, $id_f * \alpha * id_g$ is written as $f * \alpha * g$.

Definition 31 (2-functors and 2-natural transformations). Let $\underline{C}$ and $\underline{D}$ be two 2-categories. A *2-functor* $F : C \to D$ is a triple $\langle F_O, F_A, F_C\rangle$ of functions, mapping objects to objects, arrows to arrows and 2-cells to 2-cells, respectively, preserving identities and compositions of all kinds. Let $F, G : \underline{C} \to \underline{D}$ be two parallel 2-functors: a *2-natural transformation* $\eta : F \Rightarrow G$ assigns to each object $a \in C$ an arrow $\eta_a : F_O(a) \to G_O(a) \in \underline{D}$ such that

- for any arrow $f : a \to b \in \underline{C}$, $\eta_a; G_A(f) = F_A(f); \eta_b$;
- for any 2-cell $\alpha : f \Rightarrow g \in \underline{C}$, $\eta_a * F_C(\alpha) = G_C(\alpha) * \eta_b$.

If $\underline{C}$ and $\underline{D}$ are two 2-categories, the *2-functor category* $[\underline{C} \to \underline{D}]$ has 2-functors as objects and 2-natural transformations as arrows. A 2-functor $F = \langle F_O, F_A, F_C\rangle : \underline{C} \to \underline{D}$ is *2-cartesian* if for all objects a, b the canonical maps $\langle F_A(\pi_0), F_A(\pi_1)\rangle : F_O(a \times b) \to F_O(a) \times F_O(b)$ and $!_{F(0_C)} : F_O(0_C) \to 0_D$ are isomorphisms. □

Definition 32 (models of Lawvere 2-theories). If $\underline{C}$ is a 2-cartesian 2-category, a $\underline{C}$-*model* for the Lawvere 2-theory associated to a TRS $\mathcal{R}$ is a 2-cartesian 2-functor $\mathcal{M}$: $\underline{\mathbf{2\text{-}Th}(\mathcal{R})} \to \underline{C}$, while a *model morphism* is a 2-natural transformation between models. The category $\underline{C}$-$\mathbf{Mod}_{\mathcal{R}}$ of $\underline{C}$-models is the 2-cartesian 2-functor category $[\underline{\mathbf{2\text{-}Th}(\mathcal{R})} \to \underline{C}]$: its objects are $\underline{C}$-models of $\mathcal{R}$, while its arrows are model morphisms. □

Like **Set** is the paradigmatic example of category, so $\underline{\mathbf{Cat}}$, the category of small categories and functors, is the paradigmatic example of 2-category: its 2-cells are the natural transformations. Since $\underline{\mathbf{Cat}}$ is 2-cartesian, we are allowed to consider $\underline{\mathbf{Cat}}$-models of a TRS $\mathcal{R}$: they are nothing else than the $\mathcal{R}$-models.

Proposition 33 (the functor category of $\mathcal{R}$-models). *Let $\mathcal{R}$ be a* TRS*: the category of $\mathcal{R}$-models (see Definition 28) is equivalent to the category* $\underline{\mathbf{Cat}}$-$\mathbf{Mod}_{\mathcal{R}}$ *of* $\underline{\mathbf{Cat}}$*-models for $\mathcal{R}$.* □

This result was already stated in [13], and provides an elegant characterization of the $\mathcal{R}$-models. Let us now consider the **CPO**-based models for a TRS $\mathcal{R}$: according to the definition, those models should be 2-cartesian 2-functors from $\underline{\mathbf{2\text{-}Th}(\mathcal{R})}$ to a 2-cartesian 2-category. Thus the only degree of freedom is the choice of the target 2-category, and such a category should be obtained by replacing, in the models introduced so far, sets with cpo's.

It is worth stressing here in which sense we regard a $\underline{\mathbf{Cat}}$-model for $\mathcal{R}$ as a **Set**-based model, because this will hint the correct structure for the universe 2-category of **CPO**-based models. A 2-functor $\mathcal{M}$: $\underline{\mathbf{2\text{-}Th}(\mathcal{R})} \to \underline{\mathbf{Cat}}$ maps each object of $\underline{\mathbf{2\text{-}Th}(\mathcal{R})}$ (say $\underline{n}$) to a category $\mathcal{M}(\underline{n})$; in this category, the objects provide an interpretation to the n-tuples of terms, while the arrows give an interpretation to the rewrites. Since $\mathcal{M}(\underline{n})$ is a (small) category, its objects and its arrows form a set, by definition. Thus $\mathcal{M}$ is considered a **Set**-based model because terms and rewrites are interpreted in a set. By analogy, a **CPO**-model for $\mathcal{R}$ would interpret (tuples of) terms in a cpo, and similarly for the rewrites: in such cpo's we could find an interpretation also for infinite terms and for the sequents generated by the infinitary deduction rules. A model should therefore map each object of $\underline{\mathbf{2\text{-}Th}(\mathcal{R})}$ to a category having a cpo of objects and a cpo of arrows. The next section is devoted to an informal presentation of those structures, providing a result analogous to Proposition 33.

4.2 Functorial models in internal categories

To define an *internal* category simply means to consider arrows (and objects) as forming not simply a set, but a more complex class, such as a group, a monoid and so on. Informally, a category internal to **CPO**, also called a **CPO**-category, can be described as a category **C** such that its components are not just sets and functions, but are instead cpo's and strict continuous functions: so, both its collections of objects O_C and arrows A_C are cpo's, and, for example, the source and target mappings are strict continuous functions from A_C to O_C. A continuous functor is a functor such that both its components are continuous functions, while given two continuous functors $F, G : \mathbf{C} \rightarrow \mathbf{D}$, a continuous natural transformation $\alpha : F \Rightarrow G$ is a continuous function $\alpha : O_C \rightarrow A_D$ satisfying the usual conditions. As in $\underline{\mathbf{Cat}}$, also horizontal and vertical composition of continuous natural transformations can be defined.

Definition 34 (continuous $\mathcal{R}$-Models). Let $\mathcal{R} = (\Sigma, L, R)$ be a TRS. A *continuous $\mathcal{R}$-model $\mathcal{S}$* is a **CPO**-category **S** together with

- a Σ-algebraic structure: for each $f \in \Sigma_n$ a continuous functor $f_{\mathcal{S}}: \mathbf{S}^n \rightarrow \mathbf{S}$;
- for each $d : s \rightarrow t \in R$, a continuous natural transformation $\alpha_{\mathcal{S}} : s_{\mathcal{S}} \Rightarrow t_{\mathcal{S}}$.

A *continuous $\mathcal{R}$-homomorphism* $F: \mathcal{S} \rightarrow \mathcal{S}'$ is a continuous functor $F: \mathbf{S} \rightarrow \mathbf{S}'$ preserving the algebraic structure and the rewriting rules. **C$\mathcal{R}$-Mod** is the category of continuous $\mathcal{R}$-models and continuous $\mathcal{R}$-homomorphisms. □

Now, let $\underline{\mathbf{Cat(CPO)}}$ be the 2-category such that objects are **CPO**-categories, arrows are continuous functors, and 2-cells are continuous natural transformations: it is easy to check that it is 2-cartesian. Thus we can define the **CPO**-based models of a term rewriting system as 2-functors from the Lawvere 2-theory of $\mathcal{R}$ to $\underline{\mathbf{Cat(CPO)}}$. So, let $\underline{\mathbf{Cat(CPO)}}$-**Mod**$_{\mathcal{R}}$ be the category of $\underline{\mathbf{Cat(CPO)}}$-models, given by the 2-cartesian 2-functor category $[\underline{\mathbf{2\text{-}Th}(\mathcal{R})} \rightarrow \underline{\mathbf{Cat(CPO)}}]$: its objects are $\underline{\mathbf{Cat(CPO)}}$-models of $\mathcal{R}$ (i.e., 2-cartesian 2-functors), while its arrows are model morphisms (2-natural transformations). Finally, we are able to state the main result of this section, precisely relating continuous $\mathcal{R}$-models and functorial models.

Proposition 35 (the functor category of continuous $\mathcal{R}$-models). *Let $\mathcal{R}$ be a* TRS*: the category of continuous $\mathcal{R}$-models (see Definition 34) is equivalent to the category* $\underline{\mathbf{Cat(CPO)}}$-**Mod**$_{\mathcal{R}}$ *of* $\underline{\mathbf{Cat(CPO)}}$*-models for $\mathcal{R}$.* □

5 Conclusions

In this paper we introduced a new formalism for the rewriting of infinite terms, developing the rewriting logic approach originally proposed by José Meseguer. We hope we made clear the several interesting features presented by this framework: first of all, it provides a clean, algebraic presentation of infinite rewrites, exploiting the CPO-structure of the algebra of proof terms; moreover, it admits set-theoretical and categorical models; finally, it consistently includes the infinite parallel term rewriting proposal, originally presented by one of the authors ([6]).

The formalism needs however further investigations: we intend to apply our approach also to other kinds of categorical models, as those proposed by John Stell ([17]), as well as to other formalisms, in particular term graph rewriting.

References

1. J.A. Goguen, J.W. Tatcher, E.G. Wagner, J.R. Wright, *Initial Algebra Semantics and Continuous Algebras*, Journal of the ACM **24** (1), 1977, pp. 68–95.
2. G. Boudol, *Computational Semantics of Term Rewriting Systems*, chapter 8 of Algebraic Methods in Semantics, eds. M.Nivat and J. Reynolds, CUP, 1985.
3. M. Barr, C. Wells, *Category Theory for Computing Science*, Prentice Hall, 1990.
4. A. Corradini, F. Gadducci, U. Montanari, *Relating Two Categorical Models of Term Rewriting*, to appear in Proc. RTA'95, LNCS, 1995.
5. A. Corradini, F. Drewes, *(Cyclic) Term Graph Rewriting is Adequate for Rational Parallel Term Rewriting*, submitted.
6. A. Corradini, *Term Rewriting in* CT_Σ, in Proc. TAPSOFT'93 (CAAP), LNCS 668, 1993, pp. 468–484.
7. N. Dershowitz, J.-P. Jouannaud, *Rewrite Systems*, in Handbook of Theoretical Computer Science **B**, ed. J. van Leeuwen, North Holland, 1990, pp. 243–320.
8. N. Dershowitz, S. Kaplan, D.A. Plaisted, *Rewrite, Rewrite, Rewrite, Rewrite, Rewrite,...**, Journal of Theoretical Computer Science **83**, 1991, pp. 71–96.
9. G. Huet, J.J. Lévy, *Computations in Orthogonal Rewriting Systems*, chapter 11 of Computational Logic, eds. J.L. Lassez and G. Plotkin, MIT Press, 1991.
10. J.R. Kennaway, J.W. Klop, M.R. Sleep, F.J. de Vries, *On the Adequacy of Term Graph Rewriting for Simulating Term Rewriting*, ACM Transactions on Programming Languages and Systems **16** (3), 1994, pp. 493–523.
11. G.M. Kelly, R.H. Street, *Review of the Elements of 2-categories*, Lecture Notes in Mathematics 420, 1974, pp. 75–103.
12. F. W. Lawvere, *Functorial Semantics of Algebraic Theories*, in Proc. National Academy of Science **50**, 1963, pp. 869–872.
13. J. Meseguer, *Functorial Semantics of Rewrite Systems*, appendix of *Rewriting as a Unified Model of Concurrency*, SRI Technical Report CSL-93-02R, 1990.
14. J. Meseguer, *Conditional Rewriting Logic as a Unified Model of Concurrency*, Journal of Theoretical Computer Science **96**, 1992, pp. 73–155.
15. A.J. Power, *An Abstract Formulation for Rewrite Systems*, in Proc. CTCS'89, LNCS 389, 1989, pp. 300–312.
16. D.E. Rydeheard, J.G. Stell, *Foundations of Equational Deduction: A Categorical Treatment of Equational Proofs and Unification Algorithm*, in Proc. CTCS'87, LNCS 283, 1987, pp. 114–339.
17. J. G. Stell, *Modelling Term Rewriting Systems by Sesqui-categories*, Technical Report TR94-02, Keele University, 1994. To appear in Proc. CAEN.

Completeness Results for Two-sorted Metric Temporal Logics*

Angelo Montanari[1] and Maarten de Rijke[2]

[1] Dipartimento di Matematica e Informatica, Università di Udine, Via delle Scienze, 206, 33100 Udine, Italy. E-mail: montana@dimi.uniud.it
[2] Department of Software Engineering, CWI, P.O. Box 94079, 1090 GB Amsterdam, the Netherlands. E-mail: mdr@cwi.nl

Abstract. Temporal logic has been successfully used for modeling and analyzing the behavior of reactive and concurrent systems. One shortcoming of (standard) temporal logic is that it is inadequate for real-time applications, because it only deals with qualitative timing properties. This is overcome by metric temporal logics which offer a uniform logical framework in which both qualitative and quantitative timing properties can be expressed by making use of a parameterized operator of relative temporal realization. We view metric temporal logics as two-sorted formalisms having formulae ranging over time instants and parameters ranging over an (ordered) abelian group of temporal displacements.
In this paper we deal with completeness results for basic systems of metric temporal logic — such issues have largely been ignored in the literature. We first provide an axiomatization of the pure metric fragment of the logic, and prove its soundness and completeness. Then, we show how to obtain the metric temporal logic of linear orders by adding an ordering over displacements.

1 Introduction

Logic-based methods for representing and reasoning about temporal information have proved to be highly beneficial in the area of formal specifications. In this paper we consider their application to the specification of real-time systems. Timing properties play a major role in the specification of reactive and concurrent software systems that operate in real-time. They constrain the interactions between different components of the system as well as between the system and its environment, and minor changes in the precise timing of interactions may lead to radically different behaviors. Temporal logic has been successfully used for modeling and analyzing the behavior of reactive and concurrent systems (see Manna and Pnueli [8] and Ostroff [11]). It supports semantic model checking, in

* The first author was supported by a grant from the Italian Consiglio Nazionale delle Ricerche (CNR). The second author was supported by the Netherlands Organization for Scientific Research (NWO), project NF 102/62-356 'Structural and Semantic Parallels in Natural Languages and Programming Languages'. This work was carried out while the first author was visiting ILLC, University of Amsterdam.

order to verify consistency of specifications, and to check positive and negative examples of system behavior against specifications; it also supports pure syntactic deduction, in order to prove properties of systems. Unfortunately, most common representation languages in the area of formal specifications are inadequate for real-time applications, because they lack an explicit and quantitative representation of time. In recent years, some of them have been extended to cope with real-time aspects. In this paper, we focus on *metric temporal logics* which provide a uniform framework in which both qualitative and quantitative timing properties of real-time systems can be expressed.

The idea of a logic of positions (topological, or metric, logic) has originally been formulated by Rescher and Garson [12]. They defined the basic features of the logic, and showed how to give it a temporal interpretation. The logic of positions extends propositional logic with a parametrized operator P_α of positional realization. Such an operator allows one to constrain the truth value of a proposition at position α. The parameter α denotes either (i) an absolute position or (ii) a displacement with respect to the current position which is left implicit. According to interpretation (ii), $P_\alpha p$ is true at the position i if and only if p is true at a position j at distance α from i. In [12], Rescher and Garson introduced two axiomatizations of the logic of positions that differ from each other in the interpretation of parameters. Later, Rescher and Urquhart [13] proved the soundness and completeness of the axiomatization based on an absolute interpretation of parameters through a reduction to monadic quantification theory. A metric temporal logic has been independently developed by Koymans [7] to support the specification and verification of real-time systems. He extended the standard model for temporal logic based on point structures with a distance function that measures, for any pair of time points, how far they are apart in time. He provided the logic with a sound axiomatization, but no proof of completeness is given.

The main issues to confront in developing a metric temporal logic for executable specifications are:

Expressiveness (definability). Is the metric temporal logic powerful enough to express both the properties of the underlying temporal structure and the timing requirements of the specified real-time systems?

Soundness and completeness. Is the metric temporal logic provided with a sound and complete axiomatization?

Decidability. Which properties of the specified real-time system can be automatically verified? Most temporal logics for real-time systems proposed in the literature cannot be decided (see Henzinger [6]). Some of them recover decidability sacrificing completeness.

Executability. How can we prove the consistency and adequacy of specifications? In principle, decidability proof methods (e.g. via Büchi automata) outline an effective procedure to prove the satisfiability and/or validity of a formula. But as soon as certain assumptions about the nature of the temporal domain and the available set of primitive operations are relaxed, the satisfiability/validity problem becomes undecidable (Alur and Henzinger [1]). An alternative approach consists in looking at metric temporal logics as particular polymodal logics and supporting derivability by means of proof procedures for nonclassical logics or via translation in first-order theories (see D'Agostino et al [4], and Ohlbach [10]). In this case, providing the logic with a sound and complete axiomatization becomes a central issue.

The aim of this paper is to explore completeness issues of metric temporal logic; we do this by starting with a very basic system, and build on it either by adding axioms or by enriching the underlying structures. We view metric temporal logics as two-sorted logics having both formulae and parameters; formulae are evaluated at time instants while parameters take values in an (ordered) abelian group of temporal displacements. In Section 2, we define a minimal metric logic that can be seen as the metric counterpart of minimal tense logic, and we provide it with a sound and complete axiomatization. In Section 3, we characterize the class of two-sorted frames with a linearly ordered temporal domain.

2 The basic metric logic

In this section we define the minimal metric temporal logic MTL_0, and consider some of its natural extensions.

Language. We define a two-sorted temporal language for our basic calculus MTL_0. First, its algebraic part is built up from a non-empty set of *variables* X. The set of terms over X, $T(X)$, is the smallest set such that (1) $X \subseteq T(X)$, and (2) if α, $\beta \in T(X)$ then $(\alpha + \beta)$, $(-\alpha), 0 \in T(X)$. Next, the temporal part of the language is built from a non-empty set Φ of *proposition letters*. The set of MTL_0-formulae over Φ and X, $F(\Phi, X)$, is the smallest set such that (1) $\Phi \subseteq F(\Phi, X)$, and (2) if ϕ, $\psi \in F(\Phi, X)$ and $\alpha \in T(X)$, then $\neg\phi$, $\phi \wedge \psi$, $\Delta_\alpha \phi$ (and its dual $\nabla_\alpha \phi := \neg\Delta_\alpha\neg\phi$), $\bot \in F(\Phi, X)$. We will adopt the following notational conventions: p, q, ... denote proposition letters; ϕ, ψ, ... denote MTL_0-formulae; Σ, Γ, ... denote sets of MTL_0-formulae; $\alpha, \beta, \ldots$ denote algebraic terms.

Structures. We define a *two-sorted frame* to be a triple $\mathfrak{F} = (T, \mathfrak{D}; \mathrm{DIS})$, where T is the set of (time) points over which temporal formulae are evaluated, $\mathfrak{D}$ is the algebra of metric displacements in whose domain D terms take their values, and $\mathrm{DIS} \subseteq T \times D \times T$ is an accessibility relation relating pairs of points and displacements.

We require the following properties to hold for the components of two-sorted frames. First, we require $\mathfrak{D}$ to be an abelian group, that is, a 4-tuple $(D, +, -, 0)$ where $+$ is a binary function of ***displacement composition***, $-$ is a unary function of ***inverse displacement***, and 0 is the ***zero displacement*** constant, such that:

(*i*)	$\alpha + \beta = \beta + \alpha$	(commutativity of $+$)
(*ii*)	$\alpha + (\beta + \gamma) = (\alpha + \beta) + \gamma$	(associativity of $+$)
(*iii*)	$\alpha + 0 = \alpha$	(zero element of $+$)
(*iv*)	$\alpha + (-\alpha) = 0$	(inverse)

Second, we require the displacement relation DIS to respect the converse operation of the abelian group in the following sense: if DIS(i, α, j) then DIS$(j, -\alpha, i)$.

We turn a two-sorted frame $\mathfrak{F}$ into a *two-sorted model* by adding an interpretation for our algebraic terms, and a valuation for atomic temporal formulae. An *interpretation* for algebraic terms is given by a function $g : X \to D$ that is automatically extended to all terms from $T(X)$. A *valuation* is simply a function $V : \Phi \to 2^T$. Then, we say that an equation $\alpha = \beta$ is *true* in a model $\mathfrak{M} = (T, \mathfrak{D}; \text{DIS}; g, V)$ whenever $g(\alpha) = g(\beta)$. Next, *truth* of temporal formulae is defined by

$$\mathfrak{M}, i \Vdash p \text{ iff } i \in V(p)$$
$$\mathfrak{M}, i \Vdash \bot \quad \text{never}$$
$$\mathfrak{M}, i \Vdash \neg\phi \text{ iff } \mathfrak{M}, i \nVdash \phi$$
$$\mathfrak{M}, i \Vdash \phi \wedge \psi \text{ iff } \mathfrak{M}, i \Vdash \phi \text{ and } \mathfrak{M}, i \Vdash \psi$$
$$\mathfrak{M}, i \Vdash \Delta_\alpha\phi \text{ iff there exists } j \text{ such that } \text{DIS}(i, g(\alpha), j) \text{ and } \mathfrak{M}, j \Vdash \phi.$$

To avoid messy complications we only consider one-sorted consequences $\Gamma \models \phi$; for algebraic formulae '$\Gamma \models \phi$' means 'for all two-sorted models $\mathfrak{M}$, if $\mathfrak{M} \models \Gamma$, then $\mathfrak{M} \models \phi$'; for temporal formulae it means 'for all models $\mathfrak{M}$, and times instants i, if $\mathfrak{M}, i \Vdash \Gamma$, then $\mathfrak{M}, i \Vdash \phi$'.

A simple example. Even though the language of MTL_0 is very poor, it already allows us to express conditions on real-time systems. As a first example, consider a communication channel C that outputs each message with a delay δ with respect to its input time, and that neither generates nor loses messages (cf. Montanari et al [9]). C can be specified as follows:

$$out \leftrightarrow \Delta_{-\delta} in$$

This example can easily be generalized to the case of a channel C that collects messages from n different sources $S_1, \ldots, S_n$ and outputs them with a delay δ. To exclude that two input events can occur simultaneously, we add the constraint:

$$\forall i, j\, \neg(in(i) \wedge in(j) \wedge i \neq j),$$

which is shorthand for

$$\neg(in(1) \wedge in(2)) \wedge \ldots \wedge \neg(in(n-1) \wedge in(n)).$$

Then the behavior of C is specified by the formula

$$\forall i\, (out(i) \leftrightarrow \Delta_{-\delta} in(i)),$$

which is shorthand for a finite conjunction.

Notice that preventing input events from occurring simultaneously also guarantees that output events do not occur simultaneously.

Suppose now that C outputs the messages it receives from $S_1, \ldots S_n$ with a (generally different) delay $\delta_1, \ldots, \delta_n$, respectively. Constraining input events not to occur simultaneously no longer guarantees that there are no conflicts at output time. A simple strategy of conflict resolution consists in assigning a different priority to messages coming from different knowledge sources, so that, when a conflict occurs, C only outputs the message with highest priority. Accordingly, the specification of C is modified, preserving the requirement that it does not generate messages, but relaxing the requirement that it does not lose messages.

Assume that $S_1, \ldots, S_n$ are listed in decreasing order of priority. The behavior of C can be specified as follows:

$$\forall i\,(out(i) \leftrightarrow (\Delta_{-\delta_i} in(i) \wedge \neg\exists j\,(\Delta_{-\delta_j} in(j) \wedge j < i))))$$

which is a shorthand for

$$(out(1) \leftrightarrow \Delta_{-\delta_1} in(1)) \wedge (out(2) \leftrightarrow (\Delta_{-\delta_2} in(2) \wedge \neg\Delta_{-\delta_1} in(1))) \wedge \ldots \wedge$$

$$(out(n) \leftrightarrow (\Delta_{-\delta_n} in(n) \wedge (\neg\Delta_{-\delta_1} in(1) \wedge \ldots \wedge \neg\Delta_{-\delta_{n-1}} in(n-1))).$$

More realistic examples are given in the full paper.

Axioms. Our basic calculus MTL_0 has two components. On the one hand it has the usual laws of algebraic logic to deal with the displacements:

(Ref)	$\vdash \alpha = \alpha$	for all terms α (reflexivity)
(Sym)	$\vdash \alpha = \beta \implies \vdash \beta = \alpha$	(symmetry)
(Tra)	$\vdash \delta = \alpha,\ \alpha = \beta \implies \vdash \delta = \beta$	(transitivity)
(Rep)	$\vdash \alpha = \beta \implies \vdash [\alpha/x]\delta = [\beta/x]\delta$	(replacement)
(Sub)	$\vdash \alpha = \beta \implies \vdash [\delta/x]\alpha = [\delta/x]\beta$	(substitution),

as well as the above axioms (*i*)–(*iv*) for abelian groups. Here $[\alpha/x]\beta$ denotes the result of substituting α for all occurrences of x in β.

The second component of MTL_0 governs the temporal aspect of our structures; its axioms are the usual axioms of propositional logic plus

(Ax1)	$\nabla_\alpha(p \rightarrow q) \rightarrow (\nabla_\alpha p \rightarrow \nabla_\alpha q)$	(normality of ∇_α)
(Ax2)	$p \rightarrow \nabla_\alpha \Delta_{-\alpha} p,.$	

and its rules are modus ponens and

(NEC)	$\vdash \phi \implies \vdash \nabla_\alpha \phi$	(necessitation rule for ∇_α)
(SUB)	$\vdash \phi \leftrightarrow \psi \implies \vdash \chi(\phi/p) \leftrightarrow \chi(\psi/p)$	(uniform substitution)
	where (ϕ/p) denotes substitution of ϕ for the variable p	
(LIFT)	$\vdash \alpha = \beta \implies \vdash \nabla_\alpha \phi \leftrightarrow \nabla_\beta \phi$	(transfer of identities).

Axiom (Ax1) is the usual distribution axiom; axiom (Ax2) expresses that a displacement α is the converse of a displacement $-\alpha$. The rules (NEC) and (SUB) are familiar from modal logic, and the rule (LIFT) allows us to transfer provable algebraic identities from the displacement domain to the temporal domain.

A *derivation in MTL_0* is a sequence of terms and/or formulae $\sigma_1, \ldots, \sigma_n$ such that each σ_i $(1 \leq i \leq n)$ is either an axiom, or obtained from $\sigma_1, \ldots, \sigma_{n-1}$

by applying one of the derivation rules of MTL_0. We write $\vdash_{MTL_0} \sigma$ to denote that there is a derivation in MTL_0 that ends in σ.It is an immediate consequence of this definition that $\vdash_{MTL_0} \alpha = \beta$ iff $\alpha = \beta$ is provable (in algebraic logic) from the axioms of abelian groups only: whereas we can lift algebraic information from the displacement domain to the temporal domain using the (LIFT) rule, there is no way in which we can import temporal information into the displacement domain. As with consequences, we only consider one-sorted inferences '$\Gamma \vdash \phi$'.

Completeness. In this subsection we prove completeness for the basic calculus MTL_0. Our strategy will be to construct a canonical-like model by taking the free abelian group over our algebraic variables as the displacement component, by taking the familiar canonical model as the temporal component, and by linking the two in a suitable way.

The displacement domain. Recall that $T(X)$ is the collection of all algebraic terms built up from the variables in the set X. Define a congruence relation θ on $T(X)$ by taking

$$(\alpha, \beta) \in \theta \text{ iff } \vdash_{MTL_0} \alpha = \beta.$$

Then the *canonical displacement domain* $\mathfrak{D}^0$ is constructed by taking

$$\begin{aligned} D^0 &= T(X)/\theta \\ \alpha/\theta + \beta/\theta &= (\alpha + \beta)/\theta \\ -\alpha/\theta &= (-\alpha)/\theta \\ 0 &= 0/\theta. \end{aligned}$$

That $\mathfrak{D}^0$ is indeed an abelian group is easily shown using the defining axioms and rules of MTL_0. The group $\mathfrak{D}^0$ is known as the *free abelian group over* X (cf. Burris and Sankappanavar [3]).

We interpret our terms using the canonical mapping $g : T(X) \to \mathfrak{D}^0$ defined by $\alpha \mapsto \alpha/\theta$.

The temporal domain. A set of MTL_0-formulae is *maximal MTL_0-consistent* (or: an MCS) if it is MTL_0-consistent and it does not have proper MTL_0-consistent extensions. The *canonical temporal domain* T^0 is constructed by taking

$$T^0 = \{\Sigma \mid \Sigma \text{ is maximal } MTL_0\text{-consistent }\}.$$

Define a *canonical valuation* V^0 by putting $V^0(p) = \{\Sigma \mid p \in \Sigma\}$.

The canonical model for MTL_0. We almost have all the ingredients to define a canonical model for MTL_0; we only need to define a displacement relation $\mathrm{DIS}^0 \subseteq T^0 \times D^0 \times T^0$. This is done as follows:

$$\mathrm{DIS}^0(\Sigma, \alpha/\theta, \Gamma) \text{ iff for every formula } \gamma,\ \gamma \in \Gamma \text{ implies } \Delta_\alpha \gamma \in \Sigma$$
$$\text{(equivalently: for all formulae } \sigma, \text{ if } \nabla_\alpha \sigma \in \Sigma \text{ then } \sigma \in \Gamma).$$

Note that if $(\alpha, \beta) \in \theta$, then $\vdash \alpha = \beta$, hence $\vdash \nabla_\alpha \phi \leftrightarrow \nabla_\beta \phi$ by the (LIFT) rule, for all formulae ϕ. From this one easily derives that the definition of DIS^0 does not depend on the representative we take for α/θ.

Also, $\mathrm{DIS}^0(\Sigma, \alpha/\theta, \Gamma)$ implies $\mathrm{DIS}^0(\Gamma, -\alpha/\theta, \Sigma)$: if $\mathrm{DIS}^0(\Sigma, \alpha/\theta, \Gamma)$ and $\sigma \in \Sigma$, then $\nabla_\alpha \Delta_{-\alpha}\sigma \in \Sigma$ by axiom (Ax2), hence $\Delta_{-\alpha}\sigma \in \Gamma$.

Then, the *canonical model* for MTL_0 is the model $\mathfrak{M}^0 = (T^0, \mathfrak{D}^0; \mathrm{DIS}^0; g, V^0)$.

Theorem 1. *MTL_0 is sound and complete for the class of all MTL_0-frames.*

Proof. Proving soundness is left to the reader. To prove completeness we show that every consistent set of MTL_0-formulae is satisfiable in a model based on a two-sorted frame.

Let Σ be a MTL_0-consistent set of formulae; by standard techniques we can extend it to a maximal MTL_0-consistent set Σ^+ that lives somewhere in the canonical model $\mathfrak{M}^0$ for MTL_0. To complete the proof of the theorem it suffices to establish the following Truth Lemma. For all MTL_0-formulae ϕ and all $\Sigma \in T^0$:

$$\phi \in \Sigma \text{ iff } \mathfrak{M}^0, \Sigma \Vdash \phi.$$

This can be done using standard arguments from modal logic. ⊣

Imposing additional constraints. For many purposes two-sorted frames as we have studied them so far are too simple. In particular, they don't satisfy all the natural conditions one may want to impose on the displacement relation. Examples of such properties that arise in application areas such as real-time system specification include

Transitivity:	$\forall i, j, k, \alpha, \beta (\mathrm{DIS}(i, \alpha, j) \wedge \mathrm{DIS}(j, \beta, k) \rightarrow \mathrm{DIS}(i, \alpha + \beta, k))$
Quasi-functionality:	$\forall i, j, j', \alpha (\mathrm{DIS}(i, \alpha, j) \wedge \mathrm{DIS}(i, \alpha, j') \rightarrow j = j')$
Reflexivity:	$\forall i\, \mathrm{DIS}(i, 0, i)$
Antisymmetry:	$\forall i, j, \alpha (\mathrm{DIS}(i, \alpha, j) \wedge \mathrm{DIS}(j, \alpha, i) \rightarrow i = j \wedge \alpha = 0)$.

As in standard modal and temporal logic only some of the natural properties we want to impose on structures are expressible. In particular, the first three of the above properties are expressible in metric temporal logic, as follows (see Montanari et al [9]):

(Ax3)	$\nabla_{\alpha+\beta} p \rightarrow \nabla_\alpha \nabla_\beta p$	(transitivity)
(Ax4)	$\Delta_\alpha p \rightarrow \nabla_\alpha p$	(quasi-functionality w.r.t. the 3rd argument)
(Ax5)	$\nabla_0 p \rightarrow p$	(reflexivity)

In the case of Transitivity, Quasi-functionality, and Reflexivity we are able to extend the basic completeness result fairly effortlessly because each of the corresponding temporal formulae is a Sahlqvist formula. And the important feature of Sahlqvist formulae is that they are *canonical* in the sense that they are validated by the frame underlying the canonical model defined in the proof of Theorem 1 (see Goldblatt [5] for analogous arguments in standard modal and temporal logic, or De Rijke and Venema [15] for the general picture). As a consequence we have the following:

Theorem 2. *Let $X \subseteq \{\mathrm{Ax3}, \mathrm{Ax4}, \mathrm{Ax5}\}$. Then MTL_0X is complete with respect to the class of frames satisfying the properties expressed by the axioms in X.*

Further natural properties like Euclidicity ($\forall i, j, k, \alpha, \beta((\mathrm{DIS}(i, \alpha, j) \wedge \mathrm{DIS}(i, \alpha + \beta, k)) \rightarrow \mathrm{DIS}(j, \beta, k)))$ can already be derived from $MTL_0\mathrm{Ax3}$.

In the case of Antisymmetry, we have to do more work. First of all, Antisymmetry is not expressible in the basic metric language. One can use a standard unfolding argument to prove this claim (as in ordinary modal logic). Despite the undefinability of Antisymmetry, we can prove a completeness result for the class of antisymmetric two-sorted frames. Using a technique which is based on Burgess' chronicle construction (see Burgess [2]) it is indeed possible to prove the following theorem.

Theorem 3. *MTL_0 is complete with respect to the class of all antisymmetric two-sorted frames.*

3 Two-sorted frames based on ordered groups

For a variety of application purposes, our basic calculus and its semantics need to be extended with orderings. In particular, a linear order on the temporal domain is needed in many application areas; for instance, in real-time specification we want to guarantee that between any two time instants there is a unique displacement. In the following, we achieve this by adding a total ordering on the displacement domain D.

In the definition of a two-sorted frame we replace the abelian component by an *ordered* abelian group. That is, by a structure $\mathfrak{D} = (D, +, -, 0, <)$, where $(D, +, -, 0)$ is an abelian group, and $<$ is an irreflexive, asymmetric, transitive and linear relation that satisfies the comparability property (*viii*) below:

(*v*)	$\neg(\alpha < \alpha)$
(*vi*)	$\neg(\alpha < \beta \wedge \beta < \alpha)$
(*vii*)	$\alpha < \beta \wedge \beta < \gamma \rightarrow \alpha < \gamma$
(*viii*)	$\alpha < \beta \vee \alpha = \beta \vee \beta < \alpha.$

Next, there are two axioms expressing the relation between $+$ and $-$, and $<$:

(*ix*)	$\alpha < \beta \rightarrow \alpha + \gamma < \beta + \gamma$
(*x*)	$\alpha < \beta \rightarrow -\beta < -\alpha.$

One can use various languages to talk about ordered abelian groups. We do not have any clear preference, as long as the language used can be equipped with a complete axiomatization. We will simply use full first-order logic over $=$, $<$ to reason about the ordered abelian component of our two-sorted frames.

To be precise, our metric temporal language for talking about two-sorted frames based on an ordered abelian group, has a first-order component built up from terms in $T(X)$ and predicates $=$ and $<$; its temporal component is as before.

The interpretation of this language on two-sorted frames based on an ordered abelian group is fairly straightforward: the first-order component is interpreted on the group, and the temporal component on the temporal domain. Validity in this language is easily axiomatized; for the displacement component we take the axioms and rules of identity, ordered abelian groups, strict linear order together with any complete calculus for first-order logic; and for the temporal component we take the same axioms as in the case of MTL_0: axioms (Ax1), (Ax2) and the rules modus ponens, (NEC), (SUB) and (LIFT). Let MTL_1 denote the resulting two-sorted calculus.

Theorem 4. ***MTL_1 is complete with respect to the class of two-sorted frames based on ordered abelian groups.***

Proof. We can simply repeat the proof of Theorem 1 here, and replace the free algebra construction of the displacement domain by a Henkin construction for first-order logic. ⊣

3.1 Deriving a temporal ordering

Given that we have an ordering $<$ on the algebraic component of our frames, a natural definition for an ordering $\ll$ on the temporal frame suggests itself:

$$i \ll j \text{ iff for some } \alpha > 0, \mathrm{DIS}(i, \alpha, j). \tag{1}$$

So i and j are $\ll$-related if there exists a positive displacement between them. Using the relation $\ll$, we can define the qualitative operators F, P of non-metric temporal logic as follows:

$$\mathfrak{M}, i \Vdash F\phi := \exists j\,(i \ll j \wedge j \Vdash \phi) \text{ and } \mathfrak{M}, i \Vdash P\phi := \exists j\,(j \ll i \wedge j \Vdash \phi).$$

However, we will not consider this extension in this abstract.

Additional properties. The definition of $\ll$ given in (1) does not produce a temporal ordering with all the natural properties that we usually expect it to have. In particular, unless we put further restrictions on the relation of temporal displacement, $\ll$ will not be a strict linear order, and there may be time instants without a unique temporal distance between them.

To repair this situation, we assume that the displacement relation DIS satisfies the following properties: transitivity, quasi-functionality, reflexivity (as defined in Section 2), and total connectedness and quasi-functionality w.r.t. the second argument:

(xi) $\forall i, j \exists \alpha\, \mathrm{DIS}(i, \alpha, j)$ (total connectedness)
(xii) $\forall i, j, \alpha, \beta\, (\mathrm{DIS}(i, \alpha, j) \wedge \mathrm{DIS}(i, \beta, j) \rightarrow \alpha = \beta)$
(quasi-functionality w.r.t. the 2nd argument).

Given these assumptions on the displacement relation, we can show that the temporal relation $\ll$ as defined in (1) is a strict linear order. To see that $\ll$ is transitive, assume that $i \ll j \ll k$. Then there exist α, β with $\mathrm{DIS}(i, \alpha, j)$ and $\mathrm{DIS}(j, \beta, k)$. Hence $\mathrm{DIS}(i, \alpha + \beta, k)$ and $i \ll k$.

For irreflexivity, assume $i \ll i$. Then $\mathrm{DIS}(i, \alpha, i)$ for some $\alpha > 0$. By reflexivity of DIS, $\mathrm{DIS}(i, 0, i)$, hence, by quasi-functionality of the second argument, $\alpha = 0$ — a contradiction.

For asymmetry, assume $i \ll j \ll i$. Then $\mathrm{DIS}(i, \alpha, j)$ and $\mathrm{DIS}(j, \beta, i)$ for some α, $\beta > 0$. Then $\mathrm{DIS}(j, -\alpha, i)$ and so $\beta = -\alpha$, by quasi-functionality of the second argument again, which yields a contradiction.

Finally, to prove totality, take any two i, j. By total connectedness there exists α such that $\mathrm{DIS}(i, \alpha, j)$. By axiom (*viii*), $\alpha > 0 \vee \alpha = 0 \vee 0 > \alpha$. If $\alpha > 0$, then $i \ll j$. If $\alpha = 0$, then by quasi-functionality and reflexivity of DIS, $i = j$. And if $\alpha < 0$, then $-\alpha > 0$ and $\mathrm{DIS}(j, -\alpha, i)$, so $j \ll i$.

Let us call a two-sorted frame *nice* if it is transitive, reflexive, totally-connected, and quasi-functional in both the 2nd and 3rd argument of its displacement relation; a model is *nice* if it is based on an nice frame.

The next obvious question is: can we characterize the nice frames in the language of MTL_1? The answer is 'no'. To see this, we quickly adapt two truth preserving constructions from standard modal logic to the present setting. Due to space limitations we confine ourselves to frames that share the same displacement domain; however, the definitions are easily generalized to the general case.

Definition 5. Let $\mathfrak{F} = (T, \mathfrak{D}; \mathrm{DIS})$ and $\mathfrak{F}' = (T', \mathfrak{D}; \mathrm{DIS}')$ be two-sorted frames. The *disjoint union* of $\mathfrak{F}$ and $\mathfrak{F}'$ is the two-sorted frame $\mathfrak{F} \uplus \mathfrak{F}' = (T'', \mathfrak{D}, \mathrm{DIS}'')$. Here, T'' is the disjoint union of T and T', while the displacement relation DIS'' is just the disjoint union of DIS and DIS'.

Theorem 6. *Let $\mathfrak{F}$ and $\mathfrak{F}'$ be two-sorted frames, and $\mathfrak{F} \uplus \mathfrak{F}'$ their disjoint union. For all algebraic terms α, β, if $\mathfrak{F} \models \alpha = \beta$ and $\mathfrak{F}' \models \alpha = \beta$, then $\mathfrak{F} \uplus \mathfrak{F}' \models \alpha = \beta$. And, for all formulae ϕ, if $\mathfrak{F} \models \phi$ and $\mathfrak{F}' \models \phi$, then $\mathfrak{F} \uplus \mathfrak{F}' \models \phi$.*

Theorem 7. *There is no modal formula ϕ that expresses total connectedness of two-sorted frames.*

Proof. We prove the claim by showing that the existence of such a formula would violate preservation of truth under disjoint union. An intuitive account of this negative conclusion can be given noticing that disjoint unions are not totally connected frames "by definition".

Suppose that there exists a formula ϕ expressing total connectedness. By Theorem 6, it follows that ϕ is valid in the disjoint union $\mathfrak{F} \uplus \mathfrak{F}' = (T'', \mathfrak{D}; \mathrm{DIS}'')$ of any two frames $\mathfrak{F}$ and $\mathfrak{F}'$ validating ϕ. Take $i \in \mathfrak{F}$ and $j \in \mathfrak{F}'$; by definition of $\mathfrak{F} \uplus \mathfrak{F}'$, it follows that there exists no $\alpha \in \mathfrak{D}$ such that $\mathrm{DIS}''(i, \alpha, j)$. ⊣

Definition 8. Let $\mathfrak{F} = (T, \mathfrak{D}; \mathrm{DIS})$ and $\mathfrak{F}' = (T', \mathfrak{D}; \mathrm{DIS}')$ be two-sorted frames. A *bounded morphism* from $\mathfrak{F}$ to $\mathfrak{F}'$ is a mapping $f : T \to T'$ such that:

1. if $\mathrm{DIS}(i, \alpha, j)$, then $\mathrm{DIS}'(f(i), \alpha, f(j))$;
2. if $\mathrm{DIS}'(f(i), \alpha, j')$, then for some $j \in T$ both $f(j) = j'$ and $\mathrm{DIS}(i, \alpha, j)$ hold.

Theorem 9. *Let $\mathfrak{F}$ and $\mathfrak{F}'$ be two-sorted frames, and f a surjective bounded morphism from $\mathfrak{F}$ to $\mathfrak{F}'$. For all algebraic terms α, β, if $\mathfrak{F} \models \alpha = \beta$, then $\mathfrak{F}' \models \alpha = \beta$. And, for all formulae ϕ, if $\mathfrak{F} \models \phi$, then $\mathfrak{F}' \models \phi$.*

Theorem 10. *There is no modal formula ϕ that expresses quasi-functionality w.r.t. the second argument of the displacement relation.*

Proof. We prove the claim by showing that the existence of such a formula would violate preservation of truth under bounded morphisms. Suppose that there exists a formula ϕ expressing quasi-functionality with respect to the second argument of the accessibility relation.

Consider the two-sorted frames $\mathfrak{F} = (T, \mathfrak{D}; \mathrm{DIS})$ and $\mathfrak{F}' = (T', \mathfrak{D}; \mathrm{DIS}')$ such that $T = \{i_1, i_2, i_3, i_4, j_1, j_2, j_3, j_4\}$, $T' = \{i', j'\}$, DIS contains $(i_1, 1, j_1)$, $(i_1, 2, j_3)$, $(i_2, 2, j_1)$, $(i_2, 1, j_3)$, $(i_3, 1, j_2)$, $(i_3, 2, j_4)$, $(i_4, 1, j_4)$, and $(i_4, 2, j_2)$, together with the converse triplets $(j_1, -1, i_1)$, $(j_3, -2, i_1)$, and so on, while $\mathrm{DIS}' = \{(i', 1, j'), (i', 2, j'), (j', -2, i'), (j', -1, i')\}$. Clearly, $\mathfrak{F}$ satisfies the requirement of quasi-functionality, while $\mathfrak{F}'$ does not.

Now, consider the mapping $f : T \to T'$ defined by $f(i_1) = f(i_2) = f(i_3) = f(i_4) = i'$, $f(j_1) = f(j_2) = f(j_3) = f(j_4) = j'$. It is easy to verify that f is a surjective bounded morphism. Then, from $\mathfrak{F} \models \phi$ Theorem 9 allows us to infer that $\mathfrak{F}' \models \phi$, and we have a contradiction. ⊣

Enriching the language. Given that nice frames cannot be characterized in the language of MTL_1, a possible way out consists in enriching the language to make it possible to express the two properties of total connectedness and quasi-functionality of the displacement relation in its 2nd argument. We briefly show that those properties can actually be expressed by adding to the language the future and past operators F, P, the difference operator $\mathcal{D}$, and by allowing information to be lifted from the temporal domain to the displacement domain by permitting the two languages to be mixed.

First, the *difference operator* (De Rijke [14]) is a unary modal operator $\mathcal{D}$ that allows us to model unbounded jumps. Its semantic interpretation is defined as follows:

$$(\mathfrak{F}, V), i \Vdash \mathcal{D}\phi \text{ iff } \exists j (j \neq i \wedge (\mathfrak{F}, V), j \Vdash \phi)$$

with dual $\overline{\mathcal{D}}$:

$$(\mathfrak{F}, V), i \Vdash \overline{\mathcal{D}}\phi \text{ iff } \forall j (j \neq i \rightarrow (\mathfrak{F}, V), j \Vdash \phi).$$

The difference operator and its dual allow us to define three derived unary operators $\mathcal{E}$, its dual $\mathcal{A}$, and $\mathcal{U}$ that respectively model truth in at least one world, truth in all worlds, and truth in one and only one world:

$$\mathcal{E}\phi \equiv \mathcal{D}\phi \vee \phi, \ \mathcal{A}\phi \equiv \overline{\mathcal{D}}\phi \wedge \phi, \text{ and } \mathcal{U}\phi \equiv \mathcal{E}(\phi \wedge \neg\mathcal{D}\phi).$$

In a language in which the algebraic and temporal formulas may be mixed, properties *(xi)* and *(xii)* can be axiomatized by means of the qualitative operators F, P and $\mathcal{D}$, $\mathcal{E}$, and $\mathcal{U}$ as follows:

(Ax6) $\mathcal{D}p \rightarrow Fp \vee Pp$ (total connectedness of DIS)
(Ax7) $\mathcal{U}p \wedge \mathcal{U}q \rightarrow (\mathcal{E}(p \wedge \Delta_\alpha q) \wedge \mathcal{E}(p \wedge \Delta_\beta q) \rightarrow \alpha = \beta)$
(quasi-functionality of DIS w.r.t. the 2nd argument).

Details are supplied in the full paper.

However, we prefer to remain within the original language of MTL_1 and reason about nice frames there, mainly because adding the axioms Ax6 and Ax7 forces us to give up the simplicity of the basic calculus and to include non-standard derivation rules to govern the difference operator. As we will show below, the logic of nice frames can be captured in the original language.

Completeness for nice frames. Instead of increasing the expressive power of metric temporal logic, we can leave it as it stands, and prove a completeness result for nice frames in the old language. We will do this in two steps. We first prove completeness with respect to totally connected frames via some sort of generated submodel construction, and then we prove the full result.

Here's the idea for the case of total connectedness. Let $\mathfrak{F} = (T, \mathfrak{D}; \mathrm{DIS})$ be a two-sorted frame. The *master relation* on $\mathfrak{F}$ is defined by

$$(i,j) \in \mathit{Master} \text{ iff } (i,j) \in (\ll \cup \gg)^*.$$

Thus i, j are in the master relation iff there exists a zig zag path along the displacement relation from i to j in the following sense:

$$\mathrm{DIS}(i, \alpha_1, j_1), \mathrm{DIS}(j_1, \alpha_2, j_2), \ldots, \mathrm{DIS}(j_n, \alpha_{n+1}, j),$$

where $\alpha_1, \ldots, \alpha_n \in D$, and $j_1, \ldots, j_n \in T$.

A *point-generated component* of a model $\mathfrak{M} = (T, \mathfrak{D}; \mathrm{DIS}; g, V)$ is a model $(T', \mathfrak{D}; \mathrm{DIS}'; g, V')$ such that for some $i \in T$,

- $T' = \{j \in T \mid (i,j) \in \mathit{Master}\}$
- $\mathrm{DIS}' = \mathrm{DIS} \cap (T' \times D \times T')$
- $V'(p) = V(p) \cap T'$, for all p.

Proposition 11. *Let $\mathfrak{M}'$ be a point-generated component of a model $\mathfrak{M}$ based on a two-sorted frame with ordered abelian group. If $\mathfrak{M}$ has a transitive displacement relation, then $\mathfrak{M}'$ has a transitive and totally connected displacement relation.*

Lemma 12. *Let $\mathfrak{M}'$ be a point-generated component of a two-sorted model $\mathfrak{M}$. Then $\mathfrak{M}'$ satisfies exactly the same algebraic formulae as $\mathfrak{M}$. Moreover, for all $i \in T'$ and for all temporal formulae ϕ we have $\mathfrak{M}, i \Vdash \phi$ iff $\mathfrak{M}', i \Vdash \phi$.*

MTL_1Ax3 extends MTL_1 with the transitivity axiom $\nabla_{\alpha+\beta} p \rightarrow \nabla_\alpha \nabla_\beta p$.

Theorem 13. *MTL_1Ax3 is sound and complete with respect to the class of two-sorted frames based on ordered abelian groups whose displacement relation is transitive and totally connected.*

Proof. We only prove completeness, and to establish this it suffices to show that every MTL_1Ax3-consistent set of formulae is satisfiable in a model based on a frame of the right kind.

Let Γ be a MTL_1Ax3-consistent set of formulae. By a Sahlqvist style argument (see Theorem 2) it is easily seen that Γ is satisfiable in a model $\mathfrak{M}$ based on a two-sorted frame with a transitive displacement relation, say at a time instant i. Let $\mathfrak{M}'$ be a point-generated component of $\mathfrak{M}$ that contains i. By Proposition 11 $\mathfrak{M}'$ has a transitive and totally connected displacement relation, and by Lemma 12 we have $\mathfrak{M}', i \Vdash \Gamma$, as required. ⊣

To prove completeness w.r.t. the class of nice frames, we need to carry out a second construction. First, call a two-sorted frame ***almost nice*** if it is transitive, reflexive, totally-connected, and quasi-functional in the 3rd argument of its displacement relation; a model is ***almost nice*** if it is based on an almost nice frame. So a frame is nice if it is almost nice and quasi-functional in the 2nd argument of its displacement relation.

Now, to build a nice model we will take an almost nice model and carefully unfold it. To be precise, let $\mathfrak{M} = (T, \mathfrak{D}; \mathrm{DIS}; g, V)$ be an almost nice model, and let $i \in T$. The ***i-stratification of*** $\mathfrak{M}$ is the model $\mathfrak{M}' = (T', \mathfrak{D}; \mathrm{DIS}'; g, V')$ which is defined as follows:

$$\begin{aligned}
T' &= \{(0,i)\} \cup \{(\alpha, j) \mid \mathrm{DIS}(i, \alpha, j) \text{ in } \mathfrak{M}\} \\
\mathrm{DIS}_0 &= \{((0,i), \alpha, (\alpha, j)) \mid (\alpha, j) \in T'\} \cup \{((\alpha, j), -\alpha, (0,i)) \mid (\alpha, j) \in T'\} \\
\mathrm{DIS}_1 &= \{((\alpha, j), \beta - \alpha, (\beta, k)) \mid (\alpha, j), (\beta, k) \in T'\} \\
\mathrm{DIS}' &= \mathrm{DIS}_0 \cup \mathrm{DIS}_1 \\
V'(p) &= \{(\alpha, j) \in T' \mid j \in V(p)\}.
\end{aligned}$$

Observe that $\mathrm{DIS}_0 \subseteq \mathrm{DIS}_1$.

Proposition 14. *Let $\mathfrak{M}$ be an almost nice model, and let $i \in \mathfrak{M}$. The i-stratification of $\mathfrak{M}$ is nice.*

Proof. We first observe first that for any pairs (α, j), $(\gamma, k) \in T'$, and $\beta \in \mathfrak{D}$, if $\mathrm{DIS}'((\alpha, j), \beta, (\gamma, k))$ holds then $\beta = \gamma - \alpha$.

Now, to prove the proposition, we have to check the nice-ness properties. First of all, we show that $\mathrm{DIS}'((\alpha, j), \beta, (\gamma, k))$ implies $\mathrm{DIS}'((\gamma, k), -\beta, (\alpha, j))$. By the observation $\beta = \gamma - \alpha$. Also, (α, j), $(\gamma, k) \in T'$ implies $\mathrm{DIS}'((\gamma, k), \alpha - \gamma, (\alpha, j))$, that is, $\mathrm{DIS}'((\gamma, k), -\beta, (\alpha, j))$.

Next, we show that DIS' is reflexive. As $\mathfrak{M}$ is assumed to be reflexive, we have $\mathrm{DIS}(i, 0, i)$, hence $\mathrm{DIS}((0, i), 0, (0, i))$. As to other points $(\alpha, j) \in T'$, $\mathrm{DIS}_0((0, i), \alpha, (\alpha, j))$ and $\mathrm{DIS}_0((\alpha, j), -\alpha, (0, i))$ imply $\mathrm{DIS}'((\alpha, j), 0, (\alpha, j))$.

To see that DIS' is quasi-functional with respect to its 3rd argument, assume $\mathrm{DIS}'((\alpha, j), \beta, (\gamma, k))$ and $\mathrm{DIS}'((\alpha, j), \beta, (\gamma', k'))$. We need to show that $\gamma = \gamma'$

and $k = k'$. First of all, $\beta = \gamma - \alpha = \gamma' - \alpha$, hence $\gamma = \gamma'$. Therefore, DIS(i, γ, k) and DIS(i, γ, k'). So by the assumption that DIS is quasi-functional in its 3rd argument, $k = k'$.

Given that $\mathfrak{M}$ is total, the totality of its i-stratifications is immediate.

Transitivity of $\mathfrak{M}'$ may be established as follows: assume DIS$'((\alpha, j), \beta, (\gamma, k))$ and DIS$'((\gamma, k), \beta', (\delta, l))$. Then DIS$'((\alpha, j), \delta - \alpha, (\delta, l))$. As $\beta + \beta' = (\gamma - \alpha) + (\delta - \gamma)$, we are done.

Finally, to prove quasi-functionality of DIS$'$ in its 2nd argument, assume DIS$'((\alpha, j), \beta, (\gamma, k))$ and DIS$'((\alpha, j), \beta', (\gamma, k))$. It follows that $\beta = \gamma - \alpha = \beta'$. $\dashv$

Proposition 15. *Let $\mathfrak{M}$ be an almost nice model, and let $\mathfrak{M}'$ be an i-stratification of $\mathfrak{M}$. For all formulae ϕ, j in $\mathfrak{M}$, and (α, j) in $\mathfrak{M}'$, we have $\mathfrak{M}, j \Vdash \phi$ iff $\mathfrak{M}', (\alpha, j) \Vdash \phi$.*

Proof. This is by induction on ϕ. The base case and the boolean cases are trivial. So consider a temporal formula $\Delta_\gamma \psi$. Assume first that $j \Vdash \Delta_\gamma \psi$. Then there exists k with DIS(j, γ, k). Now, let α be such that $(\alpha, j) \in T'$. Then DIS(i, α, j), and hence DIS$(i, \alpha+\gamma, k)$ and $(\alpha+\gamma, k) \in T'$. By definition, DIS$_0((0, i), \alpha, (\alpha, j))$ and DIS$_0((0, i), \alpha+\gamma, (\alpha+\gamma, k))$. But then DIS$'((\alpha, j), \gamma, (\alpha+\gamma, k))$. By induction hypothesis, $(\alpha + \gamma, k) \Vdash \psi$, hence $(\alpha, j) \Vdash \Delta_\gamma \psi$.

Conversely, assume that $(\alpha, j) \Vdash \Delta_\gamma \psi$. Then there exists $(\beta, k) \in T'$ with DIS$'((\alpha, j), \gamma, (\beta, k))$ and $(\beta, k) \Vdash \psi$. Hence $\gamma = \beta - \alpha$. By construction we must have DIS(i, α, j) and DIS(i, β, k) and hence DIS$(j, \beta - \alpha, k)$. As $k \Vdash \psi$ (by induction hypothesis) and $\gamma = \beta - \alpha$, this implies $j \Vdash \Delta_\gamma \psi$, as required. $\dashv$

We are ready now for a completeness result for the class of nice frames. Let MTL_2 denote the extension of MTL_1 with axioms Ax3, Ax4 and Ax5 (expressing transitivity, quasi-functionality of DIS in its 3rd argument, and reflexivity, respectively). By an easy adaptation of the proof of Theorem 13, MTL_2 is sound and complete w.r.t. the class of almost nice frames.

Theorem 16. *MTL_2 is sound and complete with respect to the class of nice frames.*

Proof. We only show that every MTL_2-consistent set of temporal formulae is satisfiable on a nice model. Let Γ be such a set. By earlier remarks Γ is satisfiable on an almost nice model at some time instant i. Let $\mathfrak{M}'$ be the i-stratification of $\mathfrak{M}$. By Propositions 14 and 15 $\mathfrak{M}'$ is a nice model that satisfies Γ at i. $\dashv$

Conclusion

In this paper we have proved completeness results for basic systems of metric temporal logic. We started with the minimal calculus and showed how to extend it to obtain the logic of two-sorted frames with a linear temporal order in which there exists a unique temporal distance between any two time instants.

So far we have only considered simple languages that do not allow us to lift information from the temporal domain to the algebraic domain. Obviously, for application purposes they have to be extended. In particular, we are considering the possibility of a restricted form of mixing temporal and displacement formulae, so as to enable more complex ways of interaction between the two domains.

References

1. R. Alur and T.A. Henzinger. Real-time logics: complexity and expressiveness. *Information and Computation*, 104:35–77, 1993.
2. J.P. Burgess. Basic tense logic. In: D.M. Gabbay and F. Guenther (eds), *Handbook of Philosophical Logic*. Vol. 2. Dordrecht, Reidel, 1984, pages 89–134.
3. S. Burris and H.P. Sankappanavar. *A Course in Universal Algebra*. Springer, New York, 1981.
4. G. D'Agostino, A. Montanari and A. Policriti. A set-theoretic translation method for polymodal logics. In *Proc. of STACS '95*, LNCS 900, Springer, Berlin, 1995 217–228. To appear in *Journal of Automated Reasoning*.
5. R. Goldblatt. *Logics of Time and Computation*. 2nd edition. CSLI Lecture Notes No. 7, Stanford, 1992.
6. T.H. Henzinger. *The Temporal Specification and Verification of Real-Time Systems*. PhD thesis, Department of Computer Science, Stanford University, 1991.
7. R. Koymans. *Specifying Message Passing and Time-Critical Systems with Temporal Logic*. LNCS 651, Springer, Berlin, 1992. The relevant sections appeared in *Journal of Real-Time Systems*, 2:255–299, 1990.
8. Z. Manna and A. Pnueli. *The Temporal Logic of Reactive and Concurrent Systems*. Springer Verlag, 1992.
9. A. Montanari, E. Ciapessoni, E. Corsetti and P. San Pietro. Dealing with time granularity in logical specifications of real-time systems. The synchronous case. Research Report 07/92, Università di Udine, 1992. A short version appeared in *Science of Computer Programming*, 20:141–171, 1993.
10. H.J. Ohlbach. Translation methods for non-classical logics: an overview. *Bull. of the IGLP*, 1:69–89, 1993.
11. J.S. Ostroff. *Temporal Logic of Real-Time Systems*. Research Studies Press, 1990.
12. N. Rescher and J. Garson. Topological logic. *Journal of Symbolic Logic*, 33:537-548, 1968.
13. N. Rescher and A. Urquhart. *Temporal Logic*. Library of Exact Philosophy, Springer-Verlag, Berlin, 1971.
14. M. de Rijke. The modal logic of inequality. *Journal of Symbolic Logic*, 57:566–584, 1992.
15. M. de Rijke and Y. Venema. Sahlqvist's theorem for boolean algebras with operators (with an application to cylindric algebras). *Studia Logica*, 54:61–78, 1995.

On mechanizing proofs within a complete proof system for *Unity*

Naïma BROWN Abdelillah MOKKEDEM
CRIN-CNRS & INRIA-Lorraine, BP239, 54506 Vandœuvre-Lès-Nancy, France
e-mail: brown@loria.fr,mokkedem@loria.fr

Abstract. The solution proposed by Sanders in [14] consists of eliminating the need of the substitution axiom from Unity in order to eliminate the unsoundness problem caused by this axiom in Unity without loss of completeness. Sander's solution is based on the **strongest invariant** concept and provides theoretical advantages by formally capturing the effects of the initial conditions on the properties of a program. This solution is less convincing from a practical point of view because it assumes proofs of strongest invariant in the meta-level. In this paper we reconsider this solution showing that the general concept of invariant is sufficient to eliminate the substitution axiom and to provide a sound and relatively complete proof system for Unity logic. The advantage of the new solution is that proofs of invariants are mechanized inside the Unity logic itself.

1 Introduction

Unity, due to Chandy and Misra [5, 11, 9], is today highly perceived to provide a tractable methodology to formally verify and derive a wide variety of interesting parallel algorithms. The strength of *Unity* arises from its success in separating the design of an (abstract) algorithm from the control flow that occurs in any of its realizations. Issues of control flow, architectures, efficiency, etc. have no place in the initial design of an algorithm; they are later concerns. The method is completely deductive in the sense that operational reasoning is eliminated entirely; all inferences are done within a temporal logic.

On the theoretical side, however, important questions have remained controversial for quite some time. It turns out that the major of this confusion arises from an erroneous combination of the substitution axiom SA with the temporal operators *unless* and *leadsto* probably due to the informal way SA is stated. In fact, despite the very reasonable sounding justification for it as a generalization of Leibniz's rule, SA has been widely neglected and misunderstood. Sanders [14] argued that combining the substitution axiom with the temporal operators as they are defined in *Unity* [5] gives an unsound proof system. Nevertheless, if SA is omitted, then temporal operator definitions (which do have desirable properties) can be used and the proof system is sound, but incomplete [4]. To solve this problem, Sanders [14] modified the *Unity* logic in such a way that the substitution axiom is no longer needed and the logic remains complete (in the sense of Cook [6]). Her definition of the temporal operators is based on the concept of strongest invariant sst.$F.INIT$ which exists for any program F and is unique. From a theoretical point of view this solution provides a convenient result of soundness and completeness for the resulting *Unity* logic. However, from a practical point of view, when considering infinite state *Unity* programs, sst.$F.INIT$, although exists, it is not always expressible in a finitary assertional language, and its proof within a logic such as Unity is in general impossible to mechanize in practice.

In this paper we investigate a rather semantical definition for the operators *unless* and *leadsto* based on *Unity* program behaviours and strengthen rules of the logic in order to capture the effect that some states may be unreachable from the initial state. This is done by strengthen premisses of the inference rules with some *general invariant* (or simply *invariant*) which is defined independently of the operator *unless.* The resulting proof system no longer needs the substitution axiom and is sound and relatively complete for both *invariant*, *unless*, and *leadsto* operators. All proofs are mechanized within the resulting proof system for *Unity* logic. In the same way as Sanders [14], soundness and completeness are derived from an embedding of the Unity logic into Gerth-Pnueli's temporal logic defined for a class of programs called SLP (single location programs) and proved sound and relatively complete for the operators $\mathcal{U}$ (*weak until*) and $\Diamond$ (*eventually*) [8]. In order to provide a formal background for proving inter-reducibility of proofs of *unless* and $\mathcal{U}$, and *leadsto* and $\Diamond$, we define a uniform semantical framework for both Unity and SLP programs and show that for each Unity program there exists a semantically equivalent SLP program.

Our paper is organized as follows. We first give a brief introduction to the *Unity* formalism, state the problem of the substitution axiom, and recall the solution proposed by Sanders. In section 3 we define an abstract computational model and in section 4 we give a uniform representation of semantics of Unity programs and SLP programs within this computational model. In section 5 we propose a new definition for the temporal operators *invariant*, *unless* and *leadsto*, and a reformulation of rules of the Unity logic so that the effect that some states may be unreachable from the initial state becomes captured according to this definition. We close the paper with some remarks and related work.

2 Unity

Unity [5] is a formalism for specifying, designing and verifying concurrent programs. It consists of a notation for writing programs and a logic for reasoning about them. The notation provides constructs that avoid specifying sequential dependencies which are not inherent in the problem. A *Unity* program describes both the operational behavior of the program (using logical assertions) and assignment statements that describe the computation. According to *Unity* syntax, given in figure 1, we consider the **assign** section as a (finite) set of assignment statements composed with $[\!]$ (nondeterminism). Unlike sequential languages, the order of execution of the assignment statements has no relation to the order in which they are written. *Unity* programs terminate by reaching a *fixed-point.* A program reaches a fixed-point in its execution when no statement's execution changes the state of the program. In this paper, we do not use the **always** section, as every *Unity* program can be transformed into an equivalent one without **always**.

We denote a *Unity* program by a triplet $u \stackrel{def}{=} (Var_u, Init_u, Ass_u)$ where,

- Var_u is the set of variables declared in the section **declare**,
- $Init_u$ is the predicate denoting initial states in the section **initially**,
- Ass_u is the set of assignment statements contained in the section **assign**.

Problem specifications in *Unity* are written using three basic binary operators *unless*, *ensures* and *leadsto* that are defined as follows :

Program *program_name*
  declare *declare_section*
  always *always_section*
  initially *initially_section*
  assign *assign_section*
end

assign_section	$\longrightarrow$	*statement_list*
statement_list	$\longrightarrow$	*statement* $\{[\!]$ *statement*$\}^*$
statement	$\longrightarrow$	*assignment_statement* $\mid$
		$\langle [\!]$ *quantification statement_list*$\rangle$
assignment_statement	$\longrightarrow$	*assignment_component* $\{\|\|$ *assignment_component*$\}^*$
assignment_component	$\longrightarrow$	*enumerated_assignment* $\mid$
		$\langle \|\|$ *quantification assignment_statement*$\rangle$
enumerated_assignment	$\longrightarrow$	*variable_list* := *expr_list*
quantification	$\longrightarrow$	*variable_list* : *boolean_expr*

Example.

Program *Sort*
declare i : integer ; A : Array [1..N] of integer ;
assign
$\langle \|\| i \ : \ 0 \leq i < N \wedge even(i) \ :: \ A[i], A[i+1] := A[i+1], A[i] \quad if \quad A[i] > A[i+1]\rangle$
$[\!]\, \langle \|\| i \ : \ 0 \leq i < N \wedge odd(i) \ :: \ A[i], A[i+1] := A[i+1], A[i] \quad if \quad A[i] > A[i+1]\rangle$
end $\{Sort\}$

Fig. 1. Syntax of *Unity*

Chandy-Misra's Proof system : S_{CM_1}

unless operator : $p \ unless \ q \equiv \ < \forall s. \ s \in Ass_u \ :: \ \{p \wedge \neg q\} \ s \ \{p \vee q\} >$

ensures operator : $p \ ensures \ q \ \equiv \ (p \ unless \ q) \wedge < \exists s. \ s \in Ass_u \ :: \ \{p \wedge \neg q\} \ s \ \{q\} >$

leadsto operator :
$$\frac{p \ ensures \ q}{p \ leadsto \ q}$$

$$\frac{p \ leadsto \ q, \qquad q \ leadsto \ r}{p \ leadsto \ r}$$

For any set W :
$$\frac{< \forall m \in W. \ p(m) \ leadsto \ q >}{< \exists m \in W. \ p(m) > \ leadsto \ q}$$

Substitution Axiom SA :
If $(x = y)$ is an invariant of a program, then x can be replaced by y in all properties of the program. The most frequent use of the rule involves replacement of an invariant with true and inversely.

2.1 The Problem of the Substitution Axiom: Sanders's solution

Combining the substitution axiom with the defined temporal operators *unless*, *ensures* and *leadsto* as defined above gives an unsound proof system [14]. To solve this problem, Sanders [14] modifies the semantics of the operators *unless* and *ensures* in such a way that the set of reached states of a program becomes captured in the definition itself. She reformulates the Unity logic using the notion of *strongest invariant* of a program.

Sanders's Proof System: S_{SI}
$p\ unless\ q \equiv < \forall s.\ s \in Ass_u :: p \wedge \neg q \wedge SI \Rightarrow wp.s.(p \vee q) >$
$p\ ensures\ q \equiv p\ unless q \wedge < \exists s \in Ass_u :: p \wedge \neg q \wedge SI \Rightarrow wp.s.q >$
$invariant\ p \equiv [SI \Rightarrow p]$

The rules of the operator *leadsto* are similar to those of the Chandy-Misra proof system S_{CM_1}.

Using the strongest invariant in the definition of *unless* and *ensures* provides theoretical advantages by clarifying the role of the initial conditions on the properties of programs. In this sense Sander's solution provides a convenient way to prove soundness and relative completeness of the modified Unity logic. However, from a practical point of view, it does not provide an appropriate framework for a full mechanization of proofs. The temporal operators *unless* and *ensures* are defined in terms of the *strongest invariant* and thus a proof mechanization needs a mechanization of the *strongest invariant* of a program. However, for the case of infinite state programs, this predicate cannot be finitely axiomatized in the *Unity* logic in general. We will show below that a reformulation of Unity based on the general notion of *invariant* will be enough to achieve an *interactive* proof system which is sound and relatively complete. We will give a rather operational definition of the temporal operators which considers only reached states and we reformulate rules according to this new definition. The notion of general invariant provides a sufficient tool for this formulation. The resulting proof system is relatively complete and provides a convenient tool for a full mechanization of proofs.

3 A general semantical framework

We define in this section an abstract computational model that we use as a common framework to define semantics of Unity and SLP programs. The basis of this computational model is the notion of *fair transition systems.* In such a model, concurrent program behaviour can be easily modeled by all possible totally ordered execution sequences arising from (a fair) interleaving of actions in the separate 'sequential' processes of the concurrent program (interleaving semantics).

3.1 Transition systems

Definition 1. Let V be a set of variables and D be a set of values. A transition system $S(V,D)$ is defined by $S(V,D) \stackrel{def}{=} (\Gamma(V,D), A(V,D), I(V,D), FP(V,D))$, where

- $\Gamma(V, D)$ denotes the set of mappings (also called states) from V to D
- $A(V, D)$ denotes the set of actions over V and $D : A(V, D) \subseteq \mathcal{P}(\Gamma(V, D) \times \Gamma(V, D))$
- $I(V, D) \subseteq \Gamma(V, D)$ denotes the set of initial states.
- $FP(V, D)) \subseteq \Gamma(V, D)$ denotes the set of stable (or fix-point) states, such that: $\forall \gamma \in FP(V, D). \forall \alpha \in A(V, D).((\gamma, \gamma) \in \alpha)$

Remark. A stable state γ is reached when no enabled action action does change it; it is used to model the terminal state of a program.

Let $\alpha \in A(V, D)$ and γ, γ' be two states, $\gamma \xrightarrow{\alpha} \gamma'$ is equivalent to $(\gamma, \gamma') \in \alpha$ and represents a transition.

3.2 Modeling parallel transitions

Definition 2. Let α be an action in $A(V, D)$
For every transition $(\gamma, \gamma') \in \alpha$,

1. $W(\gamma, \gamma') = \{v \in V : \gamma(v) \neq \gamma'(v)\}$
   $W(\gamma, \gamma')$ denotes the set of variables modified by the transition (γ, γ').
2. $U(\gamma, \gamma') = \{v \in V : \gamma(v) = \gamma'(v)\}$
   $U(\gamma, \gamma')$ denotes the set of variables not used in the transition (γ, γ').

Definition 3. $\odot$-product. Let $(\gamma_1, {\gamma'}_1)$ et $(\gamma_2, {\gamma'}_2)$ be two transitions :

$(\gamma_1, {\gamma'}_1) \odot (\gamma_2, {\gamma'}_2)$

pre-condition

1) $\gamma_1 = \gamma_2$ (let $\gamma = \gamma_1 = \gamma_2$)
2) $v \in (U(\gamma, \gamma_1') \cap U(\gamma, \gamma_2')) \cup (W(\gamma, \gamma_1') \cap W(\gamma, \gamma_2')) \Rightarrow \gamma_1'(v) = \gamma_2'(v)$

post-condition

$(\gamma_1, {\gamma'}_1) \odot (\gamma_2, {\gamma'}_2) = (\gamma, \gamma')$ where

$$\gamma'(v) = \gamma_1'(v) \text{ if } v \in W(\gamma_1, \gamma_1')$$
$$\gamma'(v) = \gamma_2'(v) \text{ otherwise.}$$

A variable may be modified differently by two different transitions, but the definition of (the precondition of) their $\odot$-product guarantees consistency, therefore the $\odot$-product may assume an undefined value in this case. The $\odot$-product will be used to define the semantics of $\|$.

3.3 Transition system behaviour

Definition 4. Execution sequences. An execution sequence over $S(V, D)$ is an infinite sequence of transitions :

$$\gamma_0 \xrightarrow{\alpha_0} \gamma_1 \xrightarrow{\alpha_1} \dots \xrightarrow{\alpha_{i-1}} \gamma_i \xrightarrow{\alpha_i} \gamma_{i+1} \xrightarrow{\alpha_{i+1}} \dots \; s.t\ \gamma_0 \in I(V, D),\ \gamma_i \in \Gamma(V, D),\ \alpha_i \in A(V, D)$$

Let $\sigma : \gamma_0 \xrightarrow{\alpha_0} \gamma_1 \xrightarrow{\alpha_1} \ldots \xrightarrow{\alpha_{i-1}} \gamma_i \xrightarrow{\alpha_i} \gamma_{i+1} \xrightarrow{\alpha_{i+1}} \ldots$ be a sequence, $s_k.\sigma$ and $a_k.\sigma$ respectively represents the state γ_k and the action α_k in the sequence σ. The set of all possible execution sequences over $S(V, D)$ is denoted by $\Sigma(S(V, D))$.

Definition 5. Let σ, τ be two sequences. $\sigma \simeq \tau$ iff $\forall i \geq 0.\ s_i.\sigma = s_i.\tau$.
$\simeq$ is extended to set of sequences Σ_1 and Σ_2 as follows :
$\Sigma_1 \simeq \Sigma_2$ iff $\forall \sigma_1 \in \Sigma_1.\ \exists \sigma_2 \in \Sigma_2.\ \sigma_1 \simeq \sigma_2$ and $\forall \sigma_2 \in \Sigma_2.\ \exists \sigma_1 \in \Sigma_1.\ \sigma_2 \simeq \sigma_1$.

Definition 6. Justice. An execution sequence is said *unjust* if there exists an action which is continuously enabled from a state but never executed. An execution sequence which is not *unjust* is said *just*. We denote by $JUST(S(V, D))$ the set of *just* execution sequences over $S(V, D)$.

3.4 Fair transition systems

Definition 7. A fair transition system $FS(V, D)$ is a pair $(S(V, D), F(V, D))$ where, $S(V, D)$ is a transition system and $F(V, D)$ is the set of execution sequences generated under some fairness constraint F. For example, $F(V, D)$ can be $\Sigma(S(V, D))$ or $Just(S(V, D))$ or another set of execution sequences generated under a given fairness constraint.
The semantics of a concurrent program is defined in terms of a fair transition system $(S(V, D), F(V, D))$.

Definition 8. Observational equivalence. Two fair transition systems $(S_1(V, D), F_1(V, D))$ and $(S_2(V, D), F_2(V, D))$ are said to be observationally equivalent if and only if $S_1(V, D) = S_2(V, D)$ and $F_1(V, D) \simeq F_2(V, D)$.

4 A unified operational semantics for *Unity* and *SLP*

4.1 Semantics of *Unity*

Let u a *Unity* program: $u \stackrel{def}{=} (Var_u, Init_u, Ass_u)$. D_u denotes the set of data (or values) of u, $D_u = \bigcup_{x \in Var_u} D_x$ where D_x denotes the domain of the variable x.

Definition 9. Operational semantics of a Unity program. The operational semantics of u is defined by the structure $O[u] = (S_u, J_u)$ where, for every action $\alpha \in Ass_u$, $[\![\alpha]\!]$ is defined as follows :

- α is of the form $v := e$: $[\![\alpha]\!] \stackrel{def}{=} \{(\gamma, \gamma') \in \Gamma_u : \gamma' = \gamma[v \mapsto \gamma(e)]\}$
- α is of the form $v_1, \ldots, v_n := e_1, \ldots, e_n$: $[\![\alpha]\!] \stackrel{def}{=} \odot_{i \in \{1 \ldots n\}} [\![v_i := e_i]\!]$
- α is of the form $lv := le_0\ if\ b_0 \sim \ \ldots\ \sim le_n\ if\ b_n$: (lv and $le_{i_{i \in \{0, \ldots, n\}}}$ are respectively a list of variables and a list of expressions)

  $[\![\alpha]\!] \stackrel{def}{=} \{(\gamma, \gamma') \in \Gamma_u : [(\gamma, \gamma') \in [\![lv := le_0]\!] \wedge b_0(\gamma)] \vee \ \ldots\ \vee [(\gamma, \gamma') \in [\![lv := le_n]\!] \wedge b_n(\gamma)] \vee [(\gamma = \gamma') \wedge \neg(b_0 \vee \ldots \vee b_n)(\gamma)]\}$

  $\gamma(e)$ denotes valuation of e in γ and $\gamma[v \mapsto \gamma(e)]$ denotes the state in which the value of v is replaced with $\gamma(e)$.

- α is of the form $A_1 \| \ldots \| A_n$: $[\![\alpha]\!] \stackrel{def}{=} \odot_{i \in \{1 \ldots n\}} [\![A_i]\!]$

$\star$ $S_u = (\Gamma_u, A_u, I_u, FP_u)$

- $\Gamma_u = [Var_u \longrightarrow D_u]$
- $A_u = \bigcup_{\alpha \in Ass_u} [\![\alpha]\!]$
- $I_u = \{\gamma \in \Gamma_u : Init_u(\gamma)\}$
- $FP_u = \{\gamma \in \Gamma_u : \forall \alpha \in Ass_u\ (\gamma, \gamma) \in [\![\alpha]\!]\}$

$\star$ $J_u = JUST(S_u)$

4.2 Semantics of *SLP*

An *SLP* program (*Single Location Programs*) *slp* is of the form: $I : \star[A_1 \ [\!] \ \ldots \ [\!] \ A_n]$, where $A_1, \ \ldots \ A_n$ are (simple or multiple) conditional assignments and I denotes initial states. A computation step consists in executing an assignment A_i. Like *Unity*, the execution of an SLP program is infinite. The program terminates when it reaches a stable state. The fairness constraint on SLP programs is slightly weaker than Unity's fairness hypothesis. For every assignment A_i, one requires that there exists infinitely often a pair of consecutive states $(\sigma, \ \tau)$ such that $(\sigma, \ \tau) \ \in \ [\![A_i]\!]$. We don't require that every action should be executed infinitely often but only that its effect should occur infinitely often. $WJUST(V, \ D)$ denotes the set of execution sequences generated under such a fairness hypothesis, where V and D are respectively the set of program variables and the set of program data. It is thus not difficult to show that $JUST(V, \ D) \simeq WJUST(V, \ D)$ holds.

Definition 10. Operational semantics of SLP programs. Let $slp \stackrel{def}{=} (I : \star[A_1 \ [\!] \ \ldots \ [\!] \ A_n])$ be an SLP program such that Var_{slp} denotes the set of variables of *slp* and D_{slp} denotes their domain. The operational semantics of *slp* is defined by a structure $O[slp] \ = \ (S_{slp}, WJ_{slp})$ where:

- $S_{slp} = (\Gamma_{slp}, A_{slp}, I_{slp}, FP_{slp})$
  - $\Gamma_{slp} = [Var_{slp} \longrightarrow D_{slp}]$
  - $A_{slp} = \bigcup_{\alpha \in slp} [\![\alpha]\!]$
  - $I_{slp} = \{\gamma \in \Gamma_{slp} : I(\gamma_{slp})\}$
  - $FP_{slp} = \{\gamma \in \Gamma_{slp} : \forall \alpha \in slp.\ (\gamma, \gamma) \in [\![\alpha]\!]\}$
  - For every action α in *slp*, $[\![\alpha]\!]$ is defined as follows:
    * α is of the form $v := e$: $[\![\alpha]\!] = \{(\gamma, \gamma') \in \Gamma_u : \gamma' = \gamma[v \mapsto \gamma(e)]\}$
    * α is of the form $v_1, \ldots, v_n := e_1, \ldots e_n$: $[\![\alpha]\!] = \odot_{i \in \{1 \ldots n\}} [\![v_i := e_i]\!]$
    * α is of the form $b \rightarrow lv := le$: $[\![\alpha]\!] = \{(\gamma, \gamma') \in \Gamma_{slp} : [(\gamma, \gamma') \in [\![lv := le]\!] \wedge b(\gamma)] \vee [\neg b(\gamma) \wedge \gamma' = \gamma]\}$
- $WJ_{slp} = WJUST(S_{slp})$

5 A reformulation of the *Unity* logic

Interesting properties of concurrent programs are safety and liveness properties. In *Unity*, these properties are specified using the predicates *unless* (safety properties), *ensures* and *leadsto* (liveness properties). It is convenient to give an operational definition of these predicates, based on *Unity* program behaviour, in order to prove that the modified proof system we introduce in the next section is semantically complete.

5.1 Specification language

Let $u \stackrel{def}{=} < Var_u, Init_u, Ass_u >$ and J_u be the set of *just* execution sequences of the program u. Let $\sigma \in J_u$, σ is of the form $s_0 \xrightarrow{a_0} s_1 \xrightarrow{a_1} s_2 \xrightarrow{a_2} \ldots$.

In the following definitions, p and q, denote assertions in a first-order theory. The notation $s \models p$ means that p holds at the state s, and $[\![p]\!]_u \stackrel{def}{=} \{s \in \Gamma_u | \ s \models p\}$.

Two first operators, *stable* and *invariant*, are defined over states, while the second ones, *unless* and *leadsto*, are defined over behaviours.

$$[\![stable\ p]\!]_u \stackrel{def}{=} \{s \in \Gamma_u \mid \forall \alpha \in Ass_u.\ (s \models p) \wedge ((s \xrightarrow{\alpha} s') \Rightarrow s' \models p)\}$$
$$[\![invariant\ p]\!]_u \stackrel{def}{=} [\![stable\ p]\!]_u \cap [\![Init_u \Rightarrow p]\!]_u$$

$$[\![p\ unless\ q]\!]_u \stackrel{def}{=} \{\sigma \in J_u | \ \forall k :\ k \geq 0 :: (s_k.\sigma \models p \wedge \neg q) \Rightarrow (s_{k+1}.\sigma \models p \vee q)\}$$

Intuitively, $\sigma \in [\![p\ unless\ q]\!]_u$ if and only if every transition in σ transforms a state satisfying $(p \wedge \neg q)$ into a state satisfying $(p \vee q)$, which means that whenever p holds, it continues to hold unless q holds.

$$[\![p\ leadsto\ q]\!]_u \stackrel{def}{=} \{\sigma \in J_u \mid \forall k :\ k \geq 0 :: (s_k.\sigma \models p) \Rightarrow (\exists j :\ j \geq k :: s_j.\sigma \models q)\}$$

Intuitively, p *leadsto* q means that whenever p holds, q holds eventually. The specification language we consider contains the operators *unless* (for safety) and *leadsto* (for liveness). Nevertheless, a weak version of Chandy-Misra's *ensures* can be defined in terms of *unless* and *leadsto* as an additional operator to enrich the specification language:

$$p\ ensures_w\ q \stackrel{def}{=} (p\ unless\ q) \wedge (p\ leadsto\ q)$$

Unlike *ensures*, $ensures_w$ does not require the existence of *one* assignment statement that leads from $p \wedge \neg q$ to q in one step, but it requires the existence of *one or more* assignment statements that lead from $p \wedge \neg q$ to q in one or more steps.

Definition 11. Program validity. Let u be a *Unity* program and $O[u] \stackrel{def}{=} (S_u, J_u)$ its operational semantics.

- Let F be a property of the form *stable* p or *invariant* p for some predicate p.

$$\models u \textbf{ sat } F \text{ iff } \Gamma_u = [\![F]\!]_u$$

- Let F be a *unless* or *leadsto* property.

$$\models u \textbf{ sat } F \text{ iff } J_u = [\![F]\!]_u$$

5.2 The Proof System S_I

In this section we introduce the system of axioms and rules, namely S_I, dedicated to mechanize proofs within the *Unity* framework. The first level of the proof system formalizes the definition of *wp*-calculus for the actions. In the following lv denotes a list of program variables and $le, le_0, \ldots, le_n$ denotes lists of expressions. v_i and e_i denote respectively variables and expressions. Let p be an assertion, we denote by $p[e_1/v_1, \ldots, e_n/v_n]$ the ***multiple substitution*** that replaces occurrences of v_i by e_i in p.

$$wp(v_1, \ldots, v_n := e_1, \ldots, e_n, p) \stackrel{def}{=} p[e_1/v_1, \ldots, e_n/v_n]$$

$$\begin{aligned} wp((lv := le_0 \text{ if } b_0 \sim \ldots \sim le_n \text{ if } b_n), p) \stackrel{def}{=} & (b_0 \Rightarrow wp(lv := le_0, p)) \wedge \\ & \vdots \\ & (b_n \Rightarrow wp(lv := le_n, p)) \wedge \\ & (\neg(b_0 \vee \ldots \vee b_n) \Rightarrow p) \end{aligned}$$

$$wp(s, p \vee q) \stackrel{def}{=} wp(s, p) \vee wp(s, q)$$

$$wp(s, p \wedge q) \stackrel{def}{=} wp(s, p) \wedge wp(s, q)$$

$$\text{if } (p \Rightarrow q) \text{ then } wp(s, p) \Rightarrow wp(s, q)$$

Note that $\vee$-distributivity of *wp* holds because all assignment statements in *Unity* are deterministic.

Inference rules :

$$\frac{\forall s \in Ass_u.\ p \Rightarrow wp(s, p)}{u \text{ sat } (stable\ p)}$$

$$\frac{Init_u \Rightarrow p, \qquad u \text{ sat } (stable\ p)}{u \text{ sat } (invariant\ p)}$$

$$\frac{\begin{array}{l} u \text{ sat } (invariant\ I) \\ \forall s \in Ass_u.\ p \wedge \neg q \wedge I \Rightarrow wp(s, p \vee q) \end{array}}{u \text{ sat } (p\ unless\ q)}$$

$$\frac{\begin{array}{l} u \text{ sat } (p\ unless\ q), \qquad u \text{ sat } invariant\ I \\ \exists s \in Ass_u.\ p \wedge \neg q \wedge I \Rightarrow wp(s, q) \end{array}}{u \text{ sat } (p\ leadsto\ q)}$$

$$\frac{u \text{ sat } (p\ leadsto\ r), \qquad u \text{ sat } (r\ leadsto\ q)}{u \text{ sat } (p\ leadsto\ q)}$$

$$\text{for every set } W : \quad \frac{< \forall m \in W.\ u \text{ sat } (p(m)\ leadsto\ q) >}{u \text{ sat } (< \exists m \in W.\ p(m) > \ leadsto\ q)}$$

Comment. Observe that, unlike Chandy-Misra [5] and Sanders [14], the predicates *stable p* and *invariant p* (for some state formula p) are defined independently of the operator *unless*. This avoids mutual cycles in proofs of *unless* and *invariant* and gives syntactical rules characterizing all possible invariants of a program. In practice, we do not need a complete characterization of all states that are reachable from the initial state (that needs the strongest invariant) to prove some property, but just a sufficient one.

5.3 Soundness and completeness of S_I

Soundness and relative completeness of the proof system is proved according to the program semantics. The determination of a sufficient set of axioms and rules relies on a clear characterization of program properties and the role of the initial conditions on them.

To prove soundness and relative completeness of the system S_I, we explore results of Gerth and Pnueli [8] that consists of a sound and relatively complete system S_{GP} for the *SLP* programs. We proceed as follows: first, we translate *Unity* programs into *SLP*, then we define an embedding of proofs in S_I into proof in S_{GP} and conversely.

Unity programs form a subclass of SLP programs

Theorem 12. *For every Unity program, u, there exists an SLP program, slp, such that u and slp are observationally equivalent.*

Proof. To prove this theorem, we first define a translation of the *Unity* language into *SLP*, then we show that this translation preserve observational semantics.

1. The relation Tr_{u-slp} is defined to be a translation from the language *Unity* into *SLP*. A *Unity* program u of the form $< Var_u, Init_u, Ass_u >$ is translated by Tr_{u-slp} into an *SLP* program slp of the form $I_{slp} : \star[Act]$ such that:
   - $Var_{slp} \stackrel{def}{=} Var_u$
   - $D_{slp} \stackrel{def}{=} D_u$
   - $I_{slp} \stackrel{def}{=} Init_u$
   - $Act \stackrel{def}{=} Tr_{u-slp}(Ass_u)$ with :

   $Tr_{u-slp}(\alpha_1 \,[\!]\, \ldots [\!]\, \alpha_m) \stackrel{def}{=} Tr_{u-slp}(\alpha_1) \,[\!]\, \ldots [\!]\, Tr_{u-slp}(\alpha_m)$, and
   for every α in Ass_u, $Tr_{u-slp}(\alpha)$ is defined as follows. (lv and le represent respectively a list of variables and a list of expressions)

   – α is of the form $lv := le$: $Tr_{u-slp}(\alpha) \stackrel{def}{=} lv := le$

   – α is of the form $lv := le_0 \; if \; b_0 \sim \; \ldots \; \sim le_n \; if \; b_n$: $Tr_{u-slp}(\alpha) \stackrel{def}{=}$

   $$b_0 \rightarrow lv := le_0 \,[\!]\, \ldots \,[\!]\, b_n \rightarrow lv := le_n$$

2. Let $O[u]$ and $O[slp]$ be respectively the semantics of the *Unity* program u and its translation $slp = Tr_{u-slp}(u)$.

   $$O[u] \stackrel{def}{=} (S_u, J_u) \text{ where}$$
   $$S_u = (\Gamma_u, A_u, I_u, FP_u)$$
   $$O[slp] \stackrel{def}{=} (S_{slp}, WJ_{slp}) \text{ where}$$
   $$S_{slp} = (\Gamma_{slp}, A_{slp}, I_{slp}, FP_{slp})$$

A) $Var_u = Var_{slp}$ {by definition}
B) $D_u = D_{slp}$ {by definition}
C) $\Gamma_u = \Gamma_{slp}$, because $Var_u = Var_{slp}$ and $D_u = D_{slp}$
D) $I_u = I_{slp}$ {by definition}
E) $A_u = \bigcup_{\alpha \in Ass_u} [\![\alpha]\!]$ and $A_{slp} = \bigcup_{\alpha \in Tr_{u-slp}(Ass_u)} [\![\alpha]\!]$

we prove that : $\forall \alpha \in Ass_u.\ [\![\alpha]\!] \equiv [\![Tr_{u-slp}(\alpha)]\!]$.

- α is $lv := le$: obvious since $Tr_{u-slp}(\alpha) \stackrel{def}{=} lv := le$
- α is $lv := le_0\ if\ b_0 \sim \ \ldots\ \sim le_n\ if\ b_n$: $[\![Tr_{u-slp}(\alpha)]\!] \stackrel{def}{=}$

$$[\![b_0 \rightarrow lv := le_0 \ [\!] \ \ldots \ [\!] \ b_n \rightarrow lv := le_n \ ]\!]$$
$$\stackrel{def}{=} [\![b_0 \rightarrow lv := le_0 \ ]\!] \cup \ \ldots \ \cup \ [\![b_n \rightarrow lv := le_n \ ]\!]$$
$$\stackrel{def}{=} \{(\gamma,\gamma') \in \Gamma_{slp} : [(\gamma,\gamma') \in [\![lv := le_0]\!] \wedge b_0(\gamma)] \vee [\neg b_0(\gamma) \wedge \gamma' = \gamma\}$$
$$\cup \ldots \cup$$
$$\{(\gamma,\gamma') \in \Gamma_{slp} : [(\gamma,\gamma') \in [\![lv := le_n]\!] \wedge b_n(\gamma)] \vee [\neg b_n(\gamma) \wedge \gamma' = \gamma\}$$
$$\stackrel{def}{=} \{(\gamma,\gamma') \in \Gamma_{slp} : [(\gamma,\gamma') \in [\![lv := le_n]\!] \wedge b_0(\gamma)] \vee \ \ldots \ \vee$$
$$[(\gamma,\ \gamma') \in [\![lv := le_n]\!] \wedge b_n(\gamma)] \vee [(\gamma = \gamma') \wedge \neg(b_0 \vee \ldots \vee b_n)(\gamma)]\}$$

F) $FP_{slp} = \{\gamma \in \Gamma_{slp} : \forall \alpha \in A_{slp}.\ (\gamma,\gamma) \in [\![\alpha]\!]\}$
$= \{\Gamma_{slp} = \Gamma_u,\ A_{slp} = Tr_{u-slp}(A_u)\}$
$\{\gamma \in \Gamma_u : \forall \beta \in A_u.\ (\gamma,\gamma) \in [\![Tr_{u-slp}(\beta)]\!]\}$
$=$ {by E): $[\![\alpha]\!] \equiv [\![Tr_{u-slp}(\alpha)]\!]$}
$\{\gamma \in \Gamma_u : \forall \beta \in A_u.\ (\gamma,\gamma) \in [\![\beta]\!]\}$
$=$ {definition}
FP_u

G) {by E) and definition of J_u and WJ_{slp}}
$\sigma \in J_u$ iff $\exists \tau \in WJ_{slp}.\ \forall k.\ s_k.\sigma = s_k.\tau$
then {definition }
$J_u \simeq WJ_{slp}$

Embedding *Unity* into Gerth-Pnueli's logic

The temporal logic used in *SLP* framework is a first order logic extended with two temporal operators: $\Diamond$ – eventually, and $\mathcal{U}$ – weak until. Properties of *SLP* programs are described in terms of these operators according to the definition given in [8, 10]. Intuitively, $\Diamond\phi$, means that ϕ will eventually become true and $\phi\ \mathcal{U}\ \psi$, means that ϕ will continuously hold until ψ becomes true, if ever :

$$(\sigma,j) \models \Diamond p \quad \text{iff } \exists k \geq j.\ (\sigma,k) \models p$$
$$(\sigma,j) \models p\ \mathcal{U}\ q \ \text{ iff } \exists k \geq j.\ (\sigma,k) \models p \ \textit{and}\ \forall i: \ j \leq i < k :: \ (\sigma,i) \models p$$

Definition 13. Let *slp* be a *SLP* program and $O[slp] = (S_{slp}, WJ_{slp})$.

- $\models$ *slp* **sat** $\Diamond\phi$ iff $\forall \sigma \in WJ_{slp}.\ \sigma \models \Diamond\phi$.

- $\models$ *slp* **sat** $\phi\ \mathcal{U}\ \psi$ iff $\forall \sigma \in WJ_{slp}.\ \sigma \models \phi\ \mathcal{U}\ \psi$.

Remark. It is not difficult to show that :

$$p\ \textit{unless}\ q \equiv p \Rightarrow (p\ \mathcal{U}\ q)$$
$$p\ \textit{leadsto}\ q \equiv p \Rightarrow \Diamond q$$

Soundness and relative completeness of *stable* and *invariant* rules

The *stable* (resp. *invariant*) rule can be proved sound as usual (see [12]). Relative completeness relates however to the expressiveness of the assertional language. The assertional language must be sufficient to express $wp(F)$ for any program F, this condition can be satisfied by assuming *arithmetic* in the assertional language [13, 1].

Theorem 14. *Let* $u \stackrel{def}{=} < Var_u, Init_u, Ass_u >$ *be a Unity program. For every assertion* p,

1. *if* $\models u$ **sat** $stable\ p$ *then* $\vdash u$ **sat** $stable\ p$

2. *if* $\models u$ **sat** $invariant\ p$ *then* $\vdash u$ **sat** $invariant\ p$

Proof.

1.
$\models u$ **sat** $stable\ p$
iff {by definition}
$\forall s \in \Gamma_u.\ \forall \alpha \in Ass_u.\ (s \models p) \wedge (s \stackrel{\alpha}{\longrightarrow} s' \Rightarrow s' \models p)$
implies {by definition of $wp(\alpha, p)$}
$\forall s \in \Gamma_u.\ \forall \alpha \in Ass_u.\ s \models p \Rightarrow wp(\alpha, p)$
implies {We assume the underlying assertional theory to be sound and complete}
$\vdash \forall \alpha \in Ass_u.\ p \Rightarrow wp(\alpha, p)$
implies {*stable* rule}
$\vdash u$ **sat** $stable\ p$

2.
$\models u$ **sat** $invariant\ p$
iff {by definition}
$\forall s \in \Gamma_u.\ \forall \alpha \in Ass_u.\ (s \models (Init_u \Rightarrow p))\ \wedge\ (s \models p \wedge (s \stackrel{\alpha}{\longrightarrow} s' \Rightarrow s' \models p))$
implies {by definition}
$\forall s \in \Gamma_u.\ (s \models Init_u \Rightarrow p)\ \wedge\ \forall \alpha \in Ass_u.\ (s \models p \wedge (s \stackrel{\alpha}{\longrightarrow} s' \Rightarrow s' \models p))$
implies {by definition of *stable*}
$\forall s \in \Gamma_u.\ s \models Init_u \Rightarrow p\ \wedge\ u$ **sat** $stable\ p$
implies {soundness and completeness of *stable*}
$\vdash Init_u \Rightarrow p$ et $\vdash u$ **sat** $stable\ p$
implies {*invariant* rule}
$\vdash u$ **sat** $invariant\ p$

Soundness and relative completeness of *unless* rule

Let $slp \stackrel{def}{=} I : \star[A_1 \| \ldots \| A_n]$ be an *SLP* program, and p, q, r, p', q' r' be assertions. The main rule of S_{GP} to embed *unless* is:

$$Rule\,U: \quad \frac{\begin{array}{ll} \forall s \in slp.\ \{q' \wedge p'\}s\{r' \vee q'\} & \forall s \in slp.\ \{p'\}s\{p'\} \\ I \Rightarrow p' & p \wedge p' \Rightarrow q' \vee r' \\ q' \wedge p' \Rightarrow q & r' \wedge p' \Rightarrow r \end{array}}{slp\ \mathbf{sat}\ (p \Rightarrow q\ \mathcal{U}\ r)}$$

We reformulate this rule using the *wp* transformer [7] according to the following equivalence:

$$\{p\}\ s\{q\}\ \equiv\ p\ \Rightarrow\ wlp(s,\ q)$$

Recall that, in both *Unity* and *SLP*, we can substitute *wlp* by *wp* because all actions terminate.

$$\textit{Rule}\ \overline{U}\ :\ \frac{\begin{array}{ll} \forall s \in slp.\ (q' \wedge p') \Rightarrow wp(s,\ (r' \vee q')) & \forall s \in slp.\ p' \Rightarrow wp(s,\ p') \\ I \Rightarrow p' & p \wedge p' \Rightarrow q' \vee r' \\ q' \wedge p' \Rightarrow q & r' \wedge p' \Rightarrow r \end{array}}{slp\ \textbf{sat}\ (p\ \Rightarrow q\ \mathcal{U}\ r)}$$

Theorem 15. *Let* $u \stackrel{def}{=} < Var_u, Init_u, Ass_u >$ *be a Unity program.*

For every assertion p, $\models u$ **sat** p *unless* q *iff* $\vdash u$ **sat** p *unless* q

Proof. We show that:

(I) Every proof of u **sat** p *unless* q in S_I can be embedded into a proof of $Tr_{u-slp}(u)$ **sat** $p \Rightarrow p\ \mathcal{U}\ q$ in S_{GP},

(II) Every property of $Tr_{u-slp}(u)$ **sat** $p\ \Rightarrow\ p\ \mathcal{U}\ q$ in S_{GP} can be embedded into a proof of u **sat** p *unless* q in S_I.

Proof of (I):

Hypothesis: Let u be a *Unity* program and $slp = Tr_{u-slp}(u)$. Let us Assume a proof of u **sat** p *unless* q in S_I. Thus, the following premises of the *unless* rule are satisfied (there is a unique rule for deriving *unless* properties):

(**H1**) u **sat** $(invariant\ p')$
(**H2**) $\forall s \in Ass_u.\ p \wedge \neg q \wedge p'\ \Rightarrow wp(s, p \vee q)$

We substitute q by p and r by q in the rule $\overline{U}$.

$$\textit{Rule}\ \overline{U}\ :\ \frac{\begin{array}{ll} \forall s \in slp.\ q' \wedge p' \Rightarrow wp(s,\ r' \vee q') & \forall s \in slp.\ p' \Rightarrow wp(s,\ p') \\ I \Rightarrow p' & p \wedge p' \Rightarrow q' \vee r' \\ q' \wedge p' \Rightarrow p & r' \wedge p' \Rightarrow q \end{array}}{F\ \textbf{sat}\ (p\ \Rightarrow p\ \mathcal{U}\ q)}$$

Now, we replace q' by $p \wedge \neg q$ and r' by q, the rule becomes:

$$\textit{Rule}\ \overline{U}\ :\ \frac{\begin{array}{llll} \forall s \in slp.\ p \wedge \neg q \wedge p' \Rightarrow wp(s,\ q \vee (p \wedge \neg q)) & (1) & \forall s \in slp.\ p' \Rightarrow wp(s,\ p') & (2) \\ I \Rightarrow p' & (3) & p \wedge p' \Rightarrow (p \wedge \neg q) \vee q & (4) \\ p \wedge \neg q \wedge p' \Rightarrow p & (5) & q \wedge p' \Rightarrow q & (6) \end{array}}{F\ \textbf{sat}\ (p\ \Rightarrow p\ \mathcal{U}\ q)}$$

The premisses of the rule $\overline{U}$ can be deduced as follows:

- {Hypothesis (**H1**), and theorem 12}
  $\forall s \in slp.\ p' \Rightarrow wp(s,\ p')$ (2)
- {Hypothesis (**H1**), and $Tr_{u-slp}(Init_u)I$}
  $I \Rightarrow p'$ (3)
- {predicate calculus}
  $p \wedge p' \Rightarrow (p \wedge \neg q) \vee q$ (4)

- {predicate calculus}
  $p \wedge \neg q \wedge p' \Rightarrow p$ (5)
- {predicate calculus}
  $q \wedge p' \Rightarrow q$ (6)
- {Hypothesis (**H2**)}
  $\forall s \in Ass_u.\ p \wedge \neg q \wedge p' \Rightarrow wp(s,\ q \vee (p \wedge \neg q))$
  $\equiv \{\forall \alpha \in Ass_u.\ [\![\alpha]\!] = [\![Tr_{u-slp}(\alpha)]\!]$, by theorem 12$\}$
  $\forall s \in Tr_{u-slp}(u).\ p \wedge \neg q \wedge p' \Rightarrow wp(s,\ q \vee (p \wedge \neg q))$
  $\equiv \{slp = Tr_{u-slp}(u)\}$
  $\forall s \in slp.\ p \wedge \neg q \wedge p' \Rightarrow wp(s,\ q \vee (p \wedge \neg q))$

All premisses of the rule $\overline{U}$ are derived, hence *slp* **sat** $(p \Rightarrow p\ \mathcal{U}\ q)$ is deduced.

Proof of (II): Let u be a *Unity* program, $slp = Tr_{u-slp}(u)$, and a proof for *slp* **sat** $p \Rightarrow p\ \mathcal{U}\ q$ in S_{GP}.

then {the rule $\overline{U}$ is the unique rule for deriving $\mathcal{U}$}

$\forall s \in slp.\ p \wedge \neg q \wedge p' \Rightarrow wp(s,\ q \vee (p \wedge \neg q))$ (1) $\quad \forall s \in slp.\ p' \Rightarrow wp(s,\ p')$ (2)
$I \Rightarrow p'$ (3) $\quad p \wedge p' \Rightarrow (p \wedge \neg q) \vee q$ (4)
$p \wedge \neg q \wedge p' \Rightarrow p$ (5) $\quad q \wedge p' \Rightarrow q$ (6)

then {premisses (1),(2),(3)}

$\forall s \in slp.\ p \wedge \neg q \wedge p' \Rightarrow wp(s,\ q \vee (p \wedge \neg q))$
$\forall s \in slp.\ p' \Rightarrow wp(s,\ p')$
$I \Rightarrow p'$

then {predicate logic}

$\forall s \in slp.\ p \wedge \neg q \wedge p' \Rightarrow wp(s,\ (q \vee p) \wedge \neg q)$
$\forall s \in slp.\ p' \Rightarrow wp(s,\ p')$
$I \Rightarrow p'$

then {theorem 12}

$\forall s \in u.\ p \wedge \neg q \wedge p' \Rightarrow wp(s,\ (q \vee p) \wedge \neg q)$
$\forall s \in u.\ p' \Rightarrow wp(s,\ p')$
$I \Rightarrow p'$

then {*invariant* rule in S_I}

$\forall s \in u.\ p \wedge \neg q \wedge p' \Rightarrow wp(s,\ (q \vee p) \wedge \neg q)$
u **sat** *invariant* p

then {*unless* rule in S_I}

u **sat** p *unless* q

Soundness and relative completeness of *leadsto* rules

Theorem 16. *Let* $u \overset{def}{=} \langle Var_u, Init_u, Ass_u \rangle$ *be a Unity program.*

For every assertion p, $\models u$ **sat** p *leadsto* q *iff* $\vdash u$ **sat** p *leadsto* q

Proof. The proof is similar to the case of *unless*, using the following rule of *SLP* proof system to embed proofs of *leadsto* properties :

Rule $\overline{E}$:

$$\frac{\begin{array}{ll}\forall s \in slp.\ \{r'(n)\}\ s\ \{r \vee \exists m \leq n.\ r'(m)\}, & I \Rightarrow q' \\ \forall s \in slp.\ \{q'\}\ s\ \{q'\}, & q \wedge q' \wedge \neg r \Rightarrow \exists n.\ r'(n) \\ \forall s \in A_{f(n)}.\ \{r'(n)\}\ s\ \{r \vee \exists m < n.\ r'(m)\} & \end{array}}{I : \star(A_1 [] \ldots [] A_n)\ \mathbf{sat}\ (q \Rightarrow \Diamond r)}$$

for any f: $Ord \mapsto \{1, \ldots, n\}$.

with $slp = I : \star(A_1 [] \ldots [] A_n)$

In rule $\overline{E}$, f is a choice function that selects at every term the helpful action that will cause q eventually to hold. n, m are ordinals and Ord is a sufficiently[1] large set of ordinals. This rule uses the fact that every action gets executed infinitely often. Similarly to the previous proof for *unless*, we can show that any proof of u **sat** q *leadsto* r in S_I can be transformed into a proof of $Tr_{u-slp}(u)$ **sat** $q \Rightarrow \Diamond r$ in S_{GP} (and inversely) using the definition q *leadsto* $r \stackrel{def}{=} q \Rightarrow \Diamond r$.

6 Discussion

We have reconsidered Sanders' solution that eliminates the substitution axiom giving up a complete proof system for *Unity* by defining the notion of the strongest invariant of a *Unity* program. The strongest invariant has been defined semantically in [14] and no syntactic rules have been given by Sanders to mechanize it. Consequently the Sanders' proof of completeness is based on this semantic definition and assumes a complete underlying system axiomatizing the strongest invariant.

Our solution consists in showing that the general concept of invariant is sufficient to design complete rules for *unless* and *leadsto* without the need of the substitution axiom. Semantic definition of the operators and syntactic rules to derive them are clearly separated. Unlike Sanders' solution, the proof system we propose is shown sound and relatively complete without assuming an underlying complete system mechanizing the *strong invariant*. Rules for deriving general *invariant* are parts of the proposed proof system and are shown sound and relatively complete for both *finite-state* and *infinite-state Unity* programs.

We are currently working on extending a tool mechanizing proofs of *Unity* programs encoded into *B-Tool* [2] in order to capture the completeness result established in this paper. Moreover, the resulting theorem prover should provide a formal tool for developing mapping methods of *Unity* programs into parallel architectures [3].

Acknowledgements

We would like to thank Dominique Méry for useful discussions during the early stages of this work. We wish also to thank anonymous referees for their helpful comments.

[1] As large as *unbounded* nondeterminism of *Unity* can be represented.

References

1. K. R. Apt and E. R. Olderog. *Verification of sequential and concurrent programs.* Texts and Monographs in Computer Science. Springer-Verlag, 1991.
2. N. Brown and D. Mery. A Proof Environment for Concurrent Programs . In J. C. P Woodcock and P. G. Larsen, editors, *FME'93: Industrial-Strength Formal Methods.* Springer Verlag, 1993. Lecture Notes in Computer Science 670.
3. N. Brown. A Sound Mapping from Abstract Algorithms to Occam Programs. In *Transputer Research And Applications Conference.* 23-25 Octobre 1994 (University of Georgia, USA), IOS Press, pages 218-231.
4. N. Brown. Verification and Mapping of *Unity* programs into distributed architectures. *Phd. thesis,* Université de de Nancy I, France, October 1994.
5. K.M. Chandy and J. Misra. *Parallel Program Design A Foundation.* Addison-Wesley Publishing Company, 1988. ISBN 0-201-05866-9.
6. S.A. Cook. Soundness and completeness for an axiomatic system for program verification. *SIAM J. Comput.,* 7:70–90, 1978.
7. E. W. Dijkstra and C. S. Scholten. *Predicate Calculus and Program Semantics.* Texts and Monographs in Computer Science. Springer Verlag, 1990.
8. R. Gerth and A. Pnueli. Rooting UNITY. In acm, editor, *Int. Workshop on Software Specification and Design,* pages 11–19, 1989.
9. E. Knapp. An exercise in the formal derivation of parallel programs: Maximum flows in graphs. *Transactions On Programming Languages and Systems,* 12(2):203–223, 1990.
10. Z. Manna and A. Pnueli. *The Temporal Logic of Reactive and Concurrent Systems.* Springer-Verlag, 1991. ISBN 0-387-97664-7.
11. D. Mery. The \⊓ system as a development system for concurrent programs: \⊓. *Theoretical Computer Science,* 94(2):311 - 334, march 1992.
12. S. Owicki and L. Lamport. Proving liveness properties of concurrent programs. *ACM TOPLAS,* 4(3):455–495, july 1982.
13. S. Ramesh. On the completeness of modular proof systems. *Information Processing Letters,* 36:195–201, 1990.
14. B. A. Sanders. Eliminating the substitution axiom from unity logic. *Formal Aspects of Computing,* 3:189–205, 1991.

Automated reasoning about parallel algorithms using powerlists *

Deepak Kapur M. Subramaniam

Computer Science Department
State University of New York
Albany, NY 12222
kapur@cs.albany.edu, subu@cs.albany.edu

Abstract. Misra [8] recently introduced a regular data structure, called *powerlists,* using which he showed how many data parallel algorithms, including Batcher's merge sort, bitonic sort, fast Fourier transform, prefix sum, can be described concisely using recursion. The elegance of these recursive descriptions is further reflected in deducing properties of these algorithms. It is shown in this paper how such proofs can be easily automated in a theorem prover *RRL (Rewrite Rule Laboratory)* based on equational and rewriting techniques. In particular, the *cover set method* for automating proofs by induction in *RRL* generates proofs which preserve the clarity and succinctness, to a large extant, of hand proofs given in [8]. This is illustrated using a correctness proof of Batcher's merge sort algorithm. Mechanically generated proofs from specifications of powerlists and parallel algorithms using different approaches are contrasted. It is shown that one gets longer, complex proofs with many cases if powerlists are modeled as a subtype of lists. However, if powerlists are specified using a proposal by Kapur [2] in which the algebraic specification method is extended to associate *applicability conditions* with functions of a data type thus allowing constructors of a data structure to be partial, then one gets compact and elegant proofs, similar to the ones reported in [8]. Applicability conditions can be used to provide *contexts* for axioms and proofs just like type information. The effectiveness of the proposed axiomatic method becomes all the more evident while reasoning about *nested* powerlists for modeling n-dimensional arrays, for example, in specifying and reasoning about a transformation for embedding a multi-dimensional array into a hypercube such that adjacency of nodes is preserved by the transformation. A mechanically generated proof of this property of the embedding transformation, which was not proved in Misra's paper, is discussed in detail. This suggests that the proposed approach for automating reasoning about data parallel algorithms described recursively using powerlists should scale up, especially if structural properties of nested powerlists are built into *RRL.*

---

* Partially supported by the National Science Foundation Grant no. CCR-9303394 and subcontract CB0249 of SRI contract MDA904-92-C-5186 with The Maryland Procurement Office.

1 Introduction

Misra recently introduced a regular data structure called *powerlists,* and demonstrated how many data parallel algorithms including Batcher's sorting networks including merge sort and bitonic sort, fast Fourier transform, parallel prefix sum, can be described concisely using powerlists, highlighting the role of both parallelism and recursion [8]. The main strategy exploited in these descriptions is that of divide and conquer, and consequently, powerlists are (nested) lists of length 2^k of similar elements (which themselves can be powerlists). In that paper, many algebraic properties of powerlists are (hand) proved using structural induction, and these properties are then used in reasoning about parallel algorithms and doing their correctness proofs.

In this paper, we show how proofs of algebraic properties of powerlists as well as correctness proofs of parallel algorithms described using powerlists can be easily automated on a theorem prover *RRL (Rewrite Rule Laboratory).* In particular, proofs mechanically generated by *RRL* turn out to be very close to the hand-generated and carefully crafted proofs reported in Misra's paper [8]. We believe that this is due to the fact that *RRL* is suitable for reasoning based on algebraic laws expressed as equations and conditional equations which are transformed into one-directional rewrite rules, and the cover set method for automating proofs by induction and the associated heuristics in *RRL* are able to select appropriate induction schemes for generating proofs of properties of recursively defined functions.

We discuss two possibly different but related ways to axiomatize powerlists in *RRL,* and illustrate using a simple example, how different axiomatizations lead to proofs of different complexity. In particular, if powerlists are modeled as a subtype of lists, it becomes necessary to carry conditions on lists being powerlists in axioms, and many structural properties of powerlists must be explicitly proved. Proofs of algorithms become unnecessarily complicated because of many cases arising due to such structural conditions which are not relevant to the behavior of algorithms. However, if powerlists are modeled directly using constructors with applicability conditions as proposed in [2], then structural properties of powerlists can be exploited much like type information while doing proofs. This approach enables a designer to focus on relevant and interesting aspects of algorithms while formalizing concepts and doing proofs, and is thus likely to lead to better insights and enhance the designer's confidence in the formalization process. The proofs generated are elegant and compact. This contrast is first illustrated in proving a property of a simple function on powerlists and subsequently by using Batcher's merge sort algorithm.

Finally, we take a nontrivial transformation embedding n-dimensional arrays into hypercubes given in [8]. It is known that there is an embedding transformation that preserves adjacency among neighbors if an n-dimensional array is of dimensions 2^{m_0} x $\cdots$ x 2^{m_d} [9]. Modeling n-dimensional arrays as nested powerlists, a proof of this property is mechanically generated using the proposed approach on *RRL.* This property was not proved in [8], and was in fact, posed as a challenge to us by Misra in private communication for testing the effectiveness of the proposed approach. This proof exhibits the power, elegance and effectiveness of the axiomatization method for powerlists used in *RRL.*

This case study demonstrates that a theorem prover such as *RRL* developed for reasoning about algebraic specification of abstract data types and functional algorithms defined recursively on abstract data types, can be successfully used to reason about many data parallel algorithms (and hardware circuits and descriptions) if they can be described recursively. Recursive descriptions are not only concise, they can also be mechanically analyzed, and resulting proofs reflect to a large extent, the elegant structure of descriptions. By hard-wiring structural properties of powerlists, the proposed approach of mechanically reasoning about recursive descriptions on powerlists in *RRL* is likely to scale up.

2 Powerlists

Powerlists are defined in [8] to be lists of length 2^k, for some $k \geq 0$, all of whose elements are *similar*; they are built out of *scalars* such as natural numbers, binary integers 0 and 1, etc., as well as powerlists themselves. Two scalars are similar if they are of the same type and two *powerlists* are similar if they have the same length and any element of one is similar to any element of the other.[2] Powerlists can be recursively defined using the constructors, s that takes a scalar and produces a *singleton* powerlist, n to *nest* a powerlist, and *tie* for appending two similar powerlists. Instead of *tie*, *zip* could also be used as a constructor that takes two similar powerlists and produces a powerlist by successively appending the corresponding elements from the two powerlists.

If all elements of a powerlist are scalars, it is called *flat*, and can be constructed from the operators s, *tie* (or *zip*) only. Two flat powerlists are similar if their lengthes are equal, and their elements are of the same type. Powerlists whose elements are themselves powerlists, are called *nested* powerlists. They will be used in section 6 to represent matrices and multi-dimensional arrays. Until that section, powerlists are assumed to be flat.

What makes powerlists different from lists is the way they are constructed; *tie* cannot be applied on any two arbitrary powerlists; instead, the two powerlists must be similar and of the same length. However constructors of powerlists are free in the following sense:

$$A_1: \ (tie(x,y) = tie(u,v)) \equiv (x = u) \wedge (y = v), \quad A_2: \ (s(x) = s(y)) \equiv (x = y).$$

The function *zip* can be defined recursively in terms of s and *tie*; but its two arguments must also be similar and of the same length.

$$\begin{aligned} A_3: &\ zip(s(x), s(y)) = tie(s(x), s(y)), \\ A_4: &\ zip(tie(x,y), tie(u,v))) = tie(zip(x,u), zip(y,v)). \end{aligned}$$

Using this definition, it can be proved by induction that

$$A_5: \ (zip(x,y) = zip(u,v)) \equiv (x = u) \wedge (y = v).$$

Similar to functions on lists, functions on powerlists can be recursively defined. For instance, the reversal of flat powerlists *rev* can be recursively defined as:

$$1. \ rev(s(x)) \ = \ s(x), \qquad 2. \ rev(tie(x,y)) \ = \ tie(rev(y), rev(x)).$$

It is possible to prove by structural induction that

$$C_1: \ rev(zip(x,y)) \ = \ zip(rev(y), rev(x)).$$

The basis case, corresponding to x and y both being singletons leads to the subgoal, $rev(zip(s(x), s(y))) = zip(rev(s(y)), rev(s(x)))$, proved from A_3 and 1 by equational reasoning. In the induction case, the conclusion, $rev(zip(tie(x,y), tie(u,v)))$

[2] The similar relation is precisely axiomatized in section 6.

$= zip(rev(tie(u,v)), rev(tie(x,y)))$, simplifies by A_4 and 2 to: $tie(rev(zip(y,v)), rev(zip(x,u))) = tie(zip(rev(v), rev(y)), zip(rev(u), rev(x)))$, to which the following induction hypotheses apply: $rev(zip(x,u)) = zip(rev(u), rev(x))$, $rev(zip(y,v)) = zip(rev(v), rev(y))$.

3 Automating Induction by Cover Sets in RRL

We review the cover set induction method implemented in *RRL* for automatically generating induction schemes to mechanize proofs by induction. Proofs are mechanically generated by *RRL* using the cover set method. Other inference mechanisms of *RRL* used include rewriting (contextual rewriting), case analysis and linear arithmetic. For details, the reader may consult [6, 10, 4].

The cover set method generates induction schemes from complete definitions of functions given as terminating rewrite rules. In a definition of a function symbol f, the left side of each rewrite rule, $\mathtt{f}(\mathtt{t_1}, \ldots, \mathtt{t_n})$, is used to generate a subgoal of a conjecture to be proved. The recursive calls to $\mathtt{f}$, if any, in the right side of the rule generate induction hypotheses to be used in a proof. The induction scheme suggested by a term $f(x_1, \cdots, x_n)$ in a conjecture C is obtained directly from a cover set of f. The induction scheme so obtained is a finite set of induction cases of the form $< \sigma_c, cond_c, \{\theta_i\} >$, and is obtained from a cover set triple $<< s_1, \cdots, s_n >, \{\cdots, < s_1^i, \cdots, s_n^i >, \cdots\}, cond >$ as follows: $\sigma_c = \{x_1 \rightarrow s_1, \cdots, x_n \rightarrow s_n\}$, $cond_c = cond$ and , $\theta_i = \{x_1 \rightarrow s_1^i, \cdots, x_n \rightarrow s_n^i\}$. The substitution σ_c is used to generate the induction conclusion, each substitution in $\{\theta_i\}$ is used to generate an induction hypothesis, and *cond* is the condition governing the induction case. For instance, the induction scheme generated from $zip(x_1, x_2)$ using the cover set obtained from A_3 and A_4 is:

$$\mathcal{I}_1: \ \{< \{x_1 \rightarrow s(x), x_2 \rightarrow s(y)\}, \{\}, \{\} >, \ < \{x_1 \rightarrow tie(x,y), x_2 \rightarrow tie(u,v)\}, \{\}, \{\{x_1 \rightarrow x, x_2 \rightarrow u\}, \{x_1 \rightarrow y, x_2 \rightarrow v\}\} >\}.$$

To ensure that an induction scheme based on a cover set is sound, a cover set must have two properties. Firstly, it must be complete in the sense that for each induction variable, all possible values of a data type must be considered. Secondly, in the induction step, the substitutions for generating induction hypotheses must be lower in a well-founded order than the substitutions in the conclusion. The second property is automatically ensured in *RRL* since induction schemes are generated from terminating definitions.

4 Axiomatizing Powerlists

To reason about powerlists and algorithms on powerlists in an automatic theorem prover such as *RRL*, powerlists must be axiomatized. Below we discuss two different approaches for formalizing powerlists, focusing on their use in reasoning about algorithms on powerlists. The theorem C_1 presented in section 2 is used to illustrate and contrast these approaches.

4.1 Powerlists as a Subtype of Lists with s and tie

Powerlists can be modeled as a subtype of lists with free constructors s and tie where s denotes a singleton list and tie concatenates two given lists. In such a model the axioms A_1 and A_2 are automatically available and more importantly, all properties of powerlists can be established in terms of s and tie themselves. However, the data type so obtained has many extraneous elements which are not powerlists; using s and tie as free constructors one can construct almost any list (except the empty list). To restrict the elements to powerlists, *len* can be defined in terms of *s* and *tie* as:

$$len(s(x)) = 1, \quad len(tie(x,y)) = len(x) + len(y).$$

Using *len*, every usage of *tie* is constrained so that its arguments are of equal length. For instance, the definition of *zip* given by A_4 above is modified as:

$$zip(tie(x,y), tie(u,v)) = tie(zip(x,u), zip(y,v)) \text{ if } (len(x) = len(y)) \wedge (len(u) = len(v)) \wedge (len(x) = len(u)),$$

and the definition of *rev* is changed to

$$rev(s(x)) = s(x), \; rev(tie(x,y)) = tie(rev(y), rev(x)) \text{ if } len(x) = len(y).$$

The cover set method discussed in section 2 cannot be directly used to prove properties by induction using these modified definitions, since these definitions are incomplete. However, as discussed in [4], cover sets and induction schemes generated from incomplete function definitions can be used to prove conjectures that are relativized with respect to a formula ϕ with respect to which the cover set obtained from an incomplete definition is complete. For instance, $zip(x,y)$ is defined only when $len(x) = len(y)$, so the cover set generated from its definition is incomplete. But it is complete relative to $\phi = (len(x) = len(y))$. An induction scheme generated from this cover set is complete only relative to ϕ.

Let us illustrate the use of the cover set method to prove conjecture C_1. Since *zip* appears in C_1, its cover set can be used to generate the induction scheme. Since the cover set of *zip* is complete only with respect to the formula ϕ, C_1 is relativized with respect to ϕ to:

$$C_1' : \; rev(zip(x,y)) = zip(rev(y), rev(x)) \text{ if } len(x) = len(y) \; .$$

C_1' can be automatically proved in *RRL* using the properties: $len(rev(x)) = len(x)$ and $len(zip(x,y)) = len(x) + len(x) \text{ if } len(x) = len(y)$. Here is a part of the *RRL* transcript that proves C_1'.

```
Induction will be done on x, y in zip(x, y), with the scheme:
[1] P(s(x), s(y))                     [2] P(tie(x, y), tie(u, v)) if
{(len(y) = len(u)),(len(v) = len(u)),(len(x) = len(u)), P(x, u), P(y, v)}
```

The subgoal corresponding to `[1]` trivially follows from rewriting whereas the subgoal `[2]` leads to 4 subgoals. Contrast this with the single subgoal obtained in the hand proof of C_1 in section 2. Three of these subgoals are due to the constraints on *len* in C_1' above and follow by contradictory assumptions with respect to *len*. The last subgoal corresponds to the only subgoal in the induction step case of C_1 and is proved similarly. Detailed transcripts are given in [5].

4.2 Constructors with Applicability Conditions

In the approach discussed in the last subsection, constructors *s* and *tie* are total functions. Conditions expressing regularity of powerlists have to be explicitly carried as conditions governing the various definitions and properties. Such additional conditions make reasoning cumbersome and lead to unnecessarily complex proofs even in the case of very simple conjectures such as C_1 as seen above.

In [2], the algebraic specification method [1] is extended to axiomatize regular and semantic data structures. A predicate called *applicability condition* is associated with every function, including the constructors of a data type. So constructors need not be total any more. A function can be invoked only if the associated applicability condition is satisfied. If no applicability condition is explicitly specified, then it is assumed to be *true*, that is the function is applicable on any input argument satisfying the type requirements. By associating the applicability condition $len(u) = len(v)$ with $tie(u, v)$, it can be ensured that *tie* operates on powerlists of equal lengths. The same applicability condition is used for constraining *zip* as well. Powerlists are directly specified as follows.

```
[ s(x: item): powerlist ]            [ len (u: powerlist): naturals ]
[ tie(u: powerlist, v: powerlist): powerlist ]       len(u) = len(v)
[ zip (u: powerlist, v:powerlist): powerlist]        len(u) = len(v)
len(s(x)) = 1, len(tie(u,v)) = len(u) + len(v),
zip(s(x),s(y)) = tie(s(x),s(y)),
zip(tie(x, y), tie(u, v)) = tie(zip(x, u), zip(y, v)).
```

Applicability conditions on functions impose additional requirements on variables that can be substituted in an equation. For instance, the following equation, $zip(tie(x, y), tie(u, v)) = tie(zip(x, u), zip(y, v))$, really stands for:
$zip(tie(x, y), tie(u, v)) = tie(zip(x, u), zip(y, v))$ *if* $len(x) = len(y) \wedge len(x) = len(u) \wedge len(u) = len(v) \wedge len(tie(x, y)) = len(tie(u, v)) \wedge len(zip(x, u)) = len(zip(y, v))$.

The condition in the above conditional equation constitutes its *context* which can be generated mechanically from the applicability conditions of all function symbols appearing in the equation, as described in [2]. Equations with inconsistent contexts can be discarded as such equations do not convey any useful information. Henceforth, every axiom is assumed to have a consistent context.

The proposed approach is based on the fact that most axioms defining functions of a data type can be expressed as equations with implicit contexts and such equations can be used for reasoning without having to compute and store contexts for intermediate results whereas if a data type is modeled as a subsort of another data type, conditional equations would have to be used. For equations with implicit contexts, most inference rules of equality can be applied except for transitivity, i.e., given $E \vdash s = t,\ E \vdash t = u$, it is not always possible to deduce that $E \vdash s = u$. Let C_s, C_t, C_u be, respectively, the contexts of s, t, u. From $E \vdash s = t,\ E \vdash t = u$, it is not sound to deduce that $E \vdash s = u$ since t need not appear in the conclusion $s = u$. Instead, $E \vdash C_t \Rightarrow s = u$ can be deduced.

In a rewrite rule based theorem prover such as *RRL*, an equation $s = t$ is oriented either as $s \rightarrow t$ or $t \rightarrow s$ depending upon whether $s > t$ or $t > s$ in a termination ordering $>$ used to orient equations. (For a conditional equation $s = t$ *if* c, a conditional rule $s \rightarrow t$ *if* c is made if $s > t$ and $s > c$.) A finite set of axioms defining a function symbol are thus oriented into terminating rewrite rules which also guarantees that the function computation terminates.

A rewrite rule $l \rightarrow r$ is called *context preserving* if $C_l \Rightarrow C_r$. Similarly, a conditional rewrite rule $l \rightarrow r$ *if* c is context preserving if $(C_l \wedge C_c \wedge c) \Rightarrow C_r$. Of course, checking for the implication of contexts typically involves the use of definitions and properties of function symbols appearing in the applicability conditions as well as other function symbols appearing in l, r, c. We will abuse the notation and call the axiom $l = r$ context preserving also if the associated rule is context preserving. For instance, $zip(tie(x, y), tie(u, v)) \rightarrow tie(zip(x, u), zip(y, v))$ can be shown to be context preserving using the definitions and properties of *len* and *zip*. All equations and conditional equations in the specifications in this paper are context preserving when they are viewed as rewrite rules from left to right.

Given two context-preserving rewrite rules $s \rightarrow t$ and $t \rightarrow u$, it is sound to deduce that $s \rightarrow^* u$; this is so because $C_s \Rightarrow C_t$ and $C_t \Rightarrow C_u$, which means that $C_s \Rightarrow C_u$. Unlike in the case of arbitrary equations, where it would be necessary to carry the intermediate context C_t along with the deduction $s = u$, there is no need to do so for context-preserving rules. With this restriction, equations and conditional equations can be proved using classical techniques for equational reasoning as implemented in *RRL*.

The cover set method in *RRL* extends nicely to data types generated using constructors with nontrivial applicability conditions. From complete terminating definitions expressed using functions with applicability conditions, a cover set is generated from the definition in way similar to the definitions without any applicability conditions. An inductive proof of C_1 can be directly performed using this specification on *RRL*. Here is the transcript generated in *RRL*.

```
Induction will be done on x, y in zip(x, y), with the scheme:
[1] P(s(x), s(y))   [2] P(tie(x, y), tie(u, v)) if { P(x, u), P(y, v)}
```

The proof obtained in *RRL* is the same as the hand proof of C_1 presented in section 2. There is no need any more to prove unnecessary cases and discharge conditions as in the other approach resulting in longer and tedious proofs in which often, interesting aspects of the proof get hidden and submerged in details.

5 Verifying Batcher's Merge Sorting Network

Batcher's sorting networks based on even-odd merging and bitonicity of elements are well-known networks for sorting. In this section, we discuss the verification of a merge sorting network specified using powerlists as given in [8]. We discuss the correctness proof given in [8], and show how this proof can be mechanically generated in *RRL* using the cover set induction method. In fact, by associating applicability conditions with powerlists an automated proof with the same top level structure as the one in [8] is obtained in *RRL*. The lemmas employed in the automated proof are the same as those in [8] and the proof preserves the clarity and succinctness to a large extant of that in [8].

Batcher's merge operator is defined below as the function *pbm* on flat powerlists. Since Batcher's sorting networks are based on compare and swap methods, their correctness can be ensured by considering inputs made up of 0's and 1's only because of zero-one principle for sorting networks as explained on (p. 224) in [7]. The correctness proof in [8] uses the following additional functions.

```
[ minp(u: powerlist, v:powerlist): powerlist]         len(u) = len(v)
[ maxp(u: powerlist, v:powerlist): powerlist]         len(u) = len(v)
[ compare(u: powerlist, v: powerlist): powerlist ]    len(u) = len(v)
[ pbm(u: powerlist, v: powerlist): powerlist ]        len(u) = len(v)
[ zeros(u: powerlist): naturals ]   [ sort (u: powerlist): bool]

minp(s(x), s(y)) = s(x) if x <= y, minp(s(x), s(y)) = s(y) if (x > y),
minp(zip(x, y), zip(u, v)) =  zip(minp(x, u) , minp(y, v)).
maxp(s(x), s(y)) = s(y) if x <= y, maxp(s(x), s(y)) = s(x) if (x > y),
maxp(zip(x, y), zip(u, v)) =  zip(maxp(x, u) , maxp(y, v)).
compare(x, y) = zip(minp(x, y), maxp(x, y)).
```

```
pbm(s(x), s(y)) = compare(s(x), s(y)),
pbm(zip(x, y),zip(u, v)) = compare(pbm(x, v), pbm(y, u)).
zeros(s(0)) = 1, zeros(s(1)) = 0,
zeros(tie(x, y)) = zeros(x) + zeros(y).
sort(s(x)) = true,
sort(zip(x, y)) = sort(x) & sort(y) & 0 <=(zeros(x) - zeros(y))<= 1.
```

The main theorem expressing the correctness of Batcher's Merge is:

$$MT: \ sort(pbm(x,y)) \ \ if \ \ sort(x) \wedge sort(y).$$

The correctness proof in [8] uses the following lemmas.

$L_1: \ minp(x,y) = x \ if \ sort(x) \wedge sort(y) \wedge (zeros(x) >= zeros(y)),$
$L_2: \ maxp(x,y) = y \ if \ sort(x) \wedge sort(y) \wedge (zeros(x) >= zeros(y)),$
$L_3: \ zeros(compare(x,y)) = zeros(x) + zeros(y),$
$L_4: \ zeros(pbm(x,y)) = zeros(x) + zeros(y),$
$L_5: \ sort(compare(x,y)) \ if \ sort(x) \wedge sort(y) \wedge |zeros(x) - zeros(y)| \leq 1.$

MT is proved in [8] by structural induction on x and y as follows.

Proof: The basis case, $sort(compare(x,y)) \ if \ sort(x) \wedge sort(y)$ follows from lemma L_5 by case analysis. In the induction step, the conclusion is:
$sort(pbm(zip(x,y), zip(u,v)))$, with the assumptions: $\{sort(zip(x,y)),$
$sort(zip(u,v)),\ sort(pbm(x,v)) \ if \ sort(x) \wedge sort(v)$ (hypothesis),
$sort(pbm(y,u)) \ if \ sort(y) \wedge sort(u)$ (hypothesis) $\}$.

$sort(zip(x,y)) \wedge sort(zip(u,v))$
$\Rightarrow \ sort(x) \wedge sort(y) \wedge sort(u) \wedge sort(v) \wedge 0 \leq zeros(x) - zeros(y) \leq 1 \wedge$
$0 \leq zeros(u) - zeros(v) \leq 1$ {By definition of sort} (*)
$\Rightarrow \ sort(pbm(x,v)) \wedge sort(pbm(y,u)) \wedge 0 \leq zeros(x) - zeros(y) \leq 1 \wedge$
$0 \leq zeros(u) - zeros(v) \leq 1$ {By induction hypotheses }(**)
$\Rightarrow \ sort(pbm(x,v)) \wedge sort(pbm(y,u)) \wedge |(zeros(x) + zeros(v)) - (zeros(y) +$
$zeros(u))| \leq 1$ {By simplification of inequalities }
$\Rightarrow \ sort(pbm(x,v)) \wedge sort(pbm(y,u)) \wedge$
$|zeros(pbm(x,v)) - zeros(pbm(y,u))| \leq 1$ {By lemma L_1 }
$\Rightarrow \ compare(sort(pbm(x,v), pbm(y,u))$ {By lemma L_2}
$\Rightarrow \ sort(pbm(zip(x,y), zip(u,v))$ {By definition of pbm } □

5.1 Mechanically Generated Proof in RRL

A correctness proof of Batcher's merge sort was performed in *RRL* in a bottom-up fashion. The equations defining various functions are oriented into terminating rewrite rules by *RRL*. From the definitions, cover sets are generated which are used to generate appropriate induction schemes. Then lemmas L_1 to L_5 were first proved, and using these lemmas, the main theorem was finally proved. Each of the lemmas is proved using the cover set induction method, and in each case, the induction scheme is automatically obtained from the cover set generated from the definition of the functions appearing in the lemmas. Heuristics associated with the cover set method automatically selected appropriate variables to use for induction, and then appropriate induction schemes were generated.

We discuss below the proof of MT as generated mechanically by the cover set method. An induction scheme is automatically obtained from the cover set generated from the definition of *pbm* and is chosen to be the most appropriate induction scheme to attempt an induction proof of *MT*. Here is a part of the *RRL* transcript.

```
Induction will be done on x, y in pbm(x, y), with the scheme:
[1] P(s(x), s(y))     [2] P(zip(x, y), zip(u, v)) if {P(x, v), P(y, u)}
```

The basis case, `[1] sort(pbm(s(x), s(y))) = true if sort(x) and sort(y)` follows using `sort(compare(s(x), s(y))) = true` and case analysis.
The subgoal corresponding to the induction step case is as follows.

```
[2] sort(pbm(zip(x, y),zip(u, v))) if sort(zip(x, y)) & sort(zip(u, v)) &
    (sort(pbm(x, v)) if sort(x) & sort(v))) &
    (sort(pbm(y, u)) if sort(y) & sort(u)))
```

This subgoal is split into 9 intermediate subgoals. Eight of these nine subgoals are easily established due to contradictory assumptions with regard to the predicate *sort*. The last subgoal generated from [2] is:

```
[2.9]  sort(pbm(zip(x, y), zip(u, v))) if (sort(zip(x, y))) &
       (sort(zip(u, v))) & (sort(pbm(x, v))) & (sort(pbm(y, u))).
```

This is the only nontrivial subgoal and is proved using lemmas L_4 and L_5. The conclusion simplifies to `sort(compare(pbm(x, v), pbm(y, u))` which reduces to true by lemma L_5 provided the following three literals `sort(pbm(x, v))`, `sort(pbm(y, u))`, `|zeros(pbm(x, v) - zeros(pbm(y, u))| <= 1` can each be reduced to *true*. The first two literals are trivially established from the assumptions of `[2.9]`. The third literal is simplified by lemma L_4 to: `|(zeros(x) + zeros(v)) - (zeros(y) + zeros(u))|`. By definition of `sort`, `0 <= (zeros(x) - zeros(y)) <= 1` and `0 <= (zeros(u) - zeros(v)) <= 1` are assumptions. The last literal reduces to true from these assumptions since the following interesting inequality can be proved in *RRL* with the help of the linear arithmetic decision procedure [3]:

$$L_6 : \quad |(x+v)-(y+u)| \leq 1 \quad if \;\; (0 \leq (x-y) \leq 1) \wedge (0 \leq (u-v) \leq 1).$$

The proof of this subgoal as performed in *RRL* corresponds to the hand proof of Misra presented earlier in the section. In the hand proof, in order to obtain from the step (*) the subsequent step (**) it is required to discharge the conditions `sort(x)`, `sort(v)`, `sort(u)`, and `sort(v)`. This is implicitly done in the hand proof whereas in the automated proof in *RRL* this has to be be explicitly done which leads to the 8 additional subgoals discussed above.

5.2 Proof with Powerlists as a Subtype of Lists with s and tie

Now we contrast the above proof with a correctness proof of *pbm* if powerlists are modeled as a subtype of lists with *s* and *tie*. Recall from section 4.2, that while modeling powerlists as a subtype using *s* and *tie*, function definitions have to be modified to include conditions expressed using `len`. These modified definitions are not complete over lists; to prove properties using such cover sets, the properties have to be relativized with respect to a formula ϕ with respect to which the cover sets are complete. A proof of the main theorem about `pbm(x, y)` being sorted goes as follows in *RRL*:

```
MT': sort(pbm(x, y)) if  sort(x) & sort(y) & (len(x) = len(y)))
Induction will be done on x, y in pbm(x, y), with the scheme:
[1] P(s(x), s(y)) [2] P(zip(x, y), zip(u, v)) if {(len(x) = len(u)),
                (len(y) = len(u)),(len(v) = len(u)), P(x, v), P(y, u).}
```

The base case is proved the same way as in the previous subsection. The induction step case is split into 16 subgoals in contrast to the 9 generated in the previous section. Nine of these subgoals are similar to the subgoals in the previous proof with additional constraints on the arguments to `zip`. These subgoals are established in a similar way as done before.

The additional 7 subgoals are obtained due to the combinations of literals involving predicates over `len` in the hypotheses. These subgoals follow due to contradictory assumptions involving `len` predicates in the condition. However, in proving these subgoals, additional properties about the lengthes of powerlists obtained as results of different functions, such as, $len(zip(x, y)) = len(x) + len(y)$ $if\ len(x) = len(y)$, are needed. Such properties are called *length lemmas* and are proved easily by the cover set method from the function definitions and the definition of *len*. Detailed transcripts of proofs are given in [5].

A proof generated using this axiomatization properly contains the corresponding proof generated using the axiomatization based on applicability conditions as subcases. Just as we noticed in the case of a proof of a property C_1 about *rev* in section 3, there are additional cases to be considered in this approach because of the side conditions in definitions and the statement of the main theorem. The definitions and lemmas become complex because the conditions under which an equation in a definition is applicable must be explicitly included in the equation itself. Similarly, such conditions have to be explicitly associated with the lemmas also and these conditions have to be discharged additionally in order to apply the lemma. Most of the cases arising due to side conditions can be easily established though. This, however, results in longer, tedious proofs in which often, the interesting aspects of the proof get hidden in details.

6 Nested Powerlists and Multi-dimensional Arrays

So far we have discussed the use of flat powerlists whose elements are scalars. For modeling matrices (i.e. two dimensional arrays), multi-dimensional arrays, and different kinds of architectures, Misra proposed the use of nested powerlists in which the elements of a powerlist are themselves similar powerlists. Defining a nested powerlist as a subtype of nested lists is particularly not easy, so we won't even bother attempting an axiomatization of powerlists as a subtype of lists. Instead, we show below how nested powerlists can be elegantly axiomatized by associating with *tie*, an applicability condition expressed using the equivalence relation *sim* (to stand for similar relation on powerlists).

6.1 Nested Powerlists

For nested powerlists, there is an additional constructor *n* (to stand for *nest*).

```
[ s(x:item ): powerlist ]         [ n(x:powerlist): powerlist ]
[ sim(x:powerlist, y:powerlist): bool]
[ tie(x:powerlist, y:powerlist): powerlist ]          sim(x)  = sim(y)
[ zip(x:powerlist, y:powerlist): powerlist ]          sim(x)  = sim(y)

zip(s(x), s(y)) = tie(s(x) , s(y)), zip(n(x), n(y)) = tie(n(x), n(y)),
zip(tie(x, y), tie(u, v)) =  tie(zip(x, u), zip(y, v))
```

The similar predicate *sim* is axiomatized below. It can be proved in *RRL* that *sim* is an equivalence relation (i.e. it is reflexive, symmetric and transitive).

$sim(s(x), s(y)) = true$, $sim(n(x), n(y)) = sim(x, y)$, $sim(s(x), n(y)) = false$,
$sim(n(y), s(x)) = false$, $sim(s(x), tie(y, z)) = false$, $sim(tie(x, y), s(z)) = false$,
$sim(n(x), tie(y, z)) = false$, $sim(tie(x, y), n(z)) = false$,
$sim(tie(x, y), tie(u, v)) = sim(x, u) \wedge sim(y, v)$

6.2 Embedding a Multidimensional Array into a Hypercube

We now consider a transformation for embedding multi-dimensional arrays into hypercubes such that the adjacency of the elements in the array is preserved. The embedding transformation preserving adjacency is an important property since it demonstrates how most parallel algorithms can be directly implemented in a hypercube without significant change in the running time or the number of processors. This transformation is defined in Misra's paper [8], but its properties are not proved. We used *RRL* to formalize a definition of the *emb* function and adjacency relation in multi-dimensional arrays. An automated proof of correctness of this embedding transformation done on *RRL*, is briefly reviewed. The role and usefulness of applicability conditions in simplifying the proof becomes very evident in this proof.

An n-dimensional hypercube is a graph of 2^n nodes. Given an array of dimensions $2^{m_0} \times \cdots \times 2^{m_d}$, its elements can be placed at the nodes of a D-dimensional hypercube, where $D = m_0 + \cdots + m_d$ such that any two adjacent elements of the array are placed at the neighboring nodes of the hypercube. Two elements in an array are *adjacent* iff their indices differ from each other exactly in one dimension and by 1 *mod* N for a dimension of size N. (This is called *wrap-around* adjacency.)

We now specify the function *emb*; the basic idea behind *emb* can be described as: for any dimension of the given array, divide the elements of the dimension into two halves; embed these two halves into neighboring dimensions of the hypercube; the first half is embedded as it is, whereas the second half is embedded after reversal (by the function *rev*) .

$$emb(s(x)) = s(x), \qquad emb(n(x)) = emb(x),$$
$$emb(tie(x, y)) = tie(n(emb(x)), n(emb(rev(y)))).$$

6.3 Axiomatizing Adjacency in a Multidimensional Array

Below, the notion of *wrap-around* adjacency of elements in a multi-dimensional array is axiomatized.

$adj(u, v, s(y)) = false$, $adj(u, v, n(y)) = adj(u, v, y)$,
$adj(u, v, tie(s(y), s(z))) = (u = s(y) \wedge v = s(z)) \vee (u = s(z) \wedge v = s(y))$,
$adj(u, v, tie(y, z)) = true \ if \ adj(u, v, y)$, $adj(u, v, tie(y, z)) = true \ if \ adj(u, v, z)$,
$adj(u, v, tie(tie(y, z), tie(y_1, z_1))) = adj(u, v, tie(fst(y), fst(rev(z_1)))) \vee$
$\qquad adj(u, v, tie(fst(rev(z)), fst(y_1)))$,
$adj(u, v, tie(n(y), n(z))) = adj(u, v, y) \vee adj(u, v, z) \vee cadj(u, v, joinc(y, z))$

The last case above dealing with elements in neighboring dimensions is the most interesting; the other cases are self-evident. To determine the adjacency of elements in neighboring dimensions, an operation *joinc* analogous to *zip* is defined below on two powerlists producing a list whose elements are powerlists.

$$joinc(s(x), s(y)) = cons(tie(s(x), s(y)), nil),$$
$$joinc(n(x), n(y)) = cons(tie(n(x), n(y)), nil),$$
$$joinc(tie(x, y), tie(u, v)) = append(joinc(x, u), joinc(y, v)).$$

Two elements are adjacent in a list iff they are adjacent in any of its element powerlists; the predicate *cadj* is defined as:

$$cadj(u, v, nil) = false, \; cadj(u, v, cons(y, z)) = adj(u, v, y) \lor cadj(u, v, z).$$

6.4 Overview of a Mechanically Generated Proof Using RRL

Since *RRL* does not yet support the extended specification method with applicability conditions and hence, cannot generate contexts automatically, contexts must be supplied manually in certain situations. They were not needed in a correctness proof of Batcher's merge sort algorithm discussed in section 5. But for proving that *emb* preserves adjacency, it is sometimes necessary to explicitly have contexts since *emb* preserves the adjacency of elements in an multi-dimensional array only when the sizes of each of the dimensions are powers of 2. For modeling this well formedness of arrays, a predicate ***allsim*** is defined below which ensures that any element of a nested powerlist is similar to its other elements.

$$allsim(s(x)) = true, \; allsim(n(x)) = allsim(x),$$
$$allsim(tie(x, y)) = allsim(x) \land allsim(y) \land sim(x, y).$$

We would like to emphasize that once *RRL* is extended to automatically support the use of applicability conditions, ***allsim*** will not be needed. So statements of the main theorem being proved as well as required lemmas will be simpler, not explicitly requiring conditions expressed using ***allsim***. Furthermore, proofs would also be much simpler and shorter.

The correctness of the embedding transformation is stated in *RRL* as :

$$MT_1 : \; occur(x, emb(y)) = occur(x, y) \; if \; allsim(y),$$
$$MT_2 : \; adj(u, v, emb(y)) \; if \; adj(u, v, y) \land allsim(y).$$

The formula MT_1 states that the occurrences of the elements of the array are preserved by *emb* and MT_2 states that *emb* preserves the adjacency of the elements of the array. The predicate *occur* checks the occurrence of its first argument in the second. As mentioned earlier, MT_2 is not a theorem if we do not restrict the input array y to satisfy ***allsim*** since an array whose dimensions are not a power of 2 cannot be embedded into a hypercube satisfying *wrap-around* adjacency.

The proof of MT_1 is performed in *RRL* by induction on the variable y using the definition of *emb*. It leads to 3 induction subgoals corresponding to the three definition cases of *emb*. These subgoals are easily established from the definitions of *occur*, *emb* and *rev*. The proof of MT_2 in *RRL* is done by induction on y based on the definition of *adj* and the induction scheme obtained is:

```
[1] P(s(x))  [2] P(tie(s(x), s(y)))  [3] P(n(x)) if P(x)
[4] P(tie(n(x),n(y))) if {P(x),P(y)} [5] P(tie(tie(x, y),tie(x1, y1))) if
    {P(tie(fst(x), fst(rev(y1)))), P(tie(fst(rev(y)), fst(x1)))}.
```

The proofs of the induction cases `[4]` and `[5]` are the most interesting. Case `[4]` involves reasoning about elements in neighboring dimensions of the array assuming that *emb* preserves the adjacency of elements within each dimension. Case `[5]` proves that *emb* also preserves the *wrap-around* adjacency. Each of these cases leads to 9 intermediate subgoals.

Five of these intermediate subgoals have to do with checking of applicability conditions in each case. These subgoals arise due to ***allsim*** constraint analogous to the subgoals generated from the predicates over *len* in the correctness proof of Batcher's merge sort discussed in section 5. These subgoals are easily established using the definition of *allsim* and other similarity properties. None of these subgoals is directly relevant to the correctness of the embedding transformation and would not be needed if the specification method using applicability conditions is directly supported by *RRL*.

We briefly review the proofs of the above five cases as mechanically generated in *RRL*, discussing in detail, nontrivial cases, and also introduce some of the important lemmas needed in these proofs.

1. The first three cases are straightforward. In the first case, a singleton array corresponds to a 0-dimensional hypercube and the subgoal follows from the definition of *emb*. In the second case, *emb* reduces to transposing the input array and hence the adjacency is trivially preserved. The third case follows from the fact that adjacency of elements in a particular dimension of the array is independent of those in other dimensions. This is proved from the definitions of *emb* and *adj*.
2. The most interesting subcase of case [4] is when the elements are from the neighboring dimensions, i.e. they are in x and y respectively. The subgoal generated in *RRL* corresponding to this subcase is:

   ```
   [4.5] adj(u, v, emb(tie(n(x),n(y)))) if (allsim(tie(n(x), n(y)))) &
   adj(u, v, tie(n(x), n(y)))) & not(adj(u, v, x)) & not(adj(u, v, y))
   ```

   In order to prove [4.5] it is required that *emb* preserve the adjacency amongst the elements of the list obtained from the *joinc* operation, which is stated as a lemma as follows.

   $$Adj_1: \quad cadj(u,v,joinc(emb(x),emb(y)) \;\; if \;\; sim(x,y) \wedge \\ cadj(u,v,emb(x,y)) \wedge not(adj(u,v,x)) \wedge not(adj(u,v,y)).$$

   This lemma was proved by induction on the variables x and y using the definition of *joinc*. The proof leads to 9 inductive subgoals at the top level with some of these subgoals generating as many as 25 intermediate subgoals. The subgoal [4.5] simplifies by the definition of *emb*, *rev* and *adj*, and then the above lemma is applicable, thus establishing [4.5]. Three of the remaining subgoals generated in the proof of case [4] deal with the adjacency of the elements which either belong to x or y. These are easily established from the definitions of *emb* and *adj* given above and from our assumption that the adjacency is preserved for elements within a dimension. Hence these subgoals contain $adj(u,v,emb(tie(n(x),n(y))))$ as the consequent with the assumptions being either $adj(u,v,emb(x))$ or $adj(u,v,emb(y))$ or both. The remaining 5 subgoals have to do with checking of applicability conditions and are easily proved as stated earlier.
3. The above four cases establish that *emb* preserves the adjacency of elements in the array when *wrap-around* adjacency is not taken into account. Case [5] proves that *emb* also preserves *wrap-around* adjacency. The following lemmas are first proved in *RRL*.

   $$Adj_2: \; adj(u,v,emb(tie(tie(x,y),tie(x_1,y_1)))) \;\; if \\ adj(u,v,emb(tie(fst(x),fst(rev(y_1))))),$$

   $$Adj_3: \; adj(u,v,emb(tie(tie(x,y),tie(x_1,y_1)))) \;\; if \\ adj(u,v,emb(tie(fst(rev(y)),fst(x_1)))).$$

   These lemmas are then used in proving case [5]. As stated earlier the proof of case [5] leads to 9 intermediate subgoals. The proofs of 4 of these subgoals are similar and are obtained by using the adjacency lemmas Adj_1 through

Adj$_3$ given above. The remaining 5 subgoals in the proof of case [5] deal with the checking of applicability conditions and are easily proved as stated earlier. The proofs of the lemmas *Adj*$_2$ and *Adj*$_3$ are rather involved and are discussed in detail in [5].

Lemmas expressing the similarity properties of the arrays in terms of the functions *allsim* and *sim* are required in many of the above described proofs. These lemmas are analogous to the length lemmas used in the proof of Batcher's merge sort. Most of these lemmas are proved easily in *RRL* by induction using the definitions of *sim* and *allsim* functions. The proof of one of these lemmas, $allsim(tie(fst(x), fst(rev(y))))$ *if* $allsim(tie(x, y))$ particularly turned out to be cumbersome, and there were 9 top level induction subgoals. Some of these subgoals require establishing as many as 16 intermediate subgoals. The property of *sim* being an equivalence relation was extensively used.

Role of Applicability Conditions The use of applicability conditions and the advantages of supporting reasoning about them directly using contexts in a theorem prover such as *RRL* are demonstrated by the above correctness proof of the embedding transformation. It would then not be necessary to include structural conditions expressed using *sim* and *allsim* in definitions, lemmas or theorems. The well-formedness of a multi-dimensional array whose dimensions are powers of 2, is ensured using applicability conditions. Most importantly, number of intermediate subgoals to be established in lemmas and theorems would considerably be reduced because subgoals arising from different powerlists not being well-formed will not arise.

For instance, a proof of MT_2 would need 11 fewer intermediate subgoals. Similarly, the number of intermediate subgoals would reduce from 25 to 14 in a proof of *Adj*$_1$. This saving can be routinely observed in almost all the proofs of the lemmas used in this proof. Pruning such subgoals would lead to direct, shorter and elegant proofs as these subgoals are not relevant to the correctness of the embedding transformation. If nested powerlists were to be modeled as a subtype of nested lists, the number of subgoals to be considered in lemmas as well as number of additional lemmas needed in proofs are likely to go up by orders of magnitude; this is based on extrapolating our experience in doing a correctness proof of Batcher's merge sort algorithm. So we did not even bother to do a proof of the adjacency preserving property of *emb* using that approach.

Of the 25 lemmas used in the above proof, 9 were about establishing properties of the similar relation. Proofs of such structural properties constitute nearly half of the overall correctness proof in this case study and hiding such details would lead to a more direct correctness proof which is half the present size. These properties are independent of *emb* or any other application using nested powerlists, and are instead a part of infrastructure needed to use nested powerlists. They can be proved once for all to support reasoning about nested powerlists, and hard-wired into the theorem prover, much like a type-inference algorithm.

7 Conclusions

We have demonstrated how a rewrite-rule based induction theorem prover *RRL* can be used to automate proofs of properties of data parallel algorithms described recursively using powerlists. A nice feature of our methodology is that proofs mechanically generated by *RRL* mimic hand-generated proofs reported in [8]. This is because *RRL* is good for supporting reasoning using algebraic laws which can be transformed into rewrite rules, and the cover set method and the associated heuristics in *RRL* are able to select appropriate induction schemes from recursive definitions for generating proofs.

We discussed two different ways to axiomatize powerlists, and showed how these axiomatizations lead to related but different proofs. In particular, the proofs differ in the number of cases to be considered, side conditions needed, and in length and complexity. It is shown that if the information about regularity of powerlists is directly supported in an axiomatization as discussed in [2], succinct equational definitions of the primitive functions of powerlists as well as of parallel algorithms can be given. Unconditional context-preserving rewrite rules are generated from these equations simplifying proofs considerably. Contexts needed for proofs can be automatically generated from applicability conditions and axioms, and properties about contexts (such as lemmas about length and similarity relation) can be hard-wired into a theorem prover much like type-information. Including applicability conditions in a specification of a data type and a theorem prover using them whenever needed would enable a designer to focus on the behavior of definitions and algorithms. The methodology is likely to scale up and help in attempting automated proofs of larger examples. This is demonstrated in mechanically generating a proof that the embedding transformation from an n-dimensional array of dimensions 2^{m_0} x $\cdots$ x 2^{m_d} to a hypercube preserves adjacency among neighbors.

Acknowledgment: We thank Jay Misra for suggesting that we attempt proofs of algorithms in his paper on powerlists using *RRL*. He kindly answered many questions about powerlists as well as provided comments and encouragement when these proofs were attempted on *RRL*.

References

1. J.A. Bergstra, J. Heering, and P. Klint, *Algebraic Specification,* ACM Press, 1989.
2. D. Kapur, "Constructors can be partial too," Technical Report, Dept. of Computer Science, State University of New York, Albany, NY, July 1994.
3. D. Kapur and X. Nie, "Reasoning about numbers in Tecton," Proceedings of the *8th International Symposium on Methodologies for Intelligent Systems, (ISMIS'94),* Charlotte, North Carolina, October 1994, 57-70.
4. D. Kapur and M. Subramaniam, "New uses of linear arithmetic in automated theorem proving by induction," Technical Report, Dept. of Computer Science, SUNY, Albany, NY, Sept. 1994. To appear in *J. of Automated Reasoning.*
5. D. Kapur and M. Subramaniam, " Automated reasoning about parallel algorithms using powerlists," Technical Report, Dept. of Computer Science, SUNY, Albany, NY 12222, Nov. 1994. A longer version of this paper.
6. D. Kapur, and H. Zhang, "An overview of Rewrite Rule Laboratory (RRL)," *J. of Computer and Mathematics with Applications,* 29, 2, 1995, 91-114.
7. D.E. Knuth, *Sorting and Searching: The Art of Computer Programming,* Vol. 3, Addison-Wesley.
8. J. Misra, "Powerlist: A structure for parallel recursion," *A Classical Mind: Essays in Honor of C.A.R. Hoare,* Prentice Hall, Jan. 1994
9. Leighton F. T., *Introduction to Parallel Algorithms and architectures: Arrays.Trees.Hypercubes,* Morgan Kaufmann, 1992.
10. H. Zhang, D. Kapur, and M.S. Krishnamoorthy, "A mechanizable induction principle for equational specifications," Proc. *9th Intl. Conf. on Automated Deduction (CADE),* Lusk and Overbeek (eds.), LNCS 310, Springer-Verlag, 1988, 250–265.

Representing, Verifying and Applying Software Development Steps using the PVS System*

Axel Dold

Abt. Künstliche Intelligenz,
Universität Ulm,
D-89069 Ulm, Germany
dold@informatik.uni-ulm.de

Abstract. In this paper generic software development steps of different complexity are represented and verified using the (higher-order, strongly typed) specification and verification system PVS. The transformations considered in this paper include "large" powerful steps encoding general algorithmic paradigms as well as "smaller" transformations for the operationalization of a descriptive specification. The application of these transformation patterns is illustrated by means of simple examples. Furthermore, we show how to guide proofs of correctness assertions about development steps. Finally, this work serves as a case-study and test for the usefulness of the PVS system.

1 Introduction

The methodology of stepwise refinement is widely accepted in modern software engineering. The idea is to start from an abstract requirement specification of a given problem and successively apply correctness preserving transformation patterns to finally yield an executable program. These transformations can comprise development steps of different complexity. One large powerful step can be sufficient to synthesize a program while a series of smaller steps has to be applied to reach a similar result.

In this paper we focus on the representation of development steps of different complexity and kind in a rigorous formal manner. "Large" steps encode general programming knowledge which forms the basis of many algorithms. Such knowledge is frequently applied implicitly when constructing programs but even when it is explicitly described in the literature it often appears informal and lacks a rigorous formal (error-free) treatment. In a formal treatment, such development steps can be represented as schematic algorithms which, instantiated with a specific problem, synthesize a solution to this problem. These "algorithm theories" have intensively been investigated by Doug Smith [12, 13, 14] who defines,

* Part of the research reported herein has been funded by the German Federal Ministry of Research and Technology (BMFT) under contract no. 01 IS 203 K5 (KORSO).

among other things, a hierarchy of algorithm theories encoding well-known programming paradigms such as *divide-and-conquer*, *global search*, *generate-and-test* and others. However, his approach is only semi-formal, some important aspects remain informal.

The transformations developed in the CIP-project and its descendants [2, 3, 10] can be considered as "smaller" development steps since they mainly operate on the level of functions. Among them one can find transformations for optimizing functions, recursion simplification, and as well, steps which operationalize a descriptive specification.

The goal of this paper is to completely formalize and verify two selected development steps, one of each kind, and to correctly apply them to examples. In order to represent software development steps higher-order logic greatly facilitates the formalization process. Therefore, and in order to have adequate system support we choose the specification and verification system PVS [7] in which the whole process of representation and verification can be carried out. The Prototype Verification System (PVS) consists of a higher-order specification language with a rich typing system, a set of supporting tools for creating, analyzing, modifying and documenting theories and proofs, and a powerful interactive Gentzen-style theorem prover. Furthermore, a library of standard theories such as natural numbers, polymorphic sets, and lists, booleans, relations is predefined. The type system provides type constructors to form dependent and non-dependent function, tuple, record, and semantic subtypes. Specifications are realized as PVS theories which can be parameterized where the parameters can be constrained by means of *assumptions*. Detailed information about the language, prover and the usage of the system can be found in [8, 9]. A distinctive feature of the typing system is the automatic generation of proof obligations, especially when instantiating the general scheme with a specific problem.

All considered steps are represented within a parameterized PVS theory which defines the required data structures and formalizes the application conditions by means of assumptions. Applying this step to a specific situation is carried out by importing the parameterized theory where the formal theory parameters are replaced by the specific problem parameters.

Another purpose of this paper is to investigate the verification process and to provide comprehensible, reusable proof methods. We show, for example, how to utilize the use of *measure-induction* in order to prove properties about recursive functions and to use the subtyping mechanism in order to encode correctness properties. Finally, we hope that this work serves as an interesting case-study and test for the usefulness of the PVS system.

The rest of this paper is organized as follows: the next section presents a formalization and verification of the schematic algorithm *divide-and-conquer* and an application of it to a binary-search problem. The operationalization of a descriptive specification is presented in Sect. 3 and is applied to the problem of finding a minimum element in a list.

Related Work

The formalization of transformations using higher-order functions has been considered by several researchers. In [4], for example, program transformations for recursion removal are expressed as second-order patterns defined in the simply typed λ-calculus. Independently from the work described herein, my colleague Harald Rueß has formalized, among other things, the *divide-and-conquer* paradigm in his dissertation [11] using the calculus of constructions and has given a verification with the LEGO proof checker [6]. Similar work dealing with the representation of existing approaches to program synthesis, development steps, and programming paradigms as well as a library of standard theories in the context of the NUPRL system has been carried out by Christoph Kreitz [5].

2 Divide-and-Conquer

The well-known algorithmic paradigm *divide-and-conquer* is based on the principle of solving primitive problem instances directly, and large problem instances by decomposing them into 'smaller' instances, solving them independently and composing the resulting solutions. Here, we consider the decomposition of the problem into two subproblems. A general decomposition scheme would be treated analogously. Following Smith's notation of a problem specification, we parameterize theories with a descriptive problem specification described as a 4-tuple (D, R, I, O) where D denotes the problem domain, R denotes the problem range, I is a predicate constraining D to meaningful inputs, and O describes the problem as an input-/output predicate. A solution of such a problem is a function computing feasible solutions, i.e. for an input x satisfying condition I it computes a y of type R such that condition $O(x, y)$ holds. We represent this principle as a parameterized PVS theory which has a problem description (D, R, I, O) and functions *decompose*, *compose*, *dir_solve*, a predicate *primitive?*, a map *lt* from domain D to natural numbers as its parameters. PVS only allows total functions, it must be ensured that all (recursive) functions terminate. For this purpose, a *measure-function* is used. Its domain matches that of the recursive function, and its range is `Nat` or `Ordinal`. The definition of a recursive function f generates a type correctness condition (TCC) which must be discharged in order to guarantee well-definedness of f. Here, the function *lt* serves as a measure-function. Four assumptions describe the meaning of the parameters. They state that

1. the subproblems created by the decomposition operator are smaller than the original problem when applied to the measure-function *lt*.
2. if the problem is primitive enough *dir_solve* creates a solution.
3. solutions to the subproblems z_1, z_2 can be composed to build a solution to the original problem.
4. all subproblems generated by the decomposition operator satisfy the input condition I of the problem (D, R, I, O).

```
div_and_conq[D : TYPE, R : TYPE, I : [D → boolean],
             O : [D, R → boolean], decompose : [D → [D, D]],
             dir_solve : [D → R], compose : [D, R, R → R],
             primitive? : [D → boolean], lt : [D → nat]] : THEORY
 BEGIN
 ASSUMING
 x : VAR D  z1, z2 : VAR R
  ax1 :
   ASSUMPTION
    (I(x) ∧ ¬ (primitive?(x)))
       ⊃
     ((lt(proj_1(decompose(x))) < lt(x))
        ∧ (lt(proj_2(decompose(x))) < lt(x)))

  ax2 : ASSUMPTION (I(x) ∧ primitive?(x)) ⊃ O(x, dir_solve(x))

  ax3 :
   ASSUMPTION
    (I(x)
       ∧ ¬ (primitive?(x))
        ∧ O(proj_1(decompose(x)), z1) ∧ O(proj_2(decompose(x)), z2))
       ⊃ O(x, compose(x, z1, z2))

  ax4 :
   ASSUMPTION
    (I(x) ∧ ¬ (primitive?(x)))
       ⊃ (I(proj_1(decompose(x))) ∧ I(proj_2(decompose(x))))

  ENDASSUMING

 USING measure_induction[D, nat, lt, (λ (x, y : nat) : x ≤ y)]
 f_dc(x : {y : D | I(y)}) : RECURSIVE R =
   IF primitive?(x) THEN dir_solve(x)
    ELSE LET
           x1  = proj_1(decompose(x)), x2 = proj_2(decompose(x)),
           rec1  = f_dc(x1), rec2  = f_dc(x2)
      IN compose(x, rec1, rec2)
    ENDIF
  MEASURE (λ (x : {y : D | I(y)}) : lt(x))

 correct : THEOREM (∀ (x : D) : I(x) ⊃ O(x, f_dc(x)))

 END div_and_conq
```

Fig. 1. Theory of *Divide-and-conquer*

The built in selector `proj_i(x)` selects the i-th element of a tuple. The recursive function *f_dc* realizes the schematic algorithm. Its domain is specified using the subtype-mechanism of PVS. It is given by the type D such that input condition I holds. Termination is established by the given measure function *lt* for which we must show that it decreases in size for the recursive arguments. This is given immediately by the first assumption. Figure 1 shows the (LaTeX pretty-printed) PVS theory. Type-checking this theory generates the following type correctness conditions (TCC's):

```
% Subtype TCC generated (line 35) for x1
  % proved - complete
f_dc_TCC2: OBLIGATION (FORALL (x: {y: D | I(y)}):
  NOT primitive?(x) IMPLIES I(PROJ_1(decompose(x))))

% Termination TCC generated (line 35) for f_dc
  % proved - complete
f_dc_TCC3: OBLIGATION (FORALL (x: {y: D | I(y)}):
  NOT primitive?(x) IMPLIES lt(PROJ_1(decompose(x))) < lt(x))

% Subtype TCC generated (line 36) for x2
  % proved - complete
f_dc_TCC4: OBLIGATION (FORALL (x: {y: D | I(y)}):
  NOT primitive?(x) IMPLIES I(PROJ_2(decompose(x))))

% Termination TCC generated (line 36) for f_dc
  % proved - complete
f_dc_TCC5: OBLIGATION (FORALL (x: {y: D | I(y)}):
  NOT primitive?(x) IMPLIES lt(PROJ_2(decompose(x))) < lt(x))
```

`f_dc_TCC2` and `f_dc_TCC4` are generated in order to ensure that the subproblem instances satisfy the condition I while both `f_dc_TCC3` and `f_dc_TCC5` ensure termination of *f_dc*. All obligations follow immediately from the assumptions $ax1$ and $ax4$. The correctness of this schematic algorithm is stated by theorem *correct*: *f_dc* exactly calculates a solution of the given problem (D, R, I, O).

2.1 Proof of *correct*

In order to prove properties about recursive functions, a measure-induction principle is required which is predefined in PVS:

measure_induction$[T, M :$ TYPE$, m : [T \rightarrow M], \leq : (\text{well_founded?}[M])]:$ **THEORY**
  BEGIN
  measure_induction :
    LEMMA
      $(\forall (p : \text{pred}[T]) :$
        $(\forall (x : T) : (\forall (y : T) : m(y) \leq m(x) \wedge m(y) \neq m(x) \supset p(y)) \supset p(x))$
          $\supset (\forall (x : T) : p(x)))$
  END measure_induction

Measure_induction builds on well-founded induction. It allows induction over a type T for which a measure function m is defined. Here, the theory is instantiated with $D, Nat, lt, \leq_{Nat}$.

In the following we give the main ideas of the PVS proof.[2] We start with the sequence:

```
  |-------
{1}   (FORALL (x: D): I(x) IMPLIES O(x, f_dc(x)))
```

We have written a strategy which instantiates the measure-induction principle and discharges the obligation that $\leq_{Nat}$ is well-founded. In the first step we apply this strategy, expand the definition of *f_dc* and obtain:

```
{-1}    (FORALL (y: D): lt(y) <= lt(x!1) AND lt(y) /= lt(x!1)
           IMPLIES I(y) IMPLIES O(y, f_dc(y)))
{-2}    I(x!1)
  |-------
{1}   O(x!1,
            IF primitive?(x!1) THEN dir_solve(x!1) ELSE
            compose(x!1, f_dc(proj_1(decompose(x!1))),
                         f_dc(proj_2(decompose(x!1))))
            ENDIF)
```

Case analysis on `primitive?(x!1)` yields two subgoals:

```
correct.1 :

{-1}    primitive?(x!1)
[-2]    (FORALL (y: D): lt(y) <= lt(x!1) AND lt(y) /= lt(x!1)
           IMPLIES I(y) IMPLIES O(y, f_dc(y)))
[-3]    I(x!1)
  |-------
{1}   O(x!1, dir_solve(x!1))
```

Applying assumption *ax*2 completes this branch.

For the other branch where `primitive?(x!1)` is false we apply *ax*3. This yields four subgoals:[3]

---

[2] The prover maintains a proof tree. The goal is to construct a proof tree which is complete, in the sense that all of the leaves are recognized as true. Each proof goal is a sequent consisting of a sequence of antecedent formulas (indicated by negative numbers) and consequent formulas (indicated by positive numbers). The intuitive meaning of such a goal is that the conjunction of the antecedents implies the disjunction of the consequents.

[3] We omit the subgoals `correct.2.2` and `correct.2.4` since they correspond to `correct.2.1` and `correct.2.3` respectively, just substitute `proj_2` for `proj_1` in formula {1}.

```
correct.2.1 :

[-1]   (FORALL (y: D): lt(y) <= lt(x!1) AND lt(y) /= lt(x!1)
           IMPLIES I(y) IMPLIES O(y, f_dc(y)))
[-2]   I(x!1)
  |-------
{1}   O(proj_1(decompose(x!1)), f_dc(proj_1(decompose(x!1))))
[2]   primitive?(x!1)
[3]   O(x!1, compose(x!1, f_dc(proj_1(decompose(x!1))),
                          f_dc(proj_2(decompose(x!1)))))
```

```
correct.2.3 (TCC):

[-1]   (FORALL (y: D): lt(y) <= lt(x!1) AND lt(y) /= lt(x!1)
           IMPLIES I(y) IMPLIES O(y, f_dc(y)))
[-2]   I(x!1)
  |-------
{1}   I(proj_2(decompose(x!1)))
[2]   primitive?(x!1)
[3]   O(x!1, compose(x!1, f_dc(proj_1(decompose(x!1))),
                          f_dc(proj_2(decompose(x!1)))))
```

Consider the first subgoal **correct.2.1**. Automatically instantiation of the term **proj_1(decompose(x!1))** for **y** and applying assumptions *ax*1 and *ax*4 completes the proof. The second subgoal is immediately proved by applying *ax*4. □

2.2 Example: Binary Search

We apply the schematic algorithm *divide-and-conquer* to solve the following problem: Given a function f on natural numbers, a key element and two bounds i_1, i_2 denoting the interval $\{n : Nat \mid i_1 \leq n \leq i_2\}$ the problem is to check if f applied to one of the elements of the interval is equal to the key element. This problem is described by D_1, R_1, I_1, O_1 where

- D_1 is the problem domain consisting of a function f, a key element *key* and bounds i_1, i_2.
- R_1 is the type of booleans.
- I_1 constrains domain D_1 such that i_1 is positive and $i_2 \geq i_1$.
- O_1 describes the input-/output condition informally given above.

Furthermore, applying the scheme of divide-and-conquer we have to explicitly give functions and predicates *primitive1?*, *dir_solve1*, *decompose1*, and *compose1* plus a measure-function *lt1*. The idea is to use a binary-search mechanism to solve the given problem. Therefore, we choose the following *divide-and-conquer*-theory instance:

- The problem is *primitive* if the given interval is trivial, i.e. $i_1 = i_2$.
- If the problem is primitive we can *directly solve* it by checking if $f(i_1) = key$.

```
bs : THEORY
  BEGIN
  IMPORTING div
  D1 : TYPE  = [f : [nat → nat], key : nat, i1 : nat, i2 : nat]
  R1 : TYPE  = boolean

  I1(x : D1) : boolean  =
   LET i1 = proj_3(x), i2 = proj_4(x) IN (i1 > 0) ∧ (i2 ≥ i1)

  O1(x : D1 ,y : R1) : boolean  =
   LET f = proj_1(x), key  = proj_2(x), i1 = proj_3(x), i2 = proj_4(x)
   IN y = (∃ (z : {n1 : nat | i1 ≤ n1 ∧ n1 ≤ i2}) : (f(z) = key))

  primitive1?(x : D1) : boolean  =
   LET i1 = proj_3(x), i2 = proj_4(x) IN (i1 = i2)

  dir_solve1(x : D1) : R1 =
   LET f = proj_1(x), key  = proj_2(x), i1 = proj_3(x) IN (f(i1) = key)

  decompose1(x : D1) : [D1, D1] =
   LET f = proj_1(x), key  = proj_2(x), i1 = proj_3(x), i2 = proj_4(x)
   IN ((f, key, i1, div(i1 + i2, 2)), (f, key, 1 + div(i1 + i2, 2), i2))

  compose1(x : D1, y1 : R1, y2 : R1) : R1  = (y1 ∨ y2)
  lt1(x : D1) : nat  = LET i1 = proj_3(x), i2 = proj_4(x) IN abs(i2 − i1)

  IMPORTING div_and_conq[D1, R1, I1, O1,
                         decompose1, dir_solve1, compose1, primitive1?, lt1]
END bs
```

Fig. 2. A binary-search problem

- *Decomposition* is done by splitting the range given by i_1, i_2 into two parts.
- The results of the search process in both subintervals are disjunctively *composed*.
- The required measure *lt*1 is defined by the size of the interval.

All the entities are combined in the PVS theory *bs*, see Fig. 2. The PVS theory *div* defining the div-function on natural numbers together with some properties is imported. We omit this theory since it is not of great significance. All proof obligations and lemmata of *div* have been successfully discharged.

Consider the *bs* theory, we have to show that it is indeed a correct instance of the general divide-and-conquer theory. Type-checking *bs*, PVS automatically generates the four required obligations where the first one is given as

```
IMPORTING1_TCC1: OBLIGATION
  (FORALL (x: D1): (I1(x) & NOT (primitive1?(x)))
                 IMPLIES ((lt1(PROJ_1(decompose1(x))) < lt1(x)) &
                          (lt1(PROJ_2(decompose1(x))) < lt1(x))))
```

The proofs of all TCC's are established by simply expanding the definitions and using some elementary properties of *div*. Finally, having discharged all TCC's we obtain a correct solution using the instantiated algorithm *f_dc*.

3 Operationalization of a Descriptive Specification

In this section we represent a transformation called *operationalization of a choice* given in [10]. We closely follow the method described in the previous section in representing this step. However, we give another possibility to establish the correctness of such a formalization. Here, instead of using an explicit correctness theorem we encode this information into the type of the recursive function utilizing the subtyping mechanism of PVS. The transformation works as follows: Starting from a given problem (D, R, I, O), the idea is to find a predicate B such that

1. $B(x) \Rightarrow O(x, H(x))$, and
2. $\neg B(x) \Rightarrow O(x, y) = O(K(x), y)$

In the first case, whenever $B(x)$ holds $H(x)$ is a feasible solution, in the second case y is a solution to input x if and only if y is a solution to $K(x)$, where K modifies x such that it decreases w.r.t the given measure-function. A solution of the problem, i.e. a function f for which $O(x, f(x))$ holds for all x satisfying I, is then immediately obtained by the recursive function *fun* given as:

$$fun(x) = \text{IF } B(x) \text{ THEN } H(x) \text{ ELSE } fun(K(x))$$

Figure 3 shows the PVS theory. As noted above, the correctness is stated using the subtype mechanism of PVS. The range of function *fun* is of type R such that the input-/output condition O of the problem holds. Type-checking this theory automatically generates the required correctness conditions:

```
fun_TCC2: OBLIGATION
          (FORALL (x: {x1: D | I(x1)}): B(x) IMPLIES o(x, H(x)))

fun_TCC3: OBLIGATION
          (FORALL (x: {x1: D | I(x1)}): NOT B(x) IMPLIES I(K(x)))

fun_TCC4: OBLIGATION
          (FORALL (v: [x: {x1: D | I(x1)} -> {y: R | o(x, y)}]),
                  (x: {x1: D | I(x1)}):
                    NOT B(x) IMPLIES o(x, v(K(x))))
```

All but the third obligation are trivial and follow immediately from the assumptions. The only interesting obligation is the third one. The proof is established by adding type information of `v(K(x))` and applying assumptions $ax2$ and $ax3$.

```
op_of_choice_I[D : TYPE, R : TYPE, I : [D → boolean], O : [D, R → boolean],
               B : [D → boolean], H : [D → R], K : [D → D], lt : [D → nat]] :
 THEORY
  BEGIN
  ASSUMING
   x : VAR D  y : VAR R
   ax1 : ASSUMPTION (I(x) ∧ B(x)) ⊃ O(x, H(x))

   ax2 :
    ASSUMPTION (I(x) ∧ (¬ B(x))) ⊃ (O(x, y) = O(K(x), y))

   ax3 :
    ASSUMPTION
     (I(x) ∧ (¬ B(x))) ⊃ (I(K(x)) ∧ lt(K(x)) < lt(x))

   ENDASSUMING

  fun(x : {x1 : D | I(x1)}) : RECURSIVE {y : R | O(x, y)} =
     IF B(x) THEN H(x) ELSE fun(K(x)) ENDIF
   MEASURE (λ (x : {x1 : D | I(x1)}) : lt(x))

  END op_of_choice_I
```

Fig. 3. Transformation op_of_choice_I

3.1 Example: Minimum Element

Suppose we are given the problem of finding a minimum element in a given (non-empty) list of natural numbers. We formalize this problem as the following 4-tuple (D_1, R_1, I_1, O_1):

- D_1, the problem domain, is defined as a tuple type $Nat \times List(Nat)$, where the first component denotes the temporary minimum.
- R_1 is the type of natural numbers.
- I_1 is the constant true function.
- $O_1(x, y)$ is true, if y is less than or equal to x's first parameter and every element of x's second parameter (denoting the list).

In order to correctly apply the above transformation we further have to supply specific values B_1, H_1, K_1, a measure function lt_1, and have to discharge all arising proof obligations.

- B_1 is true if and only if the list is empty.
- H_1 yields the first component of the tuple.
- K_1 yields the minimum of the head element and the temporary minimum element plus the tail of the list.
- lt_1 is defined as the length of the list.

```
minel : THEORY
  BEGIN
  IMPORTING list_prop

  D1 : TYPE  = [nat, list[nat]]
  R1 : TYPE  = nat
  I1 : [D1 → boolean] = (λ (x : D1) : TRUE)

  O1(x : D1, y : R1) : boolean =
    ((y = proj_1(x) ∨ member?(y, proj_2(x)))
      ∧ (y ≤ proj_1(x)) ∧ (∀ (n : nat) : member?(n, proj_2(x)) ⊃ y ≤ n))

  B1 : [D1 → boolean] = (λ (x : D1) : null?(proj_2(x)))
  H1 : [D1 → R1] = (λ (x : D1) : proj_1(x))

  K1 : [D1 → D1] =
    (λ (x : D1) :
      (IF null?(proj_2(x)) THEN x
       ELSE
        ((IF proj_1(x) ≤ car(proj_2(x)) THEN proj_1(x)
          ELSE car(proj_2(x))
          ENDIF),
        cdr(proj_2(x)))
       ENDIF))

  lt1 : [D1 → nat] = (λ (x : D1) : length(proj_2(x)))

  IMPORTING op_of_choice_I[D1, R1, I1, O1, B1, H1, K1, lt1]

  minelf(s : {s1 : list[nat] | ¬ (null?(s1))}) : RECURSIVE nat =
    LET x = (car(s), cdr(s)) IN fun(x)
  MEASURE
    (λ (s : {s1 : list[nat] | ¬ (null?(s1))}) : LET x = (car(s), cdr(s)) IN lt1(x))

  END minel
```

Fig. 4. The minimum element problem

The function *minelf* then computes the minimum element of a given non-empty list of natural numbers using function *fun* which is called with the tuple consisting of the list's head and tail, see Fig. 4. When type-checking this theory the three required TCC's (the instantiated assumptions of *op_of_choice_I*) are generated where the second one is given as

```
(*)  IMPORTING1_TCC2: OBLIGATION
       (FORALL (x: D1), (y: R1):
        (I1(x) & (NOT B1(x))) IMPLIES (O1(x, y) = O1(K1(x), y))
```

The first and last obligation are easy to prove (simply rewrite all definitions) whereas the proof of the second obligation is lengthy and requires analyses of many cases. The complete proof script is given in the appendix. All proof obligations, lemmata and theorems have been successfully discharged, and the proof scripts are available by the author.

4 Concluding Remarks

In this paper we have demonstrated that it is possible to elegantly formalize and verify software development steps in a rigorous mathematical manner using the PVS system. We have considered steps of different complexity, both steps coding well-known algorithmic paradigms and "smaller" steps defined in the context of the CIP project. The use of higher-order logic with a rich type system greatly supported the formalization of very general transformation schemes. The method described in this paper can readily be used to represent other development steps of both kinds as we have demonstrated within the BMFT project KORSO (correct software). In another paper, for example, we have formalized the theory of *global-search* algorithms using a type-theoretic framework [1]. This framework in which all entities of the software development process can be formally represented and reasoned about has also been developed within this project. We refer to [15] for more information about the framework and the project.

Furthermore, we have shown how to synthesize a specific algorithm for some given problem simply by "filling the holes" of a general scheme. The concept of semantic subtypes has turned out to be an adequate tool to establish the correctness of the formalized steps since the type-check mechanism of PVS produces the required proof obligations automatically.

There are of course some aspects which can be improved in future versions of PVS. For example, when representing hierarchies of software development steps following the ideas of Smith [14] it is desirable to define theories which have other theories as their parameters. This is not possible in the current version. Furthermore, it is not possible to express properties about theories such as refinements between theories (theory morphisms). PVS does not allow types as parameters or results of functions. Therefore, functions like the polymorphic identity cannot be expressed directly. Finally, pattern matching could improve the readability of PVS specifications avoiding the use of projections.

Acknowledgement

I wish to thank F. W. von Henke, Harald Rueß, Martin Strecker, Detlef Schwier, and Ercüment Canver for many discussions and comments on draft versions of this paper. The constructive criticisms and suggestions provided by the anonymous referees have greatly improved the paper.

References

1. A. Dold. Formalisierung schematischer Algorithmen. Technical Report UIB-94-10, Fakultät für Informatik, Universität Ulm, January 1994.
2. The CIP Language Group. *The Munich Project CIP, Volume I: The Wide Spectrum Language CIP-L.* LNCS 183. Springer-Verlag, 1985.
3. The CIP System Group. *The Munich Project CIP - Volume II: The Program Transformation System CIP-S.* LNCS 292. Springer-Verlag, 1987.
4. G. Huet and B. Lang. Proving and Applying Program Transformations Expressed with Second-Order-Patterns. *Acta Informatica*, 11:31–55, 1978.
5. C. Kreitz. Metasynthesis - Deriving Programs that Develop Programs. Technical Report AIDA-93-03, Fachgebiet Intellektik, Technische Hochschule Darmstadt, 1993.
6. Z. Luo and R. Pollack. *LEGO Proof Development System: User's Manual.* University of Edinburgh, May 1992.
7. S. Owre, J. M. Rushby, and N. Shankar. PVS: A Prototype Verification System. In Deepak Kapur, editor, *11th International Conference on Automated Deduction (CADE)*, volume 607 of *Lecture Notes in Artificial Intelligence*, pages 748–752, Saratoga, NY, 1992. Springer-Verlag.
8. S. Owre, N. Shankar, and J.M. Rushby. *The PVS Proof Checker: A Reference Manual.* Computer Science Laboratory, SRI International, Menlo Park, CA, February 1993.
9. S. Owre, N. Shankar, and J.M. Rushby. *User Guide for the PVS Specification and Verification System.* Computer Science Laboratory, SRI International, Menlo Park, CA, February 1993.
10. H.A. Partsch. *Specification and Transformation of Programs.* Springer-Verlag, 1990.
11. H. Rueß. *Metaprogrammierung in einer typtheoretischen Umgebung.* PhD thesis, Universität Ulm, Abt. KI, to appear in 1995.
12. Douglas R. Smith. Applications of a Strategy for Designing Divide-and-Conquer-Algorithms. *Science of Computer Programming*, (8):213–229, 1987.
13. Douglas R. Smith. Structure and Design of Global Search Algorithms. Technical Report KES.U.87.12, Kestrel Institute, Palo Alto, CA, 1987.
14. Douglas R. Smith and Michael R. Lowry. Algorithm Theories and Design Tactics. *Science of Computer Programming*, (14):305–321, 1990.
15. F.W. von Henke, A. Dold, H. Rueß, D. Schwier, and M. Strecker. Construction and Deduction Methods for the Formal Development of Software. In M. Broy and S. Jähnichen, editors, *KORSO, Correct Software by Formal Methods.* Springer-Verlag, Lecture Notes in Computer Science, to appear in 1995, also available as Technical Report UIB-94-09, Fakultät für Informatik, Universität Ulm.

A Proof Script

We give the proof script of formula (*). The obligation can be proved using the prover command (TERM-TCC) which expands all relevant definitions, skolemizes by automatically generating skolem constants, tries to automatically instantiate the quantifiers, and does propositional simplification. The application of this command results in two subgoals. The boolean equality is then converted to an equivalence by (IFF). Applying (PROP) for propositional simplification to each of the subgoals yields a lot of new subgoals each of which is proved either by (TCC) or by (THEN* (INST? :SUBST ("n" "car(proj_2(x!1))")) (ASSERT)) instantiating the term car(proj_2(x!1)) for quantifier n and invoking the decision procedures. The generated proof script looks as:

```
("" (TERM-TCC)
    (("1" (IFF)
          (SPLIT)
          (("1" (FLATTEN)
                (SPLIT)
                (("1" (SPLIT)
                      (("1" (FLATTEN) (PROPAX)) ("2" (PROPAX))
                                                ("3" (TCC))))
                  ("2" (SPLIT)
                       (("1" (FLATTEN)
                             (INST? :SUBST
                                    ("n" "car(proj_2(x!1))"))
                             (ASSERT))
                         ("2" (PROPAX))
                         ("3" (TCC))))
                  ("3" (SPLIT)
                       (("1" (FLATTEN) (PROPAX)) ("2" (PROPAX))
                                                 ("3" (TCC))))))
            ("2" (FLATTEN)
                 (SPLIT)
                 (("1" (SPLIT)
                       (("1" (FLATTEN) (PROPAX)) ("2" (PROPAX))
                                                 ("3" (TCC))))
                   ("2" (SPLIT)
                        (("1" (FLATTEN) (PROPAX)) ("2" (PROPAX))
                                                  ("3" (TCC)))))))))
      ("2" (IFF)
           (SPLIT)
           (("1" (FLATTEN)
                 (SPLIT)
                 (("1" (SPLIT)
                       (("1" (FLATTEN)
                             (INST? :SUBST
                                    ("n" "car(proj_2(x!1))"))
                             (ASSERT)
                             (PROPAX))
```

```
                    ("2" (INST? :SUBST
                                ("n" "car(proj_2(x!1))"))
                         (ASSERT)
                         (PROPAX))
                    ("3" (TCC))))
               ("2" (SPLIT)
                    (("1" (FLATTEN) (PROPAX))
                      ("2" (INST? :SUBST
                                 ("n" "car(proj_2(x!1))"))
                           (ASSERT))
                      ("3" (TCC))))
               ("3" (SPLIT)
                    (("1" (FLATTEN) (PROPAX))
                      ("2" (INST? :SUBST
                                 ("n" "car(proj_2(x!1))"))
                           (ASSERT)
                           (PROPAX))
                      ("3" (TCC))))))
          ("2" (FLATTEN)
               (SPLIT)
               (("1" (SPLIT)
                     (("1" (FLATTEN) (PROPAX))
                       ("2" (INST? :SUBST
                                  ("n" "car(proj_2(x!1))"))
                            (ASSERT))
                       ("3" (TCC))))
                ("2" (SPLIT)
                     (("1" (FLATTEN) (PROPAX))
                       ("2" (INST? :SUBST
                                  ("n" "car(proj_2(x!1))"))
                            (ASSERT))
                       ("3" (TCC))))))))))
```

An Algebraic Development Technique for Information Systems*

Martin Gogolla & Rudolf Herzig

Bremen University
Department for Mathematics and Informatics
Database Systems Group
Postfach 330440, D-28334 Bremen, GERMANY
e-mail: gogolla@informatik.uni-bremen.de

Abstract. This paper reports on successful application of algebraic ideas to the formal development of software systems, in particular information systems. It describes (1) a formalism, i.e., a language, for the specification of information systems, (2) a method for the construction of specifications in this language, and (3) implemented and planned parts of a specification environment covering important phases of the software development process.

Keywords: *Algebraic and Logical Foundations* – Algebraic Methodologies for Languages and Systems; *Concurrent and Reactive Systems* – Algebraic Approaches, Object Oriented Models; *Software Technology* – Specification Languages and Tools, Integration of Pragmatic and Formal Methods.

1 Introduction

Although theory and practice of algebraic specifications have been studied by many people for about twenty years, there are opinions doubting that this line of research is well-suited for practical software development:

> *Algebraic specification techniques have been widely applied to small examples but there is little evidence yet that they are suitable for specifying large systems. It is worth trying to explain the problems of algebraic specifications with a small example. Imagine a system containing a database which we wish to update. If we model the database directly we can simply specify validity (the object to be updated exists, the type is correct, etc.). In an algebraic approach we would have to establish existence of an object by reasoning about the sequence of inputs to the system and determining whether the object had been created (successfully) since it was last deleted. This would make the specification obscure and cumbersome.*
>
> *John McDermid and Paul Rook [MR91]*

* Work reported here has been partially supported by the CEC under Grant No. 6112 (COMPASS).

We do not follow the above arguments, but propose an algebraic approach to the specification of information systems dealing with the notion of state in an implicit manner. Thus the above criticism does not hold for our approach (and other recent algebraic proposals like [GM87, DG94, EO94, GD94, MQ93]), because the notion of state for a system is (at least for our application area, namely databases and information systems) so important that we decided to incorporate it directly into our framework. The basic idea is to represent system states as well as system transitions as algebras. Therefore, although we stay on algebraic grounds, we give up the idea (present in classical algebraic specifications) that the complete system is specified as one algebra.

Thus, the main contribution of this paper is *one* attempt to explain why and how algebraic techniques can be applied successfully for practical software development. Following [ESD93], our approach to software engineering [GJM91], in particular to software development, can be visualized in Fig. 1 by a triangle picturing the tension between a formalism, a method, and an environment.

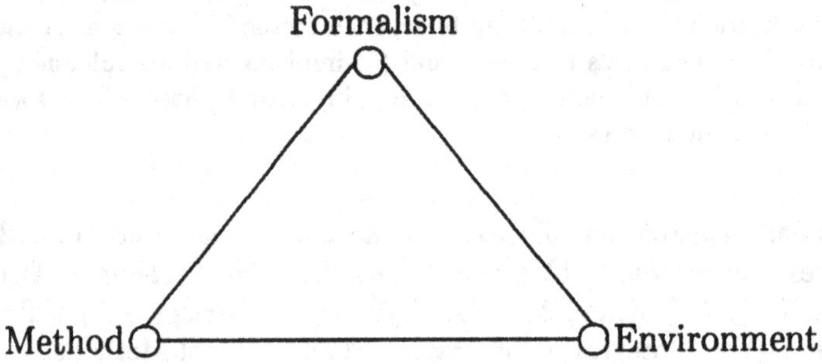

Fig. 1. Ingredients of a Software Development Approach

By *formalism* we mean a specification or programming language with well-defined syntax and semantics. However having an expressive language is not enough. Developers must have some guidelines how to proceed when they write specifications or programs. Thus, a *method* supporting the construction of software should be at hand. In order to round off the scenario, we do not expect that software development is carried out in a paper-and-pencil fashion but is assisted by a computerized development *environment* tailored to the respective formalism and method.

Let us now come back to our application area, namely information systems. Here, the aim of the conceptual modeling process is to provide a first precise description of the part of the world which is to be modeled, the so-called Universe of Discourse (UoD). By observing a certain UoD one realizes that it consists of complex structured entities with time-varying behavior. These entities exist concurrently and interact with each other. Therefore the idea of object-oriented specification is to represent the real-world entities as formal objects in a specification language. The main advantage of following the object paradigm is the fact that all relevant information concerning one object can be found within one single unit and is not distributed over a variety of locations.

Recently, a number of object specification languages like ABEL [DO91], CMSL [Wie91], GLIDER [CJO94], MONDEL [BBE+90], OCS [AAR94], OS [Bre91], and Π [Gab93] have been proposed. However, we work with the language TROLL *light*, a dialect of OBLOG [SSE87] and TROLL [JSHS91]. In TROLL *light* structure as well as behavior of real-world entities can be characterized in a direct way.
The structure of the rest of this paper mainly follows the triangle depicted in Fig. 1. In Sect. 2 we sketch the formalism we use, namely the language TROLL *light*. Section 3 explains our method for the development of information systems. In Sect. 3 we describe implemented and planed tools for supporting the method mentioned before. The last section gives some concluding remarks.

2 Formalism

2.1 Concepts and Syntax of TROLL *light*

We explain the essentials of TROLL *light* by means of an example. In Fig. 2 binary trees, where the nodes carry boolean information, are specified as object types or templates as object types are called in TROLL *light*.

```
TEMPLATE Node
  DATA TYPES   Bool;
  ATTRIBUTES   Content:bool;
  SUBOBJECTS   Left, Right:node;
  EVENTS       BIRTH create;
                     createLeft;
                     createRight;
                     update(Content:bool);
               DEATH destroy;
  VALUATION    [ create ] Content=true;
               [ update(B) ] Content=B;
  INTERACTION createLeft >> Left.create;
               createRight >> Right.create;
  BEHAVIOR     PROCESS Node =
                 ( create -> NodeLife );
               PROCESS NodeLife =
                 ( createLeft, createRight, update -> NodeLife |
                   { UNDEF(Left) AND UNDEF(Right) }
                     destroy -> Node );
END TEMPLATE
```

Fig. 2. TROLL *light* example template

DATA TYPES: In TROLL *light*, data types are employed for various purposes. Data types are assumed to be specified with a data type specification language [Wir90]. Their signature is made known in a template with the DATA

TYPES section. The basic difference between data values as instances of data types and objects as instances of object types is that values are regarded as stateless items while objects are associated with an inherent notion of state.

ATTRIBUTES: The `Node` template uses the data type `Bool` in order to define one attribute `Content` of data sort `bool`.

SUBOBJECTS: For templates (or object types) we employ the same naming convention as for data types. Template names are written capitalized, and each template (`Node`) induces a corresponding object sort written exactly as the template but with a starting lower case letter (`node`). A `Node` object is allowed to have two local sub-objects, namely `Left` and `Right`, of sort `node`.

EVENTS: The things which can happen to `Node` objects are called events. `Node` objects can be created, their left and right sub-nodes can be created, they can be updated, and they can be destroyed.

VALUATION: Valuation rules serve to define the effect of events on attributes. They specify the value an attribute has after the occurrence of an event.

INTERACTION: Interaction rules are used to describe the synchronization of events. For instance, whenever the `createLeft` event occurs in the parent object, the `create` event must occur for the left sub-object. New objects can only be created as sub-objects of existing objects by this mechanism which is called birth event calling.

BEHAVIOR: Possible life cycles, i.e., sequences of allowed events, are specified in the behavior part. For a given template there is always a process with the same name which is the initial process for objects belonging to the template. An objects' life starts with a **BIRTH** event and possibly ends with a **DEATH** event, which in this case is only allowed to occur when there are no sub-objects. In between, the events `createLeft`, `createRight`, and `update` can occur, but due to the interaction rule, for instance, `createLeft` can only occur if the called event `create` fits into the sub-objects' life cycle.

Above, we have shown a very simple example of a TROLL *light* specification for introductory purposes. For more sophisticated applications, we have to refer to [CGH92, GCH93, GHC+94]. There, you find for instance parameterized attributes and sub-object symbols, complex attributes, and non-trivial event synchronization. Several case studies demonstrating that TROLL *light* is suitable for larger specifications as well have been carried out. We mention a specification environment [GC93], a medical information system [Bec93], a production cell [HV94], and a conference information system [HH94] among other smaller TROLL *light* specifications [Ehr93].

2.2 Basic Ideas of an Algebraic Semantics

The details of the algebraic semantics of TROLL *light* can only be sketched here. The full treatment is given in [GH95]. Speaking in technical terms, object signatures consists of a data signature part (with data sorts and corresponding data operators) and object specific items like object (identifier) sorts, attribute symbols, and event symbols.

Definition 1. An **object signature** $O\Sigma = (DS, \Omega, OS, A, E)$ consists of

- a set DS of data sorts,
- a family of sets of operation symbols $\Omega = \langle \Omega_{dw,ds} \rangle_{dw \in DS^*, ds \in DS}$,
- a set OS of object sorts,
- a family of sets of attribute symbols $A = \langle A_{os,w,s} \rangle_{os \in OS, w \in S^*, s \in S}$, and
- a family of sets of event symbols $E = \langle E_{os,w} \rangle_{os \in OS, w \in S^*}$

where S refers to the union of data and object sorts: $S = DS \cup OS$.

Example 2. The `Node` template induces the following object signature.

data sorts	*bool*	
operations	$false, true, bottom :\to bool$	$(f, t, \perp)$
object sorts	*node*	
attributes	$left, right : node \to node$	(l, r)
	$content : node \to bool$	(ct)
events	$create : node$	(c)
	$createLeft, createRight : node$	(cL, cR)
	$update : node \times bool$	(u)
	$destroy : node$	(d)

We abbreviate the operation names as indicated on the right hand side above. With respect to sub-object symbols as found in `Node` we assume that these are treated just like attribute symbols. The subtle distinction is discussed in [GH95].

The interpretation of data sorts, data operations, object identifier sorts, and attribute symbols characterize the states of an object community. Event symbols are associated with state transitions. Hence we define the notions of data, attribute, and event signatures as follows.

Definition 3. Let an object signature $O\Sigma$ be given. The **data signature** induced by $O\Sigma$ consists of the data sorts and the data operations of the object signature, i.e., $D\Sigma = (DS, \Omega)$.
The **attribute signature** $A\Sigma$ induced by $O\Sigma$ consists of S, i.e., the union of data and object sorts, and the union of operation and attribute symbols, i.e., $A\Sigma = (DS \cup OS, \Omega \cup A)$.
The **event signature** $E\Sigma$ induced by $O\Sigma$ consists of the attribute signature extended by one special sort symbol $\hat{e}$ for events and an appropriate family of event symbols $\hat{E}$, i.e., $E\Sigma = (DS \cup OS \cup \{\hat{e}\}, \Omega \cup A \cup \hat{E})$ with $\hat{E} = \langle \hat{E}_{os,w,\hat{e}} \rangle_{os \in OS, w \in S^*}$ and $\hat{E}_{os,w,\hat{e}} := E_{os,w}$.

The above signatures are signatures in the classical sense. Thus, they induce corresponding classes of algebras. We will use the notions data, attribute, and event algebra to refer to them. All these algebras are partial; however, we assume that the carrier sets are explicitly completed by bottom elements ($\perp$) to express that an operation is undefined.
Besides the signature part, TROLL *light* templates further include certain axioms restricting the interpretation of object signatures.

Definition 4. Let an object signature $O\Sigma$ be given. The **axioms** for object community specification are divided into constraints, derivation rules, valuation rules, interaction rules, and behavior definitions.

- A **constraint** φ is a formula restricting the class of possible attribute algebras $ALG_{A\Sigma}$.
- A **derivation rule** is of the form $\alpha = t_\alpha$ where α is an attribute and t_α is a term with the same sort as the attribute.
- A **valuation rule** is of the form $\{\varphi\}[t_{\hat{e}}]\alpha = t_\alpha$, where φ is a formula, $t_{\hat{e}}$ is an event term, α is an attribute, and t_α is a term with the same sort as the attribute α.
- An **interaction rule** is of the form $\{\varphi\}\ t_{\hat{e}} \gg t'_{\hat{e}}$ where φ is a formula, and $t_{\hat{e}}$ and $t'_{\hat{e}}$ are event terms.
- A **behavior definition** for a template T consists of a set of process definitions of the following form:
  $\Pi_0 = (\{\varphi_1\}t_{\hat{e}_1} \to \Pi_1 \mid \{\varphi_2\}t_{\hat{e}_2} \to \Pi_2 \mid ... \mid \{\varphi_n\}t_{\hat{e}_n} \to \Pi_n)$
  Here the Π_i's are process names, the φ_i's formulas, and the $t_{\hat{e}_i}$'s event terms.

The details of the logic for formulas φ – a variation of first-order predicate calculus formulas tailored for information system purposes — can be found in [HG94].

Valuation rules, interaction rules, and behavior definitions have been mentioned before. With constraints it is possible to specify restricting conditions on object states, and with derivation rules one describes derived attributes.
Object signatures are interpreted by so-called object communities. This name was chosen because in object communities different objects co-exist and communicate with each other.

Definition 5. Let an object signature $O\Sigma$ be given. $MOD_{O\Sigma}$ denotes the class of all **$O\Sigma$-models** or **$O\Sigma$-object communities:**

$$MOD_{O\Sigma} = \{M_{O\Sigma} \mid M_{O\Sigma} \subseteq (ALG_{A\Sigma} \times ALG_{E\Sigma} \times ALG_{A\Sigma})\}$$

We may assume the interpretation of the data signature to be fixed. It is also possible to restrict the interpretation of object identifier sorts in the sense that object identifiers will be chosen from a term algebra induced by the attribute signature. Then, only attributes contribute to states. For details we have to refer to [GH95] again.
Intuitively, triples $\langle A_L, \hat{A}, A_R\rangle$ in $M_{O\Sigma}$ express that there is a state transition from attribute algebra A_L to attribute algebra A_R via the occurence of the events in $\hat{A}$. A step in an object's life can be traced along such triples: For example, an object $obj \in A_{L,os}$ for some $os \in OS$ can have for attribute α the value c_1 in A_L $[\alpha_{A_L}(obj, ...) = c_1]$; this can change in A_R to c_2 $[\alpha_{A_R}(obj, ...) = c_2]$ due to the occurrence of event e in $\hat{A}$ $[e_{\hat{A}}(obj, ...) \neq \bot]$.

Example 6 (Object community). In Fig. 3 we have sketched part of an object community for our running example. Each arrow stands for a triple $\langle A_L, \hat{A}, A_R\rangle$. Event algebras are depicted next to arrows by giving only the elements of $\hat{e}$

different from $\perp$. Let us explain some details a bit more. (1) The transition along $\{cR(i), c(r(i))\}$ stands for the event $cR(i)$ occurring in i and the create event $c(r(i))$ occurring in $r(i)$. (2) The transition along $\{d(l(i)), d(r(i))\}$ stands for the simultaneous occurence of two delete event in the respective objects. Of course much more transitions than the depicted ones are possible.

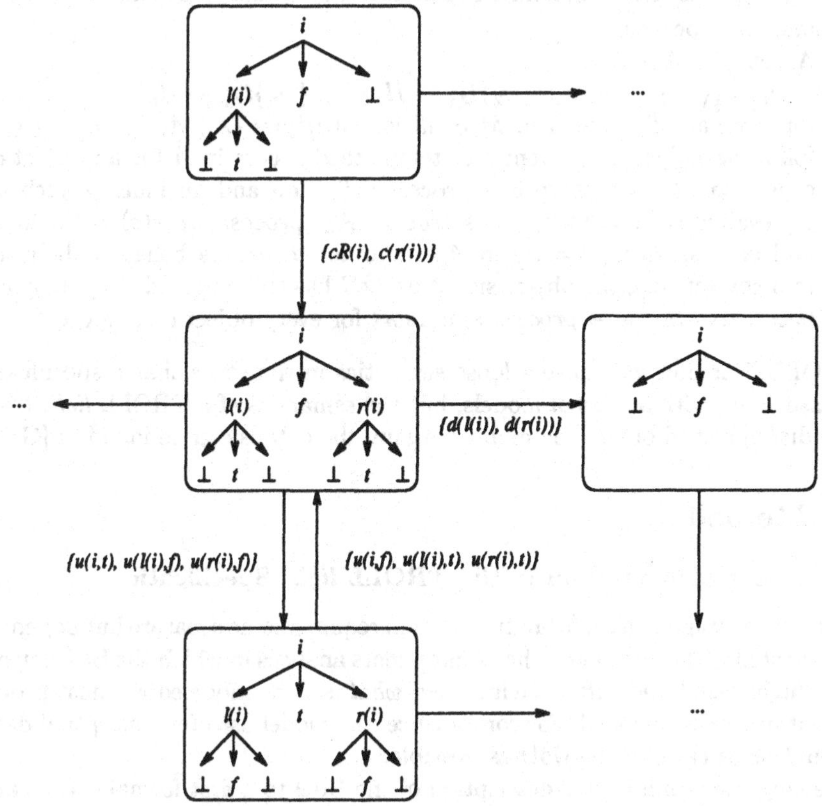

Fig. 3. Part of an example for a simple object community

In the graphical representation, left arrows represent the function $left_A$, middle arrows the function $content_A$, right arrows the function $right_A$, and the undefined object $\perp$ is duplicated as it is needed.

The possible state transition relation must obey the specified axioms. Therefore one defines the validity of axioms as follows.

Definition 7. Let an object signature $O\Sigma$, an $O\Sigma$-model $M_{O\Sigma}$ and axioms over $O\Sigma$ be given. The validity of axioms is defined as follows.

- A constraint φ is valid in $M_{O\Sigma}$ iff for all triples $\langle A_L, \hat{A}, A_R \rangle$ in $M_{O\Sigma}$ the formula φ holds in A_L and A_R.
- A derivation rule $\alpha = t_\alpha$ is valid in $M_{O\Sigma}$ iff for all triples $\langle A_L, \hat{A}, A_R \rangle$ in $M_{O\Sigma}$ attribute α evaluates in A_L and A_R to the value of t_α.

- A valuation rule $\{\varphi\}[t_{\hat{e}}]\alpha = t_\alpha$ is **valid** in $M_{O\Sigma}$ iff for all triples $\langle A_L, \hat{A}, A_R\rangle$ in $M_{O\Sigma}$ the following is true: If the formula φ holds in A_L and the event $t_{\hat{e}}$ occurs[2] in $\hat{A}$, then the attribute α evaluates in A_R to the value which t_α had in A_L.
- An interaction rule $\{\varphi\}\ t_{\hat{e}} \gg t'_{\hat{e}}$ is **valid** in $M_{O\Sigma}$ iff for all triples $\langle A_L, \hat{A}, A_R\rangle$ in $M_{O\Sigma}$ the following holds: If φ is true in A_L and $t_{\hat{e}}$ occurs in $\hat{A}$, then $t'_{\hat{e}}$ must also occur in $\hat{A}$.
- A behavior definition
  $\Pi_0 = (\{\varphi_1\}t_{\hat{e}_1} \to \Pi_1 \mid \{\varphi_2\}t_{\hat{e}_2} \to \Pi_2 \mid \dots \mid \{\varphi_n\}t_{\hat{e}_n} \to \Pi_n)$
  for template T is **valid** in $M_{O\Sigma}$ iff for all triples $\langle A_L, \hat{A}, A_R\rangle$ in $M_{O\Sigma}$ the following holds: If an event e of template T occurs in $\hat{A}$ for an object o of object sort t, then there is a process definition and an index j such that $t_{\hat{e}_j}$ evaluates in $\hat{A}$ to e, φ_j is true in A_L, $\mathit{process\_state}(o) = \Pi_0$ in A_L, and $\mathit{process\_state}(o) = \Pi_j$ in A_R. In order to handle behavior definitions in a correct way, the object signature $O\Sigma$ has to be extended by attributes $\mathit{process\_state} : os \to \mathit{process\_state\_sort}$ for every object sort $os \in OS$.

TROLL *light* does not have a loose semantics in the sense that a specification is associated with a class of models, but the semantics of a TROLL *light* text is one distinguished object community. Again, the details can be found in [GH95].

3 Method

3.1 Development of an Initial TROLL *light* Specification

The entire design of an information system requires some separate but dependent design steps. The first one is the requirements analysis in which the UoD must be thoroughly analyzed. After having fixed *what* is to be modeled we must proceed *how* it has to be modeled. For convenience, the model used for conceptual design should be as close to the UoD as possible.
The step from an informal description of the UoD to a first formal specification is difficult. Therefore we are interested in some guidelines how to obtain object descriptions which reflect the UoD as adequately as possible. We propose the following sequence of design steps.

1. **Separation of data and objects:** We will often have the choice to model a real-world entity either as a complex value, i.e., data, or as an object. In general, there are some indicators for the modeling as an object, such as object sharing, updates on objects, and cyclic object modeling [Hul89, LR89]. In summary, the first step determines what templates we have to write.
2. **Building an object hierarchy:** Objects must be organized into an object hierarchy. The topmost object of this hierarchy is a schema object. To fix sub-object relationships we can proceed either top-down, bottom-up or in yoyo-like way.

[2] We say an event e occurs in $\hat{A}$ if e evaluates to something different from $\perp$ in $\hat{A}$.

3. **Modeling object properties:** Observable properties of objects are modeled by attributes. Here a number of choices can be made: Data-valued attributes describe printable information determined by the data types; relationships between objects can be given by object-valued attributes or by object sorts as parameters for attributes.
4. **Modeling of events:** Events are abstractions of state-modifying operations on objects. Events can have data- or object-valued parameters.
5. **Specification of the effect of events:** The effect of event occurrences on attributes is specified by valuation rules. Result terms, i.e., right-hand sides of valuation rules, are evaluated in the state before the event occurs.
6. **Synchronization of events:** The synchronization of events in different objects is specified by interaction rules. This is the most important relationship between objects as far as dynamic behavior is concerned. The synchronization rules fix the event sets that cause system transitions.
7. **Determining object integrity:** Object integrity is addressed by specifying possible event sequences, by imposing integrity conditions on object states, and by giving preconditions for the application of valuation or interaction rules. The possible event sequences in an object's life starting with a `BIRTH` and possibly ending with a `DEATH` event are given as CSP-like processes. The specified valuations and interactions are only applied if their preconditions evaluate to true in the actual state; otherwise the event has no updating or calling effect.

We cannot expect that one single pass through all these steps yields a satisfactory result. Sometimes we have to go back and repeat some steps until a certain degree of correspondence to the requirements is achieved. This is supported by animating templates as described below.

3.2 Transforming Descriptive TROLL *light* Specifications

The general idea of this part of our development method is to perform several transformation steps on the initial TROLL *light* specification in order to end up with a specification which is as close as possible to an executable program. Thus, we want to achieve transformations $T_1, ..., T_n$ related to the initial TROLL *light* specification as follows: $spec_0 \xrightarrow{T_1} spec_1 \xrightarrow{T_2} ... \xrightarrow{T_n} spec_n$. In each step T_i, more and more "descriptive" TROLL *light* features should be replaced by more "operational" ones. For the concrete language elements of TROLL *light* this means the following:

1. **Constraints:** The general idea is to check the constraints as locally as possible. Thus one does not want all constraints to be checked globally for one object community state, but one should transform constraints equivalently by moving them into pre-conditions of behavior patterns. By doing so, one avoids unneccesary computations in the case that a follow-up state of an (executed) event or event set does not satisfy the constraints and, what is even more important, the applicability of events can be checked effciently.

2. **Valuations:** For the valuation rules, the pre-conditions should be transformed in order to achieve a deterministic object community so that one gets closer to conventional programs. This means that each two pre-conditions of the same event have to be checked whether they exclude each other, and if not so, they have to be transformed equivalently so that they do not hold for the same cases.
3. **Interaction and behavior:** In principle, the same strategy as for valuation rules should be applied here. Non-overlapping interaction rules and behavior patterns (non-overlapping with respect to the pre-conditions) guarantee an object community with a smaller degree of non-determinism.

We have already started to incorporate classical, semi-formal approaches to software engineering like the Entity-Relationship approach [GHC+94] into our method. But more such approaches should be covered, especially Petri Nets [Rei85] seem to be a good candidate to extend structural modelling techniques with dynamic concepts.

4 Tools

We shortly mention three tools for the development environment of TROLL *light* specifications: The animator, the proof support system, and a transformation tool. The first two (together with a context-free and -sensitive syntax checker) have been implemented, and the third one is planed.

4.1 Animating Specifications

A template describes structural and dynamic aspects of a prototypical object. Structural properties are centered around the specification of possible attribute states, dynamic properties around the specification of possible event sequences. Looking at a template with sub-object slots the prototypical object described by this template is in fact an object community where events in different objects may be synchronized by interaction rules. TROLL *light* object descriptions abstract from the causality or initiative of event occurrences. Hence making our algebraic transition system to move means to indicate certain event occurrences from the outside of the system. This is what we call *animation* of templates on the basis of their algebraic semantics. By this, a specification of a formal conceptual schema can be validated against the informal system requirements.
Animating templates. Animation of templates may help to assure that the specified behavior of objects or object communities matches the required behavior. Of course, animation of a conceptual model shows the same problems like testing an implementation. From observations that the observed behavior agrees with the requirements we would like to draw the conclusion that a conceptual schema is correct with respect to the requirements. This, however, cannot be done, because there still may be some traces in the animation which have not been tested and which may show the opposite effect. Hence animation like testing an implementation is useful to falsify informal correctness.

Requirements for a tool supporting animation. A software system with the aim to support the animation of templates should meet the following requirements. It should support (1) the exploration of actual states of an object community, (2) the specification of event occurrences for initiating state transitions, and (3) the visualization of state changes.

To be more precise exploration of states should be assisted by means to illustrate the actual global structure of an object community, to show the attribute state (and optionally the behavior state) of a single object, to traverse sub-object relationships or object-valued attributes, and eventually to formulate ad hoc queries against object states.

With respect to the specification of event occurrences it must be possible to indicate event occurrences for possibly more than one object at a time. The system should help to find event sets being closed against synchronization. Hence, when a specified event occurrence provokes a second event occurrence in another object, this event occurrence should be added automatically to the current event set.

State changes provoked by a given closed event set, as there might be insertions and deletions of objects or attribute updates, should be made visible to the user, for example by appropriate messages in a specific window. When a desired state transition cannot be carried out, for instance because the resulting state violates some integrity constraints, these conditions should also be reported to the user.

Architecture of the animation system. First of all the animation system consists of a *template dictionary* which is a persistent store for object descriptions. The template dictionary is not an exclusive part of the animation system but an integral part of the whole development environment, for instance shared by the parser and the proof support system. The second basic component of the animation system is the *object base* which is a persistent store to hold object states. In principal, the object base could also be a transient store but in the system used for implementation there was only a slight distance between transient and persistent structures so that we preferred persistent structures. Hence it is possible to stop a current animation session at one time and start it again later on.

Object descriptions contain terms and formulas of the TROLL *light* query calculus which are evaluated by the third component of the animation system, the *term evaluator*. The evaluation of terms and formulas generally depends on the current state of an object community so that the term evaluator must be able to access the object base. The *execution module* is the heart of the animation system. Its task is to compute for a given set of event occurrences a successor state or to report errors if such a state cannot be determined for different reasons. Finally the *user interface* establishes communication with the human operator.

4.2 The Proof Support System

The proof support system has the aim to verify properties of specified objects on the level of specification. This seems to be useful in order to check whether the abstract specification meets the intended properties. Hopefully, misdevelopments based on inadequate specifications are avoided by verifying such properties directly from the specification.

Verifying properties. Our verification calculus is embedded into the development environment by the following steps depicted in Fig. 4:

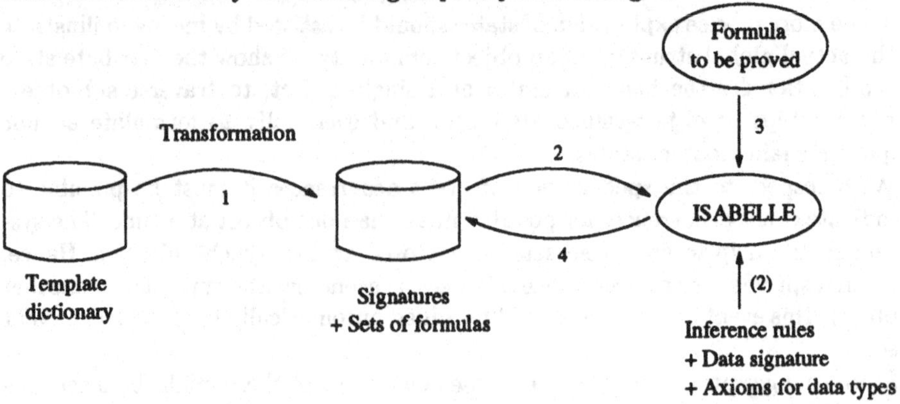

Fig. 4. Working with the proof support system.

1. Automatically transformating a TROLL *light* specification yields a corresponding template signature together with a set of formulas giving a description of the specified objects by means of the verification calculus and therefore in terms of logic. Results of these transformations are stored in a database.
2. After the transformation of a TROLL *light* specification has been completed, a generic theorem prover, in our case ISABELLE [Pau90], can be activated. In order to use this theorem prover for the verification calculus its syntax and the appropriate inference rules must be loaded into the prover (in fact we can generate an instance of ISABELLE in which these parts are already included). Next, we have to include the signature and the formulas obtained from the TROLL *light* specification to be investigated.
3. Then we can specify a property to be proven as a formula of the verification calculus. For such a formula the prover can be employed. Typically, the proof can be controlled in an interactive way, i.e., we can carry out the proof step by step by applying single inference rules or we can cause ISABELLE to automatically run parts of the proof by choosing suitable proof tactics.
4. The corresponding formula can be added to the database, if the proof of a property was successful. Proceeding this way, the formula can be reused later in proving further properties of the same object for which is was shown to be true originally.

4.3 Supporting Specification Transformation

The method described above for the translation of more "descriptive" TROLL *light* specifications into more "operational" ones can be supported by an appropriate tool. This tool should assist the developers especially in modifying formulas and process definitions. We can identify two important sub-tasks.

1. **Transformation of constraints:** In order to transform constraints into pre-conditions we must again distinguish between two cases: Treatment of attributes and treatment of sub-objects.
   Attributes: For each attribute appearing in a constraint, the tool must discover the valuation rules where the attribute is manipulated. From there one can fix the events manipulating the attribute, and the tool can point to the pre-conditions of interaction formulas and behavior definitions for the respective events. Possibly, the tool can try to integrate the constraint into these pre-conditions.
   Sub-objects: For a sub-object appearing in a constraint the tool can detect the interaction rule where the sub-object is created and fix from this point the event responsible for creation of the sub-object. The original constraint should be integrated into the pre-condition of the responsible event.
2. **Transformation of process definitions:** The tool should try to make process definitions as deterministic as possible. For example in the simple case that we have as part of a process definition $\{\varphi_1\}e \to \Pi \mid \{\varphi_2\}e \to \Pi$ this can be transformed equivalently into $\{\varphi_1 \vee \varphi_2\}e \to \Pi$. The tool should also give hints to apply rules like the above one to more involved cases.

5 Conclusion

We have explained an algebraic development technique for information systems. The technique consists of three parts: a formalism, a method, and an environment providing tools for the formalism and method. We have indicated how the relationship and dependencies between these three parts in principle look like. Of central concern was the algebraic semantics of our formalism. This semantics took deep influence on the creation of the method and the implemented and planned tools.
However, our approach is far from being perfect. Further case studies have to be carried out in order to get more insight of and criticism for our method. The existing tools can certainly be improved and new ones can be added. From such experience further improvement of our specification language TROLL *light* is expected.

References

[AAR94] R. Achuthan, V.S. Alagar, and T. Radhakrishnan. An Object-Oriented Framework for Specifying Reactive Systems. In V.S. Alagar and R. Missaoui, editors, *Proc. Colloquium on Object Orientation in Databases and Software Engineering (COODBSE'94)*, pages 18–30. Université du Quebéc à Montréal, 1994.

[BBE+90] G. v. Bochmann, M. Barbeau, M. Erradi, L. Lecomte, P. Mondain-Monval, and N. Williams. Mondel: An Object-Oriented Specification Language. Département d'Informatique et de Recherche Opérationnelle, Publication 748, Université de Montréal, 1990.

[Bec93] E. Becker. The Electronic HDMS-A Patient's Record - A TROLL *light* Case Study. Project Thesis, TU Braunschweig, 1993.

[Bre91] R. Breu. *Algebraic Specification Techniques in Object Oriented Programming Environments*. Springer, Berlin, LNCS 562, 1991.

[CGH92] S. Conrad, M. Gogolla, and R. Herzig. TROLL *light*: A Core Language for Specifying Objects. Informatik-Bericht 92–02, TU Braunschweig, 1992.

[CJO94] S. Clerici, R. Jimenez, and F. Orejas. Semantic Constructions in the Specification Language GLIDER. In H. Ehrig and F. Orejas, editors, *Recent Trends in Data Type Specification (WADT'92)*, pages 144–157. Springer, Berlin, LNCS 785, 1994.

[DG94] P. Dauchy and M.-C. Gaudel. Algebraic Specifications with Implicit State. Technical Report 887, Université de Paris-Sud, 1994.

[DO91] O.-J. Dahl and O. Owe. Formal Development with ABEL. Technical Report 159, University of Oslo, 1991.

[Ehr93] H.-D. Ehrich, editor. *Beiträge zu* KORSO- *und TROLL light-Fallstudien*. Technische Universität Braunschweig, Informatik-Bericht, 93–11, 1993.

[EO94] H. Ehrig and F. Orejas. Dynamic Abstract Data Types: An Informal Proposal. *EATCS Bulletin*, 53:162–169, 1994.

[ESD93] ESDI. *OBLOG CASE Version 1.0.* Espirito Santo Data Informatica, Lisbon, Portugal, 1993.

[Gab93] P. Gabriel. The Object-Based Specification Language Π: Concepts, Syntax, and Semantics. In M. Bidoit and C. Choppy, editors, *Proc. 8th Workshop on Abstract Data Types (ADT'91)*, pages 254–270. Springer, Berlin, LNCS 655, 1993.

[GC93] M. Gogolla and I. Claßen. An Object-Oriented Design for the ACT ONE Environment. In M. Nivat, C. Rattray, T. Rus, and G. Scollo, editors, *Proc. 3rd Int. Conf. on Algebraic Methodology and Software Technology (AMAST'93)*, pages 361–368. Springer, London, Workshops in Computing, 1993.

[GCH93] M. Gogolla, S. Conrad, and R. Herzig. Sketching Concepts and Computational Model of TROLL *light*. In A. Miola, editor, *Proc. 3rd Int. Conf. Design and Implementation of Symbolic Computation Systems (DISCO'93)*, pages 17–32. Springer, Berlin, LNCS 722, 1993.

[GD94] J.A. Goguen and R. Diaconescu. Towards an Algebraic Semantics for the Object Paradigm. In H. Ehrig and F. Orejas, editors, *Proc. 9th Workshop on Abstract Data Types (ADT'92)*, pages 1–29. Springer, Berlin, LNCS 785, 1994.

[GH95] M. Gogolla and R. Herzig. An Algebraic Semantics for the Object Specification Language TROLL *light*. In E. Astesiano, G. Reggio, and A. Tarlecki, editors, *Proc. 10th Workshop on Abstract Data Types (ADT'94)*, pages 288–304. Springer, Berlin, LNCS 906, 1995.

[GHC+94] M. Gogolla, R. Herzig, S. Conrad, G. Denker, and N. Vlachantonis. Integrating the ER Approach in an OO Environment. In R. Elmasri, V. Kouramajian, and B. Thalheim, editors, *Proc. 12th Int. Conf. on the Entity-Relationship Approach (ER'93)*, pages 376–389. Springer, Berlin, LNCS 823, 1994.

[GJM91] G. Ghezzi, M. Jazayeri, and D. Mandrioli. *Fundamentals of Software Engineering.* Prentice Hall (NJ), 1991.

[GM87] J.A. Goguen and J. Meseguer. Unifying Functional, Object-Oriented and Relational Programming with Logical Semantics. In B. Shriver and P. Wegner, editors, *Research Directions in Object-Oriented Programming*, pages 417–477. MIT Press, 1987.

[HG94] R. Herzig and M. Gogolla. A SQL-like Query Calculus for Object-Oriented Database Systems. In E. Bertino and S. Urban, editors, *Proc. Int. Symp. on Object-Oriented Methodologies and Systems (ISOOMS'94)*, pages 20–39. Springer, Berlin, LNCS 858, 1994.

[HH94] J. Hackauf and N. Hartmann. Specification of an Information System for Organizing Conferences - A TROLL *light* Case Study. Project Thesis, TU Braunschweig, 1994.

[Hul89] R. Hull. Four Views of Complex Objects: A Sophisticate's Introduction. In S. Abiteboul, P.C. Fischer, and H.J. Schek, editors, *Nested Relations and Complex Objects in Databases*, Springer, Berlin, LNCS 361, pages 87–116, 1989.

[HV94] R. Herzig and N. Vlachantonis. TROLL light — Specification with a Language for the Conceptual Modelling of Information Systems. In C. Lewerentz and T. Lindner, editors, *Case Study "Production Cell": A Comparative Study in Formal Specification and Verification*, pages 231–239. FZI-Publication 1/94, FZI, Karlsruhe (Germany), 1994.

[JSHS91] R. Jungclaus, G. Saake, T. Hartmann, and C. Sernadas. Object-Oriented Specification of Information Systems: The TROLL Language. Informatik-Bericht 91–04, Technische Universität Braunschweig, 1991.

[LR89] C. Lécluse and P. Richard. Modeling Complex Structures in Object-Oriented Databases. In *Proc. 8th ACM Symp. Principles of Database Systems*, pages 360–368, 1989.

[MQ93] J. Meseguer and X. Qian. Logic-Based Modeling of Dynamic Object Systems. In P. Bunemann and S. Jajodia, editors, *Proc. of the 1993 ACM SIGMOD Int. Conf. on Management of Data*, pages 89–98. SIGMOD Record Vol. 22, Number 2, June 1993.

[MR91] J. McDermid and P. Rook. Software Development Process Models. In J. McDermid, editor, *Software Engineer's Reference Book*, chapter 15, pages 15/1–15/36. Butterworth-Heinemann, Oxford, 1991.

[Pau90] L.C. Paulson. Isabelle: The Next 700 Theorem Provers. In P. Odifreddi, editor, *Logic and Computer Science*, pages 361–385. Academic Press, 1990.

[Rei85] W. Reisig. *Petri Nets: An Introduction.* EATCS Monographs on Theoretical Computer Science. Springer, Berlin, 1985.

[SSE87] A. Sernadas, C. Sernadas, and H.-D. Ehrich. Object-Oriented Specification of Databases: An Algebraic Approach. In P.M. Stoecker and W. Kent, editors, *Proc. 13th Int. Conf. on Very Large Databases VLDB'87*, pages 107–116. VLDB Endowment Press, Saratoga (CA), 1987.

[Wie91] R. Wieringa. Equational Specification of Dynamic Objects. In R.A. Meersman, W. Kent, and S. Khosla, editors, *Object-Oriented Databases: Analysis, Design & Construction (DS-4), Proc. IFIP WG 2.6 Working Conference, Windermere (UK) 1990*, pages 415–438. North-Holland, 1991.

[Wir90] M. Wirsing. Algebraic Specification. In J. Van Leeuwen, editor, *Handbook of Theoretical Computer Science, Vol. B*, pages 677–788. North-Holland, Amsterdam, 1990.

A Framework for Machine-Assisted User Interface Verification

Peter Bumbulis[1], P. S. C. Alencar[2], D.D. Cowan[1], C.J.P. Lucena[3]

[1] Computer Science Department, University of Waterloo, Waterloo, Ontario, Canada
[2] Departamento de Ciência da Computação, Universidade de Brasília, Brasília, Brazil
[3] Departamento de Informática, Pontifícia Universidade Católica do Rio de Janeiro, Rio de Janeiro, Brazil

Abstract. In this paper we present a formal framework for machine-assisted user interface verification. We focus on user interfaces constructed with tools that are based on a visual scripting formalism. As these tools do not provide a language for describing user interfaces (user interfaces are constructed by direct manipulation) we introduce one. Noting that user interface construction with these tools consists of "wiring" components together, we base the syntax of our language on an existing module interconnection language: in this context a user interface is described as a hierarchy of interconnected component instances. We define the semantics of user interfaces using state sequences; this allows us to reason about their ongoing behavior. We embed the semantics in higher order logic (as mechanized by the HOL system) to allow us to verify properties using formal proof.

1 Introduction

Graphical user interfaces (GUIs) are being used increasingly in security- and safety-critical applications. In this context it is important to be able to verify that they behave as intended. As empirical evidence suggests that "the only reliable method for generating quality user interfaces is to test prototypes with actual end users and modify the design based on the users' comments and performance" [28], the issue of verifying that a user interface implementation meets certain formal requirements arises[4]. In this paper we describe a framework for doing this based on (machine-assisted[5]) formal proof: we construct mathematical descriptions of both the user interface and the property to be verified and then (mechanically) prove a theorem of the form

$$\vdash \begin{array}{c}\textit{implementation}\\ \textit{description}\end{array} \ \text{satisfies}\ \begin{array}{c}\textit{property}\\ \textit{description}\end{array}$$

[4] In specification-based approaches to software development requirements verification is typically performed on specifications, not implementations.

[5] Experience has shown that machine-assisted proofs are more trustworthy than those done by hand [9].

where 'satisfies' is some *satisfaction relation*[6]. If we are to have confidence in the significance of such proofs we must ensure that the formal specification of the property to be verified accurately describes the designer's intentions, and we must ascertain that the mathematical model of the user interface truly reflects its behavior. The first issue has been well addressed by the user interface community (see [13, 3, 39], and the references contained therein) and we will not discuss it further. The second issue, on the contrary, seems to have been neglected. We address this second issue by mechanically deriving the mathematical model of the user interface from its implementation. To make this a more tractable task we restrict our attention to user interfaces constructed in a particular fashion, as described below.

1.1 Outline

Attempting to verify properties of user interface implementations presents a number of difficulties, not the least of which is that, in general, there is not a clean separation between the user interface and rest of the application [29]. However, there is a class of GUI development tools (GUIDTs) that enforce such a separation. These tools describe a user interface as a hierarchy of interconnected component instances, with primitive component instances forming the leaves of the hierarchy. We restrict our attention to user interfaces constructed with these tools. In Sect. 2 we briefly describe these tools and informally describe the behavior of user interfaces constructed with such tools.

To be able to reason formally about a user interface we need to have a description of it in a formalism with a well defined syntax and semantics. Unfortunately, while the tools just alluded to provide such descriptions, the formalisms used are proprietary and not amenable to formal reasoning. Our solution to this problem is (just as in hardware design) to construct a model of the user interface and then reason about this model. We are still left with the problem of verifying that the model is an accurate representation of the implementation, but we have reduced the problem to verifying that the representations of the primitive components and the semantics of connecting components is accurate. This only has to be done once. Section 3 sketches the syntax and semantics of the language that we use to model user interfaces.

We discuss the representation of the properties to be verified in Sect. 4, and their mechanical verification in Sect. 5. We finish with conclusions, and an outline of future work.

2 Implementations

Following Meyers [30] we use the term UIMS to denote a GUI development tool (GUIDT) that has a significant run-time component. UIMSs typically provide

[6] In this paper we describe implementations and descriptions as sets of state sequences and use subset containment as the satisfaction relation.

the user interface designer with a set of primitive components and a scripting language for connecting them. Commercial examples of UIMSs include PARTS Workbench [11], Visual Age [17], Visual Basic [26], and VX·REXX [43]. While early UIMSs only provided presentation components (menus, buttons, dialog boxes, and the like), most recent UIMSs (including the ones listed) also provide application interface primitives such as database and file accessors and facilitate the creation of new application-specific primitives. This allows for a clean separation between the user interface and the rest of the application. When developing applications with such tools the tasks of user interface designers and application programmers can be characterized as using and constructing primitive components, respectively.

Traditionally UIMSs have used a conventional programming language for scripting: specifying the behavior of a user interface essentially consists of coding callback routines. A more recent approach (as exemplarized by PARTS Workbench and Visual Age) is to provide a visual formalism for scripting: specifying the behavior of a user interface with these tools consists of "wiring" components together by direct manipulation. Fig. 1 shows one step in the construction of a (trivial) user interface using PARTS Workbench. The user interface consists of a dial and a slider connected together in such a fashion so that in the completed user interface the slider will track the motion of the dial and vice-versa. With such systems the user interface designer specifies which components are used and how they are connected but does not specify the behavior of the constituent primitive components. In this paper we will restrict our attention to user interfaces constructed with such systems.

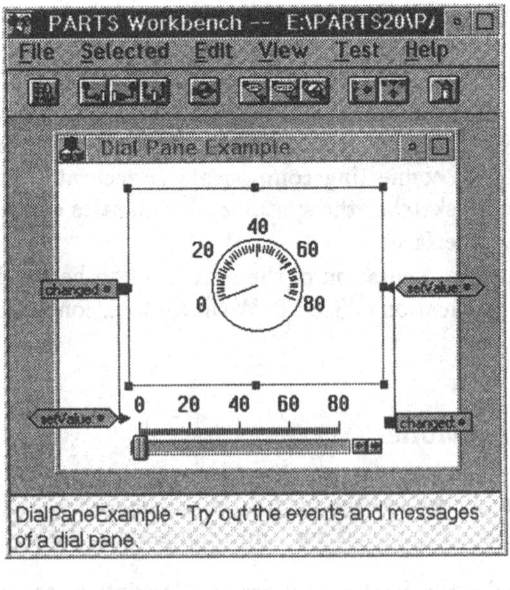

Fig. 1. Trivial PARTS Workbench example

2.1 Semantics

Wiring components together binds functions to call sites. For example, Fig. 2 gives a possible C++ [38] interpretation of the behavior of the user interface under construction in Fig. 1. Additional things to note are that presentation and application interface components also present a well defined interface to the runtime environment (in this case modeled with the functions *dispatch* and *redraw*.) The runtime environment translates events generated by the user's actions into (serialized) calls to *dispatch* which will modify state and/or invoke actions. To avoid undesirable visual artifacts and to improve performance, component implementations typically do not update the screen directly in response to a *dispatch*'ed event. Instead, they simply notify the runtime environment that the view of the component should be updated. After the call to dispatch returns, the runtime environment will call *redraw* to perform the screen update.

```
class Valuator {
  private:
          int value;
  public:
          void (*changed)(int);
          void set(int i){
             if (value != i) { value = i; /* ... */; (*changed)(i); }
          }
  //   virtual void dispatch(Event) = /* { ... set(.) ... } */ 0;
  //   virtual void redraw(Region) = 0;
};

class Dial   : public Valuator { /* ... */ } d;
class Slider : public Valuator { /* ... */ } s;

void dset(int i) { d.set(i); }
void sset(int i) { s.set(i); }

main() {
  d.changed = sset;          // bind d.changed to s.set
  s.changed = dset;          // bind s.changed to d.set
}
```

Fig. 2. Code for Fig. 1

3 The Model

3.1 Syntax

To be able to reason about a user interface, we need a description of it in some well defined formalism (language). While systems such as PARTS Workbench and Visual Age provide such a description[7], unfortunately the formalisms used are proprietary and not amenable to formal verification. As a result, we introduce our own language for describing user interfaces.

Noting that the interface construction in the context of a visual programming language typically consists of "wiring" components together by direct manipulation, we base the syntax of our language on (the static subset of) an existing module interconnection language Darwin [18]. We describe a user interface as a hierarchy of interconnected component instances, with primitive component instances forming the leaves of the hierarchy.

Each component instance has a number of *ports* available for binding. For example, Fig. 3 is a description of user interface under construction in Fig. 1[8]. Each port has a *type* (specifying the type of values that can appear at that port), *kind* (*observer* or *action*)[9], and *polarity* (specifying whether the port *requires* or *provides* values.) Only ports of the same type and kind and opposite polarity may be wired together. A port providing values can be bound to at most one port.

```
Dial changed>int set<int primitive
Slider changed>int set<int primitive

GUI {
    d:Dial s:Slider
    d.changed --> s.set
    s.changed --> d.set
}
```

Fig. 3. Syntax for Fig. 1

Work is currently under way on implementing a GUIDT that uses this language to represent user interfaces.

---

[7] Designers can save their work in files.

[8] This is an oversimplification for the purposes of exposition. Almost all components are parameterized with attributes as well. These attributes are used to specify things such as the size and position of the component, as well as the visual hierarchy and initial values.

[9] Observers are provided to inspect the state of a component; actions are provided to modify it. To simplify the following discussion we will not make any further mention of observers.

3.2 The Formal System

If we are to verify properties of user interfaces by formal proof we need to express the semantics of the user interface and the properties to be verified in some formal system. We use higher order logic[10]. Higher order logic extends first-order logic by allowing higher order variables (i.e. variables whose values are functions) and higher order functions (i.e. functions whose arguments and/or results are other functions). Using higher order logic has a number of advantages:

1. Reliable and robust proof-assistants such as HOL [14], PVS [34] and Isabelle [36] exist;
2. It will be easy to formalize the semantics of the UIMS scripting languages that we are interested in; and
3. It is simple to embed the various logics and formalisms used to express user interface specifications in higher order logic. For example, temporal logic [23, 19, 41] and Statecharts [10][11] have both been embedded.

3.3 Semantics

There is a general consensus in the user interface community that user interfaces are best described as *reactive systems* [15]; what is of interest is how they interact with their environment. As a result many of the formalisms proposed for user interface specification model behavior as a sequence of states. CSP [4], Petri nets [6], LOTOS [35], temporal logic [20] and Statecharts [24] all have been used. Similarly, we model a GUI as a 4-tuple $\langle \Pi, \Sigma, \Theta, C \rangle$ with the following components:

- Π, a set of ***state variables.***
- Σ, a set of ***states.*** Each state s in Σ is a function mapping names in Π to values.
- Θ, an initial condition. This assertion characterizes the states at which execution of the GUI can begin. A state satisfying Θ is called an ***initial state.***
- C, an assertion characterizing the possible behaviors resulting from a single action by the environment (such as a button press). Such assertions are referred to as ***commands. Behaviors*** are state sequences. Behaviors can be finite or infinite; finite behaviors must have at least one element. We denote finite behaviors as $\langle s_1, \ldots, s_n \rangle$ and infinite behaviors as $\langle s_1, \ldots \rangle$. If b is a behavior, then $\mathsf{length}\, b \in \mathbb{N} \cup \{\omega\}$ denotes the number of states in the behavior, with $\mathsf{st}\, b\, i$ denoting the ith state. We also use $\mathsf{first}\, b$ to denote the first state of the behavior and, if b is finite, $\mathsf{last}\, b$ to denote the last state.

We define the composition of two behaviors $b_1 \circ b_2$ to be b_1 if b_1 is infinite, and the sequence obtained by appending all but the first element of b_2 to b_1 if

[10] Essentially first order logic enriched with the typed lambda calculus. [5] is a good introduction.

[11] Day formalizes STATEMATE's step-dependent semantics [16]; this is quite different from the (micro-step based) semantics for Statecharts found in the literature.

b_1 is finite and $\mathsf{first}\, b_2 = \mathsf{last}\, b_1$. If b_1 is finite and $\mathsf{first}\, b_2 \neq \mathsf{last}\, b_1$ then $b_1 \circ b_2$ is undefined.

A *(little-step)* computation of a user interface $\langle \Pi, \Sigma, \Theta, C \rangle$ is defined to be any *non-extendable* composition of behaviors from C that starts from a state in Θ. A behavior b is said to be non-extendable with respect to C if b is infinite, or if b is finite and $b \circ b'$ is not defined for any b' satisfying C.

We define $\mathsf{atomic}\, b$ to be $\langle \mathsf{first}\, b, \mathsf{last}\, b \rangle$ if b is finite and $\langle \{\mathsf{first}\, b, \}^{\omega} \rangle$ if b is infinite. We extend this definition to commands (predicates of behaviors) in the usual fashion. A *big-step* computation of $\langle \Pi, \Sigma, \Theta, C \rangle$ is defined to be a computation of $\langle \Pi, \Sigma, \Theta, \mathsf{atomic}\, C \rangle$. Big-step computations are useful for reasoning about the presentation aspects of the user interface where we are not interested in intermediate states resulting from a single action by the environment. As discussed in Sect. 2.1, the display usually changes only after an event dispatch has completed.

Rather than explicitly constructing predicates for C and Θ we instead construct a predicate P that characterizes C and Θ, i.e. we construct a predicate P such that $P\, t\, c$ holds iff $t = \Theta$ and $c = C$. If we are to mechanically construct P from a description such as that found in Fig. 3, we need to be able to express P in terms of predicates describing components.

3.4 Specifying Components

We define for each type of component a predicate $\mathsf{component}\, s_0\, cmd\, p_1 \ldots p_n$ which is true if and only if cmd is a command which expresses the behavior of the component when connected as specified by the parameters and started in state s_0[12]. This relation typically gives a relation for each provided port in terms of the required ports and relation defining the behavior of the component in terms of the provided port definitions. If $p_1, \ldots, p_n$ are the required ports and $r_1, \ldots, r_m$ are the provided ports then component will have the following form (ignoring parameters dealing with state, for the moment):

$$
\begin{aligned}
\mathsf{component}\, cmd\, p_1 \ldots p_m\, r_1 \ldots r_n = \\
(p_1 = P_1\, r_1 \ldots r_n) \wedge \\
\vdots \\
(p_m = P_m\, r_1 \ldots r_n) \wedge \\
(cmd = P\, p_1 \ldots p_m)
\end{aligned}
$$

Note that this is actually a constant definition in higher order logic defining component to be the function $\lambda\, cmd\, p_1 \ldots p_m\, r_1 \ldots r_n . (\ldots) \wedge \ldots \wedge (\ldots)$[13].

We need to introduce a formalism for expressing commands (i.e. for expressing P and the P_i) to make reasoning more tractable. The subsequent development does not depend on this formalism; all we require is a means for expressing sequential composition and non-deterministic choice. We have chosen to use an

[12] A similar approach is used for hardware verification [25].

[13] In higher order logic predicates are simply boolean-valued functions.

imperative formalism since not only are mechanizations available for a variety of programming logics [27, 40], but also because of the the possibility of using symbolic execution [21] to help in constructing various proofs [8].

We choose to represent commands in an extension of Dijkstra's guarded command language [12] as defined by Nelson [31] and mechanized by Tredoux [40]. It allows for partial commands and drops the law of excluded miracle. We define each of the primitive commands as predicates on behaviors and each of the operations as predicates on commands. Figure 4 provides a partial summary of the semantics used in this paper, defining constants skip, $:=$, $\Box$, ; , $\longrightarrow$, and arb_value. Infix notation is used for the binary predicates to improve readability. We use the notation $f[y/x]$ to denote the function such that $f[y/x]\,x = y$ and $f[y/x]\,z = f\,z$ for all $z \neq x$. Figure 4 does not contain a definition of do_od: its definition is considerably more involved than the rest. Essentially do_od c is defined to be the "infinite" sequential composition of c with skip. We verify that our semantics for the guarded command language agrees with the standard semantics by deriving Dijkstra's calculus of predicate transformers.

Our mechanization is similar in approach to Tredoux's [40], but differs substantially in detail and philosophy. For example, in Tredoux's mechanization, skip is not identity for composition (;).

$$\begin{array}{l}
\mathsf{skip}\,b = \exists s.\, b = \langle s\rangle \\
(x := exp)\,b = \exists s\, b.\, b = \langle s, s[exp\ s/x]\rangle \\
(c_1 \Box c_2)\,b = c_1\,b \vee c_2\,b \\
(c_1;\, c_2)\,b = c_1\,b \wedge (\mathsf{length}\,b = \infty) \vee \exists b_1\, b_2.\, c_1\,b_1 \wedge c_2\,b_2 \wedge (b = b_1 \circ b_2) \\
(P \longrightarrow c)\,b = P\,(\mathsf{first}\,b) \wedge c\,b \\
(\mathsf{arb_value}\,c)\,b = \exists v.\, c\,v\,b
\end{array}$$

Fig. 4. State sequence semantics for the guarded command language

Given this notation, we can define the *little-step semantics* of a user interface $\langle \Pi, \Sigma, C, \Theta\rangle$ to be the command $\Theta \longrightarrow (\mathsf{do\_od}\,C)$ and the *big-step semantics* of $\langle \Pi, \Sigma, C, \Theta\rangle$ to be the command $\Theta \longrightarrow (\mathsf{do\_od}\,(\mathsf{atomic}\,C))$.

We note that although we have avoided dealing with concurrency in this stage of our work, the present approach can in principle be extended to a concurrent one through the use of another formalism for describing the commands, e.g. the temporal logic of actions (TLA) [22].

Primitive Components. We assume that the definitions for the primitive components are provided; when a new primitive is constructed, the user interface designer is presented not only with an implementation, but also with a specification of its intended behavior in this form. For example, the definition of a valuator (which we will use to model both sliders and dials) might be given as follows:

$$
\begin{aligned}
&\mathsf{valuator}\ s_0\ cmd\ id\ v_0\ set\ changed = \\
&\quad (set = (\lambda v. \\
&\qquad ((\lambda s.\, v = s\ id) \longrightarrow \mathsf{skip})\Box \\
&\qquad ((\lambda s.\, \neg(v = s\ id)) \longrightarrow ((id := (\lambda s.\, v));\ changed\ v))))\wedge \\
&\quad (v_0 = s_0\ id)\wedge \\
&\quad (cmd = \mathsf{arb\_value}\ set)
\end{aligned}
$$

This formula characterizes valuators by specifying the behavior of a valuator given the behavior of the components that it is connected to. It essentially gives the same semantics for the behavior of the valuator as described in Fig. 2. It also states that in the start state the valuator has value v_0.

Note the use of arb_value[14] in the definition of the behavior of the valuator. We model the user's actions as entering an arbitrary value. In a more realistic example there would be further constraints on the value, for example that it is a whole number in the range [0, 80].

Composite Components. Composite components are defined in terms of the constituent components. For example, the specification of the user interface of Fig. 1 would look like:

$$
\begin{aligned}
&\mathsf{dial} = \mathsf{valuator} \\
&\mathsf{slider} = \mathsf{valuator} \\
&\\
&\mathsf{gui}\ s_0\ cmd = \\
&\quad (\exists cmd_s\ cmd_d\ set_s\ set_d. \\
&\qquad \mathsf{slider}\ s_0\ cmd_s\ \text{'s'}\ 0\ set_s\ set_d\wedge \\
&\qquad \mathsf{dial}\ s_0\ cmd_d\ \text{'d'}\ 0\ set_d\ set_s\wedge \\
&\qquad (cmd = cmd_s\Box cmd_d))
\end{aligned}
$$

This does not give us a closed form expression for the behavior; instead it is specified as a system of recursive equations. Note that the actions that the user interface can perform include those that can be performed by the slider and the dial. Also note the use of existential quantification for hiding.

Note that although the primitive components must be described by hand, the composite component descriptions can be mechanically derived from the syntax of the description of the user interface.

4 Properties

In this section we present the semantics of the property representations and show how to encode these representations in higher order logic. Formally properties are assertions about behaviors: given a set of states Σ, a property $\mathcal{P}$ is an

[14] See Fig. 4 for a definition.

assertion on $\Sigma^* \cup \Sigma^\omega$. $\mathcal{P}$ is a little-step property of a GUI $\langle \Pi, \Sigma, \Theta, \mathcal{C}\rangle$ if for all behaviors $b\ \Theta \longrightarrow (\mathsf{do\_od}\ \mathcal{C})\ b \supset \mathcal{P}\ b$, and a big-step property if for all behaviors $b\ \Theta \longrightarrow (\mathsf{do\_od}\ (\mathsf{atomic}\ \mathcal{C}))\ b \supset \mathcal{P}\ b$. For example, we can express the property that the slider must track the dial as follows:

$$\mathsf{invariant}\ b\ P = \forall i.\ (i < \mathsf{length}\ b + 1) \supset P\ (\mathsf{st}\ b\ i)$$
$$\mathsf{tracks}\ c = (\forall b.\ c\ b \supset \mathsf{invariant}\ b\ (\lambda b.\ s\ \text{'s'} = s\ \text{'d'}))$$

To verify that the GUI possesses this property we have to prove a theorem of the form:

$$\forall s_0\ c.\ \mathsf{gui}\ s_0\ c \supset \mathsf{tracks}\ ((\lambda s.\ s = s_0) \longrightarrow \mathsf{do\_od}\ (\mathsf{atomic}\ c))$$

This is a trivial property; but it is instructive to verify. We note that the attempt to derive consequences from specifications not only point out flaws in the specification but also in the formalization as well.

Many different formalisms have been developed for specifying user interfaces (See the bibliographies of [3, 30], for example.) We can view specifications expressed in these formalisms as properties. We say that a GUI big-step satisfies a specification if the specification is a big-step property of the GUI and little-step satisfies a specification if the specification is a little-step property of the GUI. To be able to mechanically verify implementations against these specifications we must represent specifications as sets of state sequences, and verify that the axioms and inferences rules of the formalism are valid.

5 Mechanical Verification

In order to accomplish a machine assisted verification process we use the version of higher order logic mechanized by Gordon's HOL system [37]. The advantages of using HOL include the fact that mechanizations exist for a wide class of formalisms that have been used to specify user interfaces, including CSP [7], CCS [32], Statecharts [10], Temporal Logic [19, 40], and the Temporal Logic of Actions (TLA) [42].

The HOL system consists of a meta-language ML and a logic. The HOL logic is a natural deduction logic: assertions in the logic are sequents of the form (Γ, t), where Γ is a set of assumptions and t is the conclusion. A sequent (Γ, t) asserts that if all of the formulas in Γ are true, then so is t. A theorem is a sequent that has a proof, i.e. is either an axiom, or follows from other theorems by a rule of inference.

To ensure that the only way to get theorems is by proof, theorems are isolated in the abstract data type `thm`. There are five initial theorems corresponding to the five axioms of the HOL logic; only the (functions representing the) eight primitive inference rules of the logic have access to the internal representation of a `thm`. ML's type system ensures that the only only way to get new theorems is by using these primitive inference rules. Derived inference rules can be implemented as ML procedures; in practice most proofs use these derived rules. Goal-directed proofs are simulated with tactics: functions that, given a goal, return a list of subgoals and a procedure for combining their proofs into a proof of the original goal.

Properties are best proved using some sort of embedded logic for the formalism used to represent commands. For example, Hoare logic, or Dijkstra's predicate transformers can be used for guarded command languages. Even so, the proofs are tedious to construct and we are investigating ways of further mechanizing the process. In particular, we are investigating the use of symbolic execution as described in [8] to aid in the construction of proofs of state-invariant properties. The key step in such a proof would be the proof of a lemma of the form

$$\forall s_0\, c.\, \mathsf{gui}\, s_0\, c \supset (\forall s.\, P\, s \supset P\, (\mathsf{exec}\, c\, s))$$

where $\mathsf{exec}\, c\, s$ returns the state of the GUI after executing command c from state s. Symbolic execution (implemented as an animation conversion) would be used to prove theorem of the form:

$$\mathsf{exec}\, c\, s = s'$$

Conversions are a special class of inference rules in HOL that map terms of the logic to a theorem asserting equality of those terms to other terms. Each step of the animation is done by a systematic specification, unfolding, and simplification of the description of the command.

6 Conclusions

We have constructed a preliminary prototype of the system described in this paper using Tredoux's mechanization of execution sequence semantics as a basis [40] and have constructed several proofs using Hoare logic. This has pointed out the need for more automation. Work is currently underway to construct a more complete prototype through which we will be able to generate both an implementation (in Tk/Tcl [33]) and a specification from a particular description.

Work is proceeding on embedding a logic-based specification framework for Abstract Data Views, an object-oriented formal model for specifying interfaces. Abstract Data Views (ADVs) are Abstract Data Objects (ADOs) or objects that have been specifically augmented to support the specification of interfaces [2, 1].

We are also investigating the mechanization of ADVs (both the Temporal Logic and Statechart variants) to leverage on existing user interface specification work that we have done. The generality of the semantics adopted for the user interface descriptions (based on state sequences) ensures the feasibility of these mechanization procedures. In fact, the experimentation with these interface description styles have motivated us to attempt to construct a classification of the various properties that one would like to verify formally about user interfaces.

In contrast with a flat logical description of user interface components, the ADV formal model allows us to describe user interface abstractions structured from primitive components and to reason about local and global properties (modular verification) of the GUI components. Thus, in the ADV context, we can prove a more general class of user interface properties than the ones that can be proved within the framework of the formalism presented in this paper. To achieve this goal, we have to represent the logical ADV descriptions by state sequences and embed the ADV logic in higher order logic. Furthermore, we have to seek a suitable formalism for expressing the commands (as e.g. the Temporal Logic of Actions (TLA) [22]).

As we can in principle represent both the implementation descriptions and the ADV descriptions in this formal framework (based on the same semantics: state sequences), it should be possible to prove in some sense the equivalence between the two description levels.

References

1. P. S. C. Alencar, D. D. Cowan Cowan, C. J. P. Lucena, and L. C. M. Nova. A Formal Specification of Reusable Interface Components (to appear). Technical report, Computer Science Department, University of Waterloo, Waterloo, Ontario, Canada, 1994.
2. P.S.C. Alencar, L.M.F. Carneiro-Coffin, D. D. Cowan, and C.J.P. Lucena. Towards a Logical Theory of ADVs. In *Proceedings of the Workshop on the Logical Foundations of Object-Oriented Programming (to appear)*, August 1994.
3. Heather Alexander. *Formally-based tools and techniques for human-computer dialogues*. Ellis Horwood Limited, 1987.
4. Heather Alexander. Structuring dialogues using CSP. In M. Harrison and H. Thimbleby, editors, *Formal Methods in Human-Computer Interaction*, chapter 9, pages 273–295. Cambridge University Press, 1990.
5. Peter B. Andrews. *An introduction to mathematical logic and type theory : to truth through proof.* Academic Press, 1986.
6. Remi Bastide and Philippe Palanque. Petri net objects for the design, validation and prototyping of user-driven interfaces. In *Proceedings of IFIP INTERACT'90: Human-Computer Interaction*, Detailed Design: Construction Tools, pages 625–631, 1990.
7. A. J. Camilleri. Mechanizing CSP trace theory in higher-order logic. *IEEE Transactions on Software Engineering*, 16(9):993–1004, 1990.
8. Juanito Camilleri and Vincent Zammit. Symbolic animation as a proof tool. In Thomas F. Melham and Juanito Camilleri, editors, *Higher Order Logic Theorem Proving and Its Applications: 7th International Workshop*, volume 859 of *Lecture Notes in Computer Science*, pages 113–127, Valletta, Malta, 19–22 September 1994. Springer-Verlag.
9. Avra Cohn. The notion of proof in hardware verification. *Journal of Automated Reasoning*, 5(2):127–140, June 1989.
10. Nancy Day. A model checker for Statecharts. Technical Report TR-93-35, UBC, October 1993.
11. Digitalk. *PARTS Workbench User's Guide*, 1992.
12. Edsger W. Dijkstra. *A Discipline of Programming.* Prentice-Hall, Englewood Cliffs, New Jersey, 1976.
13. Alan Dix and Colin Runciman. Abstract models of interactive systems. In Peter Johnson and Stephen Cook, editors, *People and Computers: Designing the Interface*, pages 13–22. Cambridge University Press, September 1985.

14. Michael J.C. Gordon. HOL: A proof generating system for higher-order logic. In Graham Birtwistle and P. A. Subrahmanyam, editors, *VLSI Specification, Verification and Synthesis*, chapter 3, pages 73–128. Kluwer Academic Publishers, 1988.
15. D. Harel and A. Pnueli. On the development of reactive systems. In Krzysztof R. Apt, editor, *Logics and Models of Concurrent Systems*, volume 13 of *Series F: Computer and System Sciences*, pages 477–498. Springer-Verlag, 1985.
16. i-Logix Inc., Burlington, MA. *The Semantics of Statecharts*, January 1991.
17. IBM. *VisualAge: Concepts & Features*, 1994.
18. Imperial College of Science, Technology and Medicine. *Darwin Overview*, 1994.
19. Amit Jasuja. Temporal logic in HOL, August 3 1990. In HOL distribution: ftp://lal.cs.byu.edu/pub/hol/holsys.tar.gz.
20. C. W. Johnson. Applying temporal logic to support the specification and prototyping of concurrent multi-user interfaces. In *Proceedings of the HCI'91 Conference on People and Computers VI*, Groupware, pages 145–156, 1991.
21. Ralf Kneuper. Symbolic execution: a semantic approach. *Science of Computer Programming*, 16:207–249, October 1991.
22. Leslie Lamport. The temporal logic of actions. *ACM Transactions on Programming Languages and Systems*, 16(3):872–923, May 1994.
23. Z. Manna and A. Pnueli. Verification of concurrent programs: a temporal proof system. Technical Report CS-83-967, Stanford Univ., 1983.
24. Lynn S. Marshall. *A formal description method for user interfaces.* PhD thesis, University of Manchester, 1986.
25. Tom F. Melham. *Higher order logic and hardware verification.* Cambridge University Press, New York, 1993.
26. Microsoft Corporation. *Microsoft Visual Basic Programmer's Guide*, 1993.
27. M.J.C. Gordon. Mechanizing programming logics in higher-order logic. In Graham M. Birtwistle and P. A. Subrahmanyam, editors, *Current Trends in Hardware Verification and Automated Theorem Proving (Proceedings of the Workshop on Hardware Verification)*, pages 387–439, Banff, Canada, 1988. Springer-Verlag.
28. Brad A. Myers. *State of the Art in User Interface Software Tools*, chapter 5, pages 110–150. Ablex, Norwood, N.J., 1992.
29. Brad A. Myers. Why are human-computer interfaces difficult to design and implement? Technical Report CMU-CS-93-183, School of Computer Science, Carnegie Mellon University, July 1993.
30. Brad A. Myers. User interface software tools. Technical Report CMU-CS-94-182, School of Computer Science, Carnegie Mellon University, August 1994.
31. Greg Nelson. A generalization of Dijkstra's calculus. *ACM Transactions on Programming Languages and Systems*, 11(4):517–561, October 1989.
32. Monica Nesi. A formalization of the process algebra CCS in higher order logic. Technical Report 278, University of Cambridge Computing Laboratory, December 1992.
33. John K. Ousterhout. *Tcl and the Tk Toolkit.* Addison-Wesley, 1994.
34. S. Owre, J. M. Rushby, and N. Shankar. PVS: A prototype verification system. In Deepak Kapur, editor, *11th International Conference on Automated Deduction*, LNAI 607, pages 748–752, Saratoga Springs, New York, USA, June 15–18, 1992. Springer-Verlag.
35. F. Paternó and G. Faconti. On the use of LOTOS to describe graphical interaction. In A. Monk, D. Diaper, and M. D. Harrison, editors, *Proceedings of the HCI'92 Conference on People and Computers VII*, pages 155–173. Cambridge University Press, September 1992.

36. Lawrence C. Paulson. *Isabelle: A Generic Theorem Prover*, volume 828 of *Lecture Notes in Computer Science*. Springer-Verlag, 1994.
37. SRI International under contract to DSTO Australia, Cambridge, England. *The HOL System: Description*, 1989.
38. Bjarne Stroustrup. *The C++ Programming Language*. Addison-Wesley, second edition, 1991.
39. Kari Systä. Specifying user interfaces in DisCo. *SIGCHI Bulletin*, 26(2):53–58, 1994. Presented at a Workshop on Formal Methods for the Design of Interactive Systems, York, UK, 23rd July 1993.
40. G. Tredoux. Mechanizing execution sequence semantics in HOL. *South African Computer Journal*, 7:81–86, July 1992. Proceedings of the 7th Southern African Computer Research Symposium, Johannesburg, South Africa. Also available as part of the HOL distribution: ftp://lal.cs.byu.edu/pub/hol/holsys.tar.gz.
41. Gavan Tredoux. Mechanizing nondeterministic programming logics in higher-order logic. Technical report, Laboratory for Formal Aspects of CS, Dept Mathematics, University of Cape Town, Rondebosch 7700, South Africa, March 22 1993.
42. Joakim von Wright and Thomas Långbacka. Using a theorem prover for reasoning about concurrent algorithms. In G. v. Bochmann and D. K. Probst, editors, *Computer Aided Verification: Proceedings of the Fourth International Workshop, CAV '92*, number 663 in Lecture Notes in Computer Science, pages 56–68. Springer-Verlag, June/July 1992.
43. Watcom International Corporation, Waterloo, Ontario, Canada. *WATCOM VX·REXX for OS/2 Programmer's Guide and Reference*, 1993.

Specification of the Unix File System: A Comparative Case Study

Maritta Heisel

Technische Universität Berlin
FB Informatik – FG Softwaretechnik
Franklinstr. 28-29, Sekr. FR 5-6
D-10587 Berlin, Germany
heisel@cs.tu-berlin.de
fax: (+49-30) 314-73488

Abstract. The starting point of this investigation are two different formal specifications of the user's view of the Unix file system, one algebraic and one model-based. The different features exhibited by the specifications give rise to a discussion of desirable and undesirable properties of formal specifications.

1 Why yet another specification of the Unix file system?

The Unix file system is one of the best (or at least most) specified software systems. Several versions have been published: [1, 3, 5][1]. All of these versions are distinguished in important aspects, e.g. in the view that is considered or the executability of the specification.

We will present yet another version, and it is not even new in the above-mentioned aspects. It models the user's view of the Unix file system, just like the specification of Bidoit, Gaudel and Mauboussin [1]. In fact, it was inspired by this specification. When investigating language-independent issues of specifications [7], we thought it a nice exercise to express the given specification, written in the algebraic language PLUSS, in the model-based language Z [8], where the Z specification was to resemble the PLUSS specification as close as possible. Had this enterprise succeeded, no need would have arisen for yet another paper on the Unix file system.

The Unix file system presents itself to the user as a tree where each node has a name and an arbitrary number of successors. A specification of such trees should be present in some library for re-use, where the content of the nodes (as opposed to their names) should be a generic parameter. The first trial to define such trees indeed looked promising:

$$[NAME]$$

$$\begin{aligned} TREE[X] ::=\ & lf \langle\langle NAME \times X \rangle\rangle \\ \mid\ & node \langle\langle NAME \times \operatorname{seq} TREE[X] \rangle\rangle \end{aligned}$$

[1] Our apologies to all those not mentioned here.

It was discussed with a Z specialist (who will remain unnamed here) and approved. What an unpleasant surprise that the Z type checker rejected this specification! A look at the Z grammar showed that the checker was right (of course). Free types in connection with genericity are indeed not allowed in Z.

What began with thorough disappointment ended up in a lesson on the aesthetics of formal specifications. We were able to make a virtue out of necessity and came up with a specification that looks entirely different than the one we started out from, although it basically models "the same thing". From our point of view, the new Z specification is not inferior to the original one in PLUSS. It turned out that the strengths and weaknesses of the two versions lie in different areas, so that it is virtually impossible to prefer one over the other without reservations.

The aim of this paper is to stimulate a discussion on the various desirable features of formal specifications and how they can be achieved. In the following, we present parts of the specification in two versions[2]. Important differences are contrasted, and it is tried to distill the lessons to be learned from the example and to come to a better assessment of the qualities of formal specifications.

2 Re-Usability

That re-usability is a desirable property of specifications is undisputed. For our example, it is our intention to start out from a generic specification of named trees, and instantiate and adapt this specification to define directories.

2.1 Generic Specification of Named Trees

It seems perfectly clear that trees are defined as recursive data structures, doesn't it? The following PLUSS specification is recursive even if it does not look like it at first sight: it uses lists which are defined recursively.

```
proc NAMED_TREE(X)
   use LIST, NAME
   sorts Named_Tree
   cons <_ . _>: (Name × X) × List(Named_Tree) → Named_Tree
   func
      ...
   axiom
      ...
end NAMED_TREE(X)
```

As already stated, our trial to define trees in Z recursively was no success. Z does not support recursive definitions very well since it is part of the "Z philosophy" to define types as *sets*. It turns out that adhering to the "Z philosophy"

[2] where we do not stick literally to the specification given in [1].

yields a modeling of trees in which all the necessary functions can be expressed quite elegantly. This modeling is *not* recursive.

In Z, lists are called sequences; they are defined as finite partial functions from the natural numbers into some type X. Similarly, named trees will be finite partial functions from sequences of positive natural numbers into the Cartesian product $NAME \times X$.

$$
\begin{array}{l}
[NAME] \\
NAMED\_TREE[X] == \\
\quad \{f : \mathrm{seq}\, \mathbb{N}_1 \mathrel{+\!\!\!\!\twoheadrightarrow} NAME \times X \mid \\
\qquad \langle\rangle \in \mathrm{dom}\, f \\
\qquad \land (\forall\, path : \mathrm{seq}_1\, \mathbb{N}_1 \mid path \in \mathrm{dom}\, f \bullet \\
\qquad\qquad front\ path \in \mathrm{dom}\, f \\
\qquad\qquad \land (last\ path \neq 1 \Rightarrow front\ path \frown \langle last\ path - 1\rangle \in \mathrm{dom}\, f))\}
\end{array}
$$

This definition models trees as functions mapping "addresses" to the content of the node under the respective address. Each node consists of a name and an item of the parameter type X. The empty sequence is the address of the root. The length of an address sequence coincides with the depth of the node in the tree. The number i denotes the i-th subtree. Hence, an address can only be valid if its *front* is also a valid address. And if there is an i-th subtree for $i \geq 1$ then there must also exist an $i-1$-th subtree.

2.2 Discussion

By refraining from using free types we managed to obtain a generic tree definition in Z that may be re-used later. But this was mere luck.

Issue 1 *Is there a convincing reason why genericity in connection with free types is forbidden in Z?*

In comparison with PLUSS and other algebraic languages, Z's support for genericity and re-use is poor. In contrast, PLUSS, even offers a **param** construct that makes it possible to state restrictions on the types used as actual parameters in the instantiation of generic specifications. This feature is not too frequent in specification languages.

There might be members of the Z community who would oppose to this opinion. Wordsworth [10], p.25, for instance, states that genericity can be achieved by introducing the parameters of the specification as basic types. The following would indeed have been legal:

$$[NAME, X]$$

$$
\begin{array}{l}
TREE ::= lf\langle\!\langle NAME \times X\rangle\!\rangle \\
\qquad\quad \mid\ node\langle\!\langle NAME \times \mathrm{seq}\ TREE\rangle\!\rangle
\end{array}
$$

An instantiation of this "generic" definition would redefine X:

$$X == \ldots$$

Issue 2 *Is the use of basic types and their later redefinition a satisfactory solution for genericity?*

For example, what happens if more than one instantiation of the "generic" specification is needed in one and the same specification?

3 Descriptive vs. Recursive Specifications

Without further functions allowing one to manipulate and access named trees, the above definitions would not be of much use. We compare the different manners of specifying such functions as they are supported by the different languages.

3.1 Completing the Generic Specifications

We first give a more complete specification of named trees in PLUSS. For those familiar with algebraic languages, it bears no surprise. Lists are assumed to be defined by a constant *nil*, a constructor function "/", and selector functions *the head of_* and *the tail of_*.

```
proc NAMED_TREE(X)
   use LIST, NAME
   sorts Named_Tree
   cons <_ . _>: (Name × X) × List(Named_Tree) → Named_Tree
   func
      the name of_: Named_Tree → Name
      the content of_: Named_Tree → X
      the children of_: Named_Tree → List(Named_Tree)
      the name list of_: List(Named_Tree) → List(Name)
      the number of children of_: Named_Tree → Integer
      the child of _ named _: Named_Tree × Name → Named_Tree
   pred
      _is leaf: Named_Tree
   precond forall n:Name, t:Named_Tree
      child: the child of t named n is defined when
            n belongs to the name list of t
   axiom forall n, n':Name, t:Named_Tree, x:X, l:List(Named_Tree)
      name: the name of <(n,x) . l> = n
      cont: the content of <(n,x) . l> = x
      st: the children of <(n,x) . l>) = l
      nl1: the name list of nil = nil
      nl2: the name list of t / l = the name of t / the name list of l
      nb: the number of children of <(n,x) . l> = length(l)
      isl: <(n,x) . l> is_leaf iff l = nil
      child1: n = the name of the head of the children of t
            ⇒ the child of t named n = the head of the children of t
```

child2: n ≠ the name of the head of the children of <(n',x) . l>
    ⇒ the child of t named n
      = the child of <(n',x) . the tail of l> named n
end NAMED_TREE(X)

The corresponding definitions in Z look as follows, where we also define some auxiliary functions on *NAMED_TREE[X]* which have no counterpart in the PLUSS specification.

$$
\begin{array}{l}
\rule{1em}{0pt}[X] \\
\hline
child : NAMED\_TREE[X] \times \mathsf{N}_1 \nrightarrow NAMED\_TREE[X] \\
number\_of\_children : NAMED\_TREE[X] \longrightarrow \mathsf{N} \\
children : NAMED\_TREE[X] \longrightarrow \mathsf{P}\, NAMED\_TREE[X] \\
leafs : NAMED\_TREE[X] \longrightarrow \mathsf{P}(\mathrm{seq}\,\mathsf{N}_1) \\
\hline
\mathrm{dom}\, child = \{t : NAMED\_TREE[X];\ i : \mathsf{N}_1 \mid \langle i \rangle \in \mathrm{dom}\, t\} \\
\forall t : NAMED\_TREE[X];\ i : \mathsf{N}_1 \bullet \\
\quad \langle i \rangle \in \mathrm{dom}\, t \Rightarrow \\
\qquad child(t, i) = \{s : \mathrm{seq}\,\mathsf{N}_1;\ nx : NAME \times X \mid \\
\qquad\qquad\qquad\qquad \langle i \rangle \frown s \in \mathrm{dom}\, t \wedge nx = t(\langle i \rangle \frown s)\} \wedge \\
\quad \mathrm{dom}\, t \neq \{\langle\rangle\} \Rightarrow \\
\qquad number\_of\_children\, t = max\{k : \mathsf{N}_1 \mid \langle k \rangle \in \mathrm{dom}\, t\} \wedge \\
\quad \mathrm{dom}\, t = \{\langle\rangle\} \Rightarrow number\_of\_children\, t = 0 \wedge \\
\quad children\, t = \\
\qquad \{k : 1\,..\,number\_of\_children\, t \bullet child(t, k)\} \wedge \\
\quad leafs\, t = \{s : \mathrm{seq}\,\mathsf{N}_1 \mid s \in \mathrm{dom}\, t \wedge (\forall s1 : \mathrm{seq}\,\mathsf{N}_1 \bullet s \frown s1 \notin \mathrm{dom}\, t)\}
\end{array}
$$

With the help of these functions we can now specify:

$$
\begin{array}{l}
\rule{1em}{0pt}[X] \\
\hline
name\_of\_tree : NAMED\_TREE[X] \longrightarrow NAME \\
names : \mathsf{P}\, NAMED\_TREE[X] \nrightarrow \mathsf{P}\, NAME \\
child\_named : NAMED\_TREE[X] \times NAME \nrightarrow NAMED\_TREE[X] \\
\hline
\forall n : NAME;\ t : NAMED\_TREE[X];\ ts : \mathsf{P}\, NAMED\_TREE[X] \bullet \\
\quad name\_of\_tree\, t = first(t\langle\rangle) \wedge \\
\quad names\, ts = \{t : ts \bullet name\_of\_tree\, t\} \wedge \\
\quad n \in names(children\, t) \Rightarrow (t, n) \in \mathrm{dom}(child\_named) \wedge \\
\quad child\_named(t, n) \in children\, t \wedge \\
\qquad name\_of\_tree(child\_named(t, n)) = n
\end{array}
$$

These are not all, but the most important functions on named trees.

3.2 Discussion

It is noticeable that in the PLUSS specification, functions usually are defined as recursive equations, where the structure of the recursion follows the list constructors. The same holds true for the specification given in [1], except for the fact that there the tree and forest constructors are used as recursion schemas. This means that they are mostly executable and hence almost an implementation. In the Z specification, the functions are given as closed mathematical expressions.

Unfortunately, one confession must be made here: the above specification of *child_named* is semantically invalid in Z because in the reference manual [8] it is required that "the predicates must define the values of the constants uniquely for each value of the formal parameters." This is not the case here, because *child_named* selects an *arbitrary* child with the given name, whereas the PLUSS specification is deterministic and constructive. "Legal" possibilities would be to either define a relation instead of a function or give an unambiguous definition like in PLUSS. Since the type checker cannot find this "violation", it is hard to prevent (or even detect!) specifications like this.

However, we do not see any difficulties with a definition like the one for *child_named*. On the contrary, it has the advantage to give an implementor the greatest possible freedom: if it is more efficient to search from the back to the front instead of vice versa, it should be possible to do so. The PLUSS specification prohibits this and thus may prevent an efficient implementation. It is even questionable here if one should actually require *child_named* to be a function in the mathematical sense. One could argue that it suffices when the result it yields has the given name.

Issue 3 *Should we strive for very high-level specifications that anticipate as few implementation details as possible?*

To put it in other words: Is it satisfactory to specify functions as recursive equations along some constructors, as suggested by many algebraic specification languages?

It should be noted that this is not a language issue, but an issue of style. It would well be possible to specify functions as closed terms when using an algebraic language with a sufficiently expressive logic (i.e. more expressive than the one of PLUSS). The point is that this would be more complicated because something like the mathematical toolkit of Z had to be predefined. However, once this were done, algebraic specifications could look very much the same as in Z, as far as the use of recursive equations is concerned. It seems that the tradition to define functions with recursive equations stems from the time when algebraic specification languages had an initial semantics. In these times, there was no other possibility indeed.

4 Modularity

The next step in the specification of the user's view of the Unix file system is to specify directories, re-using the specification of named trees. It turns out that it is by far not enough to instantiate the generic parameter.

4.1 Specifying Directories

Before this can be done, the generic specification itself has to be modified in order to make it possible to use names for navigation in the tree. This can be done with *paths*. Paths are nonempty lists of names:

spec PATH **as** NONEMPTY-LIST(NAME)

$PATH == \text{seq}_1\ NAME$

The next step is to define functions working on the combination of named trees and paths. For this purpose, we have to define a predicate *is_existing_path_of* that decides if a path is valid for a given tree, and the functions *object_at_in*, *pruned_at*, and *plus_added_under* which select an item, prune the tree or add a new subtree under a given path. We only present the definition of *is_existing_path_of*.

```
proc NAMED_TREE_WITH_PATH(X)
  use NAMED_TREE(X), PATH
  pred
    _is existing path of_: Path × Named_Tree
  axiom forall n, n':Name, p: Path, t:Named_Tree, x:X, l:List(Named_Tree)
    exist1: n / nil is existing path of t iff n = the name of t
    exist2: n / n' / nil is existing path of t
        iff n = the name of t & n' belongs to the name list of t
    exist3: n' belongs to the name list of t is false
        ⇒ n / n' /p is existing path of t is false
    exist4: n' belongs to the name list of t
        ⇒ n / n' /p is existing path of t iff n = the name of t
            & n' / p is existing path of the child of t named n'
end NAMED_TREE_WITH_PATH(X)
```

It should be noted that we had the choice between two kinds of clumsiness here. Bidoit, Gaudel and Mauboussin [1] preferred to define paths from scratch. This made it possible to embed names into paths, i.e. to define every name to be a path. However, this specification is 33 lines long. We found that a bit much and preferred the one-line definition of paths as nonempty lists of names. The price for this is that we have to write "n / nil" where Bidoit, Gaudel and Mauboussin only have to write "n". This inconvenience will occur again in Section 5.1 when we define relative paths.

$$
\begin{array}{l}
[X] \\
\hline
\_is\_existing\_path\_of\_ : PATH \leftrightarrow NAMED\_TREE[X] \\
\hline
\forall t : NAMED\_TREE[X]; p : PATH \bullet \\
\quad p\ is\_existing\_path\_of\ t \Leftrightarrow \\
\qquad (head\ p = name\_of\_tree\ t \land \\
\qquad (tail\ p \neq \langle\rangle \Rightarrow (\exists\, t_1 : children\ t \bullet tail\ p\ is\_existing\_path\_of\ t_1)))
\end{array}
$$

We note that in PLUSS, a new generic specification is defined (without new constructors), whereas in Z a new global generic definition is added. Now, the actual parameters can be defined:

> **spec** UNIX-NODE **as** FILE ∪ { dir }
> **spec** DIRECTORY **as** NAMED_TREE_WITH_PATH(UNIX-NODE)

where *FILE* defines files as being either text files or binary files.

To finish the specification of directories, we must further specify some constraints related to the nodes: (i) a file may only be a leaf node; (ii) all successors of a node have different names. These constraints on the data type cannot be added to the parameter or to the generic specification but only to the whole instantiated generic specification:

> **axiom forall** d: Directory, n: Name, i,j: Integer
> file: the content of d is a file ⇒ d is leaf
> inj: n = the name of the i th element of the children of d
> & n = the name of the j th element of the children of d
> ⇒ i = j

For Z, we define:

$$UNIX\_NODE ::= dir \mid file\langle\langle FILE \rangle\rangle$$

$$DIRECTORY == NAMED\_TREE[UNIX\_NODE]$$

The global constraint that has to be added is

$$\forall d : DIRECTORY; n : NAME; i, j : \mathbb{N} \bullet$$
$$\forall p : \mathrm{dom}\, d \bullet (second(d\ p) \in \mathrm{ran}\, file \Rightarrow p \in leafs\ d) \land$$
$$\#(names(children\ d)) = \#(children\ d)$$

Requirement (ii) can be expressed somewhat more elegantly in Z because the Z specification is based on sets instead of lists. But even if sequences had been used, it would be possible to directly express that the sequence must be injective.

4.2 Discussion

Modularity is a very desirable feature for formal specifications. First, it is important for re-use. Libraries of predefined specifications should contain relatively small and self-contained modules so that they can serve as a kind of construction kit. Second, it is very hard to read, comprehend and maintain large, unstructured formal specification documents. Unfortunately this is exactly what Z forces its users to build. There is no possibility of nesting specification constructs (importing of schemas is just a shorthand for textually copying the content of the imported schema) or grouping parts of specifications together to form a new entity. Hence, the whole specification is spread out at the top-level.

In PLUSS, for instance, it is possible to import named trees without paths. In Z, you get all or nothing: once you have defined an instance of *NAMED_TREE*, you also have defined the operations dealing with paths, no matter if you want them or not.

Object oriented versions of Z are under development [6, 4] that provide better facilities for modularizing specifications. However, this is not a solution to the problem. Currently, Z is being standardized. The existence of a standard is of some importance to industry. They seem to prefer standardized products over others. If the standard will not contain better facilities for modularizing Z specifications, hundreds of industrial specifiers and programmers will have to live with the poor structuring facilities of Z.

Issue 4 *Can poor language facilities be compensated for by a better specification discipline?*

It it our experience that Z specifications can be well readable. This can be achieved by detailed comments (which are strongly advocated by the Z methodology) and by a skillful layout of the specification. But perhaps the question should be asked the other way around: why do specifiers *have* to compensate for poor language facilities?

5 Freely Generated Data Types and Z

Not only absolute, but also relative paths (relative to a given path) can be used to navigate in the directory tree. These are best specified as an abstract data type.

5.1 Defining Relative Paths

In PLUSS, relative paths, called displacements, can be defined straightforwardly:

spec RELATIVE_PATH
  use PATH
  sort Displacement
  cons
    empty_d: → Displacement
    _ : Path → Displacement
    _ / _: Displacement × Name → Displacement
    ../ : Displacement → Displacement
  func
    _ || _ : Path × Displacement → Path
  axiom forall p: Path, n: Name, dp: Displacement
    cat1: p || empty_d = p
    cat2: p || (n / nil) = p / n

cat3: p || (dp / n) = (p || dp) / n / nil
cat4: (n / nil) || ../dp = (n nil) || dp
cat5: (p / n) || ../dp = p || dp
end RELATIVE_PATH

Paths are embedded into displacements, i.e. each path is also a displacement. In Z, this is impossible. There, we have to define a function d that converts a path into a displacement (see below). The list constructor "/" is overloaded here; overloading is allowed in PLUSS. Again, we have to write "n / nil" instead of "n".

As long as no genericity is involved, we can use free types in Z:

$$
\begin{aligned}
\mathit{DISPLACEMENT} ::= \; & \mathit{empty\_d} \\
\mid \; & d\langle\!\langle \mathit{PATH} \rangle\!\rangle \\
\mid \; & (\_/\_)\langle\!\langle \mathit{DISPLACEMENT} \times \mathit{NAME} \rangle\!\rangle \\
\mid \; & ../\langle\!\langle \mathit{DISPLACEMENT} \rangle\!\rangle
\end{aligned}
$$

This definition corresponds to the **cons** part of the PLUSS specification. The function || has to be defined by an axiomatic box.

$$
\begin{array}{|l}
\_ \| \_ : \mathit{PATH} \times \mathit{DISPLACEMENT} \rightarrowtail\!\!\!\!\!\!+ \;\; \mathit{PATH} \\
\hline
\forall p : \mathit{PATH};\, n : \mathit{NAME};\, dp : \mathit{DISPLACEMENT} \bullet \\
\quad p \,\|\, \mathit{empty\_d} = p \land \\
\quad p \,\|\, d\,\langle n \rangle = p \frown \langle n \rangle \land \\
\quad p \,\|\, (dp/n) = (p \,\|\, dp) \frown \langle n \rangle \land \\
\quad \langle n \rangle \,\|\, (../dp) = \langle n \rangle \,\|\, dp \land \\
\quad (p \frown \langle n \rangle) \,\|\, ../dp = p \,\|\, dp
\end{array}
$$

Disregarding the lack of modularity of the Z specification, see Sect. 4, both specifications of relative paths are adequate. Therefore, nothing needs to be discussed and no new issues need to be raised.

6 State-Based Systems and Algebraic Languages

It is convenient to consider the user's view of the Unix file system as a *state* that can be changed by user commands.

6.1 Defining the User's View

We are now ready to define the system state: it consists of a directory, and two paths, one for the home directory and one for the working directory. In Z, such system states are easily defined by a schema:

$$\begin{array}{|l} \hline \textit{OneUserView} \\ root : DIRECTORY \\ home\_dir : PATH \\ working\_dir : PATH \\ \hline home\_dir\ is\_existing\_path\_of\ root \\ second(object\_at\_in(home\_dir, root)(\langle\rangle)) = dir \\ working\_dir\ is\_existing\_path\_of\ root \\ second(object\_at\_in(working\_dir, root)(\langle\rangle)) = dir \\ \hline \end{array}$$

As an example of a Unix command, we consider the command *cd* which changes the working directory. Since *cd* can be called with various parameters, we have to define several schemas for this operation, due to the strong typing of Z. If no argument is supplied to *cd*, the working directory is set to the home directory by default.

$$\begin{array}{|l} \hline cd\_def \\ \Delta OneUserView \\ \hline root' = root \\ home\_dir' = home\_dir \\ working\_dir' = home\_dir \\ \hline \end{array}$$

If an absolute path is supplied to *cd*, the working directory is set to this path, provided it is a legal one. Legal means that the path exists in the directory and that it leads to a directory, not to a file.

$$\begin{array}{|l} \hline cd\_abs \\ \Delta OneUserView \\ p? : PATH \\ \hline p?\ is\_existing\_path\_of\ root \\ second(object\_at\_in(p?, root)(\langle\rangle)) = dir \\ root' = root \\ home\_dir' = home\_dir \\ working\_dir' = p? \\ \hline \end{array}$$

If a displacement is supplied to *cd*, the new working directory is computed as the absolute path yielded by combining the old working directory with the given displacement.

cd_rel

$\Delta OneUserView$
$dp? : DISPLACEMENT$

$(working\_dir \parallel dp?)\ is\_existing\_path\_of\ root$
$second(object\_at\_in(working\_dir \parallel dp?, root)(\langle\rangle)) = dir$
$root' = root$
$home\_dir' = home\_dir$
$working\_dir' = working\_dir \parallel dp?$

To define state-based systems in algebraic languages, one possible "schema" (similarly to the definition of freely generated types in Z) is to first define a *data type S* (instead of a schema) modeling the global state. If the state schema consists of more than one variable, *S* has to be defined as the Cartesian product of the types of the state variables. The state invariant must be given as a global axiom on *S*. Each operation in a state-based system is specified by a function having the state before execution of the operation as an additional input parameter and the state after execution of the operation as an additional output parameter. Generally, the axioms for such a function are the conjunction of the axioms for the state definition of the "before"-state, the "after"-state and the axioms defining the operation. We are lucky: for our Unix example, it is possible to use a simpler version, although we now have the obligation to show that each *cd* function yields indeed a legal state.

```
spec ONE_USER_VIEW
   use DIRECTORY, RELATIVE_PATH
   sort User-view
   cons < -. - .- >: Directory × Path × Path → User-view
   func
      cd: User-view → User-view
      cd: User-view × Path → User-view
      cd: User-view × Displacement → User-view
      ...
   precond forall root: Directory, hd,wd, p: Path, dp: Displacement
      cd1: cd(<root.hd.wd>,p) is defined when
           p is an existing path of root
           & the object at p in root is a Directory
      cd2: cd(<root.hd.wd>,dp) is defined when
           cd(<root.hd.wd>, wd || dp) is defined
   axioms forall root: Directory, hd,wd, p: Path, dp: Displacement,
           uv:User-view
      state: uv = <root.hd.wd>
        ⇒ hd is an existing path of root
          & the object at hd in root is a Directory
          & wd is an existing path of root
```

& the object at wd in root is a Directory
cd1: cd(<root.hd.wd>) = <root.hd.hd>
cd2: cd(<root.hd.wd>,p) = <root.hd.p>
cd3: cd(<root.hd.wd>,dp) = <root.hd.wd||dp>
end ONE_USER_VIEW

We observe that, in PLUSS, there is no need to invent different names for the different versions of the *cd* command because overloading is permitted.

6.2 Discussion

The general approach to define state-based systems in algebraic languages seems to be a bit clumsy. In the worst case, it could be necessary to repeat the state invariant over and over again. We are not aware of any satisfactory solutions to this problem but would be glad to learn about them if they exist.

Issue 5 *Are there more elegant ways to deal with states in algebraic languages?*

It seems that this issue cannot be neglected by the algebraic specifications community. For implementation purposes, efficiency considerations will probably always play a major role because the impressive MIPS numbers of new computer generations are eaten up by the ever more complicated, comfortable, and large new programs that are written (and demanded!) for them. Research on functional programming languages takes this fact into account. In this area, it is even more evident that it is impractical to copy large data structures for each function call that is executed. There are approaches to avoid this and nevertheless retain referential transparency [9]. Similar ideas for algebraic specification languages would be most welcome.

7 The Moral of the Story

What do we learn of this comparative case study? The good news is that we indeed succeeded in specifying the system we wanted. The languages available today are basically useful. The bad news is that sometimes they seem to create more problems than they solve. This is irrespective of the fact that much research activity has been devoted to the design of specification languages.

There seems to be a large gap between the algebraic and the model-based communities. But in reality, in every system specification there will be parts where clean algebraic properties are wanted and other parts where a state is necessary. This means we cannot (or at least do not want) to do without one or the other of these features. Both means of expression are useful and necessary. Ignoring this fact, algebraic languages pretend there is no need for a state, and model-based languages pretend there is no need for comfortable, abstract and encapsulated data type definitions.

Additionally, we learned that a good specification discipline and style can make up for many deficiencies of today's specification languages. However, this can be no substitute for real language support.

Both stylistic and language issues come into play when we ask ourselves what makes up a "good" specification. Some of the properties discussed in the preceding sections are not disputed, e.g. re-usability or modularity. At most the best ways to achieve them are subject of discussions. Concerning the call for a very high level of abstraction in formal specifications, this is probably different. Some people strongly argue in favor of executable specifications because of their prototyping potential. This is an issue where it is hard to choose between conflicting goals: it is our strong conviction that specifications should not enforce premature implementation decisions ; on the other hand, animating specifications is very important because – as Brooks [2] put it – "For the truth is, the client does not know what he wants". It is a fact of life that system requirements cannot be fixed from the beginning and will continue to change during the project[3]. In the end, it is like everywhere: easy solutions cannot be expected.

Acknowledgment. We would like to thank Thomas Santen for many stimulating discussions and for his comments on this work.

References

1. M. Bidoit, M.-C. Gaudel, and A. Mauboussin. How to make algebraic specifications more understandable? In M. Wirsing and J.A. Bergstra, editors, *Algebraic Methods: Theory, Tools and Applications*, number 394 in LNCS, pages 31 – 67. Springer-Verlag, 1989.
2. Frederick P. Brooks. No silver bullet – essence and accidents of software engineering. *Computer*, pages 10–19, April 1987.
3. O. Declerfayt, B. Demeuse, F. Wautier, P. Y. Schobbens, and E. Milgrom. Precise standards through formal specifications: A case study: the unix file system. In *Proceedings EUUG Autumn Conference, Cascais, Portugal*, 1988.
4. Kevin Lano and Howard Haughton. Specifying a concept-recognition system in Z++. In Kevin Lano and Howard Haughton, editors, *Object-Oriented Specification Case Studies*, chapter 7, pages 137–157. Prentice Hall, 1988.
5. Carroll Morgan and Bernard Sufrin. Specification of the UNIX Filing System. In Ian Hayes, editor, *Specification Case Studies*. Prentic-Hall, 1987.
6. Gordon Rose and Roger Duke. An object-Z specification of a mobile phone system. In Kevin Lano and Howard Haughton, editors, *Object-Oriented Specification Case Studies*, chapter 5, pages 110–129. Prentice Hall, 1993.
7. Jeanine Souquières and Maritta Heisel. How to manage formal specifications? Submitted for publication, 1994.
8. J. M. Spivey. *The Z Notation – A Reference Manual*. Prentice Hall, 2nd edition, 1992.
9. Phil Wadler. Monads for functional programming. In M. Broy, editor, *Program Design Calculi*, volume 118 of *Computer and Systems Sciences*, pages 233–264. Springer-Verlag, 1993.
10. J. B. Wordsworth. *Software Development with Z*. Addison-Wesley, Wokingham, 1992.

[3] In this respect our case study was unrealistic because it specified an already existing system.

A Calculus of Countable Broadcasting Systems

Yoshinao ISOBE, Yutaka SATO, Kazuhito OHMAKI

Computer Science Division, Electrotechnical Laboratory
1-1-4 Umezono, Tsukuba, Ibaraki 305, Japan
E-mail:{ isobe|ysato|ohmaki }@etl.go.jp

Abstract. In this paper we propose a process algebra named *CCB* (a Calculus of Countable Broadcasting Systems). We define an observational congruence relation in CCB after basic definitions of CCB, and give a sound and complete axiom system for the congruence relation of finite agents.
CCB is developed for analyzing a *multi-agent model* with broadcast communication. The most important property of CCB is that a broadcaster of a message can know the number of receivers of the message after broadcasting. The property is not easily described in the other process algebras.
The multi-agent model is useful for constructing extensible systems. A disadvantage of the multi-agent model is that agents must be designed very carefully because unexpected behavior may arise by interactions between the agents. Therefore we want to analyze behavior of the agents.

1 Introduction

A design of software should be divided into several program components called *agents* which may be executed concurrently and communicate with each other through *events*, in order to develop and refine them independently thus efficiently. This approach is called a *multi-agent model*[1]. The advantages of the multi-agent model can be summarized as follows:

1. *Choice of different description languages:* Each agent can be programmed by its appropriate language and communicate with other agents through some standardized protocols modeled as events.
2. *Software reusability:* If each agent is carefully designed, it can be shared and reused. The interfaces (i.e., protocols) between agents must be designed simple and flexible to achieve this goal.
3. *Machine independence:* Machine independence should be realized in recent computer systems which contain a wide variety of machines in one network. The multi-agent model will also fit to realize this machine independence.

In order to flexibly connect agents, one of the authors has developed a mechanism named VIABUS[13] which has a software bus architecture with *broadcast communication* based on the multi-agent model. The advantage of broadcast communication is that a broadcaster of an event need not be modified when new receivers of the event are attached to VIABUS, because destinations of the event are not specified. Thus broadcast communication is useful for constructing extensible software systems. For example, VIABUS has been used for an extensible User Interface Management System named VIAUIMS[13] as shown in Fig.1.

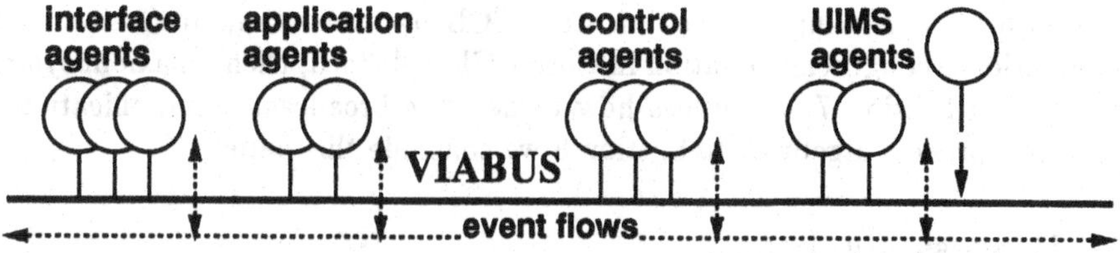

Fig. 1. The architecture of VIAUIMS

Though the multi-agent model is useful for efficient development and refinement of software, agents should be designed very carefully in order to prevent unexpected behavior such as deadlocks by interactions between agents. Thus we want a tool to analyze behavior of agents in the multi-agent model, where the tool should satisfy the following four requirements:

1. *Broadcast communication can be described.* A broadcasted event is received by *all* agents which require the event.
2. *A broadcaster of an event can know the number of receivers of the event after broadcasting.* Though a broadcaster does not specify the number of receivers of the event when broadcasting, it sometimes wants to know the number after broadcasting.
3. *Agents can dynamically change receivable events.* Agents may change their receivable events as reaction of events from the other agents.
4. *Multicast communication can be also described.* An agent sometimes wants to multicast an event to the specific number of receivers for synchronization.

We attempt to adopt a process algebra as a basic tool for analyzing agents in the multi-agent model. Many various process algebras have been proposed such as CCS[2], π-calculus[3], SCCS[4], and CBS[7, 8]. The above requirements 1 and 3 are satisfied by CBS, and the requirement 4 is satisfied by SCCS. However it is difficult for existing process algebras to satisfy the requirement 2.

If users can attach and detach agents in a system, then the number of receivers of an event can not be fixed. Furthermore one agent can change its receivable events. Therefore it is important to dynamically count the number of receivers of an event. For example, we want to describe the following behavior:

- An agent P_1 broadcasts an event *st*, and starts (triggers) the other agents $Q_{1,\cdots,n}$ which receive *st*, where P_1 must wait to continue its task until *all* the triggered agents $Q_{1,\cdots,n}$ finish their tasks.

In this case P_1 must know the number n of the triggered agents.

In this paper we propose a process algebra named *CCB* (a Calculus of Countable Broadcasting Systems). The above four requirements are satisfied by CCB. We define an observational congruence relation in CCB, and give a sound and complete axiom system for the congruence relation of finite agents.

The outline of this paper is as follows: In Section 2 we introduce process algebras, and point out some problems of existing process algebras for broadcasting systems. In Section 3 we informally introduce CCB. In Section 4 the syntax and the semantics of Core-CCB which is a base of CCB are defined. In Section 5 CCB is defined based on Core-CCB by translation. In Section 6 an observational congruence relation in Core-CCB is defined, then an axiom system is given. In Section 7 we discuss how to describe broadcast communication in the other process algebras. In Section 8 we conclude this paper.

2 Process algebras

Process algebras are well known as mathematical tools to describe and analyze concurrent and communicating systems. Behavior of agents is described as *(agent) expressions* in a process algebra, then equality of behavior of two agents is proven by rewriting their expressions according to algebraic laws in the process algebra. In general, messages are carried by *events* with *names*. When an event with a name and messages has been broadcasted, each agent decides by monitoring the name whether to read the messages in the event, or discard.

Many various process algebras have been proposed. In the rest of this section, we classify them into three groups by their communication styles, and point out some problems of existing process algebras for analyzing broadcasting systems.

2.1 Point-to-point communication

When an agent sends an event, only one agent can receive the event in CCS[2], π-calculus[3], and so on. In order to simulate broadcast communication, an event must be sent the same times as the number of agents which require the event, but it is very difficult to dynamically know the number. The following example shows that an agent $C(i)$ attempts to count the number of agents which can receive an event a but fails it, using CCS.

$$P_1 \stackrel{\mathrm{def}}{=} a.\mathbf{0}, \quad P_2 \stackrel{\mathrm{def}}{=} b.\mathbf{0}, \quad P_3 \stackrel{\mathrm{def}}{=} a.\mathbf{0}, \quad C(i) \stackrel{\mathrm{def}}{=} \overline{a}.C(i+1) + \overline{out}(i).\mathbf{0}$$
$$SYS \stackrel{\mathrm{def}}{=} (C(0)|P_1|P_2|P_3)\backslash\{a,b\}$$

The agent $C(i)$ increases the local variable i by 1, when it can send the event a (in other words, an agent can receive the event a). The agent $C(i)$ also sends out an event *out* with the value of the variable i as a message. At the initial state, two agents P_1 and P_3 can receive the event a. Therefore it is expected that the value of i sent out is 2, but it can not be predicted at all which value 0, 1, or 2 is sent out, as shown the following equation[1].

$$SYS \sim \overline{out}(0).\mathbf{0} + \tau.(\overline{out}(1).\mathbf{0} + \tau.\overline{out}(2).\mathbf{0})$$

The problem is caused by possibility that $C(i)$ sends the event *out* before sending the event a to both the agents P_1 and P_3. This problem is not solved even though the event a has higher priority[11] than the event *out*. Because SYS falls into an infinite loop if P_1 is defined as $P_1 \stackrel{\mathrm{def}}{=} a.P_1$.

[1] $\sim$ is strong equivalence in CCS[2].

2.2 Multicast communication

When an agent multicasts an event to n agents, just n agents can receive the event in SCCS[4], Meije[5], and so on. For example, in SCCS multicast communication is described as follows[2]:

$$P_0 \times P_1 \times P_2 \xrightarrow{1} P_0' \times P_1' \times P_2'$$

where each component is defined as follows:

$$P_0 \stackrel{\text{def}}{=} a^{-2}:P_0' \qquad P_1 \stackrel{\text{def}}{=} a^1:P_1' \qquad P_2 \stackrel{\text{def}}{=} a^1:P_2'$$

In this example an agent P_0 multicasts an event a to explicitly *two* agents[3]. Therefore if a new agent which require the event a is attached to this example, then P_0 must be slightly modified as $P_0 \stackrel{\text{def}}{=} a^{-3}:P_0'$.

In order to simulate broadcast communication, a multicaster of an event must count the number of receivers of the event, but it is very difficult by the same reason as shown in Subsection 2.1.

2.3 Broadcast communication

When an agent broadcasts an event, all agents which require it can receive it in CSP[6], CBS[7], and so on. Especially CBS has been developed for broadcasting systems. For example, in CBS broadcasting is described as follows:

$$(P_0|P_1|P_2|P_3)\backslash\{a\} \xrightarrow{\tau} (P_0'|P_1'|P_2|P_3')\backslash\{a\}$$

where each component is defined as follows:

$$P_0 \stackrel{\text{def}}{=} a!.P_0', \quad P_1 \stackrel{\text{def}}{=} a?.P_1' + b?.P_1'', \quad P_2 \stackrel{\text{def}}{=} b?.P_2' + c?.P_2'', \quad P_3 \stackrel{\text{def}}{=} a?.P_3'$$

Attributes ! and ? of events means transmitting and receiving, respectively. P_0 broadcasts an event a, then both P_1 and P_3 receive it. P_2 does not receive the event a, because it has no transitions like $\xrightarrow{a?}$, thus it dose not require a. Unfortunately, CBS is not interested in the number of receivers of an event.

3 Introduction of CCB

We propose a process algebra *CCB* which is an extension of CCS. The four requirements mentioned in Section 1 are satisfied by CCB. The requirement 2 is the most important, because the other requirements are satisfied also by the other process algebras.

An event in CCB has a form $a\theta\langle x\rangle(y)$, where a is a name, θ is an attribute, x is the number of receivers of this event (We call x the *received number*), and y means a message passed by this event. We often omit a received number $\langle 1\rangle$ and an empty message ( ). CCB has four attributes $\{!, ?, \mathbf{!}, \mathbf{?}\}$ explained as follows:

---

[2] : and × are a prefix and a composition combinators in SCCS, respectively.

[3] -2 in a^{-2} of P_0 means sending to two agents.

1. $a!\langle n\rangle(m)$ is used for multicasting. n is a non-negative integer (including zero). m is a message carried by this multicasting. Especially if n is zero then $a!\langle 0\rangle(m)$ is called a *silent event*. The silent event can be considered as an invisible and uncontrolled event like τ in CCS. In observational analysis, the silent event should be ignored as far as possible.
2. $a?\langle n\rangle(y)$ is used for receiving a multicasted event with the same event name a. n is a positive integer (not including zero). y is a variable bound to a message carried by the multicasted event.
3. $a\mathbf{!}\langle x\rangle(m)$ is used for broadcasting. x is a variable bound to the number of receivers of this event. m is a message carried by this broadcasting.
4. $a\mathbf{?}\langle n\rangle(y)$ is used for receiving a broadcasted event with the same event name a. n is a positive integer. y is a variable bound to a message.

For example, in CCB broadcasting is described as follows:

$$(P_0|P_1|P_2|P_3)\backslash\{a\} \xrightarrow{a!\langle 0\rangle} (P_0'(2)|P_1'|P_2|P_3')\backslash\{a\}$$

where each component is defined as follows:

$$P_0 \stackrel{\text{def}}{=} a\mathbf{!}\langle x\rangle.P_0'(x),\quad P_1 \stackrel{\text{def}}{=} a\mathbf{?}.P_1' + b\mathbf{?}.P_1'',\quad P_2 \stackrel{\text{def}}{=} b\mathbf{?}.P_2' + c\mathbf{?}.P_2'',\quad P_3 \stackrel{\text{def}}{=} a\mathbf{?}.P_3'$$

P_0 broadcasts an event a, then both P_1 and P_3 receive it. x is bound to the number 2 of the receivers as shown in $P_0'(2)$ after broadcasting.

4 The definition of Core-CCB

In this section, we define the syntax and the semantics of *Core-CCB* which is the base of CCB. Core-CCB is a process algebra with no *value variables*[4]. The definition of CCB is given based on Core-CCB in Section 5.

4.1 The syntax of Core-CCB

First, we assume that an infinite set of names, $\mathcal{N} = a, b, c, \cdots$, is given. The set of event *Event* is defined by using $\mathcal{N}$ as follows:

$$Event = \{a\theta^n : a \in \mathcal{N}, \theta \in \mathcal{T}, n \in \mathcal{I}\} - \{a?^0, a\mathbf{?}^0 : a \in \mathcal{N}\}$$

where $\mathcal{I}$ is a set of non-negative integer $\{0, 1, 2, \cdots\}$, and $\mathcal{T}$ ranged over by θ is a set of attributes $\{!, ?, \mathbf{!}, \mathbf{?}\}$. Especially $a!^0$ is a *silent event* for any a. n of $a\theta^n$ is the received number of this event, and must be a constant even for broadcasting ($\theta = \mathbf{!}$). We show how to describe broadcast communication of CCB by Core-CCB in Section 5.

Furthermore we introduce a set of *agent variables*, $\mathcal{X}$ ranged over by $X, Y, \cdots$, a set of *agent constants*, $\mathcal{K}$ ranged over by $A, B, \cdots$, and a set of renaming functions, $\mathcal{F}$ ranged over by $S : \mathcal{N} \to \mathcal{N}$.

We define the set $\mathcal{E}$ of Core-CCB expressions as follows:

[4] The value variables are used for received numbers and messages in events of CCB like x, y in $a\theta\langle x\rangle(y)$.

Definition 4.1 *The set of agent expressions, $\mathcal{E}$ ranged over by $E, F, \cdots$, is defined by the following BNF expression:*

$$E ::= X \mid A \mid \mathbf{0} \mid a\theta^n.E \mid E + E \mid E|E \mid E[S] \mid E\backslash L$$

where we take $X \in \mathcal{X}$, $A \in \mathcal{K}$, $a\theta^n \in Event$, $S \in \mathcal{F}$, and $L \subseteq \mathcal{N}$. ∎

An *agent* is an agent expression with no agent variables. The set of agents is denoted by $\mathcal{P}$ ranged over by $P, Q, \cdots$. An *agent constant* is an agent whose meaning is given by a defining equation. In fact, we assume that for every agent constant A there is a defining equation of the following form: $A \stackrel{\text{def}}{=} P$ $(P \in \mathcal{P})$, where we also assume that A is *weakly guarded*[2] in P. The weak guard is defined as follows:

Definition 4.2 *X is weakly guarded in E if each occurrence of X is within some subexpression of E of form $a\theta^n.F$.* ∎

Agent constants which are not weakly guarded make a calculus more complex, and the behavior is indefinite. Practically, we are interested in weakly guarded agent constants in CCB. (also in CBS.)

4.2 The semantics of Core-CCB

The semantics of Core-CCB is defined by the following labelled transition system like one of CCS:

$$(\mathcal{E}, Event, \{\xrightarrow{a\theta^n} : a\theta^n \in Event\})$$

For example, $E \xrightarrow{a\theta^n} E'$ $(E, E' \in \mathcal{E})$ indicates that the agent expression E may perform the event $a\theta^n$ and thereafter become the agent expression E'. The semantics for agent expressions $\mathcal{E}$ consists in the definition of each transition relation $\xrightarrow{a\theta^n}$ over $\mathcal{E}$.

Before defining the semantics, we define a function $mon(E)$ named *monitor function* which takes an agent expression and produces a subset of names.

Definition 4.3 *For each agent expression E, we inductively define monitor function $mon : \mathcal{E} \to 2^{\mathcal{N}}$ as follows:*

$$\begin{array}{r@{\;}l|r@{\;}l}
mon(\mathbf{0}) &= \emptyset & mon(E[S]) &= \{S(a) : a \in mon(E)\} \\
mon(a\theta^n.E) &= \begin{cases} \{a\} & (\theta = ?) \\ \emptyset & (otherwise) \end{cases} & mon(E\backslash L) &= mon(E) - L \\
mon(E+F) &= mon(E) \cup mon(F) & mon(A) &= mon(P) \quad (A \stackrel{\text{def}}{=} P) \\
mon(E|F) &= mon(E) \cup mon(F) & mon(X) &= \emptyset
\end{array}$$

□

Since agent constants must be weakly guarded, $mon(E)$ can be effectively evaluated. Intuitively, if an agent expression E *now* requires receiving a broadcasted event, then the event name is included in $mon(E)$.

Then the semantics of Core-CCB is defined.

Definition 4.4 *The transition relation $\xrightarrow{a\theta^n}$ over agent expressions is the smallest relation satisfying the following inference rules. Each rule is to be read as follows: if the transition relations above the line are inferred and the side conditions are satisfied, then the transition relation below the line can be also inferred.*

$$\mathbf{Event}\ \frac{}{a\theta^n.E \xrightarrow{a\theta^n} E} \qquad \mathbf{Con}\ \frac{P \xrightarrow{a\theta^n} P'}{A \xrightarrow{a\theta^n} P'}\ (A \stackrel{\mathrm{def}}{=} P)$$

$$\mathbf{Choice_1}\ \frac{E \xrightarrow{a\theta^n} E'}{E + F \xrightarrow{a\theta^n} E'} \qquad \mathbf{Para_1}\ \frac{E \xrightarrow{a\theta^n} E'}{E|F \xrightarrow{a\theta^n} E'|F}\ \begin{pmatrix} \theta \in \{!, ?\} \text{ or} \\ a \notin mon(F) \end{pmatrix}$$

$$\mathbf{Choice_2}\ \frac{F \xrightarrow{a\theta^n} F'}{E + F \xrightarrow{a\theta^n} F'} \qquad \mathbf{Para_2}\ \frac{F \xrightarrow{a\theta^n} F'}{E|F \xrightarrow{a\theta^n} E|F'}\ \begin{pmatrix} \theta \in \{!, ?\} \text{ or} \\ a \notin mon(E) \end{pmatrix}$$

$$\mathbf{Ren}\ \frac{E \xrightarrow{a\theta^n} E'}{E[S] \xrightarrow{S(a)\theta^n} E'[S]} \qquad \mathbf{Para_3}\ \frac{E \xrightarrow{a\theta^m} E' \qquad F \xrightarrow{a\phi^n} F'}{E|F \xrightarrow{a\theta^{(m-n)}} E'|F'}\ \begin{pmatrix} \theta \in \{!, !!\}, \\ \phi = @(\theta), \\ m \geq n \end{pmatrix}$$

$$\mathbf{Res_1}\ \frac{E \xrightarrow{a\theta^n} E'}{E \backslash L \xrightarrow{a\theta^n} E' \backslash L}\ (a \notin L) \qquad \mathbf{Para_4}\ \frac{E \xrightarrow{a\phi^n} E' \qquad F \xrightarrow{a\theta^m} F'}{E|F \xrightarrow{a\theta^{(m-n)}} E'|F'}\ \begin{pmatrix} \theta \in \{!, !!\}, \\ \phi = @(\theta), \\ m \geq n \end{pmatrix}$$

$$\mathbf{Res_2}\ \frac{E \xrightarrow{a\theta^0} E'}{E \backslash L \xrightarrow{a!^0} E' \backslash L}\ \begin{pmatrix} \theta \in \{!, !!\}, \\ a \in L \end{pmatrix} \qquad \mathbf{Para_5}\ \frac{E \xrightarrow{a\theta^m} E' \qquad F \xrightarrow{a\theta^n} F'}{E|F \xrightarrow{a\theta^{(m+n)}} E'|F'}\ (\theta \in \{?, ??\})$$

where @ in the conditions is a function from $\mathcal{T}$ to $\mathcal{T}$ defined as follows:

$$@(!) = ?, \quad @(?) = !, \quad @(!!) = ??, \quad @(??) = !!$$

□

Then the following proposition for monitor function is proven.

Proposition 4.1 $E \overset{a??^n}{\not\longrightarrow}$ *for any n* **iff** $a \notin mon(E)$ □

We intuitively explain $\mathbf{Para_i}$ rules.

1. By $\mathbf{Para_1}$ E multicasts an event or receives a multicasted event without respect to F. While E broadcasts an events or receives a broadcasted event only if the event name is not included in $mon(F)$, thus F does not require the event. If F requires the event, then $\mathbf{Para_3}$ or $\mathbf{Para_5}$ msut be used instead of $\mathbf{Para_1}$. This restriction relates to the requirement 1 stated in Section 1. $\mathbf{Para_2}$ is symmetric to $\mathbf{Para_1}$.
2. By $\mathbf{Para_3}$ E multicasts or broadcasts an events and F receives the event, if the number of receivers within F is not greater than the received number specified by E. If the number $(m - n)$ in the conclusion of $\mathbf{Para_3}$ is zero, then $E|F$ can no longer communicate with the other agents. This calculation of the received number relates to the requirement 2 stated in Section 1. $\mathbf{Para_4}$ is symmetric to $\mathbf{Para_3}$.
3. By $\mathbf{Para_5}$ to receive an event can be synchronized. We have no rules for synchronization of transmitting.

Then we explain an event of form $a!!^0$ which is not a silent event. In order to compare $a!^0$ with $a!!^0$, we show the following transitions:

$$(1) \qquad ((a!^2.\mathbf{0}|a?^1.\mathbf{0})|a?^1.\mathbf{0})|a?^1.\mathbf{0} \xrightarrow{a!^0} ((\mathbf{0}|\mathbf{0})|\mathbf{0})|a?^1.\mathbf{0}$$

$$(2) \qquad ((a!!^2.\mathbf{0}|a??^1.\mathbf{0})|a??^1.\mathbf{0})|a??^1.\mathbf{0} \stackrel{a!!^0}{\not\longrightarrow}$$

The transition (1) is inferred by **Para$_3$** and **Para$_1$**. However **Para$_1$** can not be used for the transition (2), because $a \in mon(a??^1.\mathbf{0})$. The following transitions by broadcasting are possible:

$$(3) \qquad ((a!!^3.\mathbf{0}|a??^1.\mathbf{0})|a??^1.\mathbf{0})|a??^1.\mathbf{0} \xrightarrow{a!!^0} ((\mathbf{0}|\mathbf{0})|\mathbf{0})|\mathbf{0}$$

$$(4) \qquad ((a!!^2.\mathbf{0}|a??^1.\mathbf{0})|a??^1.\mathbf{0})\backslash\{a\}|a??^1.\mathbf{0} \xrightarrow{a!^0} ((\mathbf{0}|\mathbf{0})|\mathbf{0})\backslash\{a\}|a??^1.\mathbf{0}$$

In the transition (3), the number specified by the broadcaster is equal to the number of receivers. In the transition (4), the scop of the event a is restricted and $a!!^0$ is changed to $a!^0$ by **Res$_2$**.

5 The definition of CCB

In this section, we define the syntax and the semantics of CCB, and give a small example. The semantics of CCB is given by translation to Core-CCB.

5.1 The syntax of CCB

The syntax of CCB is defined as follows:

Definition 5.1 *The set of agent expressions, $\mathcal{E}^+$ ranged over by $E, F, \cdots$, is defined by the following BNF expression:*

$$E ::= X \mid A(e_1, \cdots, e_n) \mid \mathbf{0} \mid a[e_1]!\langle c_0\rangle(e_2).E \mid a[e_1]!!\langle x\rangle(e_2).E \mid a[e]?\langle c_1\rangle(x).E$$
$$\mid a[e]??\langle c_1\rangle(x).E \mid E+E \mid E|E \mid E[S] \mid E\backslash L \mid [\![b]\!]E$$

where we take $X \in \mathcal{X}$, $A \in \mathcal{K}$, $a \in \mathcal{N}$, $S \in \mathcal{F}$, and $L \subseteq \mathcal{N}$. x is a bound value variable, b is a boolean expression, and e_i is an expression. c_i is also an expression, where the range of c_0 must be non-negative integer and the range of c_1 must be positive integer. All value variables in expressions b, e_i, and c_i must be bound by their occurrences on their left. □

The unconventional combinator $[\![b]\!]$ is a conditional combinator. It intuitively means that if b is true then $[\![b]\!]E$ behaves like E, otherwise it stops.

Each agent constant A with arity n has a defining equation $A(x_1, \cdots, x_n) \stackrel{\text{def}}{=} E$, where the right-hand side E may contain no agent variables and no free value variables except $x_1, \cdots, x_n$.

An event name $a[e]$ consists of a base event name a and a postfix $[e]$. We sometimes omit a postfix [0]. Intuitively an event $a[x]!$ will be received by $a[e]?$ for some e. In order to avoid ambiguous communication, x must be bound before

$a[x]!$ is multicasted. For example, the following agent INC is a procedure which takes a value y and return a value $y+1$.

$$INC \stackrel{\text{def}}{=} inc?(x).((in[x]?(y).out[x]!(y+1).\mathbf{0})|INC)$$

INC is duplicated just after called by a caller in order to prepare for the next caller. INC binds x to a process-id of a caller and certainly return $y+1$ to *the* caller through $out[x]$. For example a caller may be defined as follows:

$$CALLER23 \stackrel{\text{def}}{=} inc!(23).in[23]!(7).out[23]?(z).TASK23(z)$$

$TASK23(z)$ depends on z, and z will be bound to 8. This strategy has been used in a parallel programming language $\mathcal{M}$[2] defined in CCS.

5.2 The semantics of CCB

The semantics of CCB is given by translation. Each expression in CCB without free value variables is translated into an expression in Core-CCB by a function $\mathcal{B} : \mathcal{E}^+ \to \mathcal{E}$. We introduce a operator $\ddot{}$. It takes an expression and produce a fixed value which the expression evaluates. For example, if $e = 2+3$ then $\ddot{e} = 5$.

Definition 5.2 *The function $\mathcal{B}$ from $\mathcal{E}^+$ to $\mathcal{E}$ is defined as follows*[5]*:*

$$\begin{aligned}
\mathcal{B}(X) &= X \\
\mathcal{B}(A(e_1,\cdots,e_n)) &= A_{e_1,\cdots,e_n} \\
\mathcal{B}(\mathbf{0}) &= \mathbf{0} \\
\mathcal{B}(a[e_1]!\langle c\rangle(e_2).E) &= a_{\ddot{e}_1,\ddot{e}_2}!^{\ddot{c}}.\mathcal{B}(E) \\
\mathcal{B}(a[e_1]\mathbf{!!}\langle x\rangle(e_2).E) &= \textstyle\sum_{n\in\mathcal{I}} a_{\ddot{e}_1,\ddot{e}_2}\mathbf{!!}^{n}.\mathcal{B}(E\{n/x\}) \\
\mathcal{B}(a[e]?\langle c\rangle(x).E) &= \textstyle\sum_{v\in V} a_{\ddot{e},v}?^{\ddot{c}}.\mathcal{B}(E\{v/x\}) \\
\mathcal{B}(a[e]\mathbf{??}\langle c\rangle(x).E) &= \textstyle\sum_{v\in V} a_{\ddot{e},v}\mathbf{??}^{\ddot{c}}.\mathcal{B}(E\{v/x\})
\end{aligned}$$

$$\begin{aligned}
\mathcal{B}(E+F) &= \mathcal{B}(E)+\mathcal{B}(F) \\
\mathcal{B}(E|F) &= \mathcal{B}(E)|\mathcal{B}(F) \\
\mathcal{B}(E[S]) &= \mathcal{B}(E)[S'] \\
\mathcal{B}(E\backslash L) &= \mathcal{B}(E)\backslash L' \\
\mathcal{B}([\![b]\!]E) &= \begin{cases} \mathcal{B}(E) & \textit{if } b{=}\textit{true} \\ \mathbf{0} & \textit{otherwise} \end{cases}
\end{aligned}$$

where we assume that all values belong to the fixed value set V, and

$$S'(a_{v_1,v_2}) = S(a)_{v_1,v_2}, \qquad L' = \{a_{v_1,v_2} : a \in L, (v_1,v_2) \in V^2\}$$

where $\{v/x\}$ is a substitution of a value v into a variable x. Furthermore the single defining equation $A(x_1,\cdots,x_n) \stackrel{\text{def}}{=} E$ of an agent constant is translated into the indexed set of defining equations

$$\{A_{v_1,\cdots,v_n} \stackrel{\text{def}}{=} \mathcal{B}(E\{v_1/x_1,\cdots,v_n/x_n\}) : (v_1,\cdots,v_n) \in V^n\}. \qquad \square$$

We now give a simple example of translating CCB into Core-CCB. The following expression is described in CCB:

$$CHECK \stackrel{\text{def}}{=} a\mathbf{!!}\langle x\rangle.([\![x=0]\!]zero!\langle 1\rangle.\mathbf{0} + [\![x\neq 0]\!]nonzero!\langle 1\rangle.\mathbf{0})$$

This agent $CHECK$ broadcasts an event $a\mathbf{!!}\langle x\rangle$, and check the number of agents which receive $a\mathbf{!!}\langle x\rangle$. Thereafter if the number is zero then $CHECK$ multicasts

[5] It $I \neq \emptyset$, then $\sum_{i\in I} P_i$ is a syntactic shorthand of $P_1 + P_2 + \cdots$, otherwise it is a $\mathbf{0}$.

an event $zero!\langle 1\rangle$, otherwise multicasts an event $nonzero!\langle 1\rangle$. This expression in CCB is translated into an expression in Core-CCB by $\mathcal{B}$ as follows:

$$\begin{aligned}
\mathcal{B}(CHECK) &= \mathcal{B}(a!\langle x\rangle.([\![x=0]\!]zero!\langle 1\rangle.0 + [\![x \neq 0]\!]nonzero!\langle 1\rangle.0))\\
&= \textstyle\sum_{n\in\mathcal{I}} a!^{n}.\mathcal{B}([\![n=0]\!]zero!\langle 1\rangle.0 + [\![n\neq 0]\!]nonzero!\langle 1\rangle.0)\\
&= \textstyle\sum_{n\in\mathcal{I}} a!^{n}.(\mathcal{B}([\![n=0]\!]zero!\langle 1\rangle.0) + \mathcal{B}([\![n\neq 0]\!]nonzero!\langle 1\rangle.0))\\
&= a!^{0}.(\mathcal{B}(zero!\langle 1\rangle.0)+0) + \textstyle\sum_{n\in\mathcal{I}_1} a!^{n}.(0+\mathcal{B}(nonzero!\langle 1\rangle.0))\\
&= a!^{0}.(zero!^{1}.0+0) + \textstyle\sum_{n\in\mathcal{I}_1} a!^{n}.(0+nonzero!^{1}.0)
\end{aligned}$$

where $\mathcal{I}$ is a set of non-negative integer and $\mathcal{I}_1$ is a set of positive integer.

We show another example of translating CCB into Core-CCB by using INC and $CALLER23$ defined in Subsection 5.1.

$$\begin{aligned}
\mathcal{B}(INC) &= \mathcal{B}(inc?(x).((in[x]?(y).out[x]!(y+1).0)|INC))\\
&= \textstyle\sum_{v\in\mathcal{V}} inc_v?.((\sum_{u\in\mathcal{V}} in_{\ddot{v},u}?.out_{\ddot{v},u\ddot{+}1}!.0)|\mathcal{B}(INC))\\
&= inc_1?.((in_{1,1}?.out_{1,2}!.0 + in_{1,2}?.out_{1,3}!.0+\cdots)|\mathcal{B}(INC))+\\
&\quad\ inc_2?.((in_{2,1}?.out_{2,2}!.0 + in_{2,2}?.out_{2,3}!.0+\cdots)|\mathcal{B}(INC))+\cdots
\end{aligned}$$

$$\begin{aligned}
\mathcal{B}(CALLER23) &= \mathcal{B}(inc!(23).in[23]!(7).out[23]?(z).TASK23(z))\\
&= inc_{23}!.in_{23,7}!.(\textstyle\sum_{v\in\mathcal{V}} out_{23,v}?.TASK23_v)
\end{aligned}$$

In this case, $TASK23_8$ will be choiced through an event $out_{23,8}?$.

5.3 An example in CCB

We show an example of a *distributed used-car information system.* This system consists of many agents which have information about used-cars and give the information to car-dealers, described as follows:

$$CARINFO \stackrel{\text{def}}{=} (INFO1|INFO2|\cdots|DEALER1|DEALER2|\cdots)$$

For example, information agents about PORSCHE-928 are defined as follows:

$$\begin{aligned}
INFO45 &\stackrel{\text{def}}{=} porsche928?(x).(info[x]!(\text{`1985.RED.52940km.S34'})|INFO45)\\
INFO81 &\stackrel{\text{def}}{=} porsche928?(x).(info[x]!(\text{`1991.BLUE.21380km.S16'})|INFO81)
\end{aligned}$$

A postfix $[x]$ is used for determining a destination of information as used in INC. A car-dealer wants to gather all information about a car, though he does not know the number of agents with the car information. Therefore he broadcasts an event for the car. For example, a car-dealer may be defined as follows:

$$\begin{aligned}
DEALER7 &\stackrel{\text{def}}{=} porsche928!\langle x\rangle(7).LP(x,\texttt{nil})\\
LP(n,list) &\stackrel{\text{def}}{=} [\![n\neq 0]\!]info[7]?(y).LP(n-1,y::list) + [\![n=0]\!]PRINT(list)
\end{aligned}$$

$DEALER7$ querys about PORSCHE-928 through an event $porsche928$, then gets the number of agents with information about PORSCHE-928. Two variables n and $list$ of $LP(n,list)$ are initially bound to the number of the information agents and nil, respectively. When $LP(n,list)$ receives an event $info[7]$, y is bound to car information carried by $info[7]$. Then y is concatenated into $idlist$ and n is decreased by 1. If n is zero, then $list$ is printed out by an agent $PRINT$, because $LP(n,list)$ has already gathered all information about PORSCHE-928.

6 Equivalence and congruence relations

We want observational congruence relations in Core-CCB like in CCS. It is important that *observation congruence* $=^{[2]}$ defined in CCS is *not* a congruence relation in Core-CCB as shown in the following example:

$$P_1 = P_2, \qquad P_1|Q \neq P_2|Q$$

where each component is defined as follows:

$$P_1 \stackrel{\text{def}}{=} a??^1.b!^0.P_1, \quad P_2 \stackrel{\text{def}}{=} a??^1.P_2, \quad Q \stackrel{\text{def}}{=} a!!^0.out0!^1.Q + a!!^1.out1!^1.Q$$

P_1 and P_2 are observation congruent because the silent event $b!^0$ is ignored, but $P_1|Q$ and $P_2|Q$ are not observation congruent because P_1 sometimes fails to receive the event a broadcasted by Q while P_2 can always receive it. Therefore we define a slightly stronger observational congruence relation in this paper than observation congruence. We prepare several notations for the definition.

The set $Event^*$, ranged over by $s, t, \cdots$, is a set of event sequences including an empty sequence ε, and if $t = a_1(\theta_1)^{n_1} \cdots a_k(\theta_k)^{n_k} \in Event^*$, then we write $E \stackrel{t}{\longrightarrow} E'$ if $E \stackrel{a_1(\theta_1)^{n_1}}{\longrightarrow} \cdots \stackrel{a_k(\theta_k)^{n_k}}{\longrightarrow} E'$. Especially, if $E \stackrel{\varepsilon}{\longrightarrow} E'$ then $E' \equiv E$. ($\equiv$ means syntactic identity.) About transitions by silent events, we sometimes write $E \stackrel{\tau}{\longrightarrow} E'$ if $E \stackrel{a!^0}{\longrightarrow} E'$ for *some* event name a. We now define a new transition relation as follows:

Definition 6.1 *If $t = a_1(\theta_1)^{n_1} \cdots a_k(\theta_k)^{n_k} \in Event^*$, then $E \stackrel{t}{\Longrightarrow} E'$ if*

$$E(\stackrel{\tau}{\longrightarrow})^* \stackrel{a_1(\theta_1)^{n_1}}{\longrightarrow} (\stackrel{\tau}{\longrightarrow})^* \cdots (\stackrel{\tau}{\longrightarrow})^* \stackrel{a_k(\theta_k)^{n_k}}{\longrightarrow} (\stackrel{\tau}{\longrightarrow})^* E'$$

where $(\stackrel{\tau}{\longrightarrow})^*$ means zero or more transitions by silent events. □

Then we define an observational bisimulation relation in Core-CCB like *weak bisimulation* in CCS.

Definition 6.2 *A binary relation $\mathcal{S} \subseteq \mathcal{P} \times \mathcal{P}$ over agents is a* weak monitor bisimulation *if $(P, Q) \in \mathcal{S}$ implies, for all $a\theta^n \in Event$, that*

(*i*) *whenever $P \stackrel{a\theta^n}{\longrightarrow} P'$ then for some $Q', Q \stackrel{\widehat{a\theta^n}}{\Longrightarrow} Q'$ and $(P', Q') \in \mathcal{S}$,*
(*ii*) *whenever $Q \stackrel{a\theta^n}{\longrightarrow} Q'$ then for some $P', P \stackrel{\widehat{a\theta^n}}{\Longrightarrow} P'$ and $(P', Q') \in \mathcal{S}$,*
(*iii*) *whenever $a \notin mon(P)$ then, for some Q', Q'',*
$$Q \Rightarrow Q' \Rightarrow Q'',\ a \notin mon(Q'),\ \text{and}\ (P, Q'') \in \mathcal{S},$$
(*iv*) *whenever $a \notin mon(Q)$ then, for some P', P'',*
$$P \Rightarrow P' \Rightarrow P'',\ a \notin mon(P'),\ \text{and}\ (P'', Q) \in \mathcal{S}.$$

where $\hat{t}$ is the event sequence gained by deleting all occurrences of silent events from t. □

Weak monitor bisimulations are obtained from weak bisimulations by adding the conditions (*iii*) and (*iv*). A relation is defined by using weak monitor bisimulations.

Definition 6.3 *P and Q are* weakly monitor equivalent, *written* $P \approx_m Q$, *if* $(P, Q) \in \mathcal{S}$ *for some weak monitor bisimulation* $\mathcal{S}$. □

Weak monitor equivalence is the *largest* relation preserved by a parallel combinator | and included in observation equivalence ≈ defined in CCS, as shown in the following proposition.

Proposition 6.1 $P_1 \approx_m P_2$ **iff** *for any* Q, $P_1|Q \approx P_2|Q$ □

We show an interesting equation in the following proposition.

Proposition 6.2 $P_1 + P_2 + a!^0.(P_1 + a!^0.(Q + P_2)) \approx_m P_1 + a!^0.(Q + P_2)$ □

This equation shows a property of weak monitor equivalence, because this equation does not held for *strong monitor equivalence* defined in Definition 6.5.

We notice a special case $P \equiv P' \equiv P''$ and $Q \equiv Q' \equiv Q''$ in Definition 6.2, and define strong monitor bisimulations as follows:

Definition 6.4 *A binary relation* $\mathcal{S} \subseteq \mathcal{P} \times \mathcal{P}$ *over agents is a* strong monitor bisimulation *if* $(P, Q) \in \mathcal{S}$ *implies, for all* $a\theta^n \in Event$, *that*

(*i*) *whenever* $P \xrightarrow{a\theta^n} P'$ *then for some* $Q', Q \overset{\widehat{a\theta^n}}{\Longrightarrow} Q'$ *and* $(P', Q') \in \mathcal{S}$,

(*ii*) *whenever* $Q \xrightarrow{a\theta^n} Q'$ *then for some* $P', P \overset{\widehat{a\theta^n}}{\Longrightarrow} P'$ *and* $(P', Q') \in \mathcal{S}$,

(*iii*) $mon(P) = mon(Q)$.

□

A relation is defined by using strong monitor bisimulations like weak monitor equivalence.

Definition 6.5 *P and Q are* strongly monitor equivalent, *written* $P \simeq_m Q$, *if* $(P, Q) \in \mathcal{S}$ *for some strong monitor bisimulation* $\mathcal{S}$. □

It is helpful to consider strong monitor equivalence before weak monitor equivalence, because the definition of strong monitor equivalence is simpler than one of weak monitor equivalence and $P \simeq_m Q$ implies $P \approx_m Q$. In the rest of this paper we consider only strong monitor equivalence.

Strong monitor equivalence is an equivalence relation but is not a congruence relation because it is not preserved by a choice combinator +. Therefore we define a congruence relation from strong monitor equivalence by adding the weakest conditions. Before the definition of the congruence relation, an equivalence relation $\doteq$ over events is introduced for ignoring difference between many silent events.

Definition 6.6 Event equivalence $\doteq$ *is a binary relation over events defined as the following set:* $\{(a!^0, b!^0) : a, b \in \mathcal{N}\} \cup \{(a\theta^n, a\theta^n) : a\theta^n \in Event\}$. □

Then we define *strong monitor congruence* as follows:

Definition 6.7 *P and Q are strongly monitor congruent, written $P \cong_m Q$, if for all $a\theta^n \in Event$*

(i) whenever $P \xrightarrow{a\theta^n} P'$ then for some $Q', b, a\theta^n \doteq b\theta^n, Q \stackrel{b\theta^n}{\Longrightarrow} Q'$, $P' \simeq_m Q'$,
(ii) whenever $Q \xrightarrow{a\theta^n} Q'$ then for some $P', b, a\theta^n \doteq b\theta^n, P \stackrel{b\theta^n}{\Longrightarrow} P'$, $P' \simeq_m Q'$,
(iii) $mon(P) = mon(Q)$. □

We now extend strong monitor congruence over agents to agent expressions.

Definition 6.8 *Let agent expressions E and F contain agent variables $X_1, \cdots, X_n$ (written by $\tilde{X}$) at most. Then $E \cong_m F$ if, for all indexed sets $\tilde{P}$ of agents, $E\{\tilde{P}/\tilde{X}\} \cong_m F\{\tilde{P}/\tilde{X}\}$.*[6] □

Most of propositions for observation congruence are also held for strong monitor congruence with slight modifications. Several propositions are shown below.

Proposition 6.3 *Strong monitor congruence is a congruence relation.* □

Proposition 6.4 $P \simeq_m Q$ **iff**
$$((P \cong_m Q) \text{ or } (P \cong_m a!^0.Q + Q) \text{ or } (a!^0.P + P \cong_m Q))$$ □

Proposition 6.4 is used for strengthening $\simeq_m$ to $\cong_m$, for example, in the proof of Theorem 6.6. We show two notions defined in [2].

Definition 6.9 *X is sequential in E if every subexpression of E which contains X, apart from X itself, is of the form $a\theta^n.F$ or $\sum \tilde{F}$.* □

Definition 6.10 *X is guarded in E if each occurrence of X is within some subexpression of E form $a\theta^n.F$ such that $a\theta^n$ is not a silent event.* □

Then we give a proposition which guarantees existence of unique solution P such that $P \cong_m E\{P/X\}$ under a certain condition on the agent expression E. The solution is naturally the agent A defined by $A \stackrel{\text{def}}{=} E\{A/X\}$. Thus this proposition is useful for proving whether two agents with recursive definition are strongly monitor congruent or not.

Proposition 6.5 *Let agent expressions $\tilde{E}$ contain the variables $\tilde{X}$ at most, and let $\tilde{X}$ are guarded and sequential in $\tilde{E}$. Then if $\tilde{P} \cong_m \tilde{E}\{\tilde{P}/\tilde{X}\}$ and $\tilde{Q} \cong_m \tilde{E}\{\tilde{Q}/\tilde{X}\}$ then $\tilde{P} \cong_m \tilde{Q}$.* □

In the rest of this section, we give an axiom system for finite agents which contains no agent constants (no recursions). We define an axiom system $\mathcal{A}$.

Definition 6.11 *We write $\mathcal{A} \vdash P = Q$ if the equality of two agents P and Q can be proven by equational reasoning from an axiom system $\mathcal{A}$, where the axiom system $\mathcal{A}$ consists of the following equations:*

M1 $P_1 + P_2 = P_2 + P_1$
M2 $(P_1 + P_2) + P_3 = P_1 + (P_2 + P_3)$
M3 $P = P + P$
M4 $P = P + 0$

[6] $\{\tilde{P}/\tilde{X}\}$ means a simultaneous substitution of an agent P_i into an agent variable X_i.

E1 $P|Q \equiv (\sum_{(i\in I_1)} a_i(\theta_i)^{m_i}.P_i)|(\sum_{(i\in I_2)} b_i(\phi_i)^{n_i}.Q_i)$
$= \sum_{(i\in I_1)}\{a_i(\theta_i)^{m_i}.(P_i|Q) : \theta_i \in \{!,?\} \textit{ or } a_i \notin mon(Q)\}$
$+ \sum_{(i\in I_2)}\{b_i(\phi_i)^{n_i}.(P|Q_i) : \phi_i \in \{!,?\} \textit{ or } b_i \notin mon(P)\}$
$+ \sum_{(i\in I_1)} \sum_{(j\in I_2)}\{a_i(\theta_i)^{(m_i-n_j)}.(P_i|Q_j) : a_i{=}b_j, \theta_i \in \{!,!!\}, \phi_j{=}@(\theta_i), m_i \geq n_j\}$
$+ \sum_{(i\in I_1)} \sum_{(j\in I_2)}\{b_j(\phi_j)^{(n_j-m_i)}.(P_i|Q_j) : a_i{=}b_j, \phi_j \in \{!,!!\}, \theta_i{=}@(\phi_j), n_j \geq m_i\}$
$+ \sum_{(i\in I_1)} \sum_{(j\in I_2)}\{a_i(\theta_i)^{(m_i+n_j)}.(P_i|Q_j) : a_i = b_j, \theta_i = \phi_j \in \{?,??\}\}$
E2 $(\sum_{(i\in I)} a_i(\theta_i)^{n_i}.P_i)\backslash L = \sum_{(i\in I)}\{a_i(\theta_i)^{n_i}.(P_i\backslash L) : a_i \notin L\}$
$+ \sum_{(i\in I)}\{a_i!^0.(P_i\backslash L) : a_i \in L, \theta_i \in \{!,!!\}, n_i = 0\}$
E3 $(\sum_{(i\in I)} a_i(\theta_i)^{n_i}.P_i)[S] = \sum_{(i\in I)} S(a_i)(\theta_i)^{n_i}.(P_i[S])$
T1 $a\theta^n.(b!^0.P + P) = a\theta^n.P$
T2 $P + R + a!^0.(P + Q) = R + a!^0.(P + Q)$ *if* $mon(P) \subseteq mon(R)$
T3 $a\theta^n.(P + b!^0.Q) + a\theta^n.Q = a\theta^n.(P + b!^0.Q)$
T4 $a!^0.P = b!^0.P$ □

M1-4, **E1-3**, and **T1-4** correspond to monoid laws, expansion laws, and τ laws in CCS, respectively. Finally we show a theorem which means that $\mathcal{A}$ is a sound and complete axiom system for strong monitor congruence of finite agents.

Theorem 6.6 *Let* P *and* Q *are finite. Then* $P \cong_m Q$ **iff** $\mathcal{A} \vdash P = Q$. □

7 Related work

CBS has already been proposed for broadcasting systems. To introduce broadcast communication into process algebras, negative transitions by receive-events are important in general. In CBS transitions by special events (actions) named *discards* are used instead of the negative transitions. The advantage of using the discards instead of the negative transitions is to be able to use a synchronization algebra[12].

We adopt *monitor function* instead of the negative transitions, because we want to use monitor function as conditions in equations. For example, monitor function is used in equation **T2** in the axiom system $\mathcal{A}$. Monitor function causes the same effect as the discards. The important difference between CCB and CBS is the received number. In CCB a broadcaster of an event can know the number of receivers of the event after broadcasting.

Process algebras with notion of time have been proposed, for example TCCS[10]. In TCCS broadcast communication may be simulated by using a lot of loops with timeout. Thus for *development* of programming languages with broadcast communication, such timed process algebras are very useful. On the other hand, we want to *use* such languages which have been already developed, and to *simply* analyze programs in the languages. CCB is appropriate for this purpose.

We show [9] as another study of broadcast communication by existing process algebras. In [9] a translation from CBS to SCCS is presented, and the relation between CBS and SCCS is clarified. We want to also clarify relation between CCB and SCCS.

8 Conclusion and future works

In order to flexibly connect agents, one of the authors has developed a mechanism named VIABUS which has a software bus architecture with *broadcast communication.* Then π-calculus has been introduced into VIABUS[13] as a common communicating language to control agents written in different languages, but there are problems about describing broadcast communication in π-calculus.

In this paper we have proposed a process algebra CCB. The most important property of CCB is that a broadcaster of an event can know the number of receivers of the event after broadcasting. Then we have defined an observational congruence relation named *strong monitor congruence*, and have given a sound and complete axiom system for strong monitor congruence of finite agents. We consider *weak monitor equivalence* as the next subject.

In the future we shall extend CCB with a notion of space distance.

Acknowledgement

The authors wish to express our gratitude to Dr. Kimihiro Ohta, Director of Computer Science Division, ETL. They also thank Dr. Yoshiki Kinoshita in Computer Language Section and all colleagues in Information Base Section for their helpful discussions.

References

1. J.Coutaz, "Architecture Model for Interactive Software: Failures and Trends", Proc. of the IFIP TC 2/WG 2.7 Working Conference on Engineering for Human-Computer Interaction, pp.137 - 153, 1989.
2. R.Milner, "Communication and Concurrency", Prentice-Hall, 1989.
3. R.Milner, J.Parrow and D.Walker, "A Calculus of Mobile Processes, I and II", Information and Computation, 100, pp.1 - 40 and pp.41 - 77, 1992.
4. R.Milner, "Calculi for Synchrony and Asynchrony", Journal of Theoretical Computer Science, Vol.25, pp.267 - 310, 1983.
5. R.de Simone, "Higher-level Synchronizing Devices in Meije-SCCS", Journal of Theoretical Computer Science, Vol.37, pp.245 - 267, 1985.
6. C.A.R.Hoare,"Communicating Sequential Processes", Prentice-Hall, 1985.
7. K.V.S.Prasad, "A Calculus of Broadcasting Systems", TAPSOFT'91, Vol.1:CAAP, LNCS 493, Springer-Verlag, pp.338 - 358, 1991
8. K.V.S.Prasad, "A Calculus of Value Broadcasts", PARLE'93, LNCS 694, Springer-Verlag, pp.391 - 402, 1993
9. Uno Holmer, "Interpreting Broadcast Communication in SCCS", CONCUR'93, LNCS 715, Springer-Verlag, pp.188 - 201, 1993
10. F.Moller and C.Tofts,"An overview of TCCS", Proc. of EUROMICRO'92, Athens, June 1992.
11. L.Aceto, B.Bloom, and F.Vaandrager "Turning SOS Rules into Equations", Proc. 7th Annual IEEE Symposium on Logic in Computer Science, pp.113 - 124, 1988.
12. G.Winskel, "Synchronization trees", Journal of Theoretical Computer Science, Vol.34, pp.33 - 82, 1984.
13. Y.Sato and K.Ohmaki "A Flexible Inter-Agent Connection Scheme for Interactive Software", IPSJ Technical Report, SE89-7, pp.49 - 56, 1992.

Symbolic Timing Devices

Anne Bergeron*
LACIM, Université du Québec à Montréal

and

Department of Electrical Engineering,
Princeton University, Princeton, New-Jersey, USA 08544-5263
email: anne@lacim.uqam.ca

Abstract

Timing devices such as timers, clocks, or stopwatches, are used in a vast range of processes. In computer science, the need to specify, verify and implement real-time applications had given rise to many different formalizations of timed concurrent processes. This paper is an attempt to understand the underlying ideas of many of these approaches by focusing primarily on the timing devices. Starting with an abstract definition of a *timer*, we use the formalism of synchronized products, as developed by Arnold and Nivat, to study different formal languages associated with the concurrent operation of n timers.

1. Introduction

Formal descriptions of real-time systems have been extensively investigated in the recent years. Researchers have explored several techniques of incorporating timing constraints to qualitative models of discrete systems, such as automata or Petri nets. Many early techniques are based on introducing discrete *time actions* - or *tick events* - directly into the model. By treating the timing events at the same level as other events in the system, this approach allows the definition of associated formal languages in a natural and elegant way.

However, the necessity of expressing continuous properties of timing constraints lead to models allowing for *dense time domains* [AD90, NSY93, AD94]. In these models, a subset of $\mathbb{R}^n$ - the possible *clock values* - is associated to each state of a finite automaton describing a process. Transitions in this hybrid system are constrained by both the state of the automaton, and particular clock values. When the time constants used to define the sets of clock values are rationally related, the infinite transition system admits a finite quotient that recognizes the set of behaviors of the process consistent with the timing constraints. Working with classes of processes exhibiting regularities of timing constraints placement, [LB93] shows that the restriction to rationally related values can be removed.

In this paper, we explore the possibility of describing timing constraints *without* reference to a particular process, or to specific time constants. In order to do this, we investigate the sets of behaviors of n concurrent timers t_i, each of which can be reset, or compared to a real constant d_i. The basic problem is the following: Given constants $d_1, ..., d_n$, what can be said about the rationality of the set of behaviors of n timers whose operations are consistent with $d_1, ..., d_n$? If only partial information is known about the delays, is the set still recognizable? For most problems involving two timers, the answer is positive. Partial results, both positive and negative, are given for more than two timers.

* This work was partially supported by grants from BNR Ltd., FCAR of Québec and NSERC of Canada.

The main benefit of this study is to obtain general results about timing constraints that are independent of a particular process. Furthermore, since the concurrent operation of n timers is viewed as processes represented by finite automata, synchronizing additional real processes is straightforward: we thus recover the simplicity and elegance that was lost in the transition from discrete to dense domains.

Section 2 gives the basic definition of a timer and some useful constructions associated to the operation of one timer. Section 3 introduces the synchronized product of n timers, and develops some ideas on specification and verification that can be used to design real-time processes. In section 4, we study various formal languages associated with the concurrent operation of n timers.

2. The Basic Timer

A wide range of devices are used for measuring delays between events. However, most of them share basic features with the elementary timer of Figure 1. Since we are not particularly interested in the amount of sand, or its flowing rate, we will distinguish only two states: **Inactive**, and **Running**. When the timer is inactive, we can start it by turning it upside down. This event will be denoted **S**. After a certain amount of time d, the sand will stop running. This time out event will be denoted **T**. The events **S** and **T** are obvious transitions between the states **Inactive** and **Running**. For synchronization purposes, we will also consider null events **e**, that loop on each state. Null events can be interpreted as synchronized with external events occurring while the timer is either inactive or running.

Figure 1: An Analog Timer

Definition 2.1 A *timer* $\mathbb{T}_d$ is given by a *delay* $d \in \mathbb{R}^+$, and by the following automata $\mathbb{T}$ on the event set $\Sigma = \{$ **S**, **T**, **e**$\}$.

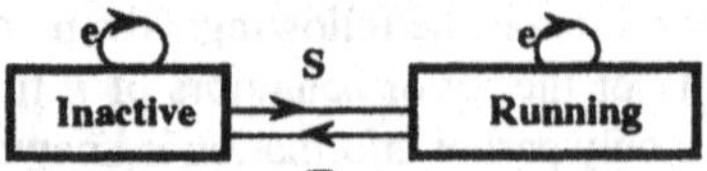

The initial state of the automata is **Inactive**, and the language recognized by the automata is denoted $\mathbb{L}$. Both states are considered to be final. □

The language $\mathbb{L}$ contains the logical sequences of actions that can executed, for example:

eSeeTeSeT

The basic intuition about timers is that if an action is synchronized with the start event **S** and if a second action is executed after the time out **T**, the delay between the two actions will be greater than the delay of the timer; and if the second action is executed while the timer is running, the delay will be less than the delay of the timer. These intuitions are captured by the following definition:

Definition 2.2 Let $y \in \mathbb{L}$, a *stamping* of y with respect to a delay d is a function S on the set $\mathcal{P}(y)$ of prefixes of the sequence y:

$$S : \mathcal{P}(y) \longrightarrow \mathbb{R}$$

such that the 3 following axioms hold:

A0)	$S(x\sigma) > S(x)$	for any event σ,
A1)	$S(x\mathbf{S}\mathbf{e}^*\mathbf{T}) - S(x\mathbf{S}) = d$,	
A2)	$S(x\mathbf{S}\mathbf{e}^*) - S(x\mathbf{S}) < d$.	□

Note that axiom **A2** is necessary to prevent the timer from staying forever in the **Running** state. A stamping of y can be described by labeling point of $\mathbb{R}$ with events in y. For example, the following labeling is a stamping of the sequence **eSeeTeSeT** with respect to delay $d = 2$:

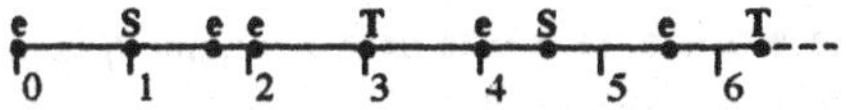

Null events have lots of flexibility concerning stamping. It is possible add to remove **e** events to a sequence without modifying its stamping in an essential way. If $y \in \mathbb{L}_D$, we define $\pi_{\mathbf{e}}(y)$ to be the sequence obtained by removing **e** events from y. Then any stamping S of y induces a stamping T of $\pi_{\mathbf{e}}(y)$ by setting $T(x') = S(x)$ where x is the shortest prefix of y such that $\pi_{\mathbf{e}}(x) = x'$. The following proposition describe how to add **e** events to a sequence:

Proposition 2.3 If S is a stamping of $\pi_{\mathbf{e}}(y)$, then there exists a stamping T of y such that S is the stamping induced by T on $\pi_{\mathbf{e}}(y)$.

Proof:
It suffices to show how to add one **e** event to a sequence y. Adding an **e** event before an event σ in y is easy: suppose that y can be written as $x\sigma z$, and that S is a stamping of y, then we can extend S to $x\mathbf{e}\sigma z$, by setting $S(x\mathbf{e}) = S(x) + \alpha[S(x\sigma) - S(x)]$ for any real value $\alpha \in (0, 1)$. Adding an **e** event to the end of the sequence y is slightly more delicate. If y is of the form $x\mathbf{T}\mathbf{e}^*$, then any value greater than $S(y)$ can be assigned to $S(y\mathbf{e})$. If y is of the form $x\mathbf{S}\mathbf{e}^*$, then we set $S(y\mathbf{e}) = S(y) + \alpha[S(x\mathbf{S}) + d - S(y)]$, for any $\alpha \in (0, 1)$. □

A concept closely related to the stamping of a sequence y, is the *reading* of a timer after the sequence of events y. The reading of a timer corresponds, in physical terms, to the amount of sand in the lowest bulb of the hourglass. When the timer is inactive, the reading is arbitrarily set to d.

Definition 2.4 If $y \in \mathbb{L}$ and if S is a stamping of y, the *reading* $R_S(y)$ after y is defined by:

if $y = x\mathbf{S}\mathbf{e}^*$ then $R_S(y) = S(x\mathbf{S}\mathbf{e}^*) - S(x\mathbf{S})$
else $R_S(y) = d$. □

3. Products of Timers

The theory of a single timer has limited interest. Complexity arises when several timers are used to synchronize different processes. To describe n timers functioning together we will record the n different states of the timers, and consider all transitions between these states that are compatible with transitions of each individual timer.

Definition 3.1 The product $\mathbb{T}^n$ is the free product of n copies of the automata $\mathbb{T}$, whose states are vectors of the form:

$$s = (s_1, ..., s_n) \text{ where } s_i \in \{ \textbf{Inactive}, \textbf{Running}\}$$

with initial state (**Inactive**, ..., **Inactive**), and whose transitions between states s and t are vectors:

$$\sigma = \begin{bmatrix} \sigma_1 \\ ... \\ \sigma_n \end{bmatrix} \text{ where } \sigma_i \text{ is a transition between } s_i \text{ and } t_i \text{ in } \mathbb{T}.$$

□

Example 3.2 The following automaton is a representation of $\mathbb{T}^2$.

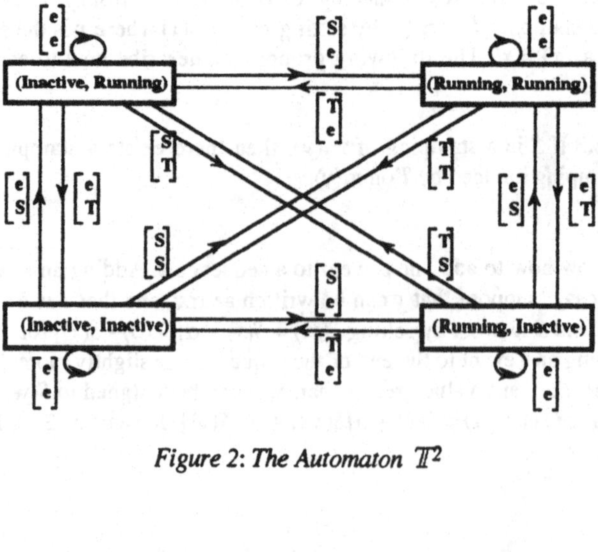

Figure 2: The Automaton $\mathbb{T}^2$

In general, the automaton $\mathbb{T}^n$ can be drawn in the n-dimensional cube. These automata are fundamental: they will be the building blocks of automata recognizing rational languages related to the concurrent functioning of n timers. □

Starting from the initial state, we define $\mathbb{L}^n$ to be the language recognized by the automaton $\mathbb{T}^n$. If y is a sequence in $\mathbb{L}^n$, then for each i, we denote by $p_i(y)$ the sequence executed by the ith timer. Let $D = (d_1, ..., d_n) \in (\mathbb{R}^+)^n$ represents the delays of the n timers, then some sequences in $\mathbb{L}^n$ are 'physically' impossible. For example, in the sequence

$$\begin{bmatrix} S \\ e \end{bmatrix}\begin{bmatrix} e \\ S \end{bmatrix}\begin{bmatrix} e \\ T \end{bmatrix}\begin{bmatrix} T \\ e \end{bmatrix}$$

timer 2 was presumably started after timer 1, and timed out before. Such a sequence of actions seems possible only if $d_2 < d_1$. On the other hand, the sequence:

$$\begin{bmatrix} S \\ e \end{bmatrix}\begin{bmatrix} e \\ S \end{bmatrix}\begin{bmatrix} e \\ T \end{bmatrix}\begin{bmatrix} T \\ e \end{bmatrix}\begin{bmatrix} e \\ S \end{bmatrix}\begin{bmatrix} S \\ e \end{bmatrix}\begin{bmatrix} T \\ e \end{bmatrix}\begin{bmatrix} e \\ T \end{bmatrix}$$

is always impossible since it would imply that $d_2 < d_1$ and $d_1 < d_2$.

Given $D = (d_1, ..., d_n)$, we will be interested in the language $\mathbb{L}_D$ of *possible* sequences.

Definition 3.3 Let $D \in (\mathbb{R}^+)^n$, and $y \in \mathbb{L}^n$. A sequence $y \in \mathbb{L}^n$ is *possible* if there exists a function S defined on the prefixes $\mathcal{P}(y)$ of y:

$$S : \mathcal{P}(y) \longrightarrow \mathbb{R}$$

such that S is a stamping with respect to d_i for each component $p_i(y)$ of y. Such a function will also be called a *stamping*. The set of possible sequences with respect to D is denoted $\mathbb{L}_D$. If S is a stamping, then each timer has a reading after $p_i(y)$ denoted $R_{S_i}(y)$. □

The basic problem is to construct automata that recognize $\mathbb{L}_D$. In order to do this, we define an equivalence relation $y \equiv y'$ on the sequences of $\mathbb{L}_D$ such that:

(1) If $y \equiv y'$ then $y\sigma$ is possible iff $y'\sigma$ is possible, and $y\sigma \equiv y'\sigma$.
(2) The number of classes is finite.

This equivalence relation will be based on the set of *possible readings* after a sequence y is executed, and before any significant event occurs. A significant event is a vector that contains at least one non-null event. For convenience, we will denote by **e**, the vector containing only e events:

$$\begin{bmatrix} e \\ e \\ \cdots \\ e \end{bmatrix}$$

Definition 3.4 Let $y \in \mathbb{L}_D$. The set $PR(y)$ of *possible readings* after y is a subset of the rectangle $(0, d_1] \times ... \times (0, d_n]$ defined by:

$$PR(y) = \{ (Rs_1(y\mathbf{e}), ..., Rs_n(y\mathbf{e})) \mid \text{for all stampings } S \text{ of } y\mathbf{e} \}.$$

We say that $y \equiv y'$ if and only if $PR(y) = PR(y')$. □

The following proposition says that null events of y are irrelevant in computing $PR(y)$:

Proposition 3.5 If $\pi_\mathbf{e}(y) = \pi_\mathbf{e}(y')$, then $y \equiv y'$.

Proof:
Let S be a stamping of $y\mathbf{e}$, then S induces a stamping T on $\pi_\mathbf{e}(y\mathbf{e})$, which is also a stamping of $\pi_\mathbf{e}(y'\mathbf{e})$. We now extend this stamping to the null events of $y'\mathbf{e}$ by setting $T(y'\mathbf{e}) = S(y\mathbf{e})$, and by applying the first construction of Proposition 2.3. Now, since $y\mathbf{e}$ and $y'\mathbf{e}$ both contain the same non-null events, $R_{T_i}(y'\mathbf{e})$ are $R_{S_i}(y\mathbf{e})$ either both d_i, or less than d_i. In the latter case, $p_i(y\mathbf{e})$ is of the form $x\mathbf{Se}^*\mathbf{e}$ and $p_i(y'\mathbf{e})$ is of the form $x'\mathbf{Se}^*\mathbf{e}$ with $\pi_\mathbf{e}(x) = \pi_\mathbf{e}(x')$. This implies that $T(x'\mathbf{S}) = S(x\mathbf{S})$ and we have

$$R_{T_i}(y'\mathbf{e}) = T(x'\mathbf{Se}^*\mathbf{e}) - T(x'\mathbf{S}) = S(x\mathbf{Se}^*\mathbf{e}) - S(x\mathbf{S}) = R_{S_i}(y\mathbf{e}).$$

□

The relation $PR(y) = PR(y')$ says that for each possible execution of the sequence of y, there exists an execution of the sequence y' that leaves the timers in the same (continuous) state. It thus seems natural that each possible event that can occur after y will also be possible after y'. This is formalized in the following:

Theorem 3.6 If y and y' are in $\mathbb{L}_D$, and if $y \equiv y'$, then
(1) $y\sigma \in \mathbb{L}_D$ if and only if $y'\sigma \in \mathbb{L}_D$,
and (2) $y\sigma \equiv y'\sigma$.

Proof:
(1) Suppose y and y' are in $\mathbb{L}_D$, and $PR(y) = PR(y')$. Let $y\sigma \in \mathbb{L}_D$, and S be a stamping of $y\sigma$. We first extend the stamping S to the sequence $y\mathbf{e}\sigma$. This stamping defines readings $(Rs_1(y\mathbf{e}), ..., Rs_n(y\mathbf{e}))$. Since $PR(y) = PR(y')$, there exists a stamping T of $y'\mathbf{e}$ that yields the same readings after $y'\mathbf{e}$. Now, we define

$$T(y'\mathbf{e}\sigma) = T(y'\mathbf{e}) + [S(y\mathbf{e}\sigma) - S(y\mathbf{e})].$$

Since T is a stamping up to $y'\mathbf{e}$, we have only to show that the new value $T(y'\mathbf{e}\sigma)$ yields a stamping for each component $p_i(y'\mathbf{e}\sigma)$.

Axiom **A0** is obvious. Suppose now that $p_i(y'\mathbf{e}\sigma)$ is of the form $x'\mathbf{Se}^*\mathbf{eT}$, then $p_i(y\mathbf{e}\sigma)$ must be of the form $x\mathbf{Se}^*\mathbf{eT}$ since the ith component of σ is the $\mathbf{T}$ event. We have, by direct computation,

$$\begin{aligned} T(x'\mathrm{Se^*eT}) - T(x'\mathrm{S}) &= [T(x'\mathrm{Se^*eT}) - T(x'\mathrm{Se^*e})] + [T(x'\mathrm{Se^*e}) - T(x'\mathrm{S})] \\ &= [S(x\mathrm{Se^*eT}) - S(x\mathrm{Se^*e})] + R_{T_i}(y'\mathrm{e}) \\ &= [S(x\mathrm{Se^*eT}) - S(x\mathrm{Se^*e})] + R_{S_i}(y\mathrm{e}) \\ &= [S(x\mathrm{Se^*eT}) - S(x\mathrm{Se^*e})] + [S(x\mathrm{Se^*e}) - S(x\mathrm{S})] \\ &= S(x\mathrm{Se^*eT}) - S(x\mathrm{S}) = d. \end{aligned}$$

If $p_i(y'\mathrm{e}\sigma)$ is of the form $x'\mathrm{Se^*ee}$, we get $T(x'\mathrm{Se^*ee}) - T(x'\mathrm{S}) = S(x\mathrm{Se^*ee}) - S(x\mathrm{S}) < d$ with a similar computation. Thus T is a stamping of $y'\mathrm{e}\sigma$, and consequently of $y'\sigma$. Note that the equality of readings is preserved since $R_{T_i}(y'\mathrm{e}\sigma) = R_{S_i}(y\mathrm{e}\sigma)$, thus $R_{T_i}(y'\sigma) = R_{S_i}(y\sigma)$.

(2) Let S be a stamping of $y\sigma\mathrm{e}$ with associated readings $(R_{S_1}(y\mathrm{e}), \ldots, R_{S_n}(y\mathrm{e}))$. Let T be the stamping of $y'\sigma$ defined in the preceding paragraph, and define:

$$T(y'\sigma\mathrm{e}) = T(y'\sigma) + [S(y\sigma\mathrm{e}) - S(y\sigma)]$$

It is again a direct computation to show that T is a stamping, and that $R_{T_i}(y'\sigma\mathrm{e}) = R_{S_i}(y\sigma\mathrm{e})$. □

Modeling timing constraints

The automaton $\mathbb{T}^n$ can be used to describe the behavior of timed processes by synchronizing actions on a process with transitions of $\mathbb{T}^n$. This is done by specifying a set V of *synchronization vectors* defining which actions of the process can occur simultaneously with transitions of $\mathbb{T}^n$. The resulting automaton is called the *synchronized product* with respect to V (see, for example, [A94]).

Consider, for example, a simple process in which actions a and b alternate.

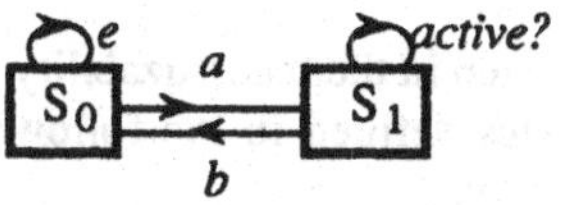

Specifying that this process is active for at least a period d can be done with synchronization with a simple timer $\mathbb{T}_d$. The first component of the synchronization vectors specify a transition of the timer, and the second, a transition of the process, yielding:

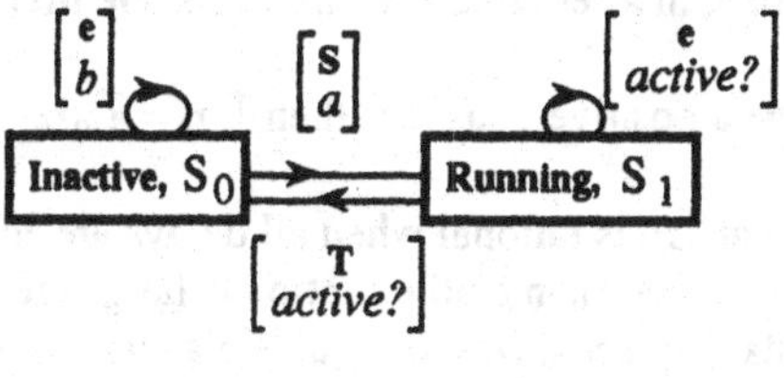

A more complex specification would be, for example, that action b occurs within d_1 of action a, and two occurrences of action a are always distant by a delay greater than d_2. This requires two timers, and can be represented by the following automaton.

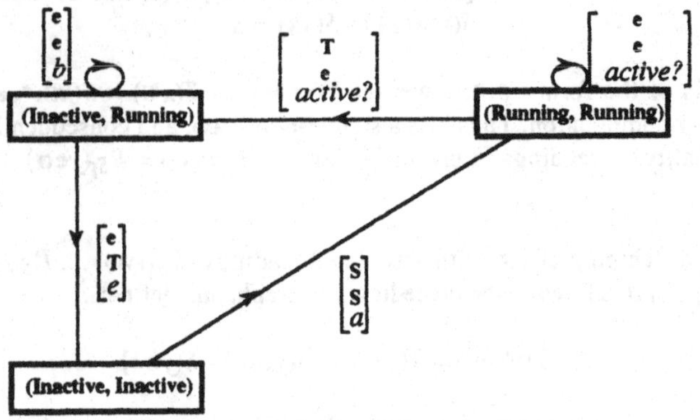

In this case, setting initial and final states to (Inactive, Inactive) yields the following language $\mathbb{L}_V$ associated with the process is the following:

$$\left(\begin{bmatrix}S\\S\end{bmatrix}\begin{bmatrix}e\\e\end{bmatrix}^*\begin{bmatrix}T\\e\end{bmatrix}\begin{bmatrix}e\\e\end{bmatrix}\begin{bmatrix}e\\T\end{bmatrix}\right)^*$$

It can be shown that any sequence in $\mathbb{L}_V$ is possible if and only if $d_1 < d_2$.

4. Timers and their Languages

In this section, we will be interested in the recognizability of $\mathbb{L}_D$, and in the more general problem of describing languages defined in the following way. Let R be a subset of $(\mathbb{R}^+)^n$, the language $\mathbb{L}_R$ is given by:

$$\mathbb{L}_R = \bigcup_{D \in R} \mathbb{L}_D$$

Such languages arise when only bounds are known for actual delays in processes, for example $R = \{(d_1, d_2) \mid 0.9 < d_1 < 1.1 \text{ and } d_2 = 2\}$, or for analyzing timing constraints given by general relations between delays, for example $R = \{(d_1, d_2) \mid d_1 < d_2\}$. A first elementary result states that, in a certain sense, time units are irrelevant to the discussion.

Proposition 4.1 Let a be a positive constant, then $\mathbb{L}_D = \mathbb{L}_{aD}$. □

The next lemma states that $\mathbb{L}_D$ is rational when all delays are equal. This is probably the most fundamental result since many other rational languages will be recognized by variations on the automata that recognizes the equal delay languages.

Lemma 4.2 If $D = (d, ..., d)$ then $\mathbb{L}_D$ is rational.

Proof:
From Theorem 3.6, it is sufficient to show that there is only a finite number of sets $PR(y)$ for $y \in \mathbb{L}_D$. Consider the set Y of sequences that contains only **s** or **e** events, that is sequences in which no timer has timed out yet. By Proposition 3.5, there is only a finite number of sets $PR(y)$ for $y \in Y$, since there is only a finite number of ways of starting n timers. For example, in the case $n = 2$, we have the six sequences:

$$\lambda \quad \begin{bmatrix} s \\ e \end{bmatrix} \quad \begin{bmatrix} e \\ s \end{bmatrix} \quad \begin{bmatrix} s \\ s \end{bmatrix} \quad \begin{bmatrix} s \\ e \end{bmatrix}\begin{bmatrix} e \\ s \end{bmatrix} \quad \begin{bmatrix} e \\ s \end{bmatrix}\begin{bmatrix} s \\ e \end{bmatrix}$$

We will show that these sets are the only possible values for $PR(y)$ for $y \in \mathbb{L}_D$.

Let $\tau y \sigma$ be a sequence without null events, such that $\tau y \in Y$, and σ contains at least one **T** event. Since all delays are equal, if the sequence $\tau y \sigma$ is possible, then the timers that time out in σ are exactly those that were started in τ. Define σ' to be the sequence obtained by replacing the **T** events of σ by **e** events, then the sequence $ey\sigma'$ is the same as $\tau y \sigma$ except for timers that time out in σ, and $ey\sigma' \in Y$. More precisely, $p_i(ey\sigma') = e^*$, if timer i timed out in σ, else $p_i(ey\sigma') = p_i(\tau y \sigma)$. We claim that $PR(\tau y \sigma) = PR(ey\sigma')$.

Let $(r_1, ..., r_n) \in PR(\tau y \sigma)$, and S be a stamping of $\tau y \sigma e$ such that $R_{S_i}(\tau y \sigma e) = r_i$. It is immediate that S is also a stamping of $ey\sigma'e$, since the projections of $ey\sigma'e$ are either equal to e^*, or to those of $\tau y \sigma e$.

If $(r_1, ..., r_n) \in PR(ey\sigma')$, and S is a stamping of $ey\sigma'e$ such that $R_{S_i}(ey\sigma'e) = r_i$, then, since there is no time out event in $ey\sigma'e$, and y has no null events, $S(ey\sigma'e) - S(e) < d$. We define the stamping S' of $\tau y \sigma e$ to be equal to S, except for $S'(\tau) = S(ey\sigma'e) - d$. The function S is increasing since $S(ey\sigma'e) - S(e) < d$, and it is easy to verify that it is a stamping for each timer. □

Theorem 4.3 Let $D = (d_1, ..., d_n)$, if d_i/d_j is rational for all i, j then $\mathbb{L}_D$ is rational

Sketch of Proof:
Suppose d_i/d_j is rational for all i, j then there exists a number d such that each d_i is a multiple of d. Let $n_i = d_i/d$, $n' = \sum n_i$, and associate to each original timer i a set $T_i = \{T_{ij} \mid 1 \le j \le n_i\}$ of n_i timers. By Lemma 4.2, there is an automata $\mathbb{A}$ that recognizes each possible sequence of the n' timers. We claim that there is a sub-automata of $\mathbb{A}$ that recognizes $\mathbb{L}_D$ with a suitable renaming of the transitions. The idea is to restrict the transitions of $\mathbb{A}$ in such a way that the timers in T_i are chained to simulate the behavior of timer i. The automaton $\mathbb{A}'$ is obtained from $\mathbb{A}$ by removing all transitions σ that does not satisfy the following:

A **S** event of timer $T_{i(j+1)}$ is synchronized with a **T** event of timer T_{ij}:

$$\sigma_{i(j+1)} = \mathbf{S} \text{ iff } \sigma_{ij} = \mathbf{T}.$$

Let σ be a vector that satisfies those conditions. We construct the vector $f(\sigma) = (\tau_1, ..., \tau_n)$ such that

$$\tau_i = \mathbf{s} \qquad \text{if } \sigma_{i1} = \mathbf{s},$$
$$\tau_i = \mathbf{T} \qquad \text{if } \sigma_{in_i} = \mathbf{T},$$
$$\tau_i = \mathbf{e} \qquad \text{otherwise}$$

Any sequence y recognized by $\mathbb{A}$ will have an image $f(y)$ in $\mathbb{L}_D$ since any stamping of y will produce a stamping of $f(y)$. On the other hand, if $y \in \mathbb{L}_D$ with stamping S, we can add n_i -1 **e** events to y at regular d intervals after each event τ such that $\tau_i = \mathbf{s}$. Each of these sequence can be transformed in an obvious way into a sequence recognized by $\mathbb{A}'$. □

When n=2, the preceding result holds without restrictions on the pair (d_1, d_2), [B94]. However, it seems unlikely that this result can be extended to higher values. One of our basic assumption was that the time spent in a state is always greater than 0, so that a timer cannot be started simultaneously as it times out. This gives an 'elastic' property to sequences of possible actions with two timers, which is lost in higher dimensions. Using, for example, two pairs of timers, it is possible to model the sequences of actions of two continuous timers. Irrationality results are then inevitable [DHKR94].

We now turn to the study of more general regions. The first result states that, for two timers, there are strong relations between the languages $\mathbb{L}_D$ when D varies in the lower – or upper part – of $(\mathbb{R}^+)^2$. Any language $\mathbb{L}_D$ with D in the shaded area will be included in $\mathbb{L}_C$.

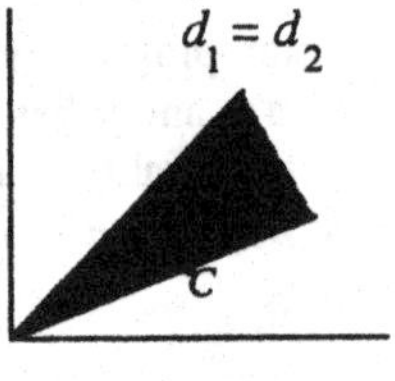

Proposition 4.4 Let $D= (d_1, d_2)$ and $C=(c_1, c_2)$, if $c_1/c_2 < d_1/d_2 < 1$ then $\mathbb{L}_D \subseteq \mathbb{L}_C$.

Sketch of proof.
Given a stamping S of y with respect to (d_1, d_2), we construct a stamping S' with respect to delays (d, d_2) with $d = (c_1/c_2)d_2$. By Proposition 4.1, and since $C = (c_2/d_2)D$, we will conclude that y is in $\mathbb{L}_C$. The idea is to shrink by $a = d/d_1$ intervals where timer 1 is running, while keeping fixed the distance between a **s** and a **T** event in timer 2. Since all intervals where timer 1 is running will be shrunk by a, this procedure will give a stamping with respect to (d, d_2).

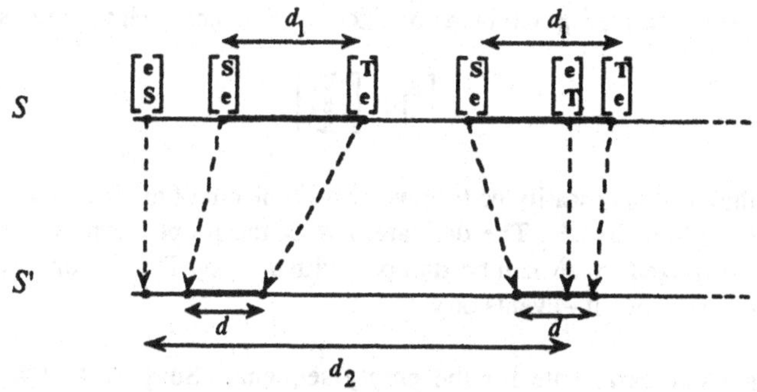

The above diagram shows the relations between a stamping S and a stamping S' where delay d_1 was shrunk to d. □

The next result shows that if $R = (\mathbb{R}^+)^2$, then the set $\mathbb{L}_R$ is rational. We will do that with the help of the following lemma that constructs an automaton recognizing any sequence that can be stamped with $d_1 < d_2$.

Lemma 4.5 Let $R = \{(d_1, d_2) \mid d_1 < d_2\}$, then $\mathbb{L}_R$ is rational.

Sketch of Proof:
The automata $\mathbb{A}$ recognizing $\mathbb{L}_R$ is the following variant on $\mathbb{T}^2$ obtained by restricting sequences where timer 1 is started before or simultaneously with timer 2 to have a necessary time out event for timer 1.

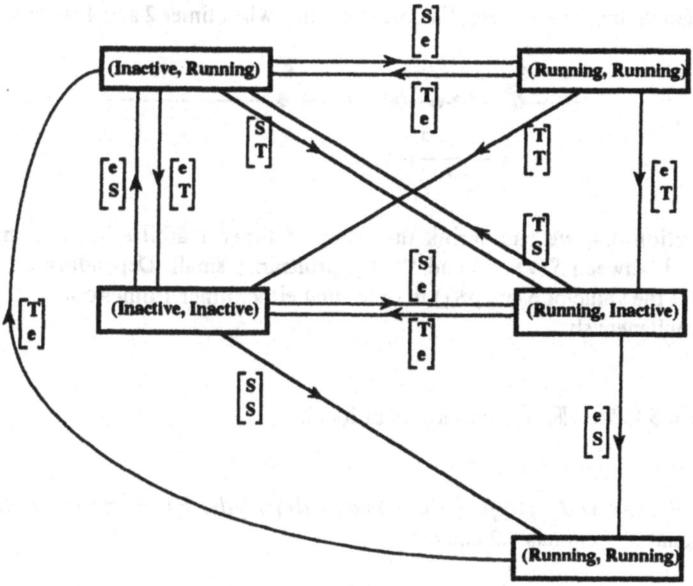

Clearly, any sequence $y \in \mathbb{L}_R$ will be recognized by $\mathbb{A}$, since each time the sequences

$$\begin{bmatrix} s \\ s \end{bmatrix} \text{ or } \begin{bmatrix} s \\ e \end{bmatrix}\begin{bmatrix} e \\ s \end{bmatrix}$$

appear in y, they will necessarily be followed by a time out of timer 1. All other sequences of $\mathbb{L}^2$ are recognized by $\mathbb{A}$. The delicate part of the proposition is to show that any sequence y recognized by $\mathbb{A}$ can be stamped with $d_1 < d_2$. This is done by induction on the number of - significant - events of y.

The statement is clearly true for the empty sequence. Suppose that a sequence y is recognized by $\mathbb{A}$, and that there exists a stamping S for y with $d_1 < d_2$. We have to show that we can construct a stamping S' for any sequence $y\sigma$ possible in $\mathbb{A}$.

If, after y, one of the timer is inactive, it is straightforward to extend a stamping S of y to $y\sigma$. If both timers are running, and if timer 1 was last started before or simultaneously with timer 2, then

$$\sigma = \begin{bmatrix} T \\ e \end{bmatrix}$$

and we set $S'(y\sigma) = S(y)+(d - R_{S_1}(y))$. It is again easy to verify that S' is a stamping with respect to (d_1, d_2). If timer 1 was last started after timer 2, then y can then be written as

$$y' s_2 x s_1$$

where s_2 and s_1 are, respectively, the last event in y when timer 2 and 1 were started.

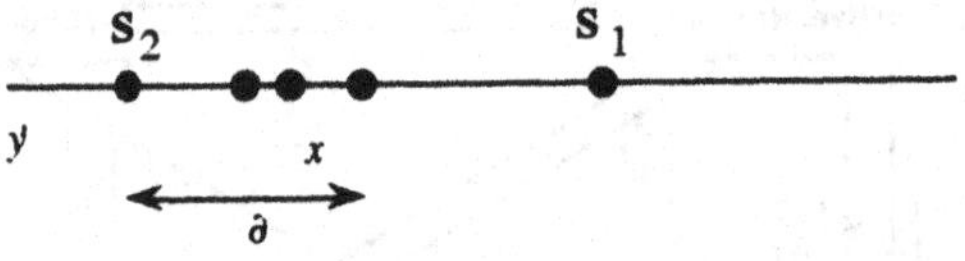

By Proposition 4.4, we can shrink the delay of timer 1 arbitrarily, thus making the difference ∂ between $S'(y's_2x)$ and $S'(y's_2)$ arbitrarily small. Depending on σ, we can then choose the value of $S'(y's_2xs_1)$ in order that either timer 1 times out first, or timer 2, or both simultaneously. □

Theorem 4.6 If $R = (\mathbb{R}^+)^2$, then $\mathbb{L}_R$ is rational.

Proof:
Since $R = \{(d_1, d_2) \mid d_1 < d_2\} \cup \{(d_1, d_2) \mid d_2 < d_1\} \cup \{(d_1, d_2) \mid d_1 = d_2\}$, we have that $\mathbb{L}_R$ is rational by Lemmas 4.2 and 4.5. □

Theorem 4.6 relies heavily on the fact the there are only two timers. In higher dimensions, languages associated to large regions are no more rational.

Proposition 4.7 Let $R = \{(d_1, d_2, d_3) \mid d_1 < d_2 < d_3\}$, then $\mathbb{L}_R$ is not rational.

Proof:
Fix $d_1 < d_2$ and consider the set of sequences defined by:

$$y_i = \begin{bmatrix} \mathbf{S} \\ \mathbf{e} \\ \mathbf{S} \end{bmatrix} x^i \begin{bmatrix} \mathbf{T} \\ \mathbf{S} \\ \mathbf{e} \end{bmatrix} \begin{bmatrix} \mathbf{e} \\ \mathbf{T} \\ \mathbf{T} \end{bmatrix}$$

where

$$x = \begin{bmatrix} \mathbf{T} \\ \mathbf{S} \\ \mathbf{e} \end{bmatrix} \begin{bmatrix} \mathbf{S} \\ \mathbf{T} \\ \mathbf{e} \end{bmatrix}$$

Each sequence y_i contains one run of timer 3 synchronized with $2i$ alternating runs timer 1 and timer 2. Thus y_i is in $\mathbb{L}_R$ if and only if $d_3 = i\,(d_1 + d_2)$, thus the sequence $y_i y_j$ is possible if and only if $i = j$. Any automata recognizing $\mathbb{L}_R$ must send y_i and y_j in different states if $i \neq j$, thus cannot be finite. □

5. Concluding Remarks

This study raises a lot of questions. The results of Section 4 show that the languages associated with two timers are relatively well behaved, and that it is always possible to find a suitable automaton to analyze a problem, even when the delays are known only within certain bounds. The same kind of constraints does not yield finite automata for higher dimensions. What kind of constraints on the delays of n timers would yield finite automata?

It is also possible to extend the basic timer with additional possibilities. If we add an operation that aborts a timer while it is still running, the theory remains almost exactly the same. This is due to the fact that **e** events that occur in the **Running** state have properties similar to an abort action with respect to stampings and readings. The possibility of suspending a timer, and resuming it later, offers a more challenging problem.

6. References

[AD90] R. Alur and D. Dill, *Automata for Modeling Real-Time Systems*, in Automata, Languages and programming, 17th International Colloquium Proceedings, Coventry, UK, July 1990, pp. 322-335.

[AD94] R. Alur and D. Dill, *The Theory of Timed Automata*, Theoretical Computer Science, 126 (1994) 183-235.

[A94] A. Arnold, Finite Transition Systems, Prentice Hall, 1994.

[B94] A. Bergeron, *A Study of Two Concurrent Timers*, 2nd AMAST Workshop on Real-Time Systems, Bordeaux, June 1995.

[DHKR94] S. Di Gennaro, C. Horn, S.R. Kulkarni and P.J. Ramadge, *Reduction of Timed Hybrid Systems*, Proceedings of the 33rd IEEE Conference on decision and Control, (1994) 4215-4220.

[LB93] W. Lam and R. Brayton, *Alternating RQ Timed Automata*, CAV'93, Lecture Notes in Computer Science, no. 697 (1993) 237-252.

[NSY93] X. Nicollin, J. Sifakis and S. Yovine, *From ATP to timed graphs and hybrid systems*, Acta Informatica, 30 (1993) 181-201.

An Algebraic Construction of the Well-Founded Model

Rajiv Bagai and Rajshekhar Sunderraman

Department of Computer Science
Wichita State University
Wichita, KS 67260-0083, U.S.A.

Abstract. An algebraic method for the construction of the well-founded model of general deductive databases is presented. The method adopts paraconsistent relations as the semantic objects associated with the predicate symbols of the database. Paraconsistent relations are a generalization of ordinary relations in that they allow manipulation of incomplete as well as inconsistent information. Algebraic operators, such as union, join, selection, are defined for paraconsistent relations. The first step in the model construction method is to transform the database clauses into paraconsistent relation definitions involving these operators. The second step is to build the well-founded model iteratively. Algorithms for both steps along with arguments for their termination and correctness are presented.

Keywords. Logic programs, Deductive databases, Relational algebra, Negative inferences, Well-founded model.

1 Introduction

The well-founded model [16] is one of the most popular semantics for logic programs with negative subgoals. It is a 3-valued semantics under which ground atoms could be assigned a truth value of *undefined*, in addition to *true* and *false*. While the well-founded model is defined for the entire class of logic programs with negative subgoals, it coincides with other semantics proposed for restricted classes of programs such as Apt, Blair and Walker's semantics [1] for *stratified* programs, and Przymusinski's "perfect model" semantics [12] for *locally stratified* programs. The well-founded semantics also strictly subsumes the semantics proposed by Fitting [8].

In this paper, we present an algebraic approach for constructing the well-founded model for general deductive databases (function-free logic programs with negative subgoals). The central idea in arriving at this model is to associate *paraconsistent relations* [4] with the predicate symbols of the given general deductive database. A paraconsistent relation essentially contains two kinds of tuples: ones for which the underlying predicate is believed to be true and ones for which it is believed to be false. Algebraic operators over paraconsistent relations that extend the standard operators, such as selection, join etc., for ordinary relations were presented in [4]. Our method for constructing the well-founded model involves two steps. In the first step, the database clauses are converted into paraconsistent relation definitions involving the operators on them. In the second step, these definitions are used to iteratively construct the model.

The approach presented in this paper lays an algebraic foundation for query processing and optimization for general deductive databases. Query processing will proceed in a bottom-up manner and will use popular rewriting strategies, such as Magic Sets [5], to focus the search for answers. Query optimization can also be achieved at the level of the paraconsistent relational algebra by making use of the laws of equalities.

The rest of this paper is organized as follows. Section 2 provides a brief overview of the well-founded model. Section 3 gives a quick introduction to paraconsistent relations and some algebraic operators over them. Section 4 presents the first part of the model construction method, namely an algorithm to convert the database clauses into algebraic equations defining paraconsistent relations. Section 5 presents the second part of the method, namely an algorithm to incrementally construct the paraconsistent relations using the equations constructed earlier. Section 6 provides a complete illustrative example of the method. Finally, Section 7 contains some concluding remarks and comparisons with related work.

2 The Well-Founded Model

In this section we give a brief overview of the well-founded model. For a detailed exposition the reader is referred to [16]. We assume an underlying language with a finite set of constant, variable, and predicate symbols, but no function symbols. A *term* is either a variable or a constant. An *atom* is of the form $p(t_1, \ldots, t_n)$, where p is a predicate symbol and the t_i's are terms. A *literal* is either a *positive literal* A or a *negative literal* $\neg A$, where A is an atom. For any literal l we let l' denote its complementary literal, i.e. if l is positive then $l' = \neg l$, otherwise $l = \neg l'$.

Definition 1. A *general deductive database* is a finite set of clauses of the form

$$a \leftarrow l_1, l_2, \ldots, l_m$$

where a is an atom, $m \geq 0$, and each l_i is a literal. □

A term, atom, literal, or clause is called *ground* if it contains no variables. The *Herbrand Universe* of the underlying language is the set of all ground terms. The *Herbrand Base* is the set of all ground atoms. A *ground instance* of a term, atom, literal, or clause Q is the term, atom, literal, or clause, respectively, obtained by replacing each variable in Q by a constant. For any general deductive database P, we let $P^\star$ denote the set of all ground instances of clauses in P. Note that since the underlying language has no function symbols, unlike logic programs, $P^\star$ is always finite.

Definition 2. A *partial interpretation* is a pair $I = \langle I^+, I^- \rangle$, where I^+ and I^- are any subsets of the Herbrand Base. □

A partial interpretation I is *consistent* if $I^+ \cap I^- = \emptyset$. For any partial interpretations I and J, we let $I \cap J$ be the partial interpretation $\langle I^+ \cap J^+, I^- \cap J^- \rangle$, and $I \cup J$ be the partial interpretation $\langle I^+ \cup J^+, I^- \cup J^- \rangle$. We also say $I \subseteq J$ whenever $I^+ \subseteq J^+$ and $I^- \subseteq J^-$.

The well-founded model of P is the least fixpoint of the immediate consequence function T_P on partial interpretations defined as follows.

Definition 3. Let I be a partial interpretation. Then $T_P(I)$ is a partial interpretation, given by

$$\begin{aligned} T_P(I)^+ = \{a \mid & \text{ for some clause } a \leftarrow l_1, \ldots, l_m \text{ in } P^\star, \text{ for each } i, 1 \le i \le m, \\ & \text{if } l_i \text{ is positive then } l_i \in I^+, \text{ and} \\ & \text{if } l_i \text{ is negative then } l_i' \in I^-\}, \\ T_P(I)^- = & \text{ the largest set } S \text{ of ground atoms, such that} \\ & \text{for each } a \in S, \text{ for every clause } a \leftarrow l_1, \ldots, l_m \text{ in } P^\star, \\ & \text{there is some } i, 1 \le i \le m, \text{ such that} \\ & \text{if } l_i \text{ is positive then } l_i \in I^- \cup S, \text{ and} \\ & \text{if } l_i \text{ is negative then } l_i' \in I^+. \; \Box \end{aligned}$$

It is easily seen that T_P is monotonic. However, an application of T_P on a consistent partial interpretation does not always result in a consistent partial interpretation. As an example, let $P = \{\mathrm{p} \leftarrow \mathrm{p}\}$ and $I = \langle\{\mathrm{p}\}, \emptyset\rangle$. Then, $T_P(I)$ is the inconsistent interpretation $\langle\{\mathrm{p}\}, \{\mathrm{p}\}\rangle$. Despite this behavior, T_P possesses a least fixpoint, which is the well-founded model of P. This least fixpoint is shown in [16] to be $T_P \uparrow \omega$, where the ordinal powers of T_P are a monotonic sequence of consistent partial interpretations defined as follows:

Definition 4. For any ordinal α,

$$T_P \uparrow \alpha = \begin{cases} \langle \emptyset, \emptyset \rangle & \text{if } \alpha = 0, \\ T_P(T_P \uparrow (\alpha - 1)) & \text{if } \alpha \text{ is a successor ordinal,} \\ \langle \bigcup_{\beta<\alpha} (T_P \uparrow \beta)^+, \bigcup_{\beta<\alpha} (T_P \uparrow \beta)^- \rangle & \text{if } \alpha \text{ is a limit ordinal.} \; \Box \end{cases}$$

In [2] the *upward closure ordinal* of the immediate consequence function is defined as the least ordinal α such that $T_P \uparrow \alpha$ is a fixpoint of T_P. The following observation for deductive databases is relevant:

Proposition 5. *For any general deductive database P, the upward closure ordinal of T_P is finite, i.e. there is a number $n \ge 0$, such that $T_P \uparrow n = T_P \uparrow \omega$.*

Proof. Immediate from the fact that the Herbrand Base is finite. □

Thus, a mechanism that "computes" the ordinal powers of T_P can be employed to construct the well-founded model of P.

3 Paraconsistent Relations

An important feature of the well-founded model is that negative information is explicitly present in it, just as the positive information. The predicate symbols of the underlying general deductive database thus need to be appropriately interpreted with explicit negative information. We achieve this by associating *paraconsistent relations* with these predicate symbols. Paraconsistent relations were recently introduced by Bagai and Sunderraman in [4]. Here we present a quick overview of these structures.

Unlike ordinary relations that can model worlds in which every tuple is known to either hold a certain underlying predicate or to not hold it, paraconsistent relations provide a framework for incomplete or even inconsistent information about tuples. They are thus extensions of ordinary relations. Some algebraic operators on paraconsistent relations, such as union, join, projection, are also defined in [4]. As expected, these operators are generalizations of their ordinary counterparts. This fact is reflected by placing a dot over an operator on ordinary relations to obtain the corresponding generalized operator on paraconsistent relations. For example, $\bowtie$ denotes the natural join among ordinary relations, and $\dot{\bowtie}$ denotes natural join between paraconsistent relations.

Let a *relation scheme* (or just *scheme*) Σ be a finite set of *attribute names*, where for any attribute $A \in \Sigma$, $dom(A)$ is a non-empty *domain* of values for A. A *tuple* on Σ is any map $t : \Sigma \rightarrow \bigcup_{A \in \Sigma} dom(A)$, such that $t(A) \in dom(A)$, for each $A \in \Sigma$. We let $\tau(\Sigma)$ denote the set of all tuples on Σ.

Definition 6. A *paraconsistent relation* on scheme Σ is a pair $R = \langle R^+, R^- \rangle$, where R^+ and R^- are any subsets of $\tau(\Sigma)$. □

Intuitively, R^+ may be considered as the set of all tuples for which R is believed to be true, and R^- the set of all tuples for which R is believed to be false. We do not assume R^+ and R^- to be mutually disjoint, though this condition holds in the well-founded model that is ultimately constructed. A non-empty overlap between R^+ and R^- is essentially contradictory information, and its possibility makes paraconsistent relations model belief systems more naturally that knowledge systems. Also, R^+ and R^- may not together cover all tuples in $\tau(\Sigma)$, giving rise to incompleteness.

Observe that a paraconsistent relation R for which $R^- = \tau(\Sigma) - R^+$ is essentially the ordinary relation R^+. Two such relations that we use in the sequel are the universal relation $\mho_\Sigma = \langle \tau(\Sigma), \emptyset \rangle$, and the empty relation $\Phi_\Sigma = \langle \emptyset, \tau(\Sigma) \rangle$. We first introduce some set-theoretic algebraic operators on paraconsistent relations:

Definition 7. Let R and S be paraconsistent relations on scheme Σ. Then,

(a) the *union* of R and S, denoted $R \mathbin{\dot{\cup}} S$, is a paraconsistent relation on scheme Σ, given by

$$(R \mathbin{\dot{\cup}} S)^+ = R^+ \cup S^+, \qquad (R \mathbin{\dot{\cup}} S)^- = R^- \cap S^-;$$

(b) the ***intersection*** of R and S, denoted $R \dot\cap S$, is a paraconsistent relation on scheme Σ, given by

$$(R \dot\cap S)^+ = R^+ \cap S^+, \qquad (R \dot\cap S)^- = R^- \cup S^-;$$

(c) the ***complement*** of R, denoted $\dot- R$, is a paraconsistent relation on scheme Σ, given by

$$(\dot- R)^+ = R^-, \qquad (\dot- R)^- = R^+. \ \square$$

An intuitive appreciation of the union operator may be obtained by interpreting relations as properties of tuples. So, $R \dot\cup S$ is the "either-R-or-S" property. Now since R^+ and S^+ are the sets of tuples for which the properties R and S, respectively, are believed to hold, the set of tuples for which the property "either-R-or-S" is believed to hold is clearly $R^+ \cup S^+$. Moreover, since R^- and S^- are the sets of tuples for which properties R and S, respectively, are believed to ***not*** hold, the set of tuples for which the property "either-R-or-S" is believed to ***not*** hold is similarly $R^- \cap S^-$.

The definitions of ***intersection***, ***complement*** and all other operators defined later can (and should) be understood in the same way.

If Σ and Δ are relation schemes such that $\Sigma \subseteq \Delta$, then for any tuple $t \in \tau(\Sigma)$, we let t^Δ denote the set $\{t' \in \tau(\Delta) \mid t'(A) = t(A), \text{ for all } A \in \Sigma\}$ of all extensions of t. We extend this notion for any $T \subseteq \tau(\Sigma)$ by defining $T^\Delta = \bigcup_{t \in T} t^\Delta$. We now define some relation-theoretic algebraic operators on paraconsistent relations.

Definition 8. Let R and S be paraconsistent relations on schemes Σ and Δ, respectively. Then, the ***natural join*** (or just ***join***) of R and S, denoted $R \dot\bowtie S$, is a paraconsistent relation on scheme $\Sigma \cup \Delta$, given by

$$(R \dot\bowtie S)^+ = R^+ \bowtie S^+, \qquad (R \dot\bowtie S)^- = (R^-)^{\Sigma \cup \Delta} \cup (S^-)^{\Sigma \cup \Delta},$$

where $\bowtie$ is the usual natural join among ordinary relations. □

It is instructive to observe that $(R \dot\bowtie S)^-$ contains all extensions of tuples in R^- and S^-, because at least one of properties R and S is believed to not hold for these extended tuples.

Definition 9. Let R be a paraconsistent relation on scheme Σ, and Δ be any scheme. Then, the ***projection*** of R onto Δ, denoted $\dot\pi_\Delta(R)$, is a paraconsistent relation on Δ, given by

$$\dot\pi_\Delta(R)^+ = \pi_\Delta((R^+)^{\Sigma \cup \Delta}), \qquad \dot\pi_\Delta(R)^- = \{t \in \tau(\Delta) \mid t^{\Sigma \cup \Delta} \subseteq (R^-)^{\Sigma \cup \Delta}\},$$

where π_Δ is the usual projection over Δ of ordinary relations. □

It should be noted that, contrary to usual practice, the above definition of projection is not just for subschemes. However, if $\Delta \subseteq \Sigma$, then it coincides with the intuitive projection operation. In this case, $\dot\pi_\Delta(R)^-$ consists of those tuples in $\tau(\Delta)$, all of whose extensions are in R^-.

A tuple $t \in \tau(\Delta)$ appears in $\dot{\pi}_\Delta(R)^+$ if any member of $t^{\Sigma \cup \Delta}$ is in $(R^+)^{\Sigma \cup \Delta}$. We also need a stronger notion of projection in which any $t \in \tau(\Delta)$ appears in the positive part of the resulting relation only if $t^{\Sigma \cup \Delta} \subseteq (R^+)^{\Sigma \cup \Delta}$.

Definition 10. Let R be a paraconsistent relation on scheme Σ, and Δ be any scheme. Then, the ***strong projection*** of R onto Δ, denoted $\dot{\mu}_\Delta(R)$, is the paraconsistent relation $\dot{-}(\dot{\pi}_\Delta(\dot{-}R))$ on scheme Δ. □

The operator $\dot{\mu}_\Delta$ is essentially a dual of $\dot{\pi}_\Delta$. We now define the last relation-theoretic operation.

Definition 11. Let R be a paraconsistent relation on scheme Σ, and let F be any logic formula involving attribute names in Σ, constant symbols (denoting values in the attribute domains), equality symbol $=$, negation symbol $\neg$, and connectives $\vee$ and $\wedge$. Then the ***selection*** of R by F, denoted $\dot{\sigma}_F(R)$, is a paraconsistent relation on scheme Σ, given by

$$\dot{\sigma}_F(R)^+ = \sigma_F(R^+), \qquad \dot{\sigma}_F(R)^- = R^- \cup \sigma_{\neg F}(\tau(\Sigma)),$$

where σ_F is the usual selection of tuples satisfying F from ordinary relations. □

Before ending this brief overview of paraconsistent relations, let us look at the identities of some operators on them, in particular $\dot{\cup}$, $\dot{\cap}$ and $\dot{\bowtie}$. If R is a paraconsistent relation on scheme Σ, then

$$R \,\dot{\cup}\, \Phi_\Sigma = \Phi_\Sigma \,\dot{\cup}\, R = R.$$

Thus, Φ_Σ is the identity of $\dot{\cup}$ for paraconsistent relations on scheme Σ. Similarly $\mho_\Sigma$ is the identity of $\dot{\cap}$ for paraconsistent relations on scheme Σ. Furthermore, for any subscheme $\Delta \subseteq \Sigma$, we have that

$$R \,\dot{\bowtie}\, \mho_\Delta = \mho_\Delta \,\dot{\bowtie}\, R = R.$$

Thus, $\mho_\Delta$ is an identity of $\dot{\bowtie}$ for any paraconsistent relation on scheme Σ, such that $\Delta \subseteq \Sigma$. Since the empty scheme $\emptyset$ is a subscheme of every scheme, it follows that $\mho_\emptyset = \langle \{()\}, \{\} \rangle$ is an unconditional identity of $\dot{\bowtie}$.

4 Generation of Algebraic Expressions

We now describe an algebraic method for constructing the well-founded model for a given general deductive database P. In this model, paraconsistent relations are the semantic objects associated with the predicate symbols occurring in P.

The method involves two steps. The first step is to obtain from P two sets, denoted $\Upsilon_1(P)$ and $\Upsilon_2(P)$, of paraconsistent relation definitions for the predicate symbols occurring in P. The set $\Upsilon_1(P)$ will be used to obtain the positive consequences, and $\Upsilon_2(P)$ for the negative consequences. The set $\Upsilon_1(P)$ contains definitions of the form

$$\mathrm{p} = D_{\mathrm{p}},$$

where p is a predicate symbol of P, and D_{p} is an algebraic expression involving predicate symbols of P and paraconsistent relation operators. The set $\Upsilon_2(P)$ contains similar definitions. The second step of our method is to iteratively build the well-founded model using the definitions in these sets. In this section we describe the construction of the definition sets $\Upsilon_1(P)$ and $\Upsilon_2(P)$ from a given general deductive database P.

A scheme Σ is a *Herbrand scheme* if *dom*(A) is the Herbrand Universe, for all $A \in \Sigma$. Let $\Gamma = \langle \nu_1, \nu_2, \ldots \rangle$ be an infinite sequence of some distinct attribute names. For any $n \geq 1$, let Γ_n be the Herbrand scheme $\{\nu_1, \ldots, \nu_n\}$. We use the following scheme renaming operators.

Definition 12. Let $\Sigma = \{A_1, \ldots, A_n\}$ be any Herbrand scheme. Then,

(a) for any paraconsistent relation R on scheme Γ_n, $R(A_1, \ldots, A_n)$ is the paraconsistent relation

$$\langle\ \{t \in \tau(\Sigma) \mid \text{for some } t' \in R^+, t(A_i) = t'(\nu_i), \text{ for all } i, 1 \leq i \leq n\},$$
$$\{t \in \tau(\Sigma) \mid \text{for some } t' \in R^-, t(A_i) = t'(\nu_i), \text{ for all } i, 1 \leq i \leq n\}\ \rangle$$

on scheme Σ, and

(b) for any paraconsistent relation R on scheme Σ, $R[A_1, \ldots, A_n]$ is the paraconsistent relation

$$\langle\ \{t \in \tau(\Gamma_n) \mid \text{for some } t' \in R^+, t(\nu_i) = t'(A_i), \text{ for all } i, 1 \leq i \leq n\},$$
$$\{t \in \tau(\Gamma_n) \mid \text{for some } t' \in R^-, t(\nu_i) = t'(A_i), \text{ for all } i, 1 \leq i \leq n\}\ \rangle$$

on scheme Γ_n. □

Before describing our method to obtain $\Upsilon_1(P)$ and $\Upsilon_2(P)$ from P, let us look at an example. Suppose the following are the only clauses in a general deductive database P with the predicate symbol p in their heads:

```
p(X) ← r(X,Y), s(a,Y), ¬p(Y)
p(Y) ← s(Y,a)
```

From these clauses the algebraic definition constructed for the symbol p in $\Upsilon_1(P)$ is

$$\begin{aligned} \mathrm{p} = {} & (\dot\pi_{\{X\}}(\mathrm{r}(X,Y) \mathbin{\dot\bowtie} \dot\pi_{\{Y\}}(\dot\sigma_{A=a}(\mathrm{s}(A,Y))) \mathbin{\dot\bowtie} \dot-\mathrm{p}(Y) \mathbin{\dot\bowtie} \mho_{\emptyset}))[X] \mathbin{\dot\cup} \\ & (\dot\pi_{\{Y\}}(\dot\pi_{\{Y\}}(\dot\sigma_{A=a}(\mathrm{s}(Y,A))) \mathbin{\dot\bowtie} \mho_{\emptyset}))[Y] \mathbin{\dot\cup} \\ & \Phi_{\Gamma_1}, \end{aligned} \quad (1)$$

and that in $\Upsilon_2(P)$ is

$$\begin{aligned} \tilde{\mathrm{p}} = {} & (\dot\pi_{\{X\}}(\dot\mu_{\{Y,A\}}(\dot\pi_{\{X,Y,A\}}(\tilde{\mathrm{r}}(X,Y)) \mathbin{\dot\cup} \\ & \dot\pi_{\{X,Y,A\}}(\dot\sigma_{A=a}(\tilde{\mathrm{s}}(A,Y))) \mathbin{\dot\cup} \dot\pi_{\{X,Y,A\}}(-\tilde{\mathrm{p}}(Y)) \mathbin{\dot\cup} \Phi_{\{X,Y,A\}})))[X] \mathbin{\dot\cap} \\ & (\dot\pi_{\{Y\}}(\dot\mu_{\{A\}}(\dot\pi_{\{Y,A\}}(\dot\sigma_{A=a}(\tilde{\mathrm{s}}(Y,A))) \mathbin{\dot\cup} \Phi_{\{Y,A\}})))[Y] \mathbin{\dot\cap} \\ & \mho_{\Gamma_1}. \end{aligned} \quad (2)$$

Such a conversion exploits the close connection between attribute names in relation schemes and variables in clauses, as pointed out by Ullman in [14]. The structure of these expressions conforms to the definition of the T_P function given in Section 2. It is interesting to note the duality among the operators occurring in the expressions (1) and (2). For example, expression (1) is essentially a union of joins, because $T_P(I)^+$ is the set of all atoms for which some ground clause is such that each of its body literal has a certain property. Similarly, expression (2) is an intersection of unions, because of the way $T_P(I)^-$ is defined. The expressions thus constructed can be used to arrive at better approximations of the paraconsistent relation p from some approximations of p, r and s. The approximations converge to the well-founded model of P as described in the next section.

Construction of $\Upsilon_1(P)$

We first describe the conversion of clauses in P to algebraic definitions in $\Upsilon_1(P)$. For any predicate symbol p in P, the expression in $\Upsilon_1(P)$ is essentially a union ($\dot{\cup}$) of the expressions obtained from each clause containing the symbol p in its head. It therefore suffices to give the following Algorithm *CONVERT1* for converting one such clause into an expression.

Algorithm CONVERT1
Input: A general deductive database clause $l_0 \leftarrow l_1, \ldots, l_m$. Let l_0 be an atom of the form $p_0(A_{01}, \ldots, A_{0k_0})$, and each l_i, $1 \leq i \leq m$, be either of the form $p_i(A_{i1}, \ldots, A_{ik_i})$, or of the form $\neg p_i(A_{i1}, \ldots, A_{ik_i})$. For any i, $0 \leq i \leq m$, let V_i be the set of all variables occurring in l_i.
Output: An algebraic expression involving paraconsistent relations.
Method: The expression is constructed by the following steps:

1. For each argument A_{ij} of literal l_i, construct argument B_{ij} and condition C_{ij} as follows:
   (a) If A_{ij} is a constant a, then B_{ij} is any brand new variable and C_{ij} is $B_{ij} = a$.
   (b) If A_{ij} is a variable, such that for each k, $1 \leq k < j$, $A_{ik} \neq A_{ij}$, then B_{ij} is A_{ij} and C_{ij} is *true*.
   (c) If A_{ij} is a variable, such that for some k, $1 \leq k < j$, $A_{ik} = A_{ij}$, then B_{ij} is a brand new variable and C_{ij} is $A_{ij} = B_{ij}$.
2. Let $\hat{l}_i$ be the atom $p_i(B_{i1}, \ldots, B_{ik_i})$, and F_i be the conjunction $C_{i1} \wedge \cdots \wedge C_{ik_i}$. If l_i is a positive literal, then let Q_i be the expression $\dot{\pi}_{V_i}(\dot{\sigma}_{F_i}(\hat{l}_i))$. Otherwise, let Q_i be the expression $\dot{-}\dot{\pi}_{V_i}(\dot{\sigma}_{F_i}(\hat{l}_i))$.
   As a syntactic optimization, if all conjuncts of F_i are *true* (i.e. all arguments of l_i are distinct variables), then both $\dot{\sigma}_{F_i}$ and $\dot{\pi}_{V_i}$ are reduced to identity operations, and are hence dropped from the expression. For example, if $l_i = \neg$p(X,Y), then $Q_i = \dot{-}$p(X,Y).

3. Let E be the expression $Q_1 \dot{\bowtie} \cdots \dot{\bowtie} Q_m \dot{\bowtie} \mho_\emptyset$. The output expression is

$$(\dot{\sigma}_{F_0}(\dot{\pi}_V(E)))[B_{01},\ldots,B_{0k_0}],$$

where V is the set of variables occurring in $\hat{l}_0$.
As in Step 2, if all conjuncts in F_0 are *true*, then $\dot{\sigma}_{F_0}$ is dropped from the output expression. However, $\dot{\pi}_V$ is never dropped, as the clause body may contain variables not in V. □

We use Algorithm *CONVERT1* to construct $\Upsilon_1(P)$ as follows:

Definition 13. For any general deductive database $P, \Upsilon_1(P)$ is a set of all equations of the form $\mathrm{p} = D_{\mathrm{p}}$, where p is a predicate symbol of P with arity n, and D_{p} is a union ($\dot{\cup}$) of Φ_{Γ_n} and all expressions obtained from Algorithm *CONVERT1* for clauses in P with symbol p in their head. □

Construction of $\Upsilon_2(P)$

For any predicate symbol p in P, the expression in $\Upsilon_2(P)$ is essentially an intersection ($\dot{\cap}$) of the expressions obtained from each clause in P containing the symbol p in its head. It therefore suffices to give the following Algorithm *CONVERT2* for converting one such clause of P into an expression.

Algorithm CONVERT2
Input: Same as the input of Algorithm *CONVERT1*.
Output: An algebraic expression involving paraconsistent relations.
Method: The expression is constructed by the following steps:

1. Same as Step 1 of Algorithm *CONVERT1*.
2. If l_i is a positive literal, then let $\hat{l}_i$ be the literal $\tilde{p}_i(B_{i1},\ldots,B_{ik_i})$. Otherwise, let $\hat{l}_i$ be the literal $\dot{-}\tilde{p}_i(B_{i1},\ldots,B_{ik_i})$.
Let F_i be the conjunction $C_{i1} \wedge \cdots \wedge C_{ik_i}$. Also, let H be the set of variables occurring in the new head $\hat{l}_0$, and B be the set of variables occurring in the new body $\{\hat{l}_1,\ldots,\hat{l}_m\}$.
Now let Q_i be the expression $\dot{\pi}_B(\dot{\sigma}_{F_i}(\hat{l}_i))$.
As a syntactic optimization, if all conjuncts of F_i are *true* (i.e. all arguments of l_i were distinct variables), then $\dot{\sigma}_{F_i}$ is reduced to the identity operation, and is hence dropped from the expression.
3. Let E be the expression $Q_1 \dot{\cup} \cdots \dot{\cup} Q_m \dot{\cup} \Phi_B$. The output expression is

$$(\dot{\sigma}_{F_0}(\dot{\pi}_H(\dot{\mu}_{B-H}(E))) \ \dot{\cup}\ \dot{\sigma}_{\neg F_0}(\mho_H))[B_{01},\ldots,B_{0k_0}].$$

As in Step 2, if all conjuncts in F_0 are *true*, then $\dot{\sigma}_{F_0}$ is dropped from the output expression, along with the term $\dot{\sigma}_{\neg F_0}(\mho_H)$. □

Now $\Upsilon_2(P)$ is defined as follows:

Definition 14. For any general deductive database $P, \Upsilon_2(P)$ is a set of all equations of the form $\tilde{\mathrm{p}} = D_{\tilde{\mathrm{p}}}$, where p is a predicate symbol of P with arity n, and $D_{\tilde{\mathrm{p}}}$ is the intersection ($\dot{\cap}$) of $\mho_{\Gamma_n}$ and all expressions obtained from Algorithm *CONVERT2* for clauses in P with symbol p in their head. □

5 Construction of the Well-Founded Model

The second and final step in our model construction process is to incrementally construct the paraconsistent relations defined by the given database P, using the definitions in $\Upsilon_1(P)$ and $\Upsilon_2(P)$. For each predicate symbol $\mathbf{p}$ in the database, the construction algorithm uses two imperative "variables", $\mathbf{p}$ and $\tilde{\mathbf{p}}$, that may contain a paraconsistent relation as value. Thus, any variable $\mathbf{p}$ has two set-valued fields, namely $\mathbf{p}^+$ and $\mathbf{p}^-$. Similarly, a variable $\tilde{\mathbf{p}}$ has the set-valued fields $\tilde{\mathbf{p}}^+$ and $\tilde{\mathbf{p}}^-$. In addition, the algorithm employs a set variable $\hat{\mathbf{p}}$.

The overall construction algorithm is rather straightforward. It is essentially an iterative bottom-up construction of the least fixpoint of the T_P function given earlier. Each execution of the loop from Step 2 to 9 in the following Algorithm *CONSTRUCT* constructs the next ordinal power of T_P and stores it in the variables of the form $\mathbf{p}$. A nested loop from Step 3 to 6 is employed to construct the negative portion of the next ordinal power of T_P. This inner loop is essentially a top-down construction of the largest set of ground atoms that qualify for being the negative inferences due to other atoms in the same set. Variables of the form $\tilde{\mathbf{p}}$ are used for that purpose.

Algorithm CONSTRUCT
Input: A general deductive database P.
Output: Paraconsistent relation values for the predicate symbols of P.
Method: The values are constructed by the following steps:

1. *(Initialization)*
   (a) Construct the sets $\Upsilon_1(P)$ and $\Upsilon_2(P)$ using Algorithms **CONVERT1** and **CONVERT2** repeatedly.
   (b) For each predicate symbol $\mathbf{p}$ in P, set $\mathbf{p}^+$ and $\mathbf{p}^-$ to $\emptyset$.
2. For each predicate symbol $\mathbf{p}$ in P, set $\hat{\mathbf{p}}$ and $\tilde{\mathbf{p}}^+$ to
$$\{\langle b_1,\ldots,b_k\rangle \mid k \text{ is the arity of } \mathbf{p}, \text{ and } \langle b_1,\ldots,b_k\rangle \notin \mathbf{p}^+\}.$$
3. For each predicate symbol $\mathbf{p}$ in P, set $\tilde{\mathbf{p}}^-$ to $\mathbf{p}^+$.
4. For each equation of the form $\tilde{\mathbf{p}} = D_{\tilde{\mathbf{p}}}$ in $\Upsilon_2(P)$, compute the expression $D_{\tilde{\mathbf{p}}}$ and set $\tilde{\mathbf{p}}$ to the resulting paraconsistent relation.
5. For each predicate symbol $\mathbf{p}$ in P, set $\tilde{\mathbf{p}}$ to $(\tilde{\mathbf{p}} \,\dot{\cup}\, \dot{-}\mathbf{p}) \,\dot{\cap}\, \hat{\mathbf{p}}$.
6. If Steps 4 and 5 involved a change in the value of some $\tilde{\mathbf{p}}^+$, goto 3.
7. For each equation of the form $\mathbf{p} = D_{\mathbf{p}}$ in $\Upsilon_1(P)$, compute the expression $D_{\mathbf{p}}$ and set $\mathbf{p}$ to the resulting paraconsistent relation.
8. For each predicate symbol $\mathbf{p}$ in P, set $\mathbf{p}^-$ to $\tilde{\mathbf{p}}^+$.
9. If Steps 7 and 8 involved a change in the value of some $\mathbf{p}$, goto 2.
10. Output the final values of all predicate symbols in P. □

Establishing the termination of the algorithms presented is quite straightforward. It follows from the fact that any database has a finite number of clauses, each clause has a finite number of literals, each of which contain only a finite number of argument terms. However, showing that the Algorithm *CONSTRUCT* constructs the well-founded model is somewhat complex, and is omitted from here.

6 An Illustrative Example

Let us now look at an example of the entire model construction method. Let P be the following general deductive database:

```
p(X, 0) ← ¬q(X)
p(1, 0) ← p(X, 1)
p(X, 1) ← p(1, 0), ¬p(0, X)
   q(0) ← ¬p(0, 0)
   q(1) ←
```

Then, $T_P \uparrow 0 = \langle \emptyset, \emptyset \rangle$, $T_P \uparrow 1 = \langle \{\mathtt{q(1)}\}, \emptyset \rangle$, and $T_P \uparrow 2 = \langle \{\mathtt{q(1)}\}, \{\mathtt{p(0,\ 1)}, \mathtt{p(1,\ 0)}, \mathtt{p(1,\ 1)}\} \rangle$. It can be verified that $T_P \uparrow 2$ is the well-founded model of P. In this model, the atom q(1) is true, while the atoms p(0, 1), p(1, 0) and p(1, 1) are false. No truth value is assigned to the remaining atoms p(0, 0) and q(0).

The set $\Upsilon_1(P)$ will have the following two definitions:

$$\begin{aligned}
\mathtt{p} = {} & (\dot{\sigma}_{\mathtt{A=0}}(\dot{\pi}_{\{\mathtt{X,A}\}}(\dot{-}\mathtt{q(X)} \ \dot{\bowtie} \ \mho_{\emptyset})))[\mathtt{X,A}] \ \dot{\cup} \\
& (\dot{\sigma}_{\mathtt{A=1} \wedge \mathtt{B=0}}(\dot{\pi}_{\{\mathtt{A,B}\}}(\dot{\pi}_{\{\mathtt{X}\}}(\dot{\sigma}_{\mathtt{C=1}}(\mathtt{p(X,C)})) \ \dot{\bowtie} \ \mho_{\emptyset})))[\mathtt{A,B}] \ \dot{\cup} \\
& (\dot{\sigma}_{\mathtt{A=1}}(\dot{\pi}_{\{\mathtt{X,A}\}}(\dot{\pi}_{\emptyset}(\dot{\sigma}_{\mathtt{B=1} \wedge \mathtt{C=0}}(\mathtt{p(B,C)})) \ \dot{\bowtie} \ \dot{-}\dot{\pi}_{\{\mathtt{X}\}}(\dot{\sigma}_{\mathtt{D=0}}(\mathtt{p(D,X)})) \ \dot{\bowtie} \\
& \ \mho_{\emptyset})))[\mathtt{X,A}] \ \dot{\cup} \\
& \Phi_{\Gamma_2} \\
\mathtt{q} = {} & (\dot{\sigma}_{\mathtt{A=0}}(\dot{\pi}_{\{\mathtt{A}\}}(\dot{-}\dot{\pi}_{\{\}}(\dot{\sigma}_{\mathtt{B=0} \wedge \mathtt{C=0}}(\mathtt{p(B,C)})) \ \dot{\bowtie} \ \mho_{\emptyset})))[\mathtt{A}] \ \dot{\cup} \\
& (\dot{\sigma}_{\mathtt{A=1}}(\dot{\pi}_{\{\mathtt{A}\}}(\mho_{\emptyset})))[\mathtt{A}] \ \dot{\cup} \\
& \Phi_{\Gamma_1}
\end{aligned}$$

and the set $\Upsilon_2(P)$ will have the following definitions:

$$\begin{aligned}
\tilde{\mathtt{p}} = {} & (\dot{\sigma}_{\mathtt{A=0}}(\dot{\pi}_{\{\mathtt{X,A}\}}(\dot{\mu}_{\emptyset}(\dot{\pi}_{\{\mathtt{X}\}}(\dot{-}\tilde{\mathtt{q}}(\mathtt{X})) \ \dot{\cup} \ \Phi_{\{\mathtt{X}\}}))) \ \dot{\cup} \ \dot{\sigma}_{\neg(\mathtt{A=0})}(\mho_{\{\mathtt{X,A}\}}))[\mathtt{X,A}] \ \dot{\cap} \\
& (\dot{\sigma}_{\mathtt{A=1} \wedge \mathtt{B=0}}(\dot{\pi}_{\{\mathtt{A,B}\}}(\dot{\mu}_{\{\mathtt{X,C}\}}(\dot{\pi}_{\{\mathtt{X,C}\}}(\dot{\sigma}_{\mathtt{C=1}}(\tilde{\mathtt{p}}(\mathtt{X,C}))) \ \dot{\cup} \ \Phi_{\{\mathtt{X,C}\}}))) \ \dot{\cup} \\
& \ \dot{\sigma}_{\neg(\mathtt{A=1} \wedge \mathtt{B=0})}(\mho_{\{\mathtt{A,B}\}}))[\mathtt{A,B}] \ \dot{\cap} \\
& (\dot{\sigma}_{\mathtt{A=1}}(\dot{\pi}_{\{\mathtt{X,A}\}}(\dot{\mu}_{\{\mathtt{B,C,D}\}}(\dot{\pi}_{\{\mathtt{X,B,C,D}\}}(\dot{\sigma}_{\mathtt{B=1} \wedge \mathtt{C=0}}(\tilde{\mathtt{p}}(\mathtt{B,C}))) \ \dot{\cup} \\
& \ (\dot{\pi}_{\{\mathtt{X,B,C,D}\}}(\dot{\sigma}_{\mathtt{D=0}}(\dot{-}\tilde{\mathtt{p}}(\mathtt{D,X}))) \ \dot{\cup} \ \Phi_{\{\mathtt{X,B,C,D}\}}))) \ \dot{\cup} \ \dot{\sigma}_{\neg(\mathtt{A=1})}(\mho_{\{\mathtt{X,A}\}}))[\mathtt{X,A}] \ \dot{\cap} \\
& \mho_{\Gamma_2} \\
\tilde{\mathtt{q}} = {} & (\dot{\sigma}_{\mathtt{A=0}}(\dot{\pi}_{\{\mathtt{A}\}}(\dot{\mu}_{\{\mathtt{B,C}\}}(\dot{\pi}_{\{\mathtt{B,C}\}}(\dot{\sigma}_{\mathtt{B=0} \wedge \mathtt{C=0}}(\dot{-}\tilde{\mathtt{p}}(\mathtt{B,C}))) \ \dot{\cup} \ \Phi_{\{\mathtt{B,C}\}}))) \ \dot{\cup} \\
& \ \dot{\sigma}_{\neg(\mathtt{A=0})}(\mho_{\{\mathtt{A}\}}))[\mathtt{A}] \ \dot{\cap} \\
& (\dot{\sigma}_{\mathtt{A=1}}(\dot{\pi}_{\{\mathtt{A}\}}(\dot{\mu}_{\emptyset}(\Phi_{\emptyset}))) \ \dot{\cup} \ \dot{\sigma}_{\neg(\mathtt{A=1})}(\mho_{\{\mathtt{A}\}}))[\mathtt{A}] \ \dot{\cap} \\
& \mho_{\Gamma_1}
\end{aligned}$$

After Step 1 of Algorithm *CONSTRUCT*, both $\mathtt{p} = \langle \emptyset, \emptyset \rangle$ and $\mathtt{q} = \langle \emptyset, \emptyset \rangle$. Thus, interpretation $T_P \uparrow 0$ is ready. It can be verified that after the loop in

Steps 3 to 6 is exited, both $\tilde{p}^+$ and $\tilde{q}^+$ are $\emptyset$. And after Step 7, $p^+ = \emptyset$, but $q^+ = \{1\}$. Now p^+ and q^+ make $(T_P \uparrow 1)^+$, whereas by Step 8, $\tilde{p}^+$ and $\tilde{q}^+$ make $(T_P \uparrow 1)^-$. A similar hand execution of the algorithm shows that at the end, the output produced by Step 10 is

$$p^+ = \emptyset, \quad p^- = \{(0,1),(1,0),(1,1)\};$$
$$q^+ = \{(1)\},\ q^- = \emptyset.$$

7 Conclusions and Related Work

We have presented an algebraic method to construct the well-founded model, proposed by van Gelder, Ross and Schlipf [16], for a general deductive database. The method is based on an extension to the relational model of data in which explicit representation of negative facts is permitted. Such relations, called *paraconsistent relations*, were introduced by Bagai and Sunderraman in [4] along with a suitable set of algebraic operators.

Our method constructs the well-founded model in a bottom-up way that mimics the T_P operator. However, since we use algebraic operators, it makes possible arriving at efficient implementations by optimizing algebraic expressions. As a consequence, the algebraic bottom-up approach is an attractive alternative to some of the other existing approaches which we discuss next.

Soon after the introduction of the well-founded semantics, van Gelder [15] proposed a constructive definition of the same (rather than an inductive one). This formed the basis for Kemp, Stuckey and Srivastava [9], who present a bottom-up operational procedure for constructing the well-founded model. However, their method works for a restricted class of deductive databases called *allowed* DATALOG programs with negation. They also consider magic-sets optimization techniques in a limited sense. Leone and Rullo [10] present a bottom-up method to construct the well-founded semantics of Datalog queries. The answers to queries are evaluated without having to construct the entire greatest unfounded set. The creation of false facts is limited to only those that contribute to derive new positive facts. Ross [13] adapts the magic-sets rewriting technique for a subclass of logic programs called *modularly stratified programs*.

There are at least two top-down approaches to this problem. Bidoit and Legay [6] present a top-down algorithm called *Well!*. This approach applies only to the class of non-floundering queries. Chen and Warren [7] present a goal-oriented method for constructing the well-founded model of general logic programs. It has the practical advantages of top-down evaluation and integration with Prolog. Our approach differs from all of these in its algebraic nature.

One possible direction for future work is the incorporation of explicit negation in disjunctive databases. We plan to add explicit negation to the generalized relational model for disjunctive databases of Liu and Sunderraman [11]. This will lay the foundation for an algebraic construction of the semantics of these databases, a subject that is almost unexplored at present.

References

1. K. R. Apt, H. A. Blair, and A. Walker. Towards a theory of declarative knowledge. In Jack Minker, editor, *Foundations of Deductive Databases and Logic Programming*, pages 89–148. Morgan Kaufmann, Los Altos, 1988.
2. R. Bagai, M. Bezem, and M. H. van Emden. On downward closure ordinals of logic programs. *Fundamenta Informaticae*, XIII(1):67–83, March 1990.
3. R. Bagai and R. Sunderraman. Bottom-up computation of the Fitting model for general deductive databases. *Journal of Intelligent Information Systems*, 1995. (To appear).
4. R. Bagai and R. Sunderraman. A paraconsistent relational data model. *International Journal of Computer Mathematics*, 55(3), 1995.
5. F. Bancilhon, D. Maier, Y. Sagiv, and J.D. Ullman. Magic sets and other strange ways to implement logic programs. In A. Silberschatz, editor, *Proceedings of the 5th Symposium on Principles of Database Systems*, pages 1–15, New York, 1986. A.C.M. SIGACT-SIGMOD.
6. N. Bidoit and P. Legay. Well!: An evaluation procedure for all logic programs. In *Proceedings of International Conference on Database Theory*, pages 335–345. Lecture Notes in Computer Science, 470, Springer-Verlag, 1990.
7. W. Chen and D. S. Warren. A goal-oriented approach to computing well founded semantics. In *Proceedings of the Joint International Conference and Symposium on Logic Programming*, Washington, D.C., 1992.
8. M. Fitting. A Kripke-Kleene semantics for logic programs. *Journal of Logic Programming*, 4:295–312, 1985.
9. D. B. Kemp, P. J. Stuckey, and D. Srivastava. Magic sets and bottom-up evaluation of well-founded models. In *Proceedings of the 1991 International Symposium on Logic Programming*, pages 337–354, San Diego, USA, 1991.
10. N. Leone and P. Rullo. The safe computation of the well-founded semantics of datalog queries. *Information Systems*, 17(1):17–31, 1992.
11. K.-C. Liu and R. Sunderraman. A generalized relational model for indefinite and maybe information. *IEEE Transactions on Knowledge and Data Engineering*, 3(1):65–77, 1991.
12. T. C. Przymusinski. Perfect model semantics. In *Proceedings of the 5th International Conference and Symposium on Logic Programming*, pages 1081–1096, Seattle, WA, August 1988.
13. K. A. Ross. Modular stratification and magic sets for datalog programs with negation. In *Proceedings of the Ninth Annual ACM Symposium on Principles of database systems*. ACM, 1990.
14. J. D. Ullman. *Principles of Database and Knowledge-Base Systems*, volume 1. Computer Science Press, 1988.
15. A. van Gelder. The alternating fixpoint of logic programs with negation. In *Proceedings of the 8th ACM Symposium on Principles of Database Systems*, pages 1–10, Philadelphia, USA, 1989. ACM Press.
16. A. van Gelder, K. A. Ross, and J. S. Schlipf. The well-founded semantics for general logic programs. *Journal of the ACM*, 38(3):621–650, 1991.

Confluence in Concurrent Constraint Programming

Moreno Falaschi
Dip. di Mat. e Inf., Univ. di Udine, Via Delle Scienze 206, Udine, Italy.
falaschi@dimi.uniud.it

Maurizio Gabbrielli
Dip. di Inf., Univ. di Pisa, Corso Italia 40, Pisa, Italy. gabbri@di.unipi.it

Kim Marriott
Dept. of Comp. Sci., Monash Univ., Clayton 3168, Victoria, Australia.
marriott@bruce.cs.monash.edu.au

Catuscia Palamidessi
DISI, Univ. di Genova, Via Benedetto XV 3, Genova, Italy. catuscia@disi.unige.it

Abstract. We investigate the subset of concurrent constraint programs (ccp) which are confluent in the sense that different process schedulings lead to the same possible outcomes. Confluence is an important and desirable property as it allows the program to be understood by considering any desired scheduling rule, rather than having to consider all possible schedulings. The subset of confluent programs is less expressive than full ccp. For example it cannot express fair merge although it can express demonic merge. We give a simple closure based denotational semantics for confluent ccp. We also study admissible programs which is a subset of confluent ccp closed under composition. We consider then applications of our results to give a framework for the efficient yet accurate analysis of full ccp. The basic idea is to approximate an arbitrary ccp program by an admissible program which is then analyzed.

1 Introduction

Concurrent constraint programming (ccp) [10, 11] is a programming paradigm which elegantly combines logical concepts and concurrency mechanisms. The computational model of ccp is based on the notion of *constraint system*, which consists of a set of constraints and an *entailment* relation. Processes interact through a common *store*. Communication is achieved by *telling* (adding) a given constraint to the store, and by *asking* (checking whether the store entails) a given constraint. Non-determinacy arises in two ways: because of a guarded choice construct and because of different process schedulings.

This work was partially supported by the ESPRIT BRA project 6707 ("Parforce"). The work of Catuscia Palamidessi is partially supported by the HCM project "Express".

Confluence, which holds when all process schedulings lead to the same results, is an important and desirable semantic property as it allows a program to be understood by considering only one scheduling. Analogous notions have been investigated for instance in lambda calculus, where it holds because of the Church-Rosser property, and in logic programming, where it holds because of the Switching Lemma. In the context of concurrency, confluence is an even more desirable property since concurrent programs are notoriously difficult to reason about and to analyze.

We investigate the subset of ccp programs which are confluent in the above sense. Note that, while in lambda-calculus confluence implies that only one result is generated, here different outcomes are still possible, due to the presence of nondeterministic choices. Crucial to our results is the monotonic evolution of the store in ccp computations.

First we investigate *confluent ccp* programs. This class is less expressive than full ccp, in fact it is not possible to define in it typical schedule-dependent processes like angelic merge [9]. However it is more expressive than determinate ccp [11], since it embodies nondeterminism. We show that confluence allows us to define a very simple denotational semantics modeling, for each input, the upward closure of the result. Our construction is based on ideas of [11], where determinate ccp agents were shown to correspond to closure operators. In the same paper an extension of such semantics was proposed for full ccp, but the resulting semantic domains and operators were much more complicated due to the non-confluence.

Our second contribution is to investigate a subclass of confluent ccp with a restricted form of choice, called *admissible programs*. Admissible, i.e. harmless choices are those which do not depend on the relative speed of other processes. For this subclass we show that the denotational semantics can be modified so to model a more refined notion of observables: the input/output relation. In this semantics, processes are modeled by sets of sets of resting points, the different sets corresponding to different local choices.

Our third contribution is to investigate the application of results to the analysis of ccp programs. The idea is to approximate an arbitrary ccp program by an admissible program, which is then analyzed. The advantage of analyzing the latter is that in the analysis we need to consider only a single process scheduling. Moreover, since confluent (and hence admissible) programs have simple denotational semantics, a simple compositional analysis can be obtained following standard methods [6]. A similar method has been investigated independently in [13], for the input/upward-closed output observables.

The application of confluence to program analysis is similar to that suggested in [4] for analyzing concurrent logic programs. However, it differs because ccp languages do not have atomic tell. This improves accuracy of the analysis itself because descriptions of the constraint store need not be "downward closed". Furthermore, the simple denotational semantics developed for confluent program provides a basis for compositional analysis. Compositional analyses for ccp based on the denotational semantics have also been investigated in [5]. However our approach here is orthogonal, in fact the approximation introduced in [5] corresponds to allowing the parallel components of a system to "restart" from scratch, whereas here it consists on generating more possibilities at the choice points.

2 Concurrent Constraint Programming

In this section we recall the definition of ccp and its operational semantics (see [11] for more details), and we give the notion of *fair* computation.

Ccp is based on the notion of cylindric constraint system:

Definition 1. A *cylindric constraint system* is a complete algebraic lattice $\mathbf{C} = \langle C, \leq, \sqcup, true, false, Var, \exists, d\rangle$, where

(i) $\sqcup$ is the lub operation (representing the logical and), and *true*, *false* are the least and the greatest elements of C, respectively[1],
(ii) for each $x \in Var$ the function $\exists_x : C \to C$ is a *cylindrification operator* [7],
(iii) for each $x, y \in Var$, $d_{xy} \in C$ is a *diagonal element* [7].

The cylindrification operators model a sort of existential quantification and are helpful when defining a hiding operator in the language. The diagonal elements are useful to model parameter passing. If C contains equality constraints, then d_{xy} is the formula $x = y$.

Ccp is parametric wrt $\mathbf{C}$ and it is described by the following syntax

$$\textit{Declarations}\ \ D ::= \epsilon \mid p(x) :\!-\ A \mid D_1, D_2$$

$$\textit{Agents}\ \ A ::= \mathbf{Stop} \mid \mathbf{tell}(c) \mid \textstyle\sum_{i=1}^{n} \mathbf{ask}(c_i) \to A_i \mid A_1 \parallel A_2 \mid \exists_x A' \mid p(x)$$

The agent **Stop** represents successful termination. In $\mathbf{ask}(c)$ and $\mathbf{tell}(c)$ c is a *finite constraint*. These actions work on a common *store* which ranges over C. The execution of $\mathbf{tell}(c)$ in the store d sets the store to $c \sqcup d$. $\mathbf{ask}(c)$ is a *guard*: it performs a test on the current store d and it is *enabled* if $c \leq d$. It does not modify the store. The *guarded choice* agent $\sum_{i=1}^{n} g_i \to A_i$ selects nondeterministically one g_i which is enabled, and then behaves like A_i. If no guards are enabled, then it *suspends*, waiting for other (parallel) agents to add information to the store. The symbol $\parallel$ represents parallel composition. With slight abuse of notation we use $\exists_x$ to indicate also an *hiding operator*; the intended meaning of $\exists_x A$ is that of an agent which behaves like A, with x considered as *local*. Finally, the agent $p(x)$ is a procedure call, where p is the name of the procedure and x is the actual parameter. The meaning of $p(x)$ is given by a procedure declaration of the form $p(y) :\!-\ A$, where y is the formal parameter. In the following, we use $+$ as a shorthand for $\sum_{i=1}^{2}$.

The operational model of ccp, informally introduced above, is described in terms of a transition system $T_D = (Conf, \longrightarrow)$ which is specified with respect to

[1] The entailment relation $\vdash$, which is commonly used in the literature, is the reverse of $\leq$. Formally: $\forall c, d \in C.\ c \vdash d \Leftrightarrow d \leq c$.

a given set of declarations (or *program*) D. The configurations in *Conf* are pairs consisting of an agent, and a constraint representing the store. Table 1 describes the rules of T_D.

R1	$\langle \mathbf{tell}(c), d\rangle \longrightarrow \langle \mathbf{Stop}, c \sqcup d\rangle$
R2	$\langle \sum_{i=1}^{n} \mathbf{ask}(c_i) \rightarrow A_i, d\rangle \longrightarrow \langle A_j, d\rangle \; j \in [1, n] \; and \; c_j \leq d$
R3	$\dfrac{\langle A, c\rangle \longrightarrow \langle A', c'\rangle}{\begin{array}{c}\langle A \parallel B, c\rangle \longrightarrow \langle A' \parallel B, c'\rangle \\ \langle B \parallel A, c\rangle \longrightarrow \langle B \parallel A', c'\rangle\end{array}}$
R4	$\dfrac{\langle A, d \sqcup \exists_x c\rangle \longrightarrow \langle B, d'\rangle}{\langle \exists_x^d A, c\rangle \longrightarrow \langle \exists_x^{d'} B, c \sqcup \exists_x d'\rangle}$
R5	$\langle p(x), c\rangle \longrightarrow \langle \Delta_y^x A, c\rangle$ $\qquad p(y)$:- A is the declaration for $p(x)$

Table 1. The transition system T_D

As we are interested in formalizing confluence, our operational semantics is given in terms of the possible computations for a particular process scheduling, which form the "computation tree". Branches in a computation tree result from different choices in a guarded choice. To formalize a process scheduling we must identify the unique "primitive" sub-agent actually reduced by a particular transition.

Let $C \longrightarrow C'$ be a transition in T_D. The *selected* agent in C is the tell, guarded choice, or procedure call agent, respectively, if the reduction is made using rule **R1**, **R2**, or **R5**. In the case of **R4** or **R3** it is the selected agent of the antecedent transition in the rule. An agent A is *enabled* in a configuration C if there is a transition from C with A as the selected agent. A *computation tree* for configuration C and declaration D is a maximal tree which has root C and each node C' is

(i) a leaf node, if no agents in C' are enabled, or
(ii) a non-leaf node, with children $C_1'', \ldots C_n''$, if for $i = 1, \ldots, n$, $C' \longrightarrow C_i''$, are transitions in T_D with a fixed selected agent A which is enabled.

A *derivation* is a maximal path in a computation tree starting from the root. We let *deriv*(T) denote the set of all derivations in the computation tree T.

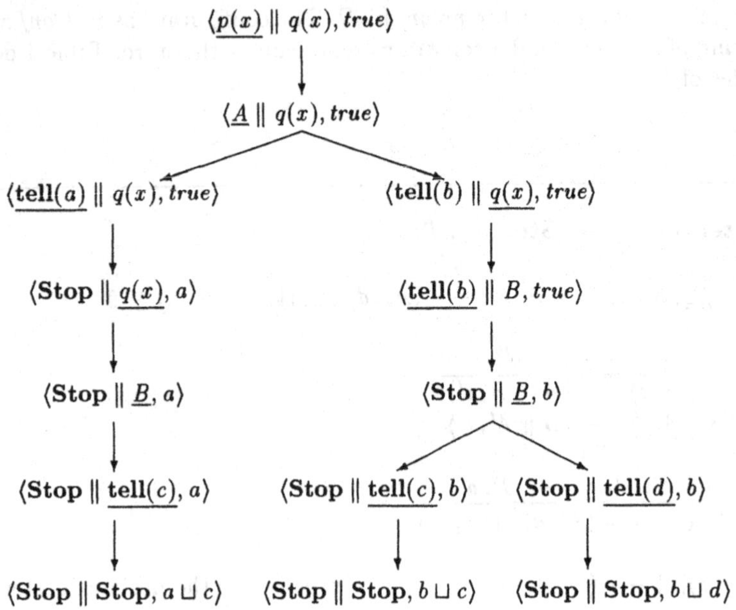

Fig. 1. A computation tree

Example 1. Consider the declarations

$$p(x) \text{ :- } (\mathbf{ask}(true) \rightarrow \mathbf{tell}(a)) + (\mathbf{ask}(true) \rightarrow \mathbf{tell}(b))$$
$$q(x) \text{ :- } (\mathbf{ask}(a) \rightarrow \mathbf{tell}(c)) + (\mathbf{ask}(b) \rightarrow \mathbf{tell}(d))$$

with the assumption $true < a < b$. In Figure 1 it is shown a computation tree for the configuration $\langle p(x) \parallel q(x), true\rangle$. In the fingure, A and B represent the bodies of $p(x)$ and $q(x)$ respectively. The selected agents are underlined and, for the sake of simplicity, the existential quantifications due to the procedure calls are omitted. Note that the agent B cannot be selected before a or b are added to the store, because it is not enabled.

As we are concerned with confluence of infinite computations some notion of "fairness" is required A derivation is *fair* if each agent enabled in some configuration in the derivation is subsequently selected. A computation tree is *fair* if all its derivations are fair. We let $ctree_D\ (C)$ denote the set of all fair computation trees for the configuration C and declarations D.

The standard notion of observables for ccp are the resulting stores computed by an agent for a given input store. Namely, given a finite or infinite derivation γ of the form $\langle B_0, c_0\rangle \longrightarrow \langle B_1, c_1\rangle \longrightarrow ... \longrightarrow \langle B_i, c_i\rangle \rightarrow ...$, the result of γ, written $store(\gamma)$, is the constraint $\sqcup\{c_0, c_1, ..., c_i, ...\}$.

Definition 2. Given a set of declarations D, the observables of an agent A, wrt an initial store c, are $\mathcal{O}(A, c) = \{store(\gamma) \mid \text{there exists } T \in ctree_D(\langle A, c\rangle) \text{ s.t. } \gamma \in deriv(T)\}$.

3 Confluent ccp

In this section we define the class of (structurally) confluent ccp programs, and we develop their denotational semantics.

Definition 3. An agent A is *confluent* if for all constraints c, and for all $T \in ctree_D(\langle A, c\rangle)$, $\mathcal{O}(A, c) = \{store(d) \mid d \in deriv(T)\}$. A is *strongly confluent* if for all constraints c the agent $A \parallel \mathbf{tell}(c)$ is confluent. A is *structurally confluent* if all its sub-agents are strongly confluent.

Not all ccp agents are confluent, in fact different process schedulings may enable different guards in the same choice. This is for instance the case with the angelic-merge process in Example 2. A simpler example is the following: Assume $a \leq b \leq c$ and consider the following declaration D

$$q(x) :\text{-}\ (\ (\mathbf{ask}(true) \rightarrow \mathbf{tell}(a)) \ + \ (\mathbf{ask}(b) \rightarrow \mathbf{tell}(c))\ ) \ \parallel \ \mathbf{tell}(b)$$

Then $\mathcal{O}(q(x), true) = \{b, c\}$ while, for $T \in ctree_D(\langle q(x), true\rangle)$ obtained by considering a leftmost selection rule, we have $\{store(d) | d \in deriv(T)\} = \{b\}$. Therefore the agent $q(x)$ is not confluent.

In the following, we show that the structurally confluent programs admit a simple compositional semantics which is correct wrt the input/upward-closed output observables.

In the following, given a set of constraints $X \subseteq \mathcal{C}$, we denote by $\uparrow X$ (upward closure of X) the set $\{d \in \mathcal{C} \mid \text{there exists } c \in X \ s.t. \ c \leq d\}$, and by $\overline{X}$ (complement of X) the set $\mathcal{C} \setminus X$. The set of all the subsets of $\mathcal{C}$ will be denoted by $\wp(\mathcal{C})$.

The input/upward-closed output observables are defined as $\mathcal{O}^u(A, c) = \uparrow \mathcal{O}(A, c)$.

We now define the denotational model: $\mathcal{D} : Agents \rightarrow \wp(\mathcal{C})$ is the least, wrt the ordering induced by $\supseteq$, among the functions which satisfy the equations in Table 2. Intuitively, $\mathcal{D}[\![A]\!]$ represents the *resting points* of the process A, i.e. those inputs stores which are not affected by a possible computation for A.

The following results show that in the case of structurally confluent agents $\mathcal{O}^u$ and $\mathcal{D}$ are equivalent, in the sense that the one can be retrieved from the other. This implies the correctness of $\mathcal{D}$ and, trivially, the so-called "full abstraction" of $\mathcal{D}$.

Proposition 4. *For any structurally confluent agent A we have $\mathcal{D}[\![A]\!] = \{c | c \in \mathcal{O}^u(A, c)\}$.*

Proposition 5. *For every structurally confluent agent A we have $\mathcal{O}^u(A, c) = \uparrow (\uparrow c \cap \mathcal{D}[\![A]\!])$*

D1	$\mathcal{D}[\![\mathbf{Stop}]\!] = \mathcal{C}$
D2	$\mathcal{D}[\![\mathbf{tell}(c)]\!] = \uparrow c$
D3	$\mathcal{D}[\![A \parallel A']\!] = \mathcal{D}[\![A]\!] \bigcap \mathcal{D}[\![A']\!]$
D4	$\mathcal{D}[\![\sum_{i=1}^{n} \mathbf{ask}(c_i) \rightarrow A_i]\!] = \bigcup_{i=1}^{n}(\uparrow c_i \cap \mathcal{D}[\![A_i]\!]) \cup \bigcap_{i=1}^{n} \overline{\uparrow c_i}$
D5	$\mathcal{D}[\![\exists_x^c A]\!] = \{c' \mid \text{there exists } c'' \in \mathcal{D}[\![A]\!] \cap \uparrow c \text{ s.t. } \exists_x c' = \exists_x c''\}$
D6	$\mathcal{D}[\![p(x)]\!] = \mathcal{D}[\![\Delta_y^x A]\!]$ where $p(y)$:- A is the declaration of p in D

Table 2. The denotational semantics for structurally confluent agents

From Propositions 4 and 5 we derive:

Theorem 6. *$\mathcal{D}$ and $\mathcal{O}^u$ induce the same equivalence relation on structurally confluent agents.*

4 Admissible agents

In this section we identify a subclass of structurally confluent programs which we call *admissible*. The interest for this class of programs comes from the fact that they can be constructed in a "bottom-up" way by using a restricted choice and the other standard operators to combine the basic agents **Stop** and **tell**(c). Moreover, this language has a simple denotational model the input/output relation, which is a notion of observables more refined than the one considered in previous section.

In the following, a *context* is an agent $A[\bullet]$ which contains a "placeholder", denoted by $[\bullet]$. We place an agent A' in this context, written $A[A']$, by replacing the placeholder by A'. The agents A and A' are *congruent*, written $A \equiv A'$, if for all contexts A'' and for any $c \in \mathcal{C}$, $\mathcal{O}(A''[A], c) = \mathcal{O}(A''[A'], c)$ holds.

Definition 7. Let G be the guarded choice $\sum_{i=1}^{n} \mathbf{ask}(c_i) \rightarrow A_i$.

(i) G is *local* if for all $i, j \in \{1, .., n\}$, $c_i = c_j$.
(ii) G is *compatible* if G is not local and for all $i, j \in \{1, .., n\}$, $\mathbf{tell}(c_i \sqcup c_j) \parallel A_i \equiv \mathbf{tell}(c_i \sqcup c_j) \parallel A_j$.

G is *admissible* if it is either local or compatible. A program D is *admissible* if any choice in D is admissible. An agent A is *admissible* (wrt a program D) if any choice in A is admissible, and D is admissible.

Composing structurally confluent agents by using admissible guarded choices preserves structural confluence. Moreover we note that the basic agents are structurally confluent and that all the other operations preserve structural confluence. Therefore we have the following result:

Proposition 8. *If the agent A is admissible then it is structurally confluent.*

In general it is not decidable if two agents are congruent. Therefore it is not decidable if an agent is admissible. However, there is an interesting subclass of admissible programs which is decidable. Define the guarded choice $\sum_{i=1}^{n} \mathbf{ask}(c_i) \to A_i$ to be *mutually exclusive* if for all $i, j \in \{1, .., n\}$, $i \neq j$ implies $c_i \sqcup c_j = \mathit{false}$ (i.e. the guards are pairwise inconsistent). Hence we have:

Proposition 9. *If each choice in the agent A and in the declaration D is either local or mutually exclusive then A and D are admissible (and hence structurally confluent).*

Note that mutually exclusive guards can be rewritten to an equivalent parallel form: $\sum_{i=1}^{n} \mathbf{ask}(c_i) \to A_i \equiv \prod_{i=1}^{n} \mathbf{ask}(c_i) \to A_i$ where $\prod_{i=1}^{n} B_i$ denotes the parallel conjunction of the n agents B_i. This is not true for local choice.

We sketch here the denotational semantics for admissible agents, which allows to retrieve the input/output relation. For technical reasons, however, we can retrieve only the final results of finite computations. The difference wrt the previous semantics is that here we associate a *set* of sets of resting points to each agent. Intuitively, each set of resting points represesents the resting points associated to a specific branch of the computation.

The denotational model is defined as follows: $\mathcal{D} : \mathit{Agents} \to \wp(\wp(\mathcal{C}))$ is the least, wrt the ordering induced by $\subseteq$, among the functions which satisfy the equations in Table 3, where equation $D5$ is used for compatible choices only.

The critical point of Table 3 is the equation **D5**, which puts together sets corresponding to alternative choices. This is not harmful since, intuitively, the contributions of different branches either correspond to incompatible initial stores or they are equivalent. On the other hand, this fusion is necessary because keeping those sets separate would generate wrong results when trying to retrieve the observables: a separate set would be interpreted as a separate computation, and this would be wrong when the corresponding branch is actually blocked.

The following result shows that this model is correct wrt the finite input/output behavior.

Theorem 10. *Let A be an admissible agent. Then*

$$\mathcal{O}_{\mathrm{Fin}}(A, c) \subseteq \bigcup \{\mathit{minimals}(\uparrow c \cap X) \mid X \in \mathcal{D}[\![A]\!]\}$$

where $\mathcal{O}_{\mathrm{Fin}}(A, c)$ denotes the subset of $\mathcal{O}(A, c)$ corresponding to finite derivations only. Moreover, if each choice in A (and in the related declaration D) is either local or mutually exclusive, and if in the constraint system there exists no c, d

D1	$\mathcal{D}[\![\mathbf{stop}]\!]=\{\mathcal{C}\}$	
D2	$\mathcal{D}[\![\mathbf{tell}(c)]\!]=\{\uparrow c\}$	
D3	$\mathcal{D}[\![A \parallel A']\!]=\{X \cap X' \mid X \in \mathcal{D}[\![A]\!],\ X' \in \mathcal{D}[\![A']\!]\}$	
D4	$\mathcal{D}[\![\sum_{i=1}^{n} \mathbf{ask}(c) \rightarrow A_i]\!]=\{(\uparrow c \cap X) \cup \overline{\uparrow c} \mid X \in \bigcup_{i=1}^{n} \mathcal{D}[\![A_i]\!]\} \cup \{\overline{\uparrow c}\}$	local c.
D5	$\mathcal{D}[\![\sum_{i=1}^{n} \mathbf{ask}(c_i) \rightarrow A_i]\!]=\{\bigcup_{i=1}^{n}(\uparrow c_i \cap X_i) \cup \bigcap_{i=1}^{n} \overline{\uparrow c_i} \mid X_i \in \mathcal{D}[\![A_i]\!]\} \cup \{\bigcap_{i=1}^{n} \overline{\uparrow c_i}\}$	comp. c.
D6	$\mathcal{D}[\![\exists_x^c A]\!]=\{\{c' \mid \text{there exists } c'' \in X \cap \uparrow c \text{ s.t. } \exists_x c' = \exists_x c''\} \mid X \in \mathcal{D}[\![A]\!]\}$	
D7	$\mathcal{D}[\![p(x)]\!]=\mathcal{D}[\![\Delta_y^x A]\!]$ where $p(y)$:- A is the declaration of p in D	

Table 3. The denotational semantics for admissible agents

such that d is finite c is not finite and $c \leq d$, then also the following inclusion holds

$$\mathcal{O}_{\text{Fin}}(A, c) \cup \{\mathit{false}\} \supseteq \bigcup\{\mathit{minimals}(\uparrow c \cap X) \mid X \in \mathcal{D}[\![A]\!]\}.$$

Finally we observe that many ccp programs are admissible. For instance, the ccp version of the following programs in [12] are admissible ccp programs: Mutual exclusion, Duplex stream protocol, Airline reservation system, MSG system, Queues, Simulator of a CP multiprocessor machine (except for the merge process), Spooler (except from the merge process), and SCAN disk-arm scheduler.

5 Application to Program Analysis

In this section we give an application of confluence for the analysis of ccp programs. The idea is the following: We first approximate an arbitrary ccp program by an admissible one, and then analyze the admissible (and hence) confluent program. The analysis can be carried out by using abstract interpretation techniques based either on the operational semantics or on a denotational one. In the first case, the advantage is that we need to consider only a single process scheduling. In the second one, a simple compositional analysis can be obtained by using as a basis the semantics presented in the previous sections.

Let A be a ccp agent. Define $\mathit{adm}(A)$ as the admissible agent obtained by replacing each non-admissible choice $\sum_{i=1}^{n} \mathbf{ask}(c_i) \rightarrow A_i$ in A by the choice

$$\mathbf{ask}(\bigvee_{1=1}^{n} c_i) \rightarrow \sum_{i=1}^{n} \mathbf{ask}(\mathit{true}) \rightarrow (A_i)$$

where $\bigvee_{1=1}^{n} c_i$ represents the intuitionistic logical disjunction, namely a constraint which is entailed by a constraint d iff there exists $i \in \{1, ..., n\}$ such that d entails c_i. In order to formalize such a disjunctive constraint we lift the constraint system **C** to the power set. A similar lifting was introduced in [2].

Given a set of constraints $\mathcal{C}$ with typical element c, properties ϕ are described by the following grammar:

$$\phi ::= c \mid \phi \wedge \psi \mid \phi \vee \psi \mid \exists_x \phi \mid D_{xy}$$

Properties are interpreted as elements of $\wp^U(\mathcal{C})$ (upward closed subsets of $\mathcal{C}$) as follows. Note that the connectives are interpreted intuitionistically, thus obtaining a Heyting-algebra:

$$\begin{array}{lll} [\![c]\!] \quad = \uparrow c & [\![\phi \wedge \psi]\!] = [\![\phi]\!] \cap [\![\psi]\!] & [\![D_{xy}]\!] = \uparrow d_{xy}. \\ [\![\phi \vee \psi]\!] = [\![\phi]\!] \cup [\![\psi]\!] & [\![\exists_x \phi]\!] \quad = \uparrow \exists_x [\![\phi]\!] & \end{array}$$

Proposition 11. *Given a cylindric constraint system* **C**, *the set of its properties, denoted by* $\mathbf{C}_\Phi$, *is a cylindric constraint system where for any* $\phi, \psi \in \mathbf{C}_\Phi$, *the entailment is defined by* $\phi \vdash \psi$ *iff* $[\![\phi]\!] \subseteq [\![\psi]\!]$.

Formally, the transformation *adm* transforms an agent A defined on a constraint system **C** into an agent $adm(A)$, specified as above, defined on a lifted constraint system $\mathbf{C}_\Phi$. The following result shows the correctness of the transformation.

Theorem 12. *Let* A *be an agent. Then for any* $c \in \mathcal{C}$, $\mathcal{O}(A, c) \subseteq \mathcal{O}(adm(A), c)$.

Note that, as observed in [13], the analysis based on the transformation *adm* can be improved by using $A_i \parallel$ **tell**(c_i) rather than A_i in the trasformed agent.

It is also worth noting that the result of computations for $adm(A)$ agents are still constraints of the original constraint system **C** since the lifted (disjunctive) constraints appear only in the ask's. Moreover, observe that $\wedge$ on properties corresponds to the *lub* ($\sqcup$) operator on the original constraints in **C**. In fact, for $c, d \in \mathcal{C}$, $[\![c \sqcup d]\!] = [\![\phi]\!] \cap [\![\psi]\!] = [\![c \wedge d]\!]$. A similar correspondence does not hold between $\vee$ and the *glb* ($\sqcap$) of the constraint system since, in general, we only have $[\![c \vee d]\!] = [\![\phi]\!] \cup [\![\psi]\!] \subseteq [\![c \sqcap d]\!]$. Using $\sqcap$ instead of $\vee$ would give an incorrect analysis.

As a consequence of Theorem 12, for any result delivered by $\langle A, c \rangle$ in D, $\langle adm(A), c \rangle$ gives the same result in some fixed computation tree for $adm(D)$. This justifies the use of the translated program to analyze properties of results.

5.1 Compositional Analysis

Abstract interpretation formalizes the notion of "approximate computation", where descriptions of data replace the data itself. An analysis is then a computation in which the program is evaluated by using a *non-standard interpretation* for data and operators in the program. According to this view, "concrete" constraints and operators are approximated by "abstract" ones. The denotational semantics presented in the previous sections can be used as the basis for a compositional analysis, following standard techniques. We consider here the semantics described in Table 2; the extension to the case of the semantics described in Table 3 is also standard.

Definition 13. A *description* $\langle \mathbf{A}, \alpha, \mathbf{C} \rangle$ consists of an *abstract domain* (a poset) $\mathbf{A} = \langle \mathcal{A}, \leq^{\mathcal{A}} \rangle$, a *concrete domain* $\mathbf{C} = \langle \mathcal{C}, \leq^{\mathcal{C}} \rangle$, and a monotonic *abstraction function* $\alpha : \mathcal{C} \rightarrow \mathcal{A}$.
Given $a \in \mathcal{A}$, $c \in \mathcal{C}$, we say that a approximates c, written $a \propto c$, iff $a \leq^{\mathcal{A}} \alpha(c)$. The approximation relation is lifted to properties (as defined in section 5), functions and sets as follows:

- For basic properties (i.e. constraints) the approximation relation is as before. For the other cases we define

  $a \propto \phi \wedge \psi$ iff $a \propto \phi$ and $a \propto \psi$
  $a \propto \phi \vee \psi$ iff $a \propto \phi$ or $a \propto \psi$

- Let $\langle \mathbf{A}_1, \alpha_1, \mathbf{C}_1 \rangle$ and $\langle \mathbf{A}_2, \alpha_2, \mathbf{C}_2 \rangle$ be descriptions, $F : \mathcal{A}_1 \rightarrow \mathcal{A}_2$ and $F' : \mathcal{C}_1 \rightarrow \mathcal{C}_2$ be functions. Then $F \propto F'$ iff $\forall d \in \mathcal{A}_1 . \forall e \in \mathcal{C}_1 . \; d \propto_1 e \Rightarrow F(d) \propto_2 F'(e)$.
- Let $\langle \mathbf{A}, \alpha, \mathbf{C} \rangle$ be a description and let $Q \in \wp(\mathcal{A})$ and $Q' \in \wp(\mathcal{C})$. Then $Q \propto Q'$ iff $\forall e \in Q' \exists d \in Q . \; d \leq^{\mathcal{A}} \alpha(e)$.

For cc languages, we are interested in descriptions of constraint systems. We give the following definition, which allows us to develop a compositional analysis based on $\mathcal{D}$.

Definition 14. Consider the cylindric constraint systems $\mathbf{C}$ and $\mathbf{A}$ with $\mathbf{C} = \langle \mathcal{C}, \leq, \sqcup, \mathit{true}, \mathit{false}, \mathit{Var}, \exists \rangle$ and $\mathbf{A} = \langle \mathcal{A}, \leq^{\mathcal{A}}, \sqcup^{\mathcal{A}}, \mathit{true}^{\mathcal{A}}, \mathit{false}^{\mathcal{A}}, \mathit{Var}, \exists^{\mathcal{A}} \rangle$. A constraint system description $\langle \mathbf{A}, \alpha, \mathbf{C} \rangle$ is a description such that

1. $\sqcup^{\mathcal{A}} \propto \sqcup$
2. $\forall x \in \mathit{Var}. \; \exists_x^{\mathcal{A}} \propto \exists_x$.
3. $\forall c \in \mathcal{C}. \; \alpha(\exists_x c) = \exists_x^{\mathcal{A}} \alpha(\exists_x c)$.
4. $\forall x, y \in \mathit{Var}. \; \alpha(d_{xy}) = d_{xy}^{\mathcal{A}}$.

For any constraint set X we let $\alpha(X) = \{\alpha(d) \mid d \in X\}$.
We will consider here finite computations only. Hence in the following definition we consider the partial ordering $\subseteq$.

Definition 15. The semantics $\mathcal{D}^{\mathcal{A}} : \mathit{Agents} \rightarrow \mathcal{P}(\mathcal{A})$ is the least function which satisfies the equations in Table 4.

The correctness of this approximation can be expressed as follows. Let $\mathcal{O}^{\mathcal{A}} : \mathit{Agents} \times \mathcal{A} \rightarrow \wp(\mathcal{A})$ be defined as:

$$\mathcal{O}^{\mathcal{A}}(A, a) = \uparrow \{b \mid b \in \mathcal{D}^{\mathcal{A}}[\![A]\!] \cap \uparrow a\}$$

Intuitively, $\mathcal{O}^{\mathcal{A}}$ are the "abstract observables" which can be retrieved from $\mathcal{D}^{\mathcal{A}}$. We have:

AD1	$\mathcal{D}^{\mathcal{A}}[\![\mathbf{Stop}]\!] = \mathcal{A}$
AD2	$\mathcal{D}^{\mathcal{A}}[\![\mathbf{tell}(c)]\!] = \alpha(\uparrow c)$
AD3	$\mathcal{D}^{\mathcal{A}}[\![A \| B]\!] = \mathcal{D}^{\mathcal{A}}[\![A]\!] \cap \mathcal{D}^{\mathcal{A}}[\![B]\!]$
AD4	$\mathcal{D}^{\mathcal{A}}[\![\sum_{i=1}^{n} \mathbf{ask}(c_i) \rightarrow A_i]\!] = \bigcup_{i=1}^{n} (\alpha(\uparrow c_i) \cap \mathcal{D}^{\mathcal{A}}[\![A_i]\!]) \cup \{a \mid \exists c . a \propto c$ and $(\bigvee_{i=1}^{n} c_i) \not\leq c\}$
AD5	$\mathcal{D}^{\mathcal{A}}[\![\exists_x^c A]\!] = \{a \mid a \in \mathcal{A}$ and there exists $b \in \mathcal{D}^{\mathcal{A}}[\![A]\!] \cap \alpha(\uparrow c)$ s.t. $\exists_x^{\mathcal{A}} a = \exists_x^{\mathcal{A}} b\}$
AD6	$\mathcal{D}^{\mathcal{A}}[\![p(x)]\!] = \mathcal{D}^{\mathcal{A}}[\![\Delta_y^x A]\!]$, where $p(y) :- A$ is the declaration of $p(x)$ in D

Table 4. The abstract denotational semantics $\mathcal{D}^{\mathcal{A}}$

Proposition 16. *For any agent A, and any constraint c, $\mathcal{O}^{\mathcal{A}}(A, \alpha(c)) \propto \mathcal{O}^{\mathrm{u}}_{\mathrm{Fin}}(A, c)$ holds, where $\mathcal{O}^{\mathrm{u}}_{\mathrm{Fin}}$ are the finite and upward-closed observables.*

6 An Efficient and Compositional Groundness Analysis

In this section we illustrate the abstract denotational semantics $\mathcal{D}^{\mathcal{A}}$ by an example of groundness analysis for ccp programs over term equations.

Consider the terms $t_1, t_2 \ldots$ on a signature and a set *Var* of variables. Let $\mathcal{E}$ be the set of existentially quantified conjunctions of equations, i.e. the least set $\mathcal{E}$ such that

- for any pair of terms t, t', $t = t' \in \mathcal{E}$,
- if $e \in \mathcal{E}$ then $\exists x . e \in \mathcal{E}$,
- if $e, e' \in \mathcal{E}$ then $e \sqcup e' \in \mathcal{E}$.

We denote by **Eqn** the Herbrand (cylindric) constraint system whose elements are those in $\mathcal{E}$ modulo logical equivalence with ordering $[e] \leq [e']$ iff $e' \models e$, and whose operations are the obvious ones ($\sqcup$ is the logical conjunction and $\exists_x$ is the existential quantifier).

Definition 17. An element $e \in \mathcal{E}$ is *solved* if e is of the form $\exists \mathbf{y} . x_1 = t_1 \sqcup \cdots \sqcup x_n = t_n$ where each x_i is a distinct variable not occurring in any of the terms t_i and each $y \in \mathbf{y}$ occurs in some t_j. ($\exists_{\{y_1, y_2, \ldots\}}$ is a shorthand for $\exists_{y_1} \exists_{y_2} \ldots$.)

It is well known that any satisfiable $e \in \mathcal{E}$ can be transformed into an equivalent one $Sol(e)$ which is solved. If e is not satisfiable we define $Sol(e) = \mathit{false}$.

The idea of groundness analysis is to infer statically which variables in the initial state are bound to ground terms in all possible successful computations.

A description of an element $e \in \mathcal{E}$ is a set of variables, with the intended meaning that any unifier of e binds these variables to ground terms. This description can be given by

$$\alpha(e) = \begin{cases} \{x \mid x = t \in Sol(e) \text{ and } t \text{ is } ground\} & \text{if } Sol(e) \neq \text{ false} \\ Var & \text{if } Sol(e) = \text{ false} \end{cases}$$

The abstract constraint system **A** has domain $\wp(Var)$ (i.e. $\mathcal{A} = \wp(Var)$) and operations defined as follows. For any $a, b \in \wp(Var)$ and for any $e, e' \in \mathcal{E}$,

1. $a \leq^{\mathcal{A}} b$ iff $a \subseteq b$,
2. $a \sqcup^{\mathcal{A}} b = a \cup b$,
3. $\exists_x^{\mathcal{A}} a = a \setminus \{x\}$,

It is easy to verify that $\langle \mathbf{A}, \alpha, \mathbf{Eqn} \rangle$ is a constraint system description.

In the analysis we are interested in resting points over $\wp(Var)$. However a more compact representation is by a Boolean function whose solutions are exactly the resting points.

Thus for example, if there are three variables $\{x, y, z\}$ then the resting points $\{\emptyset, \{x\}, \{y\}, \{z\}, \{x, z\}, \{y, z\}, \{x, y, z\}\}$ are represented by the function $\neg x \vee \neg y \vee z$. The abstract operations over the resting points can be efficiently performed over the Boolean function representation. Intersection of solutions corresponds to conjunction and restriction of a variable corresponds to existential quantification. Thus representations and techniques from [1] for analysis with Boolean functions can be used.

Example 2. Consider the following declaration D defining two producers $p1$ and $p2$, which respectively produce a stream of a's and b's of arbitrary length, and an agent m, which non-deterministically merges its two input streams x and y into an output stream z.

$$\begin{aligned} p1(x) :\!-\; & (\mathbf{ask}(true) \rightarrow \exists_{x'} (\mathbf{tell}(x = [a|x']) \parallel p1(x'))) \\ & + \\ & (\mathbf{ask}(true) \rightarrow \mathbf{tell}(x = [])) \end{aligned}$$

$$\begin{aligned} p2(y) :\!-\; & (\mathbf{ask}(true) \rightarrow \exists_{y'} (\mathbf{tell}(y = [b|y']) \parallel p2(y'))) \\ & + \\ & (\mathbf{ask}(true) \rightarrow \mathbf{tell}(y = [])) \end{aligned}$$

$$\begin{aligned} m(x, y, z) :\!-\; & (\mathbf{ask}(\exists_{x'}\, x = [a|x']) \rightarrow (\exists_{x'z'} (\mathbf{tell}(x = [a|x']) \parallel \mathbf{tell}(z = [a|z']) \parallel \\ & \qquad m(x', y, z')))) \\ & + \\ & (\mathbf{ask}(\exists_{y'}\, y = [b|y']) \rightarrow (\exists_{y'z'} (\mathbf{tell}(y = [b|y']) \parallel \mathbf{tell}(z = [b|z']) \parallel \\ & \qquad m(x, y', z')))) \\ & + \\ & (\mathbf{ask}(x = []) \rightarrow \mathbf{tell}(z = y)) \\ & + \\ & (\mathbf{ask}(y = []) \rightarrow \mathbf{tell}(z = x)) \end{aligned}$$

Imagine that we wish to analyze the agent $p1(x) \parallel p2(y) \parallel m(x, y, z)$, which specifies that the streams produced by $p1$ and $p2$ are merged into z. Note that the agent $p1(x) \parallel p2(y) \parallel m(x,y,z)$ is not confluent because $m(x,y,z)$ is not confluent.

For applying our techniques, we have first to transform m it into an admissible agent. The agents $p1$ and $p2$ are already admissible (since they have only local choice) and we do not need to transform them. The admissible program am resulting from m is:

$$
\begin{aligned}
am(x,y,z) :\!- \;& (\textbf{ask}(\exists_{x'}\, x = [a|x'] \vee \exists_{y'}\, y = [b|y'] \vee x = [] \vee y = []) \rightarrow \\
& (\textbf{ask}(true) \rightarrow \exists_{x'z'}\, (\textbf{tell}(x = [a|x']) \parallel \textbf{tell}(z = [a|z']) \parallel \\
& \qquad am(x', y, z')))) \\
& + \\
& (\textbf{ask}(true) \rightarrow (\exists_{y'z'}\, (\textbf{tell}(y = [b|y']) \parallel \text{tell}(z = [b|z']) \parallel \\
& \qquad am(x, y', z')))) \\
& + \\
& (\textbf{ask}(true) \rightarrow \textbf{tell}(z = y)) \\
& + \\
& (\textbf{ask}(true) \rightarrow \textbf{tell}(z = x))
\end{aligned}
$$

We can now analyze the agent $p1(x) \parallel p2(y) \parallel am(x,y,z)$.

The analysis corresponds to finding the least fixpoint of the following recursive Boolean equations.

$$
\begin{aligned}
p1(x) \quad & \leftrightarrow (false \vee \exists x'.(x \leftrightarrow x' \wedge p1(x'))) \\
& \quad \vee \\
& \quad (false \vee x) \\
p2(y) \quad & \leftrightarrow (false \vee \exists y'.(y \leftrightarrow y' \wedge p1(y'))) \\
& \quad \vee \\
& \quad (false \vee y) \\
am(x,y,z) & \leftrightarrow (\neg x \wedge \neg y) \\
& \quad \vee \\
& \quad (false \vee \exists x'z'.(x \leftrightarrow x' \wedge z \leftrightarrow z' \wedge am(x', y, z'))) \\
& \quad (false \vee \exists y'z'.(y \leftrightarrow y' \wedge z \leftrightarrow z' \wedge am(x, y', z'))) \\
& \quad (false \vee z \leftrightarrow y) \\
& \quad (false \vee z \leftrightarrow x).
\end{aligned}
$$

We can compute the least fixpoint using a Kleene sequence in the usual fashion [1]. We end up obtaining the solution:

$$
\begin{aligned}
& p1(x) && \leftrightarrow x \\
& p2(y) && \leftrightarrow y \\
& am(x,y,z) && \leftrightarrow \neg x \vee \neg y \vee z \\
& p1(x) \parallel p2(y) \parallel am(x,y,z) && \leftrightarrow x \wedge y \wedge z.
\end{aligned}
$$

Thus the analysis shows that at the end of any possible finite computation of the agent $p(x) \parallel q(y) \parallel am(x,y,z)$ for any input, all variables are ground. This, together with Theorem 12, implies that also the variables of the (non-confluent) agent $p(x) \parallel q(y) \parallel m(x,y,z)$ are ground at the end of any possible finite computation.

It is straightforward to extend this groundness analysis to an analysis for possible suspension. In the above example we can determine that no agents can possibly delay. This means that in the original (non-confluent) agent $p(x) \parallel q(y) \parallel m(x,y,z)$ there is no possibility of suspension.

References

1. T. Armstrong, K. Marriott, P. Schachte and H. Søndergaard. Boolean functions for dependency analysis: Algebraic properties and efficient representation. *Proc. Static Analysis Symposium, SAS'94*. B. Le Charlier (Ed.), Springer-Verlag, Vol. 864 in LNCS, pages 266–280, 1994.
2. F.S. de Boer, M. Gabbrielli, E. Marchiori and C. Palamidessi. Proving Concurrent Constraint Programs Correct. In *Proc. 21st ACM Symp. Principles of Programming Languages*, 1994.
3. F.S. de Boer and C. Palamidessi. A Fully Abstract Model for Concurrent Constraint Programming. In S. Abramsky and T.S.E. Maibaum, editors, *Proc. of TAPSOFT/CAAP, LNCS 493*, pages 296–319. Springer-Verlag, 1991.
4. M. Codish, M. Falaschi, K. Marriott, and W. Winsborough. Efficient Analysis of Concurrent Constraint Logic Programs. In *Proc. of ICALP*, volume 700 of *LNCS*, pages 633-644. Springer-Verlag, Berlin, 1993.
5. M. Falaschi, M. Gabbrielli, K. Marriott, and C. Palamidessi. Compositional Analysis for Concurrent Constraint Programming. In *Proc. Eighth IEEE Symp. on Logic In Computer Science*, pages 210–221. IEEE Computer Society Press, Los Alamitos, California, 1993.
6. P. Cousot and R. Cousot. Abstract Interpretation: A Unified Lattice Model for Static Analysis of Programs by Construction or Approximation of Fixpoints. In *Proc. of POPL*, pages 238–252, 1977.
7. L. Henkin, J.D. Monk, and A. Tarski. *Cylindric Algebras (Part I)*. North-Holland, 1971.
8. K. Marriott and H. Søndergaard. Abstract Interpretation of Logic Programs: the Denotational Approach. In A. Bossi, editor, *Proc. of the Italian Conf. on Logic Programming*, pages 399–425, 1990.
9. P. Panangaden and V. Shanbhogue. The expressive power of indeterminate dataflow primitives. *Information and Computation*, 98(1):99–131, 1992.
10. V.A. Saraswat. *Concurrent Constraint Programming Languages*. PhD thesis, Carnegie-Mellon University, January 1989. In ACM distinguished dissertation series, The MIT Press, 1993.
11. V.A. Saraswat, M. Rinard, and P. Panangaden. Semantics foundations of Concurrent Constraint Programming. In *Proc. of POPL*, 1991.
12. E. Shapiro. *A subset of Flat Concurrent Prolog and its interpreter*. Technical Report TR-003, Institute for New Generation Computer Technology (ICOT), Tokyo, 1983.
13. E. Zaffanella, G. Levi and R. Giacobazzi. Abstracting synchronisation in concurrent constraint programming. Proc. Sixth Int'l Symp. on Programming Languages Implementation and Logic Programming. LNCS 844, pages 57–72. Springer-Verlag, 1994.

A Generic Algebra for Data Collections Based on Constructive Logic

P.Rajagopalan [1] and C.P.Tsang [2]

[1] Dept. of Comp. Science, Curtin University of Tech., Bentley WA 6102, Australia
[2] Dept. of Comp. Science,University of Western Australia, Crawley WA 6009, Australia

Abstract. Data collections form the basis for the representation and manipulation of data in database systems. We describe an algebra for manipulating data collections. It has been developed using constructive logic, and is a generalisation of relational algebra. We have applied the proofs-as-programs paradigm of intuitionistic type theory for deriving executable functions from specifications of algebra operations. The properties of algebra operators such as associativity, commutativity and distributivity have been verified using the same formal system.

1 Introduction

The research into query languages for object-oriented databases in recent years has generated interest in the representation and manipulation of general data collections [1, 2, 3, 6]. Formal systems such as set theory and first order logic, type systems of languages like ML, list comprehensions, structural recursion, etc. have been used for this purpose by various researchers. Our work, however, is based on the intuitionistic type theory of Martin-Lof [8]. We make use of an implementation of Martin-Lof's theory as the environment for the development of a generic representation for collections, and an algebra for manipulating collections. A significant advantage of our approach is that the specification and implementation of the algebra are integrated by the proofs-as-programs paradigm of constructive logic and type theory. In contrast, other approaches reported in the literature are generally proposals requiring significant extra effort for their computational realisations. The type theory based system also allows us to formally state the properties of algebraic operators and to verify these properties using type theory.

Martin-Lof initially developed the theory of types as a formalisation of constructive mathematics. He subsequently showed that it could be used for constructing computer programs [7]. His method is based on the principle of propositions-as-types which is the link between constructive logic and type theory. A proposition can be considered as a type whose members are the proofs of the proposition. Proving a proposition to be true is then equivalent to showing that the type corresponding to the proposition is inhabited. If a proposition can be considered as a program specification, there is a direct correspondence between a constructive proof of the proposition and a program that implements the specification. Martin-Lof's approach combines a specification language, a programming language and a formal method of deriving programs from their specifications under a single uniform theory. Systems that implement type theory support automatic extraction of the program from the proof of a specification. A number of such systems were developed in the 1980s [4, 5]. Type theory has been used in the past for the solution of computationally interesting problems in formal mathematics and for circuit design. Our work represents the first attempt at using constructive logic for the manipulation of constructs in database models.

Data models may be viewed as abstract representations of database systems. A data model is defined [13] as a mathematical formalism with two parts: (a) a notation for describing data, and (b) a set of operators used to manipulate the data. The computational paradigm supported by database systems is based on the representation and manipulation of data collections. The type of collections, however, vary among the different database models. For example, in the relational model the collections are sets of tuples, where each tuple is made up of atomic values. In a nested relational model, on the other hand, a tuple may contain values that are themselves collections. The object-oriented databases contain collections of objects and their properties (attributes). In this paper, we denote collections by a generic type expression. A collection consists of elements of some type that satisfy a constraint. In the generic type expression, the type of elements of the collection is denoted by a variable that ranges over the members of a universe of types. The constraint in the expression is specified as a predicate function that maps a collection to a Boolean value. By substituting for the parameters of this type expression, we can get different collection types such as lists, sets and bags containing elements such as flat tuples, nested tuples, etc.

An algebra for the generic collection type is specified by a set of operators to manipulate the collections. Each operator is specified by a proposition (theorem) in constructive logic. By proving the theorem in a type theory system, a executable function corresponding to the theorem is extracted automatically. By substituting for the parameters of the function, we can get operators for specific collection types. The algebraic properties of the generic algebra are specified as theorems that can be proved to verify the properties. Transformation rules for the algebra are derived from these theorems.

In Section 2, we give an overview of constructive logic and type theory. The generic representation of collections is discussed in Section 3. The algebra for generic collections is described in Section 4. The transformation rules based on the properties of algebraic operators are discussed in Section 5. Section 6 is the conclusion of the paper and also contains some pointers for further research.

2 An Overview of Constructive Logic and Type Theory

The type theory is a formal system for reasoning about types. A type is defined as a collection of objects exhibiting a common structure. Type theory consists of a system for defining types and the notion of equality both between the members of a type, and between types. It includes a collection of inference rules which specify how types are constructed, as well as how to reason about them. Given a collection of atomic types, we can generate a bigger collection of types by combining them in various ways. If A and B are types, the following constructors can be used:

Cartesian Product: A term of type $A \times B$ consists of a pair (a,b), where a is of type A and b is of type B.

Disjoint union: A term of type $A|B$ consists of a term of type A or a term of type B, together with an indication of which of A or B it is.

Function space: A term of type A->B consists of a functional term which applied to a term of type A returns a term of type B.

Dependent product: A term of type $x{:}A{\times}B$ consists of a pair (a,b) where a is in type A and b is in type $B[a/x]$. The type $B[a/x]$ is evaluated by substituting all occurrences of x by a in B.

Dependent function: A term of type $x{:}A\text{->}B$ is a function $\lambda x.b$ where for all a of type A, $b[a/x]$ is of type $B[a/x]$.

In addition to the common type constructors given above, a number of other type constructors are also supported by systems based on type theory. We give below, two such constructors that are used in this paper:

Subset type: A term of type $\{x{:}A \mid B\}$ is an element a of type A such that the type $B[a/x]$ is inhabited.

List: A term of type *A list* can either be an empty list or a list containing elements of type A.

The Nuprl system [4, 10] is an implementation of Martin-Lof's type theory, and was developed by a group of researchers at Cornell University to provide an environment for the solution of formal problems in mathematics. It consists of an environment for specifying theorems and constructing formal proofs, and also a computation system for executing programs extracted from the proofs. It provides a collection of tools for the development, explanation, and manipulation of formal arguments that is intended to support large-scale development of formal mathematics. Nuprl has been used as the type theory environment for the research reported in this paper.

There is a precise correspondence between the type constructors and the interpretation given to the logical operators $\wedge$, $\vee$, =>, $\forall$ and $\exists$ in constructive logic. It is based on the idea that a proposition is true if and only if the corresponding type is inhabited. This is known as the Curry-Howard polymorphism [5].

Assuming that P and Q are types (propositions), and x is a member of P, the correspondences are given in Fig.1. Type theory based systems support a number of primitive types such as integers, character strings, etc. An empty type is denoted by *void* (in Nuprl). A proposition P is taken to be an empty type if it is false and a type with at least one element if it is true; so *false* corresponds to the type *void*. Detailed justifications for the type denotations of propositions are contained in [4].

Proposition	Type	Type name
$P \Rightarrow Q$	$P \text{->} Q$	Function space
$P \wedge Q$	$P \times Q$	Cartesian product
$P \vee Q$	$P \mid Q$	Disjoint union
$\forall x{:}P\ \ Q$	$x{:}P \text{->} Q$	Dependent function
$\exists x{:}P\ \ Q$	$x{:}P \times Q$	Dependent product
$\neg P$	$P \text{->}$ *void*	
$P \Leftrightarrow Q$	$(P \text{->} Q)\times (Q \text{->} P)$	

Fig. 1. Types corresponding to logical connectives and quantifiers

In this paper, we use list as a recursive type constructor for representing collections. This is justified by the fact that the list construct is commonly available in most type theory based systems. However, we could have chosen any general recursive construct that can be used to represent structures such as list, set and bag. For example, Constable et al. [4] have proposed a general inductive type constructor that extends the type system of Nuprl, and also makes the list type constructor redundant. But this general recursive constructor has not been fully implemented in Nuprl.

3 Representation of Generic Collections

Central to our representation scheme is a type expression for collections of data objects. In this representation, a generic type structure is chosen for collections. It is based on the subtype constructor described in Section 2. A subtype has the general form, $\{x{:}S \mid P\}$ and it consists of all members s of type S such that the predicate $P[s/x]$ is true (provable).

We denote a collection of data objects by the type expression, $\{r{:}A \text{ list} \mid P(r)\}$, where A is a type that denotes the structure of the objects, and P is a Boolean function of the type, (A list-> Bool). The type expression, $\{r{:}A \text{ list} \mid P(r)\}$ denotes every collection r of type (A list) that satisfies the predicate $P(r)$. The list type is used as a simple recursive type constructor (a general recursive type constructor is available as a sound extension of type theory [9], but has not yet been fully implemented in the Nuprl system). We may denote the generic type expression for collections by a notation of the form gColl(A,P), where A is the type of elements in the collection, and P is a constraint on the collections. P is of type Coll(A) -> Bool, where Coll(A) denotes a basic recursive construct such as a list containing elements of type A. The expression gColl(A,P) represents a subset type as defined below. We use Coll(A) rather than (A list) in this definition to indicate that any suitable recursive construct could be used.

gColl(A:Type, P:Coll(A)->Bool) == {r: Coll(A) | P(r)}

This definition associates the notation gColl(A,P) with the subset type $\{r{:}\text{Coll}(A) \mid P(r)\}$. It denotes the subset of all collections of type Coll(A) such that each collection r in the subset satisfies the constraint predicate $P(r)$. Specific collection types can be defined using this generic notation, by choosing appropriate values for the parameters A and P. Representations of lists, sets and bags using this notation are discussed in [11, 12].

Example 3.1

Let a predicate function Pset of type (A -> Coll(A) -> Bool), be defined such that it returns the value true if and only if a given collection has no duplicate elements in it. Then the expression gColl(A,Pset(A)) denotes the type of collections that contain only distinct elements of some type A. The definition of Pset can be given as:

Pset(A:Type)(r:Coll(A)) == $(\forall$x:A (x$\in$r) => ($\exists$s,t:Coll(A) ((s@(x.t))=r) $\wedge$ $\neg$(x $\in$ (s@t))))

where s@t denotes the concatenation of collections t and s, and (x.t) denotes the concatenation of the element x and the collection t.

Example 3.2

Let Rel be defined as Rel(A:Type) == gColl(A,Pset(A)). Assume that a type named *part* is defined as the product of types, (*pno*×*pname*×*price*), where *pno*, *pname*, and *price* are types of atomic values. Then *Rel*(*part*) denotes the type of extensions of a relation schema in the relational data model where the tuples of the relation are of type *part*.

Example 3.3

Assume that *c_name*, and *grade* denote atomic types of course names and grades. Let *course* be defined as given below.

course == (*c_name*×*grade*).

Then Rel(*course*) is the type of relations that contain tuples of type, (*c_name*×*grade*). If *s_name* denotes the type of student names, and *student* is defined as

student == (*s_name*×Rel(*course*)),

then Rel(*student*), is the type of a nested relation.

4 An Algebra for Generic Collections

The operators of the algebra to manipulate collections are specified as propositions (theorems) in constructive logic. These theorems formally describe the basic data manipulations on collections. They are proved using a type theory based system such as Nuprl. The executable functions corresponding to the proofs of theorems are extracted. In Nuprl, the extracted functions can be executed in an interactive computational environment. In this section, we describe the development of an algebraic operator in some detail, and then briefly describe two other operators that have been developed similarly. The operators discussed in this paper are select, project and join, which are generalisations of corresponding relational algebra operators. These operators and the generic counterparts of other relational algebraic operators are described in detail in [11].

4.1 Generic Select

Generic select is a generalisation of the select operator of relational algebra. Instead of relations, however, the generic select is specified for the generic collection construct introduced in Section 3. The select operator may be specified using constructive logic as follows:

∀A:Type ∀P:Coll(A)->Bool ∀Q:A->Bool ∀s:gColl(A,P) ∃r:gColl(A,P) ∀y:A (Q(y) ∧ y∈s)<=> (y∈r)

The generic select operation retrieves from a collection all the elements that satisfy a predicate. The result of the select operation on a collection *s* is specified as a collection *r*, which consists of all elements of *s* that satisfy a selection condition. If *A* is the type of elements of the collection *s*, then the selection condition *Q* is specified as a function of type, *A* ->Bool. However, as only the type specifications of the predicate functions P and Q are known, we will need a number of assumptions to complete the proof of the theorem. The assumptions used for proving this theorem are given below.

(a) ∀x:A Q(x) ∨ ¬(Q(x))

(b) P(nil)

(c) ∀h:A ∀r:gColl(A,P) P(h.r) ∨ (h∈r)

The assumption (a) is required as the law of the excluded middle is not valid in constructive logic. Given any element *x* of type *A*, the assumption states that we can prove either Q(*x*) or ¬(Q(*x*)). It may be viewed as the specification of a procedure to decide whether Q(*x*) is true or false, that is required for performing the select operation.

The proof of the theorem is derived using induction on the structure of collections. The assumption (b) is required to prove the base case of induction. As the function P is not fully specified (except for its type), we need to assume that it is possible to prove P(nil) (which means the constraint P is true for the empty collection). The assumption (c) is used to prove the step case of induction. It states that adding an element h to a collection r of type gColl(A,P) will not violate the constraint P or h is already an element of r. This assumption has been formulated based on the proof obligations that arise in the step case.

In general, assumptions may be viewed as preconditions of the function that would be extracted from the proof of a theorem. However, the assumptions of the theorem influence the course of its proof as well as the function extracted from the proof. Assumptions regarding the predicate *P* may be specified with varying degrees of generality in theorems of generic algebra. We experimented with different sets of assumptions before choosing those included in the theorems described in this section. Though we have done some work on the choice of assumptions and how they affect the functions derived from proofs, more research is needed to arrive at general conclusions. As we are not focusing on proof strategies in this paper, this part of our research is not described in detail.

In Nuprl, the relevant assumptions are included in the theorem specification itself as shown below. The assumptions correspond to proof obligations that arise when P and Q are replaced by specific functions.

THM gSelect
∀A:Type ∀P:Coll(A)->Bool ∀Q:A->Bool
(∀x:A Q(x) ∨ ¬(Q(x)) ∧ P(nil) ∧ ∀h:A ∀r:gColl(A,P) P(h.r) ∨ (h∈r)) =>
∀s:gColl(A,P) ∃r:gColl(A,P) ∀y:A (Q(y) ∧ y∈s)<=> (y∈r))

The main steps in the proof of this theorem are given in the Appendix to illustrate the nature of proofs. Once the theorem is proved, it can be used for deriving select operators for specific collection types as illustrated in Example 4.1.

Example 4.1

We consider the select operation on collections of type gColl(*A*,*Pset*(*A*)), which has a specific constraint function *Pset*. The function *Pset* was defined in Section 3. The type gColl(*A*,*Pset*(*A*)) denotes collections that contain only distinct elements. The select function is described by the theorem rSelect given below.

THM rSelect
∀A:Type ∀Q:A->Bool (∀x:A Q(x) ∨ ¬(Q(x)) ∧ ∀x,y:A (x=y) ∨ ¬(x=y)) =>
(∀s:gColl(A,Pset(A)) ∃r:gColl(A,Pset(A)) ∀y:A (Q(y) ∧ (y∈ s)) <=> (y∈ r))

An assumption that equality of any two elements of type *A* is decidable, has been included in this theorem. It is required for dealing with proof obligations arising from the use of *Pset*(*A*) as the constraint function in the type expression for collections. The proof of theorem rSelect is given in the Appendix.

An executable function is extracted from the proof of rSelect. The extracted function is also shown in the Appendix. This function can be used in the evaluation environment of Nuprl for retrieving data as shown in Example 4.2.

Example 4.2

Let *part* be defined as the type (*pno*×*pname*×*price*), as in Example 3.2 where *pno*, *pname*, and *price* are types containing only atomic values. Then *part* denotes the tuples of a relation. Consider a query to find all parts that have price greater than 100. This query can be formulated as a selection operation using the function derived from theorem rSelect. To execute the function, it is necessary to provide values for all its input parameters. The parameters correspond to the variables associated with the universal quantifiers and the assumptions of the theorem. They are obtained by viewing the theorem as a term built from the type constructors corresponding to the connectives and quantifiers of constructive logic. The theorem rSelect may be viewed as the following term:

A:Type ->Q:(A->Bool) -> (U ×V) -> s: gColl(A,Pset(A)) -> (r:gColl(A,Pset(A))×W)
where U ==(∀x:A Q(x) ∨ ¬(Q(x))), V == (∀x,y:A (x=y) ∨ ¬(x=y)),
and W == (∀y:A (Q(y) ∧ (y∈s)) <=> (y∈r)).

The corresponding input parameters are as follows:

(1) the type of elements *A* of the collection,
(2) the selection condition as a predicate function *Q*,
(3) a function to decide whether or not *Q*(*x*) is true for any *x* of type *A*,
(4) a function to decide whether or not any two elements of type *A* are equal, and
(5) the input collection *s* on which the select operation is to be performed.

The output of the function is a pair (an instance of a Cartesian product) the first component of which is the collection r that corresponds to the result of the operation. In our example query, the values corresponding to these parameters are as follows:

(1) the type *part* which denotes the tuples of the relation,
(2) a selection predicate (function) to test if the price is greater than 100,
(3) a function to decide whether the selection predicate is true for a given tuple,
(4) a function to decide whether or not any two elements of type *part* are equal, and
(5) an instance of the relation on which the selection is to be performed.

```
Qprice (z:part) ==(λz 100 < thd(z))
THM Qprice_decidable
 ∀x:part Qprice(x) ∨ ¬(Qprice(x))
THM part_eq_decidable
 ∀x,y:part x=y in (part) ∨ ¬(x=y in (part))
Qprice_dec = term_of(Qprice_decidable)
Parteq=term_of(part_eq_decidable)
assn=<Qprice_dec,Parteq>
p1=λx fst(x)
q1=p1(term_of(rSelect)(part)(Qprice)(assn)(PART))
```

Fig. 2. Functions and related objects of Select example

The formulation of the query as a function that can be executed in the computational environment of Nuprl is shown in Fig. 2. In the specification of the predicate Qprice, thd(z) is a function to extract the third component of a tuple that is an instance of the Cartesian product of 3 components. A function to decide whether Qprice is true for a given tuple is declared as Qprice_dec. The theorem part_eq_decidable is used to extract a function to determine if two elements of type *part* are equal. The need for providing these two functions reflects the rudimentary nature of the computation system and contributes to the extra effort involved in using the functions extracted from the proof system. The inputs corresponding to the two assumptions of rSelect are combined as assn, which denotes a pair of functions. As the assumptions are combined in the theorem rSelect by an '∧' which corresponds to the product type constructor '×', an instance of the assumptions is a pair. The function p1 extracts the first component of a pair. Assuming that PART is an extension of the given relation schema, the query is answered by executing the functional expression q1 in the computational environment of Nuprl.

If PART contains the tuples, (<1,"bolt",120>.<2,"nut",100>. <3,"cam",140>. <4,"flange",123>. <5,"ring",95>. <6,"pin",150>.nil), then the result of the query q1 is obtained as (<1,"bolt",120>. <3,"cam",140>. <4,"flange",123>. <6,"pin",150>. nil).

4.2 Generic Projection

In relational algebra, projection operation is used to retrieve the values corresponding to a subset of attributes from a relation. We generalise the projection operation to generic collections of data. Let the input to the projection be a collection *s* of type gColl(*A*,*P1*), and the result of the projection a collection *r* of type gColl(*B*,*P2*). If *T* is a function of type *A* -> *B*, then *r* consists of elements obtained by applying the function *T* to every element *x* of *s*. *A* and *B* can be arbitrary types that may represent flat or nested structures, and unlike the relational projection, type *B* need not consist of a subset of the components of *A*. *P1* and *P2* in the type expressions of the input and output collections can also be different predicates. For example, the input could be specified as a set and the output as a bag.

The generic project operation is specified by the theorem gProject given below. The assumptions required for completing the proof are included in the statement of the theorem. This theorem may be used for proving other theorems that specify the project operation on specific collection types such as lists, sets, or bags.

THM gProject
∀A,B:Type ∀T:A->B ∀P1:(Coll(A)->Bool) ∀P2:(Coll(B)->Bool)
(P2(nil) ∧ ∀h:B. ∀r:gColl(B,P2). P2(h.r) ∨ (h∈r)) =>
(∀s:gColl((A),P1) ∃r:gColl(B,P2)
(∀x:A x∈s => (T(x))∈r) ∧ (∀y:B y∈r => ∃x:A x∈s ∧ y=T(x)))

4.3 Generic Join

Similar to other generic operators described so far, generic join is a generalisation of relational join. The generic join is specified as an operation on two generic collections *s1* and *s2* of types gColl(*A*,*P1*) and gColl(*B*,*P2*) respectively. The result of the join is a collection *r* of type gColl(*C*,*P3*). *A*, *B*, and *C* are some types in the universe of types. The functions *P1*,*P2*, and *P3* are specified similar to the corresponding functions in the definition of Cartesian product. The join condition for matching the elements of *s1* and *s2* is specified as a predicate function *Q* of type *A*->*B*->Bool. The elements of the result r are obtained by applying a transformation function *T* of type *A*->*B*->*C* to each pair of elements (*x*,*y*) from *s1* and *s2* that satisfy the condition *Q*(*x*)(*y*).

The theorem gJoin specifies the generic join operator. The assumptions of this theorem are similar to those of the generic select operator discussed earlier. The theorem states that
(1) for every pair of elements *a* of *s1* and *b* of *s2*, if *Q*(*a*)(*b*) is true then *T*(*a*)(*b*) is an element of the collection *r*, and
(2) for every element *c* in collection *r* there exists a pair of elements *a* in the collection *s1* and *b* in collection *s2* such that *Q*(*a*)(*b*) is true and *c*=*T*(*a*)(*b*).

THM gJoin
∀A,B,C:Type ∀P1:(Coll(A)->Bool) ∀P2:(Coll(B)->Bool) ∀P3:(Coll(C)->Bool)
∀Q:A->B->Bool ∀T:A->B->C
(∀x:A ∀y:B Q(x)(y) ∨ ¬(Q(x)(y)) ∧ P3(nil) ∧ ∀h:C ∀r:gColl(C,P3) P3(h.r) ∨ h∈r)
=> ∀s1:gColl(A,P1) ∀s2:gColl(B,P2) ∃r:gColl(C,P3)
(∀a:A ∀b:B Q(a)(b) ∧ a∈s1 ∧ b∈s2 => (T(a)(b))∈r) ∧
(∀c:C c∈r => (∃a:A ∃b:B Q(a)(b) ∧ a∈s1 ∧ b∈s2 ∧ c=T(a)(b)))

5 Transformation Rules of Generic Algebra

In this section we discuss the transformation rules of generic algebra. They may be considered as generalisations of the rules of relational algebra. These rules are useful for transforming expressions into equivalent forms that are computationally more efficient. The rewrite rules of generic algebra have been verified using the Nuprl system [11]. The proofs are omitted in this paper to conserve space. A simpler functional notation is used to denote algebraic operators so that the rules can be stated more concisely. The symbol ≡ is used to denote equivalence of expressions.

5.1 Idempotence of Generic Select and Project

Let s be a generic collection of type, gColl(A,P). Let Q be a predicate function of type (A->Bool). We denote the generic select operation on s using a selection function Q, as Select(Q)(s). If Q1,Q2,Q3 are selection functions of the type, (A->Bool), then the idempotence of generic select operation can be expressed as

Select(Q1)(Select(Q2)(s)) ≡ Select(Q3)(s)
if ∀x:A Q1(x) ∧ Q2(x) <=> Q3(x).

Let A,B,C be types of elements of collections. Let s be a collection of type gColl(A,P), and let T1, T2, and T3 be the projection functions of types A->B, B->C, and A->C respectively. The projection operation on a generic collection s, using a projection function T is denoted by Project(T)(s). The rule for project operation may be stated as

Project(T2)(Project(T1)(s)) ≡ Project(T3)(s)
if ∀a:A. T2(T1(a))=T3(a) in C.

5.2 Commutativity of Generic Select and Project

The commutativity of generic select and project operations is given by the following rules:

(1) Commutativity of select

Select(Q1)(Select(Q2)(s)) ≡ Select(Q2)(Select(Q1)(s))
where s:gColl(A,P), Q1:A->Bool, and Q2:A->Bool.

(2) Commutativity of select and project

Project (T)(Select (Q1)(s)) ≡ Select (Q2)(Project(T)(s))
if ∀a:A. Q1(a) <=> Q2(T(a)),
where s:gColl(A,P), T:A->B, Q1:A->Bool, and Q2:B->Bool.

5.3 Distributivity of Unary Operations over Binary Operations

The distributivity of the generic unary operators select and project over the binary operator join is given by the following rules.

(1) Distributing selection over join

Select(Q1)(Join(Q)(T)(s1)(s2)) ≡ Join(Q)(T)(Select(Q2)(s1))(Select(Q3)(s2))
if ∀a:A.∀b:B. Q1(T(a)(b)) <=> Q2(a) ∧ Q3(b),
where s1:gColl(A,P1), s2:gColl(B,P2), Q:A->B->Bool, T:A->B->C, Q1:C->Bool, Q2:A->Bool, and Q3:B->Bool.

(2) Distributing projection over join

Project(T1)(Join(Q1)(T2)(s1)(s2))≡
Join(Q2)(T3)(Project(T4)(s1))(Project(T5)(s2))
if ∀a:A.∀b:B. Q1(a)(b) <=> Q2(T4(a))(T5(b)),
and ∀a:A.∀b:B. T1(T2(a)(b))=T3(T4(a))(T5(b)) in C,
where s1:gColl(A,P1), s2:gColl(B,P2), Q1:A->B->Bool, Q2:D->E->Bool,
T1:C->D, T2:A->B->C, T3:D->E->C, T4:A->D, and T5:B->E.

6 Conclusions and further Research

We have developed a general representation for database structures and operations that embodies the characteristics of many data models. Data collections form the basic structures on which computations are specified in database systems. The characteristics of collections, however, may vary among data models. A generic representation of collections was developed which can be used to denote structures that are commonly found in data models.

A generic algebra has been developed based on the generic representation of collections. The operators of the algebra were formally specified using constructive logic, and executable functions derived from the constructive proofs of these specifications. Examples of transformation rules based on the idempotence, commutativity and distributivity of the generic operators were given. The generic algebra can be used to derive algebras to manipulate the structures of various data models. Similarly, the generic rewrite rules can be used to derive transformation rules of specific algebras. As the generic rewrite rules generalise the transformations of relational algebra, they provide a mechanism for extending the relational rewrite rules to other data models.

Our research can be generally considered as a method of prototyping database structures and operations. It is based on formal specification of concepts, which are then formally verified and implemented as executable functions. This contrasts with the traditional development of prototypes by conventional programming. An advantage of our approach is that it takes significantly less time and effort to derive implementations of interesting concepts, compared to developing conventional prototypes for the same purpose. While a conventional prototype usually implements a particular data model, our approach is independent of specific data models. In traditional database research, the theory formulation and its implementation are separate activities which may give rise to errors and inconsistencies. In our approach, the concepts are formally specified, and the process of development ensures that the programs exactly meet the specifications. Therefore this research provides a promising new approach to the development of database theory by an integrated process of formal specification, verification and prototyping.

We view this work as establishing the feasibility of using type theory for database research. Nuprl and similar systems are designed mainly for developing formal mathematics. However, the importance of type systems is now widely recognised in database systems, though intuitionistic type theory has not been used before. Developing a type theory based system to support database research would be an interesting and useful project to undertake. Defining new primitive types such as date and money, and primitive type constructors such as arrays, sets, and trees would be useful for database research. Specialised tactics could be embedded in the system to facilitate the verification of specifications. An important requirement of such as system would be the ability to interact with other software development environments so that the functions constructed in the type theory based system can be executed in other environments.

References

1. Beeri, C., and Kornatzky, Y. Algebraic Optimization of Object-Oriented Query Languages. In Abiteboul, S., and Kanellakis, P.C.(Eds.), *ICDT'90*, LNCS, Springer-Verlag, 1990, pp. 72-88.

2. Breazu-Tannen, V., and Subrahmanyam,R. Logical and Computational Aspects of Programming with Sets/Bags/Lists. In LNCS 510: *Proc. of 18th International Colloquium on Automata, Languages, and Programming*, Springer Verlag, 1991, pp 60-75.

3. Buneman, P., Libkin, L., Suciu, D., Tannen, V., and Wong, L. Comprehension Syntax. *SIGMOD Record*, Vol.23, No.1, March 1994, pp 87-96.

4. Constable,R.L., et al. *Implementing Mathematics with the Nuprl Proof Development System*, Prentice-Hall, 1986.

5. Constable, R.L. Type Theory as a Foundation for Computer Science. In Ito, T., and Meyer, A.R. (Eds.), *Theoretical Aspects of Computer Software, Proc. Intl. Conf.*, Sendai, Japan, Sept 1991, Springer-Verlag, pp 226-243.

6. Kanellakis, P., and Schmidt, J.W. (Eds.). *Database Programming Languages: Bulk Types & Persistent Data, The Third International Workshop Proc.*, Aug 1991, Nafplion, Greece.

7. Martin-Lof, P. Constructive Mathematics and Computer Programming. In *Sixth International Congress for Logic, Methodology, and Philosophy of Science*, pp. 153-175, North-Holland, 1982.

8. Martin-Lof, P. Intuitionistic Type Theory. In *Studies in Proof Theory Lecture Notes*, BIBLIOPOLIS, Napoli, 1984.

9. Mendler, P. *Inductive Definition in Type Theory*. PhD thesis, Cornell University, Ithaca, NY, 1988.

10. The Nuprl Group. *The Nuprl Proof Development System, Version 3.2 Reference Manual and User's Guide*. Dept of Computer Science, Cornell University, September 1991.

11. Rajagopalan, P. *A Type Theory Approach to the Specification and Synthesis of Database models*. Ph.D. thesis, University of Western Australia, December 1993.

12. Rajagopalan, P., and Tsang, C.P. A Type Theory Approach to the Development of Database Concepts. In *Proc. International Symp. on Advanced Database Technologies and Their Integration (ADTI'94)*, Nara, Japan, October 1994.

13. Ullman, J.D. *Principles of Database and Knowledge-base Systems.* Computer Science Press, 1988.

Appendix

A.1 Main Steps in the Proof of Theorem gSelect

The main proof steps are shown in Fig. A1-A5 as they would appear in the Nuprl window with minor changes to enhance readability. The goal is written to the right of the turnstile symbol '>>', and the hypotheses to the left. However, the main goal of the theorem is stated without any hypothesis. Refinement (inference) rules are applied to each goal that is to be proved. Applying inference rules to a goal may produce a number of subgoals as further proof obligations. If no subgoals are generated then the proof step has been completed. The refinement rules used in the actual proof are tactics. A tactic is a function written in ML that can be applied in place of an inference rule. Tactics often combine a number of inference rules thereby simplifying the derivation of proofs. In this appendix, we have replaced the tactics with brief descriptions.

The proof is constructed in the form of a tree with the theorem to be proved as the root (denoted by 'top'). The subgoals generated by applying refinement rules represent the child nodes. The subgoals are labelled by the path from the root ('top') of the proof tree. The hypothesis lists of some child nodes have been elided when they can be read from previous figures by following the path indicated by the label.

---

top
>> ∀A:Type ∀P:(A list)->Bool ∀Q:A->Bool ∀x:A Q(x) ∨ ¬(Q(x)) ∧ P(nil) ∧ ∀h:A ∀r :gColl(A,P) P(h.r) ∨ (h:A)∈r => ∀s:gColl(A,P) ∃r:gColl(A,P) ∀y:A Q(y) ∧(y:A)∈s <=> (y:A)∈r

By Introduction rules and then by simplification
Subgoal 1:
1. A: U1
2. P: A list->Bool
3. Q: A->Bool
4. ∀x:A Q(x) ∨ ¬(Q(x))
5. P(nil)
6. ∀h:A ∀r:gColl(A,P) P(h.r) ∨ (h:A)∈r
7. s: gColl(A,P)
>> ∃r:gColl(A,P) ∀y:A Q(y) ∧ (y:A)∈s <=> (y:A)∈r

By induction on s
Subgoal 1:
7. s: A list
[8]. P(s)
>> ∃r:gColl(A,P) ∀y:A Q(y) ∧ (y:A)∈nil <=> (y:A)∈r

Subgoal 2:
7. s: A list
[8]. P(s)
9. h2: A
10. t3: A list
11. ∃r:gColl(A,P) ∀y:A Q(y) ∧ (y:A)∈t3 <=> (y:A)∈r
>> ∃r:gColl(A,P) ∀y:A Q(y) ∧ (y:A)∈(h2.t3) <=> (y:A)∈r

Fig. A1. Applying introduction rules followed by induction on the collection

top 1 2

...

11. ∃r:gColl(A,P) ∀y:A Q(y) ∧ (y:A)∈ t3 <=> (y:A)∈ r

>> ∃r:gColl(A,P) ∀y:A Q(y) ∧ (y:A)∈ (h2.t3) <=> (y:A)∈ r

By eliminating the existential quantifier in hypothesis 11

subgoal 1:

11. r1: gColl(A,P)

12. ∀y:A Q(y) ∧ (y:A)∈ t3 <=> (y:A)∈ r1

>> ∃r:gColl(A,P) ∀y:A Q(y) ∧ (y:A)∈ (h2.t3) <=> (y:A)∈ r

By case analysis of the selection condition

Subgoal 1:

>> Q(h2) ∨ ¬(Q(h2))

Subgoal 2:

13. Q(h2)

>> ∃r:gColl(A,P) ∀y:A Q(y) ∧ (y:A)∈ (h2.t3) <=> (y:A)∈ r

Subgoal 3:

13. ¬(Q(h2))

>> ∃r:gColl(A,P) ∀y:A Q(y) ∧ (y:A)∈ (h2.t3) <=> (y:A)∈ r

Fig. A2. Eliminating existential quantifier in hypothesis followed by case analysis of selection predicate

top 1 2 1 2

...

6. ∀h:A ∀r:gColl(A,P) P(h.r) ∨ (h:A)∈ r

...

>> ∃r:gColl(A,P) ∀y:A Q(y) ∧ (y:A)∈ (h2.t3) <=> (y:A)∈ r

By using hypothesis 6 with h2 and r1

Subgoal 1:

14. ∀r:gColl(A,P) P(h2.r) ∨ (h2:A)∈ r

15. P(h2.r1) ∨ (h2:A)∈ r1

16. P(h2.r1)

>> ∃r:gColl(A,P) ∀y:A Q(y) ∧ (y:A)∈ (h2.t3) <=> (y:A)∈ r

Subgoal 2:

14. ∀r:gColl(A,P) P(h2.r) ∨ (h2:A)∈ r

15. P(h2.r1) ∨ (h2:A)∈ r1

16. (h2:A)∈ r1

>> ∃r:gColl(A,P) ∀y:A Q(y) ∧ (y:A)∈ (h2.t3) <=> (y:A)∈ r

Fig. A3. Using the assumption for the step case of induction

```
top 1 2 1 2 1
...
>> ∃r:gColl(A,P) ∀y:A Q(y) ∧ (y:A)∈(h2.t3) <=> (y:A)∈r
By introducing 'h2.r1' as existential witness in the goal
Subgoal 1:
  17. y: A
  18. Q(y)
  19. (y:A)∈(h2.t3)
  >> (y:A)∈(h2.r1)
Subgoal 2:
  17. y: A
  18. Q(y) ∧ (y:A)∈(h2.t3) => (y:A)∈(h2.r1)
  19. (y:A)∈(h2.r1)
  >> Q(y)
Subgoal 3:
  17. y: A
  18. Q(y) ∧ (y:A)∈(h2.t3) => (y:A)∈(h2.r1)
  19. (y:A)∈(h2.r1)
  20. Q(y)
  >> (y:A)∈(h2.t3)
```

Fig. A4. Adding an element to the result (collection)

```
top 1 2 1 3
...
13. ¬(Q(h2))
>> ∃r:gColl(A,P) ∀y:A Q(y) & (y:A)∈(h2.t3) <=> (y:A)∈r
By introducing 'r1' as existential witness in the goal
(Subgoals similar to those in Fig. A4 are generated here. They are omitted to save space)
```

Fig. A5. When an element does not satisfy the selection predicate

A2. Proof of Theorem rSelect

The proof of rSelect is shown in Fig. A6 to A8. The first step of the proof restates the theorem in sequent form with all the hypotheses to the left of the turnstile symbol, '>>' and the goal to be proved to its right. Then the theorem gSelect is used as a lemma to prove the main goal of the theorem as shown in Fig. A7. The subgoal generated by this lemma application corresponds to the assumptions of theorem gSelect, when the predicate function *Pset*(*A*) is substituted for the parameter *P*. This subgoal is proved using a theorem named PsetLemma, as shown in Fig. A8. The theorem PsetLemma is given below. It was proved separately before being used as a lemma in the proof of theorem rSelect.
THM PsetLemma

```
∀A:Type Pset(A)(nil) ∧(∀x,y:A x=y in A ∨ ¬(x=y in A)) =>
        ∀h:A ∀r:gColl(A,Pset(A)) Pset(A)(h.r) ∨ (h:A)∈r)
```

The assumption ∀x,y:A x=y in A ∨ ¬(x=y in A), is required to prove this theorem because *A* is only a type variable. Since this theorem is used as a lemma in the proof of rSelect, the same assumption is also included in rSelect.

```
THM rSelect
top
>>∀A:Type ∀Q:A->Bool ∀x:A Q(x) ∨ ¬(Q(x)) ∧ ∀x,y:A x=y in A ∨ ¬(x=y in A) =>
∀s:gColl(A,Pset(A)).∃r:gColl(A,Pset(A)).∀y:A Q(y) ∧ (y:A)∈s <=> (y:A)∈r
By introduction rules
Subgoal 1:
1. A: Type
2. Q: A->Bool
3. ∀x:A Q(x) ∨ ¬(Q(x))
4. ∀x,y:A x=y in A ∨ ¬(x=y in A)
5. s: gColl(A,Pset(A))
>> ∃r:gColl(A,Pset(A)) ∀y:A Q(y) ∧ (y:A)∈s <=> (y:A)∈r
```

Fig. A6. Restating the theorem in sequent form

```
top 1
...
>> ∃r:gColl(A,Pset(A)) ∀y:A Q(y) ∧ (y:A)∈s <=> (y:A)∈r
By Lemma 'gSelect'
Subgoal 1:
  6. ∀x:A Q(x) ∨ ¬(Q(x))
  7. Pset(A)(nil)
  8. h: A
  9. r: gColl(A,(Pset(A)))
  >> (Pset(A))(h.r) ∨ (h:A)∈r
```

Fig. A7. Proving the main goal of the theorem using gSelect

```
top 1 1
...
>> (Pset(A))(h.r) ∨ (h:A)∈r
By Lemma 'PsetLemma'
```

Fig. A8. Proving the subgoal using PsetLemma

A3. Function extracted from the proof of rSelect

An example of the extraction from a proof is given below. Displaying the extractions of theorems is similar to viewing the object code of conventional programs.

```
λA.λQ.λv0.spread(v0;v1,v2.λs.(λv3.(λv4.(λv5.(λv6.(λv7.(λv19.v19)(v7(s)))(v6((λv8.(λv9.<v
8,v9>)((λv10.(λv11.<v10,v11>)(λh.λr.(λv12.(λv13.spread(v13;v14,v15.(λv16.(λv17.(λv18.v1
8)(v17(r)))(v16(h)))(v15(v2))))(v12(A)))(term_of(PsetLemma))))(axiom)))(v1))))(v5(Q)))(v4(
Pset(A))))(v3(A)))(term_of(gSelect)))
```

Partial Order Programming (*Revisited*)[1]

Bharat Jayaraman
Mauricio Osorio
Kyonghee Moon

Department of Computer Science
State University of New York at Buffalo
Buffalo, NY 14260
U.S.A.

E-Mail: {bharat,osorio,kmoon}@cs.buffalo.edu

Abstract

This paper shows the use of partial-order program clauses and lattice domains for functional and logic programming. We illustrate the paradigm using a variety of examples: graph problems, program analysis, and database querying. These applications are characterized by a need to solve circular constraints and perform aggregate operations, a capability that is very clearly and efficiently provided by partial-order clauses. We present a novel approach to their model-theoretic and operational semantics. The least Herbrand model for any function is not the intersection of all models, but the *glb/lub* of the respective terms defined for this function in the different models. The operational semantics combines top-down goal reduction with *monotonic memo-tables.* In general, when functions are defined circularly in terms of one another through *monotonic* functions, a memoized entry may have to monotonically updated until the least (or greatest) fixed-point is reached. This partial-order programming paradigm has been implemented and all examples shown in this paper have been tested using this implementation.

1 Introduction

Equational programming and equational reasoning lie at the heart of functional programming, and the development of modern functional languages (ML, Miranda, Haskell, etc.) has been strongly influenced by these principles. In this paper, we describe a functional language whose principal building blocks are *partial order* clauses and lattice data types. The use of partial orders and lattices in a declarative language should not be surprising, since these concepts are fundamental to the semantics of declarative languages. The motivation for our work, however, is more practical in nature: We show that partial-order clauses

[1] This research was supported by grants from the National Science Foundation and Xerox Foundation.

and lattices help obtain clear, concise, and efficient formulations of problems requiring the ability to take transitive closures, solve circular constraints, and perform aggregate operations. Applications requiring these capabilities include program analysis, database querying, etc.

There are two basic forms of a partial-order clause:

f(terms) $\geq$ *expression*
f(terms) $\leq$ *expression*

where each variable in *expression* also occurs in *terms*. Terms are made up of constants, variables, and data constructors, while expressions are in addition made up of user-defined functions, i.e., those that appear at the head of the left-hand sides of partial-order clauses. Informally, the declarative meaning of a partial-order clause is that, for all its ground instantiations (i.e., replacing variables by ground terms), the function *f* applied to argument terms is $\geq$ (respectively, $\leq$) the ground term denoted by the expression on the right-hand side. In general, multiple partial-order clauses may be used in defining some function *f*. We define the meaning of a ground expression *f(terms)* to be equal to the *least-upper bound* (respectively, *greatest-lower bound*) of the resulting terms defined by the different partial-order clauses for *f*.

We provide model-theoretic and operational semantics for a large class of partial-order clauses. The model-theoretic semantics uses a novel iterated least-model construction: A program has a unique least model if we can stratify all program clauses into several levels such that all function calls at any given level depend upon others at the same level through *monotonic* functions (in the appropriate partial order), but may depend upon calls at lower levels through non-monotonic functions. Our operational semantics is also novel in that it combines top-down goal reduction with *monotonic memo-tables. Memo tables* are used in traditional functional languages to detect dynamic common subexpressions, and they are semantically transparent. In the partial-order programming framework, however, memoization is needed in order to detect circular constraints. In fact, in general we need more than simple memoization when functions are defined circularly in terms of one another through *monotonic* functions. In such cases, a memoized entry may have to be monotonically updated in order to progress towards the least (or greatest) fixed-point.

We show that partial-order clauses help render clear and concise formulations to problems involving aggregate operations and recursion in database querying. This has been a topic of considerable interest in the literature recently [KS91, RS92, Van92, SSRB93]. An aggregate operation is a function that maps a set to some value, e.g., the maximum or minimum in the set, the cardinality of this set, the summation of all its members, etc. In considering the problems with various semantic approaches, Van Gelder [Van92] notes that, for many problems in which the use of aggregates has been proposed, the concept of *subset* is what is really necessary. Our proposed paradigm provides a natural and efficient realization of the concept of monotonic aggregation [RS92]. In order to couple an extensional database of relations with partial-order clauses, we introduce the class of *conditional partial-order clauses*:

$f(terms) \geq expression$:- *condition*
$f(terms) \leq expression$:- *condition*

where each variable in *expression* occurs either in *terms* or in *condition*, and *condition* is a conjunction of goals, each of which may be of the form *p*(*terms*), ¬*p*(*terms*), or $f(terms) = term$. The semantics of the resulting programs are a straightforward generalization of those of unconditional partial-order clauses. In this setting, we show how various examples recently discussed in the deductive database literature can be clearly and concisely formulated.

Finally, we note that our concept of partial order programming is closely related to and inspired by that of Stott Parker's [Par89]. Essentially, in his paradigm, a program is a set of clauses of the form $u_i \sqsupseteq f_i(\bar{v})$, for $i = 1 \ldots n$, where each f_i is continuous, and the goal is to minimize u_j, for some j. At a high level, that is essentially what we are proposing. There are, however, several important differences: we use partial-order clauses to *define* functions; they can be conditional; and they can use non-monotonic functions (modulo stratification). These are important features for solving problems involving circular constraints and aggregation, and, to the best of our understanding, they are not discussed in Parker's framework.

The rest of this paper is organized as follows: section 2 gives the syntax of terms and expressions, and explains using examples the informal meaning of partial-order clauses; section 3 gives the formal model-theoretic and operational semantics for stratified partial-order clauses; section 4 shows the use of conditional partial order clauses for problems involving aggregate operations; finally, section 5 presents conclusions and comparisons with related work.

2 Partial Order Clauses: An Informal Introduction

2.1 Syntax

We first discuss unconditional partial-order clauses, which have the form

$f(terms) \geq expression$
$f(terms) \leq expression$

where each variable in *expression* also occurs in *terms*. (We discuss conditional partial-order clauses in section 4). For simplicity of presentation in this paper, we assume that every function *f* is defined either with $\geq$ or $\leq$ clauses, but not both—this restriction has been easy to meet in all the examples we have considered. The syntax of *terms* and *expr* is as follows:

term ::= *variable* | *constant* | *c*(*terms*)
terms ::= *term* | *term* , *terms*
expr ::= *term* | *c*(*exprs*) | *f*(*exprs*)
exprs ::= *expression* | *expression* , *exprs*

Our lexical convention in this paper is to begin constants with lowercase letters and variables with uppercase letters. The symbol *c* stands for a constructor symbol whereas *f* stands for a non-constructor function symbol (also called user-defined function symbol). Terms are built up from constructors and stand for data objects of the language. The constructors in this language framework may be constrained by an *equational theory*; we only require that matching a ground term against a pattern (i.e., non-ground term) produces a finite number of matches. In the general case, when multiple partial-order clauses define a function *f*, all matches of a ground goal *f(terms)* against the left-hand sides of all clauses defining *f* will be used in instantiating the corresponding right-hand side expressions; and, depending upon whether the partial-order clauses are $\geq$ or $\leq$, the *lub* or the *glb* respectively of all the resulting terms is taken as the result. In case none of the clauses match the goal, the result will respectively be $\perp$ or $\top$ of the lattice.

We only consider complete lattices of *finite* terms in our language framework. Of special interest to us is the complete lattice of *finite* sets under the partial orderings subset and superset: union and intersection stand for the *lub* and *glb* respectively, and the empty set (ϕ) is the least element. In order to meet the requirements of a complete lattice, a special element $\top$ is introduced as the greatest element. We use the notation {X\T} to match a set S such that X $\in$ S and T = S - {X}, i.e., the set S with X removed. For example, matching {a,b,c} against the pattern {X\T} yields three different substitutions: {X $\leftarrow$ a, T $\leftarrow$ {b, c}}, {X $\leftarrow$ b, T $\leftarrow$ {a, c}}, and {X $\leftarrow$ c, T $\leftarrow$ {a, b}}. When used on the left-hand sides of program clauses, {X\T} allows one to decompose a set into *strictly smaller* sets.

The definition below shows a simple use of multiple partial-order clauses to define the *lub* and *glb* of two elements:

```
lub(X,Y) ≥ X          glb(X,Y) ≤ X
lub(X,Y) ≥ Y          glb(X,Y) ≤ Y
```

The definition of set-intersection shows how set patterns can finesse iteration over sets (the result is ϕ if any of the input sets is ϕ, as desired):

```
intersect({X\_},{X\_}) ≥ {X}
```

This function works as follows: For a goal intersect({1,2,3},{2,3,4}), we have intersect({1,2,3},{2,3,4}) $\geq$ {2} and intersect({1,2,3},{2,3,4}) $\geq$ {3}. Taking the *lub* of {2} and {3}, we get intersect({1,2,3},{2,3,4}) = {2,3}.

The use of remainder sets in set-matching is illustrated by the following function definition, which takes as input a collection of *propositional clauses*, i.e., a set of set of literals, and returns the set of all resolvents (note that {A,B_} is an abbreviation for {A\{B_}}).

```
resolvents({{X\S1}, {not(X)\S2} \_}) ≥ {lub(S1,S2)}
```

2.2 Memo Tables

The definition of transitive closures using partial-order clauses brings up a crucial need for memoization. As an illustration, consider the function reach below which takes a set of nodes as input and finds the set of reachable nodes from this set (we give another formulation in section 4 using a predicate for edge).

```
reach(S) ≥ S
reach({X\_}) ≥ reach(edge(X))
edge(1) ≥ {2}
edge(2) ≥ {1}
```

To illustrate top-down goal reduction with memoization, we first *flatten* the expression `reach(edge(X))` as `edge(X) = T1, reach(T1) = S1`—see section 3 for a more precise definition of flattening. Similarly, a top-level query `reach({1})` is flattened as `reach({1}) = Ans`.

Goal Sequence	Substitution	Memo Table
reach({1}) = Ans		ϕ
edge(1) = T1 reach(T1) = S1	Ans ← {1} ∪ S1	{reach({1}) = {1} ∪ S1}
reach({2}) = S1	T1 ← {2}	{reach({1})= {1} ∪ S1}
edge(2) = T2 reach(T2) = S2	S1 ← {2} ∪ S2	{reach({1}) = {1,2} ∪ S2, reach({2}) = {2} ∪ S2}
reach({1}) = S2	T2 ← {1}	{reach({1}) = {1,2} ∪ S2, reach({2}) = {2} ∪ S2}
[]	S2 ← {1,2}	{reach({1}) = {1,2} reach({2}) = {1,2}}

Note that memo-table look-up occurs in the next to last step. The binding for the variable `S2` is the smallest solution to the equation

$$S2 = \{1,2\} \cup S2.$$

The binding for the variable `Ans` in the top-level query is obtained by composing the substitutions at each step, and is easily seen to be $\{1,2\}$. Strictly speaking, the calls on `edge` in the above derivation should be memoized, but we omit doing so here since it is not necessary to memoize calls on non-recursively defined functions. As a further optimization, in this example, one need not even maintain the partial bindings for the various function calls; it suffices to record just the function calls, in order to detect the loop. When a loop is detected, one simply returns the least element, ϕ. Such a strategy can be proven to be sound in this example. However, this simple strategy does not suffice for examples to be discussed subsequently. Finally, it should be noted that no subsumption checks are needed in accessing the memo-table since all function calls have ground arguments.

2.3 Monotonic Memo Tables

In the above example, we need to solve during a look-up step an equation of the form $v = t \sqcup v$, where v is a variable and t is some term. The smallest solution to this equation is: $v \leftarrow t$. However, for more general programs involving $\geq$ clauses, it becomes necessary to solve an equation of the form $p(\ldots, v \sqcup t, \ldots) = v$, where v is a variable, t is a possibly nonground term, and p is *monotonic* in the argument shown, i.e., $t_1 \leq t_2 \Rightarrow p(\ldots, t_1, \ldots) \leq p(\ldots, t_2, \ldots)$. A symmetric situation arises with $\leq$ clauses. Consider for example:

`g(X)` $\geq$ `{10}`	`h(X)` $\geq$ `{20}`	`p(S)` $\geq$ `S`
`g(X)` $\geq$ `h(X)`	`h(X)` $\geq$ `p(g(X))`	`p(S)` $\geq$ `{30}`

The derivation from the top-level query `g(100)` is as follows.

Goal Sequence	**Substitution**	**Memo Table**
`g(100) = Ans`		ϕ
`h(100) = S1`	`Ans` $\leftarrow$ `{10}`$\cup$`S1`	`{g(100) = {10}`$\cup$`S1}`
`g(100) = T1,` `p(T1) = S2`	`S1` $\leftarrow$ `{20}`$\cup$`S2`	`{g(100) = {10,20}`$\cup$`S2` `h(100) = {20}`$\cup$`S2}`
`p({10,20}`$\cup$`S2) = S2`	`T1` $\leftarrow$ `{10,20}`$\cup$`S2`	`{g(100) = {10,20}`$\cup$`S2,` `h(100) = {20}`$\cup$`S2}`

At the last step above, we find that the argument to the function `p` is nonground, since the variable `S2` is still undetermined. At this stage, we provisionally assume that `S2` $= \phi$—the least element of the lattice—and proceed with the computation, but reconsider this assumption later. Now evaluating the goal `p({10,20}` $\cup$ `S2)` with `S2` $= \phi$ yields `p({10,20})` $=$ `{10,20,30}`. Thus the revised estimate for `S2` is `{10,20,30}`. When a variable such as `S2` has its estimate revised, the goal that used a provisional value of this variable is re-evaluated using the new estimate. Re-evaluating `p({10,20}` $\cup$ `S2)` with `S2` $=$ `{10,20,30}` yields `S2` $=$ `{10,20,30}`. Since `S2` has not changed, the least fixed-point has been reached, and the toplevel query successfully terminates, the answer being `{10,20,30}`. The successive approximations for variable `S2` are recorded by updating the corresponding location for `S2` in the memo-table. In general several variables might be nonground when attempting to reduce some monotonic function. All of them are assumed to be $\perp$ initially (resp. $\top$ for $\leq$ clauses), and their estimates are monotonically updated until a fixed-point is reached. This example also illustrates the need for monotonic functions: if `p` were not monotonic, the progressive iteration is not guaranteed to reach a fixed-point.

For a more realistic example, the program below defines the *reaching definitions* and *busy expressions* in a program flow graph, which is computed by a compiler during its optimization phase [AU77].

```
reach_out(B) ≥ reach_in(pred(B)) - kill(B)
reach_out(B) ≥ gen(B)
reach_in({B\_}) ≥ reach_out(B)
busy_out(B) ≤ busy_in(succ(B)) - def(B)
busy_out(B) ≤ use(B)
busy_in({B\_}) ≤ busy_out(B)
```

where **kill(B)**, **gen(B)**, **pred(B)**, **def(B)**, **use(B)**, and **succ(B)** are predefined set-valued functions specifying the relevant information for a given program flow graph and basic block B. The set-difference operator (-) is monotonic in its first argument, and hence its use in the bodies of the functions **reach_out** and **busy_out** is legal. Because the **reach_in** and **reach_out** functions are defined circularly (as are **busy_in** and **busy_out**), memoization is needed to avoid the infinite loop that could result when the underlying program flow-graph has cycles. Furthermore, since these functions are defined in terms of the set-difference operator, a monotonic memo-table is needed to compute the answer.

3 Semantics of Partial Order Clauses

We provide in this section model-theoretic and operational semantics of partial-order clauses. For simplicity of presentation, we consider only $\geq$ clauses in this section; the treatment of $\leq$ clauses is symmetric. In preparation for the semantics, we use the flattened form for all clauses and goals.

Definition 3.1: There are two *flattened forms* of a partial-order ($\geq$) clause:

Head

Head :- *Body*

where *Head* is $\mathtt{f}(\bar{t}) \geq u$, where $\bar{t}$ is a sequence of terms, u is a term, and *Body* is of the form $E_1, \ldots, E_n$, where each E_i is $\mathtt{f}_i(\bar{t}_i) = x_i$, where $\mathtt{f}_i$ is a user-defined (non-constructor) function, $\bar{t}_i$ is a sequence of terms, and x_i is a new variable not present on the left-hand side of :-.

For example, the flattened form of a clause f(X,Y,S) ≥ g(h(X),k(Y,S)), where g, h, and k are non-constructors, is:

```
f(X,Y,S) ≥ S2  :- h(X)=T1, k(Y,S)=S1, g(T1,S1)=S2
```

The flattened form of a *ground query expression* is similar to that of *Body*. Henceforth we assume that all programs and queries are in their flattened forms.

We will work with Herbrand Interpretations, where the Herbrand Universe consists of ground terms and the Herbrand Base consists of ground equality atoms of the form $f(t) = u$, where f is a user-defined function, t is a ground term, and u is a ground term belonging to some lattice domain[2]. In every interpretation

[2] Strictly speaking, we should work with equivalence classes of terms and atoms, due to the equality theory of the constructors. However, we will talk of terms, instead of equivalence classes of terms, for simplicity of presentation.

I, every user-defined function f is interpreted as a total function, i.e., (i) for every ground term t, there is ground term u such that $f(t) = u \in$ I; and (ii) if $f(t) = t_1 \in$ I and $f(t) = t_2 \in$ I, then $t_1 = t_2$.

3.1 Model-Theoretic Semantics

We first informally motivate our approach. Basically, we define the semantics for a ground expression $f(t)$, where f is a user-defined function and t a ground term, as the *glb* of all terms defined for $f(t)$ in the different Herbrand models for f. To see the need for taking such *glbs*, consider the following trivial program:

$\mathtt{f}(\mathtt{X}) \geq \{\mathtt{1}\}$

Each Herbrand model of this program interprets $\mathtt{f}$ as a constant function:

$\{\mathtt{f}(t) = \{1\} \cup s : t \text{ is a ground term}\}$

where s is a different ground set in each different model. The intended model for function $\mathtt{f}$, namely, $\{ \mathtt{f}(t) = \{\mathtt{1}\} : t \text{ is a ground term } \}$ is obtained not by intersecting all models but by intersecting the respective sets defined for $\mathtt{f}(t)$ in the different models. In order to obtain a computable semantics, we *stratify* or partition the program into several levels, as follows.

Definition 3.2 (strongly stratified programs):
(a) Level 1 clauses have the following syntax (f and g are at level 1):

$f(terms) \geq g(terms)$
$f(terms) \geq term$

(b) For level $j > 1$, clauses have the following syntax (f and g are at level j):

$f(terms) \geq g(lexprs)$
$f(terms) \geq lexpr$

(c) A program is partitioned into the smallest number of levels satisfying the above conditions.

In the first two cases, f is not necessarily different from g. In the second case, *lexpr* is either a term or an expression composed of functions from levels 1, ..., $j-1$, and *lexprs* is a sequence of zero or more *lexpr*. In the `reach` program shown in section 2.2, the function `edge` would be at level one, and the function `reach` would be at level two. Note that the above definition of stratification is very strong; it requires a function at any level to be directly defined in terms of other functions at the same level. We relax this requirement in section 3.3. We introduce strongly stratified programs first because their operational semantics requires simple memo-tables, but *not* monotonically updatable memo-tables.

Definition 3.3: Let P be a set of program clauses. We define P_k as those clauses of P such that the user-defined function symbols on the left-hand sides have level $\leq k$.

Definition 3.4: Given two interpretations I and J, we define $I \sqsubseteq J$ if for every $\mathtt{f}(t) = t_1 \in I$ there exists $\mathtt{f}(t) = t_2 \in J$ such that $t_1 \leq t_2$.

Definition 3.5: Let P be a stratified program. An interpretation M is a model of P, denoted by $M \models P$, if for every ground instance, $\mathtt{f}(t) \geq t_1$:- $\mathrm{E}_1 \ldots \mathrm{E}_k$, of a $\geq$ clause in P, if $\{\mathrm{E}_1 \ldots \mathrm{E}_k\} \subseteq M$ then $\{\mathtt{f}(t) = t_1\} \sqsubseteq M$.

Since we will construct the model-theoretic semantics of a stratified program level by level, in defining models at some level $j > 1$, all functions from levels $< j$ will have their models uniquely specified. Hence, all interpretations of clauses at some level j will contain the same atoms for every function from a level $< j$. For this reason we introduce the notation *level*(A) to refer to the level of the head function symbol of atom A:

Definition 3.6: For any interpretation I, $I_k := \{\ \mathrm{A} : \mathrm{A} \in I \wedge \mathit{level}(A) \leq k\}$.

Definition 3.7: For any two interpretations I and J,
$I \sqcap J := \{f(t) = u \sqcap u' \ :\ f(t) = u \in I,\ f(t) = u' \in J\}$

Definition 3.8: For any set X of interpretations, $\sqcap X$ is the natural generalization of the previous definition.

Proposition 3.9: Let X be a set of models for a program P with j levels such that for any $I \in X$ and $J \in X$, $I_{j-1} = J_{j-1}$. Then $\sqcap X$ is also a model.

Definition 3.10: Given a program P with j levels, we define the *model-theoretic semantics* of P as:
for $j = 1$, $\mathcal{M}(P_1) := \sqcap\{M : M \models P_1\}$, and
for $j > 1$, $\mathcal{M}(P_j) := \sqcap\{M : M \supseteq \mathcal{M}(P_{j-1})$ and $M \models P_j\}$.

Definition 3.11: Given a program P with j levels and a query G, we say that substitution θ is a *correct answer* for G if $\mathcal{M}(P_j) \models G\theta$.

3.2 Operational Semantics

In preparation for the operational semantics of strongly stratified programs, we first define the *lub reduction* of a ground query expression G with respect to a program P starting from their respective flattened forms, in which the order of equalities reflects the *leftmost-innermost order* of reducing expressions.

Definition 3.12: An extended goal is of the form <G, T> where G is a goal-sequence and T is a memo-table, i.e., a set of assertions of the form $\mathtt{f}(t) = u$, where $\mathtt{f}$ is a function, t is a ground term, but u may be non-ground.

Definition 3.13: Given variants of subset clauses, $\mathtt{f}(t_1) \geq s_1$:- B_1, ..., $\mathtt{f}(t_n) \geq s_n$:- B_n, in which variables have been suitably renamed, and given a query expression, G := [g_1, ..., g_m], where g_1 is $\mathtt{f}(t) = x$, t is a ground term, and x is a variable, we define the *lub reduction* relation G $\rightarrow$ G' as follows:

(a) if matching t with $t_1 \ldots t_n$ yields respectively the (finitely many) substitutions $\theta_{11}, \ldots, \theta_{1k_1}, \ldots, \theta_{n1}, \ldots, \theta_{nk_n}$, then

G' := [$B_1\ \theta_{11}, \ldots, B_1\ \theta_{1k_1}, \ldots, B_n\ \theta_{n1}, \ldots, B_n\ \theta_{nk_n}, g_2\ \sigma, \ldots, g_m\ \sigma$],
where $\sigma := \{x \leftarrow \sqcup_{i=1,n} \sqcup_{j=1,k_i} (s_i\theta_{ij})\}$;

(b) if there are no matches between t and any t_i,
G' := [$g_2\ \sigma, \ldots, g_n\ \sigma$], where $\sigma = \{x \leftarrow \bot\}$.

Definition 3.14: Given an extended goal <G, T>, let the first goal, g_1, in G be $\mathtt{f}(t) = v$. The relation <G, T> → <G', T'> is defined as follows.

(a) If there is no assertion of the form $\mathtt{f}(t) = w$ in T, then we reduce G → G' by a *lub reduction*, and T' := (T ∪ $\{\mathtt{f}(t) = v\}$) σ, where σ is the substitution for v in deriving G'.

(b) If there is an assertion of the form $\mathtt{f}(t) = w$ in T, then G' := [$g_2\ \sigma, \ldots, g_n\ \sigma$] and T' := T σ, where $\sigma := \{v \leftarrow w'\}$ and w' is the smallest solution to the equation $v = w$.

Definition 3.15: Given a program P and an extended goal G^e := <G, T>, we say that θ restricted to variables in G is the computed answer for P and G^e if there is a derivation $G^e = G^e_1 \rightarrow \ldots \rightarrow G^e_k$ = <[],T_k >, where θ_i is the substitution used in reducing G^e_i, [] is the empty goal, and $\theta = \theta_1 \ldots \theta_k$.

Theorem 3.16 (Soundness): Given a program P and an extended goal G^e := $< f(t) = x, \phi >$, the computed answer for G^e is correct.

This theorem is proved in [JOM93]; the proof is omitted here due to lack of space. The following clause shows that we do not have *completeness*: $\mathtt{f(X)} \geq \mathtt{f(\{X\})}$. Notice that a query $\mathtt{f(1)} = \mathtt{Z}$ does not have a computed answer—the function call is nonterminating—but it has a correct answer $\theta = \{\mathtt{Z} \leftarrow \phi\}$, because all functions are interpreted as total functions in the declarative semantics. Since incompleteness arises only because of this kind of nontermination, it has not been a practical problem in using this paradigm.

3.3 General Stratified Programs

The *strongly stratified language* defined in section 3.1 permits the definition of one function directly in terms of another function at the same level. However, the *general stratified language* defined below permits the definition of one function in terms of another function at the same level using *monotonic* functions.

Definition 3.17: A function f is monotonic in its ith argument if $t_1 \leq t_2 \Rightarrow f(\ldots, t_1, \ldots) \leq f(\ldots, t_2, \ldots)$, where the ith argument is the one shown and all other arguments remain unchanged in $f(\ldots, t_1, \ldots)$ and $f(\ldots, t_2, \ldots)$.

Definition 3.18 (general stratified programs):
(a) Level 1 clauses have the following syntax (where f and g are at level 1).

$f(terms) \geq g(terms)$
$f(terms) \geq term$

(b) For each level $j > 1$, clauses have the following syntax (where f and g are at level j).

$$f(terms) \geq p(lexprs, g(lexprs), lexprs)$$
$$f(terms) \geq g(lexprs)$$
$$f(terms) \geq lexpr$$

(c) A program is partitioned into the smallest number of levels satisfying the above conditions.

In the above cases, f and g are not necessarily different, and *lexpr* and *lexprs* are as defined earlier. The function p is *monotonic* in the argument where g appears. Thus, non-monotonic "dependence" occurs only with respect to lower-level functions. One can permit more liberal definitions than the one given above: First, since a composition of monotonic functions is monotonic, the function p in the above syntax can also be replaced by a composition of monotonic functions. Second, it suffices if the *ground instances* of program clauses are stratified in the above manner. This idea is, of course, analogous to that of *local stratification* [Prz88]. However, our definition is more liberal since we do permit certain forms of cyclic dependencies to occur among function calls. Hence we use the term *generalized local stratification* to refer to our definition. It is straightforward to show that the presence of monotonic functions does not call for any alteration of the model-theoretic semantics. The operational semantics, however, must be modified to incorporate monotonically updatable memo-tables. Space limitations preclude a full treatment of this topic in this paper.

4 Monotonic Aggregation

We first introduce conditional partial-order clauses:

$$f(terms) \geq expr \ \text{:-} \ condition$$
$$f(terms) \leq expr \ \text{:-} \ condition$$

where each variable in *expr* occurs either in *terms* or in *condition*, and *condition* is in general a conjunction of relational or equational goals defined as follows.

$$condition \ ::= goal \ | \ goal, condition$$
$$goal \ ::= p(terms) \ | \neg \ p(terms) \ | \ f(terms) = term$$

A well-formed program is one that satisfies the generalized local stratification condition of section 3.3. Declaratively speaking, the meaning of a conditional clause is that, for all its ground instantiations, the partial-order is asserted to be true if the *condition* is true. Procedurally, *condition* is processed first before *expr* is evaluated. When new variables appear in *condition* (i.e., those that are not on the left-hand side), the goals in *condition* are processed in such an order so that all functional goals ($f(terms)$) and all negated goals ($\neg \ p(terms)$) are invoked with ground arguments—note that negation-as-failure may be unsound for nonground negated goals. The predicates appearing in *p(terms)* may be defined through definite clauses, possibly extended with negation-as-failure. In all our examples in this paper, however, these predicates refer to extensional database relations.

We now present a few examples to explain the use of conditional partial-order clauses. The following table summarizes the various forms of clauses to be used in these examples.

Type of Partial Order	*Least/Greatest Element*	*LUB/GLB*
$\geq$	ϕ ($\perp$)	$\cup$ (*lub*)
$\leq$	`max_int` ($\top$)	`min2` (*glb*)
$\geq$	`false` ($\perp$)	`or` (*lub*)

Our implemented language is flexible in that a programmer can declare, for any given function definition, what should be the least/greatest element [JM95]. Thus `max_int` in the above table is chosen by the programmer to suit the problem at hand. It is also possible in principle to let the user specify the definitions of the *lub/glb* operations. It may be seen that specifying the least/greatest element is similar to the notion of *defaults* in the terminology of Sudarshan *et al* [SSRB93], while specifying the *lub/glb* corresponds to the notion of *first-order* aggregate operations in the sense of Van Gelder [Van92]. Furthermore, the inductive aggregates are user-definable; that is, we are not restricted to a fixed set of built-in aggregate operations.

Reachable Nodes:

```
reach(X) ≥ {X}
reach(X) ≥ reach(Y) :- edge(X,Y)
```

The above program is a reformulation of the `reach` function of section 2.2 using an `edge(X,Y)` relation (the extensional database). This definition is amenable to more efficient memoization since the argument for `reach` is a constant rather than a set. Except for this difference, the execution of a top-level query against this program is essentially identical to that of the program in section 2.2.

Shortest Distance:

```
short(X,Y) ≤ C :- edge(X,Y,C)
short(X,Y) ≤ C+short(Z,Y) :- edge(X,Z,C)
```

This definition for `short` is very similar to that for `reach`, except that the aggregate operation here `min2` (instead of $\cup$). The relation `edge(X,Y,C)` means that there is a directed edge from `X` to `Y` with distance `C` which is non-negative. The default distance between any two nodes is `max_int`. The `+` operator is monotonic with respect to the numeric ordering, and hence the program is well-defined. The *logic* of the shortest-distance problem is very clearly specified in the above program. And our computational model (reduction + monotonically updatable memo-tables) provides better efficiency than a dynamic programming algorithm because top-down control avoids solving any unnecessary subproblems. Still, this is *not* the best control strategy for the shortest-distance problem. By specifying that the underlying lattice ordering is a *total* ordering and that `min2` distributes over +, it is possible to mimic a Dijkstra-style shortest-path algorithm. While annotations for distribution are discussed in [Jay92] and is supported by our implementation, we do not yet support annotations that specify total-ordering.

Company Controls [RS92]:

controls(X,Y) $\geq$ gt(sum(holdings(X,Y)), 50)
holdings(X,Y) $\geq$ {N} :- shares(X,Y,N)
holdings(X,Y) $\geq$ {N} :- shares(Z,Y,N), controls(X,Z) = true

This example illustrates the use of an inductive aggregate operation, `sum`. The function `controls(X,Y)` returns `true` if company `X` controls `Y`, and `false` otherwise. The relation `shares(X,Y,N)` means that company `X` holds `N` % of the shares of company `Y`. Cyclic holdings are possible, i.e., company `X` may have directly holdings in company `Y`, and *vice versa*. Here we see recursion over aggregation: a company `X` controls `Y` if the sum of `X`'s ownership in `Y` together with the ownership in `Y` of all companies `Z` controlled by `X` exceeds 50%.[3] Since percentages are non-negative, `sum` is monotonic with respect to the subset ordering. The function `gt(X,Y)` stands for numeric greater-than, and is monotonic in its first argument with respect to the ordering `false <= true`. Hence the conditions are satisfied for a well-defined semantics. Note that the default value for `controls(X,Y)` is `false`. (With reference to the syntax of programs given in section 3.3, we are making use of the fact that a composition of monotonic functions is monotonic. Hence the clause `controls(X,Y)` $\geq$ `gt(sum(holdings(X,Y)), 50)` is legal, because `gt` and `sum` are monotonic.)

5 Conclusions and Related Work

We have demonstrated that partial-order clauses and lattice domains provide a very natural and flexible means for programming a wide range of problems involving circular constraints and aggregation. The contribution of this paper is primarily in providing a novel conceptual framework for programming. While the language of *unconditional* partial-order clauses is a purely functional language, the provision of *conditional* clauses shows how these clauses can be integrated with a logic programming language. When the predicates in a conditional clause refer to an extensional database of relations, the resulting language can be seen as a *functional query language*. The elegance of this framework is attested to by its simple model-theoretic and operational semantics. The computational model

[3] It would be more appropriate in this example to build a multiset (instead of a set) as the argument to `sum`. However, one can re-program this problem retaining the use of sets, by tagging each percentage value with the corresponding company-name.

for partial-order programs combines top-down goal reduction with monotonic memo-tables, and has been proved to be sound [JOM93]. The implementation of partial-order clauses was carried out by Kyonghee Moon, and all program examples in this paper were tested out using this implementation.

Partial-order clauses are a generalization and an extension of the concept of *subset clauses* described in our previous papers [Jay92,JM95,JP87,JP89]. The significance of generalizing subset clauses to partial order clauses is that it provides a simple and efficient way of programming aggregate operations. This paper extends our previous papers by treating monotonic aggregation and monotonic memo-tables. Our use of sets is related to recent work on structural recursion on sets [BBN91], and also to the CLP formulation of sets [DR93]. A distinguishing feature of our set constructor $\{X\backslash T\}$ is that it matches a set S such that $X \in S$ and $T = S - \{X\}$. The ability to form the *remainder set* T is unique to our approach and is crucial in writing recursive definitions. Basically, the work on structural recursion on sets provides a typed approach to combining relational algebra and equation-based functional programming. In contrast, we try to combine relational definite clauses with partial-order-based functional programming. In comparison with the CLP approach to finite sets [DR93], we note that their goal was more to formalize the *semantics* of finite sets in the CLP framework, and therefore they do not address the formulation of aggregate operations—aggregation being a meta-level concept with respect to the standard CLP framework.

Finally, our language has the flavor of a functional-logic language [Han94], but there are two important differences: partial-order clauses are used instead of equational clauses; functional expressions are reduced (by matching), and not narrowed (by unification). The provision of partial-order clauses and monotonic memo-tables are crucial for formulating monotonic aggregation, and these features are not so easily simulated in functional-logic languages without nontrivial changes to their semantics.

References

[AU77] A. Aho and J.D. Ullman, "Symp. on Principles of Compiler Design," Addison-Wesley, 1977.

[BBN91] V. Breazu-Tannen, P. Buneman, and S. Naqvi, "Structural Recursion as a Query Language," *Proc. 3rd Intl. Workshop on Database Programming Languages*, 1991.

[DR93] A. Dovier and G. Rossi, "Embedding Extensional Finite Sets in CLP," *Proc. Intl. Symp. on Logic Programming*, pp. 540–556, MIT Press, 1993.

[Han94] M. Hanus, "The Integration of Functions into Logic Programming: From Theory to Practice," *J. of Logic Programming*, (19/20):583–628, 1994.

[Jay92] B. Jayaraman, "Implementation of Subset-Equational Programs," *J. of Logic Programming*, 12(4):299-324, 1992.

[JM95] B. Jayaraman and K. Moon, "Implementation of Subset-Logic Programs," Submitted for publication.

[JP87] Jayaraman, B. and D. A. Plaisted, "Functional Programming with Sets," *Proc. Third Intl. Conf. on Functional Programming and Computer Architecture*, pp. 194-210, Springer-Verlag, 1987.

[JP89] Jayaraman, B. and D. A. Plaisted, "Programming with Equations, Subsets, and Relations," *Proc. N. American Conf. on Logic Programming*, pp. 1051-1068, MIT Press, 1989.

[KS91] D.B. Kemp and P.J. Stuckey, "Semantics of Logic Programs with Aggregates," *Proc. Intl. Symp. on Logic Programming*, pp. 387-401, MIT Press, 1991.

[Llo87] J.W. Lloyd, "Foundations of Logic Programming," (2 ed.) Springer-Verlag, 1987.

[JOM93] B. Jayaraman, M. Osorio and K. Moon, "Partial Order Logic Programming," Technical Report 93-040, Department of Computer Science, SUNY-Buffalo, November 1993.

[Prz88] T. Przymusinski, "On the Declarative Semantics of Stratified Deductive Databases and Logic Programs," *Proc. Foundations of Deductive Databases and Logic Programming*, J. Minker (ed.), pp. 193-216, Morgan-Kaufmann, 1988.

[Par89] S. Parker, "Partial Order Programming," *Proc. 16th Symp. on Principles of Programming Languages*, pp. 260-266, ACM Press, 1989.

[RS92] K.A. Ross and Y. Sagiv, "Monotonic Aggregation in Deductive Databases," *Proc. 11th Symp. on Principles of Database Systems*, pp. 114-126, ACM Press, 1992.

[SSRB93] S. Sudarshan, D. Srivastava, R. Ramakrishnan, and C. Beeri, "Extending the Well-Founded and Valid Semantics for Aggregation," *Proc. Intl. Symp. on Logic Programming*, pp. 590-608, MIT Press, 1993.

[Van92] A. Van Gelder, "The Well-Founded Semantics of Aggregation," *Proc. 11th Symp. on Principles of Database Systems*, pp. 127-138, ACM Press, 1992.

SPIKE: a system for automatic inductive proofs

Adel Bouhoula and Michaël Rusinowitch

INRIA-Lorraine & CRIN
BP 101, 54600 Villers-lès-Nancy, France
email: {bouhoula,rusi}@loria.fr

The SPIKE system is an automatic theorem prover in theories presented by conditional equations. SPIKE was written in Caml Light©, a functional language of the ML family. The program is provided with a graphic interface written in TCL/TK (X11 toolkit) that allows for interaction through the mouse and menus. The principal functions of SPIKE are proof by induction and an aid in the construction of correct specifications.

In contrast to the majority of current proof systems that construct their proofs step by step and require frequent user intervention, not to say a great expertise on the part of the user, SPIKE is meant to reduce the number of interactions due to the automatisation of numerous routine tasks.

The SPIKE system belongs to the family of program verification tools. The development of a program demand a certain number of proof obligations. First of all, it must be verified that the program meets the original specifications. Similarly, the transformation steps that lead to more efficient programs must be formally justified to avoid any divergence from the initial specifications. In general, the necessary proofs are tedious and verification by hand becomes rapidly unreliable, if not possible. This shows the importance of automatic proof systems such as SPIKE, for eliminating an important percentage of such computations.

References

[Bouhoula and Rusinowitch, 1993] Adel Bouhoula and Michaël Rusinowitch. Automatic case analysis in proof by induction. In Ruzena Bajcsy, editor, *Proceedings 13th International Joint Conference on Artificial Intelligence, Chambéry (France)*, volume 1, pages 88–94. Morgan Kaufmann, August 1993.

[Bouhoula and Rusinowitch, 1994] Adel Bouhoula and Michaël Rusinowitch. Implicit induction in conditional theories, 1994. Accepted for publication by *Journal of Automated Reasoning*, to appear. Also available as INRIA report 2045, 1993.

[Bouhoula *et al.*, 1995] Adel Bouhoula, Emmanuel Kounalis, and Michaël Rusinowitch. Automated mathematical induction. 1995. Accepted for publication by *Journal of Logic and Computation*, to appear. Also available as INRIA report 1636, 1992.

[Bouhoula, 1994a] Adel Bouhoula. The challenge of mutual recursion and mutual simplification in implicit induction. Invited talk at the International Workshop on the Automation of Proof by Mathematical Induction, Nancy (France), June 1994.

[Bouhoula, 1994b] Adel Bouhoula. *Preuves Automatiques par Récurrence dans les Théories Conditionnelles*. PhD thesis, Université de Nancy I, Mars 1994.

[Bouhoula, 1994c] Adel Bouhoula. Spike: a system for sufficient completeness and parameterized inductive proof. In A. Bundy, editor, *Proceedings 12th International*

Conference on Automated Deduction, Nancy (France), volume 814 of *Lecture Notes in Artificial Intelligence*, pages 836–840. Springer-Verlag, June 1994.

[Bouhoula, 1994d] Adel Bouhoula. Sufficient completeness and parameterized proofs by induction. In *Proceedings Fourth International Conference on Algebraic and Logic Programming, Madrid (Spain)*, volume 850 of *Lecture Notes in Computer Science*, pages 23–40. Springer-Verlag, September 1994.

[Bouhoula, 1995] Adel Bouhoula. Fundamental Results on Automated Theorem Proving by Test Set Induction. Technical Report 2478, INRIA, 1995.

SEAMLESS: Knowledge Based Evolutionary System Synthesis

Jutta Eusterbrock

GMD, Institute for Tele - Cooperation Technology
Rheinstr. 75, 64295 Darmstadt, Germany
eusterbr@darmstadt.gmd.de

1 Motivation

The intent of the SEAMLESS approach is to provide a knowledge based synthesis environment, supporting cyclic phases of Specification, Experimentation, Abstraction, Modification, Logical Validation and Extraction for System Synthesis, conceiving system synthesis as a non-linear, incremental knowledge acquisition process. The formal foundation is a multi-layer framework for knowledge representation within the logic programming paradigm. It allows to apply a uniform specification language and a coherent concept for describing various useful forms of knowledge, eg. domain data or software components, and supporting reasoning tasks, eg. knowledge acquisition, software construction or program optimisation. Further - from a practical point of view - it is highly desirable to have synthesis systems which assure the re-usability of components, flexibility and evolution. Hence SEAMLESS is implemented as an easily extensible workbench of re-usable encapsulated theories, which are hierarchically organized and interact by well-defined interfaces.

2 Multi-Layer Knowledge Representation

Three abstraction layers - each of them is subdivided into data type and rule specifications in the format of typed horn clauses - are introduced (cf. [Eus94]). The layers refer to the modeling of domain knowledge; generic abstractions of domain theories and strategic knowledge; and the classification, control and transformation of knowledge sources, eg. program components, respectively. For formal knowledge representation we employ a "Gödel"- like typed logic language, suitably adapted to the specific representation tasks, extended by abstract data types, metalogic features and features of "Quantified Dynamic Logic" to model knowledge transformation tactics. We assume that this approach enables efficient and terminating proof procedures. The representation framework is based on declarative model theoretic semantics and a modal logic semantics. Typed logic programs provide a means to connect data and logic. They consist of data type definitions and extended typed first order theories. According to the different abstraction layers, the data types are assigned sets of terms, formulas, and theories as universes. Especially, theories and program components are considered as typed objects, which can be constructed dynamically. Furtheron - in the given approach - the formulas of a theory are divided into defining axioms and derived knowledge, which are called generic theories and generic tactics on the higher layers. The interaction between object- and metalayers is realized by views, which instantiate generic theories by application theories. Views allow generic tactics to be specialized for a variety of applications. Tasks to be solved are submitted as goals, ie. theorems or metatheorems to be proven. Solutions are generated by computation, ie. constructive theorem proving.

3 Implemented Components

Within this framework correct and operational specifications of data types, optimisation rules for logic programs, mathematical domain knowledge, and generic tactics were derived. Based on the specified data types and tactics, the current implementation[1] provides - among others - the following features:

- canonical terms and term rewrite rules efficiently implementing directed acylic graphs, classes of isomorphic graphs and graph operations (cf. [Eus95]);
- a structured knowledge base, entailing extensive knowledge on partial-order sorting problems, their complexities and algorithms for solving subproblems;
- a generic tactic to constructively solve algorithmic problems defined over extended divide-and-conquer theories. The metainterpreter deductively evaluates proof processes, automatically generates and extends libraries with knowledge about solved specifications, derived proofs or control knowledge, allows to re-use generated knowledge and employs a three-valued logic;
- metalogic extensions, which allow to deal with programs and theories as typed data in horn clause programs, providing mechanism for dynamic binding, operations to transform and construct theories, feature terms to structure aggregated theories;
- an incremental generic tactic for theory construction from positive and negative ground atoms;
- a generic tactic to transform proof terms into executable programs.

4 Applications

SEAMLESS allows to develop new generic tactics by devising generic theories and deriving tactics as metatheorems. A SEAMLESS user develops a problem specification by instantiating generic theories through views and employing already defined data types. Complex problems are solved in an incremental process by creating, extending and refining knowledge bases.

SEAMLESS applications comprise the transformation of inefficient generate-and-test programs into more efficient ones (cf. [Eus92b]) and the discovery of a previously unknown, complex mathematical algorithm (cf. [Eus92a]). Search

[1] SEAMLESS is built in Prolog and uses a graphical interface written in C++.

complexities were tackled by extended strategic knowledge, which was implemented as annotated expert knowledge and automatically augmented at runtime by machine learning procedures. Currently, the system is extended to provide a toolkit supporting the configuration of workstation/PC-based telecommunication services which are customized to meet application specific requirements. System configurations are composed of hard- or software components, and combine services that connect to telecommunication networks, eg. ISDN or cellular networks. The requirements taken into account consist of constraints on individual components and services as well as specifications of data, temporal, or preference dependencies between components.

References

[Eus92a] J. Eusterbrock. Errata to "Selecting the top three elements" by M. Aigner: A Result of a computer assisted proof search. *Discrete Applied Mathematics*, 41:131–137, 1992.

[Eus92b] J. Eusterbrock. Speed-up transformations of logic programs by abstraction and learning. In K-K. Lau und T. Clement, editor, *Logic Program Synthesis and Transformation*, pages 167–182. Springer, 1992.

[Eus94] J. Eusterbrock. Knowledge modeling for evolutionary program synthesis. In *Proceedings of the 8th International Symposium on Methodologies for Intelligent Systems*, pages 17–30. Oak Ridge National Laboratory, N00014-94-1-0799, 1994.

[Eus95] J. Eusterbrock. Efficient reasoning with transitive DAG's, 1995. Submitted.

An Object-Oriented Front-end for Deductive Databases*

Hasan M. Jamil† **Laks V. S. Lakshmanan**

Department of Computer Science
Concordia University, Montreal, Canada
e-mail: {jamil,laks}@cs.concordia.ca

Abstract: We present the Orlog deductive object-oriented database system prototype. The implementation of the system relies on the idea of reducing inheritance to deduction and giving a relational interpretation to every Orlog database. The prototype is a user transparent front-end for CORAL deductive database system and provides a full-fledged programming environment in Orlog.

1 Introduction

In [2], we proposed a deductive query language called Orlog for object-oriented databases. Orlog provides facilities for defining methods, signatures, objects, and classes. Behavioral inheritance, conflict resolution and code sharing are made part of the semantics in an elegant way. Interested readers are referred to [2, 3, 4] for a complete discussion on the Orlog language and the theoretical basis of its implementation, and to [5] for additional readings on logic based object-oriented languages.

Several well-known systems and research prototypes are known to have been built using other existing systems as back-ends. Following the same direction, we also proposed a scheme in [4] for implementing Orlog in CORAL [6] using a translational approach. In this approach every Orlog database is encoded into a CORAL database. The encoded databases thus have a relational interpretation where every object is viewed as a theory consisting of a collection of Horn clauses. The encoding scheme relies on the key idea of reducing inheritance to deduction since CORAL is only capable of deduction. Since the idea is to allow the user to program in a declarative way, query optimization then becomes a system level concern. However, the research into query optimization in deductive object-oriented databases is still in its infancy. While sophisticated optimization techniques are being researched, we try to take advantage of what has been successfully utilized in deductive databases for query optimization. This motivates our translation based approach where users perceive their applications naturally in an object-oriented way and never go through the mental exercise of

*This research was supported in part by grants from the Natural Sciences and Engineering Research Council of Canada and the Fonds Pour Formation De Chercheurs Et L'Aide À La Recherche of Quebec.

†This author's research was additionally supported in part by grants from the Canadian Commonwealth Scholarship and Fellowship Plan and the University of Dhaka, Bangladesh.

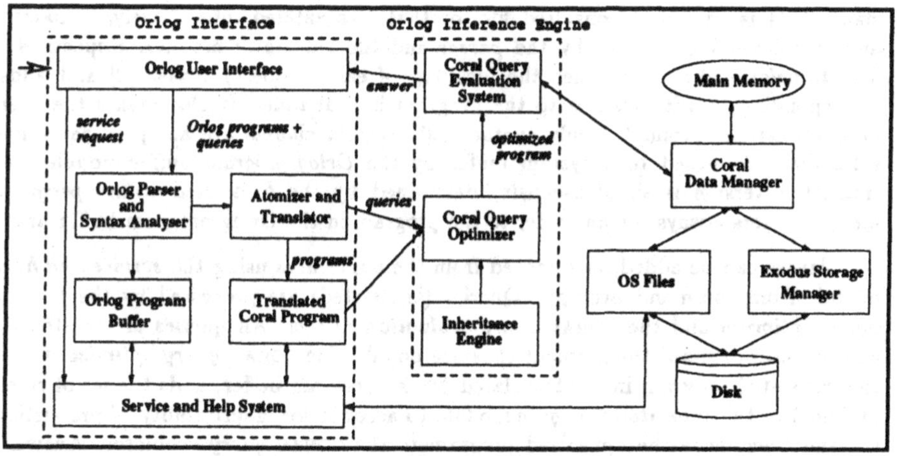

Figure 1: Orlog systems architecture.

mapping them into a non-object-oriented model. But still take advantage of the superior query optimization techniques available in the deductive paradigm.

The primary goal of this demonstration is two-fold. First, to illustrate important features of our system and to exhibit a range of programs that amply demonstrate the inheritance model of Orlog and its strengths compared to other similar languages. Secondly, to demonstrate that applications in Orlog can be built right now using our prototype. However, in this first prototype our focus has been in demonstrating the ease and elegance with which object-oriented database applications can be modeled and programmed in a declarative manner. As an aside, we advocate through this demonstration that rapid prototyping using existing deductive systems like CORAL, $\mathcal{LDL}$ [1], etc. is convenient and efficient. We have, however, plans for the near future to directly implement a full-fledged Orlog database system.

2 The Orlog System

The prototype we present is an interface which provides a complete programming environment in Orlog by translating user programs into equivalent predicate logic programs. It uses the CORAL deductive database system as the backbone inference engine in a user transparent way. The interface has been designed to serve as an on-line interactive command interpreter as well as an Orlog interpreter. It allows users to write programs in Orlog, and answer queries. It also provides facilities and basic system services like viewing, editing and printing program files, saving in-memory programs, running a stored program, getting on-line help, etc.

In Figure 1, the architecture of the system and its components are shown. The *user interface* module is an interactive shell and a command interpreter. Users are expected to interact with the system through this interface. At the command prompt users may request services, write programs interactively, or both. ASCII stored programs can also be brought into memory and executed using the consult command facility. Every clause that is part of a consulted file, or that was entered interactively, is parsed and translated into CORAL by the *parser* and the *translator* modules respectively. The *translator* also determines the locality[1] of each property clause [2] and adds corresponding locality clauses to the program in real time. It also makes the user program clauses atomic[2]. Only syntactically correct clauses are accepted as active rules that are added to a system buffer by the *Orlog program buffer* module. A translated version is simultaneously maintained by the *translated* CORAL *program* module. Users always see and refer to the program in the Orlog program buffer area.

Clauses can be added, or removed from the buffer area using the *service and help* module from command prompt. Queries (translated) are processed by the CORAL *query optimizer* and the CORAL *query evaluation system*. All queries are treated as ad hoc and processed the moment they are posed. The CORAL *query optimizer* takes the translated program in the translated CORAL program buffer, adds the set of rules (axioms) in the *inheritance engine* module to account for inheritability of properties [2], and then sends the optimized program to the CORAL *query evaluator*. Answers are then forwarded to the *user interface* by the CORAL *query evaluator* as a set of bindings to the free variables in the query.

2.1 Platform

The system is currently developed on Sun Workstation under Unix. It requires G^{++} (an extension of C^{++}) and CORAL deductive database system.

2.2 Efficiency Issues

A considerable amount of time is being spent by the interface every time it calls CORAL to evaluate a single query. The program files are being closed and opened several times, CORAL environment is being created, data structures are built for the sole purpose of a single query and destroyed subsequently. Another source of inefficiency in our system is the modular stratification in CORAL owing to our use of negation in implementing the inheritability function ∇ introduced in [2]. ∇ is implemented as a set of axioms in *inheritance engine*. These axioms account for inheritance in CORAL, and we found that the use of negation in the function implementation is unavoidable.

2.3 Future Enhancements

Despite several of its efficiency related drawbacks, the Orlog system compares very nicely with contemporary research prototypes in its class. Most of the translation based implementations have similar drawbacks, if not serious ones. However, the Orlog experience offers guidelines for addressing several important issues and suggests

[1] A clause is local to an object o if it is defined in o [2]. For example, the clause $john[age \rightarrow 20].$ is defined in $john$, and hence is local to $john$. Whereas, $X[car \rightarrow ford] \leftarrow X : employee.$ is defined in every object X such that X is an instance of the $employee$ object, and hence is local to all such Xs.

[2] Similar to F-logic [5], we also allow the users to write *molecular* formulas as a syntactic sugar. However, the molecules are broken down to atoms by the *atomizer* module, and subsequently all atomized clauses are converted to the usual clausal form (if not already).

that enhancements are possible on the present system. First, the time spent by the interface in CORAL calls can be saved by managing to run CORAL in a dedicated mode and make CORAL talk to the interface. In that way, we need to invoke CORAL only once from the interface, use shared memory space and data structures between CORAL and the interface, and spend minimal amount of time in transferring control from the interface to CORAL, and vice-versa. This can be improved in yet another potentially involved way. We can change the CORAL interface to accept Orlog code, and then translate the code internally to CORAL's native language.

Secondly, one of our goals is to eventually exploit CORAL features such as modules, execution control through meta-annotations, etc. This can be achieved very easily just by adding a simple identity function in the translator and passing the declarations to CORAL. Finally, in its current form, Orlog does not support persistent objects. This essential functionality needs to be added. A possible way to add this feature would require to extend Exodus storage manager used in CORAL with object management capabilities. A graphical user interface is also being planned for the prototype.

3 Orlog Demonstration

We show an implementation of a small object-oriented database in Orlog, several queries are executed and the answers presented. The demo example is a synthetic object-oriented aircraft design database with classes, instances, methods, signatures and object hierarchies. We show that the inheritance and code sharing work as expected, and that the resolution of inheritance conflicts can be achieved either via automatic detection or by user specification. We also show that the notion of withdrawal allows selective inheritance and thus allows the user to choose a prefered model. We also display the versatility of our system by allowing the viewers to interactively develop toy databases in Orlog and process ad hoc queries.

References

[1] D. Chimenti et. al. The $\mathcal{LDL}$ system prototype. *IEEE Journal on Data and Knowledge Engineering*, 2(1):76–90, 1990.

[2] H. M. Jamil. *Semantics of Behavioral Inheritance in Deductive Object-Oriented Databases*. PhD Thesis (in preparation), Department of Computer Science, Concordia University, Canada, 1995.

[3] H. M. Jamil and L. V. S. Lakshmanan. ORLOG: A Logic for Semantic Object-Oriented Models. In *Proc. 1st Int. Conference on Knowledge and Information Management*, pages 584–592, 1992.

[4] H. M. Jamil and L. V. S. Lakshmanan. Reducing inheritance to deduction by completion. Technical report, Department of Computer Science, Concordia University, Montreal, Canada, March 1995. In Preparation.

[5] M. Kifer, G. Lausen, and J. Wu. Logical Foundations for Object-Oriented and Frame-Based Languages. Technical Report TR-93/06, Department of Computer Science, SUNY at Stony Brook, 1993. (accepted to Journal of ACM).

[6] R. Ramakrishnan, D. Srivastava, and S. Sudarshan. CORAL : Control, Relations and Logic. In *Proc. of 18th VLDB Conference*, 1992.

The SuRE Programming Framework[1]

Bharat Jayaraman
Kyonghee Moon

Department of Computer Science
State University of New York at Buffalo
Buffalo, NY 14260
U.S.A.

E-Mail: {`bharat,kmoon`}`@cs.buffalo.edu`

Abstract

We illustrate the use of a declarative programming paradigm based upon three kinds of program clauses: equational, subset, and general relational clauses. The implemented language is called *SuRE*, which is an acronym for *Su*bsets, *R*elations, and *E*quations. (To wit, *SuRE* is the affirmative answer to the question: Can programming be declarative and practical?) Subset clauses have many uses in the logic programming context: They serve as a declarative alternative to Prolog's `mode` declarations as well as those uses of `assert` and `retract` that correspond to implementations of memo-tables or collection of results from alternative search paths, as in the `setof` construct. By re-formulating a relation as a set-valued function, one not only specifies mode information declaratively, but also gains the flexibility of operating on the resulting set incrementally (by a membership goal) or collectively (by an equational goal). In the latter case, one further has the flexibility of working lazily or eagerly. We show that lazy enumeration of solutions is the key to declaratively pruning the search space in generate-and-test problems. Subset clauses and, more generally, partial-order clauses are also useful in the deductive database context, especially for defining setof operations, transitive closures and monotonic aggregation—the concepts of subset, aggregation and monotonicity are more naturally expressed in terms of functions than predicates. A central feature in the implementation of *SuRE* programs is that of a *monotonic memo-table*, i.e., a memo-table whose entries can monotonically grow or shrink in an appropriate partial order. The implementation of *SuRE* was developed as an extension of the well-known Warren Abstract Machine (WAM) for Prolog. Even though *SuRE* differs substantially from Prolog, few additional instructions, registers, and storage structures are needed and these blend in well with the overall design of the WAM.

[1] This research was supported by a grant from the National Science Foundation.

A Declarative System for Multi-database Interoperability*

Laks V.S. Lakshmanan Iyer N. Subramanian
Despina Papoulis Nematollaah Shiri

Department of Computer Science, Concordia University, Montreal, Quebec
e-mail: {laks, subbu, papouli, shiri}@cs.concordia.ca

1 Introduction

Interoperability among database systems involves reconciling heterogeneities among datamodels, communication processing protocols, query processing systems, concurrency control protocols, consistency management, and other aspects of database systems. Though considerable amount of research has been done in the area of database interoperability, most of it has resulted in solutions that are ad-hoc and procedural. We have developed a declarative environment in which multiple heterogeneous databases interoperate by *sharing, interpreting,* and *manipulating* information, in a uniform way.

Interacting with multiple databases calls for the ability to query them in a manner independent of the discrepancies in their structure and data semantics. In this implementation, we demonstrate how we can query across multiple databases which store semantically similar data using heterogeneous schema.

2 The Language

We believe that *declarativity* is a key requirement for interoperability among multiple databases. A logic based approach for interoperability would bring the advantages of clear foundations, sound formalism, and proof procedures thus providing for a truly declarative environment. Conventional database query languages are based on predicate calculus and are useful for querying the data in a database. But as seen in many applications, interoperability necessitates a functionality to query not only the data in a database but also its schema or

*This research was supported in part by grants from the Natural Sciences and Engineering Research Council of Canada and the Fonds Pour Formation De Chercheurs Et L'Aide À La Recherche of Quebec.

meta-data. This calls for a higher-order language which treats "components" of such meta-data as "first class" entities in its semantic structure. In such a framework, queries that manipulate data as well as their schema "in the same breath" could be naturally formulated. In this vein, in [LSS93] we have proposed a language called *SchemaLog* that is syntactically higher-order but semantically first-order, and facilitates multidatabase interoperation. The language has its foundations in logic and has been studied in detail in [LSS95]. It also has syntactic features for explicitly referring to database, relation, and attribute names. In our implementation, users can write programs in *SchemaLog* to interoperate among multiple databases.

3 System Architecture

In this section, we outline the implementation strategy adopted for realizing our system. The system presently facilitates interoperability among multiple INGRES relational databases that contain semantically similar, but schematically dissimilar information.

Two aspects of our language that are of most significance are *(i)* higher-order features to access the schema information from multiple databases, and *(ii)* deduction. Our implementation handles these two aspects separately: The INGRES embedded SQL (ESQL) is used to manipulate the meta-information, and the deductive database CORAL is used for deduction. Fig 1 refers to the architecture of our implementation. Since user programs can involve complex interactions between schema manipulation and deduction, our implementation integrates these two functionalities from ESQL and CORAL closely.

In Phase 1 of the implementation the following activities are performed. The input *SchemaLog* program is rewritten to a first-order form using rewrite rules. The meta-information queried in the program is fetched and stored in appropriate intermediate tables using ESQL. During this process, various optimization strategies are employed to minimize the cost of fetching the meta-information as well as to reduce the amount of information that needs to be stored in the intermediate tables which in turn would affect the cost of (recursive) query processing by CORAL [Pap94]. Phase 2 of the implementation uses CORAL to process the rewritten program along with the meta-information fetched in Phase 1 and the answers are produced to the user.

The system sports a pleasant user interface capable, among other things, of a schema browsing facility.

4 Future Work

Various extensions to the existing implementation are currently under investigation. An important future work would involve using the database storage manager EXODUS for storing the intermediate relations, making a tighter coupling possible. We expect this to yield an impressive gain in performance, as

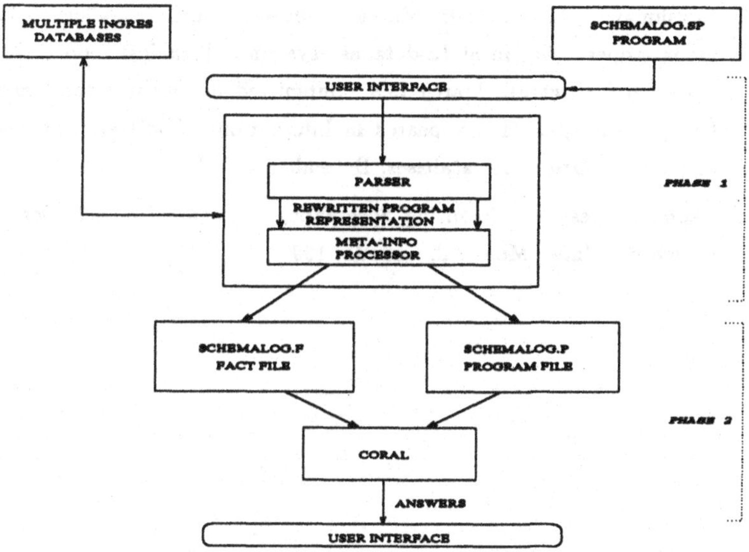

Figure 1: System Architecture

CORAL has a direct interface to EXODUS for storing and manipulating persistent relations. Other planned extensions involve incorporating disparate DBMS like Oracle, Sybase, and O_2 in the current implementation.

Language	ESQL, C
DBMS	INGRES
Deductive System	CORAL
User Interface	UIM/X
Platform	Sun 3/50 workstations Under UNIX

Figure 2: Implementation Particulars

References

[LSS93] Lakshmanan, L.V.S., Sadri, F., and Subramanian, I. N. On the logical foundations of schema integration and evolution in heterogeneous database systems. In *Proc. 3rd International Conference on Deductive and Object-Oriented Databases (DOOD '93)*. Springer-Verlag, LNCS-760, December 1993.

[LSS95] Lakshmanan, L.V.S., Sadri, F., and Subramanian, I. N. A logical language for interoperability in multi-database systems. Technical report, Concordia University, Montreal, March 1995. Submitted for publication, March 1995. (A preliminary version appeared in International Conference on Deductive and Object Oriented Databases, December 1993.).

[Pap94] Papoulis, Despina. Realizing SchemaLog. *Technical report,* Dept. of CS, Concordia Univ., Montreal, Canada, 1994.

The METAGEN System

Houari A. Sahraoui

LAFORIA - IBP, Box 169, 4 place Jussieu
75 252 Paris Cedex 05 - France
Phone + 33 1 44 27 37 17
Fax + 33 1 44 27 70 00
email sahraoui@laforia.ibp.fr

With the advent of object oriented database systems, there is an urgent need to define methodologies and tools for mapping conceptual schemes into object oriented ones. The evolution and the diversity of OO models on the one hand, and the increase of user needs on the other hand, are the two factors that make these tools require important flexibility properties, that is to say easiness in maintenance and extension.

In this context, we present a technique for rapid prototyping of such tools in order to experiment there associated methodologies. Our technique is based on metamodeling principles. It relies heavily on OO technology and on rule based programming. It borrows part of its inspiration from the AI subfield of Knowledge Acquisition.

Our technique, instanciated for OODB, can be summarized by the following steps:

- The description of a semantic formalism. This description is called a UMM (User MetaModel).
- The description of a Standard OODB formalism (e.g. ODMG, SQL3). This description is called a IMM (Implementor MetaModel).
- The definition of two transformation mechanisms:
  1. a GTM (Generic transformation mechanism) that maps a UM (User Model) into an IM (Implementor Model) and
  2. a STM (specific transformation mechanism) that translates an IM into code for a specific DBMS.

The technique is implemented as a prototype system called MetaGen. MetaGen was developed by the MetaFor group (G. Blain, J-F. Perrot, N. Revault and H.A. Sahraoui) at LAFORIA-IBP (Paris 6 Univ., France). As developped with Smalltalk 80, MetaGen runs on any host supporting the object oriented environment VisualWorks 1.0 or ObjectWorks 4.1 (i.e. mainly SparcStations, PC/Windows, Mac, HP).

For the demonstration, we propose an application of MetaGen, consisting on a re-implementation of InterSem. InterSem is an object-oriented semantic modeling tool developed by the LRBDDI group, UQAM Canada.

References

[Bla94] Blain G., Revault N., Sahraoui H.A., and Perrot J-F. A MetaModeling Technique - OOPSLA'94 addendum to the proceedings: Workshop on AI and OO Software Engineering. Portland, 1994.

[Rev95] Revault N., Sahraoui H.A., Blain G., and Perrot J-F. A MetaModeling Technique: The MetaGen System. TOOLS Europe '95 proceedings. Versailles, 1995. To appear.

[Pru92] Prud'homme B., Missaoui R., and Godin R. *INTERSEM : une interface semantique orientee-objet.* ICO, Vol. 4, N. 1-2, pp. 47-56, 1992.

[Sah94] Sahraoui H.A., and Revault N. *Modelisation Conceptuelle des Bases de Donnees: Techniques de Meta-Modelisation.* Actes du CARI'94. Ouagadougou, 1994.

Lecture Notes in Computer Science

For information about Vols. 1–865
please contact your bookseller or Springer-Verlag

Vol. 866: Y. Davidor, H.-P. Schwefel, R. Männer (Eds.), Parallel Problem Solving from Nature - PPSN III. Proceedings, 1994. XV, 642 pages. 1994.

Vol 867: L. Steels, G. Schreiber, W. Van de Velde (Eds.), A Future for Knowledge Acquisition. Proceedings, 1994. XII, 414 pages. 1994. (Subseries LNAI).

Vol. 868: R. Steinmetz (Ed.), Multimedia: Advanced Teleservices and High-Speed Communication Architectures. Proceedings, 1994. IX, 451 pages. 1994.

Vol. 869: Z. W. Raś, Zemankova (Eds.), Methodologies for Intelligent Systems. Proceedings, 1994. X, 613 pages. 1994. (Subseries LNAI).

Vol. 870: J. S. Greenfield, Distributed Programming Paradigms with Cryptography Applications. XI, 182 pages. 1994.

Vol. 871: J. P. Lee, G. G. Grinstein (Eds.), Database Issues for Data Visualization. Proceedings, 1993. XIV, 229 pages. 1994.

Vol. 872: S Arikawa, K. P. Jantke (Eds.), Algorithmic Learning Theory. Proceedings, 1994. XIV, 575 pages. 1994.

Vol. 873: M. Naftalin, T. Denvir, M. Bertran (Eds.), FME '94: Industrial Benefit of Formal Methods. Proceedings, 1994. XI, 723 pages. 1994.

Vol. 874: A. Borning (Ed.), Principles and Practice of Constraint Programming. Proceedings, 1994. IX, 361 pages. 1994.

Vol. 875: D. Gollmann (Ed.), Computer Security – ESORICS 94. Proceedings, 1994. XI, 469 pages. 1994.

Vol. 876: B. Blumenthal, J. Gornostaev, C. Unger (Eds.), Human-Computer Interaction. Proceedings, 1994. IX, 239 pages. 1994.

Vol. 877: L. M. Adleman, M.-D. Huang (Eds.), Algorithmic Number Theory. Proceedings, 1994. IX, 323 pages. 1994.

Vol. 878: T. Ishida; Parallel, Distributed and Multiagent Production Systems. XVII, 166 pages. 1994. (Subseries LNAI).

Vol. 879: J. Dongarra, J. Waśniewski (Eds.), Parallel Scientific Computing. Proceedings, 1994. XI, 566 pages. 1994.

Vol. 880: P. S. Thiagarajan (Ed.), Foundations of Software Technology and Theoretical Computer Science. Proceedings, 1994. XI, 451 pages. 1994.

Vol. 881: P. Loucopoulos (Ed.), Entity-Relationship Approach – ER'94. Proceedings, 1994. XIII, 579 pages. 1994.

Vol. 882: D. Hutchison, A. Danthine, H. Leopold, G. Coulson (Eds.), Multimedia Transport and Teleservices. Proceedings, 1994. XI, 380 pages. 1994.

Vol. 883: L. Fribourg, F. Turini (Eds.), Logic Program Synthesis and Transformation – Meta-Programming in Logic. Proceedings, 1994. IX, 451 pages. 1994.

Vol. 884: J. Nievergelt, T. Roos, H.-J. Schek, P. Widmayer (Eds.), IGIS '94: Geographic Information Systems. Proceedings, 1994. VIII, 292 pages. 19944.

Vol. 885: R. C. Veltkamp, Closed Objects Boundaries from Scattered Points. VIII, 144 pages. 1994.

Vol. 886: M. M. Veloso, Planning and Learning by Analogical Reasoning. XIII, 181 pages. 1994. (Subseries LNAI).

Vol. 887: M. Toussaint (Ed.), Ada in Europe. Proceedings, 1994. XII, 521 pages. 1994.

Vol. 888: S. A. Andersson (Ed.), Analysis of Dynamical and Cognitive Systems. Proceedings, 1993. VII, 260 pages. 1995.

Vol. 889: H. P. Lubich, Towards a CSCW Framework for Scientific Cooperation in Europe. X, 268 pages. 1995.

Vol. 890: M. J. Wooldridge, N. R. Jennings (Eds.), Intelligent Agents. Proceedings, 1994. VIII, 407 pages. 1995. (Subseries LNAI).

Vol. 891: C. Lewerentz, T. Lindner (Eds.), Formal Development of Reactive Systems. XI, 394 pages. 1995.

Vol. 892: K. Pingali, U. Banerjee, D. Gelernter, A. Nicolau, D. Padua (Eds.), Languages and Compilers for Parallel Computing. Proceedings, 1994. XI, 496 pages. 1995.

Vol. 893: G. Gottlob, M. Y. Vardi (Eds.), Database Theory – ICDT '95. Proceedings, 1995. XI, 454 pages. 1995.

Vol. 894: R. Tamassia, I. G. Tollis (Eds.), Graph Drawing. Proceedings, 1994. X, 471 pages. 1995.

Vol. 895: R. L. Ibrahim (Ed.), Software Engineering Education. Proceedings, 1995. XII, 449 pages. 1995.

Vol. 896: R. N. Taylor, J. Coutaz (Eds.), Software Engineering and Human-Computer Interaction. Proceedings, 1994. X, 281 pages. 1995.

Vol. 897: M. Fisher, R. Owens (Eds.), Executable Modal and Temporal Logics. Proceedings, 1993. VII, 180 pages. 1995. (Subseries LNAI).

Vol. 898: P. Steffens (Ed.), Machine Translation and the Lexicon. Proceedings, 1993. X, 251 pages. 1995. (Subseries LNAI).

Vol. 899: W. Banzhaf, F. H. Eeckman (Eds.), Evolution and Biocomputation. VII, 277 pages. 1995.

Vol. 900: E. W. Mayr, C. Puech (Eds.), STACS 95. Proceedings, 1995. XIII, 654 pages. 1995.

Vol. 901: R. Kumar, T. Kropf (Eds.), Theorem Provers in Circuit Design. Proceedings, 1994. VIII, 303 pages. 1995.

Vol. 902: M. Dezani-Ciancaglini, G. Plotkin (eds.), Typed Lambda Calculi and Applications. Proceedings, 1995. VIII, 443 pages. 1995

Vol. 903: E. W. Mayr, G. Schmidt, G. Tinhofer (Eds.), Graph-Theoretic Concepts in Computer Science. Proceedings, 1994. IX, 414 pages. 1995.

Vol. 904: P. Vitányi (Ed.), Computational Learning Theory. EuroCOLT'95. Proceedings, 1995. XVII, 415 pages. 1995. (Subseries LNAI).

Vol. 905: N. Ayache (Ed.), Computer Vision, Virtual Reality and Robotics in Medicine. Proceedings, 1995. XIV, 567 pages. 1995.

Vol. 906: E. Astesiano, G. Reggio, A. Tarlecki (Eds.), Recent Trends in Data Type Specification. Proceedings, 1995. VIII, 523 pages. 1995.

Vol. 907: T. Ito, A. Yonezawa (Eds.), Theory and Practice of Parallel Programming. Proceedings, 1995. VIII, 485 pages. 1995.

Vol. 908: J. R. Rao Extensions of the UNITY Methodology: Compositionality, Fairness and Probability in Parallelism. XI, 178 pages. 1995.

Vol. 909: H. Comon, J.-P. Jouannaud (Eds.), Term Rewriting. Proceedings, 1993. VIII, 221 pages. 1995.

Vol. 910: A. Podelski (Ed.), Constraint Programming: Basics and Trends. Proceedings, 1995. XI, 315 pages. 1995.

Vol. 911: R. Baeza-Yates, E. Goles, P. V. Poblete (Eds.), LATIN '95: Theoretical Informatics. Proceedings, 1995. IX, 525 pages. 1995.

Vol. 912: N. Lavrac, S. Wrobel (Eds.), Machine Learning: ECML – 95. Proceedings, 1995. XI, 370 pages. 1995. (Subseries LNAI).

Vol. 913: W. Schäfer (Ed.), Software Process Technology. Proceedings, 1995. IX, 261 pages. 1995.

Vol. 914: J. Hsiang (Ed.), Rewriting Techniques and Applications. Proceedings, 1995. XII, 473 pages. 1995.

Vol. 915: P. D. Mosses, M. Nielsen, M. I. Schwartzbach (Eds.), TAPSOFT '95: Theory and Practice of Software Development. Proceedings, 1995. XV, 810 pages. 1995.

Vol. 916: N. R. Adam, B. K. Bhargava, Y. Yesha (Eds.), Digital Libraries. Proceedings, 1994. XIII, 321 pages. 1995.

Vol. 917: J. Pieprzyk, R. Safavi-Naini (Eds.), Advances in Cryptology - ASIACRYPT '94. Proceedings, 1994. XII, 431 pages. 1995.

Vol. 918: P. Baumgartner, R. Hähnle, J. Posegga (Eds.), Theorem Proving with Analytic Tableaux and Related Methods. Proceedings, 1995. X, 352 pages. 1995. (Subseries LNAI).

Vol. 919: B. Hertzberger, G. Serazzi (Eds.), High-Performance Computing and Networking. Proceedings, 1995. XXIV, 957 pages. 1995.

Vol. 920: E. Balas, J. Clausen (Eds.), Integer Programming and Combinatorial Optimization. Proceedings, 1995. IX, 436 pages. 1995.

Vol. 921: L. C. Guillou, J.-J. Quisquater (Eds.), Advances in Cryptology – EUROCRYPT '95. Proceedings, 1995. XIV, 417 pages. 1995.

Vol. 922: H. Dörr, Efficient Graph Rewriting and Its Implementation. IX, 266 pages. 1995.

Vol. 923: M. Meyer (Ed.), Constraint Processing. IV, 289 pages. 1995.

Vol. 924: P. Ciancarini, O. Nierstrasz, A. Yonezawa (Eds.), Object-Based Models and Languages for Concurrent Systems. Proceedings, 1994. VII, 193 pages. 1995.

Vol. 925: J. Jeuring, E. Meijer (Eds.), Advanced Functional Programming. Proceedings, 1995. VII, 331 pages. 1995.

Vol. 926: P. Nesi (Ed.), Objective Software Quality. Proceedings, 1995. VIII, 249 pages. 1995.

Vol. 927: J. Dix, L. Moniz Pereira, T. C. Przymusinski (Eds.), Non-Monotonic Extensions of Logic Programming. Proceedings, 1994. IX, 229 pages. 1995. (Subseries LNAI).

Vol. 928: V.W. Marek, A. Nerode, M. Truszczynski (Eds.), Logic Programming and Nonmonotonic Reasoning. Proceedings, 1995. VIII, 417 pages. 1995. (Subseries LNAI).

Vol. 929: F. Morán, A. Moreno, J.J. Merelo, P. Chacón (Eds.), Advances in Artificial Life. Proceedings, 1995. XIII, 960 pages. 1995 (Subseries LNAI).

Vol. 930: J. Mira, F. Sandoval (Eds.), From Natural to Artificial Neural Computation. Proceedings, 1995. XVIII, 1150 pages. 1995.

Vol. 931: P.J. Braspenning, F. Thuijsman, A.J.M.M. Weijters (Eds.), Artificial Neural Networks. IX, 295 pages. 1995.

Vol. 932: J. Iivari, K. Lyytinen, M. Rossi (Eds.), Advanced Information Systems Engineering. Proceedings, 1995. XI, 388 pages. 1995.

Vol. 933: L. Pacholski, J. Tiuryn (Eds.), Computer Science Logic. Proceedings, 1994. IX, 543 pages. 1995.

Vol. 934: P. Barahona, M. Stefanelli, J. Wyatt (Eds.), Artificial Intelligence in Medicine. Proceedings, 1995. XI, 449 pages. 1995. (Subseries LNAI).

Vol. 935: G. De Michelis, M. Diaz (Eds.), Application and Theory of Petri Nets 1995. Proceedings, 1995. VIII, 511 pages. 1995.

Vol. 936: V.S. Alagar, M. Nivat (Eds.), Algebraic Methodology and Software Technology. Proceedings, 1995. XIV, 591 pages. 1995.

Vol. 937: Z. Galil, E. Ukkonen (Eds.), Combinatorial Pattern Matching. Proceedings, 1995. VIII, 409 pages. 1995.

Vol. 938: K.P. Birman, F. Mattern, A. Schiper (Eds.), Theory and Practice in Distributed Systems. Proceedings,1994. X, 263 pages. 1995.

Vol. 939: P. Wolper (Ed.), Computer Aided Verification. Proceedings, 1995. X, 451 pages. 1995.

Vol. 941: M. Cadoli, Tractable Reasoning in Artificial Intelligence. XVII, 247 pages. 1995. (Subseries LNAI).

Vol. 942: G. Böckle, Exploitation of Fine-Grain Parallelism. IX, 188 pages. 1995.

Vol. 943: W. Klas, M. Schrefl, Metaclasses and Their Application. IX, 201 pages. 1995.